GOLD NANOPARTICLES

FOR

PHYSICS, CHEMISTRY AND BIOLOGY

SECOND EDITION

GOLD NANOPARTICLES

FOR

PHYSICS, CHEMISTRY AND BIOLOGY

SECOND EDITION

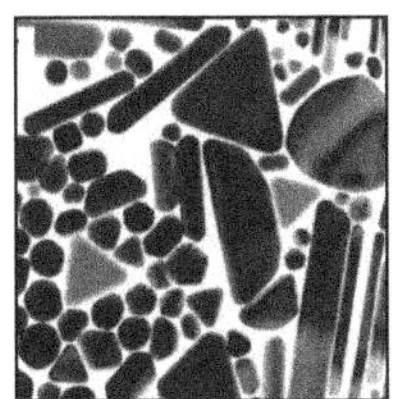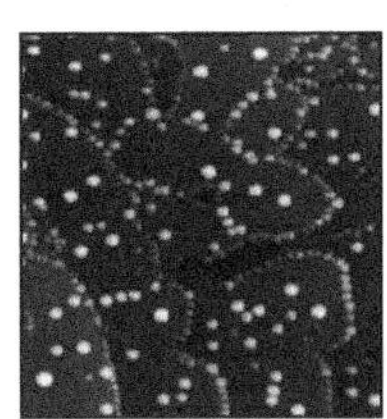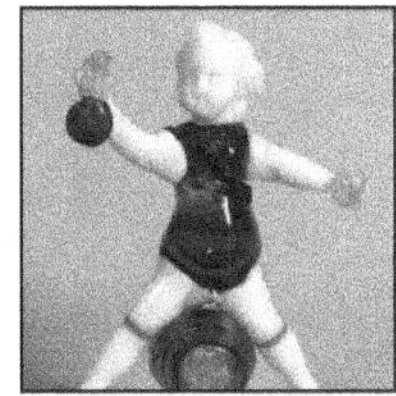

Editors

CATHERINE LOUIS
OLIVIER PLUCHERY

Université Pierre et Marie Curie, France

World Scientific

NEW JERSEY · LONDON · SINGAPORE · BEIJING · SHANGHAI · HONG KONG · TAIPEI · CHENNAI · TOKYO

Published by

World Scientific Publishing Europe Ltd.

57 Shelton Street, Covent Garden, London WC2H 9HE

Head office: 5 Toh Tuck Link, Singapore 596224

USA office: 27 Warren Street, Suite 401-402, Hackensack, NJ 07601

Library of Congress Cataloging-in-Publication Data
Names: Louis, Catherine (Chemist) | Pluchery, Olivier.
Title: Gold nanoparticles for physics, chemistry and biology / Catherine Louis (Université Pierre et
    Marie Curie, France), Olivier Pluchery (Université Pierre et Marie Curie, France).
Description: 2nd edition. | New Jersey : World Scientific, 2017. |
    Includes bibliographical references.
Identifiers: LCCN 2016034787 | ISBN 9781786341242 (hc : alk. paper)
Subjects: LCSH: Nanoparticles. | Gold.
Classification: LCC TA418.9.N35 L68 2017 | DDC 669/.22--dc23
LC record available at https://lccn.loc.gov/2016034787

British Library Cataloguing-in-Publication Data
A catalogue record for this book is available from the British Library.

First published 2017 (hardcover)
Reprinted 2024 (in paperback edition)
ISBN 9781800616332 (pbk)

Desk Editors: Herbert Moses/Mary Simpson

Typeset by Stallion Press
Email: enquiries@stallionpress.com

Contents

Contents

Contents

About the Authors

Chapters 1 and 8

Catherine Louis is a Research Director at the Laboratoire de Réactivité de Surface of the University Pierre et Marie Curie. She has been with the academic group since 1982, when she was appointed by the French National Centre of Research (CNRS). She received her PhD in Chemistry in 1985 (prepared under the direction of Prof. Michel Che). From 1986 to 1988, she was a post-doctoral fellow at the University of Berkeley with Prof. Alex Bell. She is a specialist in catalyst preparation and has worked on gold-based monometallic and bimetallic catalysts since 2000. She has authored around 140 publications. She co-authored *Catalysis by Gold* (Imperial College Press, 2006) with Geoffrey C. Bond and David T. Thompson. She is also the author of seven book chapters on synthesis of supported metal catalysts and CO oxidation of gold nanoparticles. From 2006 to 2013, she was the Director of Or-Nano (www.or-nano.com), a CNRS network gathering around 500 French researchers (physics, chemists and biologists) working with gold nanoparticles.

Chapter 2

Pekka Pyykkö was an Associate Professor of Quantum Chemistry at Åbo Akademi University (1974–1984) and Professor of Chemistry at the University of Helsinki (1984–2009). Since November 2009, he is enjoying research there as Professor Emeritus. He currently has published about 320 papers. He identified the chemical difference between silver and gold as a relativistic effect (1976, with Jean-Paul Desclaux), pointed out the importance of electron correlation, or dispersion, effects in aurophilicity 1991 (with Zhao Yongfang and later co-workers) and wrote the reviews *Theoretical Chemistry of Gold I–III* (2003–2008). He chaired the European Science Foundation programme Relativistic Effects in Heavy-Element Chemistry and Physics (REHE) during which the Hanau conference on the Science and Technology of Gold was held in 1996.

Geoffrey C. Bond held academic positions at the Universities of Leeds and Hull before being appointed Head of the Johnson Matthey Research Group on Catalysis (1962–1970). He then became Professor of Applied Chemistry at Brunel University, Uxbridge, where he held various posts (Head of the Chemistry Department, Dean of the Faculty of Science, Vice-Principal) until his retirement in 1992. His research has mainly concerned supported metal catalysts for hydrogenation and hydrogenolysis and supported oxides for selective oxidation. He has published more than 250 scientific papers and review articles. Since retirement, he has worked on gold catalysts, and has co-authored several review articles as well as the book *Catalysis by Gold* published by Imperial College Press. Earlier books include *Catalysis by Metals* (1962), *Heterogeneous Catalysis, Principles and Applications* (2nd edn., 1987) and *Metal-Catalysed Reactions of Hydrocarbons* (2005).

Chapter 3

Olivier Pluchery graduated from the Ecole Normale Supérieure de Cachan (Paris, France) in 1997 with a specialisation in laser physics. He obtained his PhD in chemical physics from University Paris-Sud in 2000 and was interested in the investigation of the electrochemical reactions on a gold interface, monitored with sum frequency generation, a nonlinear optical spectroscopy. In 2001, he joined Yves Chabal's team at Bell Labs (USA) to work on semiconductor interfaces. In 2002, he obtained a position as Associate Professor at University Pierre et Marie Curie (Paris) where he developed several research programmes dealing with the control of the adsorption of organic molecules on silicon for molecular electronics and the use of gold nanoparticles for nanoelectronics. He is the founder with Catherine Louis of the research network Or-Nano (www.or-nano.com).

Chapter 4

Bruno Palpant is a Professor at Centrale-Supélec in Paris region. He leads research activities in the Quantum and Molecular Photonics Laboratory (LPQM, belonging to CNRS, CentraleSupélec and Ecole Normale Supérieure de Cachan). He is in charge of a group devoted to the study and application of the ultrafast transient optical and thermal responses of plasmonic nanoparticles. He got his PhD in 1998 from University of Lyon (France) about quantum size effects in the optical properties of noble metal nanoparticles, before joining Keio University (Japan) for one year. Assistant professor in the Institut des NanoSciences de Paris (CNRS-UPMC) for 10 years, he has been interested in the linear and non-linear optical responses of noble metal nanoparticles as well as their link with thermal transport at small space and time scales. More recently, his activities have focused on the applications in biology, chemistry and photonics of nanoscale energy conversion processes.

Chapter 5

Changhyoup Lee is a Postdoctoral Fellow in Karlsruhe Institute of Technology (Germany). He received his PhD for quantum plasmonics in 2011 from Hanyang University (Korea). Then he has worked as a post-doctoral fellow at Hanyang University (Korea), and Centre for Quantum Technologies, National University of Singapore (Singapore) until he joined Karlsruhe Institute of Technology in 2016. His recent research focuses on the use of quantum metrology techniques in plasmonic or nanophotonic sensing platforms.

Mark Tame is an Associate Professor at the University of KwaZulu-Natal in South Africa. He leads research activities in quantum photonics at the Centre for Quantum Technology. He is in charge of a group focused on experimental and theoretical research into nanophotonic systems operating in the quantum regime. He received his PhD in 2007 from Queen's University, Belfast (UK) on the topic of optical quantum information processing. He has held research fellowship positions at Imperial College London (EPSRC) and Osaka University in Japan (JSPS). He is currently a visiting professor at Osaka University. His main research interests are in the application of quantum optics to nanoscale systems such as plasmonic nanoparticles and metamaterials, with the main goal of developing new devices that can be used in quantum information processing. This includes single-photon sources and switches, quantum sensing and entanglement generation.

Chapter 6

Woong Choi received his BSc in Chemistry at the Korea Advanced Institute of Science and Technology (KAIST) in 2013. He is currently a candidate of combined master's and doctorate programme in chemistry under the direction of Professor Hyunjoon Song at KAIST. His research has focused on the fabrication of complex nanostructures with multiple components based on noble metal nanoparticles and their applications for photocatalysts and SPR sensors.

Chapters 6 and 7

Hyunjoon Song received his B.S., M.S. and PhD degrees from the Department of Chemistry at KAIST in 1994, 1996 and 2000, under the direction of Prof. Joon T. Park. After postdoctoral works at KAIST and the University of California at Berkeley (with Prof. Peidong Yang), he was appointed as an Assistant Professor in 2005, and was promoted to an Associate and to a Full Professor in 2008 and 2014 in the Department of Chemistry at KAIST. He was appointed as a KAIST-endowed chair professor in 2015. His research interests are morphology control of metal hybrid nanostructures and their applications for surface plasmon monitoring and catalysis in organic and photochemical reactions.

Chapter 7

Souhir Boujday is an Associate Professor at University Pierre and Marie Curie (UPMC), Sorbonne Universities, Laboratory of Surface Reactivity. She is currently a Visiting Professor at Nanyang Technological University in Singapore. She graduated from UPMC with a PhD in Chemistry in 2002 on the interfacial molecular recognition between transition metal complexes and silica surface during catalyst preparation. After post-doctoral studies on photocatalysis (consortium CNRS, Rhodia, Lapeyre, Arch-Coating and Hahn Meitner Institute (Berlin, Germany)), she joined UPMC as an Assistant Professor in 2004. She was involved in the creation of a new research group on biointerfaces where she was responsible for the biosensors theme of research. Her research activity is focused on the optimisation of the molecular approach of material surface modification to ensure the highest efficiency of the functional material. She defended her habilitation to research supervision thesis on 'Surface nanostructuration and molecular recognition at the solid/liquid interface' in 2012.

Atul N. Parikh is a Professor and member of the faculty in the Departments of Biomedical Engineering and Materials Science and Engineering at the University of California, Davis. Since 2012, he is also serving as a Visiting Professor in the school of Materials Science and Engineering at Nanyang Technological University in Singapore. He studied Chemical Engineering at the Department of Chemical Technology (UDCT) University of Bombay (B. Chem. Eng., 1987) and Materials

Science (Specialisation: Polymer Science) at the Pennsylvania State University (PhD 1993). Between 1996 and 2001, as post-doctoral scholar and then technical staff member in the Chemical Science and Bioscience divisions at Los Alamos National Laboratory (LANL), he explored design of biologically inspired materials and biosensors. His research interests include surface chemistry, soft matter and membrane biophysics.

Bo Liedberg is a Full Professor of Materials Science at the School of Materials Science and Engineering, Nanyang Technological University (NTU), Singapore where he is leading a university-wide initiative on biomimetic sensor science. Liedberg is also serving as the Dean of the Interdisciplinary Graduate School, NTU. He received his PhD in applied physics from Linköping University (LiU), Sweden, in 1986. After an industrial post-doc and several years abroad, he obtained a full professorship in sensor science (2000) and later in molecular physics (2004) also at LiU. He has a long experience in surface vibrational spectroscopy, in particular for the analysis of thin molecular films and self-assembled architectures on solid supports. He has also been the principal investigator for a successful activity in biochemical and chemical sensing at LiU. This work started in early 1980s when he and his colleagues developed the surface plasmon resonance-based detection principle, which today is one of the cornerstones in the Biospecific Interaction Analysis (BIAcore) system advertised by GE Healthcare. He has published more than 270 papers/reviews in peer-reviewed international journals and magazines. Liedberg was in 2005 the recipient of the Distinguished Scientist Award for his contributions in the field of molecular physics and sensor science awarded by the Research Council of Italy (CNR).

Chapter 9

Evgeny (Eugene) Beletskiy received his PhD from the University of Minnesota under the supervision of Prof. Steven Kass investigating novel hydrogen bond catalysts, phosphate ion receptors and an organometallic catalysis methodology. He then studied gold nanoparticle and tin Lewis acid catalysts as a postdoctoral fellow with Prof. Harold Kung at the Northwestern University. He is now working on next generation oxidation catalysts as a Research Chemist at the Scientific Design Company in Little Ferry, New Jersey.

Mayfair C. Kung is a Research Associate Professor in the Chemical and Biological Engineering Department at the Northwestern University. She received her B.Sc. in biochemistry from the University of Wisconsin, Madison, PhD in chemistry from Northwestern University and her post-doctoral training at University of Pennsylvania. Her research interests include selective alkane oxidation, water purification and synthesis of organosilicon compounds as catalytic structures.

Harold H. Kung is a Walter P. Murphy Professor of Chemical and Biological Engineering at the Northwestern University. He received his B.S. from the University of Wisconsin and PhD from Northwestern University, and has been on the editorial team of *Applied Catalysis A: General* since 1996. His research interest focuses on heterogeneous catalysis, but includes energy materials, synthesis of nanostructured materials, global energy supply and consumption, and sustainability. He is the author of *Transition Metal Oxides: Surface Chemistry and Catalysis* (1989, Elsevier Science, holds six patents (one pending), has published over 270 journal articles in catalysis and energy storage and an editor of five monographs. Recently, his work was given the R.H. Wilhelm Award of the American Institute of Chemical Engineers and the Gabor A. Somorjai Award of the American Chemical Society.

Chapter 10

Ewa Kowalska is an Associate Professor and a leader of Research Cluster for Plasmonic Photocatalysis in Institute for Catalysis, Hokkaido University. She received her PhD in chemical technology from Gdansk University of Technology, Poland, in 2004. After completing JSPS fellowship (2005–2007), GCOE post-doctoral fellowships (2007–2009) in Japan and Marie Sklodowska-Curie fellowships in France (2002–2003) and in Germany (2009–2012), she joined Institute for Catalysis as an Associate Professor in 2012. Her current work focuses on heterogeneous photocatalysis, environmental protection, plasmonic nanomaterials and anti-microbial properties on nanomaterials.

Chapter 11

Eric Le Moal is a CNRS junior researcher at the Institute of Molecular Sciences of Orsay (ISMO), France. He received his PhD in Physics in 2007 from Pierre and Marie Curie Paris VI University, France, on the fluorescence enhancement of dye molecules on plasmonic nanostructures. He is a former post-doctoral fellow of the Alexander-von-Humboldt Foundation (2007–2009) in the Organic Films group of M. Sokolowski in Bonn, Germany. Before joining CNRS in 2011, he spent two post-doctoral years at the Fresnel Institute in Marseille, France, where he co-invented new optical microscopy techniques. His research interests include plasmonics, organic semiconductors, surface science and instrumentation in optics. He currently conducts experimental work on electrical nanosources of light and surface plasmons, based on techniques combining scanning tunnelling microscopy with optical microscopy.

Gérald Dujardin is a Directeur de Recherche Emérite at CNRS. He started working on 'Manipulation of single molecules with the STM' at IBM (Yorktown, USA) in 1991 with Phaedon Avouris. He studied molecular nano-machines and their electronic and optical control. His current research interests are focused on the electrical excitation of hybrid plasmon-exciton optical devices.

Elizabeth Boer-Duchemin is a Junior Professor at the University of South Paris and the Orsay Institute of Molecular Sciences (ISMO). She received her PhD in applied physics from the California Institute of Technology in 2001, under the supervision of Harry Atwater. From 2001 to 2003, she was a Research Engineer at Alcatel Opto+, Marcoussis, France, working on high-power semiconductor pump lasers for optical amplifiers, before spending a year at Thales Research and Technology on quantum cascade lasers. Her more recent interests are the development and exploitation of electrical plasmon nanosources and their integration, as well as novel uses of scanning probe microscopies.

Chapter 12

Shamil Shaikhutdinov has received his PhD (1986) in physics at the Moscow Institute of Physics and Technology. Then he joined the Boreskov Institute of Catalysis at Novosibirsk to carry out surface science studies of catalytic systems. In addition, he has been working as a post-doctoral fellow in several research centres in Germany and France. Since 2004, he is leading the group 'Structure and Reactivity' in the Department of Chemical Physics of the Fritz-Haber Institute at Berlin. His research interest is focused on an understanding of the atomic structure and surface chemistry of functional materials.

Chapter 13

Hannu Häkkinen has PhD in physics in 1991 at the University of Jyväskylä, Finland. After PhD, he worked for several years as post-doctoral researcher, senior research scientist and Academy of Finland Research Fellow at Georgia Institute of Technology, Atlanta and in University of Jyväskylä. Since 2007, he is a professor in computational nanoscience in University of Jyväskylä in a joint position at Physics and Chemistry Departments and at the Nanoscience Center. He is currently the Scientific Director of the Jyväskylä University Nanoscience Center and has been nominated as the Academy Professor for 2016–2020 by the Academy of Finland. He leads a group of about 10 researchers focusing on computational studies of electronic, optical, magnetic and catalytic properties of various metal nanoclusters, nanostructures and cluster–bionanoparticle hybrids. His teaching curriculum includes solid state physics, physical chemistry and computer simulation methods. He has co-authored about 200 peer-reviewed articles (http://users.jyu.fi/~hahakkin/).

Chapter 14

Romain Quidant received his PhD in Physics in 2002 from the University of Dijon (France). Since then, he has worked in Barcelona at ICFO in the field of nanoplasmonics. In 2006, he was appointed Junior Professor and group leader of the Plasmon NanoOptics group at ICFO. In 2009, he became ICREA Professor and tenure group leader at ICFO. Quidant carries out his research at Institut de Ciencies Fotoniques in Barcelona (ICFO) — where he leads the plasmon nanooptics group. His research focuses on the study of the optical properties of metal nanostructures, known as *nanoplasmonics*. The activities of his group cover both fundamental and applied research. The fundamental part of his work is mainly directed towards enhanced light–matter interaction for quantum optics. From a more applied viewpoint, his group investigates new strategies to control light and heat at the nanometre scale for biomedical applications, including early detection and hyperthermia.

Chapter 15

Fred Currell is a Professor of Physics at the School of Mathematics and Physics and is the Director of the Centre for Advanced and Interdisciplinary Radiation Research (CAIRR), both part of Queen's University, Belfast. He received his PhD from Manchester University in 1987. His research involves joint experimental and modelling approaches to understand interactions of radiation and matter. These studies range from fundamental physics and chemistry studies to health care applications (predominantly cancer care). He is the editor of *Cancer Nanotechnology*, and open access Springer journal.

Balder Villagomez-Bernabe is a Research Fellow at the School of Mathematics and Physics at Queen's University, Belfast, since 2014. He received his PhD from the National Optics Institute in Puebla City, Mexico in 2013. His PhD involved a stay at the Stanford Linear Accelerator in USA in order to implement Monte Carlo simulations of the Raman scattering process for optical photons. His current research comprises the performance of Monte Carlo simulations to calculate the dose enhancement in tumour cells by the use of high-density nanoparticles as radiosensitisers during radiotherapy cancer treatments.

Chapter 16

Christian Villiers is a Member of the Research Institute INSERM (National Institute for Health and Medical Research) and Vice Director of the Institute for Advanced Biosciences located in Grenoble (France). He graduated in biochemistry from a high-tech Education Establishment (INSA — Lyon, France) and received his PhD in biology and immunology from the University of Grenoble in 1984. He worked as a post-doctoral fellow at the MRC Centre of Cambridge (UK) in 1985. His current interest relates to the modification of the immune response resulting from inflammation with a special focus on the assessment of potential interferences of nanoparticles with the cellular behaviour and the immune system. He is the author of more than 90 original articles.

Chapter 17

Marie Carrière is a Senior Research Scientist at the Atomic Energy Commission (CEA) in Grenoble, France. She joined the Nucleic Acid Lesions Laboratory in the CEA Nanoscience and Cryogeny Institute (INAC) in 2011 after studying metal and nanoparticle toxicity in Laboratoire Pierre Süe, CEA Saclay, for seven years. She received her PhD from the National Institute of Agronomics at Paris in February 2002 for studying the efficiency, cell distribution and metabolisation of lipidic vectors developed for gene therapy applications. She then did post-doctoral work at CEA Saclay centred on the study of toxicological impact of heavy metals on cultured animal and human cell lines. Her current research interests are toxicology, ecotoxicology and bioavailabillity of metal and metal oxide nanoparticles as well as carbon nanotubes. She also participated in the development of innovative therapeutic and diagnostic approaches using nanoparticles.

Chapter 18

Michael Cortie is a Professor of Nanotechnology and Director of the Institute for Nanoscale Technology at the University of Technology, Sydney, in Australia. He has a BSc (Engineering) degree in physical metallurgy from the University of the Witwatersrand, South Africa (1978), a Masters in Engineering degree on from the University of Pretoria, South Africa (1983), and a PhD on metal fatigue at high temperatures from the University of the Witwatersrand (1987). He joined Mintek, the South African National Minerals and Metals Research Organisation in 1987. He was a Senior Engineer there before becoming head of its Physical Metallurgy Division between 1997 and 2002. While at Mintek, he consulted widely to the international precious metals industry in the areas of nanotechnology, catalysis and physical metallurgy. He relocated to the University of Technology Sydney in July 2002. Michael's main research interests are the nanoparticles and intermetallic compounds of the precious metals, especially as these relate to optical properties.

Chapter 1
Gold Nanoparticles in the Past: Before the Nanotechnology Era

Catherine Louis

Laboratoire de Réactivité de Surface, UPMC-CNRS, Paris, France

1.1 The First Usage of Gold

The role played by gold in history relies on its outstanding qualities among metals, making it exceptionally valuable from the earliest civilisations until the present day. As quoted by Auric Goldfinger in a James Bond movie, gold is attractive due to 'its brilliance, its colour, its divine heaviness', and also due to its incorruptibility and scarcity. Its great malleability makes gold one of the easiest metal to work with. Moreover, it often occurs naturally in a fairly pure state.

The first uses of gold were linked to deities and royalty in early civilisations. The word 'gold' exists in all old languages, often connected with the image of the Sun, with light and life-giving warmth, growth and hence power. In cultures like ancient Egypt, which deified the Sun, gold represented its earthly form. In fact, nothing has changed throughout history, and the same thinking about gold exists (golden crown of the kings, gold medals, wedding rings, cult objects, gold ingots).

1.1.1 *Quest for Gold and Gold Production*

The earliest signs of crude metallurgy occurred sometime from 9000 to 7000 BCE (before the Common Era). For instance, in Ali Kosh in Iran and Cayönü Tepesi, which is close to Ergani in Anatoly, humans first began using native copper and gold, meteoric iron, silver and tin to create tools and possibly jewellery ornamentation. Gold was most probably discovered as shining, yellow nuggets. Although it can be easily worked with because of its ductility, it is not clear whether it was worked with before copper.[a]

It is known that the Egyptians mined gold before 3500 BCE in the eastern desert of Egypt and in Nubia.[1] The *Turin Papyrus* drawn during the reign of Ramesses IV (1151–1145 BCE) is the earliest known topographic and geological map.[2] Along with specifics of the geology and topography, it shows an ancient gold-working settlement, gold-bearing quartz veins in Wadi Hammamat, a dry river bed in Egypt's eastern desert. Large mines were also present across the Red Sea in present Saudi Arabia. By 325 BCE, the Greeks had mined in areas from Gibraltar to Asia Minor and Egypt. The Romans mined gold extensively throughout the empire, developing the technology of mining to new levels of sophistication. For example, they would divert streams of water in order to mine hydraulically, and even pioneered 'roasting', the technique of separating gold from rock.

Occasional passages on mining and metallurgy of metals can be found in the works of Theophrastus (Greek, 372–288 BCE), Vitruvius (Roman 90–20 BCE), Strabo (63/64 BCE–c. 24 CE), Pliny the Elder (Roman, 23–79 CE) and Discorides (Greek, 40–90 CE). One important surviving document is the *Leyden Papyrus X* of the Museum of Antiquities in the Netherlands: it is the working notebook of a goldsmith and jeweller, probably written in the early years of the fourth century. It gathers 111 recipes of refining, alloying and working of gold; some of them are reported in Hunt's paper.[3] Details on the first techniques of gold metallurgy and gold thin film technology can be found in Greene's paper.[4]

Another important date for the history of gold is 1492, with the discovery of America and the beginning of massive expeditions and exploration with the quest for the El Dorado, and the encounter with Native American people of Central America and South America and their extensive displays of gold ornaments. The Aztecs regarded gold as literally the product of the gods, calling it *'the sweat of the sun'*.

Two hundred years later, in 1700, gold was discovered in Minas Gerais in Brazil, which became the largest producer by 1720, responsible for nearly

---

[a]One can read in some websites that the earliest traces of gold dated back to the Paleolithic period 40,000–10,000 BC and were found in Spanish caves of Maltravieso; this is wrong according to Dr Antoni Canals y Salomó (Universidad de Tarragona), a paleontolongist, and specialist of this cave.

two-thirds of the world's gold output, but the production was in rapid decline by 1760. 1799 is the year of the first discovery of gold in the United States when a 17-pound nugget was found in North Carolina. For the next 25 years, North Carolina supplied all the domestic gold coined for currency by the US. In 1848, John Marshall found flakes of gold near Sacramento in California, triggering the California Gold Rush. In 1850, E. H. Hargraves, returning to Australia from California, found gold in his home country within a week. 1868 saw the next major discovery, in South Africa, where G. Harrison uncovered gold while digging up stones to build a house, and in 1898, South Africa became the world's top gold producer with a quarter of the world production until 2006.

Up to now, it is estimated that a total of 161,000 tonnes of gold have been mined in human history; this corresponds to the volume of a single cube 20 m on a side (equivalent to 8000 m^3). Seventy-five per cent of all gold ever produced has been extracted since 1910. The typical annual production in recent years has been around 2800 tonnes per year (2860 tonnes in 2014). In 2014, the largest producers were China (15.7%), then Australia (9.4%), Russia (8.6%), United States (5.6%), Canada (5.5%), Peru and South Africa (5.2% each). India and China are the world's largest consumers of gold (1066 and 975 tonnes of gold in 2013, respectively), and the largest importer.

1.1.2 *The First Gold Jewels and Artefacts*

The most ancient gold artefacts were found in necropolis, but not in Mesopotamia or Egypt as is often believed. The history of gold starts long before the invention of writing and the establishment of the first cities of Mesopotamia and Egypt (circa 2800 BCE). It starts around 4500 BCE with 'Old Europe' civilisation in south-eastern Europe that was at that time among the most sophisticated and technologically advanced regions in the world. A necropolis with 294 graves dating to 4600–4200 BCE was discovered in 1972 in Varna on the Black Sea coast, which is located in modern-day Bulgaria. The graves contained some 300 objects made of pure gold, sceptres, axes, bracelets, other decorative pieces and bull-shaped plates. These objects attest to the high-level skill of goldsmithing. They can be seen at the Varna Archaeological Museum and at the National Historical Museum in Sofia.[5]

Three important discoveries of gold artefacts were found in tombs dated to circa 2500 BCE in three different geographical areas:

— The tomb of Djer at Abydos in Egypt. He was probably the third king of the First Dynasty (circa 2800 BCE). Although the tomb had been robbed, a human arm was discovered near the entrance, still wearing four golden bracelets.[b]

— The tomb of Queen Pu-Abi in southern Iraq. She was an important figure who lived around 2600–2500 BCE, during the First Dynasty of Ur of the Sumer civilisation. Among other excavations of the *Royal Cemetery of Ur*, by Sir Leonard Woolley between 1922 and 1934 it was discovered that, her tomb had been untouched by looters. It revealed several gold ornaments and a profusion of gold tablewares, golden beads for necklaces and belts and golden rings and bracelets.[c] The treasure was split between the British Museum in London, the Penn State Museum in Philadelphia and the National Museum in Baghdad.

— The so-called *Gold of Troy* treasure hoard, also called the *Treasure of Priam* by Heinrich Schliemann who excavated it in 1873, is on the ancient site of Troy in the area of the city of Çanakkale in Turkey. Dated to 2600–2450 BCE (i.e. 1000 years before the Trojan war!), it showed a range of gold-work from jewellery to a gold 'gravy boat' weighing 600 g.[d] Most of the treasure, which was first in Berlin, is now in the Pushkin Museum in Moscow.

A millennium later (1200 BCE), probably the most famous hoard of gold was found in the tomb of Tutankhamun in Egypt (1333–1324 BCE). It contained the largest discovered collection of gold and jewellery, including a gold coffin. At the same period, pre-Columbian goldsmiths started producing gold items in South America. Their art reached its zenith during the Moche civilisation in the northeast part of Peru between the first and eighth centuries.[e]

---

[b]Images visible on website images: four golden bracelets tomb of Djer.
[c]Images visible on website images: gold Queen Pu-Abi.
[d]Images visible on website images: gold of Troy gravy boat.
[e]Images visible on website images: gold Moche artifacts.

1.1.3 *Gold for Monetary Exchanges and the Gold Standard*

Gold has also been widely used throughout the world, as a vehicle for monetary exchange, even before the establishment of a gold standard, a monetary system in which the standard economic unit of account is a fixed weight of gold.

Egyptian Pharaohs began to commission gold tokens around 2700 BCE, but these tokens of variable purity were used as gifts, not for commerce. Much later, circa 600 BCE, the first gold coins known were minted by King Alyattes in Lydia (present-day Turkey). As a matter of fact, they were made of electrum, a natural alloy of gold and silver arising from alluvial deposits of the river running through Sardis, the Lydian capital. At the same period, 600–500 BCE, another gold coin, the Ying Yuan, was used in the kingdom of Chu in China.

Gold and silver coins were used in most of the great ancient empires, but after the ending of the Byzantine Empire in the fifteenth century, the occidental world tended to use silver coins. Paper money was first introduced in China between the seventh and fifteenth centuries, and much later in Europe in the seventeenth century. Initially, it was a promissory note, i.e. a receipt redeemable for gold and/or silver coins. In 1816, England ended its policy of bimetallic standard (gold and silver) and adopted a single gold standard while the rest of Europe remained on a silver or bimetallic standard. Between 1872 and 1900, most other major countries abandoned silver or bimetallic systems and achieved gold convertibility. At the beginning of the First World War, the gold standard was at its pinnacle, with 59 countries having adopted this standard.

However, during the First World War, governments had to face the huge war effort and boosted banknote printing, while international trade dropped dramatically. At the end of the war, all the countries had left the gold standard. However, England returned to the gold standard from 1925 to 1931, and France was the last country to abandon the convertibility in 1936. After the Second World War, the Bretton Woods Agreements (22 July 1944) created a system of fixed exchange rates, and gold was replaced by the US dollar. Nevertheless, nowadays, gold remains a safe investment.

1.1.4 *Gold for Human Well-being: Food, Drinks and Medicine*

Pure metallic gold is non-toxic and non-irritating when it is ingested. Metallic gold has been approved as a food additive by the EU (E175 in the *Codex Alimentarius*). Gold leaves are sometimes used as food decoration (for instance in France on *'palet d'or'* chocolate) and as a component of alcoholic drinks, such as *Goldschläger*, *Gold Strike* and *Goldwasser*.

Since the discovery of gold, people have thought of it as having an immortal nature and have associated it with longevity, probably because of its resistance to chemical corrosion. Many ancient cultures, such as those in India and Egypt, used gold in medicine but for its magico-religious power. An exception is China with the earliest application of gold as a therapeutic agent back in 2500 BCE. Much later, Pliny the elder, in the first century, reported that gold can be used to heal fistulas and haemorrhoids. The uses of gold were limited because at that time, people did not know how to dissolve it and make it soluble. Success in this was achieved in the eighth century by the Persian alchemist, Jābir ibn Hayyān (721–815), sometimes identified as 'Geber' in European historical records, who produced gold chloride by dissolving gold in *aqua regia*, a mixture of hydrochloric and nitric acids. The know-how was transmitted to the European alchemists[f] (two manuscripts from the ninth and tenth centuries still exist), and it is with the medieval period that gold became a prominent medicinal element, with the idea that the elixir of life, *Aurum potabile*, can restore youth. *Aurum potabile* was therefore closely related with the use of *aqua regia*, the 'royal' solvent of gold. Recently, a group of Spanish researchers has been able to reproduce a recipe of Aurum potabile of the eighteenth century and to probe the presence of gold nanoparticles in the solution.[6] A gold cordial was advocated in the seventeenth century for the treatment of ailments caused by a decrease in the vital spirits, such as melancholy, fainting, fevers and falling sickness. Later in the nineteenth century, a mixture of gold chloride and sodium chloride was used to treat syphilis.

The use of gold compounds in modern medicine began with the discovery in 1890 by the German bacteriologist Robert Koch that gold cyanide

---

[f]Note that it is during the nineteenth century that a distinction is made between alchemists and chemists.

K[Au(CN)$_2$] was bacteriostatic towards the tubercle bacillus. Gold therapy for tuberculosis was subsequently introduced in the 1920s, but soon proved to be ineffective. In contrast, gold therapy proved to be effective against rheumatoid arthritis. Since that time gold drugs have also been used to treat a variety of other rheumatic diseases such as juvenile arthritis, palindromic rheumatism and various inflammatory skin disorders such as pemphigus, urticaria and psoriasis.

Today, in allopathic medicine, in addition to organogold compounds, only salts and radioisotopes of gold are of pharmacological value, as metallic gold is inert. Some injectable gold salts do have anti-inflammatory properties and are used as pharmaceuticals in the treatment of rheumatoid arthritis. However, some forms of alternative or traditional medicine assign metallic gold a healing power. The ayurvedic medicine in India dated back to thousands of years and related to the medical use of metals and minerals, involves metallic gold in such medicines. For instance, *Swarna Bhasma*, comprises globular gold particles with an average size of about 60 nm. Gold is considered to be a rejuvenator and is taken as such by millions of Indians each year. A typical daily dose corresponds to one or two milligrams of gold incorporated into a mixture of herbs.

Metallic gold shows also a renewed potential in 'modern' medicine as nanoparticles for imaging, diagnostics, drug delivery or radiotherapy (Chapters 14–16). The malleability and resistance to corrosion make gold perfect for dental use, although its softness requires that it is alloyed, most commonly with platinum, silver or copper. So gold in alloys is also used in tooth restorations, such as crowns and permanent bridges. There are examples of its use by the Phoenicians, the Etruscans and the Romans for restoration and also for aesthetics reasons.

For more information on gold in medicine in the past and now, the reader can refer to Refs. 7–16 from which most of the information above has been drawn.

1.1.5 *Gilding Gold and Gold-like Lustre*

The use of gilded films of gold on oxide substrates to decorate glass, ceramic and mosaics may be dated from the Roman period circa first century, as reported by Pliny the Elder, but wider use of it dates from the twelfth century. The first composite materials fabricated were gold leaf *tesserae*, the small squares used as building blocks for mosaics. They were obtained by cutting 'cakes' consisting of three layers fixed at high temperatures: a layer of poured glass (the support, less than 10 mm thick), a gold leaf (beaten down to a thickness of 0.3–1.0 μm), and a thin layer of blown glass (less than 1 mm thick). The earliest examples of gold leaf *tesserae* date back to the first century CE in Rome, in the *Nymphaeum of Lucullus* and the *Domus Aurea*.[17] The peak of production of gilded gold films was in the thirteenth and fourteenth centuries with the Mamelouk production in Egypt and Syria, and also in the nineteenth century.

Gilded gold films must be distinguished from lustre of ceramics, which is a surface layer with a metallic appearance applied on glazed ceramics, i.e. on a surface of terracotta covered by a glassy layer. Lustre exhibits various colours, from gold to brown or red. However, in spite of the appearance, it does not contain any gold, but contains silver and copper metal particles in various sizes and proportions, dispersed in a glassy matrix with a gradient of size and concentration.[18–21]

1.2 The First Uses of Gold Nanoparticles

1.2.1 *Introduction*

The first use of gold nanoparticles is intimately related to the history of red-coloured glass. The production of red glass (opaque) started with the beginning of glassmaking in Egypt and Mesopotamia back in 1400–1300 BCE.[22] The colour of this red glass was given by the addition of copper. The origin of the red colour is debated, with some scientists stating that it is due to metal copper nanoparticles, while others state that it is due to cuprous oxide (cuprite) nanoparticles or some say it is due to both. The origin of the colouration depends on the sites and dates of production, the method of preparation and components of glass.[23] The production of copper red glass was a real challenge from a technological point of view because it requires a reducing atmosphere; for this reason, red glasses are less frequent than other colours.

Another way of making red glass involves the use of gold nanoparticles. According to most of the textbooks and technical encyclopaedias on gold, glass and ceramics, the production of the so-called gold ruby glass did not take place until the end of the seventeenth century. The discovery is attributed to Johann Kunckel (circa 1637–1703, Brandenburg), which is related to the former discovery by Andreas Cassius of Leyden in 1685 of the so-called *Purple of Cassius*, which is a gold preparation that is added to melted glass.[24] This preparation is a precipitate obtained from the dissolution of gold metal in *aqua regia* followed by the precipitation of metallic gold nanoparticles by a mixture of stannous and stannic chloride (see Section 1.3.2).

As a matter of fact, the history of gold ruby glass and of gold nanoparticles begins long before, without interruption until the peak of their production at the end of the seventeenth century.

1.2.2 *The Lycurgus Cup*

Hence, the first milestone in the history of gold ruby glass is a Roman opaque glass cup dated to the fourth century, the *Lycurgus cup*, which is exhibited at the British Museum in London (Figure 1.1).[25] The carved decoration

(a)   (b)

Figure 1.1   The Lycurgus cup, late Roman, fourth century CE, probably made in Rome (from the British Museum free image service). (a) Illuminated from outside and (b) illuminated from inside.

depicts the triumph of Dionysus over Lycurgus, a king of the Thracians (circa 800 BCE): One of Dionysus 'maenads, Ambrosia, transformed into a vine by Mother Earth, holds Lycurgus captive while Dionysus instructs his followers to kill him.

This cup shows a green jade colour due to the diffusion of light when it is illuminated from outside (Figure 1.1(a)) and a deep ruby-red one in transmission when it is illuminated from inside (Figure 1.1(b)) (see also Section 1.3.1). A detailed analysis of the *Lycurgus Cup*, first published in 1965 by Brill,[26] revealed the presence of minute amounts of gold (about 40 ppm) and silver (about 300 ppm) in glass. In 1980, a further analysis by Barber and Freestone.[27] attested the presence of nanoparticles of 50–100 nm in diameter by electron microscopy, composed silver–gold alloy, with a ratio of silver to gold of about 70:30. Later on, Hornyak *et al.*[28] confirmed through a theoretical study that the deep red colour of the Lycurgus cup due to light absorption being around 515 nm, is consistent with the presence of silver–gold alloy with Ag:Au of 70:30. The British Museum experts believe that the colouring of glass using gold and silver was far from routine during the Roman period since only a limited number of other Roman-period glasses appear to have been coloured by gold.[29] Moreover, no other glass of this period replicates the dichroic optical effect of the Lycurgus cup. They conclude that the technology seems to have been very restricted and did not outlast the fourth century.

However, a recent study by Verità and Santopadre[30] reports the chemical analyses of nine flesh-tone glass *tesserae* of mosaics, arising from nine important churches in Rome of the fourth to twelfth centuries. All of them reveal that the flesh colour originates from the presence of 10–30 ppm of gold or gold–silver alloy particles. Since a considerable number of flesh-coloured glass *tesserae* were employed in mosaics of these churches, the authors conclude that the colour was obtained routinely rather than by chance, and that the Roman glassmakers mastered this complex colouration process. Since there is no evidence that the Romans were able to produce *aqua regia* to prepare gold chloride during that period, the authors propose that the Roman glassmakers may have used silver slags without knowing that they also contained gold, thereby not knowing that gold was the actual colourant of glass; they also propose that the colour arises from the local dissolution of gold leaves and the formation of 'droplets' of gold ruby glass

(a)

(b)

(c)

Figure 1.2   Detail of a mosaic in basilica Saint Pudenziana in Rome (fourth century) (a), showing flesh-coloured tesserae (b); gold-foil tessera of a mosaic at Saint Sabina in Rome showing characteristic ruby 'droplets' at the glass–gold interface (c); from Ref. 30 Photo: M. Verità and P. Santopadre, All rights reserved.

since these droplets are commonly found in the gold-foil *tesserae* of Roman mosaics (Figure 1.2). Very recently, Verità *et al.* have observed experimentally that the dissolution of gold leaf and the formation of gold ruby droplets occurred only on mosaic tesserae whose glass was decolourised with antimony and not in those made with glass decolourised with manganese.[30]

1.2.3 *Medieval Period*

There are written evidences that the alchemists of the Middle Ages knew how to produce red coloured glass with gold, although samples of such glass are yet to be found.[29] It is important to mention that textbooks and websites as well as many introductions of scientific talks on gold nanoparticles very often state that the red colour of stained glasses in medieval church windows is due to the presence of gold nanoparticles. This belief propagated by Pierre Le Vieil in 1774 was at the origin of a proposition of the Convention (1792–1795) during the French Revolution, to melt the stained glasses of French churches to extract gold. Fortunately, Jean Darcet, chemist and physicist, in charge of chemical analyses, found copper and not gold.[31] This is how the French stained glasses have been preserved. Moreover, in all cases analysed so far, the dye found to be responsible for the red colour of stained glasses of churches is copper-based and contains no gold[32]: 132 examples of medieval red window glass covering the period from the twelfth to the sixteenth centuries have been examined, most of them originating from English churches, but also from windows in France, Germany, the Netherlands and Spain; none of them contain gold.

According to a paper by Sheybany from 1967,[33] the earliest known written account of a gold ruby glass comes from a treatise *Secrets of Secrets* written by Al Razi (865–925), a Persian scholar, philosopher and alchemist. The instruction is to heat a very finely powdered batch of different elements including gold powder for three days in a closed furnace fuelled with very hard wood. According to Sheybany, this may allow temperatures of 800–1000°C to be reached in a reducing atmosphere. Al Razi believed he had fulfilled the objective of the transmutation of metals; in his treatise, he stated that this glass attracted gold and silver like a magnet and that it could convert 1000 times its weight into gold.[33]

The main goal of the medieval alchemists was the making of the philosopher's stone. In alchemical writings, the philosopher's stone is often described as a red substance, which is supposed to be the key to transmutation of 'impure' base metals into gold, the unique pure metal.

1.2.4 *Fifteenth and Sixteenth Centuries*

In the Bologna manuscript *Segreti per colori* written in the first half of the fifteenth century, three recipes of gold ruby glass are described. However, according to Zecchin's paper,[34] they are inconsistent. Later on, between 1458 and 1464, Antonio Averlino, also called Filarete, provided some technical information on glass colouration in his *Trattato di Architettura*, and wrote: 'It is also said that gold makes colour'.[34]

Georgius Agricola (1494–1555, Saxony), who is considered the founder of geology, is supposed to have described the preparation of gold ruby glass in *De natura fossilium* published in 1546[24,35]: 'A famous variety of dyeing glass is made from gold and this is used to tint the glass clear ruby red'. As a matter of fact, according to Zecchin[34] and von Kerssenbrock-Krosigk,[36] this sentence is wrong and results from a mistake in the first translation from Latin to English. However, there are several other writings that refer to gold ruby glass during the sixteenth century. Benvenuto Cellini (1500–1571), a famous sculptor and goldsmith in Florence, referred to a transparent red enamel discovered by an alchemist who was also a goldsmith.[37] Later, Andreas Libavius (1540–1616), a German chemist and physician, mentioned the red colour of gold dissolved in a liquid to make red crystal in *Alchemia* published in 1597. According to Polak,[38] Andreas Libavius based himself on two earlier 'distillers', the Neopolitan Giambattista Porta (1535–1615), author of *Magiae Naturalis* (1588) and Gerhard Dorn (ca. 1530–1584), the German author of *Clavis Totius philosophiae chymistica* (1567).

1.2.5 *Seventeenth Century*

L'Arte Vetraria is the first print book exclusively devoted to glassmaking. It was published in 1612 by Antonio Neri (1576–1614), a Florentine priest, son of a physician. In Book 7, Chapter 129, one recipe mentions the use of gold to produce red glass. In short, the recipe, which is entirely reported in Franck's paper,[35] involves the calcination of gold with *aqua regia* in a furnace, which forms a red powder, which is then added to glass. The recipe attests that the potential of using gold as a red colourant was fully understood in early seventeenth century.[39] However, the only known gold ruby vessels of Italian origin of that period are a series of ribbed bowls, ewers and bottles that King Frederick IV of Denmark brought back from a trip to Venice in 1708–1709. These artefacts are visible in the Rosenborg castle at Copenhagen.

The Antonio Neri's book was then translated into English in 1662 by Christopher Merrett (1614/5–1695); he added 147 pages of his own, from other authors and his own observations. In 1679, the first German edition of the Neri–Merrett book appeared, translated with further extensive addition by the famous Johann Kunckel (see Section 1.2.1) under the title *Ars Vitraria Experimentalis*.

Other written sources were recently found by Zecchin in Murano archives.[34] A manuscript written by Giovanni Darduin (1585–1654), a glassmaker of Murano, provides a recipe of gold ruby glass among other glass recipes of his and of his father who died in 1599. Two other recipes of gold ruby glass were provided by Giusto Darduin (1661–1700) and one by Antonio dalla Rivetta (1628–1695). Zecchin could not establish the existence of a relationship between the Italian branch and the German one and Kunckel.[34] However, he suggests that a relationship may have existed with Bernard Perrot in France (see Section 1.2.5.3).

1.2.5.1 *Purple of Cassius*

As mentioned in Section 1.2.1, the paternity of the purple gold precipitate used for colouring glass, the so-called *Purple of Cassius*, has been attributed to Andrea Cassius. As described earlier, the preparation involves gold being dissolved in *aqua regia*, then its precipitation as metallic gold nanoparticles by a mixture of stannic and stannous chloride.

As a matter of fact, there were two Andreas Cassiuses, father (born circa 1605 in Schleswig and died in 1673 in Hamburg) and son (born in 1645 in Hamburg circa 1700 in Lübeck), both of whom were physicians. The son wrote *De Auro* published in 1685, in which he gave his father's recipe of the *Purple of Cassius*, obtained by reducing a gold(III) chloride aqueous solution with stannous chloride; the entire translation of the recipe can be found in Hunt's paper.[24] In a short book published in 1684 *Sole Sine Vest* (*Gold Unclothed*), Johann Christian Orschall, who was a metallurgist and also interested in gold ruby glass, reported the anecdote that Cassius, the son, succeeded in making a very fine ruby flux and sold the secret in various places.[24] On the other hand, Cassius, the son, was aware that the formula of the preparation had been used before his father and that he may have been influenced by the work of Johann Rudolf Glauber. Johann Kunckel also mentioned that Cassius was not the true inventor of the *Purple of Cassius*, and that perhaps Glauber may have given him the idea.

Johann Rudolf Glauber (1604–1670), a native of Bavaria who settled in Amsterdam, was a pharmacist, living off the sales of his medicinal preparations, which was exceptional at this time. His writing in Part IV of *Prosperitatis Germaniae* published in 1659, i.e. a quarter of a century before the publication of Cassius, is considered as the first report that mentions that gold can be precipitated with a solution of tin compound. However, there is no evidence that Glauber made use of the purple precipitate for colouring glass.[24]

It is important to stress that the seventeenth century is still a period in which alchemists were obsessed not only with attempts to unlock the secrets of nature by simulating natural processes in laboratory conditions, but also with attempts to manufacture metals for mystical purposes. They believed that the colour of metals indicated their 'souls' or essence, and that if the colour could be extracted, the extract would possess the spirit of the metal and could perform alchemical transmutation. Great scientists such as Robert

Boyle (1627–1691) and Isaac Newton (1642–1727) firmly believed in this principle. Alchemists also invested considerable efforts in making glass imitations of gemstones, and new methods of colouring glass and mixing batches were invented.[40,41] Coming back to Glauber, although he can be regarded as one of the founders of the chemical industry, he also related the production of gold ruby glass to alchemy. He claimed that the soul of gold is captured in the red colour of gold ruby glass, and he regarded the making of gold ruby glass as akin to the process of alchemical transmutation, in the sense that the substance turned red before it was transformed into gold. He also believed that this was a demonstration of the multiplication of gold, because only a small amount of gold was required to colour a large amount of glass.[40,41]

1.2.5.2 *Kunckel glass*

As already mentioned in Section 1.2.1, it is widely reported in textbooks that Johann Kunckel (1637–1703) is the first important maker of gold ruby glass. If Neri and his predecessors had managed to produce gold ruby glass in small quantities, maybe for the purpose of imitating natural stone, Kunckel is recognised as the first glassmaker to have been successful in producing gold ruby glass on a rather large scale. He was the son of an alchemist glassmaker, and himself was first an alchemist and apothecary; he taught at the University of Wittenberg in Saxony for about ten years, then moved to Postdam in Brandenburg in 1678, where the great Elector, Friedrich Wilhelm, commissioned him to take charge of a glass factory. He started developing the production of gold ruby glass vessels by 1684, but it is not known if or for how long he continued to work on it after the death of the great Elector in 1688. Four remaining vessels produced in Brandenburg before 1700 correspond to the period when Kunckel might have been the glassmaker, but none of them can be unambiguously attributed to him.[39] One feature of these vessels is the amount of cur decoration as if they would have sculpted them out of solid stone (Figure 1.3(a)).

From *Ars Vitraria Experimentalis* published in 1679, it is clear that Kunckel was unwilling to describe his recipe of gold ruby glass. Moreover, his factory was located at an isolated site, the Pfaucninsel, or Peacock island, between Berlin and Potsdam. His secret of fabrication was revealed

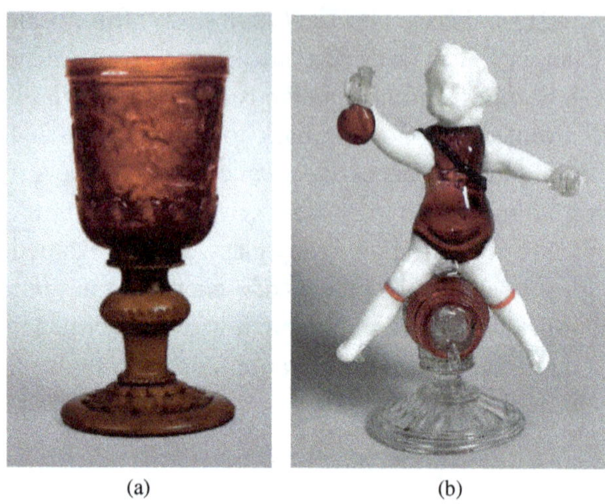

Figure 1.3   (a) Goblet, Germany, Potsdam, engraved in manner of Gottfried Spiller, about 1700, gold ruby glass; blown, cut, engraved. H. 24.1 cm. Collection of The Corning Museum of Glass, Corning, NY, gift of The Ruth Bryan Strauss Memorial Foundation (79.3.258). With permission of Corning Museum of Glass, Corning, NY, USA. (b) Bernard Perrot, end of seventeenth/beginning of eighteenth century, Orleans, Element of a table centrepiece showing child Bacchus riding a barrel, gold ruby glass and porcelain glass. H. 18.5 cm. Photo: Musée de la Céramique, Rouen. All rights reserved.

later in *Laboratorium Chymicum* published posthumously in 1716. How Kunckel managed to produce gold ruby glass on a rather large scale is still an enigma.[39] It is known that Daniel Crafft (1624–1697), who had worked as Glauber's assistant glassmaker, worked with Johann Kunckel in Dresden after 1673[41] and that Kunckel had known Antonio Neri's book *L'Arte Vetraria* (1612), since he translated it into German and published it in 1679.

According to von Kerssenbrock-Krosigk,[39] enthusiasm for gold ruby glass was at its peak in Europe between 1685 and 1705, and almost every central European sovereign owned such glass vessels. At that time, gold ruby glass was considered as a genuinely new material and a decorative folly; a way of imitating semiprecious gemstone, like crystal whose fabrication had been discovered at the same period.

1.2.5.3 *Perrot glass*

What is not reported in the previous mentioned textbooks is that 16 years before Kunckel started producing gold ruby glass, Bernard Perrot was producing glass artefacts containing gold ruby glass in France

(Figure 1.3(b)).[42] Bernardo Perrotto (1640–1709), an Italian glassmaker from Altare (Ligury), opened a glass workshop in Orleans (France), and became Bernard Perrot. In 1668 he obtained the royal privilege from Louis XIV to colour glass in red. An exhibition dedicated to his glass work was held in Orleans in 2010. Chemical analysis of various samples of glass revealed his gold ruby glass was not produced from the *Purple of Cassius*; no tin but arsenic (0.6–2.9 wt.%) was identified in the presence of 23–284 ppm gold.[43] It is not certain whether Bernard Perrot produced gold ruby glass himself; it is possible that the glass or the recipe arose from Galaup de Chastueil, an alchemist who later was involved in the famous *affair of the poisons*![44] It is also possible that gold ruby glass came from the Italian branch, because the recipes provided by Giusto Darduin and Antonio dalla Rivetta (cited at the beginning of Section 1.2.5) also involve arsenic, and that according to Zecchin,[34] a relationship between the two may have been established.

1.2.6 *Gold Ruby Glass in the Eighteenth Century*

It is also widely reported in textbooks that the art of making gold ruby glass was lost on Kunckel's death and rediscovered during the nineteenth century. This is illustrated by the Werner Herzog's movie dating to 1976, *Heart of Glass* (in German: *Herz aus Glas*): in a Bavarian village in the late eighteenth century, a glassmaker dies and takes to his grave the secret of his ruby glass. The glass factory owner goes mad trying to unearth the formula. As he goes mad, so does the village.

During the eighteenth century, even though the peak interest for gold ruby glass was over in Germany, it persisted in some regions and occasionally arose in others.[39] In Brandenburg, gold ruby glass continued to be produced, but the kings of Prussia and other statesmen in Germany favoured the hard-paste porcelain. The alchemist Johann Friedrich Böttger (1682–1719), who first became famous for producing gold 'transmutation', played a crucial role in the discovery of the hard porcelain in Europe and in the development of the first porcelain manufacture in 1707–1709 in Meissen in Saxony. Around 1713 he also experimented with gold ruby glass.

Already during Kunckel's lifetime, knowledge on gold ruby glass spread especially to Bavaria and Bohemia.[35,45] A connection between Kunckel and

Hans Christoph Fidler (1677–1702), the crystal maker of the electorate of Bavaria, who experimented with gold ruby glass in 1686–1688 at the Zákupy glasswork in Northern Bohemia, is well-documented. The production started in 1683 in Southern Bohemia and around 1700 in Silesia.[45] Several rare pieces dated from the very beginning of the eighteenth century are preserved at the Museum of Applied Art in Prague. The use of gold ruby for windows is also reported in the records of William Peckitt, a leading glassmaker and producer of stained glass during the eighteenth century in Yorkshire.[35] The interest in gold ruby glass throughout the eighteenth century is also attested by the incorporation of part of Neri, Merrett and Kunckel's books in major publications of different languages, such as the Encyclopedia Britannica.[35]

1.2.7 *Gold Ruby Glass and Cranberry Glass in the Nineteenth Century*

Hence, the seventeenth and eighteenth centuries laid the foundations for the great practical and theoretical interest in coloured glass, which took place during the nineteenth century.[35]

In the first half of the nineteenth century during the Biedermeier period (1820–1850), gold ruby glass was mostly produced in Bohemia. The fashion for these glass artefacts swept across Europe and to the United States, first with exports from Bohemia then with development of local production.

For instance, in France, a competition was organised in 1837 to find a more reliable process of colouration. The production of gold ruby glass reached its peak in England during the Victorian era (1837–1901), particularly around Stourbridge (West Midlands) and Bristol. Molineaux Webb & Co. (1826–1928) used gold ruby glass to produce window and vessel glass on quite an extensive scale. Several English companies and Baccarat, a French one, exhibited pieces of ruby glass at the first world's fair of 1851. The colour came in a variety of less saturated tints, close to pale pink, called *cranberry*, that was obtained by decreasing the gold concentration in glass. In the US, gold was also used to produce new types of glass, burmese and rose amber glasses. These are opaque glasses, ranging from yellow to pink, obtained from uranium oxide (that gives a soft yellow colour) and from gold (that gives the pink blush).

1.2.8 *Pink Enamel Porcelain: Rose Pompadour and Famille Rose*

Although each material was known from 3000 BCE, enamel combination of glass and metal was not found until the twelfth century BCE, when the Myceneans succeeded in making enamels on a gold base.[46] According to Garner's paper,[46] there is no doubt that practical means of making gold ruby enamels were known from the time of Benvenuto Cellini (1500–1571), i.e. earlier than Cassius and Glauber (see Section 1.2.4), and possibly even earlier still. However, painted enamels with a good deal of pink did not appear before 1667 in south Germany and before the end of the seventeenth century in France.

By 1719, pink enamel was prepared and used in the first porcelain factory to be established in Europe, at Meissen (see Section 1.2.6), using the *Purple of Cassius*.[47] In 1738, the Vincennes porcelain workshop in France, which became the well-known *Manufacture Royale de Sèvres* in 1756 under King Louis XV, produced colours from bright red to violet from gold-based purples.[48] The chemist Jean Hellot succeeded in producing the so-called *Rose Pompadour* in 1757 also based on the use of *Purple of Cassius* (Figure 1.4). The recipe was based on the preparation of a colloidal sol of gold, a slightly

Figure 1.4   Rose Pompadour; eighteenth century, sugar pot of the Calabre tableware in soft paste porcelain, base cover and bird cartel, from Sèvres-Cité de la Céramique, France. Photograph: Gérard Jonca, Sèvres-Cité de la Céramique.

different recipe to the *Purple of Cassius*. It was added to a powdered flux, which was then dried and ground to fine powder. Once suspended in turpentine, the enamel was then used in the decoration of porcelains and then fired at temperatures up to 880°C. These pink enamels were soon introduced in England at Worcester and Chelsea among other early porcelain factories.[47]

Meanwhile by 1723, the recipe of the *Purple of Cassius* had reached China, probably conveyed by Jesuits, and was successfully used in the production of enamel on copper and then on porcelain around 1735, which is designated as *Famille Rose* porcelain (the 'rose family').[46]

1.3 Scientific Approach of the Preparation of the Gold Ruby Colour

1.3.1 *Elucidation of the Constitution of the Purple of Cassius in the Nineteenth Century*

The elucidation of the nature and constitution of the *Purple of Cassius* remained puzzling throughout the whole of the nineteenth century in spite of many studies performed by the most famous scientists of this period.[47] In 1866, J.C. Fischer of Munich was able to draw up a list of twelve distinguished chemists who held that gold was present as an oxide and of six other ones who believed that it was in metallic form. Among the partisans of the gold oxide, there were Louis-Nicolas Vauquelin in 1811, Jöns Jacob Berzelius in 1831 and Louis Gay-Lussac in 1832. Among those of the metallic form, there were Alphonse Buisson, the assistant of Alexandre Brongniart, the director of Sèvres factory, in 1830, and Michael Faraday who conjectured in 1857 that gold was present in solution in a 'finely divided metallic state'.[49] The closest to the truth was Henry Debray, lecturer at the École Polytechnique in Paris who proposed in 1872 that the *Purple of Cassius* consisted of finely divided gold adsorbed on stannic acid. However, it is only at the turn of the century in 1905 that the true nature of the *Purple of Cassius* was finally elucidated by Richard Adolf Zsigmondy (1865–1929). Thanks to the development of a slit ultramicroscope (based on light scattering) that he developed with an optical physicist, Heinrich Siedentopf, he was able to observe finely divided gold particles on colloidal stannic acid in 1898. For these investigations, he was awarded the Nobel Prize in Chemistry in 1925. More details and references can be found in Carbert's paper.[47] Zsigmondy also confirmed the presence of colloidal particles in the ruby glass.[50] Then in 1908 Gustav Mie predicted the optical properties of homogenous spherical metal particles.[51] For a spherical nanoparticle much smaller than the wavelength of light (diameter $d \ll \lambda$), an electromagnetic field at a certain frequency (v) induces a resonant, coherent oscillation of the metal free electrons across the nanoparticle. This oscillation is known as the localised surface plasmon resonance (LSPR).[52,53] The plasmon oscillation results in a strong enhancement of absorption and scattering of electromagnetic radiation in the UV or visible range, providing intense colours and interesting optical properties to the nanoparticles when the plasmon occurs in the visible (around 520 nm for gold nanoparticles). All these concepts are more extensively developed in Chapter 3. The ratio of scattering to absorption increases with nanoparticle volume, and the dichroism of the Lycurgus cup can be briefly explained as follows. When the cup is illuminated from outside (Figure 1.1(a)), the green colour of the cup results from the scattering contribution of the large gold particles (50–100 nm) contained in the cup. In contrast, when the cup is illuminated from inside (Figure 1.1(b)), the red colour results from the absorption contribution: the green wavelength around 520 nm is absorbed, and the light that goes through the glass appears with the complementary colour, which is red.

1.3.2 *Chemical Approach to the Formation of the Purple of Cassius*

Two stages are involved in the preparation of the *Purple of Cassius*. The first one is the formation of a gold sol, and the second one its stabilisation. The first stage involves a redox reaction between stannous chloride and auric chloride leading to the formation of metallic gold:

$$2Au^{3+} + 3Sn^{2+} \longrightarrow 2Au^0 + 3Sn^{4+}.$$

The solution of stannous chloride (Sn^{2+}) also contains stannic ions (Sn^{4+}). In the past, this solution was obtained from the dissolution of tin in *aqua regia*, and the resultant stannic chloride was reduced to produce the required stannous to stannic ratio by a further addition of tin metal. The second stage is the hydrolysis of the stannic chloride into tin hydroxide that flocculates and precipitates. According to Weyl,[50] both processes, the precipitation of tin hydroxide and the formation of metallic gold occur simultaneously. When tin flocculates and precipitates, the absorbed gold particles precipitate with it and remain in dispersed form. The gold sol is therefore stabilised by absorption onto a colloidal tin hydroxide. Further operations are filtration, milling in wet conditions and drying to recover the precipitate. More details can be found in Ref. 47. The particle growth must be controlled, and 'good quality' *Purple of Cassius* requires gold particles of 10–15 nm, but a large number of factors can cause variations in the particle size and therefore in the hue and strength of the colour. The result is that despite the many years over which it has been known and studied, this preparation remains difficult to control. According to Weyl,[50] the *Purple of Cassius* did not surprise its discoverers for the colour but rather for the stability of the colour at high temperatures. It offered a new red pigment, which could be introduced into glazes and into glass. It is still in use nowadays to colour special glasses and enamels. Gold has such a strong colouring ability that only a minute amount is required even for the deepest colours: 100–1000 ppm is sufficient to produce deep pink colour glass whereas the red *sang de boeuf* colour provided by copper requires a concentration a hundred times as high as gold.

1.3.3 *Chemical Approach to the Preparation of Gold Ruby Glass*

If one keeps in mind all that has been told about the preparation of gold ruby glass so far, it appears that the addition of *Purple of Cassius* to melt glass is not the only way to prepare gold ruby glass. Calcination of gold is proposed in Darduin's manuscript (1585–1654) (Section 1.2.5): layers of gold leaves and sodium chloride are calcined several times in a furnace until the gold leaves become crumbly. The same procedure is reported by Orschall in his treatise of 1684, in which he noticed that after a calcination of eight hours, the salt turns purple.[34] Neri in *L'Arte Vetraria* (1612) reports the calcination of gold with *aqua regia* followed by the calcination of the powder, which turns red and is then added to melt glass.[34] Several experimental studies performed during the nineteenth century were gathered by

Weyl, together with his own experiments in another outstanding book after *L'Arte Vetraria*, titled *Coloured Glasses* and published in 1951.[50] Weyl confirms that there are different ways to prepare gold ruby glass. He writes that metallic gold can be directly added to molten glass and dissolves with reasonable speed, but that it is more effective to introduce gold in the form of *Purple of Cassius* or of gold chloride prepared by dissolving gold in *aqua regia*.

Practically, the preparation of gold ruby glass is based on three consecutive steps: (i) the addition of gold (several hundred ppm), most often as gold chloride or in the form of *Purple of Cassius*, in melt glass around 1400°C; at this stage, glass is colourless; (ii) a step of rapid quenching to room temperature and (iii) usually a step of reheating or annealing (also called striking step) around 500–650°C during which the red colour appears.

A simplified scheme of the principle of gold ruby glass formation gathering the different steps associated to physico-chemical phenomena and colours drawn from the scheme proposed by Weyl[50] is reported in (Figure 1.5). During the cooling step, there is oversaturation of the 'atomic gold solution' and the formation of gold nuclei. At this point, two extreme cases can be distinguished depending on the initial composition

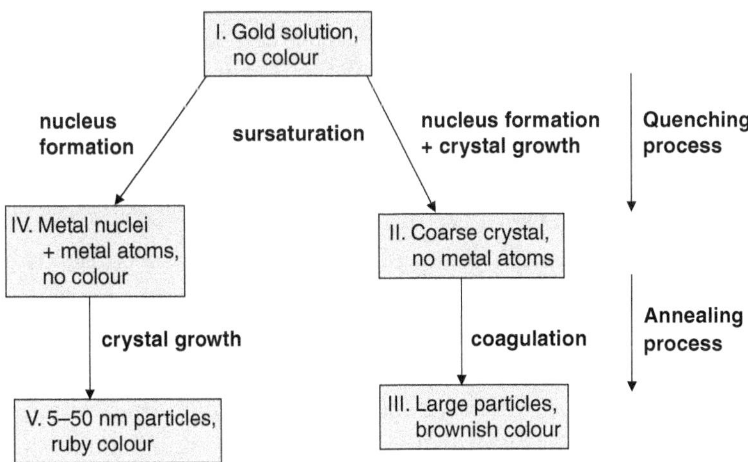

Figure 1.5　Simplified scheme of the principle of gold ruby glass formation gathering the different steps associated to physico-chemical phenomena and colours; drawn from the scheme proposed by Weyl.[50]

of glass: (i) glasses with a steep solubility curve such as sodium silicate or borax glass, lead to the growth of the nuclei and the formation of large gold particles and a brownish colour, i.e. to bad spoiled glass (Type II then Type III); (ii) the presence of lead, tin or bismuth ions in glasses enhances gold solubility, so cooling produces the nuclei (Type IV), but a substantial part of gold still remains in atomic dispersion available for the nourishment of the nuclei; the glass is still colourless at this stage. Reheating to the striking temperature causes the nuclei to grow and brings about the red colour (Type V). Striking can be defined as the step of nucleation and growth of gold atoms in glass. By adjusting the gold content, the heat treatments (temperature and duration of cooling and reheating) and the temperature coefficient solubility (controlled by the addition of lead, antimony or tin oxide), it is possible to produce nuclei in sufficient quantity at relatively high temperature and achieve the desirable hue. One can see how complex the preparation of gold ruby glass is and better understand the difficulties encountered by the glassmakers in the past to produce gold ruby glass and achieve reproducible preparations. Note that some glasses strike on cooling, so the nuclei have the chance to grow during the initial cooling. In such a case, a gold ruby glass is directly obtained; this may happen for instance when tin oxide has been added. Also note that nucleation can be produced with several other agents such as antimony oxides or using ultraviolet, X or γ radiation.[54]

There are disagreements in the literature on whether gold dissolves in glass as gold atoms or ions (Type I).[55,56] According to a ^{197}Au Mössbauer study performed on quenched colourless glasses (Type IV),[57] most of the gold is in the oxidation state I at this stage whether the gold precursor introduced in the glass was $HAu^{III}Cl_4$ or $KAu^{I}(CN)_2$; after annealing when glasses are coloured (Type V), the main species is Au^0. In another study of the same group,[58] ^{119}Sn and ^{197}Au Mössbauer spectroscopy gives an indication on the role that tin plays in the formation of much smaller but more numerous gold particles than in the absence of tin; tin provides nuclei for gold nanoparticles, and also acts as a surfactant at the surface of the gold nanoparticles, which stabilises the small gold metal particles and accelerates the kinetics of formation of gold metal through a redox mechanism similar to that occurring during the formation of *Purple of Cassius* (see Section 1.3.2):

$$Sn^{2+} + 2Au^+ \rightarrow Sn^{4+} + 2Au^0.$$

Weyl[50] states that the particles in glass, which strike to purple are between 5 and 50 nm (22 nm) and those, which lead to livery ruby are larger than 100 nm. As mentioned above, the development of the gold ruby colour is enhanced by the presence of some ions in the base glass. Lead-based glasses produce the best ruby-coloured glasses because the higher the lead content, the higher the gold solubility. The deepest colours can be obtained with 1000 ppm of gold in heavy lead glass. In soda-lime-silica glass, deep red colours can be obtained with 100–300 ppm of gold.[30]

A fragment of seventeenth-century ruby red glass found in the remains of Kunckel's factory at Peacock island (see Section 1.2.5.2) was studied by Fredrickx *et al.*[59] Gold concentration was 160 ppm, that of tin oxide was 525 ppm, and the gold particles displayed a cubo-octahedral morphology and had the right sizes (∼40 nm) to provoke the proper red colour through the phenomenon of surface plasmon resonance (Chapter 3). Iron-containing particles, mostly α-Fe_2O_3 were abundantly found in the glass matrix and were supposed to have an influence on the colour.

1.4 Conclusion

The history of gold nanoparticles, already covering centuries, through their use for the colouration of glass and ceramic, is far from over. Gold nanoparticles are still in use to make ruby glass, even though other red colourants based on copper and selenium are much more often used. Hence, gold ruby and pink glass artefacts are still produced but at a rather small scale because of the price. For instance, in France, Saint Gobain produces decorative gold pink glasses as stain glasses, and Baccarat has developed a series of gold ruby crystal glasses (Vega collection) (Figure 1.6), and edits crystal artefacts containing gold ruby crystal designed by artists. In parallel, researches are still in progress to better understand the striking mechanism and improve control of this crucial step of ruby glass fabrication.[56,60]

In the case of red enamel for porcelain, there are also alternative and less expensive red colours based on chrome-tin, but they do not offer the same range of hues and give a more opaque finish than the translucent effect obtained with the gold-based enamels. Gold-based enamels also withstand

Figure 1.6 'Cristal Rubis', ruby crystal, wine glass, From the Baccarat Vega Martini collection. Copyright ©Baccarat.

a higher temperature during the firing of the colours than cheaper base metal enamels.[47] For instance, Sèvres Manufacture in France still produces gold-based red enamels, and is still conducting research for improving the quality and the reproducibility of the hues.

The history of gold nanoparticles is now diverging with the advent of nanosciences and nanotechnologies. Researches on synthesis, properties and applications of gold nanoparticles involve now the many fields of chemistry, biology and physics that are the topics in this book.

Acknowledgements

The author deeply thanks Marco Verita (Venezia), Ian Freestone (Cardiff), Jeannine Geyssant (Paris) and Philippe Colomban (Paris), for fruitful discussions.

References

1. D. Klemm, R. Klemm, A. Murr, *J. Afr. Earth Sci.* **33**, 643 (2001).
2. J. A. Harrell, V. M. Brown, *J. Am. Res. Center Egypt* **29**, 81–105 (1992).
3. L. B. Hunt, *Gold Bull.* **9**, 24–31 (1976).
4. J. E. Greene, *Appl. Phys. Rev.* **1**, 041302 (2014).

5. V. Leusch, B. Armbruster, E. Pernicka, V. Slavčev, *Cambridge Archeol. J.* **25**, 353–376 (2015).

6. A. Mayoral, J. Agúndez, I. Miguel Pascual-Valderrama, J. Pérez-Pariente, *Gold Bull.* **47**, 161–165 (2014).

7. G. J. Higby, *Gold Bull.* **15**, 130–140 (1982).

8. R. V. Parish, S. M. Cotrill, *Gold Bull.* **20**, 3–12 (1987).

9. S. P. Fricker, *Gold Bull.* **29**, 53–60 (1996).

10. Z. Huaizhi, N. Yuantao, *Gold Bull.* **34**, 24–29 (2001).

11. E. R. T. Tiekink, *Gold Bull.* **36**, 117–124 (2003).

12. H. Knosp, R. J. Holliday, C. W. Corti, *Gold Bull.* **36**, 93–102 (2003).

13. C. L. Brown, G. Bushell, M. W. Whitehouse, D. S. Agrawal, S. G. Tupe, K. M. Paknikar, E. R. T. Tiekink, *Gold Bull.* **40**, 246–249 (2007).

14. S. Norton, *Mol. Interventions* **8**, 120–123 (2008).

15. S. J. Berners-Price, A. Filipovska, *Metallomics* **3**, 863–873 (2011).

16. R. R. Arvizo, S.Bhattacharyya, R. A. Kudgus, K. Giri, R. Bhattacharya, *Chem. Soc. Rev.* **41**, 2943–2970 (2012).

17. E. Neri, M. Verità, *J. Archaeol. Sci.* **40**, 4596–4606 (2013).

18. J. Pérez-Arantegui, J. Molera, A. Larrea, T. Pradell, M. Vendrell-Saz, *J. Am. Ceram. Soc.* **84**, 442–446 (2001).

19. E. Darque-Ceretti, D. Hélary, A. Bouquillon, M. Aucouturier, *Surf. Eng.* **21**, 352–357 (2005).

20. P. Colomban, T. Calligaro, C. Vibert-Guigue, N. Q. Liem, H. G. M. Edwards, *Archeo-Sciences* **29**, 7–20 (2005).

21. P. Colomban, *J. Nano Research* **8**, 109–132 (2009).

22. R. H. Brill, N. D. Cahill, *J. Glass Study* **30**, 16–27 (1988).

23. D. J. Barber, I. C. Freestone, K. M. Moulding. In *From Mine to Microscope: Advances in the Study of Ancient Technology*, A. J. Shortland, I. C. Freestone, T. Rehren (eds.), Oxbow Books, Barnsley, UK, pp. 115–127 (2009).

24. L. B. Hunt, *Gold Bull.* **9**, 134–139 (1976).

25. British Museum, http://www.britishmuseum.org/explore/online_tours/museu_and_exhibition/the_art_of_glass/the_lycurgus_cup.aspx.

26. R. H. Brill, *Proc 7th Int. Cong. Glass, Bruxelles, Section B, Paper* **223**, 1–13 (1965).

27. D. J. Barber, I. C. Freestone, *Archaeometry* **32**, 33–45 (1990).

28. G. L. Hornyak, C. J. Patrissi, E. B. Oberhauser, C. R. Martin, J.-C. Valmalette, L. Lemaire, J. Dutta, H. Hofmann, *Nanostruct. Mater.* **9**, 571–574 (1997).

29. I. Freestone, N. Meeks, M. Sax, C. Higgitt, *Gold Bull.* **40**, 270–277 (2007).

30. M. Verità, P. Santopadre, *J. Glass Studies* **52**, 1–14 (2010).

31. J. Geyssant, J. Lefrancq, In *Rouges & Noirs*, Vol. 67, J. Toussaint (ed.), Société Archéologique de Namur, Namur (2015).

32. J. J. Kunicki-Goldfinger, I. C. Freestone, I. McDonald, J. A. Hobot, H. Gilderdale-Scott, T. Ayers, *J. Archaeol. Sci.* **41**, 89–105 (2014).

33. H. A. Sheybany, *Glastech. Ber.* **40**, 481–484 (1967).

34. P. Zecchin, *J. Glass Stud.* **52**, 1–9 (2010).

35. S. Franck, *Glass Technol.* **25**, 47–50 (1984).

36. D. von Kerssenbrock-Krosigk, *Rubinglas des ausgehenden 17. und des 18. Jahrhunderts*, Philipp von Zabern, Mainz am Rhein (2001).

37. M. Bimson, I. C. Freestone, in *Annales du 9ème congrès de l'Association Internationale pour l'Histoire du Verre*, Nancy 1983, Liège 1985, pp. 209–222 (1985).

38. A. Polak, *Glass: Its Maker and Its Public*. Weidenfeld and Nicolson, London (1975).

39. D. von Kerssenbrock-Krosigk, In *Glass of the Alchemists*. The Corning Museum of Glass, New York, pp. 123–137 (2008).

40. P. H. Smith, In *Glass of the Alchemists*. The Corning Museum of Glass, New York, pp. 23–33 (2008).

41. W. Loibl, In *Glass of the Alchemists* (The Corning Museum of Glass, New York, 2008, pp. 63–73).

42. J. Geyssant, In *Bernard Perrot 1640–1709, secrets et chefs d'oeuvre des verreries royales d'Orléans*, Somogy Edition d'art, Paris, Musée des beaux-arts d'Orléans, Orléans, pp. 51–54 (2010).

43. I. Biron, B. Gratuze, S. Pistre, In *Bernard Perrot 1640–1709, secrets et chefs d'oeuvre des verreries royales d'Orléans*, Somogy Edition d'art, Paris, Musée des beaux-arts d'Orléans, Orléans (2010).

44. C. de Valence, *Bull. Soc. Archeol. Hist. Orléanais* **20**, 3–67 (2010).

45. O. Drahotova, *Glass Rev.* **28**, 8–11 (1973).

46. H. Garner, *Trans. Orient. Ceram. Soc.* **37**, 1–16 (1967–69).

47. J. Carbert, *Gold Bull.* **13**, 144–149 (1980).

48. O. Dargaud, L. Stievano, X. Faurel, *Gold Bull.* **40**, 283–290 (2007).

49. M. Faraday, *Trans. Roy. Soc. London* **147**, 145 (1857).

50. W. A. Weyl, *"Coloured Glasses"* The Society of Glass Technology, Sheffield, UK (1951).

51. G. Mie, *Ann. Phys.* **25**, 377–445 (1908).

52. K. L. Kelly, E. Coronado, L. L. Zhao, G. C. Schatz, *J. Phys. Chem. B* **107**, 668–677 (2003).

53. P. K. Jain, I. H. El-Sayed, M. A. El-Sayed, *Nanotoday* **2**, 18–29 (2007).

54. A. Ruivo, C. Gomes, A. Lima, M. L. Botelho, R. Melo, A. Belchior, A. Pires de Matos, *J. Cult. Herit.* **9**, e134–e137 (2008).

55. J. A. Williams, G. E. Rindone, H. A. McKinstry, *J. Am. Ceram. Soc.* **641**, 709–713 (1981).

56. B. Yin, C. Guo, C. Lu, *J. Non-Cryst. Solids* **95 & 96**, 725–732 (1987).

57. F. E. Wagner, S. Haslbeck, L. Stievano, S. Calogero, Q. A. Pankhurst, K.-P. Martinek, *Nature* **407**, 691–692 (2000).

58. S. Haslbeck, K.-P. Martinek, L. Stievano, F. E. Wagner, *Hyperfine Interact.* **165**, 89–94 (2005).

59. P. Fredrickx, D. Schryvers, K. Janssens, *Phys. Chem. Glasses* **43**, 176–183 (2002).

60. T. Jitwatcharakomol, E. Meechoowa, M. Jiarawattananon, S. Jiemsiriler, *Procedia Eng.* **32**, 584–589 (2012).

Chapter 2
Introduction to the Physical and Chemical Properties of Gold

Pekka Pyykkö* and Geoffrey C. Bond[†]

*Department of Chemistry, University of Helsinki, Finland
[†]Brunel University, Uxbridge, UK

2.1 Introduction

Gold possesses a unique combination of physical and chemical properties in both the macroscopic and microscopic states; on the macroscopic scale, gold is known for its unique yellow colour, chemical stability and high redox potential. They are the consequence of its electronic structure, the understanding of which originates with quantum chemistry coupled to Einstein's theory of relativity. On the nanoscale, the unusual electronic configuration combines with other effects due to the extremely small dimensions and (i) a high ratio of surface atoms to bulk atoms, so that overall properties are dictated by the surface atoms, (ii) electromagnetic confinement when an optical wave interacts with a gold nanoparticle, giving rise to their specific colours through a localised plasmon resonance (see Chapter 3) and (iii) quantum effects that explain the change from metallic to semiconducting character of very small particles.

Gold is the third member of Group 11 of the Periodic classification, lying below copper and silver, but its physical and chemical properties are not predictable on the basis of trends observed in other members of the group; this is evident by its bright metallic yellow colour, which resembles that of copper, but not that of silver. One could say that copper is anomalous because its 3d shell is so compact, compared to the 4d shell of silver (see the Appendix 2A). Gold is, in turn, anomalous having both a 4f shell and strong relativistic effects. The only 'normal' coinage metal is silver, which has an underlying d shell and does not yet have strong relativistic effects. The latter increase as Z^2 with the nuclear charge, Z.

Gold occupies a position at one extreme of a range of metallic properties, having excellent resistance to corrosion, considerable malleability and high density; these properties ensure its natural occurrence as metallic nuggets and powders, and its suitability for making jewellery and objects of devotion (as discussed in Chapter 1). Its lack of reactivity is demonstrated by its inability to interact with components of the atmosphere and to corrode with the formation of oxides and sulphides in the manner of copper and silver, or to dissolve in common acids. Gold artefacts are recovered unchanged from burial after many centuries and this 'nobility' also gives gold its preferred place as a coinage metal and as a form for securing wealth against all risks, except theft. The corollary of its inertness is the thermal instability and ease of reduction of compounds such as Au_2O_3, $AuO(OH)$, $Au(OH)_3$ and Au_2S_3; this follows from the small difference in electronegativity of the component atoms.

These unique physical and chemical properties are responsible for its widespread applications in both the macroscopic state[1] and in the microscopic or nanoparticulate state, as described in the following chapters of this book.

The present chapter is based on the information given in an earlier book.[2] For a general overview on the properties and uses of gold, we cite the volume edited by Schmidbaur.[3] For gold chemistry, see the monograph of Puddephatt.[4]

2.2 Physical Properties of Massive Gold

2.2.1 Crystal Structure

Gold crystallises in the face-centred cubic (fcc) structure, its metallic radius being comparable with that of silver. The single-bond covalent radius in molecules is smaller for Au than for Ag[5] (see Table 2.1). This fcc structure is responsible for its malleability. One gram can be moulded into a foil of area $1\,m^2$, the thickness of which is less than 250 atomic diameters. The same amount can also be drawn into $165\,m$ of wire that is $20\,\mu m$ in diameter.[6] These characteristics, together with many others, were discussed in detail

Table 2.1   Physical properties of gold compared to those of copper and silver (The Group 11 elements).

Property	Cu	Ag	Au
Atomic number	29	47	79
Atomic mass (amu)	63.55	107.868	196.9665
Electronic configuration	$[Ar]3d^{10}4s^1$	$[Kr]4d^{10}5s^1$	$[Xe]4f^{14}5d^{10}6s^1$
Structure	fcc	fcc	fcc
Metallic radius (pm)	128	144.47	144.20
Covalent radius (pm)[5]	112	128	124
Density (g cm^{-3})	8.95	10.49	19.32
Melting temperature (K)	1356	1234	1337
Sublimation enthalpy (kJ mol$^-$)[1]	337 ± 6	285 ± 4	343 ± 11
First ionisation energy/kJ mol^{-1})	745	731	890
Electrical resistivity at 293 K ($\mu\Omega$ cm)	1.67	1.59	2.35
Interband transition	3d 4s	4d 5s	5d 6s
Threshold energy in eV	2.1	3.9	1.84
Interband wavelength in nanometre	590	318	674

in a lengthy but fascinating paper by Michael Faraday in 1857.[7] Gold also forms alloys and intermetallic compounds with many other elements,[2,8] but it has no apparent ability to dissolve or occlude simple gases, although there is indirect evidence that hydrogen atoms can diffuse through it if formed on its surface by dissociation of molecules.[9] In jewellery, pure gold is very malleable and has to be alloyed to make it harder. The gold content is measured in carats, 24 carat gold being pure gold.

2.2.2 *Density*

While in some respects its properties reflect its greater atomic mass compared to copper and silver (e.g. density), in many cases the trend is reversed;

thus its melting point and heat of sublimation are almost the same as that of copper (see Table 2.1), the greater strength of the Au–Au bond being related to its shorter than expected length. There is only one naturally occurring isotope of gold, so the atomic mass is known very precisely (196.9665 amu, atomic mass units).

2.2.3 *Magnetic and Electrical Properties*

This isotope has a non-zero nuclear spin quantum number ($I = 3/2$) and its nucleus is therefore 'magnetic', but its receptivity relative to the proton is only 2.77×10^{-5}, so it is a hard nucleus to study.[10] Moreover it has a large quadrupole moment, which leads to line broadening in liquid-state nuclear magnetic resonance (NMR), so very refined equipment is needed for its study, and the consequential absence of hyperfine structure means that the NMR of gold is of limited diagnostic use to chemists.

Its optoelectronic properties are also unpredictable by extrapolation from its antecedents in Group 11. The electrical resistivity is greater than that of silver (see Table 2.1).

2.2.4 *Theoretical Calculations on Metallic Gold*

The yellow colour of gold has a similar origin as that of copper. Both arise from transitions from the d band to the Fermi level inside the s band (see Section 2.3). The explanation of this peculiar appearance can be traced back to the electronic properties of the gold or copper atoms themselves, and is linked to the relativistic effects for gold, and perhaps to the large electron repulsion in the compact 3d shell for copper.

Historically, the role of relativity on the band structure of gold was shown by Christensen[11] while the optical dielectric constants were explicitly calculated quite recently.[12,13] The connection to visual impressions was emphasized in Ref. 14. The connection to the Periodic Table in the sense of comparing the relativistic to non-relativistic (briefly called 'R/NR') properties of silver and gold (Ag/Au) goes back to Ref. 15. For details, see the review.[16]

In practice, the colour of gold can be changed by alloying. The addition of copper produces the 'rose gold', which is soft pink in colour; addition of aluminium leads to purple coloration, of indium to blue, and cobalt to black.

A crude idea of the band structure of metallic gold is given below.

2.2.5 *Cohesive Properties*

A R/NR comparison for silver and gold was published by Elsässer *et al.*[17] The NR lattice parameters a_0 were much larger for Au than Ag. The relativistic contraction made them comparable for the two metals. Inversely, the NR cohesive energy (E_{coh}) and bulk modulus B_0 were comparable, while the relativistic ones were much larger for gold than silver.

2.3 Relativistic Effects on the Properties of Gold

2.3.1 *Why Relativity?*

The reason for considering relativistic properties is that in heavy atoms or, more precisely, the inner parts of them, the electron speeds approach the speed of light.

The real thing. A good approach is to use a treatment based on the Dirac equation for the individual electrons. Dirac published in 1928 an equation for an electron moving in an external potential.[18,19] His equation is linear in both (x, y, z) and t coordinates, as required by special relativity. It fulfills the Lorentz invariance. Even for a many-electron atom, it turns out to be reasonable to describe each electron with its own Dirac equation. For the repulsion between electron 1 and electron 2, an acceptable starting point is the Coulomb interaction $+1/r_{12}$. It can be corrected by the Breit interaction,[20] describing both the retardation and the magnetic electron–electron interaction. This Dirac–Fock–Breit model seems to be about '101% correct'; the further quantum electrodynamical corrections, such as vacuum polarisation or self-energy, are estimated to be about -1% of the Dirac-level relativistic contributions. For a recent review, see Ref. 21. For the modern formulation of the Dirac–Fock mean-field theory for atoms, see Refs. 22 and 23.

The Dutch physicist, H.A. Kramers, also developed an equivalent treatment at about the same time.[24]

Ironically, Dirac[25] thought himself that his equation would have no importance in chemistry, because in the valence region, the electrons move slowly. He failed to notice that also the valence ns electrons make excursions to the nuclear vicinity and, while there, move fast. Moreover, there are indirect relativistic effects between the various shells.

A simple plausibility argument, published in Ref. 26 and now part of many reviews and inorganic chemistry textbooks, is the following.

Consider the innermost, 1s shell in a NR hydrogen-like atom with nuclear charge Z. (We here use the atomic units (a.u.) where the electron mass m_e, the electron charge e, the Planck constant $\hbar = h/2$ and the dielectric constant $4\pi\epsilon_0$ are all put equal to 1.) In these handy units, the average electron velocity in a 1s shell will equal the nuclear charge

$$\langle v \rangle_{1s} = Z. \tag{2.1}$$

In the same units the speed of light c equals 137.036. For Au, $Z = 79$. Compared with the rest mass m_0, the electron mass then grows to

$$m = \frac{m_0}{\sqrt{1 - (v/c)^2}}. \tag{2.2}$$

This would result in a corresponding decrease of the atomic radius from the non-relativistic Bohr radius

$$a_0 = \hbar^2/me^2. \tag{2.3}$$

The exact solutions of the Dirac equation give percental contractions of the average orbital radius of this order, not only for the lowest, 1s states but also for the higher ns states, up to the valence shell. The effects for $p_{1/2}$ shells are comparable. This is the relativistic radial contraction and concomitant energetic stabilisation of s and $p_{1/2}$ shells. (In this X_j notation the X is a letter, giving the orbital angular momentum, $X = s, p, d, f, \ldots$ for $l = 0, 1, 2, 3, \ldots$ and $j = l \pm 1$ is the total angular momentum of the electron.) Because these contracted shells in a many-electron atom now screen more strongly the nuclear attractive potential, the d and f shells will have an opposite, indirect effect, a radial expansion and energetic destabilisation. Note that these electrons are repelled from the nucleus by the centrifugal potential $+l(l + 1)/r^2$, l being the azimuthal quantum number and r the radius. For the valence shells all these relativistic effects grow as Z^2. The qualitative

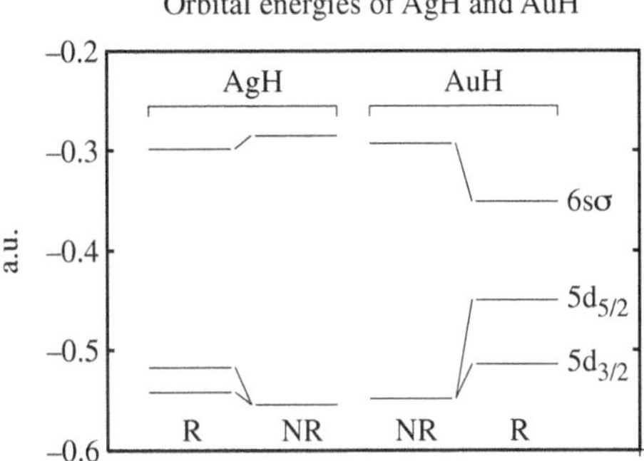

Figure 2.1   Comparison of the influence of the relativistic effect between gold and silver by calculating the orbital energies (in a.u.) of diatomic AgH and AuH molecules. Within the Koopmans theorem, these correspond to the electron binding energies. Note the similarity of the two 'NR' results, the stabilisation of the ns-based levels, the destabilisation of the $(n - 1)d$ levels and the relativistic closing of the gap between the two. Data from Ref. 15. Figure reprinted with permission from Ref. 27. Copyright 2004 Wiley-VCH.

consequences for the diatomic metal hydrides AgH and AuH are shown in Figure 2.1.

This qualitatively fits with all that we know about the chemistry of gold. The 6s relativistic stabilisation provides a natural explanation for the high electronegativity related to the high first ionisation potential IP_1 and to the high electron affinity EA. This explains the nobility of gold and the existence of auride chemistry. The easier access to the higher oxidation states $+$III and $+$V is related to the 5d relativistic destabilisation (for a confirmation of this point, see Ref. 28). Finally, the colour of gold is related to the smaller energy gap for a 5d-to-6s excitation. Everything fits.

There is a considerable dependence of the relativistic effects on the column (or 'Group'), in addition to the Z^2 trend. Figure 2.2 plots the ratio of the relativistic (R) and non-relativistic (NR) 6s radii. If the relativistic effect were insignificant, the ratio R/NR would be 1. Among the different elements of the Periodic Table (at least until $Z = 100$), the relativistic contraction of the 6s shell has a very sharp extremum at platinum and gold, where the R/NR contraction is about 0.82 (Figure 2.2).

The greater similarity between the elements in the Second and Third Transition Series compared to those in the First Series (Table 2.1) was

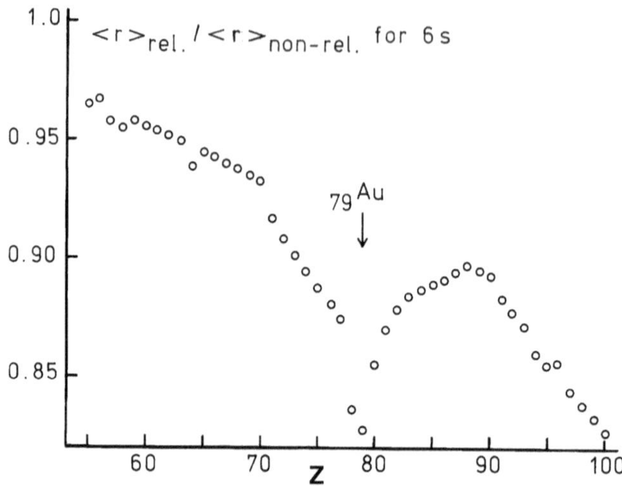

Figure 2.2   Illustration of the relativistic effect on the radius of the 6s shell. The plot of the ratio of the value calculated taking into account relativity (Dirac equation) over the non-relativistic value is always smaller than 1 (contraction). This 'gold maximum' at Group 11 is dramatic in the case of gold. Reprinted with permission from Ref. 14. Copyright 1979 American Chemical Society.

formerly ascribed solely to the lanthanide contraction, caused by the occupation of the 4f shell. This shell only partially screens the increased nuclear charge, which contracts the 6s shell. It is now realised that the lanthanide contraction and the relativistic effects are both there and act in the same direction.

2.3.2 *Optical Properties, Interband Transitions and Relativistic Effect*

The optical absorption of gold in the visible region of the spectrum is due to the relativistic decrease of the gap between the 5d band and the Fermi level (see Table 2.1). A good indication of this relativistic effect is the very low value of the interband transition energy for gold compared to silver. The interband threshold as given in Table 2.1 is the energy required to excite electrons from the top of the 5d band into the 6sp conduction band. In the case of gold, its value is 1.84 eV, which means that an optical wave corresponding to a red wavelength is able to excite this transition. In the case of silver, the interband transition is in the UV range so that the visible light is almost not affected after reflecting on a silver surface. For detailed studies of relativistic effects on these bulk properties, see Refs. 12 and 13 or the review.[16]

The exact attribution of the optical properties of gold nanoparticles to atomic level shifts, surface ('surface plasmon') effects or size and shape effects cannot be discussed here (see Chapter 3).

There are further ways to use X-ray processes to study gold band structure. A weak white line on the leading edge of the L_{III} X-ray absorption edge signifies a small number of vacancies in the d-band caused by d–s hybridisation.[29]

2.4 Chemical Properties of Gold in Relation to its Neighbours

The chemistry of gold ($5d^{10}6s^1$) is determined by (i) the easy activation of the 5d electrons and (ii) its propensity for acquiring a further electron to complete the $6s^2$ level and not to lose the one it has.[30,31] This latter effect gives it a much greater electron affinity and higher first ionisation potential than those of copper or silver, as shown in Table 2.1, and accounts for the ready formation of the Au^{-I} state. This auride chemistry has been discussed by Jansen in the Schmidbaur book,[32] or in the mini-review by the present author.[33] The 5d destabilisation obviously explains the predominance of the Au^{III} state, which in a free atom has the $5d^8$ configuration (even the Au^V state ($5d^6$) is accessible as in AuF_5), the Au^I state being of lesser importance and the Au^{II} state being unknown except in a few unusual complexes,[34] notably some diamagnetic Au^{II} dimers.

The Au(VII) compound AuF_7 had been claimed but was shown by Riedel and coworkers[35] to be a complex of AuF_5 and F_2.

For a sketch of the electronic energy levels of gold particles of various sizes, see Figure 2.3.

Gold's electronegativity (2.4 in Pauling units) equals that of selenium and approaches that of sulphur and iodine (2.5).[6] Therefore it is frequently said to have some of the properties of a halogen. Its electrode potential ($E^0 = +1.691$ V) is also extremely high for a metal. Its electronic structure determines its nobility, and its inability in the massive form to interact with oxygen or sulphur compounds, i.e. to tarnish as

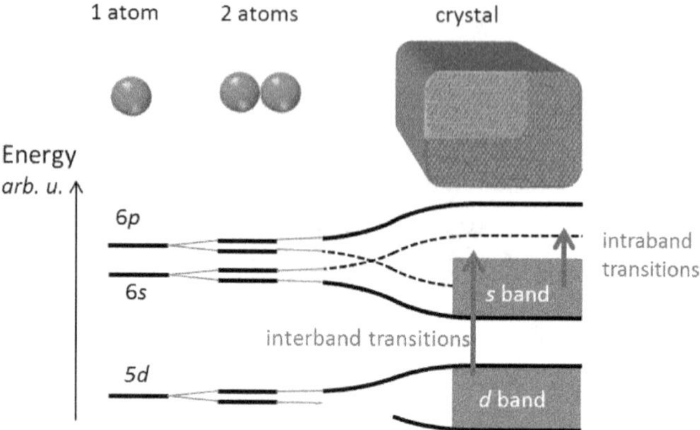

Figure 2.3   Sketch of the evolution of the electronic structure of gold. For a single atom, the electronic levels are discrete, as shown above. For a gold dimer Au–Au, the levels tend to split. For a crystal, this lifting of degeneracy widens and forms a continuum of levels and the occupied d band emerges from the *d* electrons of all the gold atoms and is completely filled with electrons. The conduction band is formed from the 6s and 6p orbitals and is partially filled (conduction band). With this structure, light can excite two kinds of transitions, the intraband transitions and the interband transitions.

silver and copper do, is in line with the instability of its oxide Au_2O_3, which decomposes at about 433 K and has a positive heat of formation. The sulphides Au_2S and Au_2S_3 are also known, but are of limited stability and importance.[6]

Gold dissolves in solutions of the heavier alkali metals in liquid ammonia, and the auride ion Au^- is formed; the electrical conductivity of cesium–gold alloys at 873 K shows a very sharp minimum at the 1:1 ratio, and the solid CsAu is regarded as a semiconductor; it has the NaCl structure. Tetramethylammonium auride is isostructural with the bromide, and the deep blue addition compound $CsAu \cdot NH_3$ has recently been prepared and characterised,[36] as also discussed in the mini-review.[33]

The dissolution of gold requires both an oxidant and a ligand to stabilise the resulting cation. Thus it dissolves in *aqua regia* (conc HCl: conc $HNO_3 = 3{:}1$) to form $AuCl_4^-$, and in the presence of oxygen in aqueous CN^- to form the $[Au(CN)_2]^-$ anion. The majority of the world's gold production is based on the cyanide method using this Au(I) dicyanide ion. In electrochemical gold plating, the gold can be introduced as AuCN or alkaline gold cyanide, depending on the type of layer desired.

Finally, we may note the existence of compounds of gold, which cannot be prepared and stored in a bottle, but whose ephemeral character may imitate transient species formed in catalytic processes. These include the hydrides AuH_3 (i.e. $HAu(H_2)$) and AuH_5 (i.e. $H_3Au(H_2)$), which have been seen in low-temperature matrices, and $AuXe^+$ and $AuXe^+$, which were theoretically predicted and subsequently detected by mass spectrometry.[37] Quite unexpectedly, bulk compounds of the cation $[Au^{II}Xe_4]^{2+}$ were also synthesised, such as the compound $[AuXe_4][Sb_2F_{11}]_2$.[38] One could say that bulk gold is quite unreactive, surfaces and small particles are more reactive and the single atoms even more so. As regards cations, an Au^I can activate one or two xenon atoms and finally an Au^{II} can activate four.

2.5 More on Gold Chemistry

As a general reference, we point the reader to the Schmidbaur volume.[3] There is one family of inorganic compounds that occurs as a mineral, that of *calaverite*, $AuTe_2$. It has historically been a minor source of gold. Its properties are interesting. The crystal structure is not simple and both Au^I and Au^{III} occur in it. It has a metallic lustre. For a theoretical band structure of the trigonal phase of $AuTe_2$, see Caracas and Gonze.[39] If a part of the gold atoms in calaverite is replaced by silver, one gets sylvanite. The silver substitutions can stabilise the incommensurate modulations of calaverite (see Bindi *et al.*[40]). In addition to gold ditelluride, there are alloys of gold, notably with silver and copper, occurring as minerals.

The rich organometallic chemistry of gold has, in addition to the eponymous Au–C bonds, also Au–X (X = halide), Au–P (often in phosphines), and Au–E (E = chalcogenide, especially S) bonds. Note here the medical uses of certain gold complexes (see the contribution by C.F. Shaw III[41,42]). One application is in combating arthritis. An example is *auranofin*, which has both a thiolate and a phosphine, bound to an Au^I.

2.6 Surface Science and Cluster Studies

The surface science or cluster studies where the gold surface is bonded to thiolate sulphur would require an entire chapter. A particular case is the clusters where an inner part is metal-only, often obeying a magic number, such as 8, 18, . . . and the outer part is a protecting shell, consisting of staples, such as –SR–Au–RS– or –SR–Au–SR–Au–RS–. A 'magic number' here means a closed electron shell structure, followed by a gap. See for these compounds the Chapter 13 by Häkkinen in the present volume.

The size, structure, substrate and reaction specificity of catalysis by gold clusters would demand its own treatise. We refer the reader to Refs. 43 and 44 and to Chapter 12.

2.7 The Aurophilic Attraction

In numerous complexes containing two or more Au^I ions, it is generally observed that the Au^I...Au^I distances between pairs of such ions are unusually short (275–350 pm), and that some form of attraction must therefore exist between them, although both are closed-shell and carry a formal positive charge.[31,34,45] The effect was termed by Schmidbaur[46] as aurophilic attraction or aurophilicity. The effect was also observed when the Au^I ions are in different molecules that pack closely together in the solid state, and when they are located at opposite sides of the same ring, formed with bidentate ligands (transannular attraction, see for an example Figure 2.4). There is an enormous literature on gold complexes in which the effect occurs, and this has been reviewed.[47,48] Focussing on the gold atoms, we can have dimers, oligomers, infinite chains or infinite

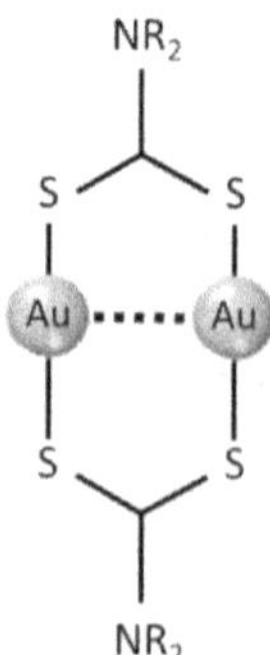

Figure 2.4   An example on a doubly bridged aurophilic bond between two gold atoms in a dimeric Au^I dithiocarbamate.

sheets. An obvious generalisation of the term is metallophilic attraction.[49] The corresponding argentophilic, $Ag^I...Ag^I$ attraction can be equally frequent[50,51] and actually equally strong, but has perhaps received less emphasis.

Aurophilicity has also been extensively studied by theoretical methods. It appears that the bond is mainly due to dispersion forces of the type that hold molecules together in a liquid or solid, but very much stronger than other van der Waals forces; it has the same kind of strength as the hydrogen bond in water and alcohols and takes values between 10 and $100\,kJ\,mol^{-1}$, depending on the separation between the atoms. This picture does not support the entirely possible idea of 5d–6s–6p hybridisation[52,53] in the systems so far studied.

While the structural evidence for metallophilic attraction is abundant, much less quantitative information is available on the bonding strength. Six values for the aurophilic interaction from temperature-dependent NMR are given in Table 5 of Ref. 54.

More information is available via the M...M' force constants, determined by Raman spectroscopy. For one summary, see Harvey.[55] The comparisons in Schmidbaur and Schier[51] suggest qualitatively stronger argentophilic than aurophilic interactions. Systematic comparisons exist for the structures of $[(NHC)MCl]_2$ (M = Ag, Au); NHC = N-heterocyclic carbene, see Ref. 56.

2.8 Dependence of Physical and Chemical Properties of Gold on Particle Size

When the size of a gold particle is progressively decreased, significant changes in physical properties and chemical reactivity are observed; they become especially noticeable when size falls below about 10 nm. Such particles, often named nanoparticles (although the term lacks quantitative definition), may be obtained by controlled growth of atomic species in various ways: (i) by condensation of vapourised metal atoms under UHV conditions to form gaseous 'clusters' having relatively few atoms (<20)[57]; (ii) by reduction of a solution of a gold compound, usually $HAuCl_4$, to form a colloidal dispersion (Chapter 6) and (iii) by deposition of a gold compound followed by decomposition or reduction to the metallic state (Chapter 7) or vapourised gold atoms (Chapter 12) onto a support. The numerous actual and potential applications of nanoparticulate gold, which are the subject of this book, have given rise to numerous studies of their physical properties by a great variety of techniques (Chapters 3 and 4), together with many theoretical studies (Chapter 13). The following paragraphs provide a very brief account of what has been observed.

The most significant consequence of decreasing the particle size is the increase in the surface/volume ratio; the rise in the fraction of surface atoms is responsible for some changes in the structural character. As a rough guide, surface atoms of a 2 nm particle constitute about 60% of the total. Notable changes include (i) a decrease in the melting temperature (2 nm particles melt at 500 K compared to 1337 K for massive gold) and (ii) a lowering of the interatomic separation from 0.288 to 0.245 nm.[59] These effects arise because surface atoms experience a resulting force acting inwards, which is not compensated by atoms above them; it causes a surface energy akin to the surface tension of liquids, and thus to a decrease of interatomic spacing throughout the particle. At the same time, the decrease in the mean number of near neighbours allows atoms a greater freedom to vibrate around their normal locations, thus accounting for a lowering of melting temperature. Theoretical studies indicate that surface atoms having low coordination number, such as are found at edges and corners of all particle shapes, are the seat of increased chemical reactivity (see Figure 2.5), and are therefore the preferred locus for the chemisorption of molecules such as CO and H_2 (Chapter 12). For example, for nanoparticles larger than 2 nm, the majority of surface atoms belong to a (111) facet. For a 10 nm cubooctahedral nanoparticle, a negligible amount of atoms are corner atoms, whereas 8% populate the edges,[60] the (100) facets and the other ones are part of the (111) surfaces. See for illustration Figure 2.5.

Changes in the electronic structure also affect the optical response of particles and the reduction of size down to the nanometre scale account for the various colours exhibited by colloidal dispersions of gold (Chapter 3) and by gold nanoparticles supported on surfaces[64] or in matrices. Changes to the X-ray photoelectron spectra (XPS) of gold nanoparticles have often been seen; the binding energy of the $4f_{7/2}$ core level typically increases by

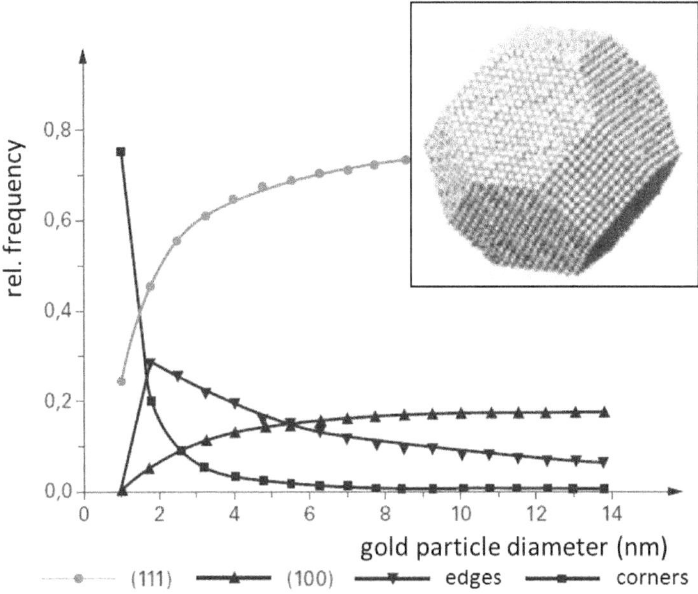

Figure 2.5   Dependence of relative amounts of surface sites on particle diameter of gold particles, based on cuboctahedron model. Reprinted with permission from Ref. 58. Copyright 2001 Science Reviews Ltd.

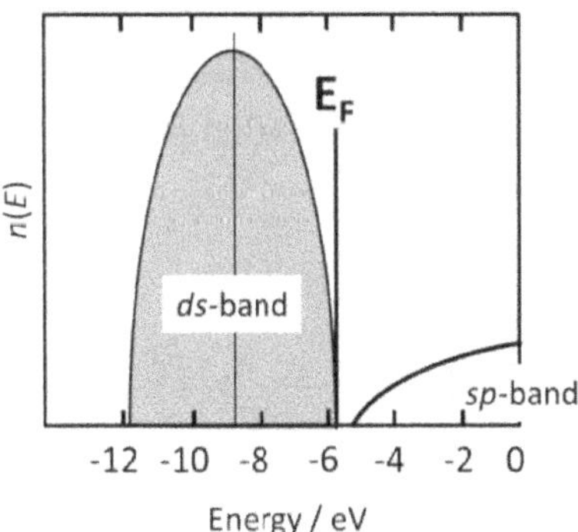

Figure 2.6   Distribution of energy levels for a 2 nm particle; there is a small band gap above the Fermi energy E_F.[61] For modern calculations of the band structure of bulk gold, see Ref. 62.

0.8–1.0 eV when size falls below 5 nm. Now metallic character is due to the broadening of electron energy levels into the overlapping bands, but with very small particles (<2 nm) the bands become narrower, and a gap appears between them (Figure 2.6).[65]

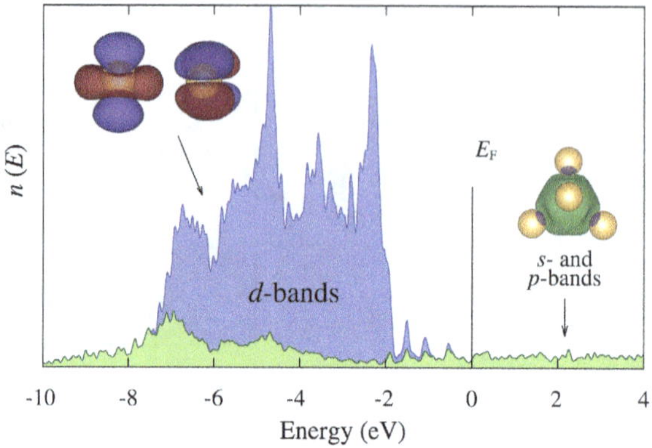

Figure 2.7   A density-of-states histogram for bulk gold (Dr. Tomatiuh Rangel, private communication). The figure is based on the scalar-relativistic (spin-orbit averaged) work of Rangel *et al.*[62] The figure projects the Bloch orbitals on a set of maximally localised Wannier functions (MLWFs)[63] with 5d character and with mixed s and p character, shown in the figure. The blue and red isosurfaces represent positive and negative isovalues of the d MLWFs. The green and light-blue isosurface represent positive and negative isovalues of the s and p like MLWFs. Using the same colour scheme; the blue area is the PDOS density-of-states on the d-bands and the green area is the PDOS on the s- and p-bands.

The metallic behaviour is then lost, and this is thought to cause a marked rise in catalytic activity for CO oxidation and other reactions (Chapter 9). This may be because nearly all the surface atoms have low coordination numbers characteristic of edges and corners on larger particles.

A recent calculation of the densities-of-states for bulk gold is shown in Figure 2.7.

2.9 Conclusion

Gold has many chemical and physical properties that are unique and not predictable by extrapolation of those of copper and silver; these include its high electronegativity and its colours, both in fine gold and its alloys and gold dispersions. All of these characteristics find their origin in the important role that relativistic effects play in the electronic structure of the element and its compounds, and they, in turn, determine its suitability for the many applications that form the subject of this book.

Acknowledgements

We express our sincere thanks to Drs R.J.F. Berger, O. Pluchery and H. Rabaâ for helpful comments on drafts of this chapter. Dr T. Rangel kindly provided Figure 2.7.

Appendix 2A: Further Theoretical Details

Commenting first on the unusually compact size of the innermost shell of every orbital angular momentum in an atom (1s, 2p, 3d and 4f) compared to their higher analogues, it has been related to their nodelessness. The lowest lying orbital of a given l has no nodes (zeros) in its radial wave function. The more the nodes are the higher the kinetic energy. For further references, see Ref. 21.

The rest of this appendix tries to discuss the present level of understanding of aurophilicity. The concepts and acronyms are well known to those in the field of chemistry. A previous mini-review was referred in the Introduction to Ref. 66. In the theoretical understanding of aurophilicity, some steps along the road were the following. First, it was found that electron correlation was needed to obtain any attraction.[67] At Hartree–Fock level, none was seen for a model dimer [(Cl–Au–PH$_3$)$_2$]. At large distances, the attraction behaved as R^{-6}, suggesting a dispersion (van der Waals) mechanism.[68,69] This picture was also basically confirmed using local orbitals.[70] After the (dipole–dipole) dispersion, the next most important mechanism was then found to be at small distances a virtual-charge-transfer contribution, decaying exponentially at large R.[70]

Concerning ligand trends, for the perpendicular dimer of the model systems [(X–Au–L)$_2$], the softer halogens X give the stronger aurophilic attraction.[49] As to the neutral ligands L, N-heterocyclic carbenes and certain phosphines give the strongest interactions.[66] That paper also gives a compact summary on metallophilicity.

An interesting twist was the dependence of the conclusion on the theoretical level used. At the lowest physically sound level which is the 'MP2', the previous results for the Ag-to-Au comparison were supported by O'Grady and Kaltsoyannis.[71] The gold–gold interaction was stronger than the silver–silver one. At the higher level, 'CCSD' or 'CCSD(T)' levels, however, the opposite was true. Thus one cannot yet claim that the aurophilic interaction was unique, nor that this were due to relativistic effects. In fact, many examples exist on the similar *argentophilic* attraction; see for reviews Jansen[50] or Schmidbaur and Schier.[51] Recall here that non-relativistic gold would be quite similar to silver, so a comparison of the two elements would also be a rough estimate on relativistic effects.

Among the Group 11 'coinage' metals, copper is here the 'odd element out'. It can also show strong and short metallophilic interactions but only when the two interacting units are coupled with one or more bridges. The reasons are not well understood.[54]

In the wave function theories (WFTs), one can have systematic sequences of theory, such as HF, MP2, MP3, MP4, CCSD, CCSD(T). It was found that the obtained aurophilic attraction strongly oscillates along this series.[68] The current best standard is CCSD(T), extrapolated to an infinite basis set.[72] The experimentally known Au...Au distances of the 'A-frames', such as $E(AuPR_3)_2$, $E = S$, Se could be used for experimental verification of the WFT models.[73–76] Empirically, the SCS-MP2 level gives results close to CCSD(T) at lower cost.[77,78]

Density functional theory (DFT) is an excellent tool in inorganic and organometallic chemistry. In its basic form it has, however, difficulties in describing dispersion-type interactions. One device, developed by the group of Grimme,[79] is the 'DFT-D3', where semiempirical correction expressions, based on monoatomic polarisabilities are added.

A 'range-separated hybrid' approach has been tested by Ángyán and co-workers with WFT at large distances and DFT at short distances.[80] For a study on the A-frames, see Alam and Fromager.[76] They used the trick of splitting the electron–electron r_{12}^{-1} interaction to short-distance and long-distance parts, handled by DFT and WFT, respectively.

A larger family of both applications and methods was tested by Andrejić and Mata.[78] In particular, they tested the 'D3' dispersion corrections from Grimme and co-workers. A short-distance damping correction had to be included. With the localised ('L') orbital analysis, individual contributions could be seen. Both electrostatics and ligand–ligand terms could exist.

Another device is the LC-ωPBE-XDM developed by Becke's and Scuseria's groups. An even larger set of systems and theoretical models was studied by Otero-de-la-Roza *et al.*[81] One further theoretical idea by them was that relativistic changes between alternative structures would be driven by the relativistic increase of the IP, which would disfavour the ionic alternatives for gold, and leave whatever remains.

References

1. C. Corti, R. Holliday (eds.), *Gold: Science and Applications* (CRC Press, Boca Raton, FL, 2010).
2. G. C. Bond, C. Louis, D. T. Thompson, *Catalysis by Gold* (Imperial College Press, London, 2006, p. 366).
3. H. Schmidbaur (ed.), *Gold. Progress in Chemistry, Biochemistry and Technology* (Wiley, Chichester, 1999, p. 894).
4. R. Puddephatt, *The Chemistry of Gold* (Elsevier, Amsterdam, 1980).
5. P. Pyykkö, *J. Phys. Chem. A* **119**, 2326–2337 (2015).
6. N. N. Greenwood, A. Earnshaw, *Chemistry of the Elements* (Butterworth Heinemann, Oxford, 1997, pp. 599, 1180, 1266, 1274).
7. M. Faraday, *Phil. Trans.* **147**, 145–181 (1857); see also W. D. Mogerman, *Gold Bull.* **7**, 22–24 (1974).
8. W. Rapson, *Gold Bull.* **29**, 141–142 (1996).
9. R. S. Yolles, B. J. Wood, H. Wise, *J. Catal.* **21**, 66–69 (1971).
10. M. Tokita, E. Haga, *J. Phys. Soc. Japan* **50**, 482–489 (1981).
11. N. E. Christensen, B. O. Seraphin, *Phys. Rev. B* **4**, 3321–3344 (1971).
12. P. Romaniello, P. L. de Boeij, *J. Chem. Phys.* **122**, 164–303 (2005).
13. K. Glantschnig, C. Ambrosch-Draxl, *New J. Phys.* **12**, 103048 (2010).
14. P. Pyykkö, J. P. Desclaux, *Acc. Chem. Res.* **12**, 276–281 (1979).
15. J. P. Desclaux, P. Pyykkö, *Chem. Phys. Lett.* **39**, 300–303 (1976).
16. P. Pyykkö, *Ann. Rev. Phys. Chem.* **63**, 45–64 (2012).
17. C. Elsässer, N. Takeuchi, K. M. Ho, C. T. Chan, P. Braun, M. Fähnle, *J. Phys: Condens. Matter* **2**, 4371–4394 (1990).
18. P. A. M. Dirac, *Proc. Roy. Soc. (London) A* **117**, 610–624 (1928).
19. P. A. M. Dirac, *Proc. Roy. Soc. (London) A* **118**, 351–361 (1928).
20. G. Breit, *Phys. Rev.* **39**, 616–624 (1932).
21. P. Pyykkö, *Chem. Rev.* **112**, 371–384 (2012).
22. I. P. Grant, *Adv. Phys.* **19**, 747–811 (1970).
23. I. P. Grant, *Relativistic Quantum Theory of Atoms and Molecules: Theory and Computation* (Springer, New York, 2007, p. 797).
24. D. ter Haar, *Masters of Modern Physics: The Scientific Contributions of H. A. Kramers* (Princeton University Press, Princeton, NJ, 1998).
25. P. A. M. Dirac, *Proc. Roy. Soc. (London) A* **123**, 714–733 (1929).

26. P. Pyykkö, *Adv. Quantum Chem.* **11**, 353–409 (1978).

27. P. Pyykkö, *Angew. Chem. Int. Ed.* **43**, 4412–4456 (2004), *Angew. Chem.* **116**, 4512–4557 (2004).

28. K. Theilacker, B. Schlegel, M. Kaupp, P. Schwerdtfeger, *Inorg. Chem.* **54**, 9869–9875 (2015).

29. J. Chevrier, L. Huang, P. Zeppenfeld, G. Comsa, *Surf. Sci.* **355**, 1–12 (1996).

30. G. C. Bond, D. T. Thompson, *Catal. Rev. Sci. Eng.* **41**, 319–388 (1999).

31. P. Pyykkö, *Gold Bull.* **37**, 136 (2004).

32. M. Jansen, A. V. Mudring, In *Gold. Progress in Chemistry, Biochemistry and Technology*, H. Schmidbaur (ed.), Wiley, New York, pp. 747–793 (1999).

33. P. Pyykkö, *Angew. Chem. Int. Ed.* **41**, 3573–3578 (2002), *Angew. Chem.* **114**, 3723–3728 (2002).

34. M. C. Gimeno, A. Laguna, *Gold Bull.* **36**, 83–92 (2003).

35. D. Himmel, S. Riedel, *Inorg. Chem.* **46**, 5338–5342 (2007).

36. A.-V. Mudring, M. Jansen, J. Daniels, S. Krämer, M. Mehring, J. P. Prates Ramalho, H. Romero, M. Parrinello, *Angew. Chem. Int. Ed.* **41**, 120–124 (2002), German version, *Angew. Chem.* **114**, 3723–3728 (2002).

37. D. Schröder, H. Schwarz, J. Hrušák, P. Pyykkö, *Inorg. Chem.* **37**, 624–632 (1998).

38. S. Seidel, K. Seppelt, *Science* **290**, 117–118 (2000).

39. R. Caracas, X. Gonze, *Phys. Rev. A* **69**, 144114/1–144114/7 (2004).

40. L. Bindi, A. Arakcheeva, G. Chapuis, *Am. Miner.* **94**, 728–736 (2009).

41. C. F. Shaw III, In *Gold. Progress in Chemistry, Biochemistry and Technology*, H. Schmidbaur (ed.), Wiley, New York, pp. 259–308 (1999).

42. C. F. Shaw III, *Chem. Rev.* **99**, 2589–2600 (1999).

43. G. Bond, *Gold Bull.* **41**, 235–241 (2008).

44. A. Taketoshi, M. Haruta, *Chem. Lett.* **43**, 380–387 (2014).

45. F. Mendizabal, P. Pyykkö, *Phys. Chem. Chem. Phys.* **6**, 900–905 (2004).

46. F. Scherbaum, A. Grohmann, B. Huber, C. Krüger, H. Schmidbaur, *Angew. Chem.* **100**, 1602–1604 (1988); *Angew. Chem. Int. Ed.* **27**, 1544–1546 (1988).

47. H. Schmidbaur, *Gold Bull.* **23**, 11–21 (1990).

48. H. Schmidbaur, A. Schier, *Chem. Soc. Rev.* **41**, 370–412 (2012).

49. P. Pyykkö, J. Li, N. Runeberg, *Chem. Phys. Lett.* **218**, 133–138 (1994).

50. M. Jansen, *Angew. Chem.* **99**, 1136–1149 (1987). Version in English: *Angew. Chem. Int. Ed. Engl.* **26**, 1098–1110 (1987).

51. H. Schmidbaur, A. Schier, *Ang. Chem. Int. Ed.* **54**, 746–784 (2015).

52. P. K. Mehrotra, R. Hoffmann, *Inorg. Chem.* **17**, 2187–2189 (1978).

53. A. Dedieu, R. Hoffmann, *J. Am. Chem. Soc.* **100**, 2074–2079 (1978).

54. P. Pyykkö, *Chem. Rev.* **97**, 597–636 (1997).

55. P. D. Harvey, *Coord. Chem. Rev.* **153**, 175–198 (1996). T. F. Carlson and J. P. Fackler, Jr., *J. Organomet. Chem.* **596**, 237–241 (2000), suggest that the Au^{II}-Au^{II} case yields two separate R versus v curves for $[Au(CH_2)_2PPh_2]X_2$ and $(AuMTP)_2X_2$, as function of X=Cl-I.

56. L. Ray, M. M. Shaikh, P. Ghosh, *Inorg. Chem.* **47**, 230–240 (2008).

57. R. Meyer, C. Lemire, Sh. K. Shaikhutdinov, H.-J. Freund, *Gold Bull.* **37**, 72–124 (2004).

58. C. Mohr, P. Claus, *Science Progr.* **84**(4), 311–334 (2001).

59. J. T. Miller, A. J. Kropf, Y. Zha, J. R. Regalbuto, L. Delannoy, C. Louis, E. Bus, J. A. van Bokhoven, *J. Catal.* **240**, 222–234 (2006).

60. N. Bartlett, *Gold Bull.* **31**(1), 22–25 (1998).

61. G. C. Bond, *Faraday Disc.* **152**, 277–291 (2011).

62. T. Rangel, D. Kecik, P. E. Trevisanutto, G.-M. Rignanese, H. Van Swygenhoven, V. Olevano, *Phys. Rev. B* **86**, 125125/1–125125/9 (2012).

63. N. Marzari, A. A. Mostofi, J. R. Yates, I. Souza, D. Vanderbilt, *Rev. Mod. Phys.* **84**, 1419–1475 (2012).

64. G. C. Bond, P. A. Sermon, *Gold Bull.* **6**, 102–105 (1973).

65. K. Okazaki, S. Ichikawa, Y. Maeda, M. Haruta, M. Kohyama, *Appl. Catal. A: General* **291**, 45–54 (2005).

66. J. Muñiz, C. Wang, P. Pyykkö, *Chem. Eur. J.* **17**, 368–377 (2011).

67. P. Pyykkö, Y.-F. Zhao, *Angew. Chem.* **103**, 622–623 (1991), Engl. transl. in *Angew. Chem. Int. Ed. Engl. 30*, 604–605 (1991).

68. P. Pyykkö, N. Runeberg, F. Mendizabal, *Chem. Eur. J.* **3**, 1451–1457 (1997).

69. P. Pyykkö, F. Mendizabal, *Chem. Eur. J.* **3**, 1458–1465 (1997).

70. N. Runeberg, M. Schütz, H.-J. Werner, *J. Chem. Phys.* **110**, 7210–7215 (1999).

71. E. O'Grady, N. Kaltsoyannis, *Phys. Chem. Chem. Phys.* **6**, 680–687 (2004).

72. P. Pyykkö, P. Zaleski-Ejgierd, *J. Chem. Phys.* **128**, 124 309 (2008).

73. J. Li, P. Pyykkö, *Chem. Phys. Lett.* **197**, 586–590 (1992).

74. P. Pyykkö, T. Tamm, *Organometallics* **17**, 4842–4852 (1998).

75. S. Riedel, P. Pyykkö, R. A. Mata, H.-J. Werner, *Chem. Phys. Lett.* **405**, 148–152 (2005).

76. M. M. Alam, E. Fromager, *Chem. Phys. Lett.* **554**, 37–42 (2012).

77. P. Pyykkö, X.-G. Xiong, J. Li, *Faraday Disc.* **152**, 169–178 (2011).

78. M. Andrejić, R. A. Mata, *Phys. Chem. Chem. Phys.* **15**, 18115–18122 (2013).

79. S. Grimme, J. Antony, S. Ehrlich, H. Krieg, *J. Chem. Phys.* **132**, 154104/1–154104/19 (2010).

80. R.-F. Liu, C. A. Franzese, R. Malek, P. S. Żuchowski, J. G. Ángyán, M. Szczęśniak, G. Chałasiński, *J. Chem. Theory Comp.* **7**, 2399–2407 (2011).

81. A. Otero-de-la-Roza, J. D. Mallory, E. R. Johnson, *J. Chem. Phys.* **140**, 18A504 (2014).

Chapter 3
Optical Properties of Gold Nanoparticles

Olivier Pluchery

Institut des NanoSciences de Paris, Université Pierre et Marie Curie-CNRS, Paris, France

3.1 Introduction

Among the remarkable properties of gold nanoparticles (AuNPs) discussed in the present book, their optical appearance is something dramatic, since gold loses its usual yellow colour when dispersed in the form of gold nanoparticles in solution, and adopts a ruby-red look. Chapter 1 provided the historical overview as well as a review of the artistic developments related to this unexpected colour. Colour is also a macroscopic manifestation of the phenomena that occur at the nanoscale where the collective behaviour of the conduction electrons of gold gives rise to the plasmon resonance. Plasmonics emerged from the plasma physics, which deals with gases where electric charges are free to move under the influence of electromagnetic or gravitational forces. In the case of metals, the conduction electrons play the role of free charges since they are detached from their ionic core and can be excited by an electromagnetic wave such as an optical beam. The oscillation of the electrons and the oscillation of the electromagnetic field are intrinsically linked. We will describe some cases when they give rise to resonant modes. These modes are sometimes termed by their corresponding quasi-particles: polaritons when the phenomenon is considered from the point of view of electrical charges and plasmon when it is studied from the electromagnetic point of view. Therefore, plasmon waves correspond to the coupling of two waves: the mechanical oscillations of charges and the electromagnetic oscillations of the electric field. Plasmonics has given photonics the ability to go to the nanoscale and this is now an established research field, with devoted journals and conferences.

3.1.1 *A Brief History of Plasmonics*

David Pines (born 1924) was interested in his PhD work (1950) in the collective description of electron interactions. He and his PhD adviser, David Bohm (1917–1992), signed a series of articles focusing on the plasma oscillation in metals and they derived the formalism for treating the electron gas either classically[1] or with the quantum physics.[2] The so-called plasma oscillations were already mentioned by Tonks and Langmuir in 1929 when they studied the plasma in gaseous discharge[3] and this phenomenon corresponds to the volume plasmon. It was Pines who first used the name 'plasmon' in 1956 to describe the quantum of energy associated with the eigenfrequency of the collective oscillation of electrons in a metal.[4] The application of the Pines and Bohm formalism to surface waves was done by Ritchie in 1957 and he predicted a collective oscillation of the electrons in thin metallic foils at an energy below the plasma frequency.[5] This was the birth of surface plasmon, or plasmon polariton, which are the concepts we will explain in the present chapter. At this time, the excitation of surface plasmon waves was achieved with low-energy electron beams (10–25 eV)[6] and the coupling with light was not investigated. Interestingly, this way of exciting the plasmon has recently been revisited with very promising perspective thanks to the progress in nanooptics (see Chapter 11). The coupling of a plasmon wave with light was discovered by A. Otto in 1968, who used an evanescent optical wave generated by a glass prism illuminated in total internal reflection.[7] Simultaneously Raether and Kretschmann succeeded in exciting the surface plasmon wave on a thin metallic film evaporated on a glass prism.[8] This was the beginning of the modern plasmonics. These phenomena raised a supplemental interest when Liedberg, Nylander and Lundstrom proposed to use the high sensitivity of the plasmon waves to the local environment for detecting biomolecules and gases in 1983.[9]

In parallel, the interest for the plasmon generated by metallic nanoparticles came into focus after the discovery of surface-enhanced Raman spectroscopy (SERS) by Van Duyne where the incredible sensitivity of the Raman signal to the molecular vibrations was assigned to the antenna effect generated by surface roughness.[10] Although the theory of interaction of light with minute particles was developed by Mie in 1908,[11] the first experimental observation of the localised plasmon with light was reported by Smithard in 1973[12] who did not interpret his results in terms of plasmon resonance. A few years later in 1978, Hansma and Broida used electrons to excite the plasmon resonance of particles[13] and their experiments were soon interpreted by Rendell, Scalapino and Mühlschlegel.[14] The complete derivation of the localised plasmon resonance in nanoparticles was later developed by Kreibig in 1985.[15,16] However, the real start of plasmonics occurred in the 1990s. The number of scientific articles referring to plasmon in their title or abstract climbed from 130 in the year 1990, to 1370 in 2002, and has reached 7874 in 2015.[a] Plasmonics has turned into one of the hottest fields in photonics nowadays.

3.1.2 *What Is the Ambition of the Present Chapter?*

This phenomenon occurring at the nanoscale shows up immediately at our scale with very specific colours: a solution containing spherical nanoparticles has a red-purple colour. This is called the localised surface plasmon resonance (LSPR) and is the consequence of the confinement of the electric field within a small metallic sphere. In other words, LSPR results from the Maxwell equations and the boundary conditions imposed on the electric field in spheres whose radius is much smaller than the wavelength. This optical resonance is the origin of many other properties that have made the AuNPs famous: it explains the red-purple colour of spherical nanoparticles, and the slight change of colour when their shape or the surrounding medium are modified; it is responsible for a strong electric field enhancement within a distance of a few times the particle diameter when this resonance is excited by light (near field exaltation), which explains the peculiar nonlinear optical properties of AuNPs and their strong Raman activities; it also results in the absorption of part of the energy of the impinging light beam and a heating of the nanoparticle (see Chapter 4).

---

[a]Data obtained through the database provided by ISI Web of Knowledge, searching 'plasmon' in the topics of the published articles (January 2016).

The present chapter focuses on the optical properties associated with plasmons and begins by clarifying the distinction between propagating plasmon waves called surface plasmon resonance (SPR) and the localised plasmon of nanoparticles traditionally called LSPR. We will derive an analytical model for describing the response of spherical metallic particles with the so-called electrostatic model and show that it gives a reasonable agreement with experiments in many cases. Then we will discuss the improvements needed to treat more complex cases when particles are not spherical or have a diameter larger than 60 nm.

Researchers entering the field of gold nanoparticles may find numerous review articles on how the plasmon resonance is used to enhance sensitivity of spectroscopic techniques or how the shape of nanoparticles influences the optical spectra, but it is more difficult to find an accessible electromagnetic description of the optical properties of ordinary nanoparticles.[17] Ordinary nanoparticles are spherical and diluted gold nanoparticles with sizes between 5 and 60 nm and their optical response can be described by the so-called electrostatic model. Such a model is less sophisticated than the very popular Mie theory, which relies on a somewhat arduous mathematical formalism. The electrostatic model, however, is a good start to understand the various effects that are in play when an electromagnetic wave interacts with the electron clouds of the metallic sphere. This chapter offers a derivation of this model as well as some simple improvements for describing the optical response of AuNP.

3.2 Distinction between Localised Surface Plasmon Resonance and Surface Plasmon Resonance

In dealing with plasmon resonance, a certain confusion reigns in the terms used. Two distinct phenomena are called plasmon resonances: the surface plasmon resonance where an evanescent wave crawls along a metallic surface and the localised plasmon, which is an oscillation of the conduction electrons within a metallic nanoobject. We have to make this distinction clear.

3.2.1 *Optical Properties of Metals*

Metals are characterised by their quasi-free electrons in the ground state, which are not bound to a single atom anymore but can freely move through the crystalline structure of the metal. These free electrons are responsible for the main properties of metals: high conductivity and high optical reflectivity. When an electromagnetic wave characterised by the electric field $\vec{E}(\vec{r}, \omega)$ interacts with a metal, it tends to make the electron cloud oscillate and this creates a dynamic polarisation $\vec{P}(\vec{r}, \omega)$. This quantity expresses how far the electric field succeeds in displacing the electrons relative to the core atoms. The polarised atoms are the sources of a depolarising field and all these effects are combined into the electric displacement $\vec{D}(r, \omega)$ linked to the excitation electric field by the relationship[b] (see Ref. 18 for a good introduction to the optical properties of condensed matter).

$$\vec{D}(\vec{r}, \omega) = \varepsilon_0 \varepsilon(\omega) \vec{E}(\vec{r}, \omega), \tag{3.1}$$

where $\varepsilon(\omega)$ is the dielectric function of the metal and it captures the entire response of a metal to the electromagnetic excitation wave for the whole frequency spectrum, starting from radio frequencies up to X-ray including, of course, optical frequencies. If the medium is vacuum, $\varepsilon(\omega) = 1$ and the displacement $\vec{D}$ is simply proportional to $\vec{E}$, but ε accounts for any kind of materials (metals, insulators, transparent or opaque media) and in the general case, it is a complex function expressed as:

$$\varepsilon(\omega) = \varepsilon_1(\omega) + i\varepsilon_2(\omega). \tag{3.2}$$

For metals, ε is dominated by its real part, which is a negative function. ε is linked to the optical complex index $\tilde{n}$ by

$$\varepsilon = \tilde{n}^2 = [n + ik]^2 = n^2 - k^2 + 2nki, \tag{3.3}$$

where ε and $\tilde{n}$ are functions of the angular frequency ω (linked to the wavelength λ) and this dependence will not be expressly indicated in the following to simplify notations, but it should be kept in mind since the plasmon resonance is directly linked to it.

---

[b]The SI system of units is used for all the equations of this chapter.

The oscillator model is the simplest analytical way to describe the polarisation and gives rise to the Drude dielectric function[18–20] expressed as follows:

$$\varepsilon_{\text{Drude}} = 1 - \frac{\omega_p^2}{\omega^2 + i\Gamma\omega} \approx 1 - \frac{\omega_p^2}{\omega^2},$$

where ω_p is the plasma frequency and Γ the damping constant.

3.2.2 *The Dielectric Function of Gold*

As explained in the previous section, the Drude model is a good start for describing the dielectric functions of metals. The formula given above is slightly modified by replacing the factor 1 by a constant ε_∞ that accounts for transitions that do no need to be explicitly expressed in the visible range.

$$\varepsilon_{\text{Drude}} = \varepsilon_\infty - \frac{\omega_p^2}{\omega^2 + i\Gamma\omega}. \tag{3.4}$$

In the case of gold, the plasma frequency and the damping constant are given by[21]

$$\begin{cases} \hbar\omega_p = 8.95 \text{ eV} \quad \text{i.e. } \omega_p = 1.36 \times 10^{16} \text{ rad} \cdot \text{s}^{-1} \\ \hbar\Gamma = 65.8 \text{ meV} \quad \text{i.e. } \Gamma = 1.0 \times 10^{14} \text{ rad} \cdot \text{s}^{-1} \\ \varepsilon_\infty = 9.5. \end{cases} \tag{3.5}$$

These values are obtained by fitting experimental values with equation (3.4) in an energy range where the free electrons are the major contributions to the dielectric function[22,23] and these values fluctuate from one author to another.

However, when a precise model is needed, the Drude model is far too simplified because it only takes into account the free electrons (intraband transitions) and completely dismisses the contribution from the bound electrons (interband transitions). The latter plays an important role for gold. This is illustrated in Figure 3.1 where the complex dielectric function is plotted. The accurate measurement of this function is crucial for reasonably modelling the optical response of AuNP. One of the preferred measurements was performed by Johnson and Christy in 1972[22] and an approximated analytical model has been published by Etchegoin.[24] The measures provided by the *Handbook of Optical Constants* have some problems near the interband threshold for gold and must be used with great care.[25] In the present

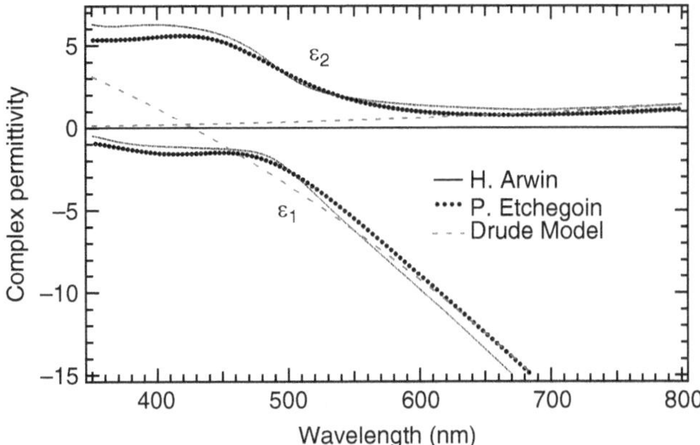

Figure 3.1 The complex dielectric function of gold $\varepsilon = \varepsilon_1 + i\varepsilon_2$ obtained from an ellipsometric measurement[26,27] (thin line) and from an analytical model[24] (large dots) based on the Johnson and Christy's data.[22] The dotted line shows the plot of the Drude model discussed in the text. This model accounts well for the free electron contribution but not for the bound electrons (intraband transitions). This discrepancy is clearly visible in the range of 300–530 nm.

chapter, we are using values measured by H. Arwin from the University of Linköping (Sweden)[26,27] because they cover a wide range of wavelength.

3.2.3 *Plasmon Resonance at Surfaces (SPR)*

An electromagnetic wave in the visible range is rapidly screened by a metal. It does not penetrate into metal much farther than the so-called skin depth given by $\delta = \sqrt{2\rho/\mu_0\mu_r\omega}$.[18,20] In the case of light beam impinging on gold with a wavelength of 500 nm, the skin depth is $\delta = 20$ nm and strongly depends on the wavelength. Therefore, the conduction electrons are excited by the electromagnetic field within an ultrathin sheet of metal close to the surface. This occurs in two cases: in the case of a flat and infinite interface and in the case of structures whose size is of the order of magnitude of the skin depth. We will quickly consider the first case and then focus our attention on the second one.

In the case of a flat interface between a metal and an insulating medium whose dielectric functions are respectively ε and $\varepsilon_{\mathrm{diel}}$, it is possible to launch a surface wave that stays confined very close to the interface. This wave is a charge density wave with a longitudinal structure (unlike light waves that propagate in vacuum with a transverse structure). The charge wave is called polariton wave and is coupled to an electromagnetic wave, which is the surface plasmon wave. This is the reason why this SPR is sometimes called surface plasmon polariton (SPP).[21] This propagation is characterised by the following dispersion relation:

$$k_x = \frac{\omega}{c} \cdot \sqrt{\frac{\varepsilon(\omega) \cdot \varepsilon_{\text{diel}}\omega}{\varepsilon(\omega) + \varepsilon_{\text{diel}}\omega}}. \tag{3.6}$$

The dispersion relation plays a key role in electromagnetism because it controls the propagation of the wave. It is an expression of the link between fundamental quantities in physics: energy ($\hbar\omega$) and momentum ($\hbar\vec{k}$). For a monochromatic light beam of photon energy ω, only waves whose wave vector has a component parallel to the interface k_x given by relation (6), can be excited. In this relationship, the number in the square root is greater than one for metals, meaning that the wave vector of the plasmon wave is greater than that of any wave travelling in free space. As a consequence, surface plasmons can only be launched with special set-ups. The most usual way is to excite the SP with an evanescent wave resulting from a total internal reflection from a prism in the so-called Kretschmann configuration depicted in Figure 3.2.[28]

When the geometrical conditions are fulfilled to excite the surface plasmon, this is the surface plasmon resonance (SPR). It shows up as a dip in the reflection spectrum when the incident beam impinges on the gold film with a given angle (see Figure 3.2(b)).

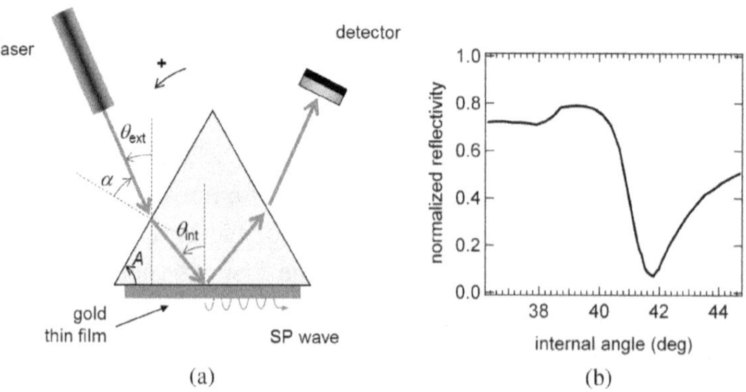

Figure 3.2   Surface plasmon resonance: (a) Sketch of the prism for coupling an excitation wave of a laser to the surface plasmon (SP) wave. The coupling is controlled by the incidence angle. (b) When the coupling is achieved, the reflected beam undergoes a strong intensity drop measured at $\theta_{\text{int}} = 41.8°$ in the present case.

Since SPR only occurs on flat surfaces approximated to planes of infinite extension, we will not discuss this kind of plasmon resonance further and restrict ourselves to charge oscillations within particles of nanometre dimensions in three-dimensional space.

3.2.4 *Localised Surface Plasmon Resonance in Nanoparticles*

In nanoparticles, the charge oscillation is different from the SPR described above. Qualitatively, if particles have sizes much smaller than the wavelength of light and smaller than the penetration depth of the field (i.e. particle size around 20 nm), the electron cloud of the particle is entirely probed by the electric field. The whole assembly of electrons is polarised, and this creates surface charges that accumulate alternately on opposite ends of the particle. This oscillating polarisation of the particles creates an electric field opposed to the excitation field and results in a restoring force. This oscillation is partially damped. The damping occurs through two channels: creation of heat and light scattering. All this can be described as a dipolar oscillator characterised by a resonance frequency $\omega_{plasmon}$, which will be discussed and used throughout this book.

In the following, this resonance will be called LSPR. This denomination is generally accepted although this localised plasmon oscillation is not primarily a surface effect, but a bulk effect taking place in the very small and confined volume of metallic nanoobjects. Thus, some authors prefer using other denominations such as nanoparticle plasmon (NPP)[29] and surface plasmon on metal nanoparticles (SPN).[30]

The main properties of LSPR can be understood within the dipolar model and can be summarised as follows:

— Spectrally, the plasmon resonance appears in the visible or near-infrared range for gold or silver nanoparticles. A light beam passing through an assembly of homogenous nanoparticles is partially absorbed at the plasmon resonance frequency so that the emerging beam displays a spectrum with a sharp absorption at $\omega_{plasmon}$. At the same time, the nanoparticles exhibit light scattering with a cross-section much larger than conventional dye.[31,32]

— The LSPR strongly depends on the environment close to the particle surface. For example, if the layer of molecules adsorbed on the NP changes, the SPR shifts. The AuNP typical shift for a protein interaction is of the order of magnitude of 10 nm (see Chapter 16). This effect is the basis of plasmonic biosensing as reviewed by Unser *et al.* in 2015.[33]

— When excited at the resonance, the dipole radiates a near field electromagnetic wave, whose amplitude can be enhanced by a factor up to 10. This plasmon amplification, also called plasmonic antenna,[34] is widely used for enhancing the sensitivity of biosensors (Chapter 14).

3.3 Theoretical Description of the Localised Plasmon Resonance

3.3.1 *About Mie Theory*

An exact theory in case of spherical particles of any size is provided by Mie theory.[11] It gives the exact solution of a plane wave interacting with a metallic sphere. The electromagnetic fields are expanded in multipole contributions and the expansion coefficients are found by applying the correct boundary conditions for electromagnetic fields at the interface between the metallic nanoparticle and its surroundings. For small particles (<60 nm), it is sufficient to restrain the multipole expansion to its first term, which is dipolar. This approximation is the dipolar approximation, also called the quasi-static or Rayleigh limit. We will develop this approximation, which is sufficient to grasp the working principles of plasmonics; readers interested in the full Mie theory should refer to textbooks such as Born and Wolf[20] or Bohren and Huffmann.[35]

3.3.2 *The Quasi-static Approximation for Describing the Localised Plasmon Resonance*

Let us consider a metallic spherical particle of radius R, which is subjected to an excitation electric field aligned along the x-axis: $\vec{E} = E(\vec{r}, t)\vec{e}_x$ (see Figure 3.3). In the quasi-static limit, the electronic polarisation is exactly in phase with the excitation field (no retardation effect) and the electrons are displaced as a whole. Therefore the charge distribution in the particle

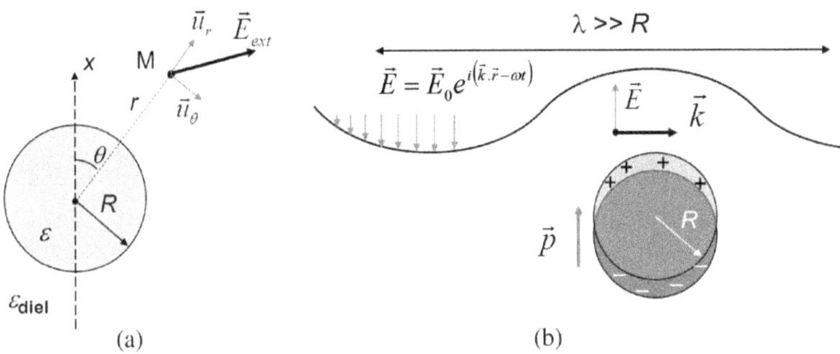

Figure 3.3   (a) Sketch of the metallic sphere and the coordinates used in the electrostatic model to calculate the external electric field. (b) Under the influence of an excitation wave with a wavelength greater than the dimension of the sphere, the electrons oscillate as a whole and the system can be handled as an oscillating dipole.

can be treated as if it were a static distribution.[21,35] The electric field should obey the Laplace equation:

$$\Delta V = 0, \tag{3.7}$$

where V is the electric potential linked to the electric field by $\vec{E} = -\vec{\nabla} V$.

Due to the symmetry of the problem, spherical coordinates are used. Moreover, the x-axis is an axis of symmetry so that the third coordinate usually required for spherical coordinates becomes useless. Therefore, Equation (3.7) written in spherical coordinates becomes

$$\frac{1}{r}\frac{\partial}{\partial r^2}(rV) + \frac{1}{r^2 \sin\theta}\frac{\partial}{\partial\theta}\left(\sin\theta\frac{\partial V}{\partial\theta}\right) = 0. \tag{3.8}$$

The solutions of this differential equation are the spherical harmonics with the following form:

$$V(r,\theta) = \sum_{n=0}^{\infty}\left(A_n r^n + \frac{B_n}{r^{n+1}}\right)P_n(\cos\theta), \tag{3.9}$$

where A_n and B_n are coefficients to be determined. P_n are the Legendre polynomials that often appear in physics problem expressed in spherical coordinates.[c]

---

[c]The Legendre polynomials are obtained as solutions of the Legendre equations or the nth element of the recurrence relation of Bonnet. The first four terms are : $P_0(X) = 1$; $P_1(X) = X$; $P_2(X) = 3/2X^2 - 1/2$; $P_3(X) = 5/2X^3 - 3/2X$.

This electric potential has two different forms $V_{int}(r,\theta)$ and $V_{ext}(r,\theta)$ inside and outside the metallic particle. Moreover it should obey the following boundary conditions:

— The electric field should be properly defined at $r = 0 : \frac{\partial V_{int}}{\partial r}|_{r=0}$ exists.
— The electric field far away from the particle should match the excitation field, which is constant in the electrostatic approximation, so that $\lim_{r \to \infty} V_{ext} = 0$.
— At the interface, the electric fields obey the continuity equation at the particle surface: $\varepsilon(\omega)E_{int}(r = R) = \varepsilon_{diel}(\omega)E_{ext}(r = R)$.
— Finally, the potential should be continuous at the particle surface: $V_{int}(r = R) = V_{ext}(r = R)$.

Applying these conditions allows us to determine the coefficients for the electric potentials inside and outside the particle. For example, for the external potential one determines that all the A_n coefficients are null except one: $A_1 = -E_0$ and similarly for the B_n coefficients, the only non-zero coefficient is given by

$$B_1 = E_0 4\pi\varepsilon_0 R^3 \frac{\varepsilon - \varepsilon_{diel}}{\varepsilon + 2\varepsilon_{diel}}. \tag{3.10}$$

Once the electric potential is obtained, the electric field is deduced by using the gradient operator: $\vec{E} = -\vec{\nabla}V$.

After some calculations, one obtains the following expression for the electric field outside of the nanoparticle, which is a sum of the incident field and a second field produced by the particle:

$$\vec{E}_{ext} = \vec{E}_0 - \alpha E_0 \left[-2\frac{\cos\theta}{r^3}\vec{u}_r - \frac{\sin\theta}{r^3}\vec{u}_\theta \right], \tag{3.11}$$

where α is the sphere polarisability given by

$$\alpha = 4\pi\varepsilon_0 R^3 \frac{\varepsilon - \varepsilon_{diel}}{\varepsilon + 2\varepsilon_{diel}}. \tag{3.12}$$

The electromagnetic response of the particle is captured in the polarisability. It is clear that the external field will go through a maximum when the polarisability is maximised. The parameter that depends on frequency is $\varepsilon = \varepsilon(\omega)$. Therefore, $|\alpha|$ is maximised when the following relationship is fulfilled:

$$|\varepsilon + 2\varepsilon_{\text{diel}}| \text{ is a minimum.} \tag{3.13}$$

As a good approximation, the dielectric permittivity of the surrounding medium, $\varepsilon_{\text{diel}}$ is a constant and a real parameter. Therefore, Equation (3.13) leads to the following condition applied to the real part of ε: $\varepsilon_1 + 2\varepsilon_{\text{diel}} = 0$. In case of a nanoparticle in water, $\varepsilon_{\text{diel}} = 1.77$, and the condition becomes $\varepsilon_1 = -2 \times 1.77 = -3.54$. From the plot of the dielectric function of gold in Figure 3.1, it is easy to check that this condition leads to a plasmon resonance at 520 nm. If the particle was in air, the plasmon resonance would be at 504 nm. These calculated values for the plasmon resonance in water or in air correspond closely to experimental values of spherical gold nanoparticles.

3.3.3 *Extinction and Scattering Cross-Sections*

However, in order to proceed one step further and calculate the intensity of light being scattered or absorbed, the electrostatic model has to be completed to take into account that light is a wave and not a static electric field. Upon interaction with the particle, light is absorbed (absorption cross-section σ_{abs}) and scattered (scattering cross-section σ_{scatt}). As a result, the beam going through a particle undergoes an extinction characterised by the extinction cross-section (σ_{ext}). The three cross-sections are linked by the simple relation

$$\sigma_{\text{abs}} = \sigma_{\text{ext}} - \sigma_{\text{scatt}}. \tag{3.14}$$

These phenomena are strongly frequency dependent and therefore time dependent. The model of the radiating dipole allows for calculating the extinction and scattering cross-sections and links them to the polarisability found in Equation (3.12):

$$\sigma_{\text{ext}} = 3\frac{2\pi}{\lambda}\sqrt{\varepsilon_{\text{diel}}}\ \text{Im}(\alpha) \tag{3.15a}$$

$$= 9\frac{2\pi}{\lambda}\varepsilon_{\text{diel}}^{\frac{3}{2}}V\frac{\text{Im}(\varepsilon)}{|\varepsilon + 2\varepsilon_{\text{diel}}|^2}, \tag{3.15b}$$

$$\sigma_{\text{scatt}} = 3\frac{(2\pi)^3}{\lambda^4}\varepsilon_{\text{diel}}^2|\alpha|^2 \tag{3.16a}$$

$$= 3\frac{(2\pi)^3}{\lambda^4}\varepsilon_{\text{diel}}^2 V^2\left|\frac{\varepsilon - \varepsilon_{\text{diel}}}{\varepsilon + 2\varepsilon_{\text{diel}}}\right|^2. \tag{3.16b}$$

This description is acceptable for homogeneous, metallic spherical nanoparticles whose diameter is roughly between 10 and 60 nm (see the following section).

For small particles, only absorption is in play with negligible scattering. A proper calculation within the discrete dipole approximation (DDA) formalism has been conducted by El-Sayed and co-workers[32] and shows that for 20 nm AuNP, no scattering occurs and the particles only absorb radiation. For diameters of 80 nm, both cross-sections are equivalent, and for larger particles, scattering dominates (see Figure 3.4). In other words, if one wants to have brilliant particles, he must choose particles with a large diameter, for example larger than 40 nm. For biomedical applications where

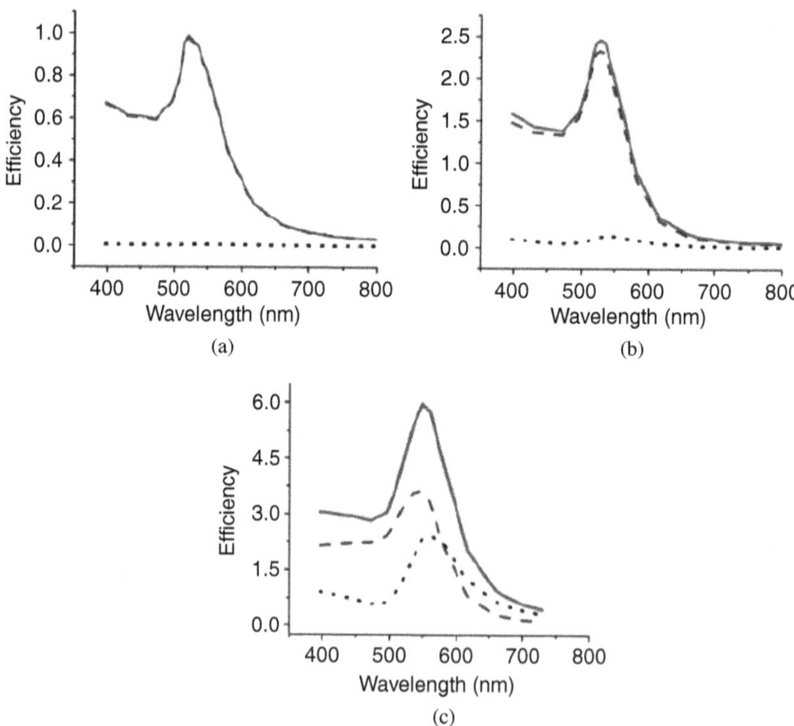

Figure 3.4   Comparison of the extinction (solid line), absorption (dashed line) and scattering (black dots) efficiencies as a function of particle diameter for gold nanoparticles. In the case of particles of 20 nm diameter (graph a), absorption is largely dominating as well as for 40 nm (graph b). For 80 nm NP, scattering and absorption are of the same order of magnitude. Calculations are made with the Mie theory. Reprinted with permission from Ref. 32. Copyright 2006, American Chemical Society.

AuNPs can be used as markers, this will be important. AuNP nanospheres with a diameter of 40 nm have a calculated absorption cross-section of 2.93×10^{-15} m^2 (thus corresponding to a molar absorption coefficient ε of 7.66×10^9 M^{-1}.cm^{-1}) at a plasmon resonance wavelength maximum λ_{max} of 528 nm. This value is five orders larger than the molar extinction coefficient for indocyanine green $\varepsilon = 1.08 \times 10^4 \text{M}^{-1} \cdot \text{cm}^{-1}$ at 778 nm), a NIR dye commonly used in laser photothermal tumour therapy (see Chapter 10).

3.3.4 *Experimental Illustrations*

The electrostatic approximation and the analytical model provided by Equation (3.15) accurately describe simple experimental situations such as spherical nanoparticles in suspensions with a concentration such that the particles stay far from each other (a few particle diameters). In this case, one usually measures the absorbance of a solution placed in the cuvette of spectrophotometer. The absorbance is linked to the extinction cross-section by

$$A = -\frac{1}{\ln 10} nb\sigma_{\text{ext}}, \tag{3.17}$$

where n is the number of nanoparticles per unit volume, b the length probed by the optical beam (thickness of the cuvette) and σ_{ext} is the cross-section given by relation (15). As an example, Figure 3.5 shows a comparison of the measured absorbance of 14.2 ± 1.3 nm gold nanoparticles in suspension in water with the calculated absorbance.

Systematic checks have been conducted and show that the plasmon resonance is measured at 519 nm in aqueous solutions and does not shift for diameters from 4 to 35 nm. The extinction coefficients are precisely measured and can be used in a Beer–Lambert law.[36] According to such a rule, the measured absorbance written as

$$A = \varepsilon bC, \tag{3.18}$$

where ε is the molar extinction coefficient, b is the cuvette thickness and C the AuNP concentration. For example, in the case of a 8.6 nm citrate-capped AuNP, $\varepsilon = 5.14 \times 10^7 \text{M}^{-1} \cdot \text{cm}^{-1}$ and for a 20.6 nm AuNP, $\varepsilon = 8.78 \times 10^8 \text{M}^{-1} \cdot \text{cm}^{-1}$.[36]

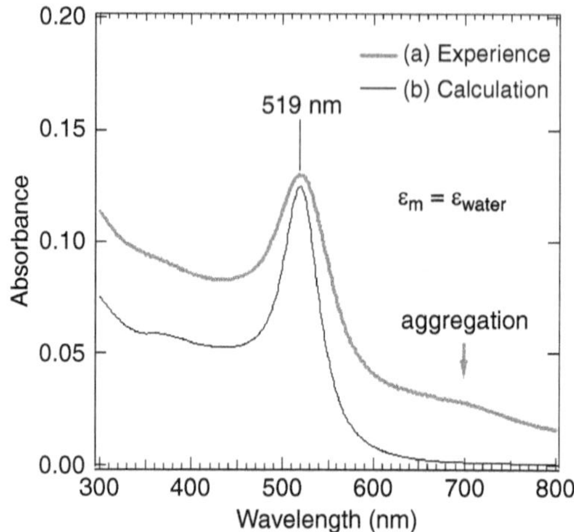

Figure 3.5   Comparison of measured and calculated extinction spectra of 14.2 nm gold nanoparticles prepared by the Turkevich method. The experimental spectrum is taken in a $b = 1$ mm thick cuvette and the density of AuNP is $n = 1.76 \times 10^{18}$ m^{-3}. A beginning of particle aggregation is visible at 700 nm on the experimental spectrum and indicated on the graph.

Aggregation of AuNPs is the most frequent cause of plasmon shift in such colloidal solutions; it causes the plasmon to redshift. An illustration can be found here.[37] The coupling between neighbouring particles is discussed in a following section.

3.3.5 *Local Field Enhancement and Nanoantennas*

From Equation (3.11), the electric field radiated by a nanoparticles can be expressed as

$$E_{out} = E_0 \frac{3\varepsilon}{\varepsilon + 2\varepsilon_{diel}}. \tag{3.19}$$

At $\lambda = 530$ nm in air, this radiated field is five times larger than the excitation field E_0. This enhancement is confined to the close vicinity of the particle (local field) and has given birth to the concept of optical nanoantennas.[21,29,34] The local field amplification generated by the plasmon is compared to the macroscopic antenna used in radio broadcasting because it enables an efficient transfer of electromagnetic energy from the near to the far-field of metal nanoparticle and vice versa. Therefore, any plasmonic nanostructures can be considered as a more or less efficient nanoantenna.

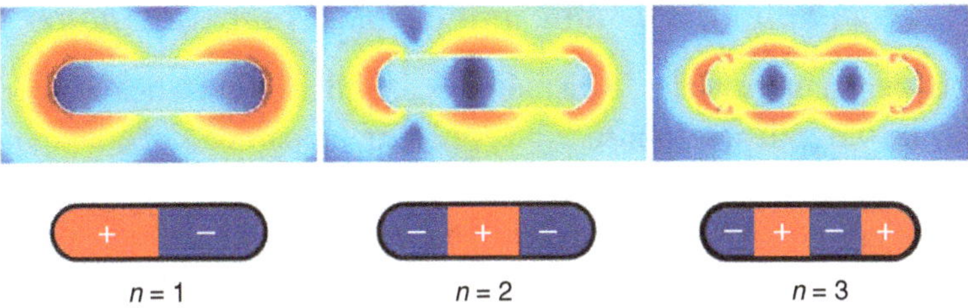

Figure 3.6   Working principles of nanoantennas illustrated in the case of a silver nanorod of 400 × 100 × 50 nm^3. The calculated mapping of the near-field generated for three optical modes at 1375, 770 and 630 nm. The nanorod is placed on top of a glass substrate and is illuminated from the top by a plane wave tilted 20° with respect the vertical direction and polarised along its long axis. The colour code for the enhancement is from 0.1 (blue) to 100 (red). Reprinted with permission from Ref. 34. Copyright 2011 American Chemical Society.

The simplest example of how local field engineering influences the far-field optical properties is the case of spherical gold nanoparticles in aqueous solution (red colour), which turns blue when the nanoparticles come in close vicinity when they aggregate. This colour change from red to blue is explained by the coupling of the near-field waves generated by the dipoles of individual particles. In general, an efficient nanoantenna should exhibit a spectrally well-defined plasmon mode and give rise to a large local field enhancement. In the case of the nanorod presented in Figure 3.6, the local electromagnetic field was calculated for three of the plasmonic modes.[34] Mode #1 is a simple dipolar charge oscillation and the local electromagnetic field enhancement is represented in false colour. The red areas at each end of the nanorod correspond to a field enhancement factor of 100. The other modes at higher energies (plasmon resonance at 770 nm for $n = 2$ and 630 nm for $n = 3$) also exhibits area with such a high enhancement. These high field areas correspond to the hot spots of the nanotenna. This figure also shows that the near field extension is of the order of magnitude of the nanostructure (50 nm in the present case).

This antenna effect can be used, for example, to enhance the optical sensitivity to a molecule. In that case, the plasmonic enhancement will be efficient if the molecule is placed in a high local field area. This local field amplification is very important for explaining the high sensitivity obtained in surface-enhanced Raman scattering (SERS),[10,38] or more complex non-linear optical spectroscopies[39] for which applications are developed in Chapter 14.

3.3.6 *Beyond the Quasi-static and Dipolar Approximations*

For spherical nanoparticles, Mie theory is an efficient but a heavy approach to describe the optical response. However, for nanoparticles of non-spherical shapes, exact solutions cannot be derived except for spheroids[40] or infinite cylinders[41] exclusively. For other shapes numerical methods have been developed and some of them are quickly reviewed in the following. A good tutorial review article was published in 2008 by Garcia de Abajo.[38]

Discrete dipole approximation (DDA) is the most popular numerical approach[42] and is used by a considerable number of groups. The nanoobject to be analysed is represented by a tridimensional cubic array of dipoles. The polarisation at each point is induced by a local electric field, which is produced by the external exciting field and the sum of the fields generated by all the other point dipoles. In case of a mapping of the nanoobject with N points, the global polarisation vector is the solution of $3N$ linear equations that are solved numerically. Quantities like extinction or scattering cross-sections are deduced for this set of dipoles in the far field as well in the near-field region. DDA can represent an object or multiple objects of arbitrary shape and composition and yields results with typically 10% accuracy.

The multiple multipole (MMP) method is a semi-analytical approach to solve Maxwell's equations in multiple connected media that have to be homogeneous, isotropic and linear.[43] The region of interest is divided into connected domains and the field inside each domain is described by a series expansion of known analytical solutions of Maxwell's equations. MMP mainly uses multipolar expansion with different origins as basis functions of the series expansion. The computational advantage of MMP is that only the boundaries, not the domains themselves, need to be discretised. The choice of suitable sets of basis functions is the most difficult task in MMP as no optimum can be defined in a unique way. The use of multipoles and not just dipoles makes this method more powerful to describe near-field effects, but increases its complexity.

With the finite difference time domain (FDTD) method, the Maxwell's curl equations are solved explicitly.[44] The equations are discretised both over time and space. The method is based on a time marching algorithm that runs over a carefully defined spatial grid. It can be used for studying both the near- and far-field electromagnetic responses for heterogeneous materials of arbitrary geometry. The time-marching aspect of the method allows one to make direct observations at any position in space (near-field and far-field) and at any time during the simulation. This last feature often brings new insight into the dynamics of the system under study.

The T-matrix method is especially suited to calculate the extinction, absorption and scattering of an ensemble of nanoparticles taking into account their multipole contribution and not just their dipolar response.[45] The transition matrix (T-matrix) establishes the relationship between the different expansion coefficients of the multipolar expansion. The averaging over the different orientations taken by the particle is made in an analytical way. Multiple scattering effects are accounted for with this method.

3.4 Factors Shifting the Plasmon Resonance of Gold Nanoparticles

The position of the plasmon resonance is basically given by relationship (Equation (3.13)) with a rather good accuracy in the case of spherical particles of moderate sizes. It is evident that this relationship does not offer any practical 'handle' to shift the plasmon resonance. In particular, relations (3.13) as well as (3.15) show that the nanoparticle size does not influence the LSPR position as long as the dipolar approximation is valid. Yet when applications are envisioned, it is often necessary to tune this plasmon wavelength. This is the case in the field of biosensors when one wants to excite nanoparticles through the human skin and needs to use a light source compatible with the biological window (650–900 nm). Another case among many others is the use of the local field enhancement at the plasmon resonance: it might be necessary to tune this nanoantenna to the wavelength of the light source.

The LSPR can be adjusted in three ways:

1. By changing the medium surrounding the particle.
2. By changing the shape of the nanoparticles: ellipsoids, cube, triangle, icosahedra, etc.
3. By using core–shell particles.

As a general behaviour, these parameters shift the LSPR to higher wavelengths compared to the case of a sphere in air ($\lambda_{\text{plasmon}} = 504$ nm). Such a redshift pushes the resonance away from the interband transitions of gold and therefore decreases their mutual coupling. Subsequently the LSPR appears as a narrower and more pronounced resonance. These effects are developed in the following three sections.

3.4.1 *What is the Dependence of the LSPR with the Nanoparticle?*

In the case of spherical AuNPs, relation (3.16) shows that the LSPR does not depend on the nanoparticle diameter. Yet, studies report a dependence of the maximum of optical absorbance of a nanoparticle suspension with their diameter. This dependence comes from the deviation from the hypothesis made in Section 3.1.3.2: actual nanoparticles are not ideal spheres and dipolar approximation does not hold anymore. Taking into account these deviations in calculations demands subtle improvements, which are beyond the scope of this chapter. Nevertheless the experimental dependence of the LSPR with the diameter of gold nanoparticles is of practical importance and is displayed in Figure 3.7. The figure shows seven nanoparticle suspensions with diameters ranging from 30 to 90 nm.[46] The colour barely changes for these nanoparticles and the plasmon shifts away from 520 nm for diameters

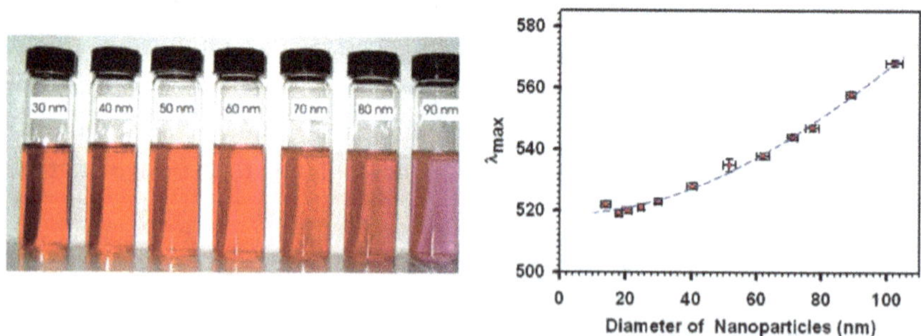

Figure 3.7   Series of suspensions of monodisperse spherical AuNPs with diameters between 20 and 100 nm. The LSPR shifts from 520 to 570 nm especially for diameter higher than 40 nm. The colour of this suspension is systematically red. Reprinted with permission from Ref. 46. Copyright 2007 American Chemical Society.

larger than 40 nm. The authors propose a practical fit with their experimental data and they link $\lambda_{\max}$ (in nm) to the diameter D (in nm).

$$\lambda_{\max} = 518.8 - 0.0172 \times D + 0.0063 \times D^2 - 1.34 \times 10^{-5} \times D^3. \quad (3.20)$$

3.4.2 *Influence of the Surrounding Medium*

From the discussion in Section 3.1.3.2, it is clear that the surrounding medium plays a role in the plasmon resonance through its optical index n linked to the dielectric permittivity: $\varepsilon_{\mathrm{diel}} = n^2$. Within the limits of the electrostatic approximation already mentioned above, the formula (3.13) provides a good accuracy between calculated and measured wavelengths. It shows that the higher the optical index, the higher the plasmon resonance. Some results are summarised in Table 3.1.

Along with the shift of the plasmon resonance to higher wavelengths, the increase of the index of the surrounding medium is accompanied by a sharp increase of the absorption cross-section. This is illustrated in Figure 3.8 where the evolution with surrounding medium of the absorbance of an assembly of gold particles of 14 nm diameter is calculated. In the present case, this assembly exhibits an absorbance of 0.04 in air, 0.18 in glass and 1.87 in titanium dioxide.

Table 3.1   Calculated plasmon resonance wavelengths in case of spherical gold nanoparticles in different surrounding media.

Plasmon resonance of spherical particles in various surrounding media					
Material	Air	Water, glycerol	Silica glass, DMSO	Alumina (Al_2O_3)	Sapphire	Titanium dioxide (TiO_2)
n	1	1.33	1.47	1.66	1.77	2.79
$\varepsilon_{\mathrm{diel}}$	1	1.77	2.16	1.77	3.13	7.78
$\lambda_{\mathrm{plasmon}}$ (nm)	**504**	**519**	**527**	**541**	**551**	**678**

Note: The electrostatic model is used with the dielectric function of gold provided by H. Arwin.[27]

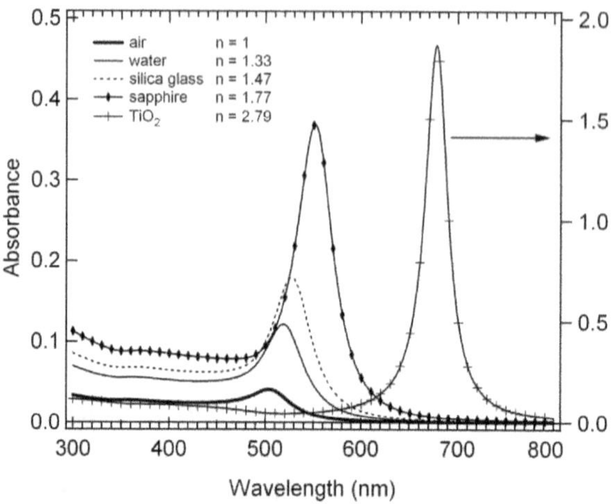

Figure 3.8 Calculated spectra of the absorbance through an assembly of gold nanoparticles when the surrounding medium is modified. This assembly is made of a sheet of particles of diameter of 14.2 ± 1.3 nm, with a density $n = 1.76 \times 10^{18}$ m^{-3} and a thickness $b = 1$ mm.

3.4.3 *Plasmon Resonance of Ellipsoids and Other Shapes*

For non-spherical particles, the absorption and scattering cross-sections can be calculated from relations (3.16a) and (3.17a) as soon as the polarisability α is known. However, an exact analytical model for calculating this polarisability is only available in the case of ellipsoids[40] and infinite cylinders.[41]

We first focus on the case of ellipsoids.[23,47] They possess three plasmon resonances corresponding to the oscillation of electrons along the three principal semi-axes, denoted a, b and c. By changing the axes lengths, the LSPR can be tuned from ca. $\lambda = 500$ nm up to the infrared range. It is obvious that the excitation of one of these plasmon resonances depends on the direction of the impinging electric field. Therefore, it depends on the direction of the light beam and its polarisation. Moreover, a crucial challenge lies in synthesising such objects (see Chapter 6): they are produced as assemblies and their size and shape are usually not homogeneous. Since they cannot be probed as individual objects except with dedicated experimental set-ups,[48,49] their optical response is an average over all the sizes and orientations present in the assembly. This may result in a smoothing or vanishing of the plasmon resonances.

For an individual metallic ellipsoid, the polarisability along the axis i ($i = x$, y or z) is given by

$$\alpha_i = \frac{V_e}{4\pi} \times \frac{\varepsilon - \varepsilon_m}{\varepsilon_m + L_i(\varepsilon - \varepsilon_m)}, \tag{3.21}$$

where ε is the dielectric function of the metal, which is a complex number and a function of the wavelength, and ε_m is the dielectric function of the surrounding medium, which can be considered as a real number, independent of the wavelength in the visible range of interest, as long as this medium is transparent with negligible absorption.

V_e is the volume of the ellipsoids given by: $V_e = \frac{4\pi}{3}abc$, with a, b and c being the semi-axes of the ellipsoids.

L_i is the depolarisation factor, which is a purely geometric factor.

For simplicity, we will consider revolution ellipsoids with two equal axes. These ellipsoids correspond to the majority of the experimental cases and belong to two families: they can be prolate (rugby ball-like with $a > b = c$) or oblate (pumpkin-like with $a = b > c$). Among the three proper modes, two of them are degenerated and two depolarisation factors are equal. They appear as functions of the eccentricity e of the corresponding ellipse along axis i. In case of prolate spheroids, the depolarisation factors are given by[23,47]:

$$L_x = \frac{1 - e^2}{2e^3}\left(\log\frac{1 + e}{1 - e} - 2e\right),$$

$$L_y = L_z = \frac{1}{2}(1 - L_x) \quad \text{and} \quad e = \sqrt{1 - \frac{b^2}{a^2}}. \tag{3.22}$$

And for oblate spheroids:

$$L_z = \frac{1 + e^2}{e^3}(e - \tan^{-1} e),$$

$$L_x = L_y = \frac{1}{2}(1 - L_z) \quad \text{and} \quad e = \sqrt{\frac{a^2}{c^2 - 1}}. \tag{3.23}$$

We label the two modes as longitudinal modes (LMs) when the electric field is along the symmetry axis and transverse modes (TMs) when it is perpendicular to this axis. Notice that this model assumes that the dipolar approximation is acceptable (nanoparticles of moderate size) and that no quadrupolar modes enter in play. For a sphere ($a = b = c$), the three resonances are degenerated and the depolarisation factors are all equal to 1/3. In this case, it is easy to check that Equation (3.21) gives the polarisability of the electrostatic model of relation (3.12).

This model allows for a fully analytical calculation of the extinction cross-section, assuming a dipolar model for the ellipsoid. Some results are presented in Figure 3.9 in the case of particles in air and compared to spherical gold particles with a radius of 7 nm whose plasmon wavelength is at 504 nm. This graph plots the extinction efficiency, which is the ratio of the extinction cross-section to the geometrical cross-section. The ellipsoid is chosen so that its volume is the same as the volume of a sphere with $r = 7$ nm. For example, in the case of the prolate ellipsoid of aspect ratio $a/c = 2$, the dimensions are the following: $a = 11.2$ nm, $b = 5.56$ nm and

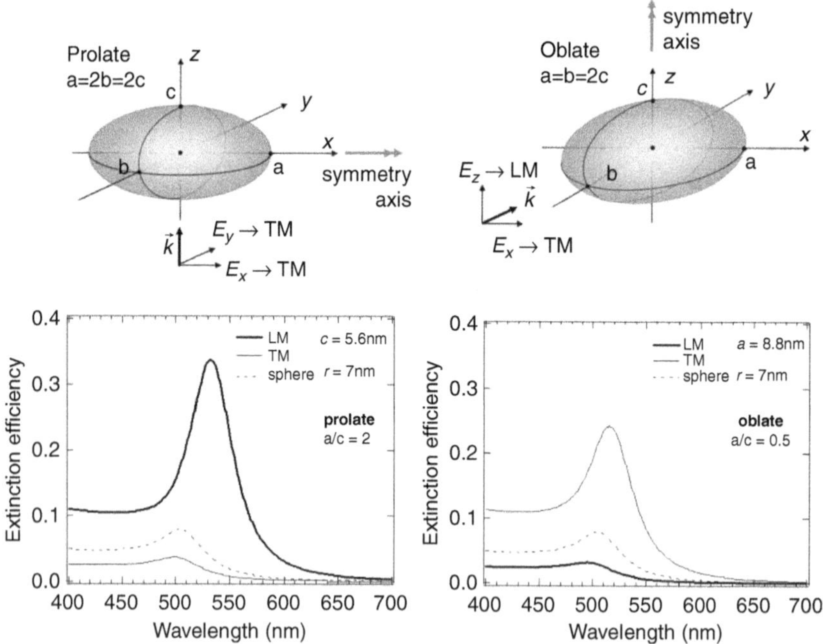

Figure 3.9 Calculated extinction efficiency of prolate and oblate ellipsoids within the dipolar approximation. Ellipsoids of aspect ratio of 2 are compared to a sphere of equal volume. The longitudinal mode corresponds to an excitation field parallel to the symmetry axis and the transverse mode is perpendicular to this axis. In case of prolate ellipsoids, LM is the most sensitive to the change of shape whereas for oblate particles TM is the most sensitive.

Table 3.2   Calculated plasmon resonance wavelengths for prolate gold spheroids ($a > b = c$) of different aspect ratios in air.

			Prolate spheroids				
Aspect ratio (a:b)	1:1	2:1	4:1	6:1	8:1	10:1	20:1
L_x	0.3333	0.1735	0.0754	0.0432	0.0284	0.0203	0.0067
$L_y = L_z$	0.3333	0.4132	0.4623	0.4784	0.4858	0.4898	0.4966
λ_{LM} (nm)	504	532	634	760	897	1037	<1700
λ_{TM} (nm)	504	500	498	496	495	495	495

$c = 5.56$ nm. In this case, the LM is shifted to 532 nm and its intensity is enhanced with a factor almost 5. TM is slightly blueshifted down to 500 nm and partially damped. By increasing the aspect ratio, this trend is amplified. For example, elongated ellipsoids of aspect ratio of 10 exhibit a longitudinal mode in the near infrared at 1037 nm whereas the transversal mode stays at 495 nm. Some other results as well as the values of the depolarisation factors are given in Table 3.2. The case of oblate ellipsoids is similar with the slight difference that the TM is now the most sensitive to the change of particle shape and shifts to higher wavelengths when the aspect ratio is increased. For example, an oblate ellipsoid of aspect ratio of 2 exhibits a transverse plasmon at 515 nm. Other values are summarised in Table 3.3.

For an electric field of random orientation relative to the symmetry axis, the extinction spectrum is a combination of the two modes and will exhibit two resonances.[23]

With the help of other calculation methods, the plasmon modes of particles of more complex shapes can be calculated.[23,50] New modes show up. Many calculations have been done to address the case of nanorods, which can be at first assimilated to ellipsoids but exhibit more complex plasmonic structures.[51] Plasmon modes of varying shapes have been calculated in the case of silver particles: they are plotted in Figure 3.10 for a cube, several truncated cubes, a cuboctahedron, an icosahedron and a sphere. Although the plasmon resonances are different for silver and for gold, the trend is similar: the number of plasmon modes increases more that the shape differs from a sphere. A cube, for example, exhibits six resonances. It is generally

Table 3.3   Calculated plasmon resonance wavelengths for oblate gold spheroids ($a = b > c$) of different aspect ratios in air.

Oblate spheroids							
Aspect ratio ($a{:}c$)	1:1	2:1	4:1	6:1	8:1	10:1	20:1
$L_x = L_y$	0.3333	0.2363	0.1482	0.1077	0.0845	0.0695	0.0369
L_z	0.3333	0.5272	0.7036	0.7846	0.8308	0.8608	0.9262
λ_{TM} (nm)	504	515	545	579	615	650	807
λ_{LM} (nm)	504	495	492	490	490	490	488

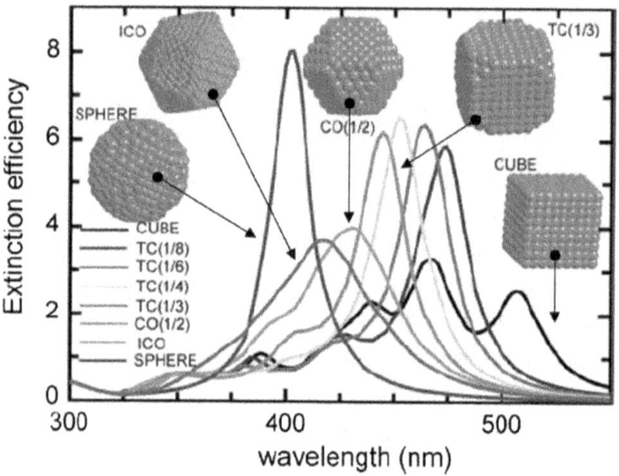

Figure 3.10   Extinction efficiencies for silver nanoparticles of different shapes. Reprinted with permission from Ref. 23. Copyright 2007 American Chemical Society.

observed that the vertices of the nanoparticles play an important role in the optical response, because the sharper they become, the greater the number of resonances. A main resonance with a dipolar character is always present along with other secondary resonances of lower intensity. It is also observed that as the nanoparticle becomes more symmetric, the main resonance is always blueshifted.[50]

3.4.4 *The Case of Very Small (Less than 5 nm) and Very Large Gold Nanoparticles (Greater than 60 nm)*

In a first approach, size does not affect the SPR position but the balance between scattering and absorption. This is true as long as the approximation that describes the particle as a radiating dipole stands. However, as soon as the particle is larger than 60 nm, multipolar effects come into play.[52] They show up as larger bands and tend to shift the dipolar contribution. The displacement of the electron cloud is no longer homogeneous over the entire particle and the oscillating charges can no longer be simply described with two opposite charges at a fixed distance (model of the dipole). The quadrupolar and higher multipolar plasmonic contributions are always located at shorter wavelengths compared to the dipolar ones, which are displaced to higher wavelengths. In the case of spherical nanoparticles, these effects can be analytically predicted using Mie theory and taking into account terms of orders higher than 2 in the multipolar development of the electromagnetic field.[17,52]

If the particle size becomes smaller than the mean free path of the free electrons (the conduction band electrons), the collisions of electrons with the particle surfaces becomes important.[16,23] It results in a slight broadening of the plasmon band for AuNPs smaller than 10 nm, and becomes really obvious below 5 nm.[53,54] For gold nanoparticles smaller than 2.5 nm, experimental studies of the UV–visible extinction coefficients show that the plasmon resonance experiences a slight blueshift and eventually vanishes as shown in Figure 3.11.[53] The theoretical description of this effect needs a semi-quantum description to take into account the interplay between the d-electrons and the s-electrons of the conduction band.[54] For large particles, this interplay between bound and free electrons also takes place but it is captured by the bulk dielectric function as already discussed above and shown in Figure 3.1.

For small nanoparticles, a typical quantum phenomenon that influences the optical response is the spill-out of the electrons. It occurs at the surface of a metal and corresponds to the fact that the density of the electrons clouds does not undergo an abrupt transition. For spherical nanoparticles, for example, the radius of the electron sphere is slightly larger than the radius of the positively charged sphere constituted with the cationic cores.

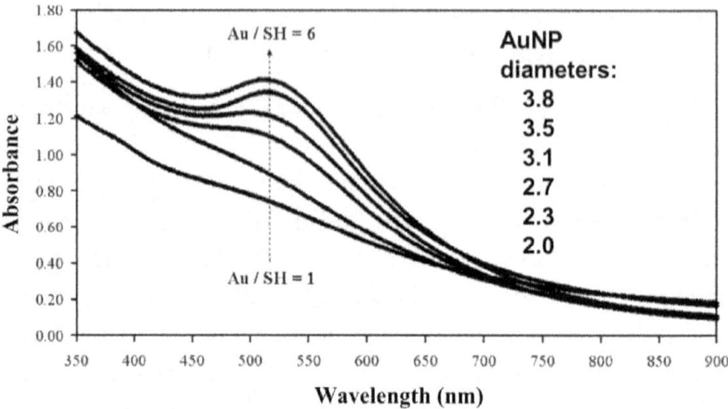

Figure 3.11   Experimental absorbance spectra of small AuNP showing the uptake of the LSPR when the nanoparticles become greater than 2.5 nm. Reprinted with permission from Ref. 53. Copyright Elsevier 2008.

The amount with which the electrons spill out is of the order of magnitude of a few atomic units (0.529 Å), which is not negligible anymore for small nanoparticles and lead to a slight blueshift in the case of gold.[55]

3.5 Optical Response of Assemblies of Nanoparticles

When measuring the optical properties of nanoobjects, one needs to systematically clarify whether the results are an average over a rather homogeneous population and if conclusions can be drawn on the optical response of a single object. Therefore the control of the sample preparation is often a crucial one.[56] To address this difficulty, the development of various techniques for measuring the spectrum of single nano-objects was decisive as detailed in several reviews.[57]

Nanoparticles are always produced in great numbers, in colloidal solutions or with other synthesis methods, and it occurs frequently that they interact with each other, especially when their density becomes high. Interactions affect the LSPR when particles are close to each other and when they are interacting with a supporting substrate. These effects modify the spectral position of the LSPR; however, a primary consequence of considering an assembly of nanoparticles is to modify the absorption cross-section. A method for taking into account assemblies is to use the effective medium approximation. All these effects will be quickly reviewed but it is clear that the next sections are just rough introductions.

3.5.1 *Supported Gold Nanoparticles*

The optical response of nanoparticles deposited on a plane substrate still exhibits a plasmon resonance but this is affected by the electromagnetic coupling with the substrate. Therefore the LSPR signature depends on the distance to the substrate and its nature (metallic or dielectric substrate).[16] Briefly, the oscillating charge distribution of the AuNP induces image charges within the substrate and the two charge distribution couple.[16] In a first approximation, this can be treated as a dipole–dipole interaction. If the exciting electric field is polarised perpendicular to the interface, it generates a charge electric distribution symmetric to the interface plane with the opposite sign (see Figure 3.12(a)). As a result, the induced dipole is aligned to the excitation dipole whereas the two dipoles are anti-parallel for an electric field polarised along the interface (see Figure 3.12(b)). The corresponding calculated spectra are presented in Figures 3.12(c) and 3.12(d) in the case of 10 nm spherical silver nanoparticles positioned at a distance d from a Al_2O_3 substrate.[23] The LSPR dependence with particle–surface distance occurs when the electric field is normal to the interface. In this case, Figure 3.12(c) shows that no interaction occurs when the particle is 100 nm away from the surface and the extinction is predicted to be identical to that of isolated nanoparticles (355 nm for silver spheres). The coupling with the surface becomes visible when the particle approaches at distances smaller than 1 nm and it shows up as a weak shoulder at 346 nm due to a quadrupolar contribution. By further approaching the particle to 0.1 nm, the quadrupolar contribution becomes almost as important as the dipolar contribution and this latter is shifted to higher wavelengths.

The case of AuNP deposited on a substrate is investigated experimentally by Wang[58] and Kooij.[59]

3.5.2 *Nanoparticle Coupling*

Coupling of nanoparticles strongly affects the LSPR but usually it occurs when a colloidal solution starts to precipitate and this coupling shows up

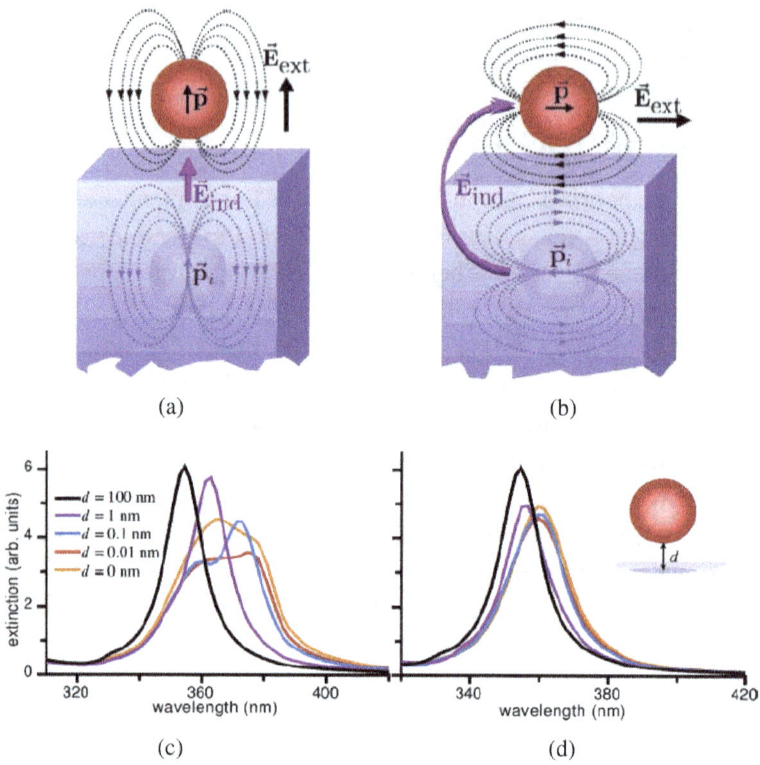

Figure 3.12   Calculation of the electromagnetic coupling of a silver nanoparticle (10 nm diameter) with an Al_2O_3 substrate. In case a, the exciting field is normal to the substrate, and the extinction spectrum (c) is strongly modified when the particle–substrate distance decreases below 1 nm. When the exciting field is parallel to the surface (b and spectra d) the dependence is not as obvious. Reprinted with permission from Ref. 23. Copyright 2007 American Chemical Society.

first as a shoulder in the infrared region and may grow as a large peak around 700 nm when particles come very close to each other.

Roughly, the optical response of a pair deviates from the mere sum of the individual particle contributions as soon as the surface-to-surface distance (d) becomes of the order of their lateral size (radius R). For a transverse excitation, the optical spectrum is not very sensitive to d (weak blueshift) and the plasmon resonance of the pair remains close to the LSPR of each particle. On the contrary, in the case of a longitudinal excitation, the LSPR is strongly shifted towards long wavelengths as d is reduced, as shown in Figure 3.13. Whatever the experimental context, these observations are quite general and

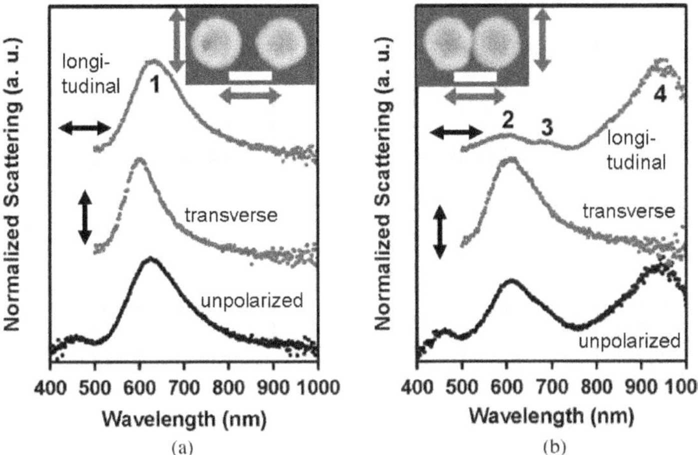

Figure 3.13   Plot of the scattering cross-sections of two interacting nanoshells made of a silica core (radius 40 nm) capped with a gold shell (radius 60 nm). In case (a), the nanoparticles are distant of 20 nm and in case (b), they are linked by a nonanedithiol molecule that keeps them separated of 1.5 nm. Reprinted with permission from Ref. 49.

are corroborated by several theoretical studies considering the ideal case of spherical particle dimers, or even more complex geometries.[49,60]

3.5.3 *Effective Medium Approximation Methods*

A medium constituted by metallic nanoparticles embedded in another material is an example of an inhomogeneous medium that can be treated by the effective medium approximations EMAs or effective medium theories (EMT). The idea of the EMA is to describe the optical response of this complex medium by a single dielectric function, which depends on three parameters: the dielectric function ε of the metallic inclusions, that of the embedding matrix ε_m and the volumic fraction of metal denoted f. The power of these approaches is that they provide with acceptable dielectric functions without specifying neither the relative positions of the particles nor their shape or sizes with a domain of validity that will be shortly discussed below. Once the dielectric function is known, it is easy to access the optical index and predict all sorts of situations such as reflection by a thin film of this complex material deposited on a substrate or its transmission coefficient.

The most commonly used EMA is the Maxwell Garnett (MG) model set up by the physicist of this name in 1904 to explain the coloration of glasses containing metallic inclusions.[61] MG theory is a local field theory that gives the same result as the Clausius Mosotti formula. It assumes that the metallic inclusions are acting like isolated electric dipoles so that the resulting polarisation of the composite material is the sum of the individual microscopic polarisability. This microscopic polarisability can be calculated if the local field acting at an individual dipole is known. This is achieved with the help of the so-called Lorentz sphere, which surrounds the dipole of interest. The Lorentz sphere delimits two regions: the outer region where the influence of the dipoles is treated globally with a mean-field theory and the inner region with neighbouring dipoles. The full derivation of this model and the different size scales cannot be discussed in details and could be found elsewhere.[16,62] However, in the case of a metal (ε) included in a matrix (ε_m) with a volumic fraction f, the resulting dielectric function is given by

$$\frac{\varepsilon_{MG} - \varepsilon_m}{\varepsilon_{MG} + 2\varepsilon_m} \times \frac{\varepsilon - \varepsilon_m}{\varepsilon + 2\varepsilon_m}, \qquad (3.24)$$

which can be solved into

$$\varepsilon_{MG} = \varepsilon_m \frac{\varepsilon(1 + 2f) + 2\varepsilon(1 - f)}{\varepsilon(1 - f) + \varepsilon_m(2 + f)}. \qquad (3.25)$$

This approach holds as long as the individual dipoles are independent. If one considers that this condition is fulfilled when one dipole feels only 1% of the electric field created by its neighbour, therefore, the maximum acceptable volumic fraction is $f_{max} = 0.1$.[62] Moreover the calculations summarised above assume that the inclusions are spherical, which explains the similarity of Equation (3.24) with Equation (3.12) and follows the choice of spherical Lorentz cavities.

Many other models of EMA exist such as the Bruggeman approximation, which considers a composite medium made by quasi-equivalent mixing of two components.[16,63]

3.6 Conclusion

This chapter has provided an overview of the basic principle of plasmonics and cannot be considered comprehensive given that plasmonics is a rapidly growing research field. However, the key features have been presented as well as the main conceptual tools so that the reader can access more specialised review articles. Some other chapters of the present book will open some complementary perspectives. Chapter 4 deals with thermo-optical properties that emerge from the plasmon resonance; Chapter 11 explains how electron beam can excite the optical plasmon and Chapters 14 and 16 will focus on some possible applications in biology, drug delivery and therapy.

Among the very various fields of modern optics, plasmonics plays a special role in a revival of optics. Plasmon amplification, enhanced sensitivity or plasmonic coupling are phenomena occurring in the near-field range. And everybody knows that the behaviour of an electromagnetic wave in the near-field range strongly differs from the usual far-field optics. Plasmonics has opened a door into sub-wavelength optics that can be overstepped even by non-specialists. Yet many non-physicists handle near-field effects, trying to engineer nanoparticles geometry in order to enhance biosensor sensitivity. However, controlling these near-field phenomena in far field is still a very difficult game. Much is still to be understood.

Finally a key for controlling the plasmonic effects probably relies on a close work between the physicists who seek to better understand how the nanoparticle shape influences its optical response and chemists who can produce more and more exotically shaped nanoparticles. This connection between chemistry and optic is occurring and some promising illustrations will be found in the following chapters of this book.

References

1. D. Pines and D. Bohm, *Phys. Rev.* **85**(2), 338–353 (1952).
2. D. Bohm and D. Pines, *Phys. Rev.* **92**(3), 609–625 (1953).

3. L. Tonks and I. Langmuir, *Phys. Rev.* **33**(2), 195–210 (1929).

4. D. Pines, Rev. *Mod. Phys.* **28**(3), 184–198 (1956).

5. R. H. Ritchie, *Phys. Rev.* **106**(5), 874–881 (1957).

6. R. H. Ritchie, *Surf. Sci.* **34**(1), 1–19 (1973).

7. A. Otto, *Zeitschrift Fur Physik* **216**(4), 398–410 (1968).

8. E. Kretschmann and H. Raether, *Zeitschrift Fur Naturforschung Part a-Astrophysik Physik Und Physikalische Chemie A* **23**(12), 2135–2136 (1968).

9. B. Liedberg, C. Nylander, and I. Lundstrom, *Sens. Actuators* **4**(2), 299–304 (1983).

10. D. L. Jeanmaire and R. P. Van Duyne, *J. Electroanal. Chem.* **84**(1), 1–20 (1977).

11. G. Mie, *Annalen der Physik* **25**, 377–445 (1908).

12. M. A. Smithard, *Solid State Comm.* **13**(2), 153–156 (1973).

13. P. K. Hansma and H. P. Broida, *Appl. Phys. Lett.* **32**(9), 545–547 (1978).

14. R. W. Rendell, D. J. Scalapino, and B. Muhlschlegel, *Phys. Rev. Lett.* **41**(25), 1746–1750 (1978).

15. U. Kreibig and L. Genzel, *Surf. Sci.* **156** (JUN), 678–700 (1985).

16. U. Kreibig and M. Vollmer, *Optical Properties of Metal Clusters* (Springer Verlag, Berlin, 1995).

17. Y. Wang, E. W. Plummer, and K. Kempa, *Adv. Phys.* **60**(5), 799–898 (2011).

18. M. Fox, *Optical Properties of Solids* (Oxford University Press, Oxford, 2001).

19. N. Ashcroft and D. Mermin, *Solid State Physics*, 1st ed. (Brooks Cole, 1976).

20. M. Born and E. Wolf, *Principles of Optics: Electromagnetic Theory of Propagation, Interference and Diffraction of Light*, 7th ed. (Cambridge University Press, 1999).

21. L. Novotny and B. Hecht, *Principles of Nano-Optics* (Cambridge University Press, Cambridge, 2006).

22. P. B. Johnson and R. W. Christy, *Phys. Rev. B* **6**(12), 4370 (1972).

23. C. Noguez, *J. Phys. Chem. C* **111**(10), 3806–3819 (2007).

24. P. G. Etchegoin, E. C. Le Ru, and M. Meyer, *J. Chem. Phys.* **125**(16), 164705–707 (2006).

25. D. W. Lynch and W. R. Hunter, edited by E. D. Palik, *Handbook of optical constants.* (Academic Press, New-York, 1985).

26. K. Johansen, H. Arwin, I. Lundstrom *et al.*, Imaging surface plasmon resonance sensor based on multiple wavelengths: Sensitivity considerations. *Rev. Sci. Instruments* **71**(9), 3530–3538 (2000).

27. H. Arwin, Dielectric function of gold measured by ellipsometry, *personal communication* (2007).

28. H. Raether, *Surface Plasmons on smooth and rough surfaces and on gratings* (Springer Verlag, 1986); O. Pluchery, R. Vayron, and K.-M. Van, *Eur. J. Phys.* **32**, 585 (2011).

29. V. M. Shalaev and S. Kawata, *Nanophotonics with Surface Plasmons* (Elsevier, Amsterdam, 2007).

30. M. L. Brongersma and P. G. Kik, *Surface Plasmon Nanophotonics* (Springer-Verlag, New York, 2006).

31. P. K. Jain, I. H. El-Sayed, and M. A. El-Sayed, Nano Today **2**(1), 18–29 (2007).

32. P. K. Jain, K. S. Lee, I. H. El-Sayed *et al.*, *J. Phys. Chem. B* **110**(14), 7238–7248 (2006).

33. S. Unser, I. Bruzas, J. He *et al.*, *Sensors* **15**(7), 15684–15716 (2015).

34. V. Giannini, A. I. Fernandez-Dominguez, S. C. Heck *et al.*, *Chem. Rev.* **111**(6), 3888–3912 (2011).

35. C. F. Bohren and D. R. Huffman, *Absorption and Scattering of Light by Small Particles* (Wiley-Interscience, New-York, 1998).

36. X. Liu, M. Atwater, J. Wang *et al.*, *Colloid Sur. B* **58**(1), 3–7 (2007).

37. S. Basu, S. K. Ghosh, S. Kundu *et al.*, *J. Colloid Interf. Sci.* **313**(2), 724–734 (2007).

38. V. Myroshnychenko, J. Rodriguez-Fernandez, I. Pastoriza-Santos *et al.*, *Chem. Soc. Rev.* **37**(9), 1792–1805 (2008).

39. O. Pluchery, C. Humbert, M. Valamanesh *et al.*, *Phys. Chem. Chem. Phys.* **11**(35), 7729–7737 (2009).

40. S. Asano and G. Yamamoto, *Appl. Opt.* **14**(1), 29–49 (1975).

41. A. C. Lind and J. M. Greenberg, *J. Appl. Phys.* **37**(8), 3195–3203 (1966).

42. G. C. Schatz, *J. Mol. Struct-Theochem* **573**(1–3), 73–80 (2001).

43. L. Novotny, R. X. Bian, and X. S. Xie, *Phys. Rev. Lett.* **79**(4), 645–648 (1997).

44. C. Oubre and P. Nordlander, *J. Phys. Chem. B* **108**(46), 17740–17747 (2004).

45. M. I. Mishchenko, D. W. Mackowski, and L. D. Travis, *Appl. Opt.* **34**(21), 4589–4599 (1995); B. N. Khlebtsov and N. G. Khlebtsov, *J. Phys. Chem. C* **111**(31), 11516–11527 (2007); N. G. Khlebtsov, *J. Quantitative Spectrosc. Radiative Transfer* **123**, 184–217 (2013).

46. P. N. Njoki, I. I. S. Lim, D. Mott *et al.*, *J. Phys. Chem. C* **111**(40), 14664–14669 (2007).

47. E. Lifshitz, M. Landau, and L. D. Pitaevskii, *Electrodynamics of Continuous Media*, 2nd ed. (Butterworth-Heinemann, Burlington, MA, 1984).

48. S. Berciaud, L. Cognet, P. Tamarat *et al.*, *Nano Lett.* **5**(3), 515–518 (2005); C. Novo, A. M. Funston, and P. Mulvaney, *Nat. Nanotechnol.* **3**(10), 598–602 (2008).

49. J. B. Lassiter, J. Aizpurua, L. I. Hernandez *et al.*, *Nano Lett.* **8**(4), 1212–1218 (2008).

50. A. L. Gonzalez and C. Noguez, *J. Comput. Theor. Nanosci.* **4**(2), 231–238 (2007).

51. J. Pérez-Juste, I. Pastoriza-Santos, L. M. Liz-Marzán *et al.*, *Coordin. Chem. Rev.* **249**(17–18), 1870–1901 (2005).

52. K. L. Kelly, E. Coronado, L. L. Zhao *et al.*, *J. Phys. Chem. B* **107**(3), 668–677 (2003).

53. G. A. Rance, D. H. Marsh, and A. N. Khlobystov, *Chem. Phys. Lett.* **460**(1 3), 230 236 (2008).

54. E. Cottancin, G. Celep, J. Lerme *et al.*, *Theor. Chem. Acc.* **116**(4–5), 514–523 (2006).

55. W. Eckardt, *Phys. Rev. B* **29**, 1558 (1984).

56. L. Dalstein, M. Ben Haddada, G. Barbillon *et al.*, *J. Phys. Chem. C* **119**(30), 17146–17155 (2015).

57. P. Zijlstra and M. Orrit, *Rep. Prog. Phys.* **74**(10) (2011); J. Olson, S. Dominguez-Medina, A. Hoggard *et al.*, *Chem. Soc. Rev.* **44**(1), 40–57 (2015).

58. D.-S. Wang and C.-W. Lin, *Opt. Lett.* **32**(15), 2128–2130 (2007).

59. E. S. Kooij, H. Wormeester, E. A. M. Brouwer *et al.*, *Langmuir* **18**(11), 4401–4413 (2002).

60. S. Marhaba, G. Bachelier, C. Bonnet *et al.*, *J. Phys. Chem. C* **113**(11), 4349–4356 (2009).

61. J. C. Maxwell Garnett, *Philos. T. Roy. Soc.* **203**, 385 (1904).

62. S. Berthier, *Optique des Milieux Continus.* (Polytechnica, Paris, 1993).

63. D. A. G. Bruggeman, *Annal. Physik* **24**, 636 (1935).

Chapter 4

Photothermal Properties of Gold Nanoparticles

Bruno Palpant

Centrale-Supélec, Laboratoire de Photonique Quantique et Moléculaire, CNRS, Ecole Normale Supérieure Paris-Saclay, Université Paris Saclay, Châtenay-Malabry, France

An outstanding number of different developments and applications have recently taken place from the ability of noble metal nanoparticles to act as nanoscale heat sources under light irradiation. The localised surface plasmon resonance (LSPR) is an efficient way to input energy in such small metallic objects, as seen in the preceding chapter, thus enhancing the yield of the nanoscale light-to-heat conversion. It has also been seen in Chapter 3 that the characteristics of the LSPR are affected by different intrinsic or extrinsic parameters, which allows us to detect different kinds of local modifications through optical monitoring. The main part of this chapter will present the photo-induced generation of thermal energy within and around gold nanoparticles, by providing the basic principles and considering cases useful for some applications. The temperature (or, more rigorously, the local thermal state) dependence of the optical properties of a medium containing gold nanoparticles will then be shortly examined. Finally, the variation of the melting temperature with AuNP size (known as the melting point depression) will be presented.

4.1 Introduction: Light to Heat Conversion at the Nanoscale

4.1.1 *Electron–Phonon Scattering in Bulk Metal*

It is more than likely that the optical response of metals confined at the nanoscale shares many common characteristics with those of the bulk metal. Hence, as the dielectric function of gold in the visible range is dominated by the contributions of both interband and intraband transitions,[1] the optical properties of AuNPs (provided their size is sufficiently large, see Chapter 2) are determined by the microscopic coupling and exchange mechanisms involved in such transitions. The plasmon resonance phenomenon — which arises from the dipolar nature of the wave-driven collective electron excitation allowed by the small NP size — is essentially a coherent set of in-phase intraband transitions. It is then also closely dependent on the energy dissipation processes involved in individual intraband transitions, which are classically accounted for in the bulk metal response by the phenomenological scattering rate Γ of the Drude model. We have seen in Chapter 3 that, if the Drude model is assumed to depict the NP dielectric function, the spectral width of the SPR is directly given by $\hbar\Gamma$. The significant contributions to Γ at room temperature are the electron–electron and electron–phonon scattering.[2] Indeed, electron scattering by defects and impurities can be neglected, except at very low temperature. Furthermore, the contribution of electron–electron collisions[3] does not exceed 15% of the total scattering rate.[4] Even at higher electron temperatures reached with ultra-short laser excitation (see below), the electron–phonon scattering contribution remains by far the dominant one. This reveals that the plasmon resonance is tightly linked with the scattering of electrons with phonons.

Heat can be seen as a broadband incoherent statistical set of vibrations and is then supposed to be mainly supported, in a solid, by phonons. Heat transport is ensured by carriers, which are elementary particles, as electrons, phonons, or even photons. In the former two cases, the mechanism involved is conduction, whereas in the latter the transfer occurs by radiation. When exposed to an incident light in (or in the vicinity of) the visible spectral domain, a gold nanoparticle can gain energy by absorbing photons through electron transitions. The main relaxation process is then the electron–phonon scattering, as stated above. Of course, an individual electron–phonon collision can result either in an energy gain, or an energy loss for the electron; but as the initial energy supplied to the NP by photon absorption is input in the form of electron excitations, the subsequent overall energy transfer progresses from electrons to lattice vibrations. This mechanism is then the cause of the photo-induced heating of a metal NP.

4.1.2 *The Localised Plasmon Resonance as an Effective Energy Input Channel*

Before going deeper into any calculation, we can foresee the great interest of the plasmon resonance for realising nanosize heat sources. Indeed, the SPR absorption is an efficient and fast way of inputting energy into a NP by macroscopic light excitation, the inner exchange mechanisms then allowing this energy to be mainly converted into heat at the nanoscale. This explains the spectacular recent raise of studies and applications of AuNPS for their photothermal ability.

4.1.3 *A Series of Energy Exchanges*

By simply depicting, in the preceding section, the link between the absorption of photons in gold NPs (at or off the plasmon resonance) and the energy dissipation through electron–phonon collisions, we have implicitly introduced the notion of dynamics. The optical properties of such nano-objects are indeed driven by a series of interlinked internal and external energy exchanges, each of them characterised by a typical time scale. It appears then relevant for getting deeper insight into these properties to study the NP impulse response, as is usual for dynamic systems in physics. Furthermore, regarding the ability of AuNPs to behave as thermal nanosources, such a study may allow (i) to understand the stationary regime of photo-induced heating and (ii) to propose new ways of controlling the heating.

Let us then imagine that a 'very short'[a] light pulse is sent onto a gold nanoparticle (see Figure 4.1). Part of the incoming light is then absorbed for inducing electron transitions. Let us stress that it is *a priori* possible to induce, whether interband transitions, or intraband ones. In the former

---

[a]By very short, we mean the duration of which is about a few electromagnetic field oscillation cycles.

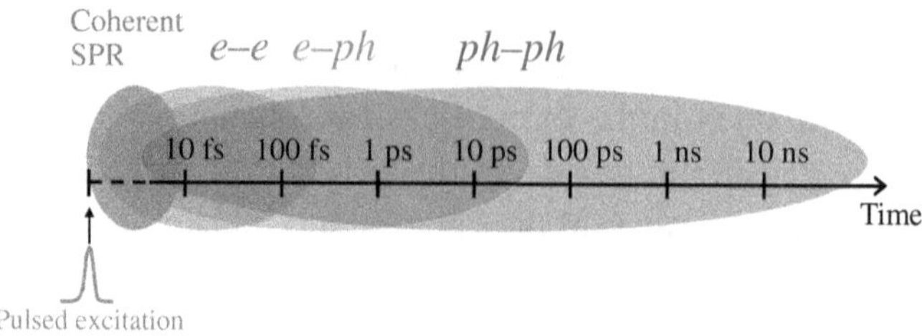

Figure 4.1 Schematic illustration of the series of energy exchanges involved in the optical impulse response of a gold nanoparticle on a logarithmetic time scale.

case, photons generate individual electron–hole pairs, while in the latter, they promote electrons up to higher levels within the conduction band. We have seen in a previous section that in spherical AuNPs both the plasmon resonance and interband transitions can be excited in the same spectral range. In this case, the SPR is damped and broadened due to Landau damping. Beyond, for sake of simplicity, we will disregard the possible excitation of interband transitions in the description of the dynamics mechanisms. Provided the ultrashort wave packet spectrum matches the SPR energy, the plasmon resonance is excited, that is, a resonant coupling with the electromagnetic wave induces a coherent set of in-phase electron excitations in the conduction band. It has been shown through optical experiments using second-harmonic generation autocorrelation, spectral hole burning or measurements of the SPR bandwidth of single NPs that the dephasing time (T_2) of the SPR is a few femtoseconds.[5] The NP is then left with an extra energy stored in the electron gas.

As a matter of fact, a few electrons have gained photon energy by absorption, the other ones remaining in the non-excited states. This puts the electron distribution out of equilibrium. Energy is then redistributed among the whole quasi-free electron gas by electron–electron collisions, leading to the recovery of an internal thermal equilibrium within the conduction band. This process occurs on a time scale which ranges from a few tens to several hundreds of femtoseconds, depending mainly on the initial energy input (the higher the proportion of excited electrons in the gas the faster the energy redistribution by collisions). At the same time, electrons scatter with phonons. Actually, there are no real collisions with such quasi-particles; rather, a quantised vibration mode of the crystal lattice (i.e. a phonon) induces a modification of the periodic potential experienced by the electrons, and then a modification of the wave function of the latter. The typical time scale of this process is about one picosecond. As for the electron–electron (e–e) scattering rate, the actual electron–phonon (e–ph) relaxation time depends on several factors as particle size and excess input energy.

Finally, as the NP is not isolated but is embedded in a medium, there is a thermal energy exchange at the interface through phonon–phonon collisions, leading to the cooling down of the NP. The dynamics of this process, as will be seen later, is very sensitive to the heat transfer properties in the host medium. It may range from a few picoseconds to nanoseconds. Let us notice that, as AuNPs are most of the time dispersed in a liquid or solid insulating medium, the heat carriers in the latter are mainly phonons (convection may be neglected at these small time scales). Of course, the NP cooling goes together with the transient heating of its close environment, which may be exploited to use gold NPs as photo-induced heat nanosources. If the neighbouring NPs are sufficiently close in the medium (that is, if the NP density is high enough), then the thermal energy released by these neighbours will affect the cooling dynamics. Let us stress that in the case of very close NPs a thermal exchange through radiative transfer is possible.[6] We won't address this particular effect in the following.

As a consequence of this series of energy exchanges, the internal energy of the electron gas subsequent to light pulse absorption undergoes (i) a sudden and strong rise, (ii) an inner redistribution within the electron gas (athermal regime), (iii) a fast decrease (e–ph scattering) and (iv) a slow return back to equilibrium (thermal transfer to the host medium).

The particle temperature is determined by the balance between the gain of energy from e–ph collisions and the heat release towards the host medium. It then presents an increase on a picosecond time scale followed by a slow decrease. The different steps described above will be examined in deeper details in Section 4.3. Before that, let's establish the simple approaches for describing the photo-heating of a AuNP.

4.2 Basic Plasmonic Photothermal Properties

In this part, we present a basic description of the light-induced heating of a gold nanoparticle and heat transfer to its environment, and provide key equations, orders of magnitude and examples for a practical use suited for many situations. Particularly, we disregard the step of light absorption by electrons and energy transfer from the electron gas to the metal ionic lattice, which will be treated in a next section of this chapter. Rather, we assume here that the energy absorbed is instantaneously converted into heat in the NP. This assumption is valid provided the typical observation time is large against the characteristic electron–phonon scattering time, of the order of a few picoseconds. Further, the heat transport in the medium surrounding the NPs will be supposed to be driven by diffusion only and then described by the Fourier law.[b] This situation corresponds to many practical applications of the plasmonic photothermal conversion and has been widely investigated by several authors, particularly by Baffou and Rigneault.[7] We will consider a dielectric homogeneous medium embedding a gold NP, but any other metal nanoparticle could be chosen rather, provided the different assumptions of the model are valid. In the different parameters and variables, which will be introduced in the following, the subscripts Au, NP and m will then refer to gold, the nanoparticle and the host medium, respectively.

4.2.1 *Power Input in Nanoparticles*

Besides the thermodynamic properties of the different media involved, and whatever the model and assumptions used, the key input parameter for calculating the temperature evolution of a nanoparticle and its environment due to photothermal conversion is the time-dependent instantaneous power density absorbed by the NP, $P_{abs}(t)$ (in W m^{-3}), which follows the time profile of the light input (pulsed or continuous). If we assume that the particle dimension in the direction of light propagation is smaller than the wave penetration depth (a few tens of nanometres for gold in the visible range), the excitation can be considered as homogeneous within the metal volume. In this condition, P_{abs} is then independent of any space coordinate and we can write at first order:

$$P_{abs}(t) = \frac{I_{inc}(t)\sigma_{abs}}{V_{NP}}, \qquad (4.1)$$

---

[b]The validity of this assumption will be discussed in Section 4.3.3.

In this equation, I_{inc} (in W m^{-2}) denotes the incoming light intensity, or incoming flux, at the NP location (supposed homogeneous), σ_{abs} (in m^2) is the NP absorption cross-section, and V_{NP} the NP volume. Equation (4.1) disregards any possible change of the absorption cross-section along the absorption–relaxation process, which could be the case under very short and/or intense laser pulse irradiation as it transiently modifies the NP optical properties. For small enough spherical NPs ($R < 25$ nm, where R is the NP radius), the light–NP interaction is dominated by absorption and σ_{abs} is given by the first-order term of the Mie expansion (see Chapter 3):

$$\sigma_{abs} = \frac{18\pi V_{NP}}{\lambda} \frac{n_m^3 \varepsilon_2}{\left[\varepsilon_1 + 2\sqrt{n_m}\right]^2 + \varepsilon_2^2}, \tag{4.2}$$

where λ is the light wavelength in vacuum, n_m the host medium refractive index and ε_1 and ε_2 the real and imaginary components of the dielectric function of metal, respectively. For more complicated NP shapes and/or for larger sizes, as well as for interacting neighbouring NPs, an alternative method should be employed to calculate the absorption cross-section, as the discrete dipole approximation[8] (DDA, see Chapter 3) or the boundary element method (BEM). As examples, σ_{abs} is worth 600 and 540 nm^2 for a AuNP with $R = 10$ nm at plasmon resonance in silica ($\lambda = 530$ nm) and water ($\lambda = 520$ nm), respectively. For a gold nanorod with the same volume and an aspect ratio of 3, $\sigma_{abs} = 7800$ nm^2 at LSPR ($\lambda = 680$ nm) in water. Let us note that the total power absorbed by the NP, $P_0 = I_{inc}\sigma_{abs}$ in Equation (4.1), can also be directly evaluated by calculating the power density dissipated in the metal, which amounts to evaluating the local electromagnetic field in the NP. This can be done by numerically solving Maxwell's equations (for instance by using the finite difference time domain approach). This approach may be employed in cases where the NP dimensions are such as the local field is not homogeneous.

Let us now come back to the simple case of homogeneity and consider different typical practical situations for evaluating $I_{inc}(t)$ and then $P_{abs}(t)$ through Equations (4.1) and (4.2):

(i) If the irradiation is continuous (cw light) with power P, having a Gaussian transverse beam profile of radius w, then the intensity varies with the radial coordinate ξ in the transverse plane as

$$I_{\text{inc}}(\xi) = \frac{2P}{\pi w^2} e^{-2\xi^2/w^2}. \tag{4.3}$$

Of course, if the NP is set at the centre of the beam, $\xi = 0$ and $I_{\text{inc }0} = 2P/\pi w^2$. For a collimated laser beam entering a microscope objective with input diameter D and focal distance (in m) $f_{\text{obj}} = 0.25/G_{\text{obj}}$, where G_{obj} is the magnification of the lens, w at the focus is worth:

$$w = \frac{2\lambda f_{\text{obj}}}{\pi D}. \tag{4.4}$$

(ii) If the irradiation is pulsed (total energy per pulse E) with a Gaussian transverse beam profile (beam radius w) and Gaussian time dependence (full width at half maximum Δt), instantaneous intensity of the incident light is written as

$$I_{\text{inc}}(\xi, t) = \frac{2E}{\pi w^2} \frac{1}{\sqrt{2\pi}\,\sigma} e^{-t^2/2\sigma^2} e^{-2\xi^2/w^2}, \tag{4.5}$$

where $\sigma = \Delta t/\sqrt{\ln 256}$. Of course, Equation (4.4) stands also for light pulses focused by a microscope lens.

(iii) In case of an ultrashort laser pulse (that is, the duration of which is smaller than the typical heat exchange from the NP to the surrounding medium), the energy surface density (in J m^{-2}) deposited in a NP set at the beam centre is worth $2E/\pi w^2$ per pulse. The total energy per pulse absorbed by the NP, ε_{abs}, is then

$$\varepsilon_{\text{abs}} = 2E\sigma_{\text{abs}}/\pi w^2. \tag{4.6}$$

4.2.2 Basic Approach: Pure Diffusion, Perfect Contact

The simplest case to be considered is the one of a metal sphere soaked in a homogeneous medium. We assume that the typical time and space scales at which the heat transfer is observed match the requirements for a pure diffusive thermal transport in the surrounding medium. The validity of this assumption will be discussed in Section 4.3.3.2. The temperature in the metal nanoparticle $T_{\text{NP}}(\mathbf{r},t)$ evolves is

$$C_{\text{Au}}\frac{\partial T_{\text{NP}}}{\partial t} = \kappa_{\text{Au}}\nabla^2 T_{\text{NP}} + P_{\text{abs}}(t). \tag{4.7}$$

where C (in J m^{-3} K^{-1}) denotes the volumetric heat capacity, with $C = \rho c$ (ρ, in kg m^{-3}, is the mass density and c, in J kg^{-1} K^{-1}, the heat capacity). κ (in W m^{-1} K^{-1}) is the thermal conductivity. ∇^2 stands for the scalar Laplacian. The host medium temperature T_m obeys the Fourier law:

$$C_m \frac{\partial T_m}{\partial t} = \kappa_m \nabla^2 T_m. \tag{4.8}$$

Considering the heat exchange at the interface (i.e. at $r = R$, the origin of distances being set at the NP centre) imposes boundary conditions. First, we equate the heat flux exiting the NP and the one getting in its environment:

$$\kappa_{\mathrm{Au}} \left. \frac{\partial T_{\mathrm{NP}}}{\partial r} \right|_{R^-} = \kappa_m \left. \frac{\partial T_m}{\partial r} \right|_{R^+}. \tag{4.9}$$

If we consider a perfect contact at the metal–host medium interface, that is, we assume the continuity of the temperature as crossing the interface, we obtain a second boundary condition:

$$T_m(R^+, t) = T_{\mathrm{NP}}(R^-, t). \tag{4.10}$$

Solving together the set of Equations (4.7) and (4.8) with boundary conditions (4.9) and (4.10) enables us to get the time and space variations of the nanoparticle and host medium temperatures. For this, a numerical approach based on the finite difference method, as described in Refs. 7, 9, or the finite element method (using, for instance, a commercial software), may be carried out.

4.2.3 *Accounting for Interface Thermal Resistance*

In the last section, we have considered a perfect contact between the metal NP and its host medium. This amounts to disregarding any interface thermal resistance, or equivalently to set the interface conductance, g (in W m^{-2} K^{-1}), to infinity. However, there are two kinds of effects which can lead to a finite value for g. First, the thermal resistance can originate from the contact itself: porosity in case of a solid surrounding matrix, stemming from the material elaboration process, poor wettability of the solvent onto the metal in the case of a colloidal solution, presence of molecules at the interface, electrical charge of the particles, which can induce a layer of solvent molecules with modified thermal properties and hydrophobicity. The result of all of this is known as 'contact thermal resistance'. Beyond, even in case of a perfect contact, the mismatch of the thermal impedances of the two media involved induces a non-perfect heat transfer from one to the other (just as the mismatch of refractive index at a dioptre does for light). As heat is an incoherent excitation of vibrations over a large spectral domain, this corresponds to a non-perfect overlap of the vibrational density of states of the two media at the interface.[10,11] This effect is known as 'Kapitza thermal resistance'. Both origins are considered together in the definition of the global interface thermal resistance, $1/g$.

The effect of the finite thermal conductance is to modify the boundary conditions (4.9) and (4.10) above. Equation (4.10) is no longer valid. Rather, a temperature jump ΔT is observed at the interface:

$$\Delta T(t) = T_{\mathrm{NP}}(R^-, t) - T_m(R^+, t). \tag{4.11}$$

Additionally, the conductance links this temperature jump with the flux balance at the interface (Equation (4.9)):

$$\kappa_{\mathrm{Au}} \left.\frac{\partial T_{\mathrm{NP}}}{\partial r}\right|_{R^-} = \kappa_m \left.\frac{\partial T_m}{\partial r}\right|_{R^+} = -g\Delta T. \tag{4.12}$$

Then:

$$\Delta T = -\ell_K \left.\frac{\partial T_m}{\partial r}\right|_{R^+}, \tag{4.13}$$

where $\ell_K = k_m/g$ is the Kapitza length. Several studies have reported values of g for different systems, as for instance: (i) $g > 20$ MW m^{-2} K^{-1} for alkanethiol-terminated AuNPs in toluene,[12] (ii) $g = 130$ MW m^{-2} K^{-1} for Au nanorods (NRs, mean aspect ratio 3.6) coated with a fully stable CTAB surfactant layer in water,[13] (iii) $g = 250$ MW m^{-2} K^{-1} for Au$_{0.92}$Pd$_{0.08}$ nanospheres stabilised by ethylene glycol in water[14] and (iv) $g = 105$ MW m^{-2} K^{-1} for gold nanospheres in glass.[10]

4.2.4 *Steady-state Photo-heating*

We will examine here the particular case of long-lasting laser pulses (for which the variations of intensity are slow against the typical energy exchange dynamics) and, of course of cw beams. This steady-state analysis has been carried out by some authors.[15,16] It is worth underlining that there are two scales for defining the stationarity of the thermal response. At the nanoparticle scale, this corresponds to the regime where, of course,

T_{NP} does not evolve with time. However, thermal energy is continuously supplied from the NP to the surrounding medium and tends to infinity; this does not represent any problem as long as the spatial extension of the medium around NP can be considered as infinite. As the total volume of inhomogeneous medium is, however, finite, a second space scale has to be considered for macroscopic times, where the stationarity now corresponds to a regime in which the total energy received by the whole medium is exactly balanced by the losses at the sample limits (at the interfaces with cuvette walls, substrate, air). We will examine the thermal behaviour at these two scales.

4.2.4.1 *Nanoparticle scale*

First, we again assume that the light excitation is homogeneous within the nanoparticle. Further, the temperature in the NP is assumed to be uniform. This is ensured by the large thermal diffusivity of gold compared with that of the host medium. G. Baffou and co-workers used a thermal extension of the green dyadic method to determine the temperature field topography in and around different kinds of gold nanostructures.[17,18] They showed that the optical near-field distribution does not always superimpose on the thermal distribution. They also demonstrated that, even if the resulting heat power density is not uniform in the NP, the high contrast between metal and surrounding medium's thermal conductivities induces a fast homogenisation of the temperature in the NP at the typical interface exchange time scale (for instance, thermal conductivity is worth 317, 1.3 and 0.6 W m^{-1} K^{-1} for gold, silica and water, respectively).[16] The NP temperature increase can then be assumed to be uniform. If the continuous light source is switched on at time $t = 0$, the typical time needed to reach the steady state is then $\tau_{st} \approx R^2/D_m$,[15] where R is still the NP radius and $D_m = \kappa_m/C_m$ is the host medium thermal diffusivity. Hence $\tau_{st} \approx 0.7$ ns for a Au nanosphere with $R = 10$ nm in water.

In the steady state, the expressions in Equations (4.7) and (4.8) are equated to zero. The temperature excess profile in the host medium in the stationary regime can be easily derived[15,16]:

$$\Delta T_m^{cw}(r) = \frac{V_{NP}P_{abs}}{4\pi\kappa_m r}. \qquad (4.14)$$

Note that for reasonably small NPs, for which Equation (4.2) is valid, it follows from Equations (4.1), (4.2) and (4.14) that $\Delta T_m^{\mathrm{cw}}(r) \sim R^3/r$ for a fixed irradiation intensity. Equation (4.14) interestingly reveals that the host medium's temperature does not depend on the interface thermal resistance, which is intuitive for the steady state. The temperature profile exhibits a $1/r$ dependence. Within the assumption of fast homogenisation of the NP temperature within the NP discussed above, the NP temperature excess reached in the steady state is written as

$$\Delta T_{\mathrm{NP}}^{\mathrm{cw}} = \frac{V_{\mathrm{NP}} P_{\mathrm{abs}}}{4\pi \kappa_m R}\left(1 + \frac{\ell_K}{R}\right). \tag{4.15}$$

The NP stationary temperature is obviously as high as the interface thermal resistance is high, as revealed by Equation (4.15). For sufficiently small NP size, σ_{abs} is proportional to V_{NP} (see Equation (4.2)). Following Equations (4.1), (4.2) and (4.15), the NP temperature excess varies as R^2 to which is added a correction term due to the finite interface resistance, proportional to R. For a gold nanorod stabilised by CTAB in water, $\ell_K \approx 5$ nm. In this case, the interface resistance correction to the expression of $\Delta T_{\mathrm{NP}}^{\mathrm{cw}}$ is significant for NP sizes ($= 2R$) lower than about 50 nm. Let us give two interesting and typical examples. We consider a AuNP with $R = 10$ nm in water, irradiated by a continuous green laser beam ($\lambda = 532$ nm). The NP is set at the beam centre. Neglecting the effect of interface resistance, the calculations based on Equations (4.2), (4.3) and (4.15) yield in the steady state:

— For a laser pointer beam, $P = 5$ mW, $D = 1$ mm: $I_0 = 13$ kW m^{-2} and $\Delta T_{\mathrm{NP}}^{\mathrm{cw}} = 90\,\mu$ K.
— For a laser beam, $P = 100$ mW, $D = 5$ mm, focused by a $\times 20$ microscope objective (Equation (4.4)): $I_0 = 90$ GW m^{-2} at focus and $\Delta T_{\mathrm{NP}}^{\mathrm{cw}} = 635$ K.

From these examples, it is obvious that the efficiency of the photothermal conversion effect strongly depends on the light source used.

4.2.4.2 *Macroscopic scale*

Photo-heating can be observed at a larger scale, that is, in a macroscopic sample containing many photothermal nanoconverters. In this case, the steady state is reached once the balance of the power absorbed and the one dissipated at the sample boundaries is ensured. Let us note that at such spatial scales, the possible pulsed nature of the irradiation has no effect, provided the time lag between two successive pulses is short against the typical diffusion time (L^2/D_m, where L is the characteristic size of the macroscopic sample). In such a case, the relevant parameter is, as for the continuous excitation, the mean power of light. By switching the light on at time origin $t = 0$, the mean temperature[c] of the whole inhomogeneous medium evolves from initial value T_0 as[19,20]:

$$T_s(t) = T_0 + \frac{A}{B}\left(1 - e^{-Bt}\right),\tag{4.16}$$

where

$$A = \frac{P_{\text{losses}}\eta}{V_s C_s}\tag{4.17}$$

is the rise slope at origin. B denotes the rate of heat dissipation towards the sample environment through the boundaries. P_{losses} is the total power lost by light ($P_{\text{losses}} = P - P_{\text{out}}$, where P_{out} is the power transmitted by the sample). η is the NP photothermal efficiency ($0 \leq \eta \leq 1$), accounting for the possible existence of other loss phenomena (scattering, luminescence, etc.). V_s and C_s are the total volume and the volumetric heat capacity of the sample; for low metal concentration, and noting that the volume heat capacities are generally of the same order of magnitude ($C_{\text{Au}} = 2.49\,\text{MJ}\,\text{m}^{-3}\,\text{K}^{-1}$, $C_{\text{water}} = 4.18\,\text{MJ}\,\text{m}^{-3}\,\text{K}^{-1}$, $C_{\text{silica}} = 1.84\,\text{MJ}\,\text{m}^{-3}\,\text{K}^{-1}$ for instance), C_s can be assimilated to C_m. Hence, measuring the evolution of the mean sample temperature allows determining the values of B and η.

4.2.5 *A Few Emblematic Applications*

The nanoscale photothermal conversion at plasmon resonance in the steady state has been used in many applications. We give a few emblematic illustrations here.

---

[c]This assumes that the temperature is homogeneous at a mesoscopic scale, which is obtained in a homogeneous medium, whether by irradiating the whole sample surface, or, for solutions, by fast steering. In other cases, the inhomogeneity of the temperature distribution has to be taken into account in the model. Note also that the energy absorbed decreases with light penetration distance as described by the Beer–Lambert's law.

Photothermal imaging

Optical microscopy does not allow for direct observation of objects smaller than the wavelength of light. However, the presence of light-to-heat nanoconverters, as gold nanoparticles can be indirectly localised by means of the thermal lensing effect. Indeed, by generating heat from an irradiated NP, a thermal gradient is created in the surrounding medium, resulting in a refractive index gradient the spatial extension of which is much larger than the NP itself. This results in a transient lens, which can modify the characteristics of a probe light beam. This principle was applied in the late 1990s to study the Brownian motion of individual silver nanoparticles as shown in Figure 4.2.[21]

This principle was developed later in a very smart way for conceiving a photothermal microscope, as reported for instance in Ref. 22. The local change in refractive index around a AuNP photo-heated by a laser beam tuned to the LSPR modifies the phase of a transmitted probe light. This change can be detected by studying the interference pattern obtained when overlapping the resulting beam with a reference one. The sensitivity of the NP detection can be further improved by modulating the heating laser

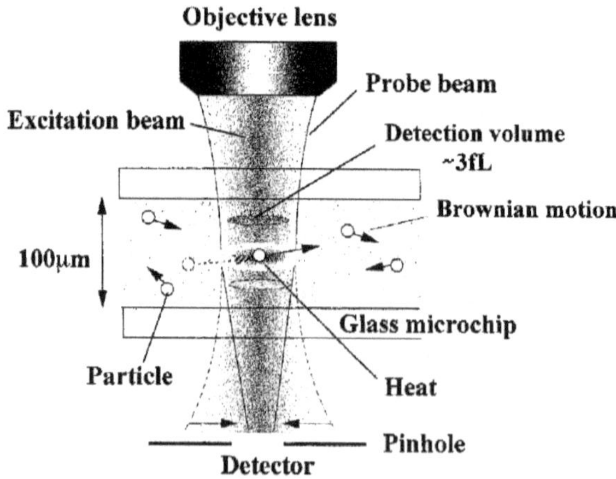

Figure 4.2   Illustration of the principle of photothermal detection of 10-nm AgNPs in water. When a NP enters the field of the excitation beam, it is heated by photothermal conversion, which creates a thermal lens around. This transient lens modifies the probe beam divergence, which is detected by setting a pinhole in front of the detector, as in the close-aperture *z*-scan method. Reprinted with permission from Ref. 21.

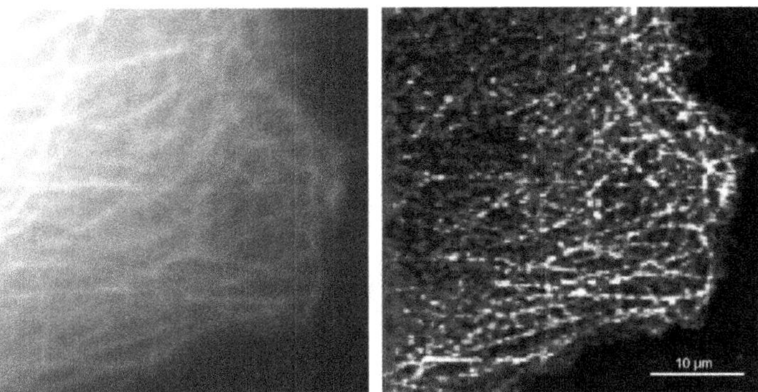

Figure 4.3   Small AuNPs can be functionalised with an antibody able to recognise different types of green fluorescent proteins (GFP) expressed by cells. In this example, the same cell, containing such nanolabels in microtubules, is observed by epi-fluorescence (left) and by photothermal imaging (right). The brightness is proportional to the local density of AuNPs, and then of GFP expressed. Reprinted with permission from Ref. 23.

intensity and using a lock-in amplifier detector. The resolution reached is much better than the one obtained by bright-field microscopy or fluorescence imaging. This photothermal imaging principle has been improved meanwhile. By grafting a proper receptor at the NP surface, the molecular selectivity can be very acute, enabling for single-molecule tracking within a complex biological environment as illustrated in Figure 4.3.[23]

Targeted therapy

The first well-known suggested application of the plasmonic photothermal conversion was to use it for local treatment for cancer tumours by hyperthermia.[24] For this, the NP LSPR has to be tuned as to match the transparency window of biological tissues (650–1320 nm).[25] This can be achieved with non-spherical AuNPs[26] or core–shell NPs.[24] Smart hybrids have more recently been proposed for drug delivery purposes.[27] The idea relies on the association of plasmonic nanostructures and thermosensitive polymers grafted on the metal surface. Such polymers in aqueous solutions exhibit a phase transition from hydrophilic to hydrophobic state when temperature increases above a certain value. Molecules like drugs can be trapped within the polymer brushes and kept in the vicinity of nanoparticles under the transition temperature. By photo-heating NPs with a sufficient power, the transition temperature is overstepped and the subsequent expelling of water from the polymer layer results in the release of the molecules. Hence, the local drug delivery can be controlled by light.

In Figure 4.4, tumour regression is efficiently induced thanks to the concomitant action of local hyperthermia and doxorubicin release.[27]

Bubble formation

The formation of solvent bubbles around light-irradiated AuNPs has been demonstrated, analysed and used in different works.[28] Cavitation effect induced by photothermal conversion around AuNPs has been suggested for therapeutic application through cell membrane breaking.[29] Other applications have been put forward as vapour production from solar light.[30]

4.3 Transient Thermal Behaviour with Pulsed-Light Irradiation

After having established, in the preceding section, the general equations driving the heat generation from light in a nanoparticle and its release towards the surrounding medium, assuming a purely diffusive heat transfer and neglecting the involvement of electrons in these exchanges, we have briefly examined the case of the steady-state heating of plasmonic gold nanostructures. We now turn to the transient case, occurring when AuNPs are irradiated by pulsed light. The exchange mechanisms have already been qualitatively described in the Introduction (see Section 4.1.3).

4.3.1 *Instantaneous Light Pulse Approximation*

With ultrashort pulsed excitation, the temperature at each point in the host medium evolves with time after each pulse. Conversely, at each time, the temperature presents a maximum value at a particular distance from the NP surface. It is then possible to track the location of this maximum along time, which defines the envelope of the temperature maximum. If the NP is sufficiently small to be considered as a point-like heat source, and if the NP heating is assumed to be instantaneous, then this envelope follows:

$$\Delta T_m^{\max}(r) = T_m^{\max}(r) - T_0 = \frac{1}{3\sqrt{3}} \frac{\varepsilon_{\text{abs}}}{C_m} \frac{1}{r^3}, \tag{4.18}$$

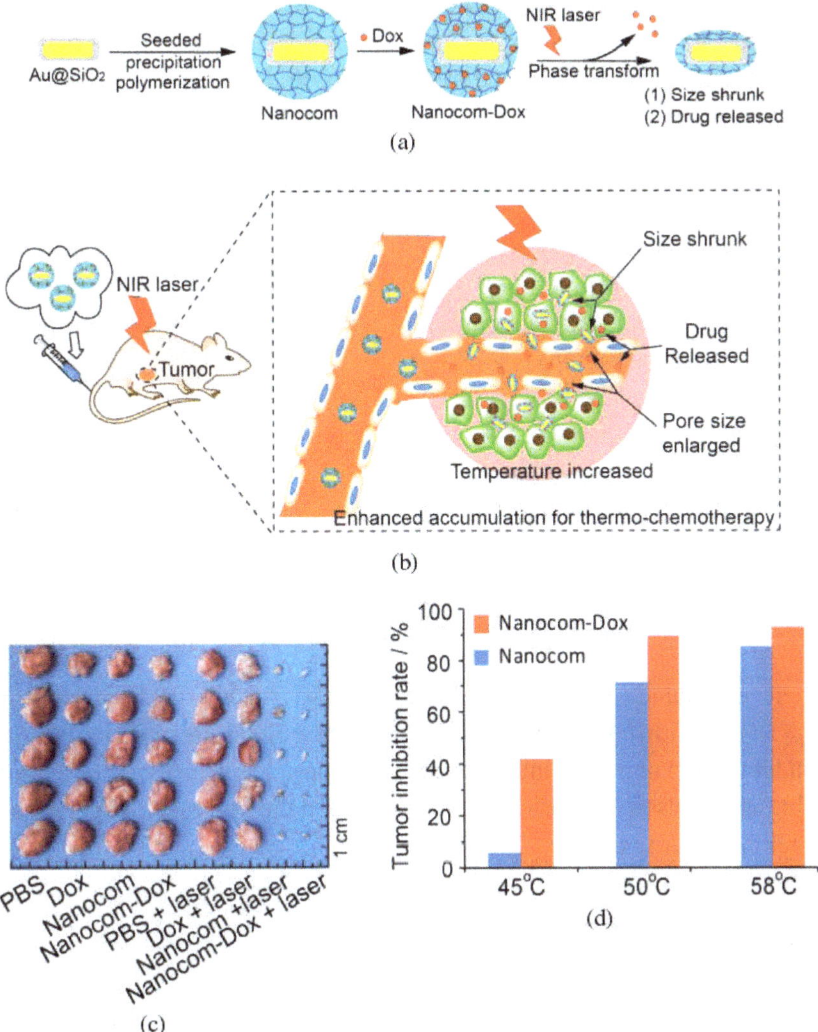

Figure 4.4   (a) Principle of the elaboration of thermosensitive plasmonic nanohybrids 'Nanocom' (pNIPAM thermosensitive polymer chains grafted on AuNR coated with SiO_2) loaded with doxorubicin ('Dox'), and of their laser-induced heating, succeeding in local temperature increase, polymer phase transition and drug release. (b) Illustration of the effect of laser irradiation of a murine breast tumour in the presence of Nanocom-Dox NPs. (c) Five tumours observed after different treatements (PBS denote the saline buffer). (d) Tumour inhibition rate after laser irradiation at three different powers, leading to three local temperatures, in the presence of Nanocom and Nanocom-Dox NPs. Reprinted with permission from Ref. 27.

where ε_{abs} (in J) is the total energy per pulse absorbed by the NP (see Equation (4.6)). If the NP is too large to be considered as a point-like heat source, it has been shown that the envelope profile follows an empirical law (stretched exponential) as[7]:

$$\Delta T_m^{max}(r) = T_m^{max}(r) - T_0 = \exp\left[-\left(\frac{r-R}{r_0}\right)^n\right], \qquad (4.19)$$

where r_0 and n are parameters fitted on the real envelope curve. If we assume no interface resistance ($\ell_K = 0$), then $n = 0.45$ and $r_0 = 0.06R$.[7]

The maximum NP temperature increase reached initially (after instantaneous pulse absorption) is given by:

$$\Delta T_{NP}^0 = \frac{\varepsilon_{abs}}{C_{Au}V_{NP}}, \qquad (4.20)$$

where V_{NP} is the NP volume.

The time dependence of the NP cooling subsequent to instantaneous heating by ultrashort pulse absorption has been found by Hu and Hartland to follow a phenomenological stretched-exponential profile as[31]:

$$\Delta T_{NP}(t) = T_{NP}(t) - T_0 = \Delta T_{NP}^0 \exp\left[-(t/\tau_{cool})^\gamma\right], \qquad (4.21)$$

where γ is found to be worth about 0.7 and the characteristic cooling time $\tau_{cool} = aR^b$ with $a \approx 0.159$ and $b = 2$ in the case of infinite interface thermal conductance. For a finite interface conductance, b has been assigned a lower value and tends to one as the heat conduction in the host medium gets much faster than across the interface.[32]

4.3.2 Athermal Regime

We have just described a simple approach where the involvement of electron excitation in the light-to-heat conversion process is disregarded. Let us now examine the photo-heating with a more realistic description of the energy exchanges, including the initial off-equilibrium state of the conduction electron gas in the metal NP when ultrashort light pulses are used. Monovalent bulk metals like alkali and noble metals exhibit quasi-free electron behaviour in the conduction band.[d] Indeed, their Fermi surface

---

[d]For noble metals, nevertheless, the lattice periodic potential opens a gap at the point L of the edge of the Brillouin zone (see Ref. 33).

in the reciprocal space is very close to a sphere, which reveals a parabolic dispersion law. This property bestows on us the right to use the quasi-free electron model for describing the conduction electron gas. Hence, at thermal equilibrium, the latter obeys a Fermi–Dirac distribution around the Fermi energy, E_F. The profile of the distribution depends on the electron temperature, T_e, defined as a means to characterise the electron internal energy at equilibrium. Before light excitation (Figure 4.5, left), T_e equals the metal lattice temperature T_l, and $T_l = T_0$ (initial temperature).[e] When sending the light pulse, part of it is absorbed to induce electron transitions the nature of which is, either intraband only if the photon energy $\hbar\omega$ is lower than the interband transition threshold, E_{ib}, or otherwise both intra- and interband. For the sake of simplicity, we will restrict our discussion to the first case only. The electron distribution then falls just below E_F and those electrons that have gained energy by photon absorption are promoted to energy levels just above E_F (Figure 4.5, middle). One is then left with an electron distribution out of thermal equilibrium; f does not follow a Fermi–Dirac distribution anymore and no electron temperature can be defined. This is the athermal regime. Note that this regime can be neglected when the excitation pulse width is long relative to the typical e–ph scattering time, as, in this case, the electron gas and the phonon bath are at every instant at quasi-thermal equilibrium.

By electron–electron collisions the energy is internally redistributed within the electron gas. This process is as efficient as the number of excited electrons is high, which explains that the duration of the athermal

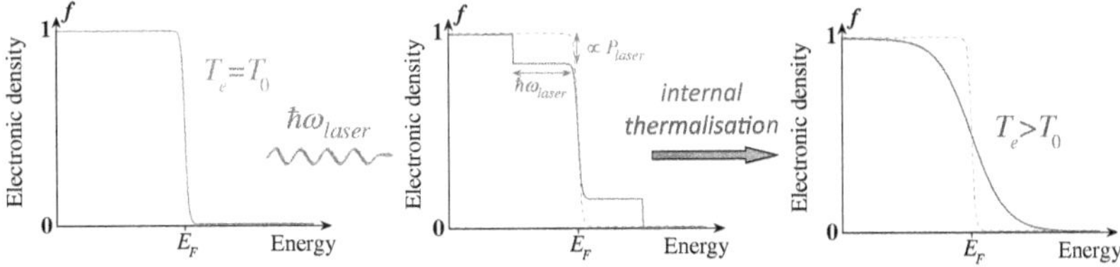

Figure 4.5 Evolution of the conduction electron occupation rate f as a function of electron energy subsequent to a light pulse absorption. Left: distribution at initial room temperature; middle: just after photon absorption; right: after internal thermalisation.

---

[e]Note that T_l has the same meaning as the classical NP temperature T_{NP} that was used in the basic approaches described in preceding sections.

regime decreases with increasing laser power. The distribution then recovers a Fermi–Dirac statistics at a temperature $T_e > T_0$ (Figure 4.5, right). This is known as Fermi smearing.

To describe the electron properties in the athermal regime, the relevant parameter then appears to be the electron distribution $f(E, t)$, which depends on electron energy E and time, the dynamics of which is governed by the Boltzmann equation:

$$\frac{\partial f(E,t)}{\partial t} = \left.\frac{\partial f(E,t)}{\partial t}\right|_{source} + \left.\frac{\partial f(E,t)}{\partial t}\right|_{e-e} + \left.\frac{\partial f(E,t)}{\partial t}\right|_{e-ph}. \tag{4.22}$$

Here, we neglect electron diffusion (assuming that the particle size is smaller than the wave penetration depth, the excitation can be considered as homogeneous) as well as the environment (the influence of which will be significant at longer times). The source term refers to the instantaneous modification of f under photon absorption. Its value at a given electron energy then depends on photon energy $\hbar\omega$ and is proportional to the number of photons absorbed per time unit, this number being proportional to the instantaneous power absorbed from the laser, $P_{abs}(t)$, as illustrated on Figure 4.5 (middle). $P_{abs}(t)$ can be evaluated from the particle absorption cross-section (or absorption coefficient of the medium knowing the NP density) and the laser pulse intensity and time profile. The second and third terms in Equation (4.22) denote the contributions of electron–electron and electron–phonon scattering, respectively, to the variation rate of f. Several approaches have been proposed to treat these contributions.[34–36] The *a priori* most rigorous one consists in including explicitly all the detailed scattering processes.[36] For the $e-e$ term, it amounts to integrating over all the wave vectors of the second electron and all the momentums exchanged in the elastic collision. For the $e-ph$ term, both the absorption and emission of phonons have to be accounted for and the integration runs over all the wave vectors of the phonons exchanged. A significant simplification can be introduced by considering the weak perturbation regime, which validates the use of the relaxation time approximation for the two scattering contributions separately:

$$\frac{\partial f(E,t)}{\partial t} = \frac{f(E,t) - f_0(E)}{\tau(E)}, \tag{4.23}$$

where f_0 is the initial equilibrium distribution and τ the typical collision time. In the case of the *e–ph* term, one has to distinguish the contribution of the spontaneous emission of phonons from the ones of phonon stimulated emission and absorption, the latter two depending on the number of states available in the phonon bath.[35] For the $e-e$ scattering term, the Landau theory of Fermi liquids allows us to express the $e-e$ mean collision time: $\tau_{e-e}(E) = \tau_0 E_F^2/(E - E_F)^2$.[37] This accounts for the fact that the scattering probability decreases as E gets closer to the Fermi level, which stems from the Pauli principle. τ_0 is a few tenths of femtosecond. In the limit of very weak perturbations, some authors have split the distribution into a major part still at thermal equilibrium and a small athermal part: $f = f_{\text{thermal}} + f_{\text{athermal}}$, with[35]

$$\int_{-\infty}^{+\infty} f_{\text{athermal}}(E)\,dE \ll \int_{-\infty}^{+\infty} f_{\text{thermal}}(E)\,dE. \qquad (4.24)$$

The model can be further refined, for instance by including the competition between transitions within the conduction band and the creation of electron–hole pairs in the valence and conduction bands.[38]

Let us notice that the only difference between the approaches for a bulk metal and a NP has up to now rested on the disappearance of any electron diffusion term in the right side of Equation (4.22) (provided that the excitation can be considered as homogeneous in the NP). In fact, some authors have shown that in the relaxation time approximation the $e-e$ and *e–ph* mean collision times exhibit a size dependence. This will be briefly evoked in Section 4.4.2.

4.3.3 *Thermal Regime*

Once the thermal equilibrium is recovered within the conduction electron gas, the energetic couplings within the nanoparticles and with their environment can be treated by more classical thermodynamics approaches as the electron temperature can be defined. Let us stress that, while the model described in the preceding section enables us to account for the off-equilibrium situation for the conduction electrons, it is restricted to short times after excitation as the heat exchange with the host medium is not accounted for. Moreover, in many cases, the athermal regime can be disregarded, as in pump–probe time-resolved experiments carried out with a high pump pulse intensity, which shortens the athermal regime duration, or for which the time scale under consideration is much larger than this duration. It can, of course, be also neglected for long pulse excitation (pulse width larger than the *e–ph* collision time) as in this case the electron gas and the crystal lattice are permanently at equilibrium.

4.3.3.1 *Analysis of the energy exchanges*

Two-temperature model

For describing the coupling between electrons and phonons, the two-temperature model (TTM) developed for bulk metals has been adapted to NPs. It consists in writing the two coupled differential equations ruling the time evolution of the electron and lattice internal energies:

$$C_e \frac{\partial T_e}{\partial t} = -G\left(T_e - T_l\right) + P_{\text{abs}}\left(t\right), \qquad (4.25)$$

$$C_l \frac{\partial T_l}{\partial t} = G\left(T_e - T_l\right). \qquad (4.26)$$

Here, C_e and C_l (in J m^{-3} K^{-1}) denote the electron gas and lattice volume heat capacities, respectively. As in noble metals the conduction electrons exhibit a quasi-free electron behaviour, as stated in Section 4.3.2, C_e can be deduced from free electron quantum statistics: $C_e = \gamma_e T_e$, where γ_e is a constant, the value of which depends on the metal (for gold, $\gamma_e = 66$ J m^{-3} K^{-2}). G is the *e–ph* coupling constant (for gold, $G = 3 \times 10^{16}$ W m^{-3} K^{-1}).[39] $C_l = 2.49 \times 10^6$ J m^{-3} K^{-1} for gold.[40] As for the athermal regime, $P_{\text{abs}}(t)$ represents the instantaneous power absorbed per metal volume unit (source term) and has the profile of the incident light pulse. Again, as the NP size is assumed to be smaller than the light penetration depth, the excitation is homogeneous and electron diffusion can be neglected.

The parameters involved in the TTM are usually taken as the bulk phase ones, but it may be easy to replace some of them in a phenomenological manner to account for finite size effects in NP. This has been done, for instance, to deduce the size-dependent G value from pump–probe experiments using Equations (4.25) and (4.26).[41] Some other authors have shown that G may vary with temperature, which itself depends on laser power.[39]

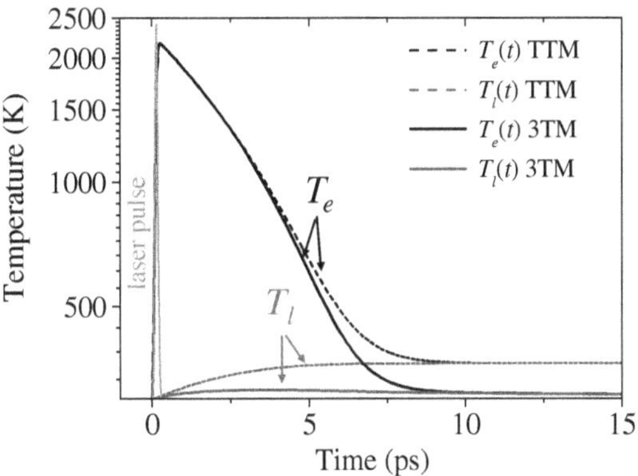

Figure 4.6 Time evolution of the electron (T_e, thick lines) and lattice (T_l, thin lines) temperatures of a AuNP in silica after light pulse absorption [$P_{\mathrm{abs}}(t)$, grey line] calculated without (TTM, dashed line) and with (3TM, solid line) considering the heat release to the host medium and a purely diffusive heat transfer in the latter. Pulse duration and peak power absorbed are worth 110 fs and 1.4×10^{21} W m^{-3}, respectively.[9]

Three-temperature model

The TTM completely neglects the thermal influence of the surrounding host medium. This might be valid as long as the heat exchange at the interface remains negligible, that is, when photo-heating with an ultrashort pulse and considering the electron temperature only during the first few picoseconds (see Figure 4.6). Of course, if the contact between the particle and the host medium is poor, if the thermal resistance at the interface is high, or if the host medium has a low thermal conductivity, its influence can be neglected over a larger time scale. In the general case, it has to be taken into account. This is the purpose of the three-temperature model (3TM). For this, Equation (4.26) has to be modified as to add the contribution of the instantaneous heat released through the interface, $H(t)$:

$$C_l \frac{\partial T_l}{\partial t} = G\left(T_e - T_l\right) - \frac{H(t)}{V_{\mathrm{NP}}}, \qquad (4.27)$$

where V_{NP} is the NP volume. The time evolution of the factor H (and subsequently of T_e and T_l) then strongly depends on the characteristics of the thermal transport through the interface and in the surrounding medium, namely, the ability of the latter to evacuate heat and then to cool down the particle. This point will be given a special attention in the next section.

Heat transfer to the host medium

As we have seen, the cooling down of a AuNP after ultrashort light pulse absorption or, by extension, the topography of the temperature field around a AuNP under light irradiation — whatever the time profile of the latter — strongly depends on the characteristics of the heat transfer in the NP host medium. For instance, the temperature of a thermally insulated NP (simulated by the TTM) is at any time higher than the temperature reached by the NP when it is in contact with a thermally conductive surrounding medium (accounted for by the 3TM), as can be seen in Figure 4.6. Beyond, it is easily understandable that, for instance, the higher the thermal conductivity of this medium, the faster the relaxation and then the lower the temperature at a given distance from the NP. This has been experimentally shown to influence the dynamics even at short times,[42] especially by comparing the optical relaxation of silver or gold NPs in silica (or glass) and alumina, the conductivities of which are roughly in the ratio of 1:30.[43,44] Beyond the only conductivity, the detailed mechanisms involved in the heat transport through the surrounding medium play a crucial role in the AuNP photo-induced thermal response. This involvement depends on the excitation conditions as well as on the observation ones as we will now show. The different approaches, which are used to describe heat transport in a medium, can be split into two categories. In the first one, known as molecular dynamics (MD), matter is described by its constituents (atoms) and the modelling consists in determining the motion of each atom by (usually) solving the classical second law of Newton once chosen the suited analytical description of the interatomic forces.[45] Once all the motions are calculated, statistical physics allows us to determine relevant quantities of the system thermodynamics as their mean values and fluctuations. Whereas, as we have seen above, molecular dynamics has allowed us to address the problem of photo-induced phase transform and partial melting of AuNPs, this powerful method has up to now been very little used to model the thermal transport in the medium surrounding a metal NP subsequent to the photo-induced heating of the latter.[46] Rather, continuous-media approaches are employed. They consider that all the media can be described by continuously varying quantities such as phonon density, energy and flux. In this category, the most general theory appropriate for this problem is the Boltzmann transport equation (BTE). It allows us to describe the time evolution of the local phonon density and is particularly suited for off-equilibrium situations. If the spectral composition of the heat transport is disregarded, namely, if a spectrally integrated phonon mean free path, Λ_{ph}, and lifetime, τ_{ph}, can be defined in a phenomenological manner, then they may be used as relevant parameters to validate successive simplifications of the BTE. First, the definition of a characteristic phonon lifetime itself may result in the use of the time relaxation approximation that we have already addressed in the section devoted to the athermal regime for the electrons. This approximation enables us to simplify the BTE. Further, it has been shown by Chen that an approach much simpler to bring into play than the BTE can nevertheless provide very similar results in the case of transient nanoscale heat transfer. This is the ballistic diffusive equations (BDE) method.[47] This takes into account the fact that at short time and space scales both ballistic and diffusive transport mechanisms can contribute to the overall heat propagation in a medium. This point will be given a more detailed development below. If the ballistic mechanism is not significant, we are then left with the Cattaneo–Vernotte model, which is a hyperbolic extension of the parabolic Fourier law for the diffusion that includes the finite lifetime of thermal phonons. Finally, the simplest approach is the Fourier law, which assumes an infinite speed for the heat transport and is then based on the assumption of a high number of anterior diffusive events to drive the thermal energy evolution at a given point of the medium and a given time. The validity of the Fourier law is then ensured when the observation time scale is much larger than τ_{ph} and the typical observation distance is much larger than Λ_{ph}. As usual examples, at room temperature, $\tau_{\mathrm{ph}} \sim 1040$ fs and $\Lambda_{\mathrm{ph}} \sim 1.6$ nm in water, $\tau_{\mathrm{ph}} \sim 130$ fs and $\Lambda_{\mathrm{ph}} \sim 0.5$ nm in silica (amorphous SiO_2) and $\tau_{\mathrm{ph}} \sim 850$ fs and $\Lambda_{\mathrm{ph}} \sim 5.4$ nm in alumina (amorphous Al_2O_3). Hence, for a AuNP in silica, the Fourier law can be used to describe thermal transport in the surrounding medium in order to determine on a few picosecond time scale the NP-transient thermal response (see Figure 4.6) or the temperature field in its vicinity — but not closer than a few nanometres to the NP surface. Note that, on the contrary, this would not be valid for a AuNP in alumina.

In that case, the interface heat exchange factor $H(t)$ of Equation (4.27) is simply worth

$$H(t) = S_{\text{NP}}\kappa_m \left.\frac{\partial T_m}{\partial r}\right|_{r=R} \tag{4.28}$$

if a perfect contact is assumed, i.e. $T_l(t) = T_m(t, R)$.[f] S_{NP} is the nanoparticle surface. The temperature field in the host medium then follows the classical diffusion law:

$$\frac{\partial T_m}{\partial t} = \frac{\kappa_m}{C_m}\Delta T_m. \tag{4.29}$$

This equation together with Equations (4.25), (4.27) and (4.28) then constitute the usual form of the 3TM.[48]

As an example, Figure 4.7 shows the mapping of the temperature field around a core–shell nanoparticle consisting of a 40 nm silica bead coated with a 0.5 nm thick gold layer (allowing to have a SPR spectral location optimised for biomedical purposes, see Chapter 14). The nanoshell is irradiated with a 7 ns Gaussian light pulse with a peak power absorbed of 5×10^{17} W m^{-3} (typical for nanosecond pulse-width lasers). The surrounding

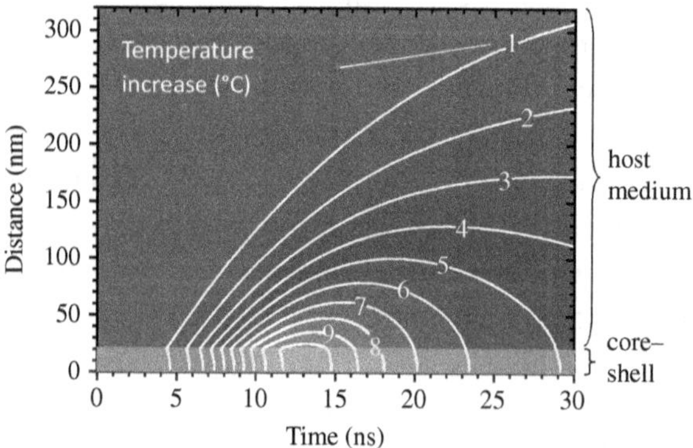

Figure 4.7   Isothermal plot of the medium (silica) surrounding a core–shell NP consisting of a 40 nm silica bead coated with a 0.5 nm thick gold layer, irradiated with a nanosecond laser pulse. The calculation is carried out in the framework of the 3TM by using Equation (4.29) for the thermal transport in silica. Pulse duration and peak power absorbed are worth 7 ns and 5×10^{17} W m^{-3}, respectively.[9]

---

[f] If an interface thermal resistance is considered, the boundary condition has to be modified according to Equation (4.12).

medium is silica, but the method could be used for any medium. Such mapping can be useful for biomedical applications, for selecting the irradiation power or the molecular spacer length between the NP heat source and the biological object to be targeted.

Whereas the method presented above can be easily applied as far as the geometry of the system remains spherical, it becomes tougher for any particle shape or for assemblies of NPs. One may then use three-dimensional finite elements methods to solve the set of coupled equations of the 3TM.

4.3.3.2 *Tuning the thermal spatial range with pulse duration*

Another interesting feature for applications lies in the dependence of the temperature field generated on the light irradiation duration. It is well known from heat physicists that the shorter the heat impulse the smaller the spatial heated range. This is nowadays used in the metal micromachining industry as laser cutting with femtosecond pulses is much more precise and creates many fewer defects than by using continuous lasers. We have already underlined the difference in the spatial extension of the temperature rise around NPs photo-heated, either by cw light, or by ultrashort laser pulses (see Equations (4.14) and (4.18)). Here, changing the pulse width can allow us to tune the effective thermal spatial range, as illustrated in Figure 4.8. This figure presents the space and time evolution of the temperature in the surrounding medium (silica again) in perfect contact with a AuNP heated by photo-absorption of a Gaussian light pulse. In (a) the pulse width is worth about 100 fs, whereas in (b), it is 30 ps and in (c) 7 ns. The respective input energies correspond to realistic experimental situations with lasers and AuNPs (peak intensity I_{00} of $11 \times 10^9, 81 \times 10^7$ and $3.25 \times 10^6 \, \mathrm{W \, cm^{-2}}$, respectively). It can be seen that despite the similar temperatures reached in the three cases the typical spatial heating range in the ultrashort regime is about 10 nm while it is about 40 nm in the picosecond regime and reaches about 100 nm in the nanosecond one.

4.3.3.3 *Cumulative thermal effect*

At the NP scale, if a train of light pulses is sent, it may be possible that the energy stored after the first pulse has not been totally released when the next pulse arrives. In this case, the NP will keep the 'thermal memory' of the

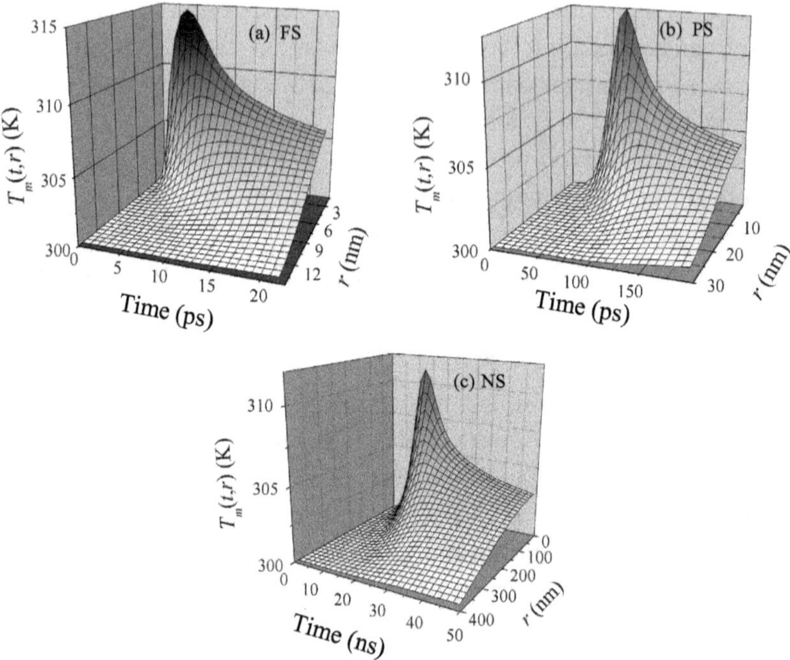

Figure 4.8   Temperature in the medium (silica) surrounding a AuNP (radius 1.3 nm) as a function of time t and distance from the NP centre, r (the back wall corresponds to the NP surface) as calculated by using the 3TM with Fourier law and perfect contact at the interface. The NP is irradiated by a Gaussian laser pulse in three different temporal regimes: (a) femtosecond, in which the conditions are those of Figure 4.6; (b) picosecond, with 30 ps pulse width and 1×10^{19} W m^{-3} peak power absorbed, and (c) nanosecond, the conditions being those of Figure 4.7.[9,48]

pulses for a given period; this results in the growth of a thermal background (hotter than the initial situation), which stabilises once the heat stored after the first pulse has been completely transferred towards the host medium. This occurs of course if the pulse repetition rate f_{pulse} is larger than the inverse of the NP thermalisation time. Thanks to the linearity of the heat exchange and conduction equations, the temperature rise above T_0 after a train of N pulses can be evaluated as[7]

$$\Delta T_{\text{NP}}(t) = \sum_{i=0}^{N-1} \Delta T_{\text{NP}}^{\text{sp}}\left(t - i/f_{\text{pulse}}\right), \tag{4.30}$$

where $\Delta T_{\text{NP}}^{\text{sp}}(t)$ denotes the time-dependent NP temperature rise above T_0, which would be obtained after a single pulse only.

Of course, at a larger space scale, heat after successive pulses is stored in the inhomogeneous medium, the mean temperature of which increases. The latter can be evaluated exactly as if the irradiation were continuous (see Section 4.2.4.2).

4.3.4 *When the Fourier Law Fails: What Occurs at Small Space and Time Scales*

The Fourier law becomes invalid at small space and time scales, and then in the cases of transient regimes at very short times (of the order of τ_{ph}), small distances (of the order of Λ_{ph}) and then at high metal concentration in a nanocomposite medium, as well as for NPs in a host medium having a high thermal conductivity. Let us imagine that an AuNP is suddenly heated by photo-absorption. At the very first instant after this excitation, the host medium in the close vicinity will be heated by thermal transfer through the interface. On the one hand, there exists no anterior diffusive event to ensure this heating. This 'phonon rarefaction' slows down the NP cooling as compared with what would be deduced from the classical diffusion law. On the other hand, if the observation time is lower than τ_{ph} and the distance from the NP smaller than Λ_{ph}, phonons emitted by the NP wall can reach the point under consideration in the surrounding medium in a ballistic (direct) way. The BDE approach introduced above is then suited to describe such phenomena. We now supply some basic elements to present the physical principle of this method, and we refer the reader to Ref. 49 and references therein for more details. As the medium is out of thermal equilibrium, temperature is no longer a good parameter to describe thermal transport. Rather, the BDE uses both internal energy $u\,(\mathbf{r}, t)$ and heat flux $\mathbf{q}\,(\mathbf{r}, t)$, which are defined locally in time and space. Then the contributions of both diffusive and ballistic phonons are added as $\mathbf{q} = \mathbf{q}_b + \mathbf{q}_d$ and $u = u_b + u_d$.

As for photons, it is possible to define a ballistic phonon intensity $I_{b\omega}\,(t, \mathbf{r}, \mathbf{\Omega})$ at point M($\mathbf{r}$) and time t, in a given direction $\mathbf{\Omega}$ and at a given phonon frequency ω (see Figure 4.9):

$$I_{b\omega}\,(t, \mathbf{r}, \mathbf{\Omega}) = I_{\mathrm{wall},\omega}\,(t - r'/|\mathbf{v}|, \mathbf{r} - r'\mathbf{\Omega}, \mathbf{\Omega})\,\exp\left(-r'/\Lambda_\omega\right), \qquad (4.31)$$

where r' is the distance from M to the NP wall in direction $\mathbf{\Omega}$ and $\mathbf{v}$ is the heat carrier speed. Λ_ω is the mean free path of phonons with angular

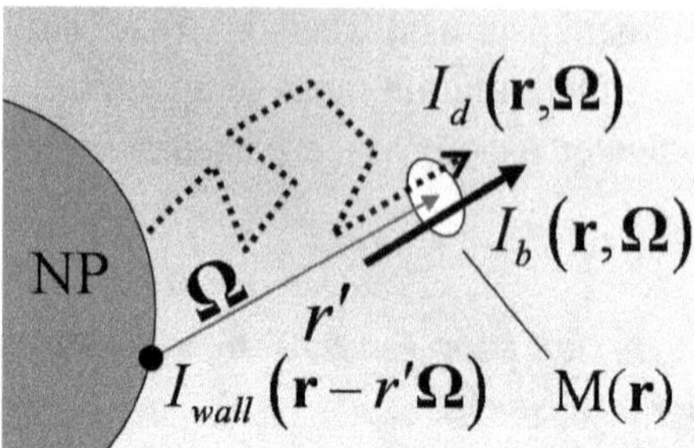

Figure 4.9 Schmatic representation of the principle of the BDE. The ballistic phonon intensity I_b at point M_r for a direction $\mathbf{\Omega}$ results from the emission of a ballistic phonon by the NP wall at point $\mathbf{r} - r'\mathbf{\Omega}$. I_d denote the diffusive intensity.

frequency ω and corresponds to the characteristic attenuation length of the phonon intensity as can be seen in the exponential of Equation (4.31). $t - r'/|\mathbf{v}|$ represents the carrier travel delay from its emission by the NP wall to point M. By definition, the NP wall emits ballistic phonons only, with a frequency-dependent intensity $I_{\text{wall},\omega}$. The ballistic flux at a given point and in a given direction is calculated by integrating the phonon intensity over all frequencies and summing over all directions. The frequency integration of the wall intensity $I_{\text{wall},\omega}$ is proportional to the temperature of the NP if any interface thermal resistance is disregarded. After some developments regarding the diffusive contribution,[49] the following equation is obtained:

$$\tau_{\text{ph}}\frac{\partial^2 u_d}{\partial t^2} + \frac{\partial u_d}{\partial t} = \frac{\kappa_m}{C_m}\Delta u_d - \nabla \cdot \mathbf{q}_b. \tag{4.32}$$

The contribution of the ballistic transport is then added to the hyperbolic Cattaneo–Vernotte expression in the form of the divergence of the ballistic flux. After imposing the suitable initial and boundary conditions, the heat exchanged at the interface in the 3TM (Equation (4.27)) is now evaluated through

$$H(t) = \int_S \mathbf{q}(\mathbf{r},t) \cdot \mathbf{n}\, ds, \tag{4.33}$$

where the integral runs over the whole NP surface and $\mathbf{n}$ is the unit vector normal to the surface.

The effect of phonon rarefaction is illustrated in Figure 4.10 where a 100-fs laser pulse heats a AuNP surrounded by an alumina shell, the thickness of which is set as the phonon mean free path (5.4 nm). The calculation

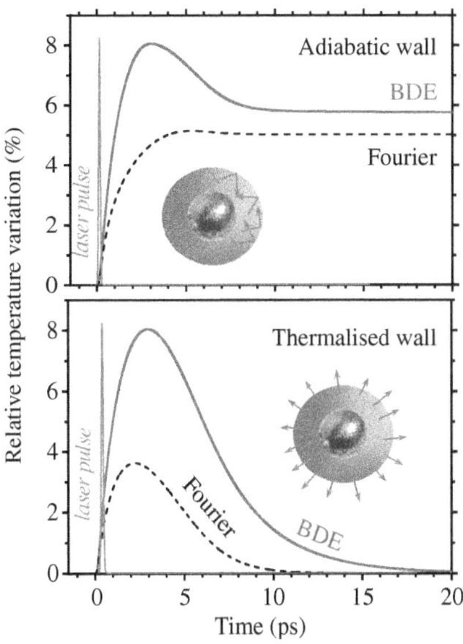

Figure 4.10   Time evolution of the relative temperature excess of a AuNP (10 nm radius) surrounded by an alumina shell after absorption of a 100-fs laser pulse. Shell thickness is of the order of the thermal phonon mean free path. The calculation is performed in a the framework of the ballistic-diffusive equation Top: adiabatic outer wall: bottom: thermalised outer wall. The result of the classical Fourier diffusion law is added for comparison.

is performed for two extreme boundary conditions: adiabatic external wall (isolated system) and external wall thermalised at room temperature. Whatever the boundary conditions imposed, the results show a slowing down of the NP cooling (and then a higher temperature reached) after energy injection by ultrashort laser pulse absorption as compared to the predictions of the classical Fourier law.[50]

4.4 Influence of Morphological Parameters

In this section, we will examine the dependence of the photothermal conversion by AuNPs on the main morphological characteristics of the inhomogeneous medium. Of course, the LSPR itself, and then the absorption cross-section, which drives the energy input during light irradiation, depend on such parameters as size, shape, spatial arrangement of the NPs in the medium, as described in Chapter 3. We won't enter into such details here but rather focus on the influence of morphology on the thermal behaviour.

4.4.1 *Nanoparticle Environment*

The refractive index of the medium around NPs has a strong influence on the spectral profile and magnitude of the LPSR, as seen in Chapter 3. Beyond this, we have already underlined that the host medium thermal conductivity, as well as the contact and Kapitza interface resistances, drive the heat exchange between NPs and their environment. The surface chemistry plays also a great role on the photothermal process, as the presence of surfactants, adsorbed or grafted molecules, modulate this exchange. For instance, a coating layer of organic molecules may degrade the conductivity at the AuNP–water interface, or on the contrary favour the overlap of the vibrational density of states and then enhance the interface conductance through better thermal impedance matching.[10]

4.4.2 *Nanoparticle Size*

From a very simple and classical consideration, it can easily be deduced that the heat release rate from a NP to its environment subsequent to light energy absorption is ruled by the competition between the energy stored in the NP, proportional to its volume, and the thermal exchange through the interface, proportional to the NP surface (see Equations (4.27) and (4.28)). Consequently, we could expect intuitively that the bigger the NP the slower the cooling down, and then the higher the temperature reached by the NP.[44,48] In the steady state, as we have seen from Equation (4.15) in Section 4.2.4.1, the NP radius dependence of the NP temperature excess is quadratic due to the conduction in the host medium, while the additional term corresponding to a finite interface resistance varies as R. The surrounding medium's temperature excess varies as R^3. We could then deduce that increasing NP size could improve the photothermal heating efficacy.[15] However, this effect is limited by the fact that light scattering, being negligible for small NPs, competes with absorption for NP a few tens of nanometres large, and dominates for larger sizes. This of course limits the efficiency of the photothermal conversion, as has been demonstrated by Jiang *et al.*[20] by macroscale

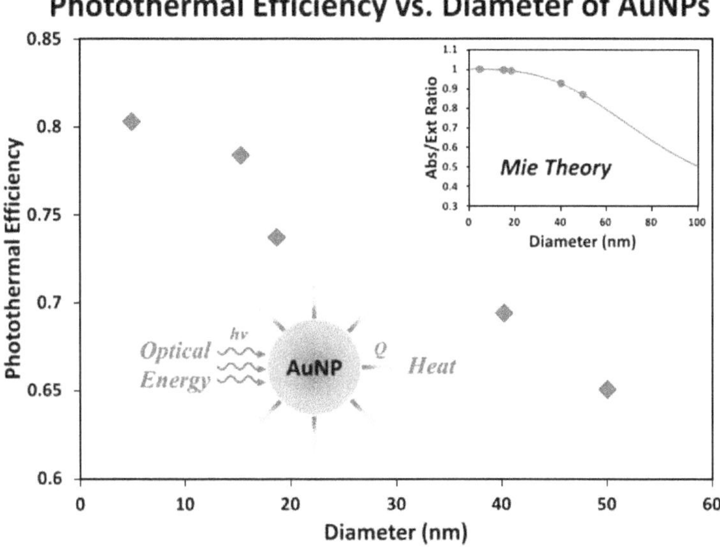

Figure 4.11 Photothermal efficiency, η, of AuNPs in water with different sizes. The insert shows the relative contribution of absorption to the total light extinction, as calculated with Mie theory. The decrease of η with increasing NP diameter is due to the rising contribution of scattering to extinction, stemming from the appearance of multipolar terms in the NP electromagnetic response, Reprinted with permission from Ref. 20.

temperature measurements such as described in Section 4.2.4.2. Their main result is reported in Figure 4.11, where the photothermal efficiency is defined as the ratio of the incident energy converted into heat and the total energy loss from the incident wave (addition of thermal and radiative losses).

In addition, due to confinement some parameters involved in the NP heating and cooling dynamics when irradiated with ultrashort pulses are likely to be modified by finite size effects. Let us merely summarise them. The increasing influence of the NP surface when decreasing particle size results in the acceleration of the global relaxation dynamics. Indeed, the $e-e$ collision time decreases due to (i) quantum size effects (electron spill-out and skin of reduced polarisability at the surface) inducing the decrease of the screening of the Coulomb interaction,[51] and (ii) electron recombination with d-holes, resonant with the SPR; this process is all the more efficient as R is small.[38] Furthermore, the *e–ph* collision time also decreases, subsequent to (i) the decrease of the Coulomb interaction screening,[52] and (ii) the appearance of new vibration modes (capillary and acoustic) in small NPs, which opens an additional relaxation channel for electrons.[53] The characteristic relaxation times τ_{e-e} and τ_{e-ph} are then reduced in a ratio of 1 to 0.4 and 1 to 0.6, respectively, from bulk gold to a AuNP with $R = 2\,\mathrm{nm}$.[51]

4.4.3 *Nanoparticle Shape*

Again, the NP shape is known to influence greatly the LSPR characteristics: splitting, spectral shift and variation of the magnitude and width of the resonance with increasing NP anisotropy (see Chapter 3). Hence, increasing the aspect ratio of Au nanorods results in the splitting of the LSPR, the longitudinal mode (excited with electric field polarised along the long nanorod axis) undergoing a rapid shift to the red as well as a magnitude increase (up to a maximum value). The resonance can then be tuned to the desired wavelength and the absorption cross-section at resonance can be much enhanced compared to the one of a sphere with same volume. Consequently, the amount of energy input for the photothermal conversion is influenced by these shape effects. The anisotropy of Au nanorods has been exploited in several works focused on the nanoscale heat photo-generation. By rotating the light polarisation relative to the nanorod orientation, provided the wavelength matches the longitudinal LSPR, enables us to modulate the heating efficiency. This has been demonstrated with single nanorods,[54] and with ensembles of nanorods aligned in the same direction.[55]

Moreover, the NP shape is likely to influence the heat transfer to the surrounding medium due to the variation of the local curvature of the interface, as well as the competition between energy storage (proportional to V_{NP}) and energy transfer through the interface (proportional to S_{NP}).[16] However, within a certain limit, the temperature profile in the host medium is not significantly affected by the shape and can be evaluated by considering an equivalent nanosphere rather than the actual nano-object, having the same volume and of course the same absorption cross-section, and using the equations that have been described in the preceding sections.[7]

4.4.4 *Nanoparticle Density*

Until now, we have considered isolated particles only. However, thermal energy exchange between neighbouring particles through the interstitial medium may affect their thermal response, either in the stationary regime or in the transient one. It can be easily understood in a qualitative manner by examining Equation (4.28). Indeed, it can be seen that the heat exchange rate crucially depends on the thermal gradient at the interface, which is a very well-known result of classical thermal transfer physics. Hence, the colder the host medium, the faster interface heat exchange (that is, the faster the NP cooling in the transient regime). Now, if a NP A has a neighbouring one B, the heat released by B induces the reduction of the thermal gradient in the vicinity of A, the cooling of which consequently slows down. This slowing down of the NP cooling has been demonstrated both experimentally[56] and theoretically[48] in nanocomposite media with varying concentrations and irradiated by light pulses. Provided A is still irradiated and goes on absorbing light energy, its temperature reaches a value higher than in the absence of B. An outcome of this is that in a nanocomposite medium the temperature of both the NPs and their close vicinity increases with NP density for a fixed absorbed power per particle.[15,48] Similarly, it has been recently shown by using the thermal extension of the DDA method that in a two-dimensional array of $N \times N$ AuNPs irradiated in the steady state the temperature at the centre increases linearly with N.[17]

4.5 Thermo-optical Properties of Gold Nanoparticles

We have seen that the specific optical properties of AuNPs can lead to make them efficient nanoscale thermal sources, thanks to an internal series of energy exchanges leading to the conversion of light into heat. Inversely, one may be interested in the way a temperature variation leads to a change in the nanoparticle optical properties. This is generally called the 'thermo-optical response' of materials. This point may be considered either in the transient case or in the stationary one. By 'transient' we design the case of ultrafast variations where the optical response, which is mainly governed in metals by the behaviour of electrons, cannot be directly linked with the particle temperature as long as the electrons and the crystal lattice are not at equilibrium. This essentially concerns the domain of ultrafast relaxation studied by pump–probe spectroscopy using femtosecond lasers. This has been largely studied for more than two decades and this is out of the scope of this chapter.

In the 'stationary' regime, a variation of temperature induces a modification of the material complex optical index, $\Delta \tilde{n}$. Provided the change is sufficiently weak around a given temperature, this variation can be linearised as

$$\Delta \tilde{n}(t) = (\partial n/\partial T + i\partial \kappa/\partial T)\, \Delta T(t) = (d_T n + id_T \kappa)\, \Delta T(t). \qquad (4.34)$$

$d_T n$ and $d_T \kappa$ are the thermo-optical coefficients for refraction and extinction, respectively.

4.5.1 *Bulk Gold*

Thanks to several experimental and theoretical investigations carried out in the 1970s on the electronic properties and band structure of noble metals, the mechanisms responsible for the optical and thermo-optical response of bulk gold are well known. Let us summarise the latter. When increasing gold temperature, the contribution of both interband and intraband electron transitions to the optical properties are modified due to (i) the increase in the $e{-}e$ and e–ph collision rates. This affects the quasi-free conduction electron contribution, and is significant at low photon energy, well below the interband transition threshold; (ii) the modification of the electron distribution around the Fermi level. This mainly modifies the contribution of interband transitions to the complex dielectric function around their onset (from $\sim$2 to 3 eV), as can be calculated through the Rosei model for the band structure[57]; (iii) the thermal expansion of the crystal lattice, which shifts the electron band energies non-uniformly and lowers the Fermi level.[58] This third effect results in the modification of the absorption spectrum above the interband transition threshold, especially around the maximum of the contribution of the L point of the Brillouin zone ($\sim$3.5–4.0 eV).[59]

The accurate analysis of different works published a few decades ago has then allowed us to extract the mean values of the thermo-optical coefficients of bulk gold in the range from 295 to 670 K.[60] Their dispersion curve reveals a monotonous decrease of the thermo-optical effect with increasing photon energy, stemming from the contribution of intraband transitions, on which superimpose strong features ascribed to the temperature dependence of the interband transitions, especially in the spectral domain of their threshold.

4.5.2 *Gold Nanoparticles*

As gold dielectric function evolves with temperature, the optical properties of AuNPs are likely to do so. In fact, the thermo-optical response of gold in a nanoparticle is strongly affected by the local electromagnetic field enhancement at the SPR.[60] This has been demonstrated experimentally on the absorption part of the optical response.[61] The consequences of heating a medium containing AuNPs on its SPR band are a quenching, a broadening and a slight red-shift. Furthermore, the refractive part of the thermo-optical response has also been shown to be significant, responsible for instance for a thermal lensing effect when using lasers.[62] As an illustration, Figure 4.12 shows the spectral variations of both $d_T n$ and $d_T \kappa$ in a Au:SiO$_2$ nanocomposite film (metal volume fraction $p = 6.6\%$; mean NP radius $R = 1.3$ nm) as deduced from ellipsometry measurements carried out at 20°C and 190°C. It is noteworthy that the profile of the curves is fully correlated to both the characteristics of interband transitions (onset around 2.0 eV, absorption at higher photon energy) and the effect of the SPR around its maximum at about 2.6 eV (denoted by the vertical line). As can be seen, in

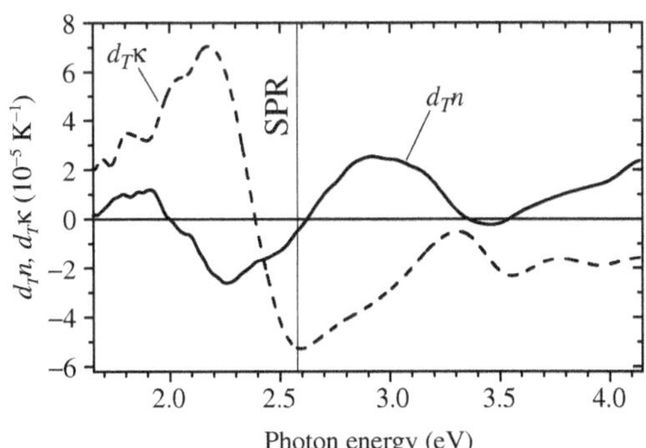

Figure 4.12   Spectral dependence of the thermo-optical coefficients $d_T n$ and $d_T \kappa$ of a thin film containing AuNPs ($p = 6.6\%_{vol}$, $R -1.3$, thickness: 150 nm). The vertical line denotes the experimental spectral location of the SPR maximum.

the visible spectral domain, the thermo-refractive coefficient may be either positive or negative depending on the wavelength, resulting in a convergent or divergent thermal lens.[63] By (analytically or numerically) differentiating relative to temperature any model describing the optical response of a nanoparticle or a nanocomposite medium from the knowledge of the metal dielectric function, it is possible to obtain the temperature dependence of this response. Of course, all the optical and morphological parameters of the model which are likely to exhibit a temperature dependence, beyond the metal index itself, may be preliminarily identified and included in the differentiation. For instance, it has been shown that the thermo-optical properties of a medium matching the conditions of validity of the Maxwell–Garnett theory (see Chapter 3) can be quite well reproduced by calculating $\partial \widetilde{\varepsilon}_{\mathit{eff}} / \partial T$.[62]

4.5.3 *Melting Point Depression in Gold Nanoparticles*

As many properties of NPs evolve with their size, the thermal properties of AuNPs are likely to present a size dependence as well. We have already briefly introduced some of these effects in the preceding sections; it appears interesting here to report such a size dependence for the melting point of AuNPs, T_{melt}. Indeed, a strong decrease of T_{melt} with decreasing NP size has been both observed and predicted. While this effect was demonstrated in some metallic nanoparticles almost six decades ago thanks to electron diffraction analysis,[64] its experimental evidence in the case of AuNPs was first reported in the 1970s.[65,66] Actually, this effect concerns all kinds of nanoparticles and is known as the 'melting point depression'. Its physical origin probably stems from the increase of the contribution of surface atoms to the total number of atoms when the particle size is decreased. As the cohesion energy of surface atoms is lower than the one of volume atoms, partial melting is favoured at the surface[67] and the NP surface melting temperature is lower than the bulk material one.[68] The classical thermodynamics approach accounting for such an effect is known as the Gibbs–Thomson equation, involving the liquid–solid interface curvature dependence of T_{melt}.[69] For a sphere of radius R, it is written as

$$T_{\text{melt}}(R) = T_{\text{melt}}(\infty)\left(1 - \frac{2\sigma_{\text{sl}}}{H_f\rho_s R}\right),$$ (4.35)

where $T_{\text{melt}}(\infty)$ denotes the bulk melting point (1337 K for gold), σ_{sl} is the solid–liquid interface energy, H_f is the bulk latent heat of fusion and ρ_s is the solid state density. The $1/R$ divergence from the bulk value has been rather well evidenced by experiments (electron diffraction analysis, calorimetric or field emission measurements), as can be observed in Figure 4.13 where the variation of T_{melt} with $1/R$ is reported from four different studies.[65,66,70,71] As several works have been devoted to the determination of the melting mechanisms and temperature of AuNPs by MD calculations,[72–75] we have also reported some of these works in Figure 4.13. The discussion about the differences in the methods and results of these studies is out of the scope of this chapter. It can be seen that the approach of Shim *et al.*[74] meets particularly well some of the experimental results within a large size range.

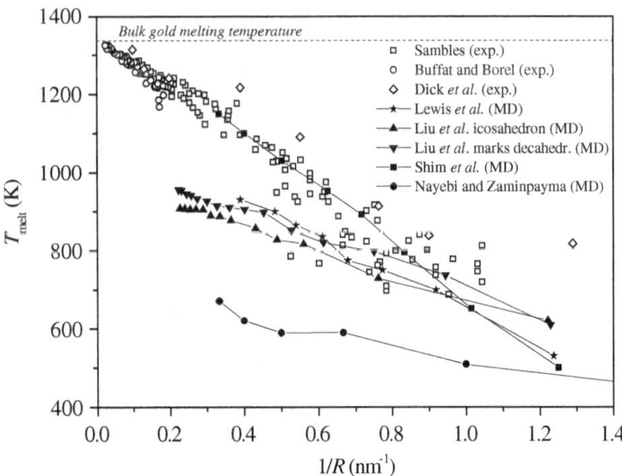

Figure 4.13   Variation of the melting temperature of AuNPs as a function of the inverse of their radius. Open symbols: experimental results from Sambles,[65] Buffat and Borel,[66] Castro *et al.*,[70] and Dick *et al.*,[71] Closed symbols: molecular dynamics calculation from Lewis *et al.*,[72] Liu *et al.*,[73] Shim *et al.*,[74] Nayebi and Zaminpayma.[75] The dashed line denotes the asymptotic value of T_{melt} for bulk gold.

The NP melting mechanisms have also been addressed through optical experiments on gold nanorods, by investigating their thermal- or photo-induced transition into the more stable spherical shape by both transmission electron microscopy and conventional optical spectrometry monitoring their SPR bands.[76,77] The role of the rigidity of the host medium has been emphasised.[76] There has been some attempts to determine the temperature of laser-induced melting of spherical AuNPs by monitoring the change in the vibration period of their low-frequency breathing acoustic mode by time-resolved pump–probe experiments; however, they have led to no clear conclusion due to the absence of any abrupt discontinuity at the phase transition.[77,78] Time-resolved X-ray scattering experiments have shown that laser-induced heating of AuNPs can result in the lattice expansion and disappearance of long-range order for sufficient laser pulse power, signing the melting phase transition.[79]

4.6 Conclusion

Many applications are being developed nowadays based on the ability of gold NPs to convert efficiently light into heat at the nanoscale. We have seen along this chapter that very basic thermodynamics considerations help in describing the photothermal behaviour of plasmonic gold nanoparticles and heat release to their environment in many usual situations. As for the optical properties themselves, the photothermal properties are determined by the NP size, shape, and spatial distribution in the medium. In addition, the characteristics of the lightning as well as the interface and host medium thermal properties play a crucial role. Hence, the duration of the light energy input drives the spatial range of the heating. Beyond, we have described the modifications of the heat transport from a plasmonic NP at short time and space scales, where the simple approaches based on a pure diffusive process are no longer valid. The dynamics of the energy exchanges has been analysed, which is particularly relevant to understand the photothermal conversion induced by ultrashort light pulses. We have then examined the dependence of the NP optical properties upon temperature variations. The effect of the local field enhancement has been emphasised. Finally, we have described and explained the strong decrease of the AuNP melting temperature with decreasing size.

Acknowledgements

B.P. warmly thanks Majid Rashidi-Huyeh, Yannick Guillet, Bruno Gallas, Yann Chalopin, Sebastian Volz and X. Wang for their contribution and enriching discussions.

References

1. M. Fox, *Optical Properties of Solids* (Oxford University Press, 2001).
2. J. M. Ziman, *Electrons and Phonons* (Clarendon Press, Oxford, 1960).
3. R. N. Gurzhi, Soviet Phys. *JETP* **8**, 673 (1959).
4. J. B. Smith and H. Ehrenreich, *Phys. Rev. B* **25**, 923 (1982).
5. B. Lamprecht, A. Leitner, and F. R. Aussenegg, *Appl. Phys. B* **64**, 269 (1997); T. Ziegler, C. Hendrich, F. Hubenthal, T. Vartanjan, and F. Träger, *Chem. Phys. Lett.* **386**, 319 (2004); C. Sonnichsen, T. Franzl, T. Wilk, G. V. Plessen, J. Feldmann, O. Wilson, and P. Mulvaney, *Phys. Rev. Lett.* **88**, 077402-1 (2002).
6. P.-O. Chapuis, M. Laroche, S. Volz, and J.-J. Greffet, *Appl. Phys. Lett.* **92**, 201906 (2008).
7. G. Baffou and H. Rigneault, *Phys. Rev. B* **84**, 1 (2011).
8. B. T. Draine and P. J. Flatau, *J. Opt. Soc. Am. A* **11**, 1491 (1994).
9. M. Rashidi-Huyeh, Influence des effets thermiques sur la réponse optique de matériaux nanocomposites métal-diélectrique, PhD thesis of Université Pierre et Marie Curie, Paris, 2006.
10. V. Juvé, M. Scardamaglia, P. Maioli, A. Crut, S. Merabia, L. Joly, N. Del Fatti, and F. Vallée, *Phys. Rev. B* **80**, 195406 (2009).
11. J. Soussi, S. Volz, B. Palpant, and Y. Chalopin, *Appl. Phys. Lett.* **106**, 093113 (2015).
12. O. M. Wilson, X. Y. Hu, D. G. Cahill, and P. V. Braun, *Phys. Rev. B* **66**, 224301 (2002).
13. A. J. Schmidt, J. D. Alper, M. Chiesa, G. Chen, S. K. Das, and K. Hamad-Schifferli, *J. Phys. Chem. C* **112**, 13320 (2008).
14. Z. Ge, D. G. Cahill, P. V. Braun, *J. Phys. Chem. B* **108**, 18870 (2004).
15. A. O. Govorov, W. Zhang, T. Skeini, H. Richardson, J. Lee, and N. A. Kotov, *Nanoscale Res. Lett.* **1**, 84 (2006).
16. G. Baffou, R. Quidant, and F. Javier García de Abajo, *ACS Nano* **4**, 709 (2010).
17. G. Baffou, R. Quidant, and C. Girard, *Phys. Rev. B* **82**, 165424 (2010).
18. G. Baffou, R. Quidant, and C. Girard, *Appl. Phys. Lett.* **94**, 153109 (2009).
19. H. H. Richardson, M. T. Carlson, P. J. Tandler, P. Hernandez, and A. O. Govorov, *Nano Lett.* **9**, 1139 (2009).
20. K. Jiang, D. A. Smith, and A. Pinchuk, *J. Phys. Chem. C* **117**, 27073 (2013).
21. K. Mawatari, T. Kitamori, and T. Sawada, *Anal. Chem.* **70**, 5037 (1998).
22. L. Cognet, C. Tardin, D. Boyer, D. Choquet, P. Tamarat, and B. Lounis, *Proc. Natl. Acad. Sci. USA.* **100**, 11350 (2003).
23. C. Leduc, S. Si, J. Gautier, M. Soto-Ribeiro, B. Wehrle-Haller, A. Gautreau, G. Giannone, L. Cognet, and B. Lounis, *Nano Lett.* **13**, 1489 (2013).

24. C. Loo, A. Lin, L. Hirsch, M.-H. Lee, J. Barton, N. Halas, J. West, and R. Drezek, *Technol. Cancer Res. Treat.* **3**, 33 (2004).

25. A. M. Smith, M. C. Mancini, and S. Nie, *Nat. Nanotechnol.* **4**, 710 (2010).

26. P. K. Jain, I. H. El-Sayed, and M. A. El-Sayed, *Nano Today* **2**, 18 (2007).

27. Z. Zhang, J. Wang, X. Nie, T. Wen, Y. Ji, X. Wu, Y. Zhao, and C. Chen, *J. Am. Chem. Soc.* **136**, 7317 (2014).

28. L. Francois, M. Mostafavi, J. Belloni, and J. A. Delaire, *Phys. Chem. Chem. Phys.* **3**, 4965 (2001); M. Hu, H. Petrova, and G. V. Hartland, *Chem. Phys. Lett.* **391**, 220 (2004); G. Baffou, J. Polleux, H. Rigneault, and S. Monneret, *J. Phys. Chem. C* **118**, 4890 (2014).

29. M. Kitz, S. Preisser, A. Wetterwald, M. Jaeger, G. N. Thalmann, and M. Frenz, *Biomed. Opt. Express* **2**, 291 (2011).

30. O. Neumann, A. S. Urban, J. Day, S. Lal, P. Nordlander, and N. J. Halas, *ACS Nano* **7**, 42–49 (2013).

31. M. Hu and G. V Hartland, *J. Phys. Chem. B* **106**, 7029 (2002).

32. A. Plech, V. Kotaidis, S. Grésillon, C. Dahmen, and G. von Plessen, *Phys. Rev. B* **70**, 195423 (2004).

33. N. W. Ashcroft and N. D. Mermin, *Solid State Physics* (HRW Int. Ed. Austin, Texas, 1976).

34. W. S. Fann, R. Storz, H. W. K. Tom, and J. Bokor, *Phys. Rev. B* **46**, 13592–13595 (1992); G. Tas and H. J. Maris, *Phys. Rev. B* **49**, 15046–15054 (1994); R. H. M. Groeneveld, R. Sprik, and A. Lagendijk, *Phys. Rev. B* **51**, 11433–11445 (1995); C. Suárez, W. E. Bron, and T. Juhasz, *Phys. Rev. Lett.* **75**, 4536–4539 (1995); V. E. Gusev and O. B. Wright, *Phys. Rev. B* **57**, 2878–2888 (1998); Y. Guillet, E. Charron, and B. Palpant, *Phys. Rev. B* **79**, 195432 (2009).

35. C.-K. Sun, F. Vallée, L. Acioli, E. P. Ippen, and J. G. Fujimoto, *Phys. Rev. B* **48**, 12365–12368 (1993); C.-K. Sun, F. Vallée, L. Acioli, E. P. Ippen, and J. G. Fujimoto, *Phys. Rev. B* **50**, 15337–15348 (1994).

36. N. Del Fatti, R. Bouffanais, F. Vallée, and C. Flytzanis, *Phys. Rev. Lett.* **81**, 922–925 (1998).

37. D. Pines and P. Nozières, *The theory of quantum liquids,* Vol. I: *Normal Fermi liquids* (W. A. Benjamin Inc., New York, 1966).

38. T. V. Shahbazyan, I. E. Perakis, and J.-Y. Bigot, *Phys. Rev. Lett.* **81**, 3120 (1998)

39. R. H. M. Groeneveld, R. Sprik, and A. Lagendijk, *Phys. Rev. B* **51**, 11433 (1995).

40. W. Benenson, J. W. Harris, H. Stocker, and H. Lutz, *Handbook of Physics* (Springer-Verlag, New York, 2000).

41. J. H. Hodak, I. Martini, and G. V. Hartland, *J. Phys. Chem. B* **102**, 6958 (1998); H. Inouye, K. Tanaka, I. Tanahashi, and K. Hirao, *Phys. Rev. B* **57**, 11334 (1998); J. Sasai and K. Hirao, *J. Appl. Phys.* **89**, 4548 (2001).

42. M. B. Mohamed, T. S. Ahmadi, S. Link, M. Braun, and Mostafa A. El-Sayed, *Chem. Phys. Lett.* **343** (2001) 55.

43. V. Halté, J. Y. Bigot, B. Palpant, M. Broyer, B. Prevel, and A. Perez, *Appl. Phys. Lett.* **75**, 3799 (1999).

44. Y. Hamanaka, J. Kuwabata, I. Tanahashi, S. Omi, and A. Nakamuka, *Phys. Rev. B* **63**, 140302 (2001).

45. D. C. Rapaport, *The Art of Molecular Dynamics Simulations*, 2nd ed. (Cambridge University Press, 2004); M. P. Allen and D. J. Tildesley, *Computer Simulation of Liquids* (Oxford University Press, 1989).

46. S. Merabia, S. Shenogin, L. Joly, P. Keblinski, and J.-L. Barrat, *PNAS* 106 (2009) 15113.

47. G. Chen, *Phys. Rev. Lett.* **86**, 2297–2300 (2001); G. Chen, *J. Heat Transfer ASME* **124**, 320–328 (2002).

48. M. Rashidi-Huyeh and B. Palpant, *J. Appl. Phys.* **96**, 4475–4482 (2004).

49. B. Palpant in *Thermal Nanosystems and Nanomaterials*, (S. Volz, ed.), series *Topics in Applied Physics*, Springer, New York, 2009, Vol. 118, p. 127.

50. M. Rashidi-Huyeh, S. Volz, and B. Palpant, *Phys. Rev. B* **78**, 125408 (2008).

51. J. Lermé G. Celep, M. Broyer, E. Cottancin, M. Pellarin, A. Arbouet, D. Christofilos, C. Guillon, P. Langot, N. Del Fatti, and F. Vallée, *Eur. Phys. J. D* **34**, 199 (2005).

52. C. Voisin, D. Christofilos, N. Del Fatti, F. Vallée, B. Prével, E. Cottancin, J. Lermé, M. Pellarin, and M. Broyer, *Phys. Rev. Lett.* **85**, 2200–2203 (2000); A. Arbouet, C. Voisin, D. Christofilos, P. Langot, N. Del Fatti, F. Vallée, J. Lermé, G. Celep, E. Cottancin, M. Gaudry, M. Pellarin, M. Broyer, M. Maillard, M.-P. Pileni, and M. Treguer, *Phys. Rev. Lett.* **90**, 177401 (2003).

53. E. D. Belotkii, S. N. Luk'yanets, and P. M. Tomchuk, *Sov. Phys. JETP* **74**, 88–94, 1992; E. D. Belotskii and P. M. Tomchuk, *Int. J. Electron.* **73**, 955 (1992).

54. H. Ma, P. Bendix, and L. Oddershede, *Nano Lett.* **12**, 3954 (2012).

55. J. Pérez-Juste, B. Rodríguez-González, P. Mulvaney, and L. M. Liz-Marzán, *Adv. Funct. Mater.* **15**, 1065 (2005); S. Maity, K. A. Kozek, W.-C. Wu, J. B. Tracy, J. R. Bochinski, and L. I. Clarke, *Part. Part. Syst. Charact.* **30**, 193 (2013).

56. H. B. Liao, R. F. Xiao, J. S. Fu, H. Wang, K. S. Wong, and G. K. L. Wong, *Opt. Lett.* **23**, 388 (1998).

57. R. Rosei, F. Antonangeli, and U. M. Grassano, *Surf. Sci.* **37**, 689 (1973).

58. P. Winsemius, H. P. Lengkeek, and F. F. van Kampen, *Physica B* **79**, 529 (1971).

59. P. Winsemius, F. F. van Kampen, H. P. Lengkeek, and C. G. Van Went, *J. Phys. F* **6**, 1583 (1976).

60. M. Rashidi-Huyeh and B. Palpant, *Phys. Rev. B* **74**, 75405 (2006).

61. R. H. Doremus, *J. Chem. Phys.* **40** (1964) 2389; R. H. Doremus, *J. Chem. Phys.* **42**, 414 (1965); U. Kreibig, *J. Phys. F: Metal Phys.* **4** (1974) 999; A. Heilmann and U. Kreibig, *Eur. Phys. J.AP* **10**, 193 (2000); S. Link and M. A. El-Sayed, *J. Phys. Chem. B* **103**, 4212 (1999); D. Dalacu and L. Martinu, *Appl. Phys. Lett.* **77**, 4283 (2000); L. M. Liz-Marzán and P. Mulvaney, *New. J. Chem.* **22**, 1285 (1998).

62. B. Palpant, M. Rashidi-Huyeh, B. Gallas, S. Chenot, and S. Fisson, *Appl. Phys. Lett.* **90**, 223105 (2007).

63. Y. Guillet, M. Rashidi-Huyeh, D. Prot, and B. Palpant, *Gold Bull.* **41**, 341 (2008).

64. M. Takagi, *J Phys. Soc. Jpn.* **9**, 359 (1954).

65. J. R. Sambles, *Proc. Roy. Soc. Lond. A* **324**, 339 (1971).

66. Ph. Buffat and J-P. Borel, *Phys. Rev. A* **13**, 2287 (1976).

67. R. Kofman, P. Cheyssac, A. Aouaj, Y. Lereah, G. Deutscher, T. Ben-David, J. M. Penisson, and A. Bourret, *Surf. Sci.* **303**, 231 (1994).
68. P. R. Couchman and W. A. Jesser. *Nature* **269**, 481 (1977).
69. L. Makkonen, *Langmuir* **16**, 7669 (2000).
70. T. Castro, R. Reifenberger, E. Choi, and R. P. Andres, *Phys. Rev. B* **42**, 8548 (1990).
71. K. Dick, T. Dhanasekaran, Z. Zhang, and D. Meisel, *J. Am. Chem. Soc.* **124**, 2312 (2002).
72. L. Lewis, P. Jensen, and J.-L. Barrat, *Phys. Rev. B* **56**, 2248 (1997).
73. H. Liu, J. A. Ascencio, M. Perez-Alvarez, and M. J. Yacaman, *Surf. Sci.* **491**, 88 (2001).
74. J. Shim, B. J. Lee, and Y. W. Cho, *Surf. Sci.* **512**, 262 (2002).
75. P. Nayebi and E. Zaminpayma, *J. Cluster Sci.* **20**, 661 (2009).
76. S.-S. Chang, C.-W. Shih, C.-D. Chen, W.-C. Lai, and C. R. C. Wang, *Langmuir* **15**, 701 (1999).
77. H. Petrova, J. P. Juste, I. Pastoriza-Santos, G. V. Hartland, L. M. Liz-Marzan, and P. Mulvaney, *Phys. Chem. Chem. Phys.* **8**, 814 (2006).
78. G. V. Hartland, M. Hu, and J. E. Sader, *J. Phys. Chem. B* **107**, 7472 (2003).
79. A. Plech, V. Kotaidis, S. Gresillon, C. Dahmen, and G. von Plessen, *Phys. Rev. B* **70**, 195423 (2004).

Chapter 5

Quantum Properties of Gold Nanoparticles

Changhyoup Lee* and Mark Tame[†]

*Karlsruhe Institute of Technology, Institute of Theoretical Solid
State Physics, Karlsruhe, Germany
[†]School of Chemistry and Physics, University of KwaZulu-Natal,
Durban, South Africa

5.1 Introduction

Metal nanoparticles provide a unique setting for controlling light at the nanoscale due to their ability to confine the electromagnetic (EM) field to regions well below the diffraction limit in a variety of configurations.[1,2] Here, the EM field is confined to the surface of a nanoparticle in the form of a localised surface plasmon (LSP) – a joint excitation of light coupled to electron charge density oscillations in the metal.[3] These plasmonic excitations and their supporting metal nanoparticles have opened up a wide range of applications based on extreme light concentration, including optical metamaterials,[4,5] biochemical sensing,[6] near-field nano-imaging and super-lensing,[7] nanoscale lasers,[8,9] nonlinear nanophotonics,[10] photovoltaics for enhanced solar cells[11] and antennas transmitting and receiving light signals at the nanoscale.[12] The recent rapid development of all these applications has been made possible by the large array of experimental tools that have become available in the past few years for nanoscale fabrication and theoretical tools for efficient EM simulation. Most recently, there has been a growing interest in a new direction for plasmonic nanoparticles, one that is focused on the exploration of their quantum properties for potential use in building novel plasmonic devices that operate faithfully at the quantum level.[13]

From a quantum mechanical perspective, LSPs supported by metal nanoparticles can be thought of as 'quasi-particles' in that they are made up of both light (photons) and matter (electrons). This makes them rather unusual, as they can have both a photonic and an electronic response, providing a hybrid optoelectronic system in the quantum regime.[14] Due to their optical field confinement, LSPs have the potential to provide strong coupling of light to matter in the form of emitter systems, such as atoms, molecules and artificial systems, e.g. quantum dots[15] and nitrogen vacancy (NV) centres.[16] The dynamics associated with these interactions brings the prospect of using metal nanoparticles to build devices for quantum technologies, such as single-photon sources,[17] switches,[18] sensors[19–21] and more complex many-body quantum simulators.[22] Although plasmonic systems provide a highly confined field, they are affected by a large amount of loss and this can impact the performance of a quantum plasmonic device. Fortunately, researchers have found several ways to deal with this in the quantum regime by introducing schemes that, for example, exploit the dissipation and use it as a resource,[23,24] providing a new way in which to control a given system's quantum dynamics. All of the above-mentioned devices and schemes are part of an exciting new field of research that considers quantum effects in plasmonic systems and is known as quantum plasmonics.

In this chapter, we review recent progress in the study of gold nanoparticles in quantum plasmonics. There are many types of material that can be used to support localised plasmons in the quantum regime, including aluminium, silver, graphene and more exotic materials like superconductors.[25] However, gold is one of the most well characterised, stable and easily fabricated materials for plasmonics, making it highly suitable for research purposes. The main physical principles and behaviours highlighted in this chapter can be observed using other metallic materials. We begin the chapter by describing the transition from the classical description of LSPs to the quantum description. We show how to quantise a single LSP and discuss in which limit a classical or quantum description should be used in order to correctly model the dynamics. We then extend the quantisation to several LSPs interacting in an array of nanoparticles and describe the new features that arise, including resonances, transfer dynamics and quantum interference. The interaction of LSPs with emitter systems is another aspect we discuss, and we summarise the two main regimes of interactions: the weak and strong coupling regimes. In these regimes, we look at recent work on developing quantum sources of light, switching at the few-photon level and using emitter–nanoparticle systems as unit cells in quantum metamaterials. We also describe how loss, rather than being a problem, can actually be utilised as a resource in emitter–nanoparticle systems using the dissipative-driven model. Finally, as nanoparticles not only have an optical response but also an electronic response, we describe how researchers have recently used a range of techniques to accurately model electronic behaviours in the quantum regime — studying phenomena such as quantum size effects and electron tunnelling. We conclude the chapter with an outlook on various challenges that remain and new directions for research into the quantum properties of metal nanoparticles and quantum plasmonics in general.

5.2 Quantum Optical Properties

A LSP is a coupled mode of an EM field and a free-electron density oscillation. The quantum nature of EM fields or electrons naturally reveals itself when either a quantum vacuum fluctuation of the EM fields is involved, the discretised/particle nature of the EM fields is present, or a 'size effect' of the supporting nanoparticle becomes important. The collective behaviour of electrons can be treated macroscopically by taking into account an appropriate dielectric constant when the size of the nanoparticle is large enough such that the energy bands of the associated electrons are continuous. In this case, which is our main interest in this section, the vacuum fluctuation of light and the discretized nature of light — key features of quantised light — are responsible for the non-classical aspects of a LSP at a nanoparticle. To describe such quantum features, the quantisation of the EM fields of the LSP is required.

5.2.1 *Single Nanoparticles — From Classical to Quantum*

The quantisation of the EM field is typically carried out by replacing the electric field vector E and magnetic field vector B by their corresponding quantum operators. It typically consists of three steps: (1) classical mode description, (2) discretisation of classical modes and (3) quantisation via the correspondence principle. Following recent work on the quantisation of

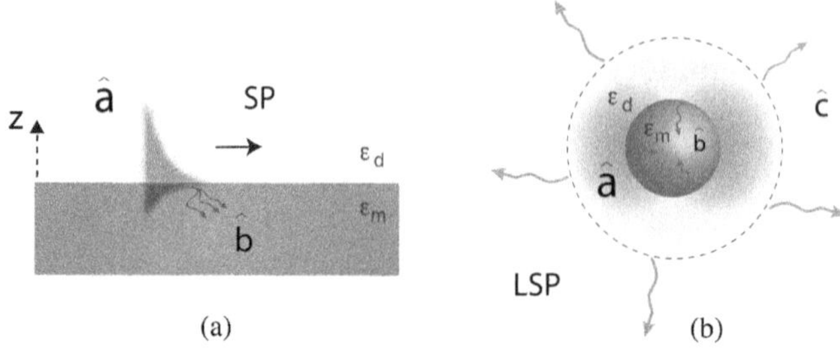

Figure 5.1 (a) A Surface plasmon at the interface between bulk gold and a dielectric, e.g. air. (b) A localised surface plasmon located in the near field of a gold nanoparticle. In both, the plasmonic quantum excitation is described by the operator $\hat{a}$, while loss is modelled by the operator $\hat{b}$ and the far-field excitation linked to the near field is given by the operator $\hat{c}$.

a LSP at a metallic nanoparticle,[26–28] we briefly present how to quantise the EM fields of a LSP.

We begin with the quantisation of a surface plasmon at the interface of a bulk metal (such as gold) and air,[13] as shown in Figure 5.1(a), after which we describe the quantisation of a LSP at a spherical metal nanoparticle, as shown in Figure 5.1(b). As already mentioned, we focus on the regime where the behaviour of the electrons in the metal can be macroscopically described by a dielectric constant $\epsilon_m(\omega)$ and initially loss is neglected. One can generally describe the total EM field in Figure 5.1(a) in terms of a vector potential, $A(r, t)$, where the electric and magnetic fields are given by $E = -\partial A/\partial t$ and $B = \nabla \times A$, and the Coulomb gauge ($\nabla \cdot A = 0$) is used. Maxwell's equations give a general form for the vector potential of the surface plasmon as

$$A(r,t) = \frac{1}{(2\pi)^2} \int d^2k \, \alpha_k u_k(r) \exp(-i\omega t) + \text{c.c.},$$

where c.c. denotes the complex conjugate, k is a real wave vector parallel to the interface and the frequency ω is linked to the wave number along the propagation direction, $k = |k|$, by the dispersion relation

$$k = \frac{\omega}{c} \sqrt{\frac{\epsilon_m(\omega)\epsilon_d}{\epsilon_d + \epsilon_m(\omega)}}.$$

The term α_k is an amplitude and the normalised mode function $u_k(r)$ is given as

$$u_k(r) = \frac{1}{\sqrt{L(\omega)}} \exp(-\kappa_j |z|) \left(\hat{k} - i\frac{k}{\kappa_j}\hat{z} \right) \exp(i\,k \cdot r),$$

where $L(\omega)$ is a normalisation length and $\kappa_j = k^2 - \epsilon_j \omega^2/c^2$ characterises the decay of the field in the z direction, with $\epsilon_1 = \epsilon_d$ and $\epsilon_2 = \epsilon_m(\omega)$, the relative permittivities for dielectric and metal, respectively. In the above, $\hat{v}$ denotes the unit vector for the vector $\boldsymbol{v}$.

For the discretisation of the surface plasmon mode functions, we assume a virtual square of area $S = L_x \times L_y$ introduced on the surface. This enables discretised values to be defined for the wavenumbers $k_x = n_x 2\pi/L_x$, and $k_y = n_y 2\pi/L_y$, where n_x and n_y are integers. By making the following substitutions

$$\frac{1}{(2\pi)^2} \int d^2k \rightarrow \frac{1}{S} \sum_k$$

$$\alpha_k \rightarrow S A_k,$$

the vector potential A can be rewritten in a discretised form as

$$A(\boldsymbol{r}, t) = \sum_k A_k \boldsymbol{u}_k(\boldsymbol{r}) \exp(-i\omega t) + \text{c.c.}$$

The total energy of the EM field in the virtual square defined as

$$U = \int dt \int d\boldsymbol{r} \left(E\frac{\partial D}{\partial t} + H\frac{\partial B}{\partial t} \right),$$

where $D = \epsilon_0 E + P = \epsilon_j E$ and $H = \mu_0 B$, can then be rewritten as

$$U = \sum_k \epsilon_0 \omega^2 S[A_k A_k^* + A_k^* A_k].$$

The above Hamiltonian is in the form of a harmonic oscillator, so that the two variables A_k and A_k^* form a pair of canonical conjugate variables that can be quantised. As performed in the quantisation of a simple harmonic oscillator, we transform the two scalar conjugate variables into quantum operators as

$$A_k \rightarrow \sqrt{\frac{\hbar}{2\epsilon_0 \omega S}} \hat{a}_k,$$

$$A_k^* \rightarrow \sqrt{\frac{\hbar}{2\epsilon_0 \omega S}} \hat{a}_k^\dagger,$$

where the bosonic field operators $\hat{a}_k$ and $\hat{a}_k^\dagger$ obey the commutation relations $[\hat{a}_k, \hat{a}_{k'}^\dagger] = \delta_{k,k'}$.[29,30] As a result, the Hamiltonian is written in a simpler form as

$$\hat{H} = \sum_k \hbar\omega_k \left(\hat{a}_k^\dagger \hat{a}_k + \frac{1}{2} \right).$$

Thus, a single quantised surface plasmon excitation with energy $\hbar\omega_k$ at a metal–dielectric interface can be written as $|1_k\rangle = \hat{a}_k^\dagger|\text{vac}\rangle$, where $|\text{vac}\rangle$ denotes the vacuum state of a surface plasmon at a bulk metal surface.

We now turn to the case of a LSP at a nanoparticle. In fact, the above quantisation procedure can be applied to more complex waveguides or the near-field of a LSP at a nanoparticle, with the only change being the mode function $u_\mathbf{k}(r)$, which includes all information about the physical structure and the classical wavelike properties of the excitation. The vector potential for the EM near-field at a nanoparticle under the dipole approximation can be written as

$$A(r, t) = \sum_i \alpha_i u_i(r)\sin(\omega_0 t),$$

where the frequency ω_0 is determined by the Fröhlich criterion, $\text{Re}[\epsilon_m(\omega)] = -2\epsilon_d$, for the nanoparticle and the mode function is given by

$$\begin{cases} u_i(r) = \hat{i}, & \text{for } r < R \\[2mm] u_i(r) = -\dfrac{R^3}{r^3}(3(\hat{i} \cdot \hat{r})\hat{r} - \hat{i}), & \text{for } r > R, \end{cases}$$

where the subscript i represents the three-dimensional coordinate ($i = x, y, z$), R is the radius of the nanoparticle and r is the radial coordinate of the position vector r, taken with respect to the centre of the nanoparticle. Using similar steps as the above, one obtains bosonic annihilation and creation operators $\hat{a}$ and $\hat{a}^\dagger$ for the LSP mode at a metal nanoparticle, where $|1\rangle = \hat{a}^\dagger|\text{vac}\rangle$ represents a single quantised LSP of energy $\hbar\omega_0$ at the nanoparticle.

During the quantisation of the surface plasmon modes at the interface between a dielectric and a bulk-metal or a spherical metal nanoparticle, we have so far ignored metallic loss that is caused by the scattering of electrons with background ions, phonons and themselves in the conduction band (Ohmic loss),[12,25] and at high frequencies by interband transitions.[12] In quantum optics, loss can be incorporated into the quantised modes of the surface plasmon or LSP by allowing it to couple to a reservoir of bath modes $\hat{b}$ whose coupling strength is phenomenologically determined from the imaginary part of $\epsilon_m(\omega)$, which is a result of the damping caused by the electron collisions. This is mathematically equivalent to the more rigorous reservoir method,[31] i.e. the LSP modes at metal nanoparticles can be effectively regarded as a leaky cavity in the quantum optics formalism.

5.2.2 *Many-nanoparticle Array*

If many nanoparticles are placed close to each other, they start to interact, leading to coherent exchange of LSP excitations. When the distance between nanoparticles is sufficiently larger than their size, a dipole excitation dominates all other modes, and the interaction can be treated in terms of a dipole–dipole interaction. A decade ago, ordered arrays of closely spaced noble-metal nanoparticles were proposed as a means to guide EM energy on scales far below the diffraction limit.[32–34] The energy transport relies on near-field coupling between LSPs of neighbouring particles,[35] with the suppression of radiative scattering into the far field.[36–38] In this section, we introduce the quantum version of this type of system, where the quantised excitations of LSPs interact in an array of nanoparticles. A key feature of the corresponding quantum version is the quantum (relative) phase assigned over different sites or states.

The Hamiltonian for a linear gold nanoparticle array system consisting of N nanoparticles (see Figure 5.2(a)) can be written as

$$\hat{H} = \sum_{i=1}^{N} \hbar\omega\hat{a}_i^{\dagger}\hat{a}_i + \sum_{i=1}^{N-1} \hbar g_i\,(\hat{a}_i^{\dagger}\hat{a}_{i+1} + \hat{a}_{i+1}^{\dagger}\hat{a}_i),$$

where ω_i is the natural resonant frequency of the field oscillation at the ith nanoparticle, and g_i is the coupling strength between the fields of the ith and $(i+1)$th nanoparticles. The interaction terms in the above Hamiltonian are a result of considering a weak-coupling approximation, where the values of the couplings $g_{i,j}$ are much less than that of the natural frequency, ω_i. This approximation also covers two other approximations that are in good agreement with a number of classical experiments: a point-dipole approximation, where multipolar interactions are negligible,[39] and a near-field approximation, where the nearest-neighbour interaction is dominant via the Förster fields.[33,40,41] The natural frequency satisfies the Fröhlich criterion,[35,36] which considers the nanoparticles to be small enough compared to the operating wavelength such that only dipole-active excitations are important.[42] For the dielectric function of gold, a Drude–Sommerfeld model can be considered, leading to a best fit to experimental data at frequencies corresponding to free space wavelengths $\lambda_0 \geq 350$ nm.[43] For simplicity, a homogeneous array of nanoparticles is considered, where all nanoparticles have the same dielectric function $\epsilon_m(\omega)$ and radius R, leading to $\omega_i = \omega_0$ for all i. Furthermore, all the couplings are assumed to be equal, i.e. $g_i = g$, and the coupling strength can be obtained from the correspondence of the quantum description of the nanoparticle array with the classical description.[35]

For the injection and extraction of quantum states of LSP excitations, tapered nanowires can be considered as input (source) and output (drain) channels at the respective ends of the array (see Figure 5.2(a)). The theoretical description of this process can be provided by quantum optics input–output theory,[44,45] where continuum modes of the input and output nanowires are coupled to the discretised modes of the nanoparticles. Metallic loss can be also incorporated in the input–output theory, where additional channels into which the excitations of the LSP can leak are added, and the coupling strength determined by the rate of damping. In the frequency domain, the boundary conditions for the source, drain and bath modes can be written as

$$\hat{a}_1(\omega) = \frac{1}{\sqrt{g_{\text{in}}}}\left(\hat{s}_{\text{in}}(\omega) + \hat{s}_{\text{out}}(\omega)\right),$$

$$\hat{a}_N(\omega) = \frac{1}{\sqrt{g_{\text{out}}}}\left(\hat{d}_{\text{in}}(\omega) + \hat{d}_{\text{out}}(\omega)\right),$$

$$\hat{a}_i(\omega) = \frac{1}{\sqrt{\Gamma_i}}\left(\hat{A}_{i,\text{in}}(\omega) + \hat{A}_{i,\text{out}}(\omega)\right) \quad \text{for } i = 1,\ldots,N,$$

where $g_{\text{in/out}}$ denotes the coupling strength to the input/output channels, and Γ_i denotes the damping rate of the ith nanoparticle. By eliminating the internal nanoparticle operators $\hat{a}_i^\dagger$ from a set of coupled equations given by the Heisenberg equations for all operators,[46] a scattering matrix is obtained that links the source, drain and bath operators, $(\hat{s}_{\text{in/out}}^\dagger(\omega), \hat{d}_{\text{in/out}}^\dagger(\omega)$, and

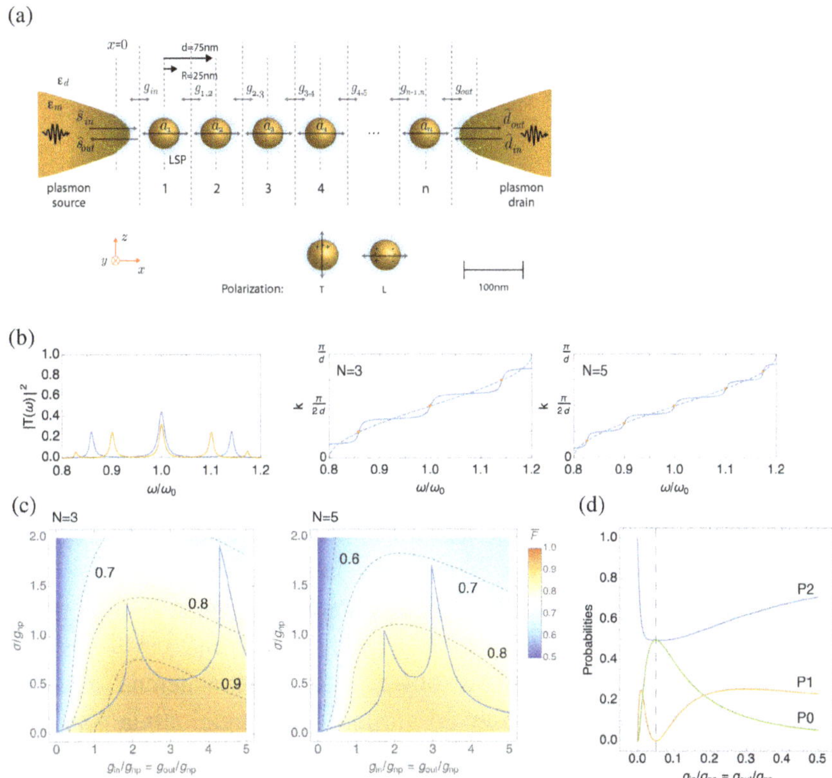

Figure 5.2 (a) Metallic gold nanoparticle array with a source nanowire (left) and a drain nanowire (right) allowing insertion and extraction of quantum states. For arrays of three and five nanoparticles, (b) presents the transmission and dispersion properties with varying frequency and (c) shows averaged fidelity with in–out coupling strength and bandwidth of initial wavepacket for a single-qubit input. (d) Survival probabilities for quantum states with different numbers of plasmonic quanta inserted into an array of three nanoparticles.

$\hat{A}^{\dagger}_{i,\text{in/out}}(\omega)$, respectively) as input–output relations:

$$\hat{s}^{\dagger}_{\text{in}}(\omega) = R_s(\omega)\,\hat{s}^{\dagger}_{\text{out}}(\omega) + T_s(\omega)\,\hat{d}^{\dagger}_{\text{out}}(\omega) + \sum_i S_{s,i}(\omega)\,\hat{A}^{\dagger}_{i,\text{out}}(\omega),$$

$$\hat{d}^{\dagger}_{\text{in}}(\omega) = T_d(\omega)\,\hat{s}^{\dagger}_{\text{out}}(\omega) + R_d(\omega)\,\hat{d}^{\dagger}_{\text{out}}(\omega) + \sum_i S_{d,i}(\omega)\,\hat{A}^{\dagger}_{i,\text{out}}(\omega),$$

where $\left|R_{s(d)}(\omega)\right|^2 + \left|T_{s(d)}(\omega)\right|^2 + \sum_i \left|S_{s,i(d,i)}(\omega)\right|^2 = 1$, and the bath modes are assumed to be initially in the vacuum state $|\text{vac}\rangle$. The transmission coefficient $T_{s(d)}(\omega)$ contains information about the transmission spectrum and the dispersion properties (see Figure 5.2(b) for the case of three and five nanoparticles in an array). In Figure 5.2, the damping rate Γ of the LSP resonance at a nanoparticle is assumed to be $\Gamma = 0.1\omega_0$ for simplicity. Then, for a given input quantum state $|\psi_{\text{in}}\rangle$ from the source, the output state $|\psi_{\text{out}}\rangle$ can be obtained by applying the above relation within the Heisenberg picture. Further details of this formalism for a nanoparticle array can be found in Ref. 46.

The formalism provided allows one to describe quantum state transfer through a metallic nanoparticle array where the input quantum state is encoded on the propagating LSP mode and is excited by a coupling from the source nanowire. From the overall scattering behaviour, the quantum state can either be transmitted or reflected. As a typical example of quantum state transfer, an arbitrary qubit state that consists of single plasmon and vacuum is considered, i.e. $|\Psi_{\text{in}}\rangle = a|\text{vac}\rangle + b|1\rangle$, where $|a|^2 + |b|^2 = 1$. A figure of merit known as the fidelity can be used to quantify the quality of the state transfer and is defined as $F = |\langle\Psi_{\text{out}}|\Psi_{\text{in}}\rangle|^2$. Its averaged quantity over all possible qubit states is shown in Figure 5.2(c), where $\bar{F} = \frac{1}{4\pi}\int_0^\pi d\theta \int_0^{2\pi} d\phi\, F \sin\theta$ and the Bloch sphere coordinates $a = \cos(\theta/2)$ and $b = e^{i\phi}\sin(\theta/2)$ are used. This shows that quantum coherence assigned to the single qubit can be preserved during its propagation via near-field coupling of LSP excitations along the array. Thus, a metallic nanoparticle array can be used for quantum communication; however, it is limited to the case of short-distance communication due to large metallic loss.

In addition to the preservation of quantum coherence during quantum state transfer, quantum interference can also be observed through the array of nanoparticles. In particular, an initial condition that a single plasmonic excitation is injected into each end of the array can be considered. In the absence of losses (or when losses are negligible), the well-known Hong–Ou–Mandel quantum interference effect[47] is predicted when $|R_{s,d}|^2 = |T_{s,d}|^2 = 1/2$, i.e. the two plasmons are always found to be in the same output mode (either the source or the drain nanowire). On the other hand, when losses are taken into account, a new interference effect occurs due to the additional excitation paths induced by the loss channels, i.e. the nanoparticle array no longer effectively serves as a two-port beam splitter, but a multi-port beam splitter.

In Figure 5.2(d), the probabilities of two, one and no plasmon survival are presented as a function of the ratio of in–out coupling to the inter-particle coupling strength. One can see that there is a particular point marked by the dashed line at which the probability for one of the plasmons to survive (or be absorbed) is zero, $P_1 \sim 0$, i.e. there is no one-plasmon absorption. In other words, either both plasmons are absorbed, $P_0 \sim 1/2$, or neither is absorbed, $P_2 \sim 1/2$, which is a nonlinear effect even though the damping in the nanoparticle array is a linear process. This effect is due to quantum interference of the plasmonic excitations associated with the input/output and loss channels, and cannot be described in terms of a classical treatment of the nanoparticle array.[46,48]

With the quantum model of metallic nanoparticles presented here, a linear array of nanoparticles is only considered; however, the theory can be applied to more complex arrangements of quantum plasmonic nanoparticles.[40,49,50]

5.2.3 *Single Nanoparticles Interacting with Emitters: Weak and Strong Coupling Regime*

In the previous section, we introduced a theoretical description of the interaction between quantum plasmonic nanoparticles, upon which two examples, quantum state transfer and quantum interference, were presented. Here we describe the interaction between a quantum emitter (QE) and the EM fields in plasmonic nanostructures. In general, the interaction between light and matter can be considered in two interesting regimes depending on the coupling strength (or coupling rate), g, of light to matter. One is the weak coupling regime, where g is less than all other time scales, for instance the decay rates of a cavity mode (supporting the light) and matter, and the resonant frequencies of the cavity mode and matter. The other regime is the strong coupling regime, where g dominates all other times scales, so that light and matter may exchange their energies before the system energies and their coherence are irreversibly lost. We will review the weak coupling and strong coupling regimes for plasmonic structures[13,51] and show some examples for nanoparticles.

In the weak coupling regime, losses dominate the coupling strength, and thus one cannot easily single out the coherent process in the composite system. Instead, the physical system that supports light modes in this regime will alter the EM local density of states (DOS) surrounding the QE, subsequently resulting in the modification of the radiative properties of the QE. In 1946, Purcell discovered the enhancement of spontaneous emission (SE) rates of QEs for the first time when he considered a QE placed in the vicinity of an EM environment — the so-called Purcell effect.[52] This was a surprising result as it had previously been thought that the emission was an inherent and unchangeable property of a QE. The first experiments of the Purcell effect were performed by Drexhage in the 1970s for dye layers near smooth metal surfaces.[53,54] Since then, there have been many efforts to investigate the modification of SE rates in various systems, including an absorbing cavity,[55] complex nanostructures,[56,57] wedge[58] and photonic crystals.[59–61] The enhancement of the SE rate is typically characterised by the so-called Purcell factor, defined as the ratio between the modified and free-space emission rates, which can be written as

$$F_Q = \frac{\Gamma'}{\Gamma_0} = \frac{3}{4\pi^2}\left(\frac{\lambda_0}{n}\right)^3 \frac{Q}{V},$$

where $\Gamma'(\Gamma_0)$ is the modified (free-space) emission rate, λ_0/n is the wavelength in the environment, and Q and V are the quality factor and mode volume of the environment, respectively. As can be seen in the formula, the Purcell factor is enhanced by either increasing the quality factor of the environment or decreasing the mode volume. In this sense, metallic structures provide a promising platform to enhance the Purcell factor due to the fact that the electrons coupled to EM fields at a metal surface enable a reduction of the mode size V despite its low quality factor limited by a significantly large dissipation inside the metal. This can be equivalently understood in terms of the enhanced local DOS in the near-field of the metallic structure, which produces a strong emission modification of a QE.[62–65] In a system supporting hybrid surface plasmon modes, Purcell factors of about 60 have been achieved.[66,67] Furthermore, for molecules in gaps between a silver surface and a silver nanowire or nanocube, Purcell factors as high as 1000 have recently been observed.[68–70] The effect of metallic nanoparticles on the Purcell enhancement has been also studied extensively.[71–76]

The Purcell effect can be explained within a semi-classical description where the energies of a QE are quantised but the EM fields in the plasmonic system are treated using classical mean-field theory. The Purcell factor can be also derived using a full quantum optical model where the surface plasmon field is also quantised.[28,77] Although the above formula provides an intuitive understanding of the Purcell factor associated with the quality factor and mode volume, one should take into account the exact calculation of the local density of states to correctly determine the quantitative emission rate of a QE near a plasmonic resonance.[78,79] Several reviews on the modification of the emission rate of a QE can be found in Refs. 17, 26 and 80.

The Purcell factor can be used to quantify the modification of a QEs fluorescence near a plasmonic structure. Enhanced fluorescence of a QE near a nanoparticle is observed as it approaches the plasmonic structure, but the fluorescence rate will drop at very short distances. The fluorescence quenching comes from the interplay of two processes; excitation and radiative emission processes. The excitation rate increases as a QE approaches the plasmonic structure, while the radiative emission rate, called the quantum yield, decreases. This results in an optimal distance at which a QE exhibits the maximal fluorescence rate. The fluorescence enhancement and quenching with respect to a given distance have been observed near a gold nanoparticle, also showing that multipole excitation of the nanoparticle is a key mechanism for the fluorescence quenching, i.e. the quenching cannot be predicted within a dipole approximation.[81–83] Modified fluorescence of a single molecule has been also used for near-field microscopy using nanoparticles[84] and nanotips.[85] Further details about the modification of single molecule fluorescence can be found in Refs. 80 and 86.

Contrary to the weak coupling regime, a hybrid mode of light and matter emerges when the coupling strength is greater than other time scales — the so-called strong coupling regime.[87,88] This regime can be achieved by either increasing the oscillator strength (e.g. the dipole moment) of the matter part, or reducing the mode volume of the light part. Thus, a plasmonic system can be a useful platform for enabling the strong coupling regime to be reached when it is placed in the vicinity of a QE, due to the EM field enhancement at the metal surface. The strong coupling enables a coherent exchange of energy between a QE and a plasmonic mode and leads to a splitting in the spectrum, corresponding to the eigen energies of the hybrid system. This is equivalently shown as a Rabi oscillation in the time domain. Although such a splitting followed by hybrid modes can be predicted within a classical treatment, where classical EM fields and a Lorentzian oscillator are considered, a quantitatively correct analysis is only provided by a full-quantum mechanical treatment, where an EM field and a QE are both quantised. A comparison among classical, semi-classical and quantum treatments with respect to the strong coupling regime can be found in Ref. 89. As an example, when a QE is coupled to a single-cavity mode at resonance, i.e. $\omega_{cav} = \omega_{em} \equiv \omega_0$, the splitting is quantified by the hybrid eigenmodes as

$$\Omega_{\pm} = \omega_0 - \frac{i}{4}\left(\gamma_{cav} + \gamma_{em}\right) \pm \sqrt{g^2 - \left(\frac{\gamma_{cav} - \gamma_{em}}{4}\right)^2},$$

where ω_0 is the natural frequency of the cavity and QE modes, $\gamma_{em}(\gamma_{cav})$ is the decay rate of the QE (cavity), and g is the coupling constant. Although large losses in the metal (corresponding to γ_{cav} in the above) can be a major obstacle to driving a plasmonic system into the strong coupling regime, the Rabi splitting between plasmonic modes and organic excitons has been demonstrated.[90–93] Furthermore, recent experiments have shown strong coupling between individual plasmonic structures and molecular J-aggregates.[94,95] Theoretical works have also focused on the strong coupling between a single QE and a metal nanoparticle,[26–28,96–98] or a dimer.[99] A strong coupling of a surface plasmon to a single QE has not yet been experimentally observed so far. However, very recently one experiment has reported vacuum Rabi splitting using a silver bowtie plasmonic cavity with a single quantum dot.[100]

In addition to a spectral splitting, Fano interferences have also been studied at the single QE and nanoparticle limit,[101,102] and a transition from Fano interference to Rabi splitting is explained under a single theoretical framework by varying the damping rate or linewidth of the plasmonic resonance mode.[103] The interplay between strong coupling and quenching near a metal nanoparticle has been also studied at the short distance limit.[104]

5.2.4 *Nanoparticle Systems as Unit Cells in Metamaterials*

Metamaterials are artificial materials composed of periodic lattices of identical subwavelength unit cell scatterers. The collective response of all the unit cells governs completely the EM properties of the entire bulk material.[4] Advancements made in fabricating metamaterials have led to the development of novel EM materials, such as those with a negative refractive index,[105,106] which in turn has opened up many exciting new applications, including the super lens,[107] transformation optics[108] and EM cloaking.[109]

Plasmonic nanostructures such as metal nanoparticles are natural electric resonators at optical frequencies and so researchers have increasingly looked towards plasmonics to design metamaterials in the optical domain. A big challenge has been developing plasmonic magnetic resonators that can be combined with the electric resonances in order to achieve a negative refractive index.[110–112] One of the ways in which a negative refractive index and other interesting optical effects may be achieved is via a quantum response of the unit cells. Recent experimental work has shown that metamaterials based on plasmonic nanostructures respond well in the quantum regime, with the observation of phenomena such as mediated quantum interference,[113] remote quantum interference[114] and quantum size effects.[115] Applications have also been experimentally demonstrated in the form of coherent absorption of single photons[116] and the distillation of photon entanglement.[117,118] On the theoretical side, researchers have looked at how to model active metamaterials using a quantum formalism,[119,120] engineer the SE of many-body quantum states,[121] detect entanglement,[122] quantify dispersion and absorption properties,[123] include quantum corrections[124] and even carry out low-loss quantum information processing.[125–127] Recent work has also explored the use of hyperbolic metamaterials for quantum state engineering.[128,129]

The extra functionality provided by quantum plasmonic structures in the unit cells of metamaterials comes in the form of their interaction with emitter systems, leading to tunability and interference effects. This has the potential to provide significant advantages over earlier classical metamaterial designs by providing more options to control the unit cell dynamics. In the following, we describe briefly such a scenario, with an example that considers two-level semiconductor quantum dot emitters integrated in a negative permeability metamaterial unit cell made of metal nanoparticles.[130] This example brings together phenomena from metamaterials and quantum plasmonics in order to achieve a negative refractive index.

The unit cell consists of a ring of metal nanoparticles, as shown in Figure 5.3(a). This configuration produces a magnetic dipole in the z-direction into the plane. The magnetic dipole response is achieved because the magnetic field that impinges on the unit cell sets up a circular displacement current in the nanoring, which then produces a magnetic response in the z-direction.[110,111] For making a three-dimensional isotropic metamaterial out of these nanorings, one could use a face-centred unit cell and have three nanorings, each in an orthogonal direction. The idea is then to incorporate quantum dots into the nanoring design and couple them to the nanoparticles of the unit cells to provide tunability. One way to obtain this tunability is via a Fano interference effect,[131,132] which can occur between a metal nanoparticle (a broad resonance) and a quantum dot (a narrow resonance) in what is collectively called a metamolecule, as shown in Figure 5.3(b). The Fano interference can be seen as the sharp resonances of the polarisability in Figure 5.3(b).

The Fano effect can then be manifested in the macroscopic response of a material made up of the nanorings with quantum dots, as seen in the real and imaginary parts of the permeability in Figure 5.4(a). By tuning the energy levels of the quantum dots via a global external field, e.g. a Stark or Zeeman shift, one can spectrally shift the Fano resonance and hence the frequency at which the permeability of the material is negative (see Figure 5.4(b)), opening up a window of opportunity in which to match the negative permeability of the nanoring with the natural negative permittivity of the individual nanoparticles in order to achieve a negative refractive index.[130]

The above example is one of several recently introduced that include quantum effects in the response of the unit cells. This is an exciting new topic in its early stages. Due to the possibility of arranging nanoparticles and quantum emitters in many different configurations and interaction regimes, there is great potential for developing quantum plasmonic metamaterials with novel optical properties. Future unit cell designs may achieve optical responses in the quantum regime that are simply not possible with classical unit cells. This may lead to new applications based on metamaterials for quantum technologies, such as single- and multi-photon sources, and single-photon switches, which are important ingredients for quantum communication and quantum computing.

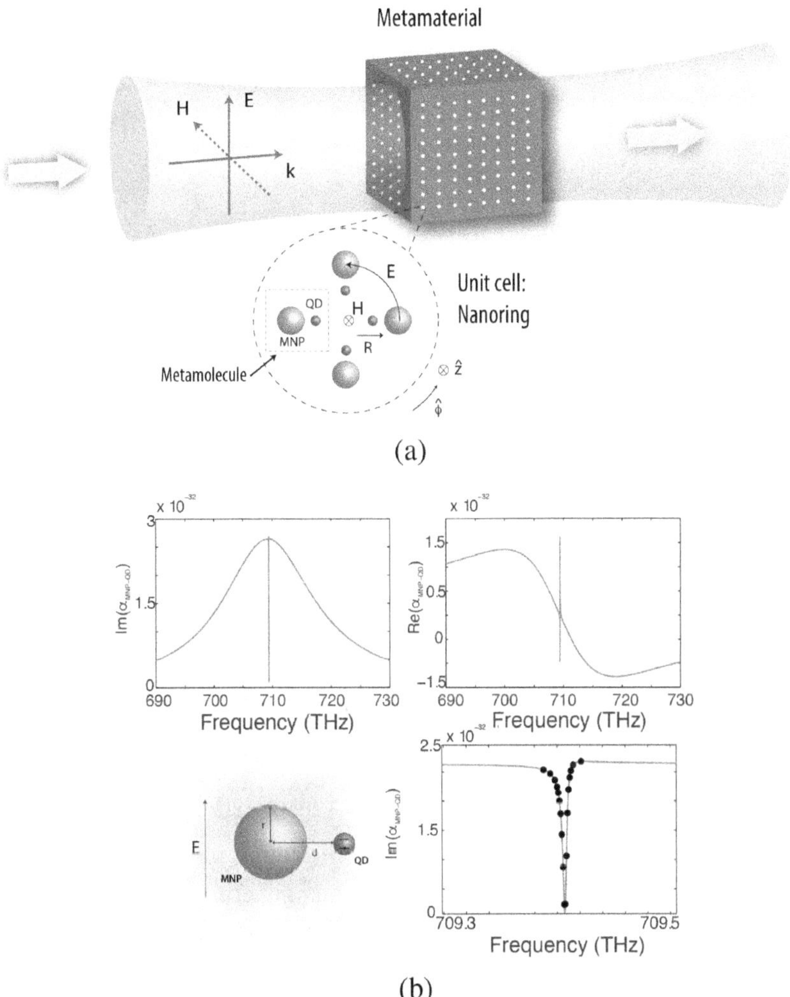

Figure 5.3   A quantum plasmonic metamaterial made from unit cells consisting of metal nanoparticles interacting with quantum dots. (a) An arbitrary optical field can be injected into the metamaterial and the figure shows an example of a transverse plane wave at the centre of a focused beam. In order to isolate the ring's magnetic dipole response, a quasi-static magnetic field H directed along the z-axis can be considered, as shown in the inset. This is a standard approach used to isolate the electric and magnetic response of a unit cell. The metal nanoparticles and the quantum dots are then excited by the induced azimuthal E field, as shown. The interaction between the quantum dots and metal nanoparticle fields at each site, or 'metamolecule', causes a Fano profile to appear in the scattered field, as seen from the polarisability shown in (b). The combination of the scattered fields from all metamolecules results in a Fano profile appearing in the total scattered magnetic field of the unit cell, which provides a Fano profile in the magnetic dipole response.

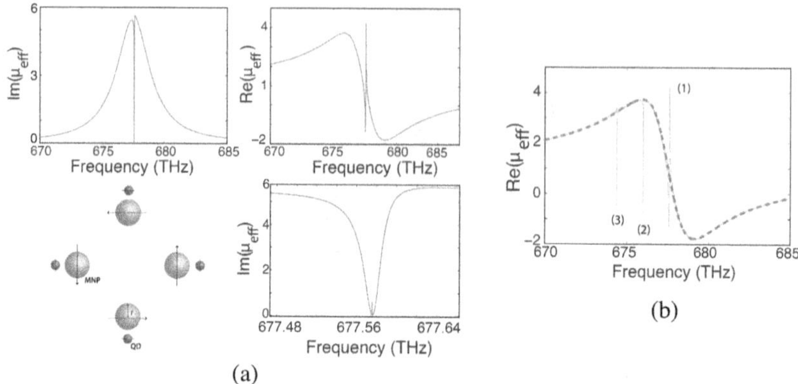

Figure 5.4   The bulk response of a metamaterial made of unit cells comprised of nanoparticles and quantum dots. (a) The magnetic permeability, where the Fano effect from each metamolecule can be seen as manifested in the bulk response. (b) By tuning the energy levels of the quantum dots (for three different splittings) one can shift the position of the Fano resonance and thus the spectral position of the negative permeability. Combined with the negative permittivity of the nanoparticles, this can give rise to a negative refractive index.

5.2.5 *Dealing with Metallic Loss*

Due to the large imaginary part of the metallic dielectric function $\epsilon_m(\omega)$ in the optical regime, loss in plasmonic systems is so significant that it is experimentally challenging to observe or keep desired quantum features for a long time. Furthermore, the induced decoherence limits the performance of a given quantum task to be performed in plasmonic systems. In this section, we briefly describe how one can generally deal with losses in quantum technology and review several interesting related studies in plasmonics, including the use of nanoparticle structures. The fundamental ideas for overcoming loss that are presented below are applicable to any kind of plasmonic structure.

The first natural way to overcome loss that occurs in a system is to correct the errors that are induced by the loss.[133–135] This approach is known as error correction and has been used for decades in classical information processing. To apply this method in the quantum realm, one has to encode quantum information on a particular type of initial state with an additional ancillary system, and then, after a time evolution or an operation, one corrects the output state depending on the result of the measurement of the ancillary system. Recently, Hanson *et al.* have shown that for graphene plasmons, the quantum error correction method enables the transfer of quantum states of light over large distances in the presence of loss.[136] A second alternative method to address the problem of loss suggests not to try to battle with the loss, but rather accept and manipulate the subsequent effects of dissipation — the so-called driven-dissipative approach.[23,24] This method enables the steady state of the system to yield a desired state in the long time limit by only engineering static degrees of freedom, such as system parameters or coupling strengths to the environment that causes the loss. The benefit of this approach is given by it being independent of any careful initial preparation of the system. This interesting method has been employed in a variety of plasmonic structures,[137–139] where the generation of entanglement between qubits (represented by QEs) is achieved via mediating LSP excitations at nanoparticles or a propagating plasmonic mode in a waveguide, where dissipation is clearly not negligible. A third approach to deal with loss is called post-selection, which allows one to simply throw away outcomes from a quantum process when unwanted results stemming from loss are obtained, then repeating the process until the desired result is picked up. This is a common scheme used in quantum information processing when an operation is generally probabilistic.[140,141] The post-selection method yields a high-fidelity performance at the expense of low efficiency (or low probability of obtaining the desired result). Using this idea, a robust-to-loss scheme for generating entanglement between two quantum dots in a plasmonic nanoparticle array has been proposed.[142]

In addition to the methods mentioned above, there have been other attempts to tackle the problem of loss in plasmonic systems by employing various types of gain media[8] or metamaterials.[125–127] Overcoming dissipation and decoherence is generally an important issue in quantum technology,[13,143] and thus a variety of remedies have so far been proposed. These facilitate the use of quantum plasmonic systems for various quantum applications that can keep their intrinsic benefits even in the presence of large metallic loss.

5.3 Quantum Electronic Properties

In contrast to quantum optical properties of nanoplasmonic structures, the quantum nature of the conduction electrons in metals was predicted decades ago for sub-nanometre metallic structures. Recent advances in experimental techniques have allowed the experimental demonstration of theoretical predictions, thus stimulating theoretical development to correctly explain observed quantum phenomena.[144] In this section, we discuss some of the main quantum electronic properties, including the recent achievements in the field.

5.3.1 *Quantum Size Effect: Analytical and Numerical*

When considering electrons in a metal nanoparticle, a classical treatment using the macroscopic classical Maxwell's equations correctly describes quantitative and qualitative behaviours of nanoparticles of size larger than about 10 nm.[145] However, this approach starts to fail when the size of the nanoparticle enters a regime where careful consideration of individual electrons is needed to understand and predict the dynamics. For example, as the size of a nanoparticle becomes as small as a few nanometres, the electronic transition between discretised energy levels of conduction electrons begins to play an important role. This was predicted long ago by Fröhlich in 1937,[146] and is known as the quantum size effect. The size effect has been treated in terms of an analytical description where the dielectric function of the metal is modified to include extra terms corresponding to the electronic transitions.[147–150] The revised particle response can be written as[151]

$$\epsilon(\omega) = \epsilon_{IB} + \omega_p^2 \sum_i \sum_j \frac{S_{if}}{\omega_{if}^2 - \omega^2 - i\gamma\omega},$$

where the term ϵ_{IB} accounts for the interband transition of d-band valence electrons, ω_p is the plasma frequency, and the electron transition frequency ω_{if} and the oscillator strength S_{if} take into account all initial to final state transitions of the electrons, with γ representing a size-dependent scattering frequency.[152,153] Such an analytical quantum mechanical treatment explains a significant blueshift and linewidth broadening of a single nanoparticle.[152]

Although the above analytical approaches have shown a quantitative agreement with some experimental data, the used assumption that the metal surface is abrupt and sharp limits its usefulness. Alternatively, *ab initio* methods, e.g. the numerical time-dependent local density approximation (TDLDA) or time-dependent density functional theory (TDDFT), have been employed to provide more faithful calculations, thus predicting a quantitatively correct response with respect to experimental observations. Prodan *et al.* have investigated the electronic and optical properties of metallic nanoshells by using the TDLDA method to include quantum effects such as the effect of a dielectric background, the electron spill-out, and discrete electronic structures. This consequently shows a size-dependent shift of plasmon modes and an excellent agreement with experimental data.[154–158] Alternatively, TDDFT has been employed as a first principles method for metallic nanoshells[159,160] and for metallic nanorods,[161] which includes the effects of the electronic structure of the particles and spatial distribution of the plasmon-induced electron surface charges. TDDFT has also enabled Townsend and Bryant to show that in quantum-sized nanospheres there can be two types of collective oscillations, quantum core plasmons in the centre and classical surface plasmons throughout the particle.[162] Quantum size effects in thin films[163,164] and graphene[165] have also been studied. In addition, size effects in nonlinear optical properties have been experimentally measured for a nanoparticle.[166] In general, TDDFT is limited to atomic-scale systems, which contain only a few electrons since it requires high-power computational resources. Alternatively, the so-called projected dipole method has recently been proposed to deal with the mesoscopic scale, which is too small to treat within classical theory, but too large for an efficient quantum treatment.[167]

Furthermore, the roles of different microscopic mechanisms such as surface screening and size quantisation (or surface spill-out) have been investigated recently, showing a blueshift or redshift of the plasmon modes depending on the material parameters.[168] The theoretical models have been compared with experimental data for spheres of nanometric size.[169–171] Theoretical studies using TDDFT have been progressing well, however further work still needs to be done in order to explain the response to other environmental factors, such as the temperature or substrate,[172] where there is a quantitative difference among the experimental data.[152,173–175]

5.3.2 *Quantum Tunnelling: Linear and Nonlinear Regimes*

Another quantum effect that occurs with nanoparticles is when they are placed from each other with a sub-nanometre gap, that is, electron tunnelling. Although the fundamental mechanism behind this is simple to understand, only very recently has this phenomena begun to be studied for nanostructures as it requires very challenging experimental techniques and high computational resources to model.

For a dimer of nanoparticles, classical models predict that the associated bond dipolar plasmon redshifts with decreasing inter-particle separation and starts to blueshift when the particles enter a touching regime allowing a conductive overlap. The latter gives rise to a charge transfer plasmon (CTP), which involves conduction electrons flowing back and forth between the nanoparticles. The classical approach assumes that the nanoparticle surface is abrupt and sharp, leading to a discontinuous transition from nontouching to touching particles.[176,177] This assumption starts to break down in the nearly touching regime since electrons are expected tunnel through a potential barrier between the nanoparticles, enabling a CTP in this regime. Furthermore, screening due to the electron density distribution may occur in the junction between the nanoparticles. This is an important effect, requiring an appropriate theoretical description to correctly explain what is happening in the system.[178–180]

With a fully quantum mechanical description using TDDFT, Zuloaga *et al.* have investigated the effect of electron tunnelling and screening for a dimer with a interparticle separation below 1 nm.[181] Quantum effects in dimers strongly modify the optical response, e.g. the EM field enhancement is reduced in contrast to the classical prediction. This quantum effect was experimentally demonstrated by Savage *et al.*[182] and Scholl *et al.*[183] where both the electrical and optical properties of two nanoparticles were investigated in the quantum regime of tunnelling plasmonics below a few nanometres. The observed behaviour in the experiments agrees well with quantum-corrected models of plasmonic systems, where quantum mechanical effects are effectively incorporated into a classical electrodynamic framework for the benefit of feasibility of theory for realistically sized quantum systems.[184] Quantum phenomena such as a strong reduction of the EM field enhancement and non-local screening have been investigated for a concentric core–shell nanoparticle, called a 'nanomatryushka' when the inter-metallic gaps becomes narrower and approach sub-nanometre widths.[185] Zhang *et al.* have recently studied the effect of atomic structure, showing that atomic structures have to be taken into account for quantitatively accurate predictions beyond the standard jellium model.[186]

Nonlinear effects in quantum tunnelling have also been investigated for moderate incident light intensities, showing that the increased electron current due to the strong field photoemission enables the neutralisation of the charge densities on the opposite sides of the junction.[187] The Fowler–Nordheim tunnelling phenomenon has also been theoretically studied in a nanoparticle dimer, where a strong field effect enables a CTP that is not via direct tunnelling, but through a field emission effect, leading to tunnelling from the conduction band of one nanoparticle into the gap.[188]

At present, a variety of experimental observations at the sub-nanometre scale for metal particles have stimulated the development of various sophisticated theoretical treatments. For example, Mortensen *et al.* have presented a generalised non-local optical response theory to include both quantum-mechanical and classical effects in a single framework.[189] Quantum effects in CTPs have also been revisited again by looking at the optical response of a metallic dimer bridged with a two-level system, demonstrating the importance of the electronic structure of the junction for the existence of CTP modes.[190] Several recent review articles provide many more interesting details of this fascinating field.[150,191–193]

5.4 Conclusion

In this chapter, we reviewed recent progress in the study of gold nanoparticles in the quantum regime. We examined the transition between the classical description of LSPs and the quantum description, and showed how to quantise a single LSP. We also discussed in which limits a classical or quantum description needs to be used in order to correctly model the dynamics. We then extended the quantisation to several LSPs interacting in an array and described the new features that arise. The interaction of LSPs with emitter systems was another aspect discussed, where we summarised the two main regimes of coupling: the weak and strong coupling regimes. We looked at recent work on developing quantum sources of light, switching at the few-photon level and using emitter-nanoparticle systems as unit cells in novel quantum plasmonic metamaterials. We also described how loss can be utilised as a resource in emitter-nanoparticle systems using the dissipative-driven model. Finally, we described how researchers are using numerical techniques to accurately model electronic behaviour in the quantum regime — studying phenomena such as quantum size effects and electron tunnelling.

The study of the quantum properties of metal nanoparticle systems is currently experiencing a huge interest from researchers. Many novel properties have been found, but there is still much more to be explored. For instance, it will be interesting to see what types of quantum information processing tasks can be performed with loss present, or how to use loss as an advantage, as in the dissipative-driven model for entanglement generation. In addition, new analytical and numerical methods are needed in order to efficiently model more complex nanoparticle-emitter systems, especially those in the unit cells of plasmonic-based metamaterials. The use of nanoparticles in plasmonic-based lasers, and single- and multi-photon sources is also an important topic for developing new types of light source for classical and quantum applications. Another interesting aspect is to combine quantum models of the optical response of nanoparticles with those of the electronic response. This has the potential to lead to novel quantum devices with hybrid quantum optoelectronic functionality. These devices may be based on the dynamics of only a few photons and electrons interacting, which would lead to exciting new physical phenomena. It is clear there is a bright future for research into the quantum properties of gold and other metallic nanoparticles.

References

1. D. K. Gramotnev and S. I. Bozhevolnyi, Plasmonics beyond the diffraction limit. *Nat. Photon.* **4**, 83 (2010).
2. J. A. Schuller, E. S. Barnard, W. Cai, Y. C. Jun, J. S. White and M. L. Brongersma, Plasmonics for extreme light concentration. *Nat. Mat.* **9**, 193 (2010).
3. S. A. Maier and H. A. Atwater, Plasmonics: Localization and guiding of electromagnetic energy in metal/dielectric structures. *J. App. Phys.* **98**, 011101 (2005).
4. W. Cai and V. Shalaev, *Optical Metamaterials: Fundamentals and Applications* (Springer, Dordrecht, 2010).
5. O. Hess, J. B. Pendry, S. A. Maier, R. F. Oulton, J. M. Hamm and K. L. Tsakmakidis, Active nanoplasmonic metamaterials. *Nat. Mat.* **11**, 573 (2012).

6. J. N. Anker, W. P. Hall, O. Lyandres, N. C. Shah, J. Zhao and R. P. Van Duyne, Biosensing with plasmonic nanosensors. *Nat. Mat.* **7**, 442 (2008).

7. S. Kawata, Y. Inouye and P. Verma, Plasmonics for near-field nano-imaging and super-lensing. *Nat. Photon.* **3**, 388 (2009).

8. P. Berini and I. De Leon, Surface plasmon-polariton amplifiers and lasers. *Nat. Photon.* **6**, 16 (2011).

9. R.-M. Ma, R. F. Oulton, V. J. Sorger and X. Zhang, Plasmon lasers: Coherent light source at molecular scales. *Laser Photonics Rev.* **7**, 1 (2012).

10. M. Kauranen and A. V. Zayats, Nonlinear plasmonics. *Nat. Photon.* **6**, 737 (2012).

11. H. A. Atwater and A. Polman, Plasmonics for improved photovoltaic devices. *Nat. Mat.* **9**, 205 (2010).

12. V. Giannini, A. I. Fernández-Dominguez, S. C. Heck and S. A. Maier, Plasmonic nanoantenas: Fundamentals and their use in controlling the radiative properties of nanoemitters. *Chem. Rev.* **111**, 3888 (2011).

13. M. S. Tame, K. R. McEnery, S. K. Ozdemir, J. Lee, S. A. Maier and M. S. Kim, Quantum plasmonics. *Nat. Phys.* **9**, 329–340 (2013).

14. E. Ozbay, Plasmonics: Merging photonics and electronics at nanoscale dimensions, *Science*, **311**, 189 (2006).

15. S. Buckley, K. Rivoire and J. Vuckovic, Engineered quantum dot single-photon sources. *Rep. Prog. Phys.* **75**, 126503 (2012).

16. I. Aharonovich, S. Castelletto, D. A. Simpson, C.-H. Su, A. D. Greentree and S. Prawer, Diamond-based single-photon emitters. *Rep. Prog. Phys.* **74**, 076501 (2011).

17. M. Pelton, Modified spontaneous emission in nanophotonic structures. *Nat. Photon.* **9**, 427 (2015).

18. D. E. Chang, V. Vuletic and M. D. Lukin, Quantum nonlinear optics — photon by photon. *Nat. Photon.* **8**, 685 (2014).

19. C. Lee, F. Dieleman, J. Lee, C. Rockstuhl, S. A. Maier and M. Tame, Quantum plasmonic sensing: Beyond the shot-noise and diffraction limit. *ACS Photonics* **3**, 992 (2016).

20. W. Fan, B. J. Lawrie and R. C. Pooser, Quantum plasmonic sensing. *Phys. Rev. A* **92**, 053812 (2015).

21. R. C. Pooser and B. Lawrie, Plasmonic trace sensing below the photon shot noise limit. *ACS Photonics* **3**, 8 (2016).

22. J. I. Cirac and P. Zoller, Goals and opportunities in quantum simulation. *Nat. Phys.* **8**, 264 (2012).

23. F. Verstraete, M. M. Wolf and J. I. Cirac, Quantum computation and quantum-state engineering driven by dissipation. *Nat. Phys.* **5**, 633 (2009).

24. B. Kraus, H. P. Büchler, S. Diehl, A. Kantian, A. Micheli and P. Zoller, Preparation of entangled states by quantum Markov processes. *Phys. Rev. A* **78**, 042307 (2008).

25. P. Tassin, T. Koschny, M. Kafesaki and C. M. Soukoulis, A Comparison of graphene, superconductors and metals as conductors for metamaterials and plasmonics. *Nat. Photon.* **6**, 259 (2012).

26. V. V. Klimov, M. Ducloy and V. S. Letokhov, Spontaneous emission of an atom in the presence of nanobodies. *Sov. J. Quantum Electron.* **31**, 569 (2001).

27. A. Trügler and U. Hohenester, Strong coupling between a metallic nanoparticle and a single molecule. *Phys. Rev. B* **77**, 115403 (2008).

28. E. Waks and D. Srindharan, Cavity QED treatment of interaction between a metal nanoparticle and a dipole emitter. *Phys. Rev. A* **82**, 043845 (2010).

29. M. S. Tame, C. Lee, J. Lee, D. Ballester, M. Paternostro, A. V. Zayats and M. Kim, Single-photon excitation of surface plasmon polaritons. *Phys. Rev. Lett.* **101**, 190504 (2008).

30. D. Ballester, M. S. Tame, C. Lee, J. Lee and M. Kim, Long-range surface plasmon polariton excitation at the quantum level. *Phys. Rev. A* **79**, 053845 (2009).

31. B. Huttner and S. M. Barnett, Quantization of the electromagnetic field in dielectrics. *Phys. Rev. A* **46**, 4306 (1992).

32. M. Quinten, A. Leitner, J. R. Krenn and F. R. Aussenegg, Electromagnetic energy transport via linear chains of silver nanoparticles. *Opt. Lett.* **23**, 1331 (1998).

33. S. A. Maier, P. G. Kik and H. A. Atwater, Optical pulse propagation in metal nanoparticle chain waveguides. *Phys. Rev. B* **67**, 205402 (2003).

34. S. A. Maier, P. G. Kik, H. A. Atwater, S. Meltzer, E. Harel, B. E. Koel and A. A. G. Requicha, Local detection of electromagnetic energy transport below the diffraction limit in metal nanoparticle plasmon waveguides. *Nat. Mat.* **2**, 229 (2003).

35. M. L. Brongersma, J. W. Hartman and H. A. Atwater, Electromagnetic energy transfer and switching in nanoparticle chain arrays below the diffraction limit. *Phys. Rev. B* **62**, R16356 (2000).

36. J. R. Krenn, A. Dereux, J. C. Weeber, E. Bourillot, Y. Lacroute, J. P. Goudonnet, G. Schider, W. Gotschy, A. Leitner, F. R. Aussenegg and C. Girard, Squeezing the Optical Near-Field Zone by Plasmon Coupling of Metallic Nanoparticles. *Phys. Rev. Lett.* **82**, 2590 (1999).

37. J. R. Krenn, M. Salerno, N. Felidj, B. Lamprecht, G. Schider, A. Leitner, F. R. Aussenegg, J. C. Weeber, A. Dereux and J. P. Goudonnet, Light field propagation by metal micro- and nanostructures. *J. Microsc.* **202**, 122 (2001).

38. S. A. Maier, *Plasmonics: Fundamentals and Applications* (Springer, New York, 2007).

39. S. Y. Park and D. Stroud, Surface-plasmon dispersion relations in chains of metallic nanoparticles: An exact quasistatic calculation. *Phys. Rev. B* **69**, 125418 (2004).

40. B. Yurke and W. Kuang, Passive linear nanoscale optical and molecular electronics device synthesis from nanoparticles. *Phys. Rev. A* **81**, 033814 (2010).

41. J. R. Krenn, A. Dereux, J. C. Weeber, E. Bourillot, Y. Lacroute, J. P. Goudonnet, G. Schider, W. Gotschy, A. Leitner, F. R. Aussenegg and C. Girard, Squeezing the optical near-field zone by plasmon coupling of metallic nanoparticles. *Phys. Rev. Lett.* **82**, 2590 (1999).

42. A. V. Zayats, I. I. Smolyaninov and A. A. Maradudin, Nano-optics of surface plasmon polaritons. *Phys. Rep.* **408**, 131 (2005).

43. P. B. Johnson and R. W. Christy, Optical constants of the noble metals. *Phys. Rev. B* **6**, 4370 (1972).

44. M. J. Collett and C. W. Gardiner, Squeezing of intracavity and traveling-wave light fields produced in parametric amplification. *Phys. Rev. A* **30**, 1386 (1984).

45. D. F. Walls and G. J. Milburn, *Quantum Optics* (Springer, Berlin, 1994).

46. C. Lee, M. Tame, J. Lim and J. Lee, Quantum plasmonics with a metal nanoparticle array, *Phys. Rev. A* **85**, 063823 (2012).

47. C. K. Hong, Z. Y. Ou and L. Mandel, Measurement of subpicosecond time intervals between two photons by interference. *Phys. Rev. Lett.* **59**, 2044 (1987).

48. S. M. Barnett, J. Jeffers, A. Gatti and R. Loudon, Quantum optics of lossy beam splitters. *Phys. Rev. A* **57**, 2134 (1998).

49. R. Baer, K. Lopata and D. Neuhauser, Properties of phase-coherent energy shuttling on the nanoscale. *J. Chem. Phys.* **126**, 014705 (2007).

50. G. Weick, C. Woollacott, W. L. Barnes, O. Hess and E. Mariani, Dirac-like plasmons in honeycomb lattices of metallic nanoparticles. *Phys. Rev. Lett.* **110**, 106801 (2013).

51. T. Hümmer, F. J. García-Vidal, L. Martín-Moreno and D. Zueco, Weak and strong coupling regimes in plasmonic QED. *Phys. Rev. B* **87**, 115419 (2013).

52. E. M. Purcell, Spontaneous emission probabilities at radio frequencies. *Phys. Rev.* **69**, 681 (1946).

53. K. H. Drexhage, Influence of a dielectric interface on fluorescence decay time. *J. Luminesc.* **1**, 693 (1970)

54. K. H. Drexhage, Interaction of light with monomolecular dye layers. *Prog. Opt.* **12**, 163 (1974).

55. M. S. Tomas, Local-field corrections to the decay rate of excited molecules in absorbing cavities: The Onsager model. *Phys. Rev. A* **63**, 053811 (2001).

56. L. A. Blanco and F. J. García de Abajo, Spontaneous light emission in complex nanostructures. *Phys. Rev. B* **69**, 205414 (2004).

57. S. Evangelou, V. Yannopapas and E. Paspalakis, Modifying free-space spontaneous emission near a plasmonic nanostructure. *Phys. Rev. A* **83**, 023819 (2011).

58. F. S. S. Rosa, T. N. C. Mendes, A. Tenorio and C. Farina, Spontaneous emission of an atom near a wedge. *Phys. Rev. A* **78**, 012105 (2008).

59. A. Kress, F. Hofbauer, N. Reinelt, M. Kaniber, H. J. Krenner, R. Meyer, G. Böhm and J. J. Finley, Manipulation of the spontaneous emission dynamics of quantum dots in two-dimensional photonic crystals. *Phys. Rev. B* **71**, 241304 (2005).

60. D. Englund, D. Fattal, E. Waks, G. Solomon, B. Zhang, T. Nakaoka, Y. Arakawa, Y. Yamamoto and J. Vucković, Controlling the spontaneous emission rate of single quantum dots in a two-dimensional photonic crystal. *Phys. Rev. Lett.* **95**, 013904 (2005).

61. P. Lodahl, A. Floris van Driel, I. S. Nikolaev, A. Irman, K. Overgaag, D. Vanmaekelbergh and W. L. Vos, Controlling the dynamics of spontaneous emission from quantum dots by photonic crystals. *Nature* **430**, 654 (2004).

62. K. H. Drexhage, H. Kuhn and F. P. Schäfer, Variation of the fluorescence decay time of a molecule in front of a mirror. *Ber. Bunsenges. Phys. Chem.* **72**, 329 (1968).

63. R. R. Chance, A. Prock and R. Silbey, Lifetime of an emitting molecule near a partially reflecting surface. *J. Chem. Phys.* **60**, 2744 (1974).

64. J. Gersten and A. Nitzan, Spectroscopic properties of molecules interacting with small dielectric particles. *J. Chem. Phys.* **75**, 1139 (1981).

65. W. L. Barnes, Electromagnetic crystals for surface plasmon polaritons and the extraction of light from emissive devices. *J. Lightw. Technol.* **17**, 2170 (1999).

66. Y. C. Jun, R. D. Kekatpure, J. S. White and M. L. Brongersma, Nonresonant enhancement of spontaneous emission in metal–dielectric–metal plasmon waveguide structures. *Phys. Rev. B* **78**, 153111 (2008).

67. V. J. Sorger, N. Pholchai, E. Cubukcu, R. F. Oulton, P. Kolchin, C. Borschel, M. Gnauck, C. Ronning and X. Zhang, Strongly enhanced molecular fluorescence inside a nanoscale waveguide gap. *Nano Lett.* **11**, 4907 (2011).

68. K. J. Russel, T.-L. Liu, S. Cui and E. L. Hu, Large spontaneous emission enhancement in plasmonic nanocavities. *Nat. Photon.* **6**, 459 (2012).

69. A. Rose, T. B. Hoang, F. McGuire, J. J. Mock, C. Ciracì, D. R. Smith and M. H. Mikkelsen, Control of radiative processes using tunable plasmonic nanopatch antennas. *Nano Lett.* **14**, 4797 (2014).

70. G. M. Akselrod, C. Argyropoulos, T. B. Hoang, C. Ciracì, C. Fang, J. Huang, D. R. Smith and M. H. Mikkelsen, Probing the mechanisms of large Purcell enhancement in plasmonic nanoantennas. *Nat. Photon.* **8**, 835 (2014).

71. E. Dulkeith, A. C. Morteani, T. Niedereichholz, T. A. Klar, J. Feldmann, S. A. Levi, F. C. J. M. van Veggel, D. N. Reinhoudt, M. Möller and D. I. Gittins, Fluorescence quenching of dye molecules near gold nanoparticles: Radiative and nonradiative effects. *Phys. Rev. Lett.* **89**, 203002 (2002).

72. D. J. Maxwell, J. R. Taylor and S. Nie, Self-assembled nanoparticle probes for recognition and detection of biomolecules. *J. Am. Chem. Soc.* **124**, 9606 (2002).

73. N. Liu, B. S. Prall and V. I. Klimov, Hybrid gold/silica/nanocrystal-quantum- dot superstructures: Synthesis and analysis of semiconductor-metal interactions. *J. Am. Chem. Soc.* **128**, 15362 (2006).

74. O. G. Tovmachenko, C. Graf, D. J. can den Heuvel, A. van Blaaderen and H. C. Gerritsen, Fluorescence enhancement by metal-core/silica-shell nanoparticles. *Adv. Mater.* **18**, 91 (2006).

75. G. Schneider, *et al.* Distance-dependent fluorescence quenching on gold nanoparticles ensheathed with layer-by-layer assembled polyelectrolytes. *Nano Lett.* **6**, 530 (2006).

76. M. Ringler, A. Schwemer, M. Wunderlich, A. Nichtl, K. Kürzinger, T. A. Klar and J. Feldmann, Shaping emission spectra of fluorescent molecules with single plasmonic nanoresonators. *Phys. Rev. Lett.* **100**, 203002 (2008).

77. A. Archambault, F. Marquier and J.-J. Greffet, Quantum theory of spontaneous and stimulated emission of surface plasmons. *Phys. Rev. B* **82**, 035411 (2010).

78. X. Zambrana-Puyalto and N. Bonod, Purcell factor of spherical Mie resonators. *Phys. Rev. B* **91**, 195422 (2015).

79. A. F. Koenderink, On the use of Purcell factors for plasmon antennas. *Opt. Lett.* **35**, 4208 (2010).

80. S. Kühn, G. Mori, M. Agio and V. Sandoghdar, Modification of single molecule fluorescence close to a nanostructure: radiation pattern, spontaneous emission and quenching. *Molec. Phys.* **106**, 893 (2008).

81. R. Carminati, J.-J. Greffet, C. Henkel and J. M. Vigoureux, Radiative and non-radiative decay of a single molecule close to a metallic nanoparticle. *Optics Commun.* **261**, 368 (2006).

82. P. Anger, P. Bharadwaj and L. Novotny, Enhancement and quenching of single-molecule fluorescence. *Phys. Rev. Lett.* **96**, 113002 (2006).

83. S. Kühn, U. Hakanson, L. Rogobete and V. Sandoghdar, Enhancement of single-molecule fluorescence using a gold nanoparticle as an optical nanoantenna. *Phys. Rev. Lett.* **97**, 017402 (2006).

84. R. X. Bian, R. C. Dunn, X. S. Xie and P. T. Leung, Single molecule emission characteristics in near-field microscopy. *Phys. Rev. Lett.* **75**, 4772 (1995).

85. E. J. Sánchez, L. Novotny and X. S. Xie, Near-field fluorescence microscopy based on two-photon excitation with metal tips. *Phys. Rev. Lett.* **82**, 4014 (1999).

86. Y. Fu and J. R. Lakowicz, Modification of single molecule fluorescence near metallic nanostructures. *Laser Photon. Rev.* **3**, 221 (2009).

87. R. J. Thompson, G. Rempe and H. J. Kimble, Observation of normal-mode splitting for an atom in an optical cavity. *Phys. Rev. Lett.* **68**, 1132 (1992).

88. M. Brune, F. Schmidt-Kaler, A. Maali, J. Dreyer, E. Hagley, J. M. Raimond and S. Haroche, Quantum rabi oscillation: A direct test of field quantization in a cavity. *Phys. Rev. Lett.* **76**, 1800 (1996).

89. P. Torma and W. L. Barnes, Strong coupling between plasmon polaritons and emitters: a review. *Rep. Prog. Phys.* **78**, 013901 (2015).

90. I. Pockrand, A. Brillante, D. Mobius, Exciton-surface plasmon coupling: An experimental investigation. *J. Chem. Phys.* **77**, 6289 (1982).

91. J. Bellessa, C. Bonnand and J. C. Plenet, Strong coupling between surface plasmons and excitons in an organic semiconductor. *Phys. Rev. Lett.* **93**, 036404 (2004).

92. T. K. Hakala, J. J. Toppari, A. Kuzyk, M. Pettersson, H. Tikkanen, H. Kunttu and P. Torma, Vacuum Rabi splitting and strong-coupling dynamics for surface plasmon polaritons and Rhodamine 6G molecules. *Phys. Rev. Lett.* **103**, 053602 (2009).

93. Y. Sugawara, Strong coupling between localized plasmons and organic excitons in metal nanovoids. *Phys. Rev. Lett.* **97**, 266808 (2006).

94. A. E. Schlather, N. Large, A. S. Urban, P. Nordlander and N. J. Halas, Near-field mediated plexcitonic coupling and giant Rabi splitting in individual metallic dimers. *Nano Lett.* **13**, 3281 (2013).

95. G. Zengin, Realizing strong light-matter interactions between single nanoparticle plasmons and molecular excitons at ambient conditions. *Phys. Rev. Lett.* **114**, 157401 (2015).

96. C. V. Vlack, P. T. Kristensen and S. Hughes, Spontaneous emission spectra and quantum light-matter interactions from a strongly coupled quantum dot metal-nanoparticle system. *Phys. Rev. B* **85**, 075303 (2012).

97. M. M. Dvoynenko and J.-K. Wang, Revisiting strong coupling between a single molecule and surface plasmons. *Opt. Lett.* **38**, 760 (2013).

98. K. V. Nerkararyan and S. I. Bozhevolnyi, Relaxation dynamics of a quantum emitter resonantly coupled to a metal nanoparticle. *Opt. Lett.* **39**, 1617 (2014).

99. S. Savasta, R. Saija, A. Ridolfo, O. Di Stefano, P. Denti and F. Borghese, Nanopolaritons: vacuum Rabi splitting with a single quantum dot in the center of a dimer nanoantenna. *ACS Nano* **4**, 6369 (2010).

100. K. Santhosh, O. Bitton, L. Chuntonov and G. Haran, Vacuum Rabi splitting in a plasmonic cavity at the single quantum emitter limit. *Nat. Commun.* **7**, 11823 (2016).

101. W. Zhang, A. O. Govorov and G. W. Bryant, Semiconductor-metal nanoparticle molecules: hybrid excitons and the nonlinear Fano effect. *Phys. Rev. Lett.* **97**, 146804 (2006).

102. A. Ridolfo, O. D. Stefano, N. Fina, R. Saija and S. Savasta, Quantum plasmonics with quantum dot-metal nanoparticle molecules: influence of the Fano effect on photon statistics. *Phys. Rev. Lett.* **105**, 263601 (2010).

103. J. A. Faucheaux, J. Fu and P. K. Jain, Unified Theoretical Framework for realizing diverse regimes of strong coupling between plasmons and electronic transitions. *J. Phys. Chem. C* **118**, 2710 (2014).

104. A. Delga, J. Feist, J. Bravo-Abad and F. J. Garcia-Vidal, Quantum emitters near a metal nanoparticle: strong coupling and quenching. *Phys. Rev. Lett.* **112**, 253601 (2014).

105. W. J. Padilla, D. N. Basov and D. R. Smith, Negative refractive index metamaterials. *Materials Today* **9**, 28 (2006).

106. M. W. McCall, A. Lakhtakia and W. Weiglhofer, The negative index of refraction demystified. *Eur. J. Phys.* **23**, 353 (2002).

107. J. B. Pendry, Negative refraction makes a perfect lens. *Phys. Rev. Lett.* **85**, 3966 (2000).

108. H. Chen, C. T. Chan and P. Sheng, Transformation optics and metamaterials. *Nat. Mat.* **9**, 387 (2010).

109. D. Schurig, J. J. Mock, B. J. Justice, S. A. Cummer, J. B. Pendry, A. F. Starr and D. R. Smith, Metamaterial electromagnetic cloak at microwave frequencies. *Science* **314**, 977 (2006).

110. A. Alu, A. Salandrino, N. Engheta, Negative effective permeability and left-handed materials at optical frequencies. *Opt. Express* **14**, 1557 (2006).

111. A. Alu and N. Engheta, Dynamical theory of artificial optical magnetism produced by rings of plasmonic nanoparticles. *Phys. Rev. B* **78**, 085112 (2008).

112. C. R. Simovski and S. A. Tretyakov, Model of isotropic resonant magnetism in the visible range based on core-shell clusters. *Phys. Rev. B* **79**, 045111 (2009).

113. S. M. Wang, S. Y. Mu, C. Zhu, Y. X. Gong, P. Xu, H. Liu, T. Li, S. N. Zhu and X. Zhang, Hong-Ou-Mandel interference mediated by the magnetic plasmon waves in a three-dimensional optical metamaterial. *Opt. Lett.* **20**, 5213 (2012).

114. P. K. Jha, X. Ni, C. Wu, Y. Wang and X. Zhang, Metasurface-Enabled Remote Quantum Interference. *Phys. Rev. Lett.* **115**, 025501 (2015).

115. J. A. Scholl, A. Garcia-Etxarri, G. Aguirregabiria, R. Esteban, T. C. Narayan, A. L. Koh, J. Aizpurua and J. A. Dionne, Evolution of plasmonic metamolecule modes in the quantum tunneling regime. *ACS Nano* **10**, 1346 (2016).

116. T. Roger, S. Vezzoli, E. Bolduc, J. Valente, J. J. F. Heitz, J. Jeffers, C. Soci, J. Leach, C. Couteau, N. I. Zheludev and D. Faccio, Coherent perfect absorption in deeply subwavelength films in the single-photon regime. *Nat. Commun.* **6**, 7031 (2015).

117. M. Asano, M. Bechu, M. Tame, S. K. Özdemir, R. Ikuta, D. Ö. Güney, T. Yamamoto, L. Yang, M. Wegener and N. Imoto, Distillation of photon entanglement using a plasmonic metamaterial. *Sci. Rep.* **5**, 18313 (2015).

118. Md. A. al Farooqui, J. Breeland, M. I. Aslam, M. Sadatgol, Ş. K. Özdemir, M. Tame, L. Yang and D. Ö. Güney, Quantum entanglement distillation with metamaterials. *Opt. Exp.* **23**, 17941 (2015).

119. S. Wuestner, A. Pusch, K. L. Tsakmakidis, J. M. Hamm and O. Hess, Overcoming losses with gain in a negative refractive index metamaterial. *Phys. Rev. Lett.* **105**, 127401 (2010).

120. A. Chipouline, S. Sugavanam, V. A. Fedotov and A. E. Nikolaenko, Analytical model for active metamaterials with quantum ingredients. *J. Opt.* **14**, 114005 (2012).

121. N. Gheeraert, S. Bera, S. Florens, Spontaneous emission of many-body Schrödinger cats in metamaterials with large fine structure constant. arXiv:1601.01545 (2016).

122. J. del Pino, J. Feist, F. J. Garcia-Vidal and J. J. Garcia-Ripoll, Entanglement detection in coupled particle plasmons. *Phys. Rev. Lett.* **112**, 216805 (2014).

123. Z. -Y. Zhou, D. -S. Ding, B. -S. Shi, X. -B. Zou and G. -C. Guo, Characterizing dispersion and absorption parameters of metamaterial using entangled photons. *Phys. Rev. A* **85**, 023841 (2012).

124. V. Yannopapas, Quantum corrections in the optical properties of three-dimensional plasmonic metamaterials with subnanometer gaps. *Solid Stat. Comm.* **222**, 18 (2015).

125. A. Kamli, S. A. Moiseev and B. C. Sanders, Coherent control of low loss surface polaritons. *Phys. Rev. Lett.* **101**, 263601 (2008).

126. S. A. Moiseev, A. Kamli and B. C. Sanders, Low-loss nonlinear polaritonics. *Phys. Rev. A* **81**, 033839 (2010).

127. M. Siomau, A. A. Kamli, A. A. Moiseev and B. C. Sanders, Entanglement creation with negative index metamaterials. *Phys. Rev. A* **85**, 050303 (2012).

128. C. L. Cortes, W. Newman, S. Molesky and Z. Jacob, Quantum nanophotonics using hyperbolic metamaterials. *J. Opt.* **16**, 129501 (2014).

129. A. N. Poddubny, I. V. Iorsh and A. A. Sukhorukov, Generation of photon-plasmon quantum states in nonlinear hyperbolic metamaterials. *Phys. Rev. Lett.* **117**, 123901 (2016).

130. K. R. McEnery, M. S. Tame, S. A. Maier, M. S. Kim, Tunable negative permeability in a quantum plasmonic metamaterial. *Phys. Rev. A* **89**, 013822 (2014).

131. B. Lukyanchuck, N. I. Zheludev, S. A. Maier, N. J. Halas, P. Nordlander, H. Giessen and C. T. Chong, The Fano resonance in plasmonic nanostructures. *Nat. Mat.* **9**, 707 (2010).

132. A. E. Miroshnichenko, S. Flach and Y. S. Kivshar, Fano resonances in nanoscale structures. *Rev. Mod. Phys.* **82**, 2257 (2010).

133. A. M. Steane, Error correcting codes in quantum theory. *Phys. Rev. Lett.* **77**, 793 (1996).

134. C. H. Bennett, D. P. DiVincenzo, J. A. Smolin and W. K. Wootters, Mixed-state entanglement and quantum error correction. *Phys. Rev. A* **54**, 3824 (1996).

135. B. M. Terhal, Quantum error correction for quantum memories. *Rev. Mod. Phys.* **87**, 307 (2015).

136. G. W. Hanson, S. Ali Hassani Gangaraj, C. Lee, D. G. Angelakis and M. Tame, Quantum plasmonic excitation in graphene and loss-insensitive propagation. *Phys. Rev. A* **92**, 013828 (2015).

137. A. Gonzalez-Tudela, D. Martin-Cano, E. Moreno, L. Martin-Moreno, C. Tejedor and F. J. Garcia-Vidal, Entanglement of two qubits mediated by one-dimensional plasmonic waveguides. *Phys. Rev. Lett.* **106**, 020501 (2011).

138. D. Martin-Cano, A. Gonzalez-Tudela, L. Martin-Moreno, F. J. Garcia-Vidal, C. Tejedor and E. Moreno, Dissipation-driven generation of two-qubit entanglement mediated by plasmonic waveguides. *Phys. Rev. B* **84**, 235306 (2011).

139. M. Gullans, T. G. Tiecke, D. E. Chang, J. Feist, J. D. Thompson, J. I. Cirac, P. Zoller and M. D. Lukin, Nanoplasmonic lattices for ultracold atoms. *Phys. Rev. Lett.* **109**, 235309 (2012).

140. E. Knill, R. Laflamme and G. J. A. Milburn, A scheme for efficient quantum computation with linear optics. *Nature* **409**, 46 (2001).

141. L. M. Duan, M. D. Lukin, J. I. Cirac and P. Zoller, Long-distance quantum communication with atomic ensembles and linear optics. *Nature* **414**, 413 (2001).

142. C. Lee, M. Tame, C. Noh, J. Lim, S. A. Maier, J. Lee and D. G. Angelakis, Robust-to-loss entanglement generation using a quantum plasmonic nanoparticle array. *New J. Phys.* **15**, 083017 (2013).

143. J. B. Khurgin, How to deal with the loss in plasmonics and metamaterials. *Nat. Nano.* **10**, 2 (2015).

144. F. J. García de Abajo, Microscopy: Plasmons go quantum. *Nature* **483**, 417 (2012).

145. G. Mie, Beiträge zur Optik trüber Medien, speziell kolloidaler Metallösungen. *Ann. Phys.* **25**, 377 (1908).

146. H. Fröhlich, Die spezifische wärme der elektronen kleiner metallteilchen bei tiefen temperaturen. *Physica* **6**, 406 (1937).

147. L. Genzel, T. P. Martin and U. Kreibig, Dielectric function and plasma resonances of small metal particles. *Z. Physik B* **21**, 339 (1975).

148. W. P. Halperin, Quantum size effects in metal particles. *Rev. Mod. Phys.* **58**, 533 (1986).

149. O. Keller, M. Xiao and S. Bozhevolnyi, Optical diamagnetic polarizability of a mesoscopic metallic sphere: Transverse self-field approach. *Opt. Commun.* **102**, 238 (1993).

150. N. J. Halas, S. Lal, W.-S. Chang, S. Link and P. Nordlander, Plasmons in strongly coupled metallic nanostructures. *Chem. Rev.* **111**, 3913 (2011).

151. W. A. Kraus and G. C. Schatz, Plasmon resonance broadening in small metal particles. *J. Chem. Phys.* **79**, 6130 (1983).

152. J. A. Scholl, A. L. Koh and J. A. Dionne, Quantum plasmon resonances of individual metallic nanoparticles. *Nature* **483**, 421 (2012).

153. R. A. de la Osa, J. M. Sanz, J. M. Saiz, F. Gonzalez and F. Moreno, Quantum optical response of metallic nanoparticles and dimers. *Opt. Lett.* **37**, 5015 (2015).

154. E. Prodan and P. Nordlander, Exchange and correlations effects in small metallic nanoshells. *Chem. Phys. Lett.* **349**, 153 (2001).

155. E. Prodan and P. Nordlander, Electronic structure and polarizability of metallic nanoshells. *Chem. Phys. Lett.* **352**, 140 (2002).

156. P. Nordlander and E. Prodan, Optical properties of metallic nanoshells. *Proc. SPIE*, **4810**, 91 (2002).

157. E. Prodan and P. Nordlander, Structural tunability of the plasmon resonances in metallic nanoshells. *Nano Lett.* **3**, 543 (2003).

158. E. Prodan, P. Nordlander and N. J. Halas, Electronic structure and optical properties of gold nanoshells. *Nano Lett.* **3**, 1411 (2003).

159. E. Prodan, A. Lee and P. Nordlander, The effect of a dielectric core and embedding medium on the polarizability of metallic nanoshells. *Chem. Phys. Lett.* **360**, 325 (2002).

160. E. Prodan, P. Nordlander and N. J. Halas, Effects of dielectric screening on the optical properties of metallic nanoshells. *Chem. Phys. Lett.* **368**, 94 (2003).

161. J. Zuloaga, E. Prodan and P. Nordlander, Quantum plasmonics: Optical properties and tenability of metallic nanorods. *ACS Nano* **4**, 5269 (2010).

162. E. Townsend and G. W. Bryant, Plasmonic properties of metallic nanoparticles: The effects of size quantization. *Nano Lett.* **12**, 429 (2012).

163. Z. Yuan and S. Gao, Linear-response study of plasmon excitation in metallic thin films: Layer-dependent hybridization and dispersion. *Phys. Rev. B* **73**, 155411 (2006).

164. R. C. Jaklevic, J. Lambe, M. Mikkor and W. C. Vassell, Observation of electron standing waves in a crystalline box. *Phys. Rev. Lett.* **26**, 88 (1971).

165. S. Thongrattanasiri, A. Manjavacas and F. J. García de Abajo, Quantum finite-size effects in graphene plasmons. *ACS Nano* **6**, 1766 (2012).

166. R. Sato, M. Ohnuma, K. Oyoshi and Y. Takeda, Experimental investigation of non-linear optical properties of Ag nanoparticles: Effect of size quantization. *Phys. Rev. B* **90**, 125417 (2014).

167. W. Yan, M. Wubs and N. A. Mortensen, Projected dipole model for quantum plasmonics. *Phys. Rev. Lett.* **115**, 137403 (2015).

168. R. C. Monreal, T. J. Antosiewicz and S. P. Apell, Competition between surface screening and size quantization for surface plasmons in nanoparticles. *New J. Phys.* **15**, 083044 (2013).

169. K. P. Charle, W. Schulze and B. Winter, The size dependent shift of the surface plasmon absorption band of small spherical metal particles. *Z. Phys. D: At., Mol. Cluster*, **12**, 471 (1989).

170. J. Lerme, B. Palpant, B. Prevel, M. Pellarin, M. Treileux, J. Vialle, A. Perez and M. Broyer, Quenching of the size effects in free and matrix-embedded silver clusters. *Phys. Rev. Lett.* **80**, 5105 (1998).

171. E. Cottancin, G. Celep, J. Lerme, M. Pellarin, J. Huntzinger, J. Vialle and M. Broyer, Optical properties of noble metal clusters as a function of the size: Comparison between experiments and a semi-quantal theory. *Theor. Chem. Acc.* **116**, 514 (2006).

172. H. Haberland, Looking from both sides. *Nature* **494**, E1 (2013).

173. J. Tiggesbaumker, L. Koller, K. H. Meiwes-Broer and A. Liebsch, Blue shift of the Mie plasma frequency in Ag clusters and particles. *Phys. Rev. A* **48**, 1749(R) (1993).

174. W. Harbich, S. Federigo and J. Buttet, The optical absorption spectra of small silver clusters (n = 8–39) embedded in rare gas matrices. *Z. Phys. D At. Mol. Cluster* **26**, 138 (1993).

175. K.-P. Charle, L. Konig, S. Nepijko, I. Rabin and W. Schulze, The surface plasmon resonance of free and embedded Ag-clusters in the size range 1.5 nm< D<30 nm. *Cryst. Res. Technol.* **33**, 1085 (1998).

176. J. B. Lassiter, J. Aizpurua, L. I. Hernandez, D. W. Brandl, I. Romero, S. Lal, J. H. Hafner, P. Nordlander and N. Halas, Close encounters between two nanoshells. *Nano Lett.* **8**, 1212 (2008).

177. I. Romero, J. Aizpurua, G. W. Bryant, F. J. García de Abajo, Plasmons in nearly touching metallic nanoparticles: singular response in the limit of touching dimers. *Opt. Exp.* **14**, 9988 (2006).

178. I. A. Larkin and M. I. Stockman, Imperfect Perfect Lens. *Nano Lett.* **5**, 339 (2005).

179. E. Prodan and R. Car, Tunneling conductance of amine linked alkyl chains. *Nano Lett.* **8**, 1771 (2008).

180. F. J. García de Abajo, Nonlocal effects in the plasmons of strongly interacting nanoparticles, dimers and waveguides. *J. Phys. Chem. C* **112**, 17983 (2008).

181. J. Zuloaga, E. Prodan and P. Nordlander, Quantum description of the plasmon resonances of a nanoparticle dimer. *Nano Lett.* **9**, 887 (2009).

182. K. J. Savage, M. M. Hawkeye, R. Esteban, A. G. Borisov, J. Aizpurua and J. J. Baumberg, Revealing the quantum regime in tunnelling plasmonics. *Nature* **491**, 574 (2012).

183. J. A. Scholl, A. Garcia-Etxarri, A. L. Koh and J. A. Dionne, Observation of quantum tunnelling between two plasmonic nanoparticles. *Nano Lett.* **13**, 564 (2012).

184. R. Esteban, A. G. Borisov, P. Nordlander and J. Aizpurua, Bridging quantum and classical plasmonics. *Nat. Commun.* **3**, 825 (2012).

185. V. Kulkarni, E. Prodan and P. Nordlander, Quantum plasmonics: optical properties of a nanomatryushka. *Nano Lett.* **13**, 5873 (2013).

186. P. Zhang, J. Feist, A. Rubio, P. Garcia-Gonzalez and F. J. Carcia-Vidal, Ab initio nanoplasmonics: The impact of atomic structure. *Phys. Rev. B* **90**, 161407(R) (2014).

187. D. C. Marinica, A. K. Kazansky, P. Nordlander, J. Aizpurua and A. G. Borisov, Quantum plasmonics: Nonlinear effects in the field enhancement of a plasmonic nanoparticle dimer. *Nano Lett.* **12**, 1333 (2012).

188. L. Wu, H. Duan, P. Bai, M. Bosman, J. K. W. Yang and E. Li, Fowler–Nordheim tunnelling induced charge transfer plasmons between nearly touching nanoparticles. *ACS Nano* **7**, 707 (2013).

189. N. A. Mortensen, S. Raza, M. Wubs, T. Sondergaard and S. I. Bozhevolnyi, A generalized non-local optical response theory for plasmonic nanostructures. *Nat. Commun.* **5**, 3809 (2014).

190. V. Kulkarni and A. Manjavacas, Quantum effects in charge transfer plasmons. *ACS Photonics* **2**, 987 (2015).

191. E. Townsend and G. W. Bryant, Which resonances in small metallic nanoparticles are plasmonic? *J. Opt.* **16**, 114022 (2014).

192. S. Raza, S. I. Bozhevolnyi, M. Wubs and N. A. Mortensen, Nonlocal optical response in metallic nanostructures. *J. Phys.: Condens. Matter* **27**, 183204 (2015).

193. W. Zhu, R. Esteban, A. G. Borisov, J. J. Baumberg, P. Nordlander, H. J. Lezec, J. Aizpurua and K. B. Crozier, Quantum mechanical effects in plasmonic structures with subnanometre gaps. *Nat. Commun.* **7**, 11495 (2016).

Chapter 6

Synthesis of Gold Nanoparticles in Liquid Phase

Woong Choi and Hyunjoon Song

Department of Chemistry, Korea Advanced Institute of Science and Technology, and Center for Nanomaterials and Chemical Reactions, Institute for Basic Science, Daejeon, 34141, Daejeon, Republic of Korea

6.1 Introduction

Nanometre-sized gold particles are of fundamental and practical interest, as their chemical, electronic and optical properties can potentially be exploited for diverse applications.[1–5] These properties are extremely dependent upon their size and shape; therefore numerous approaches to synthetically and systematically control morphology and surface compositions have been developed in recent decades. Design of a generic method for the preparation of gold nanostructures with a broad range of well-defined and controllable morphologies is needed in order to fully exploit their unique properties for practical applications. Numerous methods have been developed to prepare gold nanostructures, including electrochemical and gas and liquid phase methods. Among these, liquid phase synthesis offers many advantages such as large-scale synthesis and ready quality control. In particular, the synthesis under well-controlled reaction conditions possibly provides uniform and monodisperse nanoparticles. The liquid phase preparation is also important in biological applications because of their compatibility in environments that living cells can exist.

In this chapter, we briefly introduce chemical properties important for liquid phase synthesis (Section 6.2), and review general strategies for the synthesis, shape control and surface functionality modification of gold nanostructures in liquid phase. Generally, gold nanoparticles are synthesised via the reduction of gold(I) or gold(III) precursors in aqueous or organic media in the presence of surface stabilisers. However, a certain synthetic condition provides only a narrow-size range of the nanoparticles, and thus appropriate synthetic methods must be chosen in consideration of various factors including final particle size and size distribution, readiness of the reaction procedure, reaction scale and surface functionality suitable for specific applications (Section 6.3). Shape control of gold nanoparticles is another important and timely issue, because light scattering and surface-dependent chemical activity are known to be very sensitive to the morphology of nanostructures (Section 6.4). Bimetallic nanoparticles containing gold can widen the coverage of applications using gold nanoparticles with unique physico-chemical properties (Section 6.5). Surface functionalisation of gold nanoparticles for their chemical and bio-applications is reviewed in Chapter 7. Synthesis of gold nanoparticles on solid supports is also reviewed in upcoming chapters by chemical (Chapter 8) and physical (Chapter 12) methods.

6.2 Chemical Properties and Characterisation of Gold Nanoparticles for Liquid Phase Synthesis

6.2.1 *Structure and Size Range of Gold Nanoparticles*

A representative structure of gold nanoparticles synthesised in liquid phase comprises three parts, including inner gold atoms with a closed-packed crystal structure (central atoms), outer layers exposed on the surface (surface atoms) and surface protecting organic ligands or surfactants (Figure 6.1).[6] This structure is historically referred to as a 'monolayer protected cluster' or briefly MPC. The central gold atoms determine the crystallinity of the structure. The geometry of the surface atoms is different from that of the central atoms and forms surface facets and edges that dominate catalytic activity. The surface-protecting ligands are anchored on the surface atoms, stabilise them and sometimes provide surface functionality that influences the chemical nature of the particles in solution.

Gold nanoparticles can be classified into three types according to their size and size-dependent properties: size ranges of less than 2 nm, 2–10 nm, and 10–300 nm (Table 6.1). Particles smaller than 2 nm in diameter are

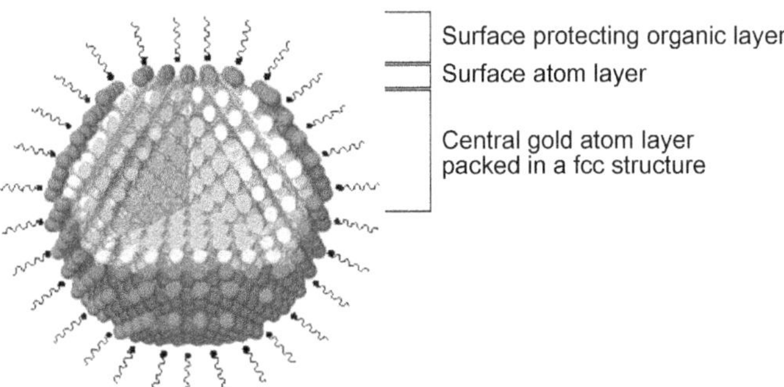

Figure 6.1   A structural model of a single gold nanoparticle.

Table 6.1   Classification of gold nanoparticles by size range.

Size range	Class	Major property	Example
<2 nm	Cluster	Molecular- to metal-like	$Au_{11}, Au_{28}, Au_{39}, Au_{55}, Au_{102}$
2–10 nm	Catalytic particle	Enhanced catalytic activity	Tiny nanoparticles
10–300 nm	Plasmonic crystal	Surface plasmon resonance	Polyhedrons, rods, wires, plates

called 'gold clusters',[7–9] which consist of a few tens to a few hundreds of gold atoms. The cluster surface is normally stabilised by organic molecules such as organothiolates, and physical properties vary from molecular-like to metal-like. Evolution of the electronic structure in gold clusters is observed in this range. Particles smaller than 10 nm in diameter are very useful in heterogeneous catalysis.[10] The surface atom fraction versus a total number of atoms (or volume) rapidly increases with reduction of the particle diameter. Surface atoms are generally more active than closed-packed atoms inside the particle, due to their outwardly exposed dangling bonds. When the particle size is less than 5 nm, surface properties exceed intrinsic features of the materials, and dominate total physical properties. For instance, the melting point of bulk gold is 1064°C, but it abruptly decreases to 380°C for 1.5-nm sized particles due to high surface energy.[11,12] Large surface area and high density of edge and kink atoms are other factors that lead to increased chemical activity compared to bulk materials. Particles with sizes of 10–300 nm strongly scatter light, mostly in the visible range, which is attributed to localised surface plasmon resonance (LSPR).[1–4] The LSPR behaviour is extremely dependent upon the size and shape of metallic objects. Gold nanoparticles with this range of diameter are ideal to adjust LSPR extinctions from visible to near infrared (NIR).

6.2.2 *Electrochemical Potentials of Gold Precursors*

Gold(III) halides are the precursors commonly used in gold nanoparticle synthesis. The gold precursors are prepared by dissolution of bulk gold in aqua regia or metal cyanide. Under well-controlled reaction conditions, reducing agents (i.e. $NaBH_4$ by Brust *et al.*[13] ascorbic acid by Murphy *et al.*[14] Tetrakis(hydroxymethyl)phosphonium chloride (THPC)-NaOH by Baiker *et al.*[15] alcohols for the polyol process,[16] and citrate for Turkevich *et al.*[17] see Section 6.3.1) are introduced in the precursor solution. However, because the reduction potentials of the gold ions are higher than those of other noble metal ions,[18] and they depend on the nature of the precursors (Table 6.2), proper choice of the reducing agent is very important to synthesise regular nanoparticles with narrow shape and size distributions.

Table 6.2   Standard reduction potentials of various gold precursors at 25°C, 1 atm.[18]

Reaction	$E°$ (V)
$Au^+ + e^- \leftrightarrow Au$	1.692
$Au^{3+} + 3e^- \leftrightarrow Au$	1.498
$AuBr_2^- + e^- \leftrightarrow Au + 2Br^-$	0.959
$AuBr_4^- + 3e^- \leftrightarrow Au + 4Br^-$	0.854
$AuCl_4^- + 3e^- \leftrightarrow Au + 4Cl^-$	1.002
$Au(OH)_3 + 3H^+ + 3e^- \leftrightarrow Au + 3H_2O$	1.450

6.2.3 *Surface Energy and Particle Morphology*

The structural total energy (E_t) using macroscopic concepts as a guide is given by

$$E_t(N) = E_B N + E\sigma N + E\gamma S,$$

where N is the number of atoms, E_B is the bulk energy per atom, $E\sigma$ is the structural strain energy per atom, $E\gamma$ is the average surface energy per unit area and S is the surface area of the particles.[19,20] When the particles are stabilised with surface-protecting ligands, the interaction between surface atoms and protecting ligands should be added as a separate term. As the particle size decreases, the surface area decreases more slowly than the number of atoms, and the surface energy dominates the total energy. In principle, a sphere is the most thermodynamically stable structure in an isotropic system, because it has the minimum number of surface atoms per volume and the minimum surface energy.[21] Taking into consideration atomic packing in a tiny nanoparticle surface, however, low index facets such as {100}, {111} and {110} are more stable than other facets due to their high density of atomic packing, and these facets tend to form the surface of polyhedral structures. Among the low-index facets, the surface energy is in the order of $\gamma\{111\} < \gamma\{100\} < \gamma\{110\}$, where the {111} facets have the largest surface density and the lowest number of dangling bonds of the surface atoms. Consequently, the particles tend to have {111} facets on the surface without any kinetic adjustment. Among three-dimensional polyhedrons, a decahedron (Figure 6.2) is one of the most stable structures, having only {111} surfaces and a nearly spherical shape, although it has strain energy originating from twin boundaries.[12,19,22] Theoretical calculations show that decahedral and icosahedral shapes with {111} surface facets are excellent models for particles with a size of less than 5 nm. However, a

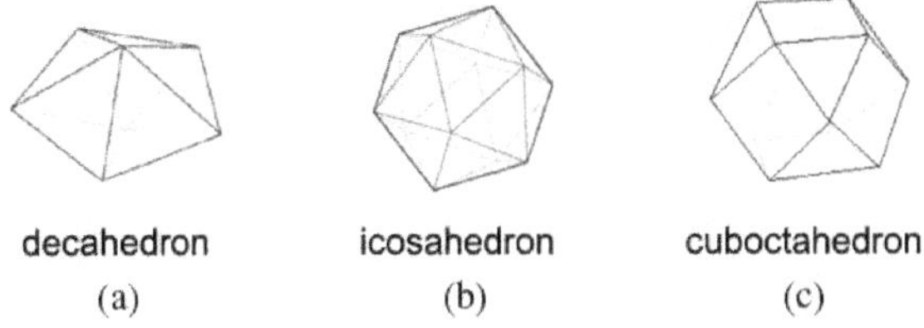

Figure 6.2   Ideal (a) decahedral, (b) icosahedral and (c) cubooctahedral structures.

single-crystalline cuboctahedron is a common structure for larger particles, because lattice strain rapidly increases with particle size enlargement. These polyhedrons are regarded as plausible morphologies of the seeds that are further grown into various structures (Section 6.4). Introduction of surface regulating agents and heterometallic species that are selectively bound to the specific facets can precisely alter relative surface energies and the resulting particle morphology.

6.2.4 *Characterisation of Nanoparticles*

6.2.4.1 *Morphology characterisation*

Electron microscopy, including transmission electron microscopy (TEM) and scanning electron microscopy (SEM), makes use of the wave character of electrons, but is analogous to optical microscopy in that an image of an object is constructed from the scattering pattern of a focused beam of electrons. The technique of high-resolution transmission electron microscopy (HRTEM) together with selected electron diffraction (SAED) and energy dispersive X-ray spectroscopy (EDX) can be utilised to analyse the crystal structure and composition of objects. X-ray diffraction (XRD) is meanwhile one of the most important techniques for the determination of structure and composition.

For surface plasmon resonance, UV-visible spectroscopy provides valuable information regarding shape, size, interparticle distance and aggregation of nanoparticles. Dynamic light scattering (DLS) analysis determines the average hydrodynamic size and size distribution profiles of the particles in solution.

6.2.4.2 *Surface characterisation*

X-ray photoelectron spectroscopy (XPS) is an essential tool to analyse surface atomic layers of nanoparticles within several nanometres from the surface. Information of atomic composition and chemical state of the surface atoms can be obtained from XPS spectra. Fourier transform infrared (FT-IR) spectroscopy is meanwhile a powerful tool to characterise functional groups of the surface protecting organic layers. Thermal analysis (TA) including thermogravimetric analysis (TGA) and differential scanning calorimetry (DSC) can be used to analyse the amount of organic residues and surface melting properties.

6.2.4.3 *Theoretical simulation*

The surface plasmon phenomenon is basically a resonance between light and conductive matter (see Chapter 3), and thus it can be comprehensively analysed by Maxwell equations.[23] The Mie theory provides an exact solution of the equations, but is limited to analysis of the morphology of spheres and spheroids. Recent advances in electrodynamic theory make it possible to determine numerically converged solutions for nanoparticles of arbitrary shapes with dimensions up to a few hundred nanometres. The discrete dipole approximation (DDA)[24,25] and finite difference time domain (FDTD)[26,27] methods are generally used for analysing scattering and absorption bands of gold nanoparticles.

6.3 Synthetic Methods of Gold Nanoparticles in Liquid Phase

Known since ancient times, the synthesis of colloidal gold was originally used as a method of staining glass (see Chapter 1). According to the demands of given applications such as chemo- and biosensors, catalysts and electric circuits, numerous synthetic processes including the use of chemical reductants in both aqueous and non-aqueous solvents, as well as photochemical, radiolytic, electrochemical, sonochemical and microwave assisted methods, have been developed.

6.3.1 *Kinetic Consideration for Highly Monodisperse Nanoparticles*

The synthesis of nanoparticles with a narrow size distribution has long been of scientific and technological interest since Faraday's production of gold sols in 1857.[28] For the preparation of monodispersed nanoparticles, the nucleation step must be separated from the growth step in order to prevent simultaneous secondary nucleation and growth. LaMer *et al.* proposed the concept of 'burst nucleation', where many nuclei were simultaneously generated, and applied their concept to synthesise a series of monodisperse nanoparticles through a temporally discrete nucleation event followed by a controlled growth on the existing nuclei.[29] It is necessary to induce a single nucleation event in the nucleation step and to prevent additional nucleation during the growth step to prepare highly uniform nanoparticles.[30]

There are two major techniques to separate nucleation and growth. The seed-mediated growth utilises preformed nanocrystal seeds followed by slow growth on their surface. The precursor concentration is kept at a low level during the growth to prevent homogeneous nucleation. This method is also useful to generate core–shell nanocrystals by introducing hetero-components[31] or to generate anisotropic structures such as nanorods by adding surface-regulating surfactants (see Section 6.4.1). The second is so-called 'hot-injection', which was originally introduced in the synthesis of semiconductor nanoparticles.[32] This method is rapid injection of a high concentration of the precursor solution into a hot surfactant solution, leading to fast and concomitant formation of nuclei. The nucleation rate rapidly decreases just after the nucleation process, and the growth occurs on the seed surface.[33]

During the growth step, the particle size distribution is generally diminished by 'size focusing' through mass-transport processes.[34] Talapin *et al.* simulated two related mechanisms for the control of size distributions.[35] If no additional nucleation occurs and the precursor concentration is relatively high, the particle growth rate is inversely proportional to its radius, and thus the particle size distribution always decreases as long as all particles are continuously growing. On the other hand, under the low precursor concentration, Ostwald ripening occurs.[36] In this process, small crystallites have higher surface free energy, and thus are less stable against the dissolution in solvent than the large crystallites. This stability gradient leads to slow diffusion of the materials from the small to large crystal surfaces, and the relative standard deviation approaches to a constant in the equilibrium state between the dissolution and reprecipitation.

6.3.2 *Chemical Reduction of Gold Precursors*

Reduction of gold salts is a simple process, which requires only mixing of the reagents under well-controlled external conditions. These conditions can affect the final morphology of the particles in subtle ways. According to the synthetic route used, different characteristics of final products are obtained (Table 6.3). Besides the strength of the reductant, the action of a stabiliser is critical in liquid phase synthesis. The reaction temperature is one of the main factors to determine particle size, because the oxidation potential and related kinetics of the reductant are normally dependent upon the temperature. Representative examples are: (1) organic phase synthesis involving a two-phase process at room temperature, with 2–10 nm range (Brust method, Figure 6.3(a)); (2) the single-phase water-based reduction of a gold salt by citrate at 100°C, which produces almost spherical particles over a tunable range of 10–20 nm diameter (Turkevich method, Figure 6.3(c)); and (3) alcohol phase synthesis at 20–300°C over a tunable range of 20–300 nm diameter (polyol process, Figure 6.3(b)).

6.3.2.1 *Chemical reduction in aqueous media*

Baiker method is the most pertinent one to prepare gold colloids with the size smaller than 2 nm in aqueous solution.[15] In this method, the gold colloids were synthesised through the reduction of gold(III) ions by THPC, which behaved also as a stabiliser.

For producing monodisperse gold nanoparticles of 10–20 nm diameter in an aqueous solution, the citrate reduction method of $HAuCl_4$ has been widely adopted and improved. This method was pioneered by Turkevich *et al.* in 1951[17] and refined by Frens in the 1970s.[38,39] The reaction kinetics have been addressed in a study from Chow and Zukoski focusing on the stabilisation mechanism.[40] In this reaction, sodium citrate behaves both as a reducing agent and as a capping agent that stabilises the nanoparticles at 100°C. Generally, this method is useful to produce modestly monodisperse spherical gold nanoparticles suspended in water in diameters of 10–20 nm with a relatively narrow size distribution (Figure 6.3(c)).[41] The reduction of the sodium citrate concentration diminishes the citrate ion concentration available for particle stabilisation, which causes aggregation of small particles into larger ones. During the synthesis, networks of gold nanowires are formed as a transient intermediate.[42]

Table 6.3   Various well-known synthetic methods of gold nanoparticles in liquid phase.

Reduction method	Reaction media	Reductant	Surface protecting agent	Particle size range (nm)	Reaction temperature (°C)	Refs.
Brust–Schiffrin method	Organic	$NaBH_4$	Organothiol	2–10	r.t.	13, 45–52
Turkevich method	Aqueous	Citrate	Citrate	10–20	100	17, 38–42
Murphy method	Aqueous	Ascorbic acid	CTAB	10–50	r.t.	14
Perrault method	Aqueous	Hydroquinone	Citrate	50–200	r.t.	43
Polyol process	Alcohol	Diols	PVP	20–200	20–300	37, 64–67

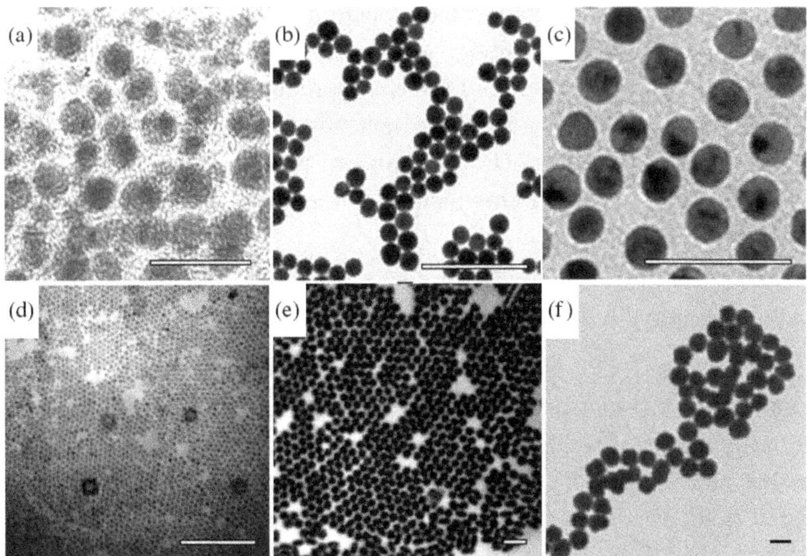

Figure 6.3   TEM images of gold nanoparticles prepared through (a) Brust–Schiffin method, (b) polyol process, (c) Turkevich method, (d) and (e) Murphy method and (f) Perrault method. Scale bar: (a) 10 nm, (b) 500 nm, (c) 50 nm and (d–f) 100 nm. (a) Modified with permission from Ref. 13. Copyright 1995 Royal Society of Chemistry, (b) Ref. 37. Copyright 2006 American Chemical Society, (c) Modified with permission from Ref. 41. Copyright 2010 American Chemical Society, (d,e) Ref. 14, Copyright 2001 American Chemical Society, (f) Ref. 43, Copyright 2009 American Chemical Society.

Other reductants such as inorganic reductants and organic amines have also been successfully utilised. Ascorbic acid is one of the most widely used reducing agents. In this reaction, the gold ion forms a complex with a surfactant molecule, such as cetyltrimethylammonium bromide (CTAB), but is not directly reduced by ascorbic acid. The addition of 3.5 nm gold seeds or a small amount of a strong reducing agent leads to the growth of gold nanostructures with different morphologies at room temperature (Figures 6.3(d) and 6(e)).[14]

For the realisation of monodisperse gold nanoparticles in the range of 50–200 nm in diameter in an aqueous solution, Perrault and Chan used hydroquinone to reduce $HAuCl_4$ in an aqueous solution that contains gold nanoparticle seeds (Figure 6.3(f)).[43] This process is similar to that used in photographic film development, wherein silver grains within the film grow through the addition of reduced silver onto their surface. Similarly, gold nanoparticles can catalyse the reduction of gold ions onto their surface. The presence of a stabiliser such as citrate results in controlled particle growth. Typically, the gold seeds are produced via citrate reduction. The hydroquinone reduction method extends the size range of monodispersed spherical particles. This method can produce particles of at least 50–200 nm, whereas the Frens method is ideal for obtaining particles of 10–20 nm diameter.

6.3.2.2 *Chemical reduction in organic media*

Stabilisation of gold nanoparticles with organothiol ligands was first reported by Mulvaney and Giersig in 1993.[44] The Brust–Schiffin method, for the synthesis of small gold clusters and nanoparticles ranging in diameter between 1.5 and 5.2 nm,[13] has since had a considerable impact on the overall field, as it offers facile synthesis of thermally and air stable gold nanoparticles of reduced dispersity and controlled size (Figure 6.4).[45–50] Indeed, these monolayer-protected clusters (MPCs) can be repeatedly isolated and redispersed in common organic solvents without irreversible aggregation or decomposition, and they can be easily handled and functionalised, similar to stable organic and molecular compounds. This procedure typically involves three steps: phase transfer of gold precursors from an aqueous solution to the organic phase using a long-chain ammonium salt (e.g. tetraoctylammonium bromide (TOAB)); reduction of Au(III) (from $AuCl_4^-$) to Au(I) by thiol through the formation of gold–thiol intermediates; and further reduction of Au(I) to Au(0) by a second reductant (e.g. $NaBH_4$). Recently, the reduction kinetics[51] and identification of the precursors[52]

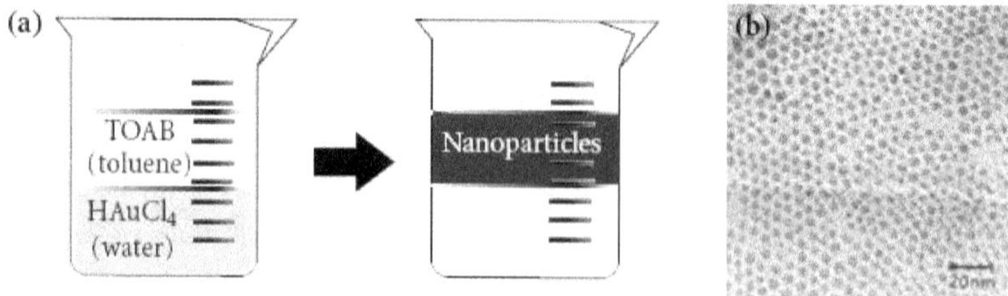

Figure 6.4   (a) Schematic illustration of Brust–Schiffin method for gold nanoparticles and (b) TEM image of resulting gold nanoparticles. Modified with permission from Ref. 49. Copyright 1997 American Chemical Society.

have been studied in detail, and an improved synthetic scheme has been developed.[16,53]

The synthesis of Au_{55} clusters is also categorised in this method. The Au_{55} clusters gain special interest because of their ultimate size of 1.4 nm and ideal cubooctahedral structure. $Au_{55}(PPh_3)_{12}Cl_6$ was successfully synthesised through the reduction of Ph_3PAuCl in benzene or toluene by gaseous B_2H_6 as a reducing agent instead of $NaBH_4$,[54] because B_2H_6 not only directly reduced Au(I) to Au(0), but also bound excess PPh_3 to form Ph_3P-BH_3, and removed them from the clusters. The resulting Au_{55} clusters can replace their surface ligands and generate either hydrophilic (with borate and monosulphonated phosphine) or hydrophobic (with silsesquioxane and aryl phosphine) characters.

6.3.2.3 *Synthesis in micelles*

The syntheses involve a two-phase system with a surfactant that causes the micelle formation maintaining a favourable nano- and micro-environment, together with the extraction of metal ions from the aqueous phase to the organic phase.[50] A micelle is an aggregate of surfactant molecules dispersed in a liquid, which have hydrophobic and hydrophilic domains. Although this method uses a similar two-phase system, it is distinct from the Brust–Schiffin method because the nanoparticle formation actually occurs in the micellar structure. Synthetic copolymers are reported to form effective micelle structures for gold nanoparticle synthesis, due to their stabilisation of the gold surface and surface functionality control.[55–57] The size and shape of the micelles could be readily tuned by variation of composition, molecular architecture and constituent block length of the copolymers. Various copolymers including two different polymer units, such as diblock,[58,59] graft[60] and star-shaped copolymers,[61,62] can form nanometre-sized micelles consisting of a soluble corona and an insoluble core in a solvent selective for one of the blocks, and these micelles were successfully employed as synthetic templates for gold nanoparticles with the size range of 0.5–50 nm.

6.3.2.4 *Polyol process*

The polyol process originally entailed the reduction of metal salts to prepare metal and metal oxide particles.[63] It is a convenient, versatile and a low-cost method for the synthesis of metal nanostructures in a large scale. A metal compound or salt is dispersed in a long-chain diol such as ethylene glycol that acts both as a solvent and as a reductant. After the reduction, the resulting particles are monodisperse and non-agglomerated. The particles with a relatively large size range, from several nano- to micro-metres, can be synthesised. The products are not spherical, but rather polyhedral or plate-like with stable surface facets. Poly(vinyl pyrrolidone) (PVP) is a representative surface regulating reagent that kinetically controls the growth rates of various facets for the formation of anisotropic gold nanostructures, on the basis of its excellent adsorption ability. Yang *et al.* synthesised the so-called platonic nanocrystals, isotropic gold polyhedrons with sizes of 100–300 nm, by a polyol process.[64] Importantly, these highly symmetric structures can provide fundamental insight into the origin of symmetry and the formation mechanism of nanoscale materials. Song *et al.* employed the polyol process to syntheside gold polyhedrons, rods and wires under various reaction conditions (see Section 6.4.3).[37,65–67]

6.3.3 *Non-chemical Reduction for Preparation of Gold Nanoparticles*

Although chemical reduction methods with various reductants are simple and convenient, the reaction conditions must be carefully adjusted for particular applications. For example, an excess amount of reducing agents, their oxidation products and the surfactants used for chemical reduction may contaminate the final nanoparticles. So, in this section, we introduce alternative synthetic processes, including electrochemical, photochemical, sonochemical and microwave-assisted methods, which do not require the use of chemical reductants at high temperatures. As shown in Table 6.4, the reduction of the gold precursors is assisted by various energy sources, with or without surface stabilising reagents.

6.3.3.1 *Photochemical and radiolytic methods*

Reduction of gold ions by photochemical and radiolytic methods has the advantage of not employing an excess of reducing agents, which may contaminate the nanoparticles. For this reason, these methods are widely applied to catalytic applications due to their simple and one-step procedures

Table 6.4   Non-chemical reduction methods for gold nanoparticle synthesis.

Preparation method	Energy source	Reduction mechanism
Chemical	Reductant (sometimes with heating)	Electrochemical potential difference
Photochemical	Ultraviolet	Reductive elimination of ligands from the metal precursors
Electrochemical	Electron	Direct reduction by applying a negative potential
Sonochemical	Ultrasonic wave	Radicals generated from solvents or other reagents
Microwave-assisted	Microwave	Organic reducing agents heated by microwave irradiation

as well as cleanliness of the nanoparticle surface. Ultraviolet (UV) is a commonly used light source to excite the gold precursor in the presence of surfactants. Irradiation of UV light led to the effective photoreduction of $AuCl_4^-$ in formamide through the dissociation and direct reduction by the photogenerated free radicals from the formamide molecules, and yielded uniform gold nanoparticles of 12 nm at room temperature in the presence of polymer surfactants.[68] The use of near-infrared laser irradiation is also reported to prepare thiol-stabilised gold nanoparticles.[69] In the radiolytic method, γ-irradiation generates hydrated electrons or active radicals, which reduce the gold precursors. In order to prevent particle aggregation, surface-protecting reagents or surfactants such as polymers and dendrimers are generally added to the particle dispersion (see Section 6.5.2). In the presence of poly(vinyl alcohol) (PVA) or PVP, γ-irradiation of aqueous solutions of $KAuCl_4$ yielded tiny gold nanoparticles with a size between 1.5 and 2.5 nm.[70] The hydrated electrons generated in the aqueous solvent reduced Au(III) to form Au(I), and after a certain period, Au(I) ions accumulated in the solution were further reduced to generate Au(0). In the mixture with methanol, the hydroxyl radicals from the radiolysis of water rapidly reacted with methanol, and the resulting hydroxymethyl radicals reduced additional Au(III).[71] Large gold nanoparticles with more than 20 nm were formed by the radiolysis of more concentrated PVA–AuCl$_4^-$ solutions.[72] Controlled enlargement of particle size is carried out by γ-irradiation of the aqueous solutions containing KAu(CN)$_2$ and methanol with gold colloids, and the stepwise radiation growth can grow gold nanoparticles to any desired size up to 120 nm.[73]

6.3.3.2 *Electrochemical methods*

The gold precursors are directly reduced via an electrochemical method without the use of any reducing agents. In the electrochemical method, current flow between two active electrodes by applying a potential can drive oxidation and reduction processes of the materials suspended in solution, generally at room temperature. The particle size is controlled by adjustment of current density. Wang *et al.* demonstrated the synthesis gold nanorods in various sizes stabilised by tetraalkylammonium halides such as CTAB via the electrochemical method.[74,75] In this reaction, a bulk gold plate is oxidised from the anode to form gold nanorods most probably either at the interfacial region of platinum cathodic surface or in electrolytic solution.

Anisotropic gold nanorods with a controllable aspect ratio are also yielded via electrodeposition in hard templates such as nanoporous aluminium oxide (see Section 6.4.4). Zande *et al.* prepared monodisperse dispersion of gold nanorods in water by electrodeposition. The length of the gold rods could be easily varied by changing deposition time, while the thickness was determined by the pore diameter of nanoporous template.[76,77] In order to disperse the gold nanorods, the templates and substrates were selectively dissolved, and PVP was added as a stabiliser.

6.3.3.3 *Sonochemical method*

Sonolysis entails the application of ultrasound to the precursor solution. The gold ions are reduced to form colloidal nanoparticles in a standing wave system generated by an ultrasonic generator. In the presence of organic compounds such as alcohol, ultrasound irradiation generates organic radical species, which reduce the gold precursors. The size of the resulting nanoparticles depends on the frequency used and the Au(III) reduction rate.[78–80] Grieser *et al.* controlled gold nanoparticle sizes from 10 to 30 nm via sonochemistry in the presence of 1-propanol.[81] It was found that the rate of $AuCl_4^-$ reduction induced by cavitation was the highest at 213 kHz in the range of 20–1062 kHz. The size of the gold nanoparticles produced was closely related to the sonochemical reduction rate of $AuCl_4^-$. Su *et al.* prepared gold nanoparticles by ultrasound irradiation without alcohols and stabilisers.[82] Instead, they found that the addition of citrates gave a drastic effect on the formation of the well-separated, monodisperse gold particles with the diameters of around 20 nm.

6.3.3.4 *Microwave-assisted methods*

Microwave dielectric heating is useful for effective and uniform heating, and thus is applied for the rapid synthesis of gold nanostructures.[83] Microwaves are in the electromagnetic spectrum with a frequency range of 300 MHz to 300 GHz. In these wavelengths, polar solvents such as water, alcohols, dimethylformamide (DMF) and ionic liquids[84] continuously orientate under an alternating electric field, and lose energy in the form of heat by molecular friction. The microwave heating is rapid and homogeneous, and can even provide superheating beyond the boiling point of solvent. Therefore, it provides uniform nucleation and growth conditions, shorter reaction times and reduced energy consumption for gold nanoparticle synthesis. Many studies have explored the synthesis of gold nanoparticles by microwave-assisted methods, and it is known that the reaction time, temperature and ramping rate are important factors in enhancement of particle monodispersity. For example, monodisperse gold nanoparticles with 20 nm in diameter could be synthesised within 15 min at 125°C, with a fast ramping rate (>20°C/min) in the presence of citrate.[85] Tsuji *et al.* employed microwave heating to the polyol process conditions for the synthesis of gold nanopolygons, where $AuCl_4^-$ was reduced in ethylene glycol in the presence of PVP.[86]

6.4 Shape Control of Gold Nanoparticles

Spherical gold nanoparticles are commonly employed as electronic reservoirs or structural supports for chemical and biofunctionalities.

Control over the morphology of gold nanoparticles in terms of size, shape and surface structures, however, can enhance catalytic, electronic and optical properties of the nanoparticles such that can be used in a broader range of applications. For example, the presence of sharp edges and tips significantly promotes electric field enhancement, which is useful for optical sensing of chemical and biomolecules (see Chapter 14).[87–90] Nanoparticle morphology also determines three-dimensionally assembled structures,[91,92] which are very important for real-device applications. Many researchers have successfully developed rational synthetic methods to control gold nanostructures, yielding, for example, polyhedrons, rods, plates and branched structures. Among a variety of nanostructures, one-dimensional nanostructures such as rods and wires have been intensively studied owing to their unique applications in mesoscopic physics and for the fabrication of nanoscale devices (see Chapters 10, 14, 16 and 18).[93] In particular, by adjusting the aspect ratio of gold nanorods, their optical properties can be tuned in a wide range from visible to NIR. In order to prepare anisotropic structures such as nanorods, a rational strategy of symmetry breaking must be developed, because gold has an isotropic crystallographic face-centred cubic (fcc) structure.

6.4.1 *Shaping Strategies with Seed-mediated Growth*

In general, the growth mechanism of metal nanoparticles involves nucleation and growth steps.[94] Seed-mediated growth, wherein tiny seeds serve as nucleation centres for surface growth, can effectively separate both steps and successfully control the dispersity of the nanoparticles (see Section 6.3.1). Another advantage of seed-mediated growth is that the original seeds can influence the growth direction. For instance, Murphy *et al.* synthesised nanorods by reduction of the gold precursors with $NaBH_4$ in the presence of 4-nm sized gold seeds. In this reaction, a decahedral shape of gold seeds led to anisotropic growth of gold nanorods by controlling the growth conditions (see Section 6.4.2).[95,96] Huang *et al.* applied this seed-mediated growth to obtain diverse morphologies of gold nanorods by changing growth solutions and pH.[97]

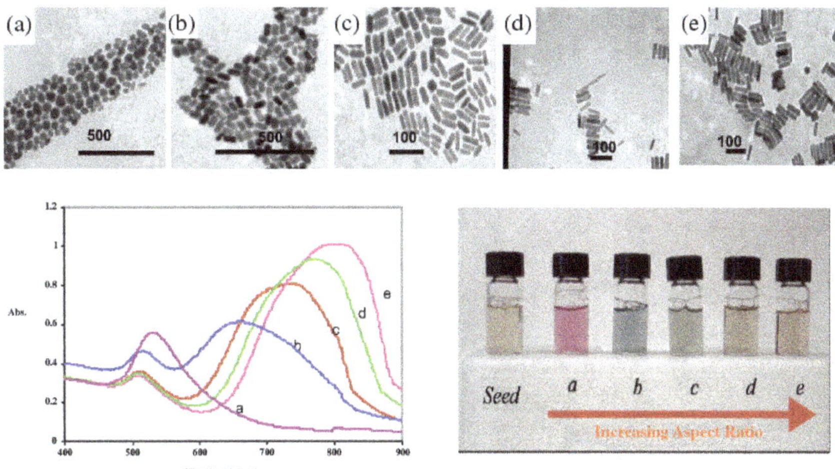

Figure 6.5   TEM images (top), UV-visible spectra (left), and photographs (right) of aqueous solutions of gold nanorods of various aspect ratios. Ref. 96. Copyright 2005 American Chemical Society.

6.4.2 *Selective Binding of Capping Reagents*

As noted earlier, Murphy *et al.* synthesised nanorods with a tunable aspect ratio using gold seeds through optimisation of the CTAB and ascorbic acid concentrations and by applying a step seeding process (Figure 6.5).[95,96] Mann *et al.* confirmed that the electron diffraction patterns of the resulting nanorods were consistent with a pentagonally twinned prism with five {100} side faces capped with five {111} faces at both ends, with growth again being observed to be along the [110] direction.[97] The formation mechanism of one-dimensional rods was proposed to entail preferential binding of the CTAB head group to {110} and {100} faces of gold existing along the sides of twinned rods, as compared to {111} faces at the tips. This is based on the finding that CTAB absorbs onto gold nanorods in a bilayer fashion with the trimethylammomnium headgroups of the first monolayer facing the gold surface.

Numerous complex gold nanostructures, such as rods,[98,99] platonic crystals (Figures 6.6(a) and 6(b)),[98] triangles (Figure 6.6(c))[100] and branched structures (Figure 6.6(d))[101–104] have been synthesised using CTAB-based synthetic methods, where the morphology and dimensions of the gold

Figure 6.6   (a, c, d) TEM, and (b, e, f, g) TEM images of gold nanostructures of various kinds of morphology synthesised with CTAB analogues as surfactants in aqueous solutions. The bars represent: (a, b, h) 100 nm; (c) 166 nm; (d) 10 nm; (e, f) 200 nm and (g) 500 nm. (a, e) Ref. 105. Copyright 2009 American Chemical Society; (b) Ref. 98. Copyright 2009 American Chemical Society; (c) Ref. 100. Copyright 2005 American Chemical Society; (d) Ref. 102. Copyright 2003 American Chemical Society; (f) Ref. 106. Copyright 2009 American Chemical Society; (g) Ref. 108. Copyright 2010 American Chemical Society; (h) Ref. 109. Copyright 2008 Wiley-VCH.

nanocrystals can be controllably varied by the manipulation of synthetic parameters. Recently, single-crystalline gold nanocrystals with high-index facets have been successfully generated by the selective coordination of CTAB analogues, including rhombic dodecahedral with {110} facets (Figure 6.6(e))[105] and regular and elongated tetrahexahedral with {037} facets (Figure 6.6(f)).[106] Concave polyhedra are another set with high-energy surfaces,[107] such as concave cubes with {720} facets (Figure 6.6(g))[108] and trisoctahedra with {221} faces (Figure 6.6(h)).[109,110]

Selective binding of the capping agents also enables to tune the morphology of gold nanoparticles in an organic media. Oleylamine is effectively bound to the {111} facets on gold seeds and generates truncated gold decahedral in tetrahydrofuran at room temperature.[111] Interestingly, three research groups simultaneously reported the formation of ultrathin gold nanowires by slow reduction of the gold precursors. The resulting nanowires show a single-crystalline nature with an average diameter of less than 2 nm, and are unidirectionally grown along ⟨111⟩.[112–114]

6.4.3 *Underpotential Deposition of Heterometallic Additives*

The presence of silver ions allows better control of the shape of gold nanorods synthesised through an electrochemical method. Murphy *et al.*[115] also reported on the effects of silver on the preparation of gold nanorods and spheroids. It was found that the addition of silver nitrate influences not only the yield and aspect ratio control of the gold nanorods, but also the mechanism of gold nanorod formation and the crystal structure. It has been hypothesised that silver adsorbs at the surface of gold, and the silver deposition takes place in an epitaxial fashion due to close matching of the lattice parameters between gold and silver.[116–119]

Yang *et al.* and Song *et al.* have prepared isotropic cubic, octahedral, truncated tetrahedral and icosahedral nanostructures by polyol processes in the presence of PVP and silver nitrate.[37,64–66] These highly symmetric structures are very important, because they can provide fundamental insight into the origin of symmetry and the formation mechanism in nanoscale materials. In these reactions, a small amount of silver species ($\sim$1/100 molar ratio than the gold precursor) is introduced as an additive with gold precursor and PVP. During the reaction, the silver species are deposited on a specific surface of the gold nanoparticles through underpotential deposition (UPD).[67] UPD is a well-known atomic layer scale deposition technique in electrochemistry, where metal overlayers are electrodeposited onto a foreign metal substrate at a potential that is less negative than the Nernst equilibrium potential.[117,120] The underpotential shift of the Au/Ag$^+$ system is particularly facilitated due to a large work function gap ($>$0.5 eV) between silver and gold.[121] The underpotential shift is very sensitive to the distinct surface structures, and theoretical calculations on the Au/Ag$^+$ system demonstrate selective island growth of Ag on Au(100) but not on Au(111) at intermediate deposition rates.[122] This selective deposition of the silver species on the gold {100} surface appears to suppress {100} surface growth and/or enhance {111} growth. Exploiting this phenomenon, subsequent growth is precisely controlled to prepare polyhedral structures[65] and pentagonal nanorods.[67] A rational combination of seed-mediated synthesis and underpotential deposition of silver ions leads to the precise adjustment of both size and shape of gold polyhedral nanocrystals from polygonal to cubic, cuboctahedral and octahedral structures (Figure 6.7).[65]

Figure 6.7   Size and shape control of gold polyhedral nanocrystals from polygonal seeds. SEM images of (a, e) polygonal seeds (average diameters: 81 and 40 nm), (b, f) cubes (average edge lengths: 116 and 67 nm), (c, g) cuboctahedra (122 and 54 nm), and (d, h) octahedral (236 and 88 nm). The bar represents 100 nm for all images. Modified with permission from Ref. 65. Copyright 2008 Wiley-VCH.

6.4.4 *Template-directed Synthesis*

Template-directed synthesis represents a straightforward route to generate One-dimensional structures. In this strategy, templates present on the surface of a solid substrate and channels in a porous membrane serve as a scaffold within which different materials are generated *in situ*. The resulting nanostructure has morphology complementary to that of the original template.[123] Many kinds of methods use template-directed preparation of the gold nanostructures. One example in the area of liquid phase preparation is electrodeposition onto step edges on the surface of a solid substrate such as highly oriented pyrolytic graphite.[124] The nanowires were found to preferentially nucleate and grow along the step edges present on a graphite surface into a two-dimensional parallel array that could be transferred onto another surface.

Channels in a porous membrane also provide one-dimensional nanostructures, an approach pioneered by Martin and other researchers.[125] Two types of porous membranes are commonly used, i.e. ion-track-etched membranes and anodic aluminium oxide (AAO) templates, because both are commercially available. Using these channels, gold ions are reduced from electrolytic solutions through the application of a negative potential, typically in a three-electrode electrochemical cell.

6.5 Synthetic Methods of Gold–Metal Bimetallic Nanoparticles in Liquid Phase

6.5.1 *Structure and Composition of Gold–Metal Bimetallic Nanoparticles*

Gold–metal bimetallic nanoparticles have been intensively investigated for the widespread use of unique optical and catalytic properties of gold nanoparticles.[126–128] For instance, a combination of gold and other metals exhibits large Raman scattering enhancements,[129,130] and shows better efficiency for CO oxidation reactions.[131–133] Due to their dual composition, the structures of gold–metal bimetallic nanoparticles have multiple forms, including intermetallic or alloy, core–shell and heterostructures (Figure 6.8). The intermetallic or alloy nanoparticles are commonly yielded when the two components are homogeneously miscible.[133–135] On the other hand, if the reduction kinetics are completely different between two compositions, either core–shell or heterostructures are generated.[136–138] In the case that lattice parameters of the secondary metal are similar to those of gold, the metal atoms are evenly deposited on the gold surface forming gold–metal core–shell nanoparticles.[139–141] But in the opposite case, the metal atoms are selectively grown on the seeds preformed on the gold surface, resulting

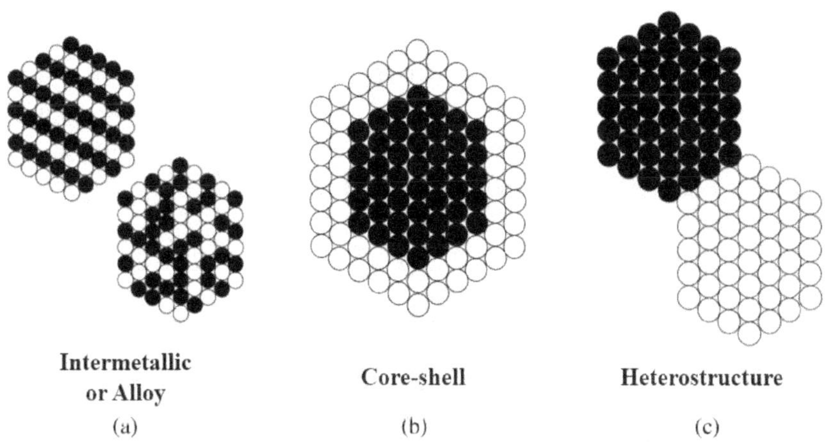

Figure 6.8   Plausible structures of gold–metal bimetallic nanoparticles.

in the formation of heterostructures, such as dumbbell-like and dimeric nanoparticles.[142–144]

For the synthesis of uniform bimetallic nanoparticles, precise control of the reaction conditions by consideration of thermodynamic and kinetic factors is highly demanded. Crystal structures and lattice parameters, bond strengths, surface energies and reduction potentials of the metal components are regarded as thermodynamic factors, which are most significant at high-temperature reactions.[127] However, kinetic factors, such as concentrations of the metal precursors and reductants, addition speeds and sequence of the reaction processes, are also critical, particularly for morphology control in complex structures.[137,138] Table 6.5 briefly summarises some important physical properties of the metal components readily forming bimetallic nanoparticles with gold. These physical properties directly influence the final morphology of gold–metal bimetallic nanoparticles. A small lattice mismatch between gold and the secondary metal determines

Table 6.5   Physical properties of representative metal components.[147,148]

Metal	Lattice constant (Å)	Reduction potential	Surface energy[a] ($J\,m^{-2}$)	Miscibility with gold in bulk
Au	4.065	$AuCl_4^- + 3e^- \leftrightarrow Au$ (1.692 V) $Au^{3+} + 3e^- \leftrightarrow Au$ (1.498 V)	1.283	—
Pt	3.912	$Pt^2 + 2e^- \leftrightarrow Pt$ (1.18 V)	2.299	Immiscible
Pd	3.859	$Pd^{2+} + 2e^- \leftrightarrow Pd$ (0.987 V)	1.920	Partially miscible
Ag	4.079	$Ag^+ + e^- \leftrightarrow Ag$ (0.7996 V)	1.172	Miscible
Cu	3.597	$Cu^+ + e^- \leftrightarrow Cu$ (0.521 V) $Cu^{2+} + 2e^- \leftrightarrow Cu$ (0.3419 V)	1.952	Miscible
Ni	3.499	$Ni^+ + 2e^- \leftrightarrow Ni$ (−0.257 V)	2.011	Partially miscible
Co	2.510 (hcp)	$Co^{2+} + 2e^- \leftrightarrow Co$ (−0.28 V)	2.775[b]	Immiscible
Fe	2.856 (bcc)	$Fe^{2+} + 2e^- \leftrightarrow Fe$ (−0.447 V)	2.222[c]	Immiscible

[a]Calculated values by full charge density theory of (111) surface of fcc structure.
[b](0001) surface energy of hcp structure.
[c](100) surface energy of bcc structure.

the preference of either an ordered intermetallic or a core–shell structure, where the metal is grown on the gold surface in an 'epitaxial' way and generates a smooth interface.[145] The reduction potential of the metal precursors is also a critical factor. Due to the high reduction potential of the gold ions, co-reduction with the metal precursors commonly yields a gold–metal core–shell-like structure (Section 6.5.2.1). The galvanic replacement reaction, which is an electrochemical process involving the oxidation of one metal by the reduction of another metal ions, is also induced by the reduction potential difference, which generates alloy hollow shell structures (Section 6.5.2.3). Surface energies, relative bond strengths and binding strengths are related to the degree of mixing or segregation of the metals on the surface.[146] Thermodynamically, a metal with lower surface energy prefers segregation into the surface in vacuum. High miscibility of gold with various metals is another factor to yield intermetallic and alloy nanoparticles.

There have been many efforts to develop synthetic strategies for gold–metal bimetallic nanoparticles in liquid phase by the rational adjustment of thermodynamic and kinetic parameters. Three representative synthetic methods that have widely been used are co-reduction, seed-mediated growth and galvanic replacement reactions.

6.5.2 Synthetic Protocols of Gold–Metal Bimetallic Nanoparticles

6.5.2.1 Co-reduction

Co-reduction is a reduction process of the mixture of gold and other metal precursors using chemical reductants in one batch. The co-reduction is a one-pot process in principle; however, in the actual mechanism during the reaction, gold ions are reduced first generating small seeds due to their high reduction potential, and then the other metals are grown on the surface. Accordingly, gold–metal alloy nanoparticles tend to be yielded with a gradual change of the composition as a core–shell-like structure, and an additional post-annealing process is required to produce atomically ordered alloy nanoparticles. Yin *et al.* treated AuAg nanospheres at a critical temperature of 930°C with the surface protection by SiO_2 layers, providing a homogeneous distribution of gold and silver in particle size of 22 nm.[149] Shaak *et al.*

developed a post-annealing process in liquid phase, so-called 'metallurgy in a beaker',[150] and applied it for disordered AuCu alloy nanoparticles in the range of 15–30 nm to generate an ordered phase.[151] To directly make an ordered intermetallic phase, a rapid co-reduction of the gold and metal precursors is required by using strong reducing agents under sufficiently high-temperature conditions. Eichhorn *et al.* used butyllithium as a strong reductant and successfully generated AuPt alloy nanoparticles with the average size of 2.5 nm.[152] Schaak *et al.* extended this approach to yield Au_3Fe, Au_3Co and Au_3Ni nanoparticles with the average particle sizes of ~10 nm, although these metals are immiscible or partially miscible in bulk (Table 6.5).[153]

During the co-reduction process, preformed gold seeds can assist the reduction of other metals at much higher reduction potentials. For example, Sun *et al.* synthesised Au–Ag alloy nanoparticles with size measured to be 8–9 nm in a various ratio of Au and Ag in the presence of oleylamine.[133] Under this condition, gold nanoparticles were prepared at 65°C, while silver nanoparticles were formed at 180°C. Interestingly, the Au–Ag alloy nanoparticles were formed at the low temperature of 120°C in the co-reduction process, indicating that the preformed gold seeds facilitated the reduction of silver ions.

Although 'co-reduction' mainly refers to chemical reduction process, similar strategies using simultaneous reduction of the metal precursor mixtures have also been employed for non-chemical reduction methods. AuCu intermetallic nanoparticles with wide size range up to hundreds nm were synthesised in a poly(vinyl alcohol) film by photolysis,[154] AuAg, AuPd and AuPt alloy nanoparticles with sizes less than 5 nm were generated by radiolysis of the aqueous solutions,[155–157] and Au–Pd core–shell nanoparticles with an average diameter of 13 nm were synthesised by a sonochemical method.[158]

6.5.2.2 *Seed-mediated growth*

A seed-mediated growth process for the synthesis of gold–metal bimetallic nanoparticles commonly indicates a mild reduction of secondary metals in the presence of gold seeds that are prepared in a separate batch. This approach is very effective for the formation of core–shell and heterostructures.[127] In general, high surface energy of the seeds can lower the free-energy barrier of a nucleation process, and make heterogeneous nucleation more favourable than homogenous nucleation during the synthesis. To generate a uniform structure, however, multiple factors, including both thermodynamic and kinetic reaction parameters, need to be considered.

As introduced in Section 6.4, the shape and crystal structure of the original gold seeds largely influence the growth of additional metal components. Heteroepitaxial growth of palladium and silver was induced on 30-nm gold octahedral cores, forming isotropic Au–Pd and Au–Ag core–shell nanocrystals with various morphologies, including cubes, octahedrons and concave polyhedrons (Figure 6.9(a)).[145] Lee *et al.* synthesised polyhedral palladium nanocrystals enclosed by high-index facets through an epitaxial growth of palladium layers on 55 nm concave gold nanocrystal seeds.[159] Song *et al.* guided the anisotropic growth of silver to form silver nanorods and nanowires using multiply twinned gold decahedra with 67 nm edges as a seed surface (Figure 6.9b).[116] In some cases, the crystal structure of the seeds could not alter the final morphology, because the gold seeds were evenly coated with the metal atoms at the initial stage, and then the growth kinetics were regulated by homogeneous nucleation. Xia *et al.* synthesised uniform silver octahedrons with an average diagonal length of 65 nm from gold nanorods with 38 nm in length,[160] and Mirkin *et al.* observed stepwise evolution of silver octahedron to tetrahedron and icosahedron on spherical gold seeds with the diameter of 25 nm.[161]

Due to the weak miscibility with gold, platinum is commonly deposited as agglomerates or dendritic shells on the gold surface. Han *et al.* reported that the addition of platinum formed thin layers with the thicknesses of 4–5 nm evenly coating various gold nanocrystal cores.[162] The addition of silver ions or the use of CTAB selectively blocked side walls of gold nanorods, and led to the preferential deposition of platinum atoms at the tip regions (Figure 6.9(c)).[142]

Gold deposition on heterometal seeds is plausible only for the metals with similarly high reduction potentials, such as platinum and palladium; otherwise galvanic replacement reaction occurs (see Section 6.5.2.3). The overgrowth of gold on platinum cubes as large as 18 nm led to the formation of gold nanorods with $\sim$185 nm length and $\sim$25 nm diameter, because the

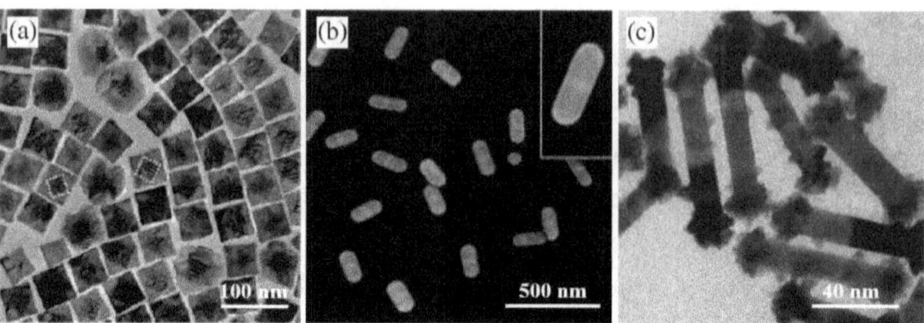

Figure 6.9   Gold–metal bimetallic nanoparticles synthesised via seed-mediated growth. (a) TEM image of Au–Ag core–shell nanopartices, (b) SEM image of Ag–Au–Ag heterometal nanorods, and (c) TEM image of Pt-tipped Au nanorods. (a) Ref. 145. Copyright 2008, American Chemical Society; (b) Ref. 116. Copyright 2008, American Chemical Society; (c) Ref. 142. Copyright 2007, American Chemical Society.

lattice mismatch with the platinum surface induced twinning defects.[143] Sun *et al.* reported that gold branches were grown on 3–7 nm platinum seeds forming pear-, peanut-, and clover-like AuPt heteronanoparticles, of which final morphology was dependent upon the size of seeding nanoparticles and solvent polarity.[144] Precise adjustment of the addition speed of the gold precursor also induced an anisotropic growth of gold domains on the surface of palladium nanocubes.[137,138] Although the deposition of gold on silver without galvanic replacement was not plausible, Qin *et al.* developed a strategy of galvanic replacement-free deposition through the concomitant addition of a strong reductant and successfully generated ultrathin gold shells of 0.6 nm thick on the surface of silver nanocubes.[130]

6.5.2.3 *Galvanic replacement*

The galvanic replacement reaction is a general and effective method for preparing metallic nanostructures by consuming more reactive components without additional reducing agents. Since the standard reduction potential of the gold precursor is higher than those of other metals such as silver and cobalt (Table 6.5), the gold precursors are reduced to Au(0) while the other metals are oxidised. When the Au(III) precursors galvanically substitute Ag(0) nanostructures, vacancies are generated inside the structures to form either hollow or shell morphologies. Using this strategy, Xia *et al.* synthesised gold nanoboxes and nanotubes from silver nanocubes and nanowires in an aqueous phase (Figure 6.10(a)).[163,164] The exterior

Figure 6.10 (a) SEM image of gold nanoboxes, and TEM images of (b) AuAg double-walled nanoboxes and (c) AuAgPd heterogeneous metal nanoparticles. (a) Ref. 164. Copyright 2007, Nature Publishing Group; (b) Modified with permission from Ref. 167. Copyright 2011, American Association of the Advancement of Science; (c) Modified with permission from Ref. 169. Copyright 2013, American Chemical Society.

shapes of the final structures tend to reproduce those of the sacrificial silver counterparts without any change to their overall morphology. Alivisatos *et al.* also demonstrated the same chemical transformation to prepare hollow structures of gold from silver particles of ~10 nm size in an organic phase.[165] Partial galvanic replacement reactions were induced by careful control of the reaction kinetics, and provided anisotropic single hollows as well as symmetric double hollows from silver–gold–silver heterometal nanorods.[166]

Galvanic replacement can couple with other nanoscale reactions to provide various complex nanostructures. Puntes *et al.* developed a route for the production of complex gold–silver hollow structures from silver nanocubes by the combination of galvanic replacement and the Kirkendall effect, which results in the formation of pores due to the difference in diffusion rates between two components (Figure 6.10(b)).[167] The addition of gold salts to palladium nanorods led to the galvanic reaction, but the deposition of the gold atoms was selectively localised to the tips, forming a Pd–Au tadpole structure.[168] Selective UPD of silver on gold octahedrons and sequential galvanic replacement with palladium could generate a large variety of complex heterogeneous metallic nanostructures (Figure 6.10(c)).[169]

6.6 Conclusion

In this chapter, we briefly reviewed the synthesis and morphology control of gold and gold–metal bimetallic nanoparticles in liquid phase. Since chemical and physical properties of gold nanoparticles critically depend on not only their size but also their shape and surface structure, synthesis of monodisperse nanoparticles is highly demanding for the improvement of their performances in versatile applications. Although precise control of morphology, sizes and surface facets have been a major topic with respect to the preparation of gold nanoparticles in liquid phase over the past decade, careful design of nanostructures with consideration of a variety of factors is still necessary to develop specific applications. Because of their outstanding properties, gold nanoparticles can find application in fields in a wide range of areas, including chemical or biological sensors, imaging, medical materials, catalysis for organic chemical synthesis and energy generation, and due to technological development, the scope will continue to broaden. Consequently, easy, rapid and large-scale synthesis, and diverse and stable surface treatments of gold nanoparticles are still required. In addition to the formation of bimetallic nanostructures, hybridisation with other materials, such as semiconductors, polymers and biomolecules, with well-designed morphology and properties is another emerging topic being explored to realise the full use of unique physical and chemical properties in gold nanoparticles.

References

1. Y. Xia and N. J. Halas, *Mat. Res. Soc. Bull.* **30**, 338 (2005).
2. C. J. Murphy, *Science* **298**, 2139 (2002).
3. P. V. Kamat, *J. Phys. Chem. B* **106**, 7729 (2002).
4. M. A. El-Sayed, *Acc. Chem. Res.* **34**, 257 (2001).
5. J. J. Mock, M. Barbic, D. R. Smith, D. A. Schultz and S. Schultz, *J. Chem. Phys.* **116**, 6755 (2002).
6. P. D. Jadzinsky, G. Calero, C. J. Ackerson, D. A. Bushnell and R. D. Kornberg, *Science* **318**, 430 (2007).
7. J. Zhao, J. Yang and J. G. Hou, *Phys. Rev. B* **67**, 85404 (2003).
8. J. Li, X. Li, H.-J. Zhai and L.-S. Wang, *Science* **299**, 964 (2003).
9. P. Pyykko, *Angew. Chem. Int. Ed.* **43**, 4412 (2004).
10. H.-G. Boyen, G. Kastle, F. Eeigl, B. Koslowski, C. Dietrich, P. Ziemann, J. P. Spatz, S. Riethmuller, C. Hartmann, M. Moller, G. Schmid, M. G. Garnier and P. Oelhafen, *Science* **297**, 1533 (2002).
11. P. Buffat and J.-P. Borel, *Phys. Rev. A* **13**, 2287 (1976).
12. K. Dick, T. Dhanasekaran, Z. Zhang and D. Meisel, *J. Am. Chem. Soc.* **124**, 2312 (2002).

13. M. Brust, J. Fink, D. Bethell, D. J. Schiffrin and C. Kiely, *J. Chem. Soc., Chem. Commun.* 1665–1666 (1995).
14. N. R. Jana, L. Gearheart and C. J. Murphy, *Langmuir* **17**, 6782 (2001).
15. D. G. Duff, A. Baiker and P. Edwards, *J. Chem. Soc., Chem. Commun.* 96 (1993).
16. F. Fievet, J. P. Lagier, B. Blin, B. Beaudoin and M. Figlarz, *Solid State Ionics* **32**, 198 (1989).
17. J. Turkevich, P. C. Stevenson and J. Hillier, *Discuss. Faraday. Soc.* **11**, 55 (1951).
18. P. Vanysek, In *CRC Handbook of Chemistry and Physics*, 86th ed., D. R. Lide (ed.), CRC Press, Taylor & Francis, Boca Raton, FL, Vol. 8, pp. 8–20 (2005).
19. M. J. Yacamán, J. A. Ascencio, H. B. Liu and J. Gardea-Torresdey, *J. Vac. Sci. Technol. B* **19**, 1091 (2001).
20. C. L. Cleveland and U. Landman, *J. Chem. Phys.* **94**, 7376 (1991).
21. Z. L. Wang, *J. Phys. Chem. B* **104**, 1153 (2000).
22. A. S. Barnard and L. A. Curtiss, *ChemPhysChem* **7**, 1544 (2006).
23. K. L. Kelly, E. Coronado, L. L. Zhao and G. C. Schatz, *J. Phys. Chem. B* **107**, 668 (2003).
24. W.-H. Yang, G. C. Schatz and R. P. Van Duyne, *J. Chem. Phys.* **103**, 869 (1995).
25. B. T. Draine and P. J. Flatau, *J. Opt. Soc. Am. A* **11**, 1491 (1994).
26. L. Novotny, R. X. Bian and X. S. Xie, *Phys. Rev. Lett.* **79**, 645 (1997).
27. R. X. Bian, R. C. Dunn, X. S. Xie and P. T. Leung, *Phys. Rev. Lett.* **75**, 4772 (1995).
28. J. T. G. Overbeek, *Adv. Colloid Interface Sci.* **15**, 251 (1982).
29. V. K. LaMer and R. H. Dinegar, *J. Am. Chem. Soc.* **72**, 4847 (1950).
30. J. Park, J. Joo, S. G. Kwon, Y. Jang and T. Hyeon, *Angew. Chem. Int. Ed.* **46**, 4630 (2007).
31. H. Yu, P. C. Gibbons, K. F. Kelton and W. E. Buhro, *J. Am. Chem. Soc.* **123**, 9198 (2001).
32. C. B. Murray, D. J. Norris and M. G. Bawendi, *J. Am. Chem. Soc.* **115**, 8706 (1993).
33. N. R. Jana and X. Peng, *J. Am. Chem. Soc.* **125**, 14280 (2003).
34. II. Rciss, *J. Chem. Phys.* **19**, 482 (1951).
35. D. V. Talapin, A. L. Rogach, M. Haase and H. Weller, *J. Phys. Chem. B* **105**, 12278 (2001).
36. A. L. Smith, *Particle Growth in Suspensions* (Academic Press, London, 1983, pp. 3–15).
37. D. Seo, J. C. Park and H. Song, *J. Am. Chem. Soc.* **128**, 14863 (2006).
38. G. Frens, *Colloid Polym. Sci.* **250**, 736 (1972).
39. G. Frens, *Nature (London), Phys. Sci.* **241**, 20 (1973).
40. M. K. Chow and C. F. Zukoski, *J. Colloid Interface Sci.* **165**, 97 (1994).
41. H. Xia, S. Bai, J. Hartmann and D. Wang, *Langmuir* **26**, 3585 (2001).
42. B.-K. Pong, H. I. Elim, J.-H. Chong, W. Ji, B. L. Trout and J.-Y. Lee, *J. Phys. Chem. C* **111**, 6281 (2007).
43. S. D. Perrault and W. C. W. Chan, *J. Am. Chem. Soc.* **131**, 17042 (2009).
44. M. Giersig and P. Mulvaney, *Langmuir* **9**, 3408 (1993).
45. M. Brust, M. Walker, D. Bethell, D. J. Schiffrin and R. Whyman, *J. Chem. Soc., Chem. Commun.* 801 (1994).

46. M. J. Hostetler, S. J. Green, J. J. Stokes and R. W. Murray, *J. Am. Chem. Soc.* **119**, 4212 (1996).

47. R. S. Ingram, M. J. Hostetler and R. W. Murray, *J. Am. Chem. Soc.* **118**, 4212 (1996).

48. A. C. Templeton, W. P. Wuelfing and R. W. Murray, *Acc. Chem. Res.* **33**, 27 (2000).

49. K. V. Sarathy, G. Raina, R. T. Yadav, G. U. Kulkarni and C. N. R. Rao, *J. Phys. Chem. B* **101**, 9876 (1997).

50. M. C. Daniel and D. Astruc, *Chem. Rev.* **104**, 293 (2004).

51. M. Zhu, E. Lanni, N. Garg, M. E. Bier and R. Jin, *J. Am. Chem. Soc.* **130**, 1138 (2008).

52. P. J. G. Goulet and R. B. Lennox, *J. Am. Chem. Soc.* **132**, 9582 (2010).

53. N. R. Jana and X. Peng, *J. Am. Chem. Soc.* **125**, 14280 (2003).

54. G. Schmid, *Chem. Soc. Rev.* **37**, 1909 (2008).

55. G. Riess, *Prog. Polym. Sci.* **28**, 1107 (2003).

56. R. K. Oreilly, M. J. Joralemon, C. J. Hawker and K. L. Wooley, *J. Polym. Sci. Part A: Polym. Chem.* **44**, 5203 (2006).

57. S. Abraham, C. S. Ha and I. Kim, *J. Polm. Sci. Part A: Polym. Chem.* **43**, 6367 (2005).

58. Y. Kang and T. A. Taton, *Angew. Chem. Int. Ed.* **44**, 409 (2005).

59. Y. Kang, K. J. Erickson and T. A. Taton, *J. Am. Chem. Soc.* **127**, 13800 (2005).

60. G. Carrot, J. C. Valmalette, C. J. G. Plummer, S. M. Scholz, J. Dutta, H. Hofmann and J. G. Hilborn, *Colloid Polym. Sci.* **276**, 853 (1998).

61. A. B. Lowe, B. S. Sumerlin, M. S. Donovan and C. L. McCormick, *J. Am. Chem. Soc.* **124**, 11562 (2002).

62. M. Filali, M. A. R. Meier, U. S. Schubert and J. F. Gohy, *Langmuir* **21**, 7995 (2005).

63. N. Zheng, J. Fan and G. D. Stucky, *J. Am. Chem. Soc.* **128**, 6550 (2006).

64. F. Kim, S. Connor, H. Song, T. Kuykendall and P. Yang, *Angew. Chem. Int. Ed.* **43**, 3673 (2004).

65. D. Seo, C. I. Yoo, J. C. Park, S. M. Park, S. Ryu and H. Song, *Angew. Chem. Int. Ed.* **47**, 763 (2008).

66. D. Seo, C. I. Yoo, I. S. Chung, S. M. Park, S. Ryu and H. Song, *J. Phys. Chem. C* **112**, 2469 (2008).

67. D. Seo, J. H. Park, J. Jung, S. M. Park, S. Ryu, J. Kwak and H. Song, *J. Phys. Chem. C* **113**, 3449 (2009).

68. M. Y. Han and C. H. Quek, *Langmuir* **16**, 362 (2000).

69. K. Mallick, Z. L. Wang and T. Pal, *J. Photochem. Photobiol.* **140**, 75 (2001).

70. J. Westerhausen, A. Henglein and J. Lilie, *Ber. Bunsen-Ges. Phys. Chem.* **85**, 182 (1981).

71. A. Henglein, *Langmuir* **15**, 6738 (1999).

72. E. Gachard, H. Remita, J. Khatouri, B. Keita, L. Nadjo and J. Belloni, *New J. Chem.* **22**, 1257 (1998).

73. A. Henglein and D. Miesel, *Langmuir* **14**, 7392 (1998).

74. Y.-Y. Yu, S.-S. Chang, C.-L. Lee and C. R. C. Wang, *J. Phys. Chem. B* **101**, 6661 (1997).

75. G.-T. Wei, F.-K. Liu and C. R. C. Wang, *Anal. Chem.* **71**, 2085 (1999).

76. B. M. I. van der Zande, M. R. Bohmer, L. G. J. Fokkink and C. Schonenberger, *J. Phys. Chem. B* **101**, 852 (1997).

77. B. M. I. van der Zande, M. R. Bohmer, L. G. J. Fokkink and C. Schonenberger, *Langmuir* **16**, 451 (2000).

78. Y. Nagata, Y. Mizukoshi, K. Okitsu and Y. Maeda, *Radiat. Res.* **146**, 333 (1996).

79. K. Okitsu, A. Yue, S. Tanaba, H. Matsumoto and Y. Yobiko, *Langmuir* **17**, 7717 (2001).

80. R. A. Caruso, M. Ashokkumar and F. Grieser, *Langmuir* **18**, 7381 (2002).

81. K. Okitsu, M. Ashokkumar, F. Grieser, *J. Phys. Chem. B* **109**, 20673 (1005).

82. C.-H. Su, P.-L. Wu and C.-S. Yeh, *J. Phys. Chem. B* **107**, 14240 (2003).

83. M. Tsuji, M. Hashimoto, Y. Nishizawa, M. Kubokawa and T. Tsuji, *Chem. Eur. J.* **11**, 440 (2005).

84. Y. Jiang and Y.-J. Zhu, *Chem. Lett.* **33**, 1390 (2004).

85. F.-K. Liu, C.-J. Ker, Y.-C. Chang, F.-H. Ko, T.-C. Chu and B.-T. Dai, *Jpn. J. Appl. Phys., Part 1* **42**, 4152 (2003).

86. M. Tsuji, N. Miyamae, M. Hashimoto, M. Nishio, S. Hikino, N. Ishigami and I. Tanaka, *Coll. Surf. A* **302**, 587 (2007).

87. K. L. Kelly, E. Coronado, L. L. Zhao and G. C. Schatz, *J. Phys. Chem. B* **107**, 668 (2003).

88. G. C. Schatz and R. P. Van Duyne. In *Handbook of Vibrational Spectroscopy*, J. M. Chalmers and P. R. Griffiths (eds.), John Wiley, Chichester, p. 1 (2002).

89. Y. Sun and Y. Xia, *Anal. Chem.* **74**, 5297 (2002).

90. T. Rindzevicius, Y. Alaverdyan, A. Dahlin, F. Hook, D. S. Sutherland and M. Kall, *Nano Lett.* **5**, 2335 (2005).

91. O. C. Compton and F. E. Osterloh, *J. Am. Chem. Soc.* **129**, 7793 (2007).

92. A. Tao, P. Sinsermsuksakul and P. Yang, *Nat. Nanotech.* **2**, 435 (2007).

93. Y. Xia, P. Yang, Y. Sun, Y. Wu, B. Mayers, B. Gates, Y. Yin, F. Kim and H. Yan, *Adv. Mat.* **15**, 353 (2003).

94. X. Peng, J. Wickham and A. P. Alivisatos *J. Am. Chem. Soc.* **120**, 5343 (1998).

95. C. J. Murphy, L. B. Thompson, A. M. Alkilany, P. N. Sisco, S. P. Boulos, S. T. Sivapalan, J. A. Yang, D. J. Chernak and J. Huang, *J. Phys. Chem. Lett.* **1**, 2867 (2010).

96. C. J. Murphy, T. K. Sau, A. M. Gole, C. J. Orendorff, J. Gao, L. Gou, S. E. Hunyadi and T. Li, *J. Phys. Chem. B* **109**, 13857 (2005).

97. C. J. Johnson, E. Dujardin, S. A. Davis, C. J. Murphy and S. Mann, *J. Mat. Chem.* **12**, 1765 (2002).

98. K. Sohn, F. Kim, K. C. Pradel, J. Wu, Y. Peng, F. Zhou and J. Huang, *ACS Nano* **3**, 2191 (2009).

99. S.-Y. Wu, W.-L. Huang and M. H. Huang, *Cryst. Growth Des.* **7**, 831 (2007).

100. J. E. Millstone, S. Park, K. L. Shuford, L. Qin, G. C. Schatz and C. A. Mirkin, *J. Am. Chem. Soc.* **127**, 5312 (2005).

101. H.-L. Wu, C.-H. Chen and H. Huang, *Chem. Mater.* **21**, 110 (2009).

102. S. Chen, Z. L. Wang, J. Ballato, S. H. Foulger and D. L. Carroll, *J. Am. Chem. Soc.* **125**, 16186 (2003).

103. T. K. Sau and C. J. Murphy, *J. Am. Chem. Soc.* **126**, 8648 (2004).

104. D. Y. Kim, T. Yu, E. C. Cho, Y. Ma, O. O. Park and Y. Xia, *Angew. Chem. Int. Ed.* **50**, 6328 (2011).

105. W. Niu, S. Zheng, D. Wang, X. Liu, H. Li, S. Han, J. Chen, Z. Tang and G. Xu, *J. Am. Chem. Soc.* **131**, 697 (2009).

106. T. Ming, W. Feng, Q. Tang, F. Wang, L. Sun, J. Wang and C. Yan, *J. Am. Chem. Soc.* **131**, 16350 (2009).

107. H. Zhang, M. Jin and Y. Xia, *Angew. Chem. Int. Ed.* **51**, 7656 (2012).

108. J. Zhang, M. R. Langille, M. L. Personick, K. Zhang, S. Li and C. A. Mirkin, *J. Am. Chem. Soc.* **132**, 14012 (2010).

109. Y. Ma, Q. Kuang, Z. Jiang, Z. Xie, R. Huang and L. Zheng, *Angew. Chem. Int. Ed.* **120**, 9033 (2008).

110. Y. Yi, Q. Zhang, X. Lu and J. Y. Lee, *J. Phys. Chem. C* **114**, 11119 (2010).

111. Y. Ma, J. Zeng, W. Li, M. Mckiernan, Z. Xie and Y. Xia, *Adv. Mater.* **22**, 1930 (2010).

112. X. Lu, M. S. Yavus, H.-Y. Tuan, B. A. Korge and Y. Xia, *J. Am. Chem. Soc.* **130**, 8900 (2008).

113. C. Wang, Y. Hu, C. M. Lieber and S. Sun, *J. Am. Chem. Soc.* **130**, 8902 (2008).

114. Z. Huo, C.-k. Tsung, W. Huang, X. Zhang and P. Yang, *Nano Lett.* **8**, 2041 (2008).

115. C. J. Orendorff and C. J. Murphy, *J. Phys. Chem. B* **110**, 3990 (2006).

116. D. Seo, C. I. Yoo, J. Jung and H. Song, *J. Am. Chem. Soc.* **130**, 2940 (2008).

117. M. C. Gimenez, M. G. Del Popolo, E. P. M. Leiva, S. G. Garcıa, D. R. Salinas, C. E. Mayer and W. J. Lorenz, *J. Electochem. Soc.* **149**, E109 (2002).

118. S. Garcia, D. Salinas, C. Mayer, E. Schmidt, G. Staikov and W. J. Lorenz, *Electrochim. Acta.* **43**, 3007 (1998).

119. A. Kuzume, E. Herrero, J. M. Feliu, R. J. Nichols and D. J. Schiffrin, *J. Electroanal. Chem.* **570**, 157 (2004).

120. M. C. Gimenez, M. G. Del Popolo and E. P. M. Leiva, *Electrochim Acta* **45**, 699 (1999).

121. L. B. Rogers, D. P. Krause, J. C. Griess, Jr and D. B. Ehrlinger, *J. Electrochem. Soc.* **95**, 33 (1949).

122. M. C. Giménez, M. G. Del Pópolo, E. P. M. Leiva, S. G. García, D. R. Salinas, C. E. Mayer and W. J. Lorenz, *J. Electrochem. Soc.* **149**, E109 (2002).

123. S. J. Hurst, E. K. Payne, L. Qin and C. A. Mirkin, *Angew. Chem. Int. Ed.* **45**, 2672 (2006).

124. E. C. Water, B. J. Murray, F. Favier, G. Kaltenpoth, M. Grunze and R. M. Penner, *J. Phys. Chem. B* **106**, 11407 (2002).

125. C. R. Martin, *Chem. Mater.* **8**, 1739 (1996).

126. R. Ferrando, J. Jellinek and R. L. Johnston, *Chem. Rev.* **108**, 845 (2008).

127. D. Wang and Y. Li, *Adv. Mater.* **23**, 1044 (2011).

128. M. B. Cortie and A. M. McDonagh, *Chem. Rev.* **111**, 3713 (2011).

129. F. Bao, J. Li, B. Ren, J.-L. Yao, R.-A. Gu and Z.-Q. Tian, *J. Phys. Chem. C.* **112**, 345 (2008).

130. Y. Yang, J. Liu, Z.-W. Fu and D. Qin, *J. Am. Chem. Soc.* **136**, 8153 (2014).

131. V. Abdelsayed, A. Aljarash, M. S. El-Shall, Z. A. Al Othman and A. H. Alghamdi, *Chem. Mater.* **21**, 2825 (2009).

132. J. C. Bauer, D. Mullins, M. Li, Z. Wu, E. A. Payzant, S. H. Overbury and S. Dai, *Phys. Chem. Chem. Phys.* **13**, 2571 (2011).

133. C. Wang, H. Yin, R. Chan, S. Peng, S. Dai and S. Sun, *Chem. Mater.* **21**, 433 (2009).

134. M. P. Mallin and C. J. Murphy, *Nano Lett.* **2**, 10 (2002).

135. M. J. Hostetler, C. J. Zhong, B. K. H. Yen, J. Anderegg, S. M. Gross, N. D. Evans, M. Porter and R. W. Murray, *J. Am. Chem. Soc.* **120**, 9396 (1998).

136. Y. Yang, W. Wang, X. Li, W. Chen, N. Fan, C. Zou, X. Chen, X. Xu, L. Zhang and S. Huang, *Chem. Mater.* **25**, 34 (2013).

137. C. Zhu, J. Zeng, J. Tao, M. C. Johnson, I. Schmidt-Krey, L. Blubaugh, Y. Zhu, Z. Gu and Y. Xia, *J. Am. Chem. Soc.* **134**, 15822 (2012).

138. S.-U. Lee, J. W. Hong, S.-I. Choi and S. W. Han, *J. Am. Chem. Soc.* **6**, 5221 (2014).

139. G. Park, D. Seo, J. Jung, S. Ryu and H. Song, *J. Phys. Chem. C.* **115**, 9417 (2011).

140. Y. Ma, W. Li, E. C. Cho, Z. Li, T. Yu, J. Zeng, Z. Xie and Y. Xia, *ACS Nano* **4**, 6725 (2010).

141. M. Tsuji, N. Miyamae, S. Lim, K. Kimura, X. Zhang, S. Hikino and M. Nishio, *Cryst. Growth Des.* **6**, 1801 (2006).

142. M. Grzelczak, J. Pérez-Juste, F. J. Garcíade Abajo and L. M. Liz-Marzán, *J. Phys. Chem. C.* **111**, 6183 (2007).

143. S. E. Habas, H. Lee, V. Radmilovic, G. A. Somorjai and P. Yang, *Nat. Mater.* **6**, 692 (2007).

144. C. Wang, W. Tian, Y. Ding, Y. Q. Ma, Z. L. Wang, N. M. Markovic, V. R. Stamenkovic, H. Daimon and S. Sun, *J. Am. Chem. Soc.* **132**, 6524 (2010).

145. F. R. Fan, D. Y. Liu, Y. Wu, S. Duan, Z. X. Xie, Z. Y. Jiang and Z. Q. Tian, *J. Am. Chem. Soc.* **130**, 6949 (2008).

146. X. Xia, Y. Wang, A. Ruditskiy and Y. Xia, *Adv. Mater.* **25**, 6313 (2013).

147. R. G. Chaudhuri and S. Paria, *Chem. Rev.* **112**, 2373 (2011).

148. L. Vitos, A. V. Ruban, H. L. Skriver and J. Kollár, *Surf. Sci.* **411**, 186 (1998).

149. C. Gao, Y. Hu, M. Wang, M. Chi and Y. Yin, *J. Am. Chem. Soc.* **136**, 7474 (2014).

150. R. E. Schaak, A. K. Sra, B. M. Leonard, R. E. Cable, J. C. Bauer, Y. F. Han, J. Means, W. Teizer, Y. Vasquez and E. S. Funck, *J. Am. Chem. Soc.* **127**, 3506 (2005).

151. A. K. Sra and R. E. Schaak, *J. Am. Chem. Soc.* **126**, 6667 (2004).

152. S. Zhou, G. S. Jackson and B. Eichhorn, *Adv. Funct. Mater.* **17**, 3099 (2007).

153. J. F. Bondi, R. Misra, X. Ke, I. T. Sines, P. Schiffer and R. E. Schaak, *Chem. Mater.* **22**, 3988 (2010).

154. M. Sakamoto, T. Tachikawa, M. Fujitsuka and T. Majima, *Adv. Funct. Mater.* **17**, 857 (2007).

155. F. Ksar, L. Ramos, B. Keita, L. Nadjo, P. Beaunier and H. Remita, *Chem. Mater.* **21**, 3677 (2009).

156. J. Belloni, M. Mostafavi, H. Remita, J.-L. Marignier and M.-O. Delcourt, *New J. Chem.* **22**, 1239 (1998).

157. H. Remita, A. Etcheberry and J. Belloni, *J. Phys. Chem. B.* **107**, 31 (2003).

158. T. Akita, T. Hiroki, S. Tanaka, T. Kojima, M. Kohyama, A. Iwase and F. Hori, *Catal. Today* **131**, 90 (2008).

159. Y. Yu, Q. Zhang, B. Liu and J. Y. Lee, *J. Am. Chem. Soc.* **132**, 18258 (2010).

160. E. C. Cho, P. H. C. Camargo and Y. Xia, *Adv. Mater.* **22**, 744 (2010).

161. M. R. Langile, J. Zhang, M. L. Personick, S. Li and C. A. Mirkin, *Science* **337**, 954 (2012).

162. Y. Kim, J. W. Hong, Y. W. Lee, M. Kim, D. Kim, W. S. Yun and S. W. Han, *Angew. Chem. Int. Ed.* **122**, 10395 (2010).
163. Y. Sun, B. Wiley, Z.-Y. Li and Y. Xia, *J. Am. Chem. Soc.* **126**, 9399 (2004).
164. S. E. Skrabalak, L. Au, X. Li and Y. Xia, *Nature Protocols* **9**, 2182 (2007).
165. Y. Yin, C. Erdonmez, S. Aloni and A. P. Alivisatos, *J. Am. Chem. Soc.* **128**, 12671 (2006).
166. D. Seo and H. Song, *J. Am. Chem. Soc.* **131**, 18210 (2009).
167. E. Gonzalez, J. Arbiol and V. F. Puntes, *Science* **334**, 1377 (2011).
168. P. H. C. Camargo, Y. Xiong, L. Ji, J. M. Zuo and Y. Xia, *J. Am. Chem. Soc.* **129**, 15452 (2007).
169. Y. Yu, Q. Zhang, Q. Yao, J. Xie and J. Y. Lee, *Chem. Mater.* **25**, 4746 (2013).

Chapter 7

Functionalisation of Gold Nanoparticles

Souhir Boujday[*,†], Atul N. Parikh[†,‡], Bo Liedberg[†] and Hyunjoon Song[§]

*Laboratoire de Réactivité de Surface, Université Pierre et Marie Curie — CNRS, Paris, France
[†]Center for Biomimetic Sensor Science,
School of Materials Science and Engineering, Singapore
[‡]Departments of Biomedical Engineering and Materials Science,
University of California, Davis, California, USA
[§]Department of Chemistry, Korea Advanced
Institute of Science and Technology, Daejeon, Republic of Korea

7.1 Introduction

Geometry dictates that the fraction of atoms at the surface of a bulk consolidated matter scales with area per volume, increasing with the reciprocal of the linear dimension, i.e. radius. As a consequence, with diminishing dimension of the bulk, in the limit of nanoparticulate matter, surface assumes a dominant role — becoming key determinant of the physical-chemical properties of the nanoparticle (e.g. solubility, stability and melting point) and, together with quantum size effects, producing emergent, size-dependent effects (e.g. electronic structure, photoluminescence and plasmonics).[1] Moreover, during the synthesis of nanoparticulate matter itself, the rapidly growing surface plays critical roles in determining size, shape and morphology of the incipient nanoparticle.[2] Thus, a control over surface physical and chemical properties of a nanoparticle is of significant fundamental and practical interest.

Surface differs from the bulk in that the atoms at the surface have fewer direct neighbours than those in the bulk. Thus, the nanoparticle, with a preponderant surface, becomes characterised by low mean coordination number and a population of unsaturated 'dangling bonds'. Saturating these dangling bonds, such as through molecularly defined surface functionalisation or 'capping', provides a means to both control the kinetic processes of nucleation and growth during synthesis and the resulting attributes, such as physical properties, chemical reactivities and biological interactions, of the nanoparticle.

For the specific case of gold nanoparticles,[3,4] this dual importance of surface functionalisation — both during synthesis and in determining the material properties of the nanoparticulate matter — is abundantly clear. Commonly synthesised by the so-called Turkevich synthesis[5] — which involves the reduction of Au ions using reducing agents such as citrate — gold nanoparticles form through a classical nucleation and growth process: The reduction of Au ions seed the nucleation sites, which grow through a rather complex series of agglomeration reactions. The growth process is terminated by the citrate ($Na_3C_6H_5O_7$), which, acting as a surface functional layer or a capping agent, controls the nanoparticle size and morphology, and acts to prevent inter-particle aggregation through electrostatic repulsion.

Although the citrate capping agent is well-suited as a native functional layer during the nanoparticle synthesis, substituting citrate, which is only bound loosely to the nanoparticle surface through weak van der Waals interactions, with a more tightly bound and well-defined capping layer (e.g. mercaptoderivatives on gold) opens door to endowing nanoparticles with a broad range of surface functionalities.[6] In what follows, we present (1) the geometric consideration of gold nanoparticles in determining structure and packing of the functional layer: why does the size matter? (2) major strategies for chemical derivatisation using organic moieties; (3) silica capping of the nanoparticles and (4) biomolecular functionalisation. The references and attributions are far from comprehensive, but should act as a guide to the vast literature focused on these topics.

7.2 Geometric Considerations: Why Does the Size Matter?

As discussed above, the large surface-to-volume ratio of gold nanoparticles makes their surfaces critically important in controlling the vast majority of their chemical and physical properties. Surfaces can be seen as a huge defect in nanocrystals. As stated by Wolfgang Pauli, 'God made the bulk; surfaces were invented by the devil.' Therefore, the dominant role of surface phenomena in nanoparticles gives rise to complex behaviour and emergent properties. For gold nanoparticles, plasmonic properties are perhaps the most extensively studied ones, together with size-related catalytic properties of gold. However, two geometric parameters, which are a common feature shared by all spherical nanoparticles, need to be considered when it comes to surface functionalisation or biofunctionalisation: the arrangement and coordination chemistry of surface atoms, and the curvature of nanoparticles. Both parameters are dependent on the surface-to-volume ratio and scale inversely with particle size. We don't aim here at addressing a complete analysis of the geometric parameters influencing the structure and properties of nanoscale interfaces; we refer to recent reviews in the field (references 7 and 8 and references therein) and to Chapter 13. Here, we simply intend to consider the above-mentioned geometric aspects from the perspective of biofunctionalisation: how nanoparticle dimension affects surface (bio)functionalisation?

7.2.1 Coordination and Arrangement of Surface Atoms

Though commonly thought as 'spherical', gold nanoparticles often have a cuboctahedral shape, exhibiting mainly (111) and (100) facets with hexagonal and square faces[9] (see Figure 7.1). When the diameter of nanoparticle is large, these facets can be seen as contiguous nanometre scale domains, upon which the adsorption of alkanethiols would lead to miniature self-assembled monolayers (SAMs).[7,10] The extensive effort on planar gold surfaces in the self-assembled monolayers field (Ref. 11) and references therein) can be extrapolated to nanoparticle surfaces. This is well documented in a number of excellent reports, which establish a direct link between planar SAMs and miniature SAM on gold nanoparticles 10. Moreover, because the nanoparticles are faceted with distinct (111) and

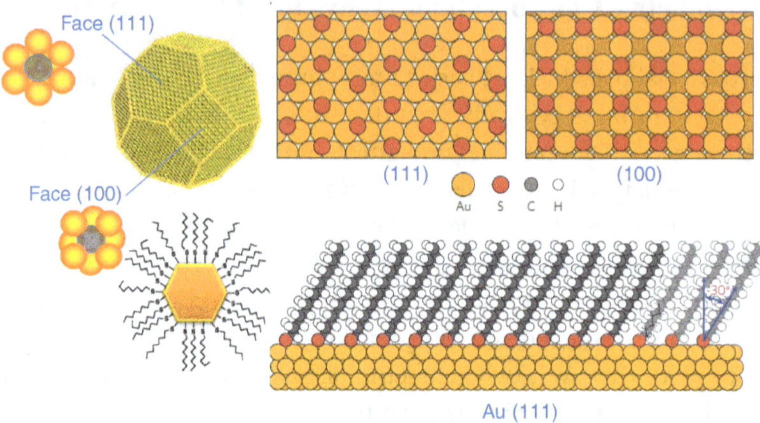

Figure 7.1   Left top: Cuboctahedral shape gold nanoparticles exhibiting Au (111) and (100) facets.[9] Bottom: Schematic representation of *n*-alkanethiolate monolayers on gold cluster illustrating the types of defects observed.[14] Right top: The binding pattern of sulphur headgroups (red spheres) on Au (111) and (100) facets. Right bottom: Sketch of self-assembled monolayer (SAM) of hexadecanethiol molecules adsorbed on Au(111) surface as an idealised picture of a nanocrystal capping layer. Hydrocarbon chains are fully extended, tilted with respect to the surface normal, and in all-*trans* configuration. Adapted with permission from Refs. 12 and 7.

(100) single crystal faces, and because thiols assemble differently on these crystallographic textures,[12,13] it is likely that thiol-based functionalisation is facet-dependent, establishing which has remained challenging because of characterisation challenges. This facet-dependent differences in molecular organisation notwithstanding, geometric constraints themselves give rise to significant differences in the structures of SAMs at nanoparticle surfaces. The first point is that these miniature SAMs can be seen as 3D-SAMs compared to the 2D-SAMs on planar surfaces.[14] The main consequence is that the packing density, defined as the percentage of surface coverage, of alkanethiolates bonded to the gold cluster surface (62–68%) is nearly twice that found on planar Au(111) (33%). In addition, despite considering the large number of different gold sites on the nanoparticles surface, the concentration of near surface defects on the 3D-SAMs is similar to that in 2D-SAMs. The second major difference is due to the radius of curvature that increases when the particle size decreases. One consequence on alkanethiol adsorption is that the units of the alkane chains become progressively less densely packed when moving farther from the core. Therefore,

the outermost functional groups exhibit greater liberty of movement, which may explain the highest coverage. As the size decreases, the concentration of sub-coordinated atoms or surface defect-like sites (corner, edges) increases. These atoms are very reactive and react with thiols upon adsorption, contributing additional issues in packing of the thiolate functional layer.

7.2.2 *Particle Curvature Influence*

Surface curvature scales inversely with the particle size: The maximum is obtained for the smallest nanoparticle and with increasing particle diameter; the curvature effects diminish adopting 2D planar structure beyond a certain threshold. Besides the effects discussed above on atom coordination and 3D-SAMs, curvature strongly influences, at the macroscopic level, especially when the functionalisation layer comprises macromolecules and proteins. Specifically, conformational changes are closely related to curvature. By combining several characterisation techniques, including NMR and circular dichroism, Lundqvist *et al.* could establish this correlation upon adsorption of human carbonic anhydrase I (HCAI) on silica nanoparticles.[15] The three particle sizes studied, i.e. 6, 9 and 15 nm, were relatively small; nevertheless their effect on the protein secondary structure, the 3D form of local segments of proteins, was different; a conformational change was observed on the biggest particles (lower curvature) while proteins retained their conformation on small ones (Figure 7.2).

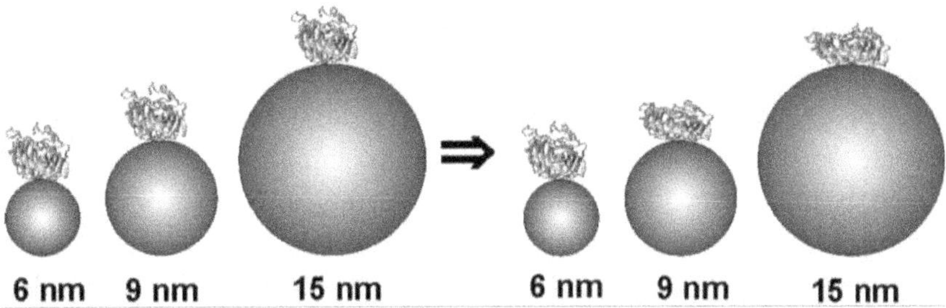

Figure 7.2   Left: Differences in size between HCAI (represented by a ribbon structure) and the different particles (represented by grey spheres) in a 2D plane. Right: Schematic illustration of the effect of the particle curvatures on the protein's secondary structure. From Ref. 15.

A similar trend was observed for a larger diameter of nanoparticles, upon the interaction of bovin serum albumin (BSA) with hydrophobic polystyrene nanoparticles and upon human serum albumin (HSA) interaction with gold nanoparticles.[16,17] The influence of surface curvature was probed by electrophoretic light scattering and UV-visible spectroscopy for BSA adsorption on polystyrene nanoparticles of 39, 82 and 220 nm in diameter.[16] BSA molecules retained their native conformation on the smallest nanoparticles and were more conformationally disturbed on the biggest ones. All the same, HSA interaction with gold nanoparticles from 5 to 100 nm in size was discussed on the basis of ζ-potential data and FTIR spectroscopy.[17] Differences in the conformations of the adsorbed protein molecules onto small (<40 nm) and large nanoparticles were observed; this effect was enhanced as the curvature of the nanoparticles decreased.

Of course, these geometric considerations need to be rationalised with additional parameters intrinsically linked to particle size and curvature such as ζ-potential whose role is prominent in electrostatic interaction. But viewing the size ratio of protein to nanoparticle is mandatory for a balanced biofunctionalisation.

7.3 Major Strategies for Organic Chemical Derivatisation

7.3.1 *Self-Assembly of Monomeric Thiol and Amine Molecules on Gold Nanoparticles*

Although capping layers derived from a variety of organic molecules via simple physical adsorption or electrostatic interactions can readily functionalise gold nanoparticles, surface modification using discrete-length chain molecules displaying higher affinity ligands is of significant interest. This is because capping layers derived from such molecules offer greater molecular-level definition in their structural order because of self-assembly-induced chain organisation and significantly improved stability against thermal, chemical and environmental assaults because of stronger or covalent bonding. Major candidates, in this regard, include chain molecules, which present thiolate (–SH), disulfide (–S–S), dithiocarbamate, amine (NH_2), carboxylate (COO–), selenide (–Se), isothiocyanate (–NC) or phosphine — all of which exhibit high affinities for binding to gold. Among them, the self-assembly of organosulphur compounds (e.g. $X(CH_2)_nSH$) on gold nanoparticles represents perhaps the most widely practiced route.[4,18] These molecules spontaneously adsorb from the solution (or vapour) phase onto gold nanoparticle surfaces[19] in a self-limiting manner producing monomolecular layers. The surface binding mercapto- or amine head-groups have a specific affinity for the arrays of single or multiple, distinct adsorption sites at the nanoparticle surface. This molecule-surface recognition landscape (S–Au, ~45 kcal/mol), together with the intermolecular tail–tail interactions between the adsorbate molecules [e.g. $(CH_2)_n|(CH_2)_n$, ~1.5 kcal/mol per CH_2] and the interfacial interactions of the exposed end-groups with the surrounding ambient phase (e.g. X-solvent interface), provides the balance of forces, which is needed for the formation of not only a strong and robust but also structurally most well-defined capping layer.[20]

Structurally, the balance of forces above produces a well-formed monomolecular thiol layer. For the case of aliphatic-chain bearing organosulphur compounds [e.g. $X-(CH_2)_n-SH$], this results in a dense packing of highly oriented aliphatic chains in predominantly all-*trans* conformations with the average alkyl chain [$-(CH_2)_n-$] is tilted from the surface normal by ~30° with S atoms arranged in an epitaxially commensurate $\sqrt{3} \times \sqrt{3}$ R30° structure with one molecule per unit cell[21]; a possible

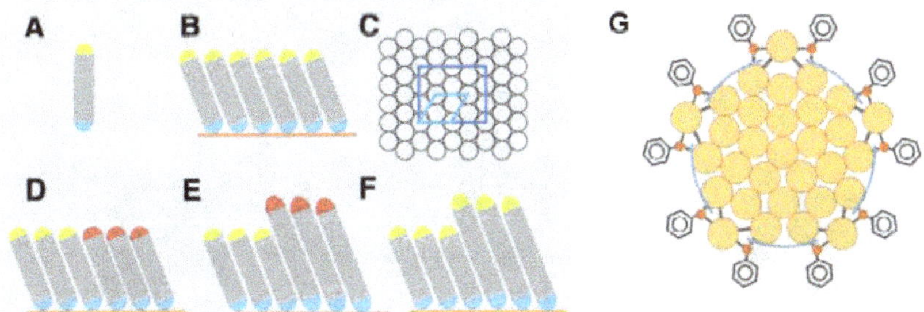

Figure 7.3 Gold nanoparticle functionalisation using thiols. Schematics of self-assembled monolayers of alkanethiols on Au(111). (A) molecular building block consisting of head-group (blue), aliphatic tail (grey) and tail-group (yellow); (B) Uniform monolayer deposited on Au surface (orange); (C) packing habits of alkanethiols on Au lattice: $\sqrt{3} \times \sqrt{3}R30°$ (light blue) and c(4 × 2) (dark blue); (D–F) Binary SAMs used to create chemical and topographic textures at monolayer surfaces; (G) Plausible disulphide bonding model in which a complex involving an Au adatom and a thiolate (Au–SR), and a polymeric chain structure where monothiolates are bridging Au adatoms at the gold–thiolate interface of thiolate-protected gold nanoclusters from X-ray single crystal studies.

superstructure in a c(4 × 2) habit with inequivalent molecule per supercell has also been proposed.[22] More recently, this standard model has been challenged, primarily by density functional theory based computations, which predict an even more extended bonding pattern: here, a disulphide bonding, consisting of a polymeric chain structure of thiolates bridging Au adatoms has been proposed (Figure 7.1).

These discrepancies regarding the details of gold–thiolate bonding notwithstanding, there is a significant consensus in stability of the gold–thiolate bond. An important consequence of strong affinity of thiols for Au surface is that the thiol adsorbates can readily displace other loosely bound, adventitious molecules from the surface.[18,23] Moreover, to incorporate specific functionalities on the surface, end-group functionalised (ω-functionalised, $X(CH_2)_n SH$) organothiols can be utilised or mixed with pre-synthesised gold nanoparticles and other surfactants[24] Figure 7.4

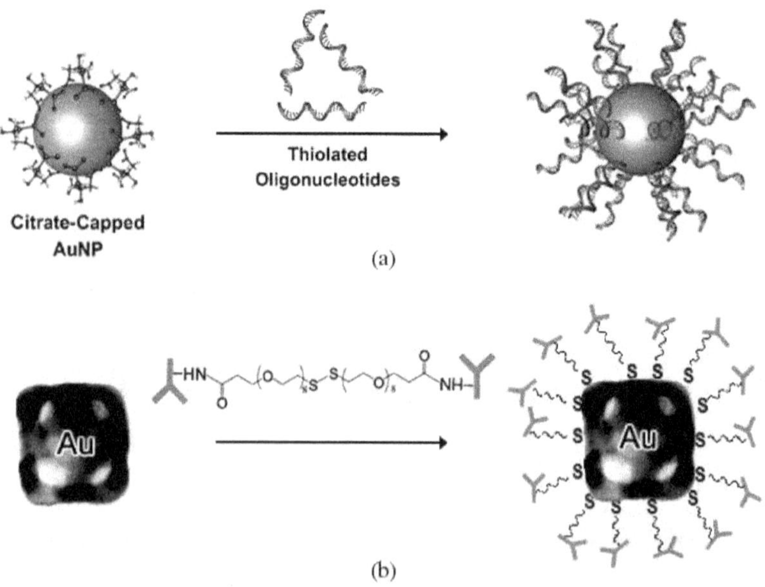

Figure 7.4   Surface functionalisation using thiol linkage, (a) oligonucleotide gold nanoconjugate using alkanethiol-terminated oligonucleotides from citrate-stabilised gold nanoparticles, (b) antibodies conjugate using disulphide molecules (succinimidyl propionyl poly(ethylene glycol) disulfide) to gold nanocages. (a) Modified with permission from Ref. 25, Copyright 2010 Wiley-VCH. (b) Ref. 26, Copyright 2005 American Chemical Society.

illustrates an application of this approach for introducing oligonucleotides and antibodies on the surface of gold nanoparticles through thiol–gold linkages. Another method to introduce surface functionality using thiol molecules is the variation of end groups through well-established substitutions by ligand exchange.

Other functional ligands such as amines and phosphines also form stable gold colloids, although their binding forces with the gold surface are weaker and the ligands displaced readily by thiolated molecules. Dodecylamine and oleylamine can behave as surface passivants, and the resulting amine-passivated gold nanoparticles are stable in organic solvents including toluene, hexane and THF.[27] Reduction of the gold precursors by $NaBH_4$ in a mixture of tri-*n*-octylphosphine oxide (TOPO) and octadecylamine at 190°C yield uniform gold nanoparticles with the diameter of 8.6 nm, which is stable in toluene for as long as a month.[28] The Au_{55} cluster can also be stabilised by triphenyl phosphine, which could be substituted by various ligands.

7.3.2 *Surface-regulating Polymers*

For a variety of applications, most notably those involving interactions with biological systems, gold nanoparticles must be stable and thoroughly dispersed in aqueous media. Gold nanoparticles functionalised with the obvious selection of cetyltrimethylammonium bromide (CTAB), quaternary amine or citrate, are, however, toxic.[29] Therefore, thiol molecules containing carboxylate[30,31] or poly(ethylene oxide) units[29] are preferred candidates of choice in the preparation of water-soluble, biocompatible gold nanoparticles. Furthermore, these thiolated gold nanoparticle conjugates can be used for molecular delivery, owing to cleavable gold–sulphur bonds by light or reductive cytoplasm.[32,33] However, relatively poor long-term stability of gold–sulphur bonds in biological media hampers their biomedical and catalytic applications. Alternative ligands of choice are polymers with polar groups such as poly(vinyl pyrrolidone) (PVP).[34,35] poly(vinyl alcohol) (PVA),[36] poly(acrylic acid) (PAA)[37] and their copolymers.

In the polyol process, a random copolymer containing a vinylpyrrolidone unit can be utilised for the synthesis of surface-functionalised nanomaterials. Song *et al.* have used a copolymer, poly(vinyl pyrrolidone-*ran*-vinyl acetate) (PVP–PVAc), which comprises both vinyl pyrrolidone for gold nanoparticle formation and vinyl acetate for functionalisation, and have successfully synthesised functionalised gold nanoparticles through a one-step reaction of the gold precursors. The resulting nanoparticles have multiple hydroxyl groups on their surface, derived by acid hydrolysis from the acetate group of PVP–PVAc during the reaction.[31]

Block copolymers have ordered structures in solution with periodic thicknesses between 10 and 100 nm. These structures can provide nano-sized domains as nanoreactors, which can be used for gold nanoparticle synthesis. Diblock copolymers such as poly(styrene)-*b*-poly(vinylpyridine) (PS-*b*-P4VP)[38] and poly(styrene)-*b*-poly(acrylic acid) (PS-*b*-PAA)[39] have been used to synthesise gold nanoparticles and assemble them into polar domains.[40]

Dendrimers, well-defined polymer molecules that are known to act as hosts for guest molecules, are another potential template for the formation of gold nanoparticles.[30,41,42] Basically, one dendrimer molecule can entrap one or more gold nanoparticles. Poly(amidoamine) (PAMAM) dendrimers are commonly used polymers for incorporating gold nanoparticles, because of their monomodal and well-defined size and shape. Basically, one dendrimer molecule can entrap one or more gold nanoparticles.[43] Poly(amidoamine) (PAMAM) dendrimers are commonly used polymers for incorporating gold nanoparticles, because of their monomodal and well-defined size and shape. Esumi *et al.* reported the reduction of gold ions by a chemical or photo-chemical reduction method in the presence of generation 4 (G4) PAMAM dendrimers (Figure 7.5). PAMAM-conjugated gold nanoparticles are func-tionalised with hydrophobic groups for solubilisation in organic media, and have been employed as pH sensors and in cell imaging.[44,45]

7.3.3 *Competitive Displacement*

The 'place exchange' method in the monomeric thiol system is a well-established route for incorporating various surface functionalities on nanoparticles. Murray *et al.* demonstrated the place exchange of a con-trolled proportion of thiol ligands by various functional thiols at the end.[6,46] Recently, other thiol-ended ligands containing organic/inorganic

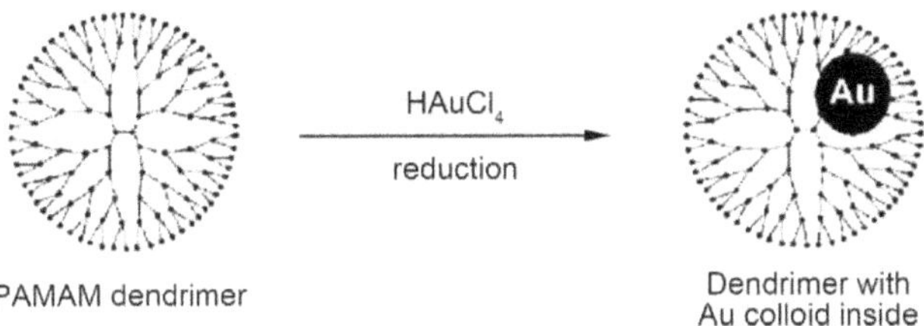

Figure 7.5   Schematic representation of dendrimer nanotemplating in an aqueous solution. Modified with permission from Ref. 42. Copyright 2000 American Chemical Society.

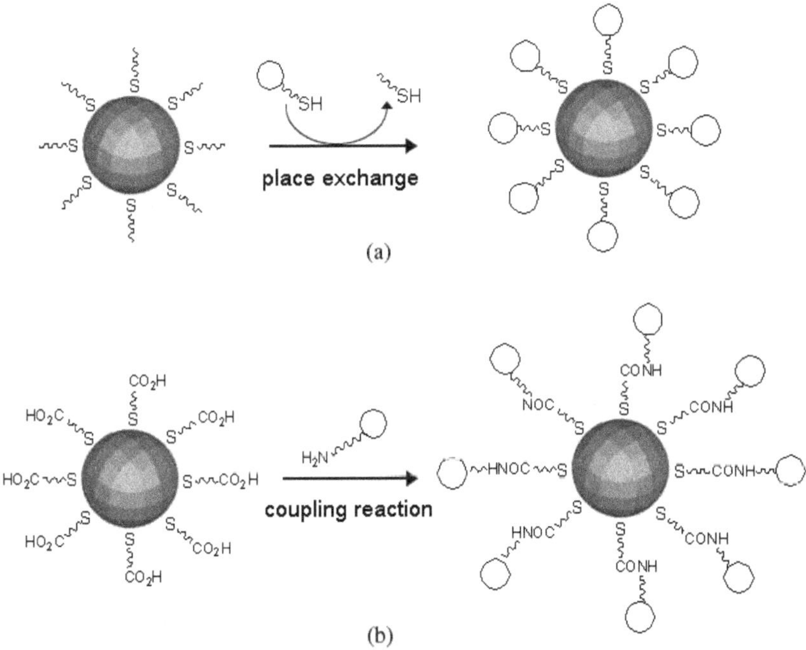

Figure 7.6   Schematic representation of surface modification method through (a) place exchange in organothiol system and (b) secondary modification of ligand end groups.

dyes,[47] smart polymers,[48] bio-molecules,[25,26,32] drug molecules[49] and other nanoparticles have been used to prepare hybrid gold nanoparticles.[50]

Competitive displacement is also commonly employed when the surface functionality on gold nanoparticles has activated groups such as carboxylates and hydroxyls (Figure 7.6). The secondary reaction on the surface is accomplished using chemical coupling,[6,51] polymerisation,[52] electrostatic interaction[53] and selective interaction between biomolecules.[25,27,54] The most well-established methods are coupling and esterification. The gold nanoparticles functionalised with carboxylic acid-terminated thiol ligands readily form amide linkages with other molecules through EDC (1-ethyl-3-(3-dimethylaminopropyl) carbodiimide) coupling. The hydroxyl group of the surface functionality can directly react with acyl chloride to generate various functionalities through esterification.[31]

7.4 Silica Capping of Gold Nanoparticles

Over the last decades, there have been considerable effort in engineering core–shell gold nanoparticles by encapsulating them into other materials, mainly oxides or polymers.[55,56] Among these coating materials, silica occupies a pre-eminent position for obvious reasons.[45,57] First, silica provides an ideal protective shell for the gold core, which is tunable in thickness, stable in various solvents, biocompatible and easily dispersible in silica–glass hosts. Silica synthesis, especially through sol–gel and/or Stöber process, has reached a level of maturity that enables a nanometric control of thickness, porosity and homogeneity of the shell (Figure 7.7). All the same, the surface chemistry of silica is well-mastered, which simplify the further chemical modification of the core–shell nanoparticles to modulate their properties for a set of applications such as biolabelling, biosensing, medicinal diagnostics and drug delivery/therapeutics.

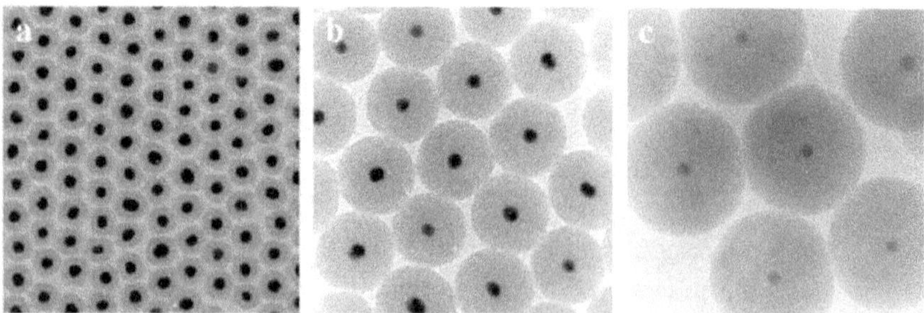

Figure 7.7   TEMs of silica-coated gold nanoparticles, and subsequently further growing the silica shell via the hydrolysis of TEOS (Stöber method) in an ethanol–water mixture. The silica shell thickness on the 15-nm gold nanoparticles was varied from approximately (a) 10 through (b) 60 to (c) 80 nm during growth of the silica shell. From Ref. 55.

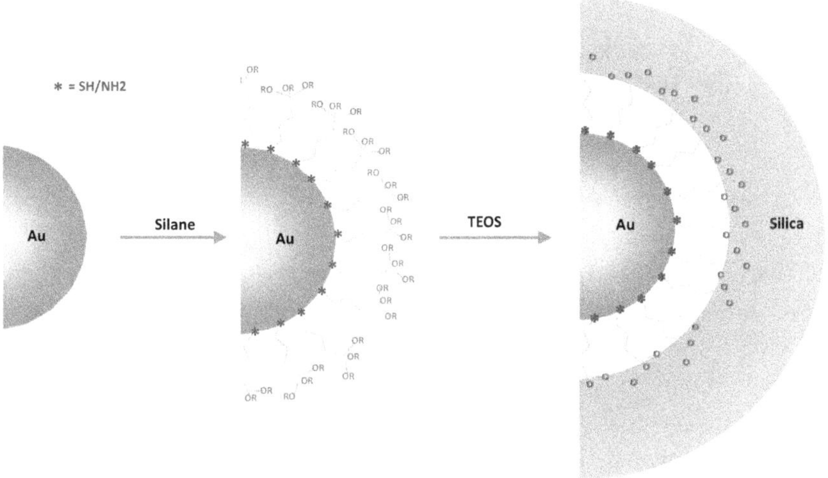

Figure 7.8   Sketch of the three-step process for silica shell growth on silane-modified citrate-stabilised gold nanoparticles. Adapted from Ref. 57.

7.4.1 *Primer-mediated Silica Coating*

In the early studies, it was believed that a surface primer was mandatory to initiate the growth of a silica shell on a gold surface due to the low chemical affinity of gold to silica. Therefore, silica-coating of citrate-stabilised gold nanoparticles has been achieved by means of protocols that make use of surface priming. Most of the described methods typically involve three main steps (Figure 7.8): first rending gold nanoparticle surfaces vitreophilic using a surface primer, second silica deposition, or nucleation on the surface, and third further growth through the polycondensation of silicon alkoxides (usually tetraethoxysilane (TEOS)) in sol–gel reactions or following the Stöber method in ethanol–ammonia–water mixtures.[58] Silanes are the most encountered linkers but not the unique ones; the non-ionic polymer poly(vinylpyrrolidone) (PVP) for example, has been used to form homogeneous silica shells.[59] PVP facilitates the transfer of gold nanoparticles into ethanol and also promotes silica grafting. Silanes were first used as surface primer by Liz-Marzan *et al.* to engineer core–shell nanoparticles with the Stöber process.[57] The commonly used silane are either amine or thiol-terminated, relying on the affinity of these groups to gold.[60] Mixture of silanes either with alcohol terminal groups or using carboxyl acid silane also lead to a stable homogeneous silica shell.[61] Silane linkers perform two functions as they 'vitreophile' nanoparticle surfaces and also initiate silica shell growth through the remaining alkoxy groups.

In this three-step process, the mechanism of silica nucleation is poorly investigated, especially the nature of the core–shell interface in SiO_2@Au nanoparticles. Poovarodom *et al.* removed the gold core via cyanide etching in core–shell particles synthesised via a 3-aminopropyltrimethoxysilane (APTMS) route as building blocks, and characterised the remaining material mainly by ^{13}C CPMAS NMR spectroscopy.[62] Their results showed a very low yield of covalent amine incorporation (less than 10%). The core−shell interface in gold@silica would therefore contain siloxane and silanol groups with randomly dispersed aminosilane in low yield. Their results question the need of primers to grow the silica shell.

7.4.2 *Direct Silica Coating*

More recently, it was found that surface primers were not mandatory to coat citrate-stabilised gold nanoparticles by silica using a slightly modified Stöber process. Lu *et al.* could directly coat the surfaces of commercial 50 nm gold nanoparticles by a silica shell grown by an ammonia-catalysed hydrolysis/condensation of TEOS.[63] The shell thickness could be changed from tens to several hundred nanometres. Mines *et al.* also succeeded in direct coating of gold nanoparticles with silica by a seeded polymerisation technique.[64] They used smaller citrate-stabilised nanoparticles (~15 nm), and could tackle the aggregation and instability issues encountered when transferring small particles into alcoholic solution by addition of TEOS and water prior to ammonia. This order was found to be critical to obtain a proper coating. The silica shell thickness was varied from 30 to 90 nm.

Ever since, other authors have reported the successful direct coating of gold nanoparticles with silica.[65] However, the mechanism of interaction is also obscure. The role of citrates is unclear, their presence in the junction gold/silica is a plausible scenario, but remains a hypothesis.

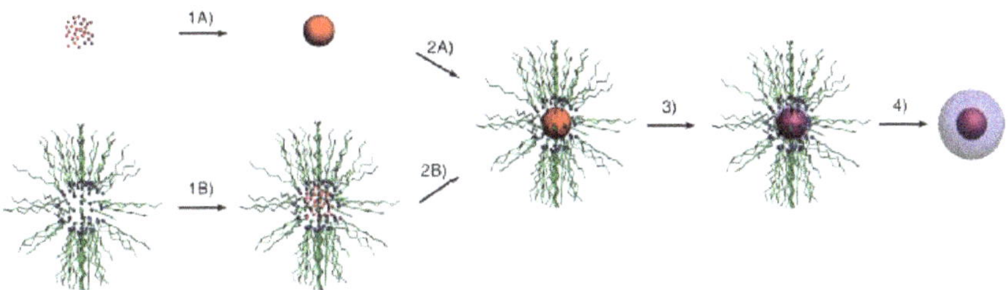

Figure 7.9   Schematic illustration of the conventional water/oil (W/O) microemulsion silica coating of nanoparticles. In the first initial route, nanoparticle cores are previously synthesised (1A) and transferred into stable W/O microemulsions prepared by mixing organic solvent, water and surfactant (2A). In the second initial approach, both solubilisation (1B) and reaction (2B) of the core precursors inside reverse micelles is followed by *in situ* hydrolysis and condensation of TEOS (3). Ageing and washing lead to the formation of cores with sizes and coating thickness that can be experimentally controlled (4). From Ref. 56.

7.4.3 *Other Protocols for Citrate-stabilised Nanoparticle Coating*

Though the vast majority of Au@SiO$_2$ nanoparticles are prepared through the Stöber process, efficient alternatives are also described, among them, the very popular water/oil microemulsion process depicted in Figure 7.9.[56–66] This process can be used for small particles (from 10 to 50 nm) for whom aggregation issues rise in water–ammonia–ethanol mixture needed for the Stöber process.

Han *et al.* reported reverse microemulsion method for preparing monodispersed silica-coated gold nanoparticles without the use of a surface primer.[67] This method shows many advantages compared to the previous ones: first, a finer adjustment of the silica thickness with nanometre precision is reached. Furthermore, the coating can be achieved on gold nanoparticles at concentrations three orders of magnitude higher than that achieved by the Stöber method. Finally, and more interestingly, various functional groups can be grafted during the synthesis step by condensing reactive silanes with TEOS, which leads to functional nanoparticles ready-to-use for further biological conjugation.

Microwave-mediated silica growth is a second alternative for rapid synthesis of core–shell nanoparticles. Bahadur *et al.* described a rapid and simple method for preparing silica-coated gold using microwave irradiation.[68] Silica coating was achieved by mixing citrate-stabilised gold

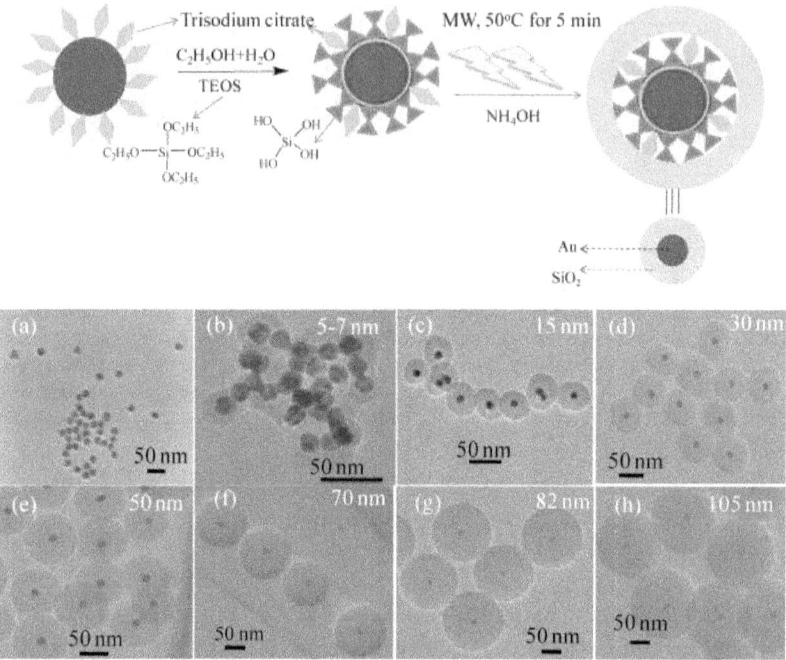

Figure 7.10   Silica-coated gold nanoparticles (Au@SiO$_2$) prepared by a fast and one-step microwave method with different silica shell thicknesses (5–105 nm). From Ref. 68.

nanoparticles ($\sim$16 nm) with TEOS and ammonia and then irradiating the mixture by microwaves (MW). The colloidal Au particles were covered with uniform silica shell within very short time (5 min) without any preliminary treatment for a wide range of silica shell thickness from 5 to 105 nm (see Figure 7.10). The size uniformity and monodispersity shown on the TEM images are impressive, especially in view of the rapidity and simplicity of the process, which makes this protocol promising for further uses.

7.4.4 *Silica-capping of CTAB-stabilised Gold Nanoparticles*

All the procedures described above are applicable to citrate or ascorbic acid stabilised nanoparticles. These capping agents are weekly bound to gold surfaces and therefore easy to exchange by the ligands mentioned above. When the application requires non-spherical nanoparticles, the most

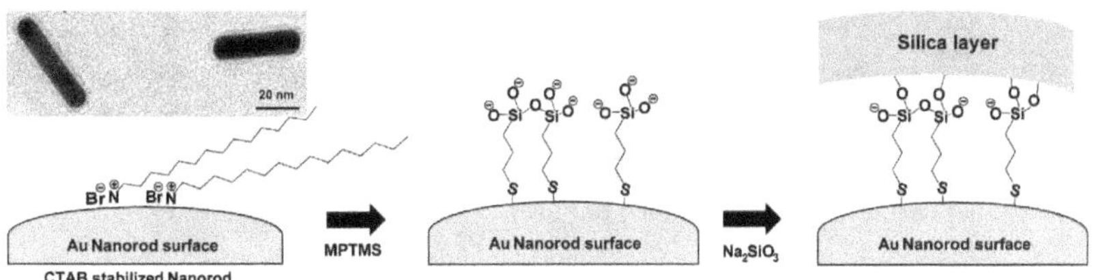

Figure 7.11 Sketch of the three-steps process for silica shell growth on silane modified CTAB-stabilised gold nanorods. From Ref. 71.

successful methods for size and shape control rely on the use of cationic surfactant CTAB as the 'shape-inducing' agent.[69] However, the strong binding of CTAB to the gold nanoparticle surface makes its displacement by thiol- or amine-terminated silanes difficult, especially at the flat sides of the nanorods. The complexity of CTAB exchange makes the silica capping of nanorods more challenging and more needed as this would allow the use of the silica surface chemistry for further tethering. Very few methods are described for a controlled and homogeneous silica coating CTAB-stabilised gold nanorods. Murphy *et al.*[70] have reported the use of mercaptopropyl-silanes for the silica coating of high-aspect-ratio gold rods with a coating thicknesses ranging from 5 to 10 nm. Their method was applied for smaller aspect ratio nanorods by Corn *et al.*[71] and a thin silica shell, 3 to 5 nm, was reached (see Figure 7.11).

Liz-Marzan *et al.* achieved thicker silica shells by combining the layer-by-layer (LBL) technique and the hydrolysis and condensation of TEOS in 2-propanol–water mixture and reached homogeneous coatings with tight control on shell thickness[35] (Figure 7.12).

Finally PVP was also used as primer for CTAB-stabilised particles.[35] In the classical three-step synthesis process, PVP helped transferring nanoparticles into ethanol but silica condensation and growth on the nanoparticles failed due to the interference of the CTAB groups.

7.5 Biofunctionalisation of Gold Nanoparticles

The ability to functionalise gold nanoparticle surfaces with biomolecules is important for the successful realisation of myriad of new and emerging

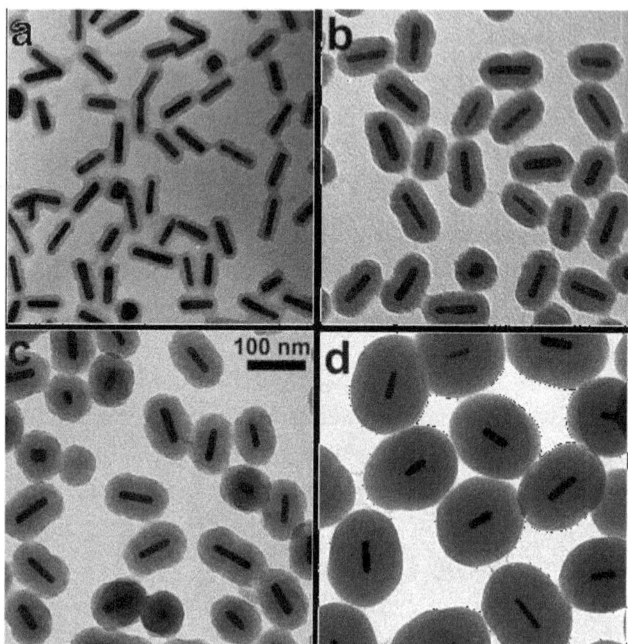

Figure 7.12   Transmission electron micrographs (TEM) of silica-coated gold nanorods, with silica shell thickness increasing from (a) to (d). The scale is the same for all images from Ref. 35.

applications across biology, biotechnology and medicine (see Figure 7.13). By affording a direct molecular-level control of the interface between the nanoparticle and the surrounding (biological) environment, surface functionalisation provides a powerful strategy to control nanoparticle inter-actions with the biological milieu, affords targeting through biospecific recognition and affinity modulation of specific biomolecules, and enables spatial localisation. These abilities can evade nanoparticle capture by the immune system, enhance cellular internalisation, reduce cytotoxicity, target delivery of any encapsulated payload, minimise non-specific adsorption of proteins and sense specific biomolecules enabling applications across fields of drug delivery, nanoscale imaging, diseases diagnostics, biosensing and therapeutics.

Multiple disparate strategies are needed for achieving these functional-ities, namely passive biocompatiblisation and active functionalisation for biospecific activities, using gold nanoparticles. These are discussed in detail in following sections.

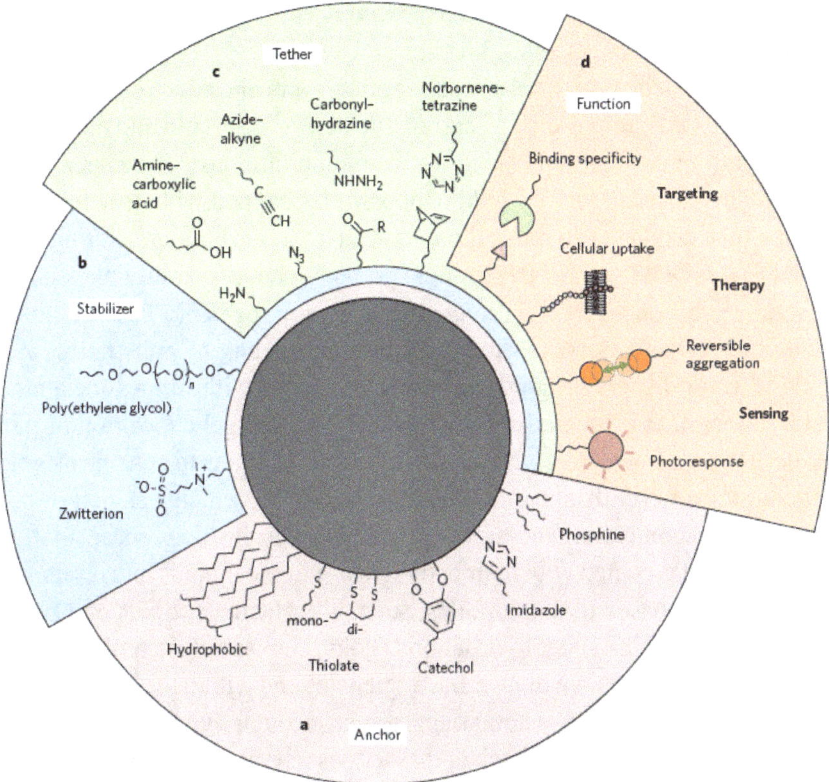

Figure 7.13   Modular design of surface ligands for biocompatible nanomaterials. (a) Anchor links capping molecules to the crystallite through hydrophobic interaction with native ligands or through atoms that bind directly to the nanocrystal surface. (b) Stabiliser region incorporates polyethylene glycol (PEG) or zwitterionic chains that strongly bind water molecules to impart hydrophilicity and prevent non-specific adsorption. (c) Capping molecules may feature reactive groups serving as a tether point for covalent conjugation to biofunctional molecules after ligand exchange. (d) The biofunctional units can offer targeting, therapeutic or sensing capability. From Ref. 7.

7.5.1 *Water-dispersible Gold Nanoparticles*

An important pre-requisite for biofunctionalisation of gold nanoparticles is to render them water dispersible. A variety of approaches are available to achieve this goal, among which the most successful approaches include (1) direct substitution of the native capping layer with capping agents displaying hydrophilic groups; (2) deposition of thin inorganic oxide layers;

and (3) the encapsulation of the nanoparticle into organised amphiphilic mesophases, such as micelles, bicelles and vesicles.

In the first approach, thiol-based capping layers end-functionalised with hydrophilic end-groups such as carboxylic acid,[30] tiopronin and coenzyme A,[72] glutathione,[73] sulphonic acid[74] and ammonium ions[75] have been used. In many instances, however, gold nanoparticles capped with only terminal groups that displayed hydrophilic character (e.g. arylthiolates terminated with OH–, COOH– and $-NH_2$ groups) proved to be less readily dispersible in water. Subsequent efforts directed at replacing mercapto-based binding by amine derivatives proved more successful, leading to preparation of a class of water-dispersible gold nanoparticles capped with amines and amino acids. Second, as discussed above (Section 1.3), sol–gel deposition of thin oxide layers provides a convenient and versatile route to rendering gold nanoparticles water dispersible. The third approach, which was first developed for semiconductor quantum-dot nanoparticles, involves encapsulation of nanoparticles coated with their native hydrophobic ligands into organised discrete mesophases of amphiphiles, such as surfactants, lipids and block copolymers. For instance, hydrophobic gold nanoparticles can be readily incorporated into the core of micellar aggregates and within the hydrophobic core of bilayer shells of vesicular aggregates through hydrophobic effects. A major benefit of this approach is that it does not alter the dangling bonds at the nanoparticle surface and introduces minimal perturbations to the quantum size effects of the nanoparticles.

7.5.2 *Non-biofouling Gold Nanoparticles*

An important component in the biofunctionalisation of nanoparticle surfaces is to endow them with an ability to resist non-specific adsorption of proteins from the biological milieu.[76] Such adsorption is undesirable for *in vivo* applications in which uncontrolled protein adsorption to the nanoparticle surface may reduce *in vivo* circulation of particles by accelerating renal clearance, elicit inadvertent immune responses, blood coagulation or microbial adhesion and mask the functions that require a pristine interface between the nanoparticle and its biological environment, such as in targeting, localising and sensing. *In vitro* bioanalytical applications of gold nanoparticles (e.g. affinity biosensors and bioanalytic devices) are also negatively affected by non-specific adsorption losing their specificity for designed biological interactions and producing 'false-positives'.[77] Rational engineering of synthetic, 'non-fouling' surfaces to minimise non-specific protein adsorption requires the suppression of all major non-specific protein adsorption pathways, which encompass one or more of attractive factors including physical (e.g. van der Waals, electrostatic, hydrophobic and entropically driven adsorption) and chemical (e.g. Lewis acid–base interactions, bond formation) processes.

One of the best strategies to confer protein binding resistance to nanoparticle surfaces involves presenting oligo- or poly(ethylene glycol) (OEG or PEG) moieties at surfaces.[78] Both long segments of PEG units and high densities of OEG are known to offer protein resistance. The thiol-based monolayers provide an elegant means to present the requisite high surface densities simply by placing PEG-ylated moieties at the end-groups without significantly compromising surface molecular densities of SAM-forming molecules.[76] Indeed, even very short, densely packed oligo(EG) (e.g. tri- or hexa-(ethylene glycol)) terminated alkanethiol monolayers on gold have been shown to confer protein resistance to surfaces.

The origin of mechanisms of protein resistance by PEG-ylation has remained a subject of vigorous debate in the literature.[79] Two main mechanistic considerations are invoked: (1) does PEG confer protein resistance or the surrounding water and (2) is the protein resistance of PEG-ylated surfaces an enthalpic effect or an entropic one. Physical models propose that the van der Waals attraction between the surface and the protein are overwhelmed by the steric repulsion originating from the penalties associated with the compression of PEG induced by protein binding.[80] Chemical models, by contrast, emphasise the importance of Lewis acid–base interactions: the electron lone pairs on the oxygen atoms of the ethylene oxide repeating unit provides a strong enthalpic contribution to the PEO–water interaction, which reorients interfacial water with the oxygen atoms pointing away from PEO.[81] There is some consensus that PEG-ylated monolayers optimise these interactions by producing amorphous PEG surface in aqueous environment.

Despite the successful protein resistance achieved with PEG-ylated SAMs, achieving long-term stability of thiolate-based SAMs has remained a persistent challenge. Specifically, defects are known to appear in SAMs terminated with OEG moieties under physiological aqueous environments compromising the structural integrity of SAMs, diminishing their protein resistance and thus continuing to hamper their practical applications. A likely cause for this loss of protein resistance is the oxidation at the Au–S interface in the aqueous environment, culminating into monolayer desorption. Recent developments of zwitterionic surfactants that interact with water molecules even more strongly than PEG have enabled further reduction of capping-layer thickness, while preserving effective adsorbate repulsion. The combination of PEG or zwitterionic ligands with multidentate anchors offers highly non-fouling nanoparticles.

7.5.3 *Active Biofunctional Gold Nanoparticles*

The introduction of specific biological activities in gold nanoparticles falls broadly into four different categories: sensing, targeting, therapy and as tools for biophysical and biochemical characterisation of biological mechanisms (Figure 7.13(d)). In each of these cases, specific functional groups are required to be immobilised onto the nanoparticle surface together with those required for water dispersal (Section 7.4.1) and reduction of non-specific protein adhesion (Section 7.4.2).

For biosensors for immunoassays, antigen- or antibody-displaying gold nanoparticles are used either via direct binding of antigen-displaying gold nanoparticles to an antibody-modified surface or the exposure of an antibody-derived surface to free antigen and then to a secondary antibody-conjugated gold nanoparticle. In this same vein, gold nanoparticles derivatised with peptides, enzymes, proteins, membranes, drugs and even viruses have been prepared. Excellent reviews of these functionalisation procedures are available in literature.[82] Synthetic peptides have been extensively employed as recognition entities in affinity assays and as substrates in enzymatic/catalytic assays. Aili *et al.*, for example, used a 42-mer helix-loop-helix motif that was modified with a small molecule ligand, benzene-sulphone amide, for colorimetric 'naked eye' affinity detection of human carbonic anhydrase at a limit of detection in the low nanometre regime.[83] A cysteine (SH-containing) residue was introduced in the inactive loop region for proper and oriented immobilisation onto gold nanoparticles. Linear peptides also have been modified with a cysteine at the terminal position for attachment to noble metal nanoparticles either directly via Au–S bond formation[84] or via attachment through a surface bound maleimide linker.[85] Liu *et al.* developed a cleavage assay for specific and sensitive detection of Botulinum serotype A (one of the most toxic substances known to man) at sub-nantometre concentrations.[86] Wang *et al.* improved the sensitivity further in into the pM regime by employing a fluorescence-based AuNP quenching assay.[87] In yet another study, Chen and co-workers[88] utilised a synthetic peptide motif possessing two dominant cleavage sites for rapid and sensitive detection of Matrilysin or metalloproteinase-7 (MMP-7) a potential biomarker that is up-regulated during many cancers, for example colorectal and salivary gland cancer. A mini-review on the modification of nanoparticles and nanomaterials using non-antibody recognition entities, including peptides, was recently reported by Chen *et al.*[89]

7.6 Conclusions

In summary, chemical functionalisation of nanoparticle surfaces has proven to be a key enabling strategy playing central roles not only in controlling how nanoparticles interact with their physical, chemical and biological environments but also in directing the nanoparticle synthesis itself. For gold nanoparticles, the strong Au–S affinity provides one of the most elegant and widely used means to achieve such functionalisation enabling the formation of robust capping layers consisting of discrete-chain alkanethiol molecules, polymers, peptides, proteins and even inorganic capping layers. Depending on the application, the capping layer can be chosen to display desired end-functional groups (or surface epitopes) in controlled densities, distributions and conformations. By affording a direct molecular-level control of the interface between the nanoparticle and the surrounding (biological) environment, surface functionalisation provides a powerful strategy to control nanoparticle interactions with the biological structures including membranes, proteins and cells. These abilities can help design nanoparticles that can evade capture by the immune system, enhance or reduce cellular internalisation, control cytotoxicity, target delivery of any encapsulated payload such as from nanoporous gold, minimise non-specific adsorption of proteins, sense specific biomolecules and spatially localise in pre-determined physiological niches enabling applications across fields of drug delivery, nanoscale imaging, diseases diagnostics, biosensing and therapeutics.

References

1. E. Roduner, *Chem. Soc. Rev.* **35** (7), 583–592 (2006); A. P. Alivisatos, *Science* **271**, 933–937 (1996).
2. A. R. Tao, S. Habas, and P. D. Yang, *Small* **4**, 310–325 (2008).
3. R. Sardar, A. M. Funston, P. Mulvaney *et al.*, *Langmuir* **25**, 13840–13851 (2009); C. N. R. Rao, G. U. Kulkarni, P. J. Thomas *et al.*, *Chem. Soc. Rev.* **29**, 27–35 (2000).
4. M. C. Daniel and D. Astruc, *Chem. Rev.* **104**, 293–346 (2004).
5. J. Turkevich, P. C. Stevenson, and J. Hillier, *Discuss. Faraday Soc.* (11), 55–75 (1951).
6. A. C. Templeton, M. P. Wuelfing, and R. W. Murray, *Accounts Chem. Res.* **33**, 27–36 (2000).
7. M. A. Boles, D. Ling, T. Hyeon *et al.*, *Nat. Mater.* **15**, 141–153 (2016).
8. H. Hakkinen, *Nat. Chem.* **4**, 443–55 (2012).
9. M. J. Yacaman, J. A. Ascencio, H. B. Liu *et al.*, *J. Vacuum Sci. Technol. B* **19**, 1091–103 (2001).
10. A. Badia, L. Demers, L. Dickinson *et al.*, *J. Am. Chem. Soc.* **119**, 11104–11105 (1997).
11. J. C. Love, L. A. Estroff, J. K. Kriebel *et al.*, *Chem. Rev.* **105**, 1103–1170 (2005).
12. L. H. Dubois, B. R. Zegarski, and R. G. Nuzzo, *J. Chem. Phys.* **98**, 678–688 (1993).
13. H. Sellers, A. Ulman, Y. Shnidman *et al.*, *J. Am. Chem. Soc.* **115**, 9389–401 (1993).
14. M. J. Hostetler, J. J. Stokes, and R. W. Murray, *Langmuir* **12**, 3604–3612 (1996).
15. M. Lundqvist, I. Sethson, and B.-H. Jonsson, *Langmuir* **20**, 10639–10647 (2004).
16. J. A. Sánchez-Pérez, A. M. Gallardo-Moreno, M. L. González-Martín *et al.*, *Appl. Surf. Sci.* **353**, 1095–1102 (2015).
17. S. Goy-López, J. Juárez, M. Alatorre-Meda *et al.*, *Langmuir* **28**, 9113–9126 (2012).
18. M. Brust, M. Walker, D. Bethell *et al.*, *J. Chem. Soc. Chem. Commun.* (7), 801–802 (1994).
19. R. G. Nuzzo and D. L. Allara, *J. Am. Chem. Soc.* **105**, 4481–4483 (1983); J. Sagiv, *J. Am. Chem. Soc.* **102**, 92–98 (1980).
20. D. K. Schwartz, *Ann. Rev. Phys. Chem.* **52**, 107–37 (2001); H. Hakkinen, *Nat. Chem.* **4**, 443–455 (2012).
21. P. E. Laibinis, G. M. Whitesides, D. L. Allara *et al.*, *J. Am. Chem. Soc.* **113**, 7152–7167 (1991).
22. A. Cossaro, R. Mazzarello, R. Rousseau *et al.*, *Science* **321**, 943–946 (2008); P. Fenter, A. Eberhardt, and P. Eisenberger, *Science* **266**, 1216–1218 (1994).
23. M. Brust, J. Fink, D. Bethell *et al.*, *J. Chem. Soc. Chem. Commun.* (16), 1655–1656 (1995).
24. P. V. Kamat, S. Barazzouk, and S. Hotchandani, *Ang. Chemie Int.* **41**, 2764–2767 (2002); M. J. Hostetler, S. J. Green, J. J. Stokes *et al.*, *J. Am. Chem. Soc.* **118**, 4212–4213 (1996).
25. D. A. Giljohann, D. S. Seferos, W. L. Daniel *et al.*, *Ang. Chem.Int.* **49**, 3280–3294 (2010).

26. J. Chen, F. Saeki, B. J. Wiley *et al.*, *Nano Lett.* **5**, 473–477 (2005).
27. D. V. Leff, L. Brandt, and J. R. Heath, *Langmuir* **12**, 4723–4730 (1996).
28. M. Green and P. O'Brien, *Chem. Commun.* (3), 183–184 (2000).
29. E. Glogowski, R. Tangirala, J. He *et al.*, *Nano Lett.* **7**, 389–393 (2007); M. Zheng, Z. G. Li, and X. Y. Huang, *Langmuir* **20**, 4226–4235 (2004); M. K. Corbierre, N. S. Cameron, and R. B. Lennox, *Langmuir* **20**, 2867–2873 (2004); W. P. Wuelfing, S. M. Gross, D. T. Miles *et al.*, *J. Am. Chem. Soc.* **120**, 12696–12697 (1998); N. F. Steinmetz and M. Manchester, *Biomacromolecules* **10**, 784–792 (2009); B. C. Mei, K. Susumu, I. L. Medintz *et al.*, *Nat. Protocols* **4**, 412–423 (2009).
30. S. H. Chen and K. Kimura, *Langmuir* **15**, 1075–1082 (1999).
31. C. I. Yoo, D. Seo, B. H. Chung *et al.*, *Chem. Mat.* **21**, 939–944 (2009).
32. G. B. Braun, A. Pallaoro, G. Wu *et al.*, *Acs Nano* **3**, 2007–2015 (2009).
33. M. Oishi, J. Nakaogami, T. Ishii *et al.*, *Chem. Lett.* **35**, 1046–1047 (2006).
34. D. Seo, J. C. Park, and H. Song, *J. Am. Chem. Soc.* **128**, 14863–14870 (2006); D. Seo, J. H. Park, J. Jung *et al.*, *J. Phys. Chem. C* **113**, 3449–3454 (2009); D. Seo, C. I. Yoo, I. S. Chung *et al.*, *J. Phys. Chem. C* **112**, 2469–2475 (2008); F. Kim, S. Connor, H. Song *et al.*, *Ang. Chem. Int.* **43**, 3673–3677 (2004).
35. A. Sanchez-Iglesias, I. Pastoriza-Santos, J. Perez-Juste *et al.*, *Adv. Mater.* **18**, 2529–2534 (2006).
36. Y. Zhou, C. Y. Wang, Y. R. Zhu *et al.*, *Chem. Mater.* **11**, 2310–2312 (1999); L. Prati and G. Martra, *Gold Bull.* **32**, 96–101 (1999); M. Comotti, C. Della Pina, R. Matarrese *et al.*, *Appl. Catal. A* **291**, 204–209 (2005).
37. H. Jans, K. Jans, L. Lagae *et al.*, *Nanotechnology* **21**, 455702 (2010); E. Gachard, H. Remita, J. Khatouri *et al.*, *New J. Chem.* **22**, 1257–1265 (1998); J. M. Petroski, Z. L. Wang, T. C. Green *et al.*, *J. Phys. Chem. B* **102**, 3316–3320 (1998); T. S. Ahmadi, Z. L. Wang, T. C. Green *et al.*, *Science* **272**, 1924–1926 (1996).
38. J. H. Youk, M. K. Park, J. Locklin *et al.*, *Langmuir* **18**, 2455–2458 (2002); J. F. Gohy, N. Willet, S. Varshney *et al.*, *Ang. Chem. Int.* **40**, 3314–3316 (2001); J. P. Spatz, S. Mossmer, C. Hartmann *et al.*, *Langmuir* **16**, 407–415 (2000).
39. Y. J. Kang and T. A. Taton, *Ang. Chem. Int.* **44**, 409–412 (2005); C.-M. Huang, K.-H. Wei, U. S. Jeng *et al.*, *Macromolecules* **40**, 5067–5074 (2007).
40. A. Aqil, C. Detrembleur, B. Gilbert *et al.*, *Chem. Mater.* **19**, 2150–2154 (2007).
41. M. Stemmler, F. D. Stefani, S. Bernhardt *et al.*, *Langmuir* **25**, 12425–12428 (2009).
42. F. Grohn, B. J. Bauer, Y. A. Akpalu *et al.*, *Macromolecules* **33**, 6042–6050 (2000).
43. X. Shi, K. Sun, and J. R. Baker, Jr., *J. Phys. Chem. C* **112**, 8251–8258 (2008).
44. K. Esumi, H. Houdatsu, and T. Yoshimura, *Langmuir* **20**, 2536–2538 (2004); K. Esumi, K. Satoh, and K. Torigoe, Langmuir **17**, 6860–6864 (2001).
45. K. Hayakawa, T. Yoshimura, and K. Esumi, *Langmuir* **19**, 5517–5521 (2003).
46. A. C. Templeton, M. J. Hostetler, C. T. Kraft *et al.*, *J. Am. Chem. Soc.* **120**, 1906–1911 (1998); M. J. Hostetler, A. C. Templeton, and R. W. Murray, *Langmuir* **15**, 3782–3789 (1999).
47. E. C. Walter, B. J. Murray, F. Favier *et al.*, *J. Phys. Chem. B* **106**, 11407–11411 (2002).
48. C. R. Martin, *Chemi. Mater.* **8**, 1739–1746 (1996).
49. P. Ghosh, G. Han, M. De *et al.*, *Adv. Drug Delivery Rev.* **60**, 1307–1315 (2008); J. D. Gibson, B. P. Khanal, and E. R. Zubarev, *J. Am. Chem. Soc.* **129**, 11653–11661 (2007).

50. K. G. Thomas and P. V. Kamat, *Accounts Chem. Res.* **36**, 888–898 (2003).
51. S. Banerjee and S. S. Wong, *Nano Lett.* **2**, 195–200 (2002); J. Liu, A. G. Rinzler, H. J. Dai *et al.*, *Science* **280**, 1253–1256 (1998).
52. T. K. Mandal, M. S. Fleming, and D. R. Walt, *Nano Lett.* **2**, 3–7 (2002).
53. K. Y. Jiang, A. Eitan, L. S. Schadler *et al.*, *Nano Lett.* **3**, 275–77 (2003); J. Kolny, A. Kornowski, and H. Weller, *Nano Lett.* **2**, 361–364 (2002); W. Cheng and E. Wang, *J. Phys. Chem. B* **108**, 24–26 (2004); Y.-M. Chen, C.-J. Yu, T.-L. Cheng *et al.*, *Langmuir* **24**, 3654–3660 (2008).
54. H. Reiss, *J. Chem. Phys.* **19**, 482–487 (1951); B. L. Frankamp, O. Uzun, F. Ilhan *et al.*, *J. Am. Chem. Soc.* **124**, 892–893 (2002).
55. F. Caruso, *Adv. Mater.* **13**, 11–22 (2001).
56. A. Guerrero-Martínez, J. Pérez-Juste, and L. M. Liz-Marzán, *Adv. Mater.* **22**, 1182–1195 (2010).
57. L. M. Liz-Marzan, M. Giersig, and P. Mulvaney, *Langmuir* **12**, 4329–4335 (1996).
58. W. Stober, A. Fink, and E. Bohn, *J. Colloid Interface Sci.* **26**, 62–69 (1968).
59. C. Graf, D. L. J. Vossen, A. Imhof *et al.*, *Langmuir* **19**, 6693–6700 (2003).
60. M. M. Y. Chen and A. Katz, *Langmuir* **18**, 8566–8572 (2002).
61. S. Liu, Z. Zhang, Y. Wang *et al.*, *Talanta* **67**, 456–461 (2005); J. Xu and C. C. Perry, *J. Non-Crystal. Solids* **353**, 1212–1215 (2007).
62. S. Poovarodom, J. D. Bass, S. J. Hwang *et al.*, *Langmuir* **21**, 12348–12356 (2005).
63. Y. Lu, Y. Yin, Z.-Y. Li *et al.*, *Nano Lett.* **2**, 785–788 (2002).
64. E. Mine, A. Yamada, Y. Kobayashi *et al.*, *J. Colloid Interface Sci.* **264**, 385–390 (2003).
65. S. H. Liu and M. Y. Han, *Adv. Func. Mater.* **15**, 961–967 (2005).
66. S. Santra, R. Tapec, N. Theodoropoulou *et al.*, *Langmuir* **17**, 2900–2906 (2001).
67. Y. Han, J. Jiang, S. S. Lee *et al.*, *Langmuir* **24**, 5842–5848 (2008).
68. N. M. Bahadur, S. Watanabe, T. Furusawa *et al.*, *Colloids Surf. A: Physicochem. Eng. Aspects* **392**, 137–144 (2011).
69. B. Nikoobakht and M. A. El-Sayed, *Chem. Mater.* **15**, 1957–1962 (2003).
70. S. O. Obare, N. R. Jana, and C. J. Murphy, *Nano Lett.* **1**, 601–603 (2001).
71. I. E. Sendroiu, M. E. Warner, and R. M. Corn, *Langmuir* **25**, 11282–11284 (2009).
72. A. C. Templeton, S. W. Chen, S. M. Gross *et al.*, *Langmuir* **15**, 66–76 (1999).
73. T. G. Schaaff, G. Knight, M. N. Shafigullin *et al.*, *J. Phys. Chem. B* **102**, 10643–10646 (1998).
74. Y. S. Shon, W. P. Wuelfing, and R. W. Murray, *Langmuir* **17**, 1255–1261 (2001).
75. D. E. Cliffel, F. P. Zamborini, S. M. Gross *et al.*, *Langmuir* **16**, 9699–9702 (2000).
76. K. L. Prime and G. M. Whitesides, *J. Am. Chem. Soc.* **115**, 10714–10721 (1993).
77. Z. Matharu, A. J. Bandodkar, V. Gupta *et al.*, *Chem. Soc. Rev.* **41**, 1363–1402 (2012).
78. K. L. Prime and G. M. Whitesides, *Science* **252**, 1164–1167 (1991).
79. E. Ostuni, R. G. Chapman, R. E. Holmlin *et al.*, *Langmuir* **17**, 5605–5620 (2001); M. Morra, *J. Biomater. Sci.-Polym. Ed.* **11**, 547–569 (2000).
80. I. Szleifer, *Physica A* **244**, 370–388 (1997); A. Halperin, *Langmuir* **15**, 2525–2533 (1999); S. I. Jeon, J. H. Lee, J. D. Andrade *et al.*, *J. Colloid Interf. Sci.* **142**, 149–158 (1991).
81. P. Harder, M. Grunze, R. Dahint *et al.*, *J. Phys. Chem. B* **102**, 426–436 (1998).

82. P. D. Howes, R. Chandrawati, and M. M. Stevens, *Science* **346**, 53 (2014); D. S. Ling, M. J. Hackett, and T. Hyeon, *Nano Today* **9**, 457–477 (2014).
83. D. Aili, R. Selegård, L. Baltzer *et al.*, *Small* **5**, 2445–2452 (2009).
84. J. Wetterö, T. Hellerstedt, P. Nygren *et al.*, *Langmuir* **24**, 6803–6811 (2008).
85. J. Su and M. Mrksich, *Ang. Chem. Int.* **41**, 4715–4718 (2002).
86. X. Liu, Y. Wang, P. Chen *et al.*, *Anal. Chem.* **86**, 2345–2352 (2014).
87. Y. Wang, X. Liu, J. Zhang *et al.*, *Chem. Sci.* **5**, 2651–2656 (2014).
88. P. Chen, R. Selegard, D. Aili *et al.*, *Nanoscale* **5**, 8973–8976 (2013).
89. H. Chen, J. Huang, A. Palaniappan *et al.*, *Analyst* **141**, 2335–2346 (2016).

Chapter 8
Chemical Synthesis of Gold Nanoparticles on Surfaces and in Matrices

Catherine Louis

Laboratoire de Réactivité de Surface, Université Pierre et Marie Curie — CNRS, Paris, France

8.1 Introduction

The size of the gold nanoparticles (NPs) that must be prepared onto supports or in matrices depends on the final use of the materials. For catalytic purposes, the gold particles must be as small as possible, i.e. smaller than 5 nm (Chapter 9), not only to reduce catalyst costs, but also because only the low coordination surface sites of the particles (or defects sites) are reactive, and the smaller the NPs, the higher the proportion of low coordination sites (Figure 2.5 in Chapter 2). For physics (Chapters 3–5 and 11) and biological applications (Chapters 14–16), larger particles can be required. Moreover, most often, for academic work in catalysis, supports are in a powder form, and essentially as oxides with or without porosity. Gold nanoparticles can also be deposited on organic supports or embedded into inorganic or organic matrices. For specific studies in physics and for 'model' catalysis (Chapter 12), gold particles must be deposited on planar surfaces (thin films or single crystals); physical techniques of deposition are often used, but also chemical ones, especially when controlled distance between the particles and ordering are needed.

In the past few years, many materials have emerged as supports for gold NPs: graphenes,[1–4] carbides,[5,6] polymers,[7–12] porous coordination polymers (PCPs) or metal–organic frameworks (MOFs),[9,13–16] and also specific oxides such as clays,[17–20] hydrotalcites[21–25] and hydroxyapatites.[26–29]

The chemical methods of preparation of gold nanoparticles onto supports are mainly based upon two principles, each being usually performed in two steps:

— Deposition of a gold precursor onto support in aqueous phase, followed by thermal or chemical reduction to produce metal NPs. This is called 'deposition–reduction' in the following. It is noteworthy that because of the instability of the Au(III) and Au(I) precursors, thermal treatments performed under oxidising atmosphere, oxygen or air, usually lead to the formation of metallic gold nanoparticles.
— Pre-formation of gold nanoparticles in liquid phase (aqueous or organic), also called gold sol or colloid, obtained by chemical reduction of a gold precursor in the presence of stabilising agents (Chapter 6) followed by the deposition onto a support. This is called 'reduction–deposition' in the following.

The deposition–reduction methods are essentially used for catalyst preparation when very small particles are needed. The reduction–deposition methods can be used for any application since the gold particle size can be adjusted within a large range depending on the method used for colloid preparation (Chapter 6); an additional step of decomposition of the stabilising agents may be required after deposition onto a support. Note that the metallic state of gold nanoparticles is easily detected with eyes, as the colour is red, pink or violet depending on the nature of the support or matrix, the gold particle shape and size and their concentration (Chapter 3).

Only a small number of gold precursors are commercially available: gold trichloride and tetrachloroauric acid or salts (Na, K) are the most commonly used, gold acetate and gold nitrate are both poorly soluble in water, and dimethyl-acetylacetonate gold(III) ($(CH_3)_2Au(acac)$) must be handled in air-free conditions. It is noteworthy that gold chloride speciation evolves as pH increases, leading to chloride hydrolysis and the formation of $[AuCl_{4-x}(OH)_x]^-$ species. A few Au(I) compounds are used, such as AuCl and $KAu(CN)_2$.

The characterisation of the gold particles mainly relies on the measurement of the particle sizes to get a distribution of size and an average particle size, and on the observation of their shape. For divided supports, various techniques of electron microscopy are used, while for planar supports, atomic force microscopy (AFM) and scanning tunnelling microscopy (STM) are used. Their electronic properties are often studied by X-ray photoelectron spectroscopy (XPS) or infrared spectroscopy (IR) coupled with probe molecule adsorption, such as CO. UV-visible spectroscopy is also used to detect the plasmon resonance.

Considering the huge amount of papers dedicated to the preparation of gold NPs on and in substrates, we had to limit the number of methods of preparation and the number of examples. Apart from a few exceptions, the limitations are the following:

— Chemical preparations of supported gold NPs in liquid phase.
— Preparations of supported gold NPs without specific shapes, i.e. no supported nanorods or other exotic shapes as shown in Chapter 6.
— Mere oxides as oxide supports and not composite oxides.

We have excluded:

— Supports whose particle sizes are close to the gold particles, so very few nanocomposites are described.
— Self-assembly monolayer particles (examples can be found in Chapter 7).

Except when this is specified, the particle sizes given correspond to the average sizes (d_{Au}), calcination treatments imply a thermal treatment under O_2 or air and thermal reductions imply the use of H_2.

In this second edition, we have included some preparations of supported gold-based bimetallic particles at the end of each main section. These preparations are more complex to describe because of the number of possible metal couples, and because some of them are miscible (Au–Pd, Au–Cu, Au–Ag) and other are not or poorly miscible (Au–Pt, Au–Ni, Au–Ru, Au–Ir, Au–Rh) (Table 6.5 in Chapter 6); it is obvious that it is more challenging to prepare bimetallic particles when the metals are poorly miscible in the bulk state. Most of the preparation methods are based on the same principle as those of monometallics, but they can involve co- or successive deposition–reduction or reduction–deposition, and in the case of successive preparations, two different methods can be used. Moreover, specific preparation methods have been developed for supported bimetallic samples (Section 8.2.4). Finally, in addition to the control of the particle size and distribution, there are at least two other challenges in the synthesis of supported bimetallic nanoparticles: the control of their bimetallic character and the determination of their structure (e.g. core–shell or alloy-type, Figure 6.8 in Chapter 6).

The formation of bimetallic particles can be attested by X-ray diffraction (XRD) if the particles are big enough and if the lattice parameters are different, but many other techniques can be used:

— Various techniques of electron microscopy, such as high-resolution electron microscopy (HRTEM) with the measurement of the d-spacing, high-angle annular dark field imaging (HAADF), energy-dispersive X-ray spectroscopy (EDX) of individual particles and electron energy loss spectroscopy (EELS) of individual particles or particle ensembles.
— Various techniques of spectroscopy, X-ray photoelectron spectroscopy (XPS) whose binding energy (BE) shifts may attest for charge transfer between the two metals, UV-visible spectroscopy showing possible shifts and shape changes of the plasmon band, IR coupled with the adsorption of probe molecules, X-ray absorption fine structure (XAFS).
— Even some catalytic reactions can be used to indirectly characterise the samples.

To avoid a too long text, the techniques used to demonstrate the bimetallic character of particles have not been reported. When the particle structures have been determined, this is specified in the text.

This chapter is divided into three main sections: preparation of gold nanoparticles on inorganic powder supports, embedded into matrices of various types and supported on planar surfaces. The section on the preparation of gold nanoparticles on powder inorganic supports is the longest one, and it is divided into three large sub-sections: deposition–reduction, reduction–deposition and in between a sub-section dedicated to the reduction in the liquid phase.

8.2 Gold Nanoparticles Supported on Powder Inorganic Supports

The powder supports are often oxide supports, such as alumina, titania and ceria, but can be also structured mesoporous supports, zeolites, nanotubes of different materials, carbons and carbenes. As mentioned in the Introduction, the choice of the supports depends on the application, and more specific supports can also be used.

8.2.1 *Deposition–Reduction (Deposition of Gold Precursor)*

These methods are mainly used for the preparation of catalysts, which means that the goal is to form metal particles as small as possible (<5 nm).

8.2.1.1 *Impregnation and related methods*

a. *Impregnation*

Impregnation is the simplest method of preparation of metal-supported catalysts, and it was the first one used for gold catalyst preparation. It consists of wetting the support with an aqueous solution containing the metal precursor, usually $HAuCl_4$. When the volume of solution is limited to the one corresponding roughly to pore filling, impregnation is called incipient wetness or dry impregnation (Figure 8.1(a)), and when the solution is in excess, it is called impregnation in excess of solution (Figure 8.1(b)). Afterwards, the sample is dried and then a thermal treatment is performed to reduce the precursor into metallic particles. The thermal treatment can

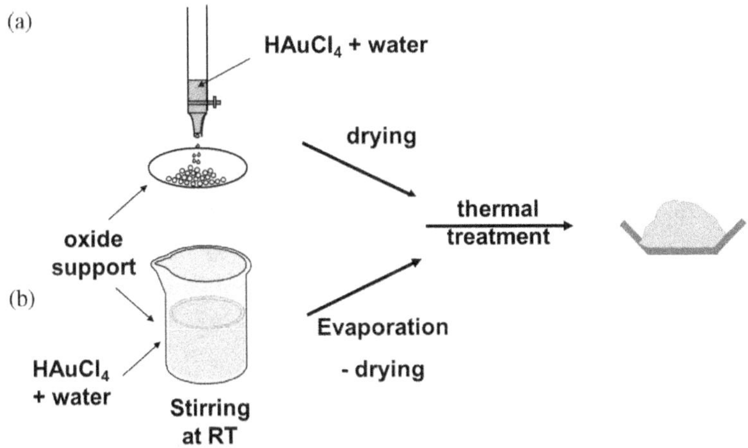

Figure 8.1   Scheme showing the various steps of the preparation of supported gold samples by (a) incipient wetnesss impregnation; (b) impregnation in excess of solution.

be performed under a reducing gas such as hydrogen or an oxidising gas such as oxygen or air since in both cases, and gold is reduced because of the instability of Au(III) compounds as mentioned above. This method has the advantage of being usable with any type of support, but the drawback is that the chlorides of the gold precursor are also deposited onto the support, which induces gold particle sintering during calcination and the formation of large gold particles (>10 nm) even for low gold loading (1–2 wt.%). This is unacceptable for catalytic applications as this leads to poorly active catalysts.[30–32] Smaller gold particles can be obtained under H_2, but the presence of residual chlorides may also be detrimental to catalytic performances[33] (Chapter 9).

Alternatively, impregnation of alkaline solution of chloride-free gold acetate (dissolved at pH 10–11 in Na_2CO_3 under boiling reflux because of its low solubility) can be performed on various supports (Al_2O_3, CeO_2, TiO_2, SiO_2 as well as silicates such as saponite clay and Y-type zeolite).[20] After solvent evaporation at 40°C and calcination at 350°C, the sample was extensively washed to remove Na ions and then dried again. The Au nanoparticles were smaller than 6 nm, and the smallest ones (2.3 nm) were obtained on saponite clay. Any metallic oxide can be applied as support, from strongly acidic (e.g. HY) to basic oxides (e.g. Al_2O_3) regardless of the PZC of the materials, provided that the support does not dissolve in basic solution.

b. Anion adsorption

Another way to deposit gold is to generate specific interaction between the gold precursor and the oxide support, so as to be able to wash the sample after deposition and remove the chlorides without leaching gold. The preparation is based on the fact that the hydroxyl groups present on the oxide surface may be protonated (OH_2^+) or deprotonated (O^-) depending on the solution pH in which the oxide is immersed. The pH value at which the total electric charge of the surface is zero is the point of zero charge (PZC) (Figure 8.2).

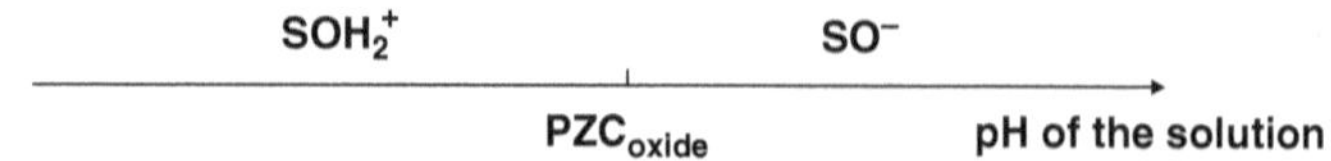

Figure 8.2   Oxide surface charge of an oxide in aqueous solution as a function of pH.

Adsorption of chloroauric anions on oxide supports is therefore in principle possible, provided that the pH of the solution is lower than the PZC of the oxide support, making the support surface positively charged; this excludes supports such as silica, the PZC of which is very low ($\sim$2). Practically, the powder is immersed into a solution containing the gold precursor at pH lower than the support PZC, stirred for a given time (usually around 1 h) and most often at room temperature. The sample is recovered after filtering or centrifugation, and washed several times with water to eliminate the chlorides as much as possible and any gold species not interacting with the support. In this type of preparation, the amount of gold deposited is limited by the capacity of adsorption of the oxide supports, which mainly depends on the PZC, the surface area and nature of the support and the pH of the solution. For usual oxide supports, the maximum gold loading is around 1–2 wt.%.[34,35] The gold particles are reasonably small after calcination $\sim$4 nm, and smaller after reduction under hydrogen, but the chlorides present in the adsorbed gold complex are not totally eliminated.[35]

c. Washing with ammonia

An efficient way to eliminate the chlorides after preparation is to wash the samples with ammonia solution (0.1–1 M). This was performed with samples prepared by incipient wetness impregnation[35–37] and anion adsorption.[38–40] Most of the gold was retained while chlorides were eliminated, which led to smaller gold particles after thermal treatment (3–4 nm). It was proposed that during ammonia washing at pH higher than the oxide PZC, gold hydroxychloride anion transformed into amino-hydroxo-aquo gold cation, $[Au(NH_3)_2(H_2O)_{2-x}(OH)_x]^{(3-x)+}$, which could interact with the support.[37] Initially performed with gold supported on alumina, silica and titania, this method has been more recently applied to other oxides, goethite and Mn- and Co-substituted goethites,[41] ceria-zirconia mixed oxides,[42] ceria doped with Fe, La or Zr[43] and La_2O_3.[44]

8.2.1.2 Deposition–precipitation and related methods

The method of deposition–precipitation is certainly the method most used for the preparation of gold catalysts since it readily leads to the formation of small gold particles (2–3 nm) with a very low amount of chloride. This method was first developed by Haruta's group, and thanks to this method, they discovered the extraordinary catalytic performances of small gold particles supported on titania in the reaction of CO oxidation at room temperature,[45,46] which is at the origin of the burst of catalysis by gold (Chapter 9). As shown below, the Haruta's deposition–precipitation method was performed at fixed basic pH, but deposition–precipitation can also be performed at increasing pH.

a. Deposition–precipitation at fixed pH

Deposition–precipitation at fixed pH involves the same procedure as for anion adsorption (Section 8.2.1.1b) except that the pH of the aqueous suspension containing both $HAuCl_4$ and the oxide support is adjusted at rather high pH, 7 to 10, by addition of a base, NaOH, KOH, Na_2CO_3, K_2CO_3 or NH_3. The mixture is usually stirred at 70–80°C for around 1 h. As for anion adsorption, the catalyst is washed with water to remove traces of the base and chlorides, dried between RT and 100°C, and usually calcined in air. This method is suitable for oxide supports, the PZC of which is higher than 5, such as MgO, TiO_2, Al_2O_3, ZrO_2, CeO_2, SnO_2, Fe_2O_3[47–51] and not for SiO_2 (PZC $\sim$2), SiO_2–Al_2O_3 (PZC $\sim$1)[48] or for activated carbon.[52]

The choice of the pH of 7–8 for titania support is the result of a compromise between the yield of gold deposition (which usually is lower than 100%, except when the targeted gold loading is low ($\sim$1 wt.%)), and the gold particle size resulting from the further thermal treatment. It is noteworthy that the conditions of preparation are such that both the support surface and the gold complex are negatively charged, which raises the question of the nature of the interaction between gold precursor and the support since it cannot be related to electrostatic interaction. As a matter of fact, the gold uptake on titania supports versus pH looks like a volcano curve with a maximum at pH 6, which is the PZC of TiO_2 (Figure 8.3).[53] An explanation for the interaction with the oxide support would be a reaction between anion gold complexes and OH group of the support, leading to the formation of a surface gold complex:

$$TiOH + AuCl(OH)_3^- \rightarrow [(TiO)Au(OH)_2Cl]^- + H_2O.$$

The amount of deposited gold is maximum when pH is close to the PZC because of the higher number of OH groups. More details on the mechanism of gold deposition can be found in Refs. 49, 54–56.

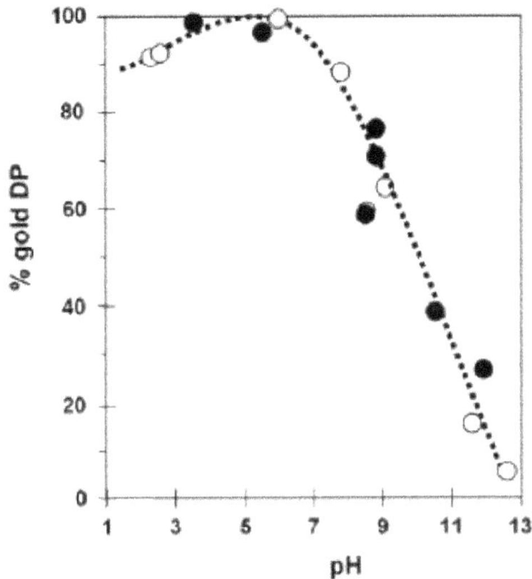

Figure 8.3   Gold uptakes on TiO$_2$ as a function of the pH upon addition of NaOH for a nominal gold loading of 1 wt.%. Adapted with permission from Ref. 51. Copyright (2007) Elsevier.

In the last few years, this method has been applied to other kinds of support, such as cobalt-modified silica,[57] carbon nitride-modified silica,[58] buckminsterfullerene(C$_{60}$)-modified silica[59] to favour Au deposition on silica, hydroxyapatite (Ca$_{10}$(PO$_4$)$_6$(OH)$_2$)[27,28,60–62] and various metal phosphates MPO$_4$ (M = Ca, Y, La, Pr, Nd, Sm, Eu, Ho, Er, Ce).[63–65]

b. Deposition–precipitation at increasing pH (with urea)

More recently, a variant based on the gradual rise in pH like in the procedure of deposition–precipitation developed first by Geus *et al.* for the preparation of metallic systems such as Ni/SiO$_2$ or Cu/SiO$_2$.[66,67] The rise in pH is obtained by the addition of urea (CO(NH$_2$)$_2$), which decomposes above 35–40°C into

$$CO(NH_2)_2 + 3H_2O \rightarrow CO_2 + 2NH_4^+ + 2OH^-.$$

Applied to the preparation of gold catalysts, the procedure consists of stirring an aqueous solution of HAuCl$_4$ and urea with the oxide support in suspension at ∼80°C. An extensive study performed by Louis *et al.*[34,49] showed that small gold particles (∼2 nm) could be obtained on various supports (TiO$_2$, Al$_2$O$_3$ and CeO$_2$) but again not on oxides with low PZC

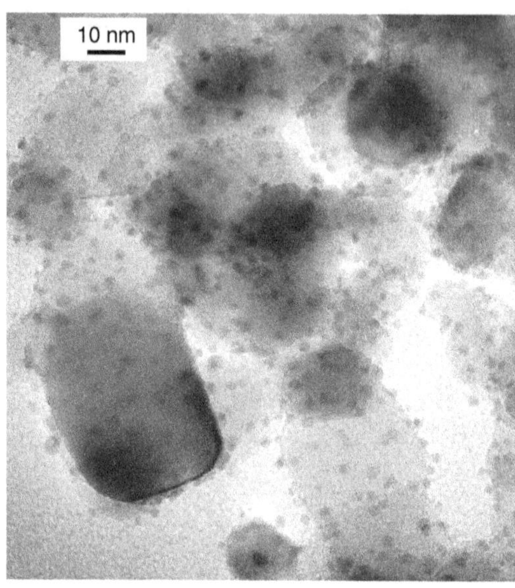

Figure 8.4   TEM image showing Au NPs around 2 nm in an 8 wt.% Au/TiO$_2$ sample prepared by DPU for 16 h and calcined at 300°C. Reprinted with permission from Ref. 34. Copyright (2002) American Chemical Society.

such as silica. In contrast with deposition–precipitation at fixed pH, all the gold of the solution (at least up to 8 wt.%, the highest Au loading studied) could be rapidly deposited onto the supports (within less than 1 h), but the gold particle size obtained after calcination at 300°C depended on the duration of the preparation (5.5, 5.2, 2.7 and 2.5 nm after 1, 2, 4 and 16 h, respectively) (Figure 8.4). Details on the mechanism of deposition–precipitation of gold with urea can be found in Refs. 35, 49, 55 and 68. When low gold loading is required ($\leq$1 wt.% Au, i.e. lower than the adsorption capacity of the support), it is possible to drastically reduce the preparation time to $\sim$1 h and to get small gold particles (3.0 and 2.2 nm after 1 and 16 h, respectively).[35]

Deposition–precipitation with urea (DPU) is applicable to the same type of supports as for deposition–precipitation at fixed pH, but with the advantage that all of the gold in solution is deposited onto the support, so the Au loading can be easily controlled *a priori*. Moreover, it is possible to prepare a set of samples with the same gold loading but with different particle sizes, varying the DP time, and not with the conditions of subsequent thermal treatment.

DPU of gold has been applied to several other oxide supports. In one method, 3 wt.% Au were deposited on ferric oxide with a yield close to 100%, and the gold particles were quite small (3–7 nm) after calcination at 350°C[69] and smaller (2 nm) for a sample with 1 wt.% Au calcined at 300°C.[70] Another study reports the use of this method to deposit gold on MgO and CaO supports.[71] After calcination at 400°C, gold particles of moderate size were obtained on magnesium oxide (8 nm for a sample with 7.5 wt.% Au) and on calcium oxide (6 nm for a sample with 4.7 wt.% Au).

Gold DPU has also been applied to more complex oxides like ceria-alumina,[72,73] ceria-zirconia,[74] ceria-gallia,[75] $Ce_xTb_yZr_zO_{2-v}$,[76] Cu-Cr oxide spinel[77] and also to more exotic supports like titano-silicalite,[78] β-MnO_2,[79] hydroxyapatite,[80,81] carbides[6] and nitrides.[82]

8.2.1.3 *Less common preparation methods*

a. Cation adsorption

Gold(III) ethylenediamine cation, $[Au(en)_2]^{3+}$ (en $= NH_2CH_2CH_2NH_2$), was used for cation exchange in zeolites[83,84] and for cation adsorption on oxide supports, TiO_2,[34,49] amorphous[85,86] and ordered mesoporous silicas,[87–89] zirconia and tungstated zirconia[90] and activated carbon.[91] $[Au(en)_2]Cl_3$ is easy to synthesise.[92] Cation adsorption requires that the oxide surface is negatively charged with O^- groups, which means that the pH of the aqueous solution containing $[Au(en)_2]Cl_3$ must be higher than the PZC of the support (Figure 8.2).

$[Au(en)_2]^{3+}$ adsorption must be performed at RT since the complex decomposes at around 60°C. As in the case of anion adsorption, the gold loading is limited by the adsorption capacity of the support, $\sim$1 wt.% on TiO_2 with gold particles of 2–3 nm after calcination at 300°C[49] and up to 9 wt.% on SBA-15 of much higher surface area with gold particles of 4 is -5 nm.[86] Again, the particles are smaller after reduction under H_2 than after calcination, but the samples retain residual organics after reduction, and it is necessary to proceed to further calcination at 400°C.[86]

Gold tetrammine nitrate, $Au(NH_3)_4](NO_3)_3$, is another precursor that can be used for cation adsorption, and it must also be synthesised. It was used with mesoporous carbon materials[93] and ETS-10 titanosilicate supports,[94] and small gold particles could be obtained after thermal treatment.

b. Deposition of organogold precursors

There are only a few examples of preparation involving organogold precursors, probably because they must be synthesised and are not easy to handle because of their toxicity and instability in ambient air.

The phosphine gold complex $[Au(PPh_3)]NO_3$ and the phosphine-stabilised gold clusters $[Au_9(PPh_3)_8](NO_3)_3$ and $[Au_6(PPh_3)_6](BF_4)_2$ and also larger ones, $Au_{101}(PPh_3)_{21}Cl_5$, were used to impregnate various oxides such as TiO_2, SiO_2 or α-Fe_2O_3 in organic solvents.[95–98] They led to rather small gold particles, between 3 and 7 nm, after thermal treatment to decompose the phosphines. Other organogold compounds were used for the impregnation of TiO_2 (1 wt.% Au): $[Au_2^I(dppm)_2](PF_6)_2$, $Au_3^I(Ph_2pz)_3$, $[Au_4^I(dppm)_2(3,5-Ph_2pz)_2](NO_2)_2$, and $Au_4^I(form)_4$, (dppm $=$bis(diphenylphosphino)methane, form $=$formamidinates, pz $=$pyrazolate), leading to gold particles of 7.7, 4.8, 3.1 and 3.0 nm, respectively.[99] The advantage of these precursors is that they do not contain chlorides, but traces of phosphorus may remain onto the support.

The thiol-ligated $Au_{38}(SC_{12}H_{25})_{24}$ cluster was also used as a precursor for TiO_2 support impregnation, and led to 3.9 nm Au particles.[100]

Another type of organogold compound is the dimethyl-acetylacetonate gold(III), $(CH_3)_2Au(acac)$; it is commercially available, but expensive, and it must also be handled in the absence of moisture and air. Impregnation in dried pentane or hexane was performed with various supports, MgO,[101] γ-alumina,[102] TiO_2,[103] La_2O_3[104] and Na–Y zeolites.[105] Chemical vapour deposition (CVD) is another method to deposit this compound on oxides and activated carbon.[106] In both cases, impregnation and CVD, gold particles around 3 nm were obtained after decomposition of the grafted complex. The mechanism of $(CH_3)_2Au(acac)$ adsorption and decomposition into gold particles was studied in Refs. 102 and 107. This compound is also used as a precursor for another type of deposition method, solid grinding, described below.

c. Solid grinding

This method has been first applied in 2008 by Haruta and co-workers to the preparation of supported gold catalysts.[9] It is very simple since it consists of grinding a mixture of powder support and dimethyl-acetylacetonate gold(III) in the absence of solvent and of a subsequent thermal treatment.

Grinding can be performed manually in a mortar or by ball-milling, in air at room temperature.

This method was developed first for the deposition of gold nanoparticles on several porous coordination polymers (PCPs) or metal–organic frameworks (MOFs)[9] (Section 8.3.4). A loading of 1 wt.% gold could be deposited, leading to small gold particles of 1.5 nm after H_2 treatment at 120°C. Several other supports have been used such as carbon, oxides and cellulose.[108–111] The amount of gold deposited depends on the nature of the support and on the establishment of bonding between the gold precursor and the OH's of the support. Moreover, depending on the support, the preparation must be performed either under low humidity levels (<50%) (mesoporous titano-silicate), under high humidity levels (cellulose) or in acetone (alumina or zirconia). After calcination treatment, particles around 2 nm were obtained except on the semiconducting supports, ceria and titania, which gave larger particles because of the uncontrolled gold reduction during grinding. More recent papers confirm that gold particle size strongly depends on the nature of the support and the loading.[112,113]

8.2.1.4 *Gold-based bimetallic catalysts prepared by deposition–reduction*

Impregnation, which is no longer the method most used nowadays for the preparation of monometallic gold supported on oxides, still remains used for the preparation of bimetallic catalysts. Several variants are possible, such as co-impregnation, sequential impregnation of one metal precursor followed by the other one, or impregnation associated with another method.

Among bimetallics, the most studied and therefore the most frequently prepared is certainly bimetallic Au–Pd. Supported Au–Pd samples have been prepared by co-impregnation of $HAuCl_4$ and $PdCl_2$ or H_2PdCl_4 on various supports, TiO_2,[114,115] Al_2O_3,[35] mesoporous silica,[116] MOF,[117,118] CeO_2,[119] activated carbon[120] and carbon nanotubes.[121] After impregnation, the samples are either calcined and reduced, or directly reduced at various temperatures. For unknown reasons, in some cases, rather good results in terms of bimetallic character and of small particle sizes are obtained,[35,119,121] while in other ones, bimodal size distributions or large size are obtained.[114,116,117,120] In spite of the fact that Au and Pd are miscible metals in any proportion, the samples containing bimodal particle size distribution also show heterogeneous composition; the large particles (up to 50 nm or more) are gold-rich while the smaller ones are Pd-rich. Attempts were made to reduce the particle sizes by addition of HCl during the preparation[115] or washing with NH_4Cl[122] (see also Section 8.2.1.1c).

Co-impregnation and successive impregnations have also been used to prepare other oxide-supported bimetallics, such as Au–Cu,[123,124] Au–Ir[125] and Au–Pt.[126,127] Again, the resulting particle sizes are at variance, and they also depend on the Au/M ratio. In the two last systems, although Au is not miscible in Ir and Pt, small bimetallic particles were obtained. In the case of Au–Pt/TiO_2,[126] Au core–Pt shell structures were identified after reduction at 250°C and alloy structures after reduction at 500°C.

Ion adsorption is barely used in the preparation of bimetallics. For instance, $PdCl_2$ and $HAuCl_4$ were co-adsorbed at pH 4 on an hydrotalcite (a Mg–Al layered double hydroxide).[128] After chemical reduction in ethanol and hydrazine (NH_2NH_2) (Section 8.2.2.1), 2.4 nm particles were obtained. Ion adsorption can also be associated with impregnation. For the preparation of Au_1–Cu_3/TiO_2, gold anion adsorption was performed first (Section 8.2.1.1b), followed by washing with ammonia and then impregnation of $Cu(NO_3)_2$.[129] After reduction at 300°C, bimetallic particles of the size 3.5 nm were formed.

The method most used for the preparation of supported bimetallics by deposition–reduction is probably deposition–precipitation (DP), co-DP or successive DP. Co-DP is in principle not suitable for the Au–Ag system (miscible) because the chlorides of the Au precursor can induce the precipitation of AgCl. However, co-DP with NH_3 was applied to the preparation of Au–Ag/CeO_2, leading to the formation of bimetallic alloy particles smaller than 3 nm, the size of which decreased as the Ag/Au ratio increased.[130] Co-DP with NaOH was used for the preparation of Au–Pd[131] and Au–Pt on TiO_2,[132] and led after thermal treatment to bimetallic particles around 7 and 5 nm, respectively. Co-DP with urea of Au–Pd on TiO_2 and Al_2O_3 with Au/Pd $\gg$ 1 led to bimetallic particles smaller than 3 nm after reduction at 500°C.[35] Co-DPU was also very efficient for the deposition of Au–Cu on TiO_2, also leading to small and bimetallic Au–Cu particles.[133,134]

Among the two-step preparations, one can find examples combining deposition–precipitation and then impregnation[135–137] and vice-versa, impregnation then deposition–precipitation[127,138] or examples of sequential deposition–precipitations.[139–142]

8.2.2 *Reduction in Liquid Phase*

Up to here, only thermal treatments have been discussed to reduce supported gold precursors into gold particles, but it is also possible to perform a chemical reduction by addition of a reducing agent, such as sodium borohydride ($NaBH_4$), in the suspension containing the powder sample. Chemical reduction can also be assisted by microwaves. Alternatively, other reduction modes can be performed in the liquid phase, such as photoreduction or sonochemical reduction; moreover, deposition and reduction can take place in a single step. The reduction mode may have an influence on the final particle size, but the extent of gold deposition is often found to be quantitative. There is an interesting paper in which four reduction methods have been applied to an Au/TiO_2 sample prepared by DP NaOH at pH 10: UV irradiation, microwave irradiation (MW), thermal reduction under H_2 at 200°C and chemical reduction with $NaBH_4$.[143] The gold loading was more or less the same in all samples (0.3 wt.%), but the Au particle size was greatly influenced by the reduction method. The largest particles ($\sim$11 nm) were obtained using UV and MW irradiations while the smallest ones (2.4 nm) were obtained after H_2 reduction. With $NaBH_4$, intermediate size particles were obtained (4.7 nm). Gold was completely reduced to the metallic state except in the UV-reduced sample.

8.2.2.1 *Chemical reduction*

One can cite three examples of chemical reduction: 1 wt.% Au/TiO_2 prepared by DP NH_4OH and reduced with $NaBH_4$, leading to particles smaller than 5 nm,[144] 1 wt.% Au/TiO_2 prepared by DPU and also reduced with $NaBH_4$, leading to particles of 3.8 nm[145] and 0.3 to 2 wt.% Au/TiO_2 prepared by DP NaOH and reduced with dimethylamine borane in ethanol, leading to particles of 3–4 nm for Au.[146] Thorough washing of the solid must be performed to ensure that samples are not contaminated by residual compounds, such as sodium, which might have an influence on the catalytic properties. This is systematically applied to all sample preparations reported in the following sections.

8.2.2.2 *Chemical reduction assisted by microwave irradiation*

Fast heating by microwaves accelerates the reduction of metal precursors in the presence of a stabiliser or a mild reducer, such as PVP, polyethylene glycol (PEG), ethylene glycol (EG) and citrate, therefore favouring the formation of small particles (Chapter 6). When MW irradiation is applied to a solution also containing a support in suspension, oxide or carbon, the metal particles formed can directly interact with the support.

MW irradiation has been mostly applied to the preparation of gold supported on carbons.[147–149] For instance, gold particles (5–10 nm) supported on carbon nanotubes (CNTs) were obtained from a suspension of CNTs in a mixture containing $HAuCl_4$, ethylene glycol, oleylamine and oleic acid; the suspension was irradiated by MW with power adjusted to set up the temperature at 180°C.[149] Active sites created on the surface of CNTs under microwaving ensured the nucleation of gold nanoparticles. Au on CNTs was prepared from CNTs introduced in a solution containing Na citrate, propylene–glycol and $HAuCl_4$, which was irradiated for 5 min.[150] The resulting sample contained 40 wt.% Au with Au particles of 10–13 nm.

8.2.2.3 *Photochemical deposition–reduction*

Photo-deposition is based on the principle that metal cations with appropriate redox potentials can be reduced by the photoelectrons generated by bandgap illumination of the semiconductor used as support with UV light. This is a one-pot one-step preparation. The mechanism of photo-reduction has been described in Ref. 151. This method is often used for the preparation of photo-catalysts, and especially recently with the extensive development of photocatalysis and assisted-plasmon catalysis (Chapter 10). This method is efficient for titania and zinc oxide supports, but not for tin and iron oxides even though they are also semiconductors.[152] One can also find examples of gold photo-deposition on composite supports such as ZnO-CNT[153] or C-TiO_2 composites.[154]

Practically, the preparation procedure starts with a solution containing the gold precursor, most often $HAuCl_4$, and the oxide support in pure water or mixtures with methanol or ethanol. The suspension is generally de-aerated and UV irradiated under an inert gas to avoid the presence of O_2, which is an electron acceptor. Additional parameters can be controlled, such as pH,[155,156] addition of citric acid[157] and temperature.[158] The irradiation time is at variance from a few minutes (2 min[159]) to several hours (24 h[160]). The final gold particle size on titania (0.5 wt.% Au) varies from 5–10 nm[161] to 25 nm.[162] The gold loadings are typically between 0.5 and 2 wt.% Au, and the deposition yield can reach 100%. Hidalgo *et al.*[161] studied the influence of the irradiation time and the light intensity. They recommend the use of low-light intensity and to play with the deposition time to get small gold particles well distributed onto the support surface with an effective deposition yield.

8.2.2.4 *Sonochemical deposition–reduction*

Like UV irradiation, sonication can be used to simultaneously deposit and reduce gold on a support (one-pot one-step preparation). Sonication is also used to synthesise gold colloids and the principle was developed in Chapter 6.

This method has been most largely developed by Gedanken and co-workers for the formation of gold nanoparticles on supports such as silica submicrospheres,[163] silica and titania[164] and mesoporous TiO_2, Fe_2O_3 and Fe_2O_3–TiO_2.[165] After sonication under Ar or Ar and H_2, washing and drying, gold particles of 4–5 nm were obtained with 100% deposition yield (5–7 wt.% Au). Gold nanoparticles on carbon ($d_{Au} = 4-8$ nm),[166] on ZnO (d_{Au} between 5 and 10 nm)[167] and on MgO ($d_{Au} - 6.5$ nm)[168] were also prepared with this technique. Sonication was also applied to the preparation of gold NPs (6.7 nm) on γ-Fe_2O_3 in the presence of polyethylene glycol monostearate (PEG-MS).[169] PEG-MS was supposed to act both as a surfactant and as a reducing agent.

Note that most often, sonication is simply used in the preparation of supported gold to disperse the powder support in solution[167] or to avoid the aggregation of gold colloids and favour their interaction with a support.[170]

8.2.2.5 *Gold-based bimetallic catalysts obtained by reduction in liquid phase*

Again different strategies explained below can be used to prepare supported bimetallic particles by reduction in the liquid phase.

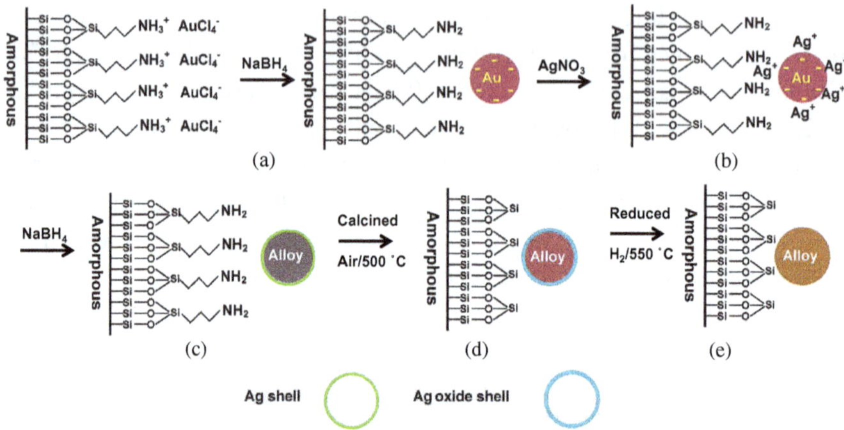

Figure 8.5   Schematic illustration of the procedure of synthesis of SiO_2-supported Au–Ag alloy nanoparticles. (a) Initial formation of Au particles; (b) Ag^+ adsorbed on the Au particles; (c) Au–Ag nanostructure with alloy core surrounded by a silver shell (green); (d) Au–Ag nanostructure with more gold-rich alloy core covered by a Ag oxide shell (blue); (e) random alloy of Au–Ag nanoparticles on SiO_2 support. Adapted with permission from Ref. 171. Copyright (2009) American Chemical Society.

Liu *et al.* used a two-step method for the preparation of Au–Ag and Au–Cu on silicas functionalised with APTES (3-aminopropyltriethoxysilane) (Figure 8.5).[171,172] $HAuCl_4$ solution was added to APTES-SBA-15 and to APTES-SiO_2, followed by reduction with $NaBH_4$. Then, the washed solid was immersed into a copper or silver nitrate solution, which was also reduced by $NaBH_4$, leading to Au core–Ag or Cu shell particles. After washing and drying, the solid was calcined then reduced at 550°C to decompose APTES, and transformed the structure of the NPs into Au–Ag or Au–Cu nanoalloy. The size of the Au–Ag particles was 3.3 nm for Au/Ag = 3/1 and 3.4 nm for Au/Ag = 1/1.[171] The same trend was observed for Au–Cu 2.4–2.8 nm in SBA-15 and 3.0–3.6 nm on silica according to the Au/Cu ratio,[172] and for both systems they were smaller than in the monometallic Au sample (5.7 nm).

Alternatively, Au–Cu supported on silica was prepared by direct adsorption of $Au(en)_2^{3+}$ cations on silica (Section 8.2.1.3), followed by reduction at 150°C, then by sample immersion in a deoxygenated solution of copper acetate in 1-octadecene containing reducing agents, oleic acid and oleylamine.[173] The resulting particles were bimetallic with a size of 4 nm and again smaller than those in the Au NPs in Au/SiO$_2$ (4.9 nm).

Alloy-type Au–Pd NPs of 4–5 nm were prepared on phosphate-modified hydrotalcite in a single step, by co-impregnation with a solution of HAuCl$_4$ and PdCl$_2$ followed by chemical reduction with lysine followed by NaBH$_4$.[25] Au–Ni (non-miscible) was also prepared in a single step on activated carbon:[174] A suspension of carbon in a ethyleneglycol solution containing HAuCl$_4$ and NiCl$_2$ was heated at 85°C and hydrazine was added. However, Au–Ni alloying was obtained by heating the sample under N$_2$ between 300 and 500°C; alloying improved with temperature, but particle size increased (>10 nm) and unalloyed Ni remained on the sample.

Photo-deposition techniques (Section 8.2.2.3) can also be used to prepare supported bimetallic particles. This was done for Au–Pt in mesoporous titania from a solution containing HAuCl$_4$, H$_2$PtCl$_6$ and the support dispersed in a methanol solution.[175] The deoxygenated mixture was subjected to UV irradiation. The Au–Pt particles were formed with a yield of 90%, and a composition close to the nominal one. They were 7.3 nm with an Au core–Pt shell structure (Figures 8.6(a) and 8.6(b)). After calcination at 350°C, the particle size did not change (7.9 nm), but the structure became an alloy (Figures 8.6(c) and 8.6(d)).

Chen *et al.*[176] described a one-pot synthesis of Au–Pd nanoparticles on graphene, which acted as a three-function agent, a reducer, a stabiliser and a support. Graphene and HAuCl$_4$ in aqueous solution were mixed first, then K$_2$PdCl$_4$ was added at RT, and bimetallic particles of 3.4 nm were directly obtained.

8.2.3 *Reduction–Deposition (Deposition of Preformed Gold Particles)*

Deposition of gold nanoparticles preformed in solution (gold colloids or gold in micelles or in dendrimers), also called sol immobilisation, is another strategy to prepare supported gold samples (Figure 8.7). In principle, the advantage of starting from colloids is that the metal particles have a controlled size with a narrow distribution (Chapter 6). A recent review on sol immobilisation can be found in Ref. 177.

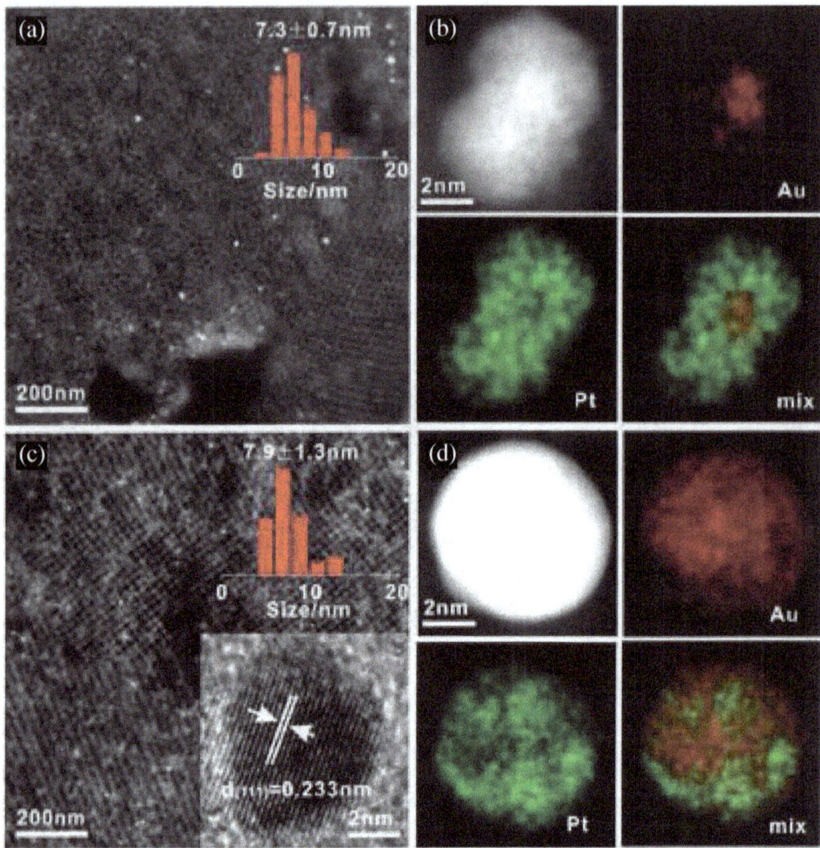

Figure 8.6 HAADF-STEM images and particle size distribution, typical EDS mappings of $Au_{50}Pt_{50}/EP$-TiO_2 after the photo-deposition (a and b) and after annealing at 350°C (c and d). Adapted with permission from Ref. 175. Copyright (2014) Royal Society Chemistry.

8.2.3.1 *Gold colloids*

As mentioned in Chapter 6, gold colloids or sols are obtained by reduction of a gold precursor, usually $HAuCl_4$, in aqueous or organic phase and in the presence of stabilising or capping agents, which can be molecules, such as CO, citrate, thiol, amine (e.g. lysine or oleyalmine) or polymers, such as polyvinylpyrrolidone (PVP) and polyvinylalcohol (PVA). The stabilisers can also be micelles formed from diblock copolymers, surfactants (Chapter 6) or dendrimers (Section 8.2.3.3). The stabilisers are added to the

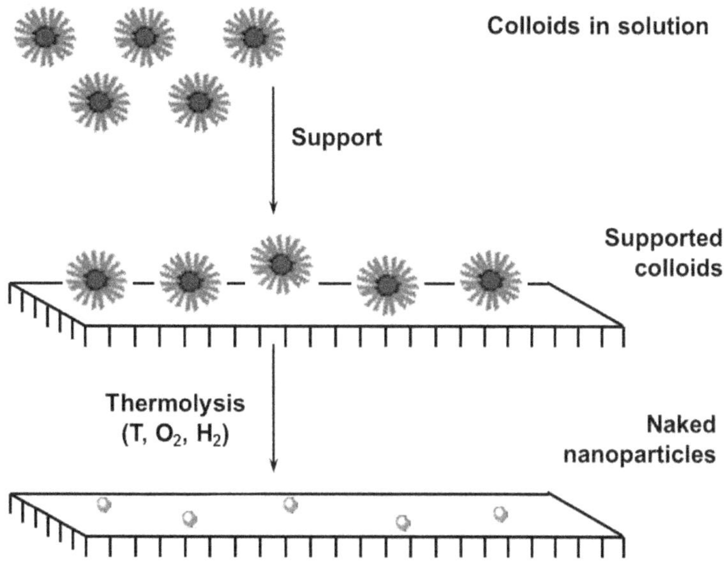

Figure 8.7   Scheme showing the principle of the reduction–deposition preparation method.

solution to control the growth of the particles during reduction and avoid aggregation and precipitation. Reduction is usually performed by addition of a chemical agent like those mentioned in Section 8.2.2, i.e. sodium boro-hydride, hydrazine or weaker reducing agents such as amine–borane com-plexes, methanol or glucose. The stabiliser can also act as a reducer; this is the case for sodium citrate and tetrakis(hydroxymethyl)phosphonium chloride (THPC). As already mentioned in Section 8.2.2, reduction can be assisted by heating, sonication,[178] radiolysis, UV[179] or microwave irra-diation.[180] The formation of a gold sol is attested by the colour change of the solution, from pale yellow or colourless to pink or red. The aver-age size and size distribution of the so-prepared gold particles, as well as the shape, strongly depend on the conditions of synthesis, i.e. temper-ature, ageing time, concentration and nature of the different constituents and reduction technique; the size may vary between a few to hundreds of nanometres.

In the case of gold colloid synthesis in aqueous medium, sol immo-bilisation is performed through colloid adsorption, by mere immersion of the support into the colloidal solution, followed by washing with water and drying. Gold colloid adsorption is attested by the decolouration of the solution and the colouration taken by the support. This method is appropriate for supports such as carbons because of their acidic carboxylic groups, which can easily interact with colloids.[108] Colloid surfaces are in general negatively charged because of the stabilising agents. They can electrostatically interact with an oxide support provided that the pH of the solution is lower than the support PZC (Figure 8.2). The deposition step is not a trivial issue in terms of deposition yield, final gold particle size and particle dispersion onto the support. For instance, the gold particles have a tendency to grow on active carbons, except when PVA is used as stabilising agent (SA).[181] The growth trend also depends on the SA/Au ratio.[177] Hence, a delicate balance must be achieved between several parameters, such as the nature and the concentration of the SA, the SA/Au ratio, the solution pH and the nature of the support. Different sizes were reported for citrate-stabilised gold particles on silica (6–8 nm),[182] titania (10–15 nm)[183] and alumina (10–30 nm).[184] Note that the particle sizes reported are those of the gold particles without the corona of stabilisers. To favour interactions between gold colloids and the oxide support, it is also possible to functionalise it.[185–187] For instance, particles of Fe_3O_4 were functionalised with APTES (3-aminopropyltriethoxysilane) to interact with negatively charged THPC-stabilised gold colloids.[185]

In the case of gold sols in organic medium, the problem is to reconcile colloid hydrophobicity with support hydrophilicity. One solution is to proceed to a mere impregnation of the support with the colloidal solution, followed by solvent evaporation: for instance, dodecanethiol-stabilised gold colloids (1.9 nm, prepared with the Brust method, see Chapter 6) in ethyl acetate were impregnated on ceria ($\sim$2 wt.% Au);[188] other colloids (1.8, 3.9 and 9.9 nm) in hexane were impregnated on silica, titania and carbon ($\sim$2 wt.% Au).[189] However, this method does not permit the control of the dispersion of the gold particles over the supports because of the absence of specific interaction. One solution proposed by Zheng and Stucky[190] is to use an aprotic solvent such as chloroform or methylene chloride. As a result, dodecanethiol-stabilised gold colloids were readily adsorbed on oxide powders, such as TiO_2, SiO_2, ZnO and Al_2O_3. Herranz *et al.*[191] used this method to adsorb hexadecanethiol-stabilised gold colloids (2, 3.7 and 4.8 nm) in chloroform on SiO_2 and TiO_2 (1–2 wt.% Au).

Depending on the final applications, stabilising agents may have to be eliminated; this is the case for further gas-phase catalytic reactions. In the case of liquid phase reactions, the density of SA must be controlled, and the presence of SA can even play a positive role in catalytic performances.[192] The removal of SA is a critical step because gold particles may sinter, and stabiliser residues may poison the catalyst.[193] Thermal treatments under air or oxygen are the most frequently used. For unknown reasons, some studies report the growth of the particles during thermal treatment[188,194–196] while others do not in spite of the rather high temperature applied in all cases ($\sim 300°C$).[170,190,191,197–199] Sintering can be prevented with a SA that decomposes at low temperature. This is the case for lysine that decomposes at $200°C$ without growth of the Au particles (<5 nm) supported on $\alpha\text{-Fe}_2\text{O}_3$.[170]

Alternatively, other types of treatments can be used: ozone treatment or plasma activation, both performed at RT. However, they require thin layers of materials, and these techniques are most often applied to planar surfaces (Section 8.4). Only in a few cases, they have been used for powder supports: the decomposition of thiolate phosphine ligands by ozone led to much smaller Au particles (1.2 nm) on TiO_2 than after calcination at $400°C$ (2.7 nm),[200] the decomposition of dodecanethiol by oxygen plasma led to gold NPs on cordierite ($Mg_2Al_4Si_5O_{18}$) much smaller (3.5 nm) than after thermal treatment at $250°C$ in He (10–20 nm),[201] and the removal of PVP by UV-ozone treatment preserved the initial gold particle size on silica (4.9 instead of 4.7 nm).[202]

Recently, new ways to remove the stabilisers have been investigated. Since many of the SA are water soluble, attempts have been made to remove them by washing, for instance under water reflux at $90°C$ in the case of PVA[203,204]: 20% of PVA could be removed and gold particles on TiO_2 only slightly increased in size. PVP and other organothiols could be removed by washings with solutions containing $NaBH_4$ and/or *tert*-butylamine followed by washing in ethanol/acetone without affecting the size of the NPs.[205–207] In some cases, acidic treatments have also been efficient.[208,209]

8.2.3.2 *Gold in micelles*

Reverse micelles as stabilisers, i.e. water-in-oil emulsion, are most often used for planar surfaces as they allow quasi-hexagonal ordering of the gold nanoparticles (Section 8.4.2), but there are a few examples involving this technique for gold particles immobilisation on powder supports. Several strategies can be adopted.

Gold particles in micelles can be synthesised before support deposition. For instance, Au particles were synthesised in a mixture of water, poly(ethyleneglycol)-dodecylether (BRIJ 30) and *n*-heptane by the reduction of $HAuCl_4$ with $NaBH_4$ in the core of the micelle then adsorption on carbon nanotubes (CNTs) followed by washing and drying.[150] Small particles of 5 nm were obtained in spite of the high Au loading of 32 wt.%.

Micelles can also be used as a medium to deposit gold precursor: Diblock copolymers dissolved in toluene formed micelles with a hydrophilic core containing $HAuCl_4$, which were deposited onto oxides by impregnation. After calcination at 300°C to reduce gold and to remove the polymers, gold particles of various sizes (8, 12 and 22 nm) were obtained on TiO_2, ZnO and ZrO_2, respectively.[210]

Micelles can also be deposited onto a support before gold introduction. This is the case for gold on yttrium hydroxide nanotubes (YNTs) with first, the adsorption of poly(ethylene oxide)-*b*-poly(4-vinylpyridine) (PEO-*b*-P4VP) micelles.[211] Then, $HAuCl_4$ was introduced and reduced with $NaBH_4$. The samples exhibited a high loading capacity (17 wt.% Au) and small Au particle size (4 nm). There was no attempt to decompose the organics in this work.

8.2.3.3 *Gold in dendrimers*

The stabilisers of the gold colloids can also be dendrimers, often polyamidoamine (PAMAM, commercially available), which are hyperbranched polymers that ramify from a single core and form a porous sphere[212,213] (Figure 8.8). Dendrimers are defined by the number of generations and the nature of the termination function, e.g. G4-OH or $G6-NH_2$. Dendrimer-encapsulated nanoparticles (DENs) are synthesised by sequestering metal ions, which are afterwards chemically reduced. When NH_2-terminated dendrimers are used, the ammine functions can reduce the Au(III) precursors.[214] DENs can be synthesised in water or alcohols. The resulting metal NPs sizes are usually small and nearly monodispersed, and they can be tuned by

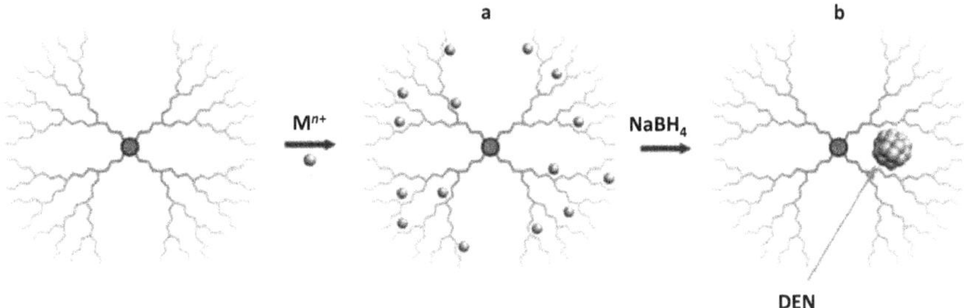

Figure 8.8   Schematic representation of a dendrimer (a) metal ions into a dendrimer (b) dendrimer encapsulated nanoparticles (DENs) after chemical reduction. Adapted with permission from Ref. 216. Copyright (2013) Springer.

varying the metal to dendrimer ratio, the dendrimer composition and size, and the pH. As in the case of colloids, DENs can be deposited onto supports or used as a templating agent for the preparation of embedded nanoparticles (Section 8.3).

As for colloids, it may be desirable to remove the dendrimers from the gold particles, and this may also lead to an increase of the metal particle size. For instance, a calcination at 500°C was required to remove PAMAM dendrimers from gold DENs impregnated on titania, which led to particle growth from ~2 to ~7 nm.[215,216] As for colloids, lower temperature routes, such as oxygen plasma treatment, must permit the retention of the particle size. Another strategy to prevent nanoparticle sintering is to extract them from dendrimers. For instance, gold nanoparticles in PAMAM were extracted by *n*-alkanethiol in toluene. The thiol-protected Au particles were then adsorbed on TiO$_2$[199] and on CNTs.[217] In the first case, the thiols were eliminated by H$_2$ treatment at 120°C, and the Au NPs were almost as small (3.4 nm) as stabilised with thiols (3 nm). In the second case, the gold particles on CNTs were 2.4 nm.

8.2.3.4   *Gold-based bimetallic catalysts prepared by reduction–deposition*

a. Bimetallic colloids

The methods used for the synthesis of bimetallic colloids described in Chapter 6 can be applied before deposition onto supports.

Preparations involving the deposition of Au–Pd colloids on activated carbon were widely developed by the Prati's group, using PVA as a stabiliser and $NaBH_4$ as a reducing agent.[218–221] Their best results in terms of single-phase bimetallic particles with small sizes were obtained by sequential reduction–deposition,[219] i.e. adsorption of PVA-gold colloids on activated carbon, followed by palladium reduction on gold nanoparticles by bubbling H_2. In fact, the second step derives from the redox method described later in Section 8.2.4.2. Small Au–Pd nanoparticles of ∼3.6 nm were obtained.[221] Konuspayeva *et al.*[222] used almost the same type of method to prepare Au–Pd and Au–Rh (a non-miscible system) on TiO_2 nanorods. They reported that a treatment under H_2 or under O_2 then H_2 at 350°C was efficient enough to remove PVA without drastic increase in particle size: from 2.5 to ∼3 nm for Au–Pd and from 3 to 3.7 nm for Au–Rh. The particles in Au–Pd/TiO_2 remained bimetallic whereas those in Au-Rh/TiO_2, initially Janus-type, tended to segregate into small Rh particles and larger Au ones. PVA was also used to stabilise Au–Ag colloids first with the synthesis of PVA–Ag colloids then the reduction of $HAuCl_4$. After adsorption on TiO_2, calcination at 400°C and then reduction at 350°C, the particles were alloy-type with sizes between 2.9 and 5.2 nm depending on the Au/Ag ratio.[223] Based on the same principle of preparation, PVA–Au–Pt colloids were adsorbed on CNTs. After treatment in N_2 at 350°C, the particles were ∼4 nm and bimetallic.[224] PVP was also used for the preparation of Au–Pd particles by co-reduction of $HAuCl_4$ and $PdCl_2$ with ethanol and water at 90°C.[225] After adsorption on silica, the particle size was 3.5 and 6.2 nm after calcination at 400°C.

Tetrakis(hydroxymethyl) phosphonium chloride (THPC), which is both a stabiliser and a reducing agent, was used to prepare Au–Pd supported on titania, alumina and silica.[226] After co-reduction of $HAuCl_4$ and Na_2PdCl_4 in a basic solution of THPC, adsorption on the supports and then reduction at 250°C, the particles were bimetallic with 3.5 nm size on alumina, 3.0 nm on TiO_2 and 5.6 nm on SiO_2. Citrate, which is also a stabiliser and a reducing agent, was used for the preparation of Au–Pd particles on TiO_2[227,228] and graphene.[3] It was also used for the preparation of Au–Ag colloids; after adsorption on TiO_2 under UV irradiation, the sample contained bimetallic particles of 10 nm size.[229]

One can find also a few examples of colloid preparation in the organic phase. For instance, Au–Cu/TiO$_2$ samples with different Au/Cu ratios were obtained from pre-formed thiol-capped nanoparticles in toluene, after phase transfer of AuCl$_4^-$ and Cu^{2+} ions from water to toluene and reduction with NaBH$_4$.[230] After adsorption on titania followed by calcination at 400°C, ~5.6 nm particles with alloy structure were obtained.

b. Bimetallics in micelles

Au–M particles (M = Fe, Co, Ni, Cu, Zn) synthesised in reverse microemulsion were supported on carbon.[231,232] Two sets of reverse micelle solutions were prepared by adding sodium bis(2-ethylhexyl)sulfosuccinate (AOT, the surfactant) in *n*-heptane (the oil phase) to an aqueous solution of HAuCl$_4$ and NiCl$_2$ and to one containing NaBH$_4$.[231] After mixing of the two solutions, the carbon support was added. After reduction at 250°C to decompose AOT, the particles were bimetallic and uniformly dispersed on carbon with a size of ~3 nm. Au–Pd particles on carbon were prepared with the same technique using Triton-X 114 as the surfactant, cyclohexane as the oil phase and hydrazine as the reducing agent;[233] alloy-type Au–Pd particles were 5–8 nm in size after drying at 120°C.

Block copolymer micelles in toluene were also used to stabilise bimetallic Au–Ag nanoparticles of 3 nm after hydrazine reduction.[234] After deposition on various supports, TiO$_2$, nanostructured TiO$_2$ (*n*-TiO$_2$) and Al$_2$O$_3$, the polymer matrix was removed by thermal treatment at 400°C. After treatment in H$_2$, rather large particles between 5 and 20 nm were obtained on TiO$_2$ whereas the particles were much smaller (3 nm) on *n*-TiO$_2$ or Al$_2$O$_3$. Using O$_2$, similar results were obtained on *n*-TiO$_2$ while they sintered on TiO$_2$ and alumina (12 nm).

Reverse micelles were also used to accommodate anionic gold and palladium complexes in their hydrophilic core. Deposition onto TiO$_2$ followed by a reduction treatment at 350°C led to bimetallic nanoparticles of 3 nm.[235]

c. Bimetallics in dendrimers

One can find studies exploring the use of dendrimer-encapsulated nanoparticles (DENs) (Section 8.2.3.3 and Figure 8.8) to synthesise supported bimetallic catalysts. These DENs can be obtained by co-complexation of both metal salts followed by reduction or through successive complexations and then reduction. Several protocols of deposition are proposed: the formation of DENs then support deposition, the use of DENs as nano-reactors to form bimetallic particles that are then extracted and deposited on a substrate and the use of DENs to template the porosity of support (Section 8.3.1).

Initially, the synthesis of bimetallic particles in dendrimers was developed to overcome the difficulty of preparing bimetallic particles with miscibility gap, such as Au–Pt, Au–Ni or Au–Ir. Hydroxy-terminated fifth generation PAMAM dendrimers (G5-OH) were used to synthesise Au–Pt DENs before deposition on silica, alumina and titania.[236,237] After Au–Pt DENs adsorption onto the support, calcination and reduction at 300°C, the nanoparticles were found bimetallic and most of them were smaller than 3 nm. For the preparation of supported Au–Ni bimetallic particles, the method was more complex.[238] Ammine-terminated G5 PAMAM dendrimers on a silica support were used to template the Au–Ni nanoparticles resulting from the co-reduction of $NiCl_2$ and $HAuCl_4$ with $NaBH_4$. These nanoparticles were subsequently extracted by toluene containing decanethiol that acted as new capping ligands. After adsorption on TiO_2, the thiol ligands were decomposed under H_2 at 300°C, leading to bimetallic nanoparticles of ∼3 nm with a Ni core–Au shell structure. In another work, four sequential synthesis routes involving G4-OH PAMAM dendrimers were tested to produce bimetallic Au−Ir particles supported on alumina.[239] The sample containing *in fine* the smallest metal particles (1.6 nm) was the one prepared from $IrCl_3$ mixed with the dendrimer solution first before $HAuCl_4$ addition. The resulting solution was used for alumina impregnation, followed by calcination at 350°C and then reduction at 400°C.

d. Radiolysis-assisted reduction

Radiolytic reduction of metal ions in aqueous solutions (Chapter 6) is a method suitable for the synthesis of mono and bimetallic nanoparticles, but in the case of supported materials, it has been mostly used for the synthesis of bimetallic NPs. The hydrated electrons and the reducing radicals produced during the radiolysis of water by irradiation with γ-rays or accelerated electrons reduce the metal ions. Depending on the dose rate, alloy-type or core–shell particles can be obtained. Reduction can be performed either before support adsorption, which requires the addition of stabilising agent, or directly in the presence of the support.

Au–Pd nanoparticles were produced by radiolysis before adsorption on an alumina support.[240] The aqueous solution containing $HAuCl_4$ and $Pd(NO_3)_2$ was irradiated under a N_2 atmosphere in the presence of PVA and of 2-propanol as scavenger of oxidising OH· radicals. Under γ irradiation, Au core–Pd shell particles of 3–4 nm were obtained whereas under electron beam, i.e. at higher dose rate, very fast reduction occurred and alloy-type nanoparticles of 2–3 nm were formed. After adsorption on alumina, reduction at 300°C induced particle reconstruction in both cases without drastic change of size ($\sim$3 nm) but with different proportions of Pd on the particle surface.

Radiolysis can also be used to overcome the difficulty of preparing bimetallic nanoparticles with a miscibility gap. Bimetallic Au–Pt nanoparticles (1:0.75) were prepared by reduction of H_2PtCl_6 and $HAuCl_4$ under γ-rays in an aqueous solution containing PVA and polyacrylic acid (PAA).[241] After adsorption onto a silica support, calcination at 400°C followed by reduction at 500°C, the metal particles were 4.6 nm in size, and the Au core–Pt shell structure after radiolysis transformed into an alloy-type structure with an Au-rich surface.

Radiolysis can also be performed directly in the presence of the support and without stabiliser. An aqueous solution containing $HAuCl_4$ and H_2PtCl_6 (1:1), 2-propanol and γ-Fe_2O_3 powder in suspension was irradiated by an electron beam.[242] The resulting particles were bimetallic with a size of 2.9 nm.

e. Microwave-assisted reduction

Several protocols of MW-assisted reduction were applied for the preparation of bimetallic catalysts: co-reduction before deposition onto the support, co-reduction in the presence of the support and two-step reduction–deposition.

In this example, Au–Cu colloids were synthesised by co-reduction of $HAuCl_4$ and $CuSO_4$ by glucose in alkaline aqueous solution containing PVP.[243] Microwaving allowed the solution to rapidly reach 90°C, and to induce rapid reduction of the metal precursors. The colloids were then immobilised onto TiO_2, leading to bimetallic particles of 4.4 nm size after drying at 120°C.

Au–Pt nanoparticles supported on carbon black (10 wt.% Pt + 10 wt.% Au) were prepared by microwave-assisted polyol reduction of $HAuCl_4$ and H_2PtCl_6 in the presence of the support at pH 10.[244] After washing and drying, the metal particle size was 3.3 nm and after treatment at 500°C under H_2, 4.7 nm. The characterisation indicated the formation of Au–Pt nanoalloy. Since Au and Pt are immiscible, the alloy formation was attributed to the fast reduction rates of both metal ions under microwave conditions.

8.2.4 Specific Methods for the Preparation of Supported Bimetallic Particles

8.2.4.1 Bimetallic clusters

Organo bimetallic precursors are sometimes used to prepare supported bimetallic particles. One can find only very few examples for gold-based bimetallics. $Pt_2Au_4(C{\equiv}CtBu)_8$ was adsorbed onto silica from hexane solution; this led to particles of $\sim$2.5 nm size after calcination at 300°C and reduction at 200°C.[245] $[NEt_4][AuFe_4(CO)_{16}]$ is another precursor, which was impregnated on titania, but after decomposition at 400°C in N_2, gold nanoparticles (from 3 to 7 nm for Au loadings from 2 to 7 wt.%) were found anchored onto the titania surface through an iron oxide interface.[246]

8.2.4.2 Surface redox methods

Methods based on surface reduction–oxidation (redox) reactions to prepare supported bimetallic catalysts were developed first by Barbier *et al.*[247–249] and then more recently by Monnier *et al.*[250,251] They are based on the principle that a second metal can be deposited on a first one already supported, thanks to a redox reaction selectively occurring on the particle surface of the first metal.

This can be achieved by the so-called direct surface redox reactions,also called galvanic replacement reaction (GRR) Figure 8.9), i.e. by a redox reaction, which occurs spontaneously when the redox potential of the second metal is higher than that of the first metal: The precursor of the second metal

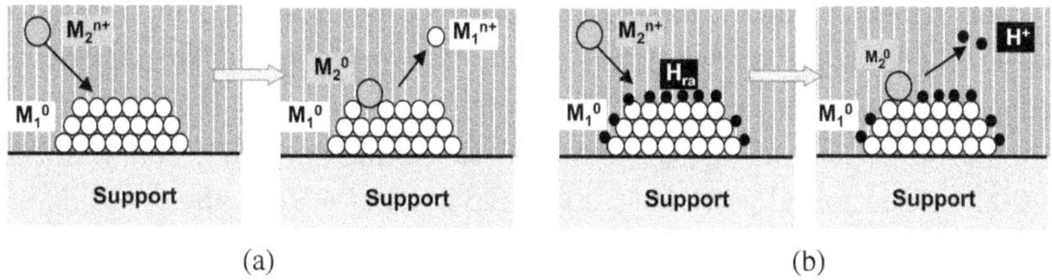

Figure 8.9 Schematic illustration of the surface redox methods: (a) direct reduction of M_2^{n+} ions by M_1^0 metal particle, also called galvanic replacement reaction; (b) reduction of M_2^{n+} ion by active H adsorbed on M_1^0 metal particle resulting from H_2 dissociation or activation of a reducing agent on M_1^0.

is reduced on the first metal and the latter is oxidised and dissolved in the solution, and therefore also acts as a sacrificial component.

Surface redox preparation can also be achieved through redox reaction assisted by a reducing agent, for instance by hydrogen dissociatively adsorbed on the first metal (Figure 8.9)[247–249] or a reducing agent activated by the first supported metal[250,251]; this method is also called electroless deposition (ED). Note that successful ED requires that the precursor of the second metal cannot reduce in solution (or that its kinetics of reduction is much slower than that of reduction on the first metal) and cannot adsorb onto the support. All these methods have been applied to the preparation of gold-based bimetallic catalysts.

Au–Ag bimetallic NPs were synthesised by GRR on microspheres consisting of a Fe_3O_4 core and a carbon shell.[252] The addition of a solution of $HAuCl_4$ to metallic Ag NPs supported on these microspheres, followed by incubation at 50°C led to the formation of the Ag–Au particles: the Ag particles were oxidised into Ag^+ ions, which were leached out, and $HAuCl_4$ was reduced on the Ag particles. Bimetallic Au–Ag NPs supported on TiO_2 were also obtained through GRR, between photo-deposited Ag/TiO_2 and aqueous $HAuCl_4$ solution heated under reflux.[253] The particle size remained almost unchanged (1.5 nm). Au–Cu/TiO_2 was synthesised by a one-pot photo-deposition-GRR method:[159] A suspension of TiO_2 in a water–ethanol solution containing Cu acetate was UV-irradiated to deposit Cu nanoparticles onto TiO_2. Then, $HAuCl_4$ solution was added to perform GRR in the dark. Small-size metal nanoparticles (<2 nm) with Au-rich core/Cu shell structure were found on TiO_2 surface.

The redox method with preadsorbed hydrogen was used to prepare Au–Pd/SiO$_2$ catalysts: HAuCl$_4$ was added to an aqueous suspension of reduced Pd/SiO$_2$ on which hydrogen was pre-adsorbed by bubbling H$_2$.[254] After washing and drying, the sample was calcined at 300°C and reduced at 200°C. The initial particle size was 3.5 nm and 4.3 nm after gold deposition and thermal treatment, and the particles were alloy-type. The same procedure was used for a titania support.[255] This redox method was also used to synthesise bimetallic nanoparticles from non-miscible metals. Au–Pt/SiO$_2$ catalysts were prepared by depositing Au (up to 2.2 wt.%) from a HAuCl$_4$ solution at pH 1 on Pt/SiO$_2$ (6.3 wt.% and $d_{Pt} = 2.7$ nm).[256] After washing and drying, the particles were bimetallic but much larger, 6.5 nm, because of sintering in the liquid phase. Au–Ru/C samples were prepared according to the same principle.[257] However, after gold deposition (0.85 wt.%) on a Ru/C catalyst (5 wt.%), the particles were only slightly larger (2.9 nm) than the Ru particles (2.6 nm). Those in the 1.5–3 nm range contained mainly Ru, while the mid-size range ones (5–10 nm) contained both Au and Ru, and the few larger ones ($>$10 nm) mainly contained Au.

To our knowledge, ED with gold has been applied only to the preparation of Au–Pd particles supported on silica[258] and on carbon.[259] Au ED was performed at pH 9 on large Pd particles of commercial Pd/SiO$_2$ catalysts using KAu^{I}(CN)$_2$ as precursor and hydrazine as the reducing agent. After ED completion, the particles were found to be bimetallic, and the coverage of Pd by Au was controlled by the amount of gold in the ED solution. For the preparation of Au–Pd/C, the same protocol was adopted with Pd/C, except that the solution was heated at 40°C and pH was adjusted to 12.

8.3 Gold Nanoparticles Embedded into a Matrix

There are several strategies for the synthesis of gold nanoparticles embedded into matrices. Such types of materials may be needed for studies of optical properties (Chapter 3), but when matrices contain open porosity and the gold particles remain accessible to reactants, they can be used for the purpose of catalysis.

8.3.1 *Gold Embedded into an Inorganic Matrix*

Most of the methods described below involve one-pot preparation for both the matrix and the gold particles. When the gold loadings are mentioned in the papers, they are usually close to the nominal ones, indicating that all the gold in solution can be precipitated or trapped into the oxide; in the works dealing with non-porous matrices, it is rarely mentioned which fraction of gold particles is located on surface and which one is embedded into the matrix.

8.3.1.1 *Monometallic gold*

One method is co-precipitation, which allows the one-pot preparation of gold particles and oxide supports, such as α-Fe_2O_3, Co_3O_4, MnO_2, CuO, CeO_2 and ZnO.[260–268] Co-precipitation is generally performed by addition of sodium carbonate to aqueous solutions containing both $HAuCl_4$ and the nitrate precursor of the oxide support at controlled pH and temperature. After ageing, the co-precipitate is washed thoroughly to remove sodium and chlorides, dried and calcined to form metallic gold particles and to transform the oxy-hydroxy-carbonate precipitate into oxide; this is a kind of deposition-reduction method. Attempts to explain the chemistry of co-precipitation can be found in Ref. 266 for the Au/Fe_2O_3 system, which is one of the most frequently system prepared with this method. Co-precipitation can lead to small gold particles (≤ 5 nm), but the size depends on several parameters, such as the affinity between gold and the support, the temperature and pH of precipitation. This method is limited to the supports that can precipitate at a rate consistent with that of gold precipitation or deposition. One can find a few studies reporting the influence of preparation parameters on the final Au particle size: timing of $HAuCl_4$ addition,[265] pH and stirring rate,[267] washing volumes[262] and drying conditions.[268] The preparation method has been reported as highly reproducible.[266]

Another one-pot preparation is the sol–gel method. It involves the controlled hydrolysis of an alkoxide precursor of the oxide support in a water–alcohol solution containing $HAuCl_4$ or other gold precursors, such as gold acetate or hydrogen tetra-nitratoaurate. Various alkoxide precursors are used, tetra-ethoxysilane (TEOS),[269] tetrabutoxy titanate,[270–272] aluminium butoxide,[273] or aluminium triacetate.[274] Most of the preparations involve a gelation step performed at controlled pH and temperature and sometimes in an autoclave, followed by washing and drying steps (xerogel), and finally a thermal treatment to remove the organics, transform the hydroxide into oxide (aerogel) and reduce gold. The average gold particle size can vary within a large range depending on the preparation conditions.

Gold can also be introduced as a colloid into the solution mixture used for the sol–gel synthesis of the oxide, and this is a kind of reduction-deposition method. For instance, Au NPs in mesoporous TiO_2 microspheres were obtained after a hydrothermal treatment at 180°C of the precipitate resulting from mixing tetrabutyl-titanate into a water–ethanol solution containing citrate-stabilised gold colloids; this led to 6–7 nm gold particles into the TiO_2 microspheres (1.2–1.5 μm).[275] Gold–titania aerogels were synthesised from alkanethiol-stabilised gold particles added to a titania sol.[276] After calcination at 425°C, gold particles of 5–10 nm were found between the titania nanoparticles (10–12 nm), creating multiple points of contact between Au and TiO_2 particles.

There are also strategies for generating an open porosity in the oxides, so as to keep the gold particles accessible to reactants for catalytic reactions and to limit their sintering at high temperature. For instance, 2 nm gold particles in PAMAM dendrimers were incorporated into the sol–gel matrix of titania.[277] After calcination at 500°C, they barely increased in size (2.7 nm) and were found located in the 4.5 nm pores arising from the imprint of the PAMAM dendrimers. Budroni *et al.*[278] introduced gold colloids (2 nm) capped with both 1-dodecanethiol and 3-mercaptopropyltrimethoxysilane (MPMS) into a solution of ethanol containing TEOS whose hydrolysis was catalysed by NH_4F. The alkoxysilane groups of MPMS promoted the hydrolysis and condensation of TEOS around the gold particles, providing a link between them and the silica after calcination. Calcination at 450°C resulted in 3.5 nm gold particles in 9 nm pores (Figure 8.10). This strategy of synthesis has also been applied by the same group to the synthesis of gold in mesoporous titania; rigid carboxylic acids with different lengths were used as spacers to tune the porosity.[279]

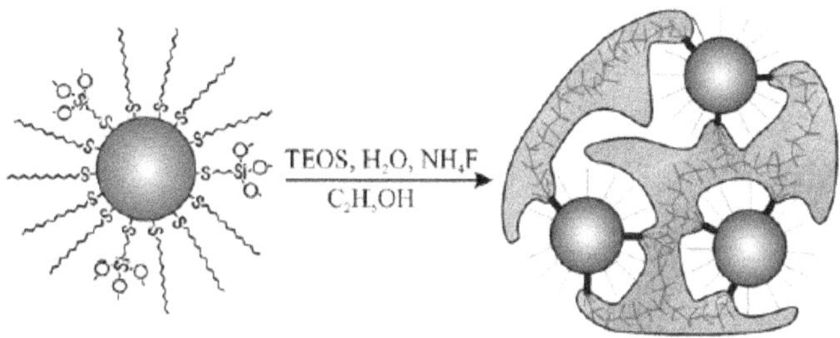

Figure 8.10   Schematic representation of gold nanoparticles capped with 1-dodecanethiol (DT) and MPMS and the AuNP-organic-SiO$_2$ structure. MPMS (thick lines) on the right-hand side of the figure provides the organic links between the metal particles and the inorganic mesoporous silica. Reprinted with permission from Ref. 278. Copyright (2006) John Wiley & Sons.

8.3.1.2 *Bimetallic systems*

Co-precipitation is barely used for the preparation of bimetallic particles in matrices. However, one can cite an example with the preparation of Au–Pt/CeO$_2$ from an aqueous solution containing H$_2$PtCl$_6$ and HAuCl$_4$, Ce(NO$_3$)$_3$ and urea, which was aged at 100°C.[280] After calcination at 500°C, the metal particles were 5–10 nm and bimetallic.

Metal precursors or bimetallic colloids can be introduced in sol–gel preparation. One can first cite the case of the synthesis of Au–Cu particles embedded into SiO$_2$, ZrO$_2$ and TiO$_2$ matrices; they were obtained by introduction of HAuCl$_4$ and CuCl$_2$ into the oxide precursor sols arising from alkoxides.[281] After drying, calcination and reduction at 500°C, the particles were bimetallic of alloy-type with sizes between 25 and 38 nm depending on the matrix.

PVP-stabilised Au–Pd colloids with several Au/Pd ratios and synthesised by co-reduction or sequential reduction, were added to a sol–gel precursor mixture of alumina.[282] After calcination and reduction at 300°C, the particles resulting from co-reduction had a slightly Au-rich core and Pd-rich shell, with a size (5.5 nm) slightly larger than in colloids (4.1 nm). In contrast, those arising from sequentially reduction maintained their initial Pd core–Au shell structure with almost the same size as the colloids (5.5 nm instead of 5.2 nm).

Au–Ni nanoparticles of the size 3–4 nm were embedded in 15 nm silica nanospheres.[283] They were synthesised in reverse-micelle preparation

(Section 8.2.3.4b) to which ammonia solution and then TEOS were added for the sol–gel synthesis.

Finally, we can cite the use of dendrimers to encapsulate bimetallic nanoparticles (Section 8.2.3.4c) and template the porosity of a support during sol–gel synthesis. Au–Pd nanoparticles in G4-PAMAM dendrimers were added to a solution of titanium isopropoxide in methanol, which was then hydrolysed.[284] After ageing and drying, the metal particle size was 1.8 nm, and increased to 3.2 nm after calcination and reduction at 500°C. The particles were bimetallic.

8.3.2 *Gold in an Inorganic Matrix with Ordered Porosity*

Gold can be introduced during the preparation of the matrix, and the main difference with the sol–gel methods previously described is that a surfactant is added to the sol to generate and structure the porosity of the oxide. Gold can be also introduced after the synthesis of the mesoporous materials and also in some cases, after pore functionalisation.

Hence, gold colloids can be introduced into the matrix sol or during the gelation step. For instance, gold particles smaller than the pore size were incorporated into mesoporous silicas, such as MCM-41, MCM-48 or SBA-15.[285–288] After calcination at 550°C, the gold particles were slightly larger than in the colloidal solution (for instance 4 nm in MCM-41 instead of 2 nm), but were homogeneously distributed into the mesoporous materials.

Gold precursor can be added to the matrix sol. For instance, $HAuCl_4$ was added to the solution mixture used for the synthesis of MCM-41, i.e. cetylpyridinium bromide (CPBr), HCl and TEOS[289] or of mesoporous titanium oxide, i.e. triblock copolymer and Ti tetrapropoxide.[290] To better ensure that the gold particles are located in the porosity of the matrix and are of controlled size, the mesoporous materials can be functionalised during or after synthesis. For instance, bifunctional molecules such as amino-organosilane (AAPTS) were added to the matrix sol.[291] The amino groups of AAPTS complex the gold anions while the organosilanes bind to the porous MCM-41 silica matrix. Removal of the templating surfactants via ion-exchange reaction and thermal reduction of gold led to ordered meso-porous materials containing uniform gold nanoparticles of size 2–5 nm in the pores with a gold loading as high as 7 wt.%. In this example, 1,4-bis(triethoxysilyl)propane tetrasulphide was added to the sol used for the synthesis of mesoporous silica thin films.[292] It co-condensed with TEOS, and after gelation and washing, $[Au(en)_2]^{3+}$ cations were introduced and adsorbed on the SO_3^- groups. After reduction at 300°C, the gold nanoparticles were located in the mesopores with a 3–7 nm size range.

In this example, functionalisation was performed after synthesis: SBA-15 was functionalised with PAMAM dendrimers in which $HAuCl_4$ was introduced and reduced with $NaBH_4$, leading to gold particles not exceeding 5 nm.[293] Several structured mesoporous silicas, HMS, MSU and PMO, were also functionalised after their synthesis with bifunctional amine ligands in such a way that they covalently bonded to silica via siloxane groups and they complexed the AuIII precursor via the amine functional groups.[294] Gold nanoparticles were smaller than 5 nm, and located inside the mesoporous materials structures. Bimetallic Au–Pd nanoparticles of size 1.5 nm were also synthesised in G4-PAMAM dendrimers built into the channels of SBA-15.[295] They were synthesised by co-complexation of K_2PdCl_4 and $HAuCl_4$ by dendrimers, followed by chemical reduction with $NaBH_4$.

8.3.3 *Gold on/in Organic Materials*

Gold nanoparticles can also be deposited on organic polymers or in the porosity, but the number of reports is much less numerous than for deposition on oxides. The resulting hybrid inorganic–organic materials are useful for applications in catalysis, life science or analytical science. As in the case of the inorganic supports, it is possible to proceed via deposition–reduction or reduction–deposition, using most often the same type of preparation methods. Preparation of gold in/on organic–inorganic nanocomposites is reported in Section 8.3.4 with the case of gold in MOF.

Gold nanoparticles can be generated within the interconnected macropores of a poly(styrene-*co*-divinylbenzene) matrix (polyHIPE) by forcing a water–acetone solution containing $HAuCl_4$ to fill up the void spaces by applying and releasing a vacuum.[296] Spontaneous reduction led to gold particles around 50–200 nm located in the porosity (200–300 nm). Other protocols require specific functions in polymers to allow adsorption of gold complexes before reduction or to stabilise gold nanoparticles after reduction. In the first case, for instance, solid polystyrene core onto which long linear cationic polyelectrolyte chains (poly(2-aminoethyl methacrylate hydrochloride [PAEMH]) were grafted, were used to immobilise $AuCl_4^-$ ions in water[297] while thioether groups of gel-type polyacrylic resin[298] and of related cross-linked co-polymers[299] were used as coordination sites for $AuCl_3$ precursor in acetonitrile. In all cases, the gold nanoparticles were smaller than the pore size of the polymers, indicating their location into the pores. In the second case, poly(propyleneimine)-G2 dendrimers grafted on a cross-linked poly(4-vinylpyridine) matrix of polymer beads[10] or benzene rings of polystyrene derivatives[300] was used to stabilise gold nanoparticles after the reduction of $AuClPPh_3$ with $NaBH_4$. Thermal grinding is another way to form gold NPs on polymers; this has been described in Section 8.2.1.3c.

Preparations based on reduction–deposition are also efficient. For instance, Au(THPC) colloids (2.5 nm) were adsorbed on a conducting polymer, polyaniline (PANI).[301] Au NPs of different average sizes (8 to 55 nm) stabilised by citrate were immobilised on resin beads of polystyrene-based anion exchangers by stepwise replacement of the chlorides on the resin.[302] Adsorption of Au(citrate) colloids of 20 nm size on polystyrene beads (107 μm) containing positively charged NH_3^+ functions led to a uniform coating.[303]

One can also find a few papers reporting the one-step synthesis of gold NPs in polymers, for instance, by UV photoreduction at RT of a gold precursor in the presence of a urethane methacrylate carrying a quaternary ammonium end that can interact with gold particles.[304]

Finally, one can find examples of synthesis of gold-based bimetallic particles in organic matrices using the same type of preparation methods as for monometallic particles[305,306] or using galvanic replacement reaction[307] (Section 8.2.4.2).

8.3.4 *Gold on/in Inorganic–Organic Materials*

This section essentially concerns metal–organic frameworks (MOFs), also called porous coordination polymers (PCPs); these are 3D structures of an organometallic polymer containing metal cations centres linked by organic ligands. The very first method used to introduce gold was the chemical vapour deposition of $(CH_3)Au(PMe_3)$, which led to broad size distribution, 5–20 nm, due to the absence of interaction with the substrate.[308] Solid grinding (Section 8.2.1.3c) led to much smaller Au particles (<5 nm) in various types of MOF.[9,309]

Au nanoparticles were also prepared in IRMOF-3 (Zn_4O clusters linked by 2-amino-1,4-benzendicarboxylic acid (NH_2–BDC)) according to two methods: a one-pot synthesis and a post-covalent modification.[310] The second method, which involved the addition of salicylaldehyde to functionalise the MOFs with pendant amino groups, favoured gold reduction and interaction with MOF, leading to gold particles of 3.3 nm. The first one was more simple as it resulted from mixing of NH_2–BDC, $Zn(NO_3)_2$, salicylaldehyde and AuCl or $NaAuCl_4$ at 100°C, and led to smaller particles from 1.7 to 2.5 nm depending on the loading. A new type of pore functionalisation of a MOF (consisting of layers of ZnO_4N_2 octahedra linked by dicarboxylate-type ligands, which are connected by bipyridine pillars and forming a 3D structure) with terminal alkyne moieties allowed the adsorption of up to 50 wt.% Au from $HAuCl_4$ precursor owing to the alkynophilicity of the Au^{3+} ions.[311] As a result, after reduction by $NaBH_4$, small Au particles of 1.9 nm were mainly located inside the pores of the MOF in spite of the high Au loading.

Deposition–precipitation with urea is also an efficient method to introduce gold in MOFs. This was done with MIL-101(Cr) (a 3D structure forming mesoporous cages built with trimers of Cr(III) oxide octahedra and dicarboxylate linkers); reduction at 200°C led to the formation of 1.8 nm Au particles consistent with the size of the MOF cages (2.9 and 3.4 nm).[312]

Chen *et al.*[4] explored the use of Au(PVP) colloids around 15 nm as seeds to grow zeolitic imidazolate framework-8 and selectively synthesise Au in ZIF-8 ($Zn(MeIM)_2$, MeIM = 2-methylimidazole), with single- or multi-gold cores in ZIF shell structures. The size distribution of Au@ZIF-8 structures ranges from 135 to 200 nm, and there was no apparent change in the diameter of the Au NPs.

One can also find examples of synthesis of bimetallic NPs in MOF, mostly of Au–Pd, either by adsorption of colloids such as Au–Pd(PVP)[313] or by introduction of Au and Pd precursors that are reduced thermally[117] or chemically.[314,315]

8.4 Gold Nanoparticles on Planar Surfaces

Materials based on gold nanoparticles supported on planar surfaces can be used as model systems for surface science studies (gold on thin films or single crystals, explained in Chapter 12, as biosensors, as materials for nanophotonics, memory devices, surface-enhanced Raman spectroscopy (SERS) and also as modified electrodes). In addition to the control of the particle size and shape, periodic arrangements of the gold particles on these planar surfaces may be desirable, for instance for nanoelectronic applications or model catalytic studies. This can be done using various physical techniques with top-down or bottom-up approaches (Chapters 7 and 12) and also using chemical methods. As for divided supports, these are based on reduction–deposition, deposition–reduction and also on other techniques more adapted to planar surfaces.

8.4.1 *Non-ordered Deposition*

The simplest chemical method consists of depositing droplets of colloidal solution onto these surfaces, followed by water evaporation or spinning. The excess particles can be removed by mere rinsing with water or using sonication. Such type of preparations was performed on thin films of indium tin oxide (ITO)[316] or polyelectrolyte,[317] using gold(citrate) colloids larger than 15 nm. Smaller gold nanoparticles functionalised with thiols were deposited on thin films of organic semiconductor matrix by spin-coating, which resulted on the formation of 2.3-nm particle monolayer.[318] Even smaller gold entities $\sim$1 nm high were formed onto a $TiO_2(110)$ single crystal by deposition of gold phosphine clusters, $[Au_6(PPh_3)_6][BF_4]_2$, followed by electron beam surface irradiation to decompose the cluster.

The substrate can be functionalised; Au(citrate) colloids of various sizes were adsorbed on ITO-coated glass pre-functionalised with (3-amino-propyl)tri-methoxysilane (APTMS)[319] and on ITO and Pt electrodes functionalised with 1,12-dodecanediamine (DDA).[320]

With the procedure of deposition–reduction, Au particles were obtained on layered double hydroxides (LDH) films on glass after thin-film immersion in a solution of $KAuCl_4$ in formamide to which ascorbic acid was added for gold reduction[321] or on ITO glass after its immersion in a basic solution containing $HAuCl_4$ to which eugenol (2-methoxy-4-allyl-phenol) was added.[322] Gold particles in the size range of 30–80 nm were obtained.

Another method, specific for planar surface is the electrodeposition. A deposition potential (0 V vs. SCE) is applied to an ITO substrate in the presence of a solution of $HAuCl_4$ and KCl.[320] The size of the gold particles depends on the electrodeposition duration, 150 nm for 15 s and 525 nm for 30 min. Another method allows the formation of gold nanoparticles on semi-conductive substrates by galvanic displacement, i.e. by reduction of gold ions in aqueous solution by the electrons arising from the semiconductor itself: immersion of an InP(100) wafer in an aqueous solution of $HAuCl_4$ at RT resulted in the rapid deposition of strongly adhering gold nanoparticles of several tens of nanometres.[323] Note that these methods do not require the use of organics to stabilise the gold particles.

8.4.2 *Ordered Deposition*

Self-organisation of diblock copolymers into micellar structures in an appropriate solvent is an efficient chemical way to end up with well-ordered arrays of metal nanoparticles on planar surfaces with narrow distributions in particle size and interparticle spacing.

The method based on the deposition of micelles containing gold was developed by Spatz *et al.*[324] Reversed micelles were generated by dissolution of amphiphilic diblock copolymers in toluene to which $HAuCl_4$ was added. Reduction by hydrazine resulted in the formation of metal particles in micelles. Dip coating of planar substrates (mica, thin glass plates, n^+-doped Si wafers, carbon grids) led to self-assembled monolayers of micelles whereby the particles arrange in a mesoscopic quasi-hexagonal two-dimensional lattice (Figure 8.11). Oxygen plasma treatment yielded naked gold particles of identical size and in the same quasi-hexagonal order as the micellar film. The size of the clusters could vary between 1 and 15 nm depending on the metal salt concentration, and the length of the blocks, i.e. the interparticle distance between 30 and 140 nm.[324] This type of preparation has been extended to other planar substrates from conducting to insulating substrates: natively oxidised *n*-doped Si(001), Ti on *n*-doped Si(001), Ti foil and ITO-coated glass,[325] glassy carbon[326] and also to single-crystal

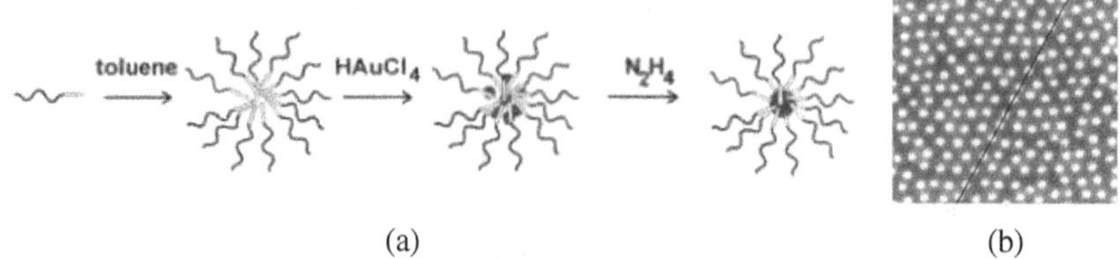

Figure 8.11 (a) Schematic drawing of the micelle formation of poly(styrene)-*block*-poly(2-vinylpyridine) (PS-*b*-P2VP) block copolymers in toluene; (b) Scanning force microscopy topography image of a monomicellar film cast from a PS(1700)-*b*-P[2VP(HAuCl$_4$)$_{0.3}$(450)] solution onto a glass substrate. Reprinted with permission from Ref. 324. Copyright (2000) American Chemical Society.

surfaces, such as rutile TiO$_2$(110).[327–330] In these last examples, the oxygen or argon plasma treatments were performed to decompose the micelles, and also to reduce gold.

Another strategy is to self-assemble diblock-copolymer micelles on planar surfaces, then to fill them with gold ions and reduce them. Semi-conductive surfaces of doped Si(100), Ge(100), GaAs(100) and InP(100) coated with a self-assembled monolayer of poly(styrene)-*block*-poly(2-vinylpyridine) (PS-*b*-P2VP) or poly(styrene)-*block*-poly(4-vinylpyridine) (PS-*b*-P4VP) were immersed in an aqueous solution of HAuCl$_4$ before reduction by galvanic displacement.[331] Micelle elimination was performed by dissolution in toluene. According to the authors, this strategy led to better patterning of the gold particles than the one described previously. Depending on the molecular weight of the polymer, the gold particle size varies between 8 and 60 nm and the interparticle distance between 90 and 170 nm. Laskar *et al.*[332] employed the same strategy for a fluorinated surface of silicium wafer, on which diblock copolymers, PEO$_m$-*b*-PMA(Az)$_n$, formed ordered arrays of spherical micelles. After immersion in an aqueous solution containing gold ions, excimer vacuum UV radiation led to gold reduction, micelles elimination and formation of regular hexagonal arrays of gold particles of 3.8 nm in height.

8.5 Conclusion

The actual trend in metal-based nanomaterial synthesis is to make them more and more complex. The supports can themselves be of nanosize; it is also possible to prepare nanosized oxide-core-bimetallic shell particles or metallic particles embedded in oxide shells or in nano-boxes or nano-bells. Gold is more and more often associated to another metal to form bimetallic particles, for instance, for catalysis purposes or optical properties. Another trend is to play with the metal particle shape, and to develop more exotic shapes as already shown in Chapter 6 for non-supported particles.

References

1. Y. Li, X. Fan, J. Qi, J. Ji, S. Wang, G. Zhang, F. Zhang, *Mater. Res. Bull.* **45**, 1413–1418 (2010).
2. Y. Choi, H. S. Bae, E. Seo, S. Jang, K. H. Park, B.-S. Kim, *J. Mater. Chem.* **21**, 15431–15436 (2011).
3. H. Chen, Y. Li, F. Zhang, G. Zhang, X. Fan, *J. Mater. Chem.* **21**, 17658–17661 (2011).
4. L. Chen, Y. Peng, H. Wang, Z. Gua, C. Duan, *Chem. Commun.* **50**, 8651–8654 (2014).
5. J. A. Rodriguez, P. Liu, Y. Takahashi, F. Vines, L. Feria, E. Florez, K. Nakamura, F. Illas, *Catal. Today* **166**, 2–9 (2011).
6. N. Perret, X. Wang, L. Delannoy, C. Potvin, C. Louis, M. A. Keane, *J. Catal.* **286**, 172–183 (2012).
7. F. Shi, Q. Zhang, Y. Ma, Y. He, Y. Deng, *J. Am. Chem. Soc.* **127**, 4182–4183 (2005).
8. T. Ishida, M. Haruta, *Ang. Chem. Int. Ed.* **46**, 7154–7156 (2007).
9. T. Ishida, M. Nagaoka, T. Akita, M. Haruta, *Chem. Eur. J.* **14**, 8456–8460 (2008).
10. E. Murugan, R. Rangasamy, *J. Polym. Sci. A* **48**, 2525–2532 (2010).
11. H. Tsunoyama, Y. Liu, T. Akita, N. Ichikuni, H. Sakurai, S. Xie, T. Tsukuda, *Catal. Surv. Asia* **15**, 230–239 (2011).
12. A. Azetsu, H. Koga, A. Isogai, T. Kitaoka, *Catalysts* **1**, 83–96 (2011).
13. T. Ishida, N. Kawakita, T. Akita, M. Haruta, *Gold Bull.* **42**, 267–274 (2009).
14. H. Liu, Y. Liu, Y. Li, Z. Tang, H. Jiang, *J. Phys. Chem. C* **114**, 13362–13369 (2010).
15. A. Corma, M. Iglesias, F. X. Llabres i Xamena, F. Sanchez, *Chemistry* **16**, 9789–9795 (2010).
16. L. Liu, X. Zhang, J. Gao, C. Xu, *Green Chem.* **14**, 1710–1720 (2012).
17. V. Belova, H. Mohwald, D. G. Shchukin, *Langmuir* **24**, 9747–9753 (2008).
18. T. L. M. Martinez, M. I. Dominguez, N. Sanabria, W. Y. Hernandez, S. Moreno, R. Molina, J. A. Odriozola, M. A. Centeno, *Appl. Catal. A* **364**, 166–173 (2009).
19. L. Zhu, S. Letaief, Y. Liu, F. Gervais, C. Detellier, *Appl. Clay Sci.* **43**, 439–446 (2009).
20. H. Sakurai, K. Koga, Y. Iizuka, M. Kiuchi, *Appl. Catal. A* **462– 463**, 236–246 (2013).
21. I. Dobrosz, K. Jiratova, V. Pitchon, J. M. Rynkowski, *J. Mol. Catal. A* **234**, 187–197 (2005).
22. C.-T. Chang, B.-J. Liaw, C.-T. Huang, Y.-Z. Chen, *Appl. Catal. A* **332**, 216–224 (2007).
23. A. Noujima, T. Mitsudome, T. Mizugaki, K. Jitsukawa, K. Kaneda, *Ang. Chem. Int. Ed.* **50**, 2986–2989 (2011).
24. A. Takagaki, A. Tsuji, S. Nishimura, K. Ebitani, *Chem. Lett.* **40**, 150–152 (2011).

25. Q. Xiao, Z. Liu, A. Bo, S. Zavahir, S. Sarina, S. Bottle, J. D. Riches, H. Zhu, *J. Am. Chem. Soc.* **137**, 1956–1966 (2015).

26. A. Venugopal, M. S. Scurrell, *Appl. Catal. A* **245**, 137–147 (2003).

27. N. Phonthammachai, Z. Ziyi, G. Jun, F. Y. Han, T. J. White, *Gold Bull.* **41**, 42–50 (2008).

28. M. I. Dominguez, F. Romero-Sarria, M. A. Centeno, J. A. Odriozola, *Appl. Catal. B* **87**, 245–251 (2009).

29. J. Huang, L.-C. Wang, Y.-M. Liu, Y. Cao, H.-Y. He, K.-N. Fan, *Appl. Catal. B* **101**, 560–569 (2011).

30. S. Lin, M. A. Vannice, *Catal. Lett.* **10**, 47–62 (1991).

31. M. Haruta, S. Tsubota, T. Kobayashi, H. Kageyama, M. J. Genet, B. Delmon, *J. Catal.* **144**, 175–192 (1993).

32. H. S. Oh, J. H. Yang, C. K. Costello, Y. M. Wang, S. R. Bare, H. H. Kung, M. C. Kung, *J. Catal.* **210**, 375–386 (2002).

33. G. C. Bond, C. Louis, D. Thompson, *Catalysis by Gold* (Imperial College Press, London, 2006, Vol. 6).

34. R. Zanella, S. Giorgio, C. R. Henry, C. Louis, *J. Phys. Chem. B* **106**, 7634–7642 (2002).

35. A. Hugon, L. Delannoy, J.-M. Krafft, C. Louis, *J. Phys. Chem. C* **114**, 10823–10835 (2010).

36. Q. Xu, K. C. C. Kharas, A. K. Datye, *Catal. Lett.* **85**, 229–235 (2003).

37. L. Delannoy, N. El Hassan, A. Musi, N. Nguyen Le To, J.-M. Krafft, C. Louis, *J. Phys. Chem. B* **110**, 22471–22478 (2006).

38. S. Ivanova, V. Pitchon, C. Petit, *J. Mol. Catal. A* **256**, 278–283 (2006).

39. S. Ivanova, V. Pitchon, Y. Zimmermann, C. Petit, *Appl. Catal. A* **298**, 57–64 (2006).

40. Y. Azizi, V. Pitchon, C. Petit, *Appl. Catal. A* **385**, 170–177 (2010).

41. B. C. Campo, O. Rosseler, M. Alvarez, E. H. Rueda, M. A. Volpe, *Mater. Chem. Phys.* **109**, 448–454 (2008).

42. I. Dobrosz-Gomez, I. Kocemba, J. M. Rynkowski, *Appl. Catal. B* **88**, 83–97 (2009).

43. P. Sudarsanam, B. Mallesham, P. S. Reddy, D. Großmann, W. Grunert, B. M. Reddy, *Appl. Catal. B* **144**, 900–908 (2014).

44. J. D. Lessard, I. Valsamakisz, M. Flytzani-Stephanopoulos, *Chem. Commun.* **48**, 4857–4859 (2012).

45. S. Tsubota, D. A. H. Cunningham, Y. Bando, M. Haruta, *Stud. Surf. Sci. Catal.* **91**, 227–235 (1995).

46. G. R. Bamwenda, S. Tsubota, T. Nakamura, M. Haruta, *Catal. Lett.* **44**, 83–97 (1997).

47. A. Wolf, F. Schüth, *Appl. Catal. A* **226**, 1–13 (2002).

48. M. Haruta, *Cattech* **6**, 102–115 (2002).

49. R. Zanella, L. Delannoy, C. Louis, *Appl. Catal. A* **291**, 62–72 (2005).

50. F. Moreau, G. C. Bond, *Catal. Today* **114**, 362–368 (2006).

51. F. Moreau, G. C. Bond, *Catal. Today* **122**, 215–221 (2007).

52. L. Prati, G. Martra, *Gold Bull.* **32**, 96101 (1999).

53. F. Moreau, G. C. Bond, *Catal. Today* **122**, 260–265 (2007).

54. F. Moreau, G. C. Bond, A. O. Taylor, *J. Catal.* **231**, 105–114 (2005).

55. J. Radnik, L. Wilde, M. Schneider, M.-M. Pohl, D. Herein, *J. Phys. Chem. B* **110**, 23688–23693 (2006).

56. C. Louis, In *Gold catalysis: preparation, characterization and Applications*, L. Prati, A. Villa (eds.), Pan Stanford Publishing, Singapore, pp. 1–38 (2016).

57. B. Solsona, M. Perez-Cabero, I. Vazquez, A. Dejoz, T. García, J. Alvarez-Rodríguez, J. El-Haskourib, D. Beltran, P. Amoros, *Chem. Eng. J.* **187**, 391–400 (2012).

58. A. J. Binder, Z.-A. Qiao, G. M. Veith, S. Dai, *Catal. Lett.* **143**, 1339–1345 (2013).

59. K. Qian, L. Luo, C. Chen, S. Yang, W. Huang, *ChemCatChem* **3**, 161–166 (2011).

60. E. L. Reddy, A. Prabhakharn, J. Karuppiah, *Intern. J. Nanosci.* **11**, 12400041–12400047 (2012).

61. K. Zhao, B. Qiao, J. Wang, Y. Zhang, T. Zhang, *Chem. Commun.* **47**, 1779–1781 (2011).

62. Y.-F. Han, N. Phonthamamchai, K. Ramesh, Z. Hong, T. White, *Environ. Sci. Technol.* **42**, 908–912 (2008).

63. W. Yan, S. Brown, Z. Pan, S. M. Mahurin, S. H. Overbury, S. Dai, *Angew. Chem. Int. Ed.* **45**, 3614–3618 (2006).

64. Z. Ma, H. Yin, S. H. Overbury, S. Dai, *Catal. Lett.* **126**, 20–30 (2008).

65. C. Tian, S.-H. Chai, X. Zhu, Z. W. A. Binder, J. C. Bauer, S. Brwon, M. Chi, G. M. Veith, Y. Guo, S. Dai, *J. Mater. Chem.* **22**, 25227–25235 (2012).

66. L. A. Hermans, J. W. Geus, *Stud. Surf. Sci. Catal.* **4**, 113 (1979).

67. J. A. van Dillen, J. W. Geus, L. A. Hermans, J. van der Meijden, in *Proc. 6th Intern. Congr. Catal., London, 1976*, G. C. Bond, P. B. Wells, F. C. Tompkins (eds.), The Chemical Society, London, p. 677 (1977).

68. C. Baatz, N. Thielecke, U. Prüße, *Appl. Catal. B* **70**, 653–660 (2007).

69. M. Khoudiakov, M.-C. Gupta, S. Deevi, *Appl. Catal. A* **291**, 151–161 (2005).

70. Y. Guo, D. Gu, Z. Jin, P.-P. Du, R. Si, I. Tao, W.-Q. Xu, Y.-Y. Huang, S. Senanayake, Q.-S. Song, C.-J. Jia, F. Schüth, *Nanoscale* **7**, 4920 (2015).

71. N. S. Patil, B. S. Uphade, P. Jana, S. K. Bharagava, V. R. Choudhary, *J. Catal.* **223**, 236–239 (2004).

72. P. Lakshmanan, L. Delannoy, V. Richard, C. Méthivier, C. Potvin, C. Louis, *Appl. Catal. B* **96**, 117–125 (2010).

73. E. Smolentseva, A. Simakov, S. Beloshapkin, M. Estrada, E. Vargas, V. Sobolev, R. Kenzhin, S. Fuentes, *Appl. Catal. B* **115–116**, 117–128 (2012).

74. E. del Rio, G. Blanco, S. Collins, M. Lopez-Haro, X. Chen, J. J. Delgado, J. J. Calvino, S. Bernal, *Top. Catal.* **54**, 931–940 (2011).

75. J. Vecchietti, S. Collins, J. J. Delgado, M. Malecka, E. Rio, X. Chen, S. Bernal, A. Bonivardi, *Top. Catal.* **54**, 201–209 (2011).

76. E. del Rio, M. Lopez-Haro, J. M. Cies, J. J. Delgado, J. J. Calvino, S. Trasobares, G. Blanco, M. A. Cauqui, S. Bernal, *Chem. Commun.* **49**, 6722–6724 (2013).

77. I. Sobczak, K. Szrama, R. Wojcieszak, E. Gaigneaux, M. Ziolek, *Catal. Today* **187**, 48–55 (2012).

78. X. Lu, G. Zhao, Y. Lu, *Catal. Sci. Technol.* **3**, 2906–2909 (2013).

79. Q. Ye, J. Zhao, F. Huo, D. Wang, S. Cheng, T. Kang, H. Dai, *Microp. Mesop. Mater.* **172**, 20–29 (2013).

80. J. Huang, L.-C. Wang, Y.-M. Liu, Y. Cao, H.-Y. He, K.-N. Fan, *Appl. Catal. B* **101**, 560–569 (2011).

81. H. Sun, F.-Z. Su, J. Ni, Y. Cao, H.-Y. He, K.-N. Fan, *Angew. Chem. Int. Ed.* **48**, 4390–4393 (2009).

82. N. Perret, F. Cardenas-Lizana, D. Lamey, V. Laporte, L. Kiwi-Minsker, M. A. Keane, *Top. Catal.* **55**, 955–968 (2012).

83. D. Guillemot, V. Y. Borovskov, V. B. Kazansky, M. Polisset-Thfoin, J. Fraissard, *J. Chem. Soc. Faraday Trans.* **93**, 3587–3591 (1997).

84. D. Guillemot, M. Polisset-Thfoin, J. Fraissard, *Catal. Lett.* **41**, 143–148 (1996).

85. R. Zanella, A. Sandoval, P. Santiago, V. A. Basiuk, J. M. Saniger, *J. Phys. Chem. B* **110**, 8559–8565 (2006).

86. H. Zhu, Z. Ma, J. C. Clark, Z. Pan, S. H. Overbury, S. Dai, *Appl. Catal. A* **326**, 89–99 (2007).

87. H. Zhu, C. Liang, W. Yan, S. H. Overbury, S. Dai, *J. Phys. Chem. B* **110**, 10842–10848 (2006).

88. Y. Guan, E. J. M. Hensen, *Appl. Catal. A* **361**, 49–56 (2009).

89. L.-F. Gutiérrez, S. Hamoudia, K. Belkacemi, *Appl. Catal. A* **425–426**, 213–223 (2012).

90. M. Kantcheva, M. Milanova, I. Avramova, S. Mametsheripov, *Catal. today* **187**, 39–47 (2012).

91. D. A. Bulushev, I. Yuranov, E. I. Suvorova, P. A. Buffat, L. Kiwi-Minsker, *J. Catal.* **224**, 8–17 (2004).

92. B. P. Block, J. J. C. Bailar, *J. Am. Chem. Soc.* **73**, 4722–4725 (1951).

93. P. A. Pyryaev, B. L. Moroz, D. A. Zyuzin, A. V. Nartova, V. I. Bukhtiyarov, *Kinet. Catal.* **51**, 885–892 (2010).

94. J. Xu, Y. Liu, H. Wu, X. Li, M. He, P. Wu, *Catal. Lett.* **141**, 860–865 (2011).

95. Y. Yuan, A. P. Kozlova, K. Asakura, H. Wan, K. Tsai, Y. Iwasawa, *J. Catal.* **170**, 191–199 (1997).

96. A. I. Kozlov, A. P. Kozlova, H. Liu, Y. Iwasawa, *Appl. Catal. A* **182**, 9–28 (1999).

97. T. V. Choudhary, C. Sivadinarayana, C. C. Chusuei, A. K. Datye, J. P. Fackler Jr., D. W. Goodman, *J. Catal.* **207**, 247–255 (2002).

98. R. H. Adnan, G. G. Andersson, M. I. J. Polson, G. F. Metha, V. B. Golovko, *Catal. Sci. Technol.* **5**, 1323–1333 (2015).

99. Z. Yan, S. Chinta, A. A. Mohamed, J. P. Fackler Jr., D. W. Goodman, *Catal. Lett.* **111**, 15–18 (2006).

100. S. Gaur, H. Wu, G. G. Stanley, K. More, C. S. S. R. Kumar, J. J. Spivey, *Catal. Today* **208**, 72–81 (2013).

101. J. Guzman, B. C. Gates, *Nano Lett.* **1**, 689–692 (2001).

102. J. Guzman, B. C. Gates, *Langmuir* **19**, 3897–3903 (2003).

103. J. C. Fierro-Gonzalez, B. C. Gates, *J. Phys. Chem. B* **109**, 7275–7279 (2005).

104. J. C. Fierro-Gonzalez, V. A. Bhirud, B. C. Gates, *Chem. Commun.* 5275–5277 (2005).

105. J. C. Fierro-Gonzalez, B. C. Gates, *J. Phys. Chem. B* **108**, 16999–17002 (2004).

106. M. Okumura, S. Tsubota, M. Haruta, *J. Mol. Catal. A* **199**, 73–84 (2003).

107. M. Hisamoto, R. C. Nelson, M.-Y. Lee, J. Eckert, S. L. Scott, *J. Phys. Chem. C* **113**, 8794–8805 (2009).

108. H. Okatsu, N. Kinoshita, T. Akita, T. Ishida, M. Haruta, *Appl. Catal. A* **369**, 8–14 (2009).

109. J. Huang, T. Takei, T. Akita, H. Ohashi, M. Haruta, *Appl. Catal. B* **95**, 430–438 (2010).

110. T. Ishida, H. Watanabe, T. Bebeko, T. Akita, M. Haruta, *Appl. Catal. A* **377**, 42–46 (2010).

111. T. Ishida, N. Kawakita, T. Akita, M. Haruta, *Stud. Surf. Sci. Catal.* **175**, 839–842 (2010).

112. Y. Maeda, T. Akita, M. Kohyama, *Catal. Lett.* **144**, 2086–2090 (2014).

113. T. Akita, Y. Maeda, M. Kohyama, *J. Catal.* **324**, 127–132 (2015).

114. J. K. Edwards, B. E. Solsona, P. Landon, A. F. Carley, A. Herzing, C. J. Kiely, G. J. Hutchings, *J. Catal.* **236**, 69–79 (2005).

115. M. Sankar, Q. He, M. Morad, J. Pritchard, S. J. Freakley, J. K. Edwards, S. H. Taylor, D. J. Morgan, A. F. Carley, D. W. Knight, C. J. Kiely, G. J. Hutching, *ACS Nano* **6**, 6600–6613 (2012).

116. X. Yang, C. Huang, Z. Fu, H. Song, S. Liao, Y. Sud, L. Du, X. Li, *Appl. Catal. B* **140–141**, 419–425 (2013).

117. X. Gu, Z.-H. Lu, H.-L. Jiang, T. Akita, Q. Xu, *J. Am. Chem. Soc.* **133**, 11822–11825 (2011).

118. J. Long, H. Liu, S. Wu, S. Liao, Y. Li, *ACS Catal.* **3**, 647–654 (2013).

119. J. K. Edwards, J. Pritchard, L. Lu, M. Piccinini, G. Shaw, A. F. Carley, D. J. Morgan, C. J. Kiely, G. J. Hutchings, *Angew. Chem. Int. Ed.* **53**, 2381–2384 (2014).

120. D. Gudarzi, W. Ratchananusorn, I. Turunen, M. Heinonen, T. Salmi, *Catal. Today* **248**, 58–68 (2015).

121. Y. Hao, G.-P. Hao, D.-C. Guo, C.-Z. Guo, W.-C. Li, M.-R. Li, A.-H. Lu, *ChemCatChem* **4**, 1595–1602 (2012).

122. F. Menegazzo, M. Manzoli, M. Signoretto, F. Pinna, G. Strukul, *Catal. Today* **248**, 18–27 (2015).

123. R. J. Chimentao, F. Medina, J. L. G. Fierro, J. Llorca, J. E. Sueiras, Y. Cesteros, P. Salagre, *J. Mol. Catal. A* **274**, 159–168 (2007).

124. S. Belin, C. L. Bracey, V. Briois, P. R. Ellis, G. J. Hutchings, T. I. Hyde, G. Sankar, *Catal. Sci. Technol.* **3**, 2944–2957 (2013).

125. Y. Guan, E. J. M. Hensen, *J. Catal.* **305**, 135–145 (2013).

126. A. Gallo, T. Montini, M. Marelli, A. Minguzzi, V. Gombac, R. Psaro, P. Fornasiero, V. D. Santo, *ChemSusChem* **5**, 1800–1811 (2012).

127. H. Na, T. Zhu, Z. Liu, *Catal. Sci. Technol.* **4**, 2051–2057 (2014).

128. C. Chen, H. Yang, J. Chen, R. Zhang, L. Guo, Huimei Gan, B. Song, W. Zhu, L. Hua, Z. Hou, *Catal. Commun.* **47**, 49–53 (2014).

129. X. Liao, W. Chu, X. Dai, V. Pitchon, *Appl. Catal. B*, **142–143**, 25–37 (2013).

130. N. Sasirekha, P. Sangeetha, Y.-W. Chen, *J. Phys. Chem. C* **118**, 15226–15233 (2014).

131. N. Mimura, N. Hiyoshi, M. Date, T. Fujitani, F. Dumeignil, *Catal. Lett.* **144**, 2167–2175 (2014).

132. H. Na, T. Zhu, Z. Liu, *Catal. Sci. Technol.* **4**, 2051–2057 (2014).

133. L. Li, C. Wang, X. Ma, Z. Yang, X. Lu, *Chin. J. Catal.* **33**, 1778–1782 (2012).

134. L. Delannoy, G. Thrimurthulu, P. S. Reddy, C. Méthivier, J. Nelayah, B. M. Reddy, C. Ricolleau, C. Louis, *Phys. Chem. Chem. Phys.* **16**, 26514–26527 (2014).

135. P. M. More, D. L. Nguyen, M. K. Dongare, S. B. Umbarkar, N. Nuns, J.-S. Girardon, C. Dujardin, C. Lancelot, A.-S. Mamede, P. Granger, *Appl. Catal. B* **162**, 11–20 (2015).

136. Z. Zhang, Y. Wang, X. Li, W. L. Dai, *Chin. J. Catal.* **35**, 1846–1857 (2014).

137. L. E. Chinchilla, C. M. Olmos, L. Villa, A. Carlsson, L. Prati, X. Chen, G. Blanco, J. J. Calvino, A. B. Hungría, *Catal. Today* **253**, 178–189 (2015).

138. P. Kittisakmontree, B. Pongthawornsakun, H. Yoshida, S.-i. Fujita, M. Arai, J. Panpranot, *J. Catal.* **297**, 155–164 (2013).

139. A. Sandoval, A. Aguilar, C. Louis, A. Traverse, R. Zanella, *J. Catal.* **281**, 40–49 (2011).

140. A. Sandoval, C. Louis, R. Zanella, *Appl. Catal. B* **140–141**, 363–377 (2013).

141. A. Sandoval, L. Delannoy, C. Méthivier, C. Louis, R. Zanella, *Appl. Catal. A* **504**, 287–294 (2015).

142. A. Aguilar-Tapia, R. Zanella, C. Calers, C. Louis, L. Delannoy, *Phys. Chem. Chem. Phys.* **17**, 28022–28032 (2015).

143. M. Meire, P. Tack, K. de Keukeleere, L. Balcaen, G. Pollefeyt, F. Vanhaecke, L. Vincze, P. van Der Voort, I. van Driessche, P. Lommens, *Spectrochim. Acta B* **110**, 45–50 (2015).

144. K. Mallick, M. J. Witcomb, M. S. Scurell, *Appl. Catal. A* **259**, 163–168 (2004).

145. J. A. Lopez-Sanchez, N. Dimitratos, N. Glanville, L. Kesavan, C. Hammond, J. K. Edwards, A. F. Carley, C. J. Kiely, G. J. Hutchings, *Appl. Catal. A* **391**, 400–406 (2011).

146. R. Isono, T. Yoshimura, K. Esumi, *J. Coll. Inter. Sci.* **288**, 177–183 (2005).

147. P. Zhang, B. Zhang, R. Shi, *Front. Environ. Sci. Eng. China* **3**, 281–288 (2009).

148. T. Suprabha, H. G. Roy, S. Mathew, *Sci. Adv. Mater.* **2**, 107–114 (2010).

149. Q. Hu, Z. Gan, X. Zheng, Q. Lin, B. Xu, A. Zhao, X. Zhang, *Superlat. Microstr.* **49**, 537–542 (2011).

150. S. A. Neto, T. S. Almeida, D. M. Belnap, S. D. Minteer, A. R. de Andrade, *J. Power Ress.* **273**, 1065–1072 (2015).

151. A. Fernandez, A. Caballero, A. R. Gonzalez-Elipe, J.-H. Herrmann, H. Dexpert, F. Villain, *J. Phys. Chem.* **99**, 3303–3309 (1995).

152. D. Li, J. T. McCann, M. Gratt, Y. Xia, *Chem. Phys. Lett.* **394**, 387–391 (2004).

153. J. Khanderi, R. C. Hoffmann, J. Engstler, J. J. Schneider, J. Arras, P. Claus, G. Cherkashinin, *Chem. Eur. J.* **16**, 2300–2308 (2010).

154. N. R. de Tacconi, K. Rajeshwar, W. Chanmanee, V. Valluri, W. A. Wampler, W.-Y. Lin, L. Nikiel, *J. Electrochem. Soc.* **157**, B147–B153 (2010).

155. X. Wang, G. Wu, L. Li, N. Guan, *Adv. Mater. Res.* **148–149**, 1258–1263 (2011).

156. S. Dulnee, A. Luengnaruemitchai, R. Wanchanthuek, *Int. J. Hydr. Ener.* **39**, 6443–6453 (2014).

157. H. Kominami, A. Tanaka, K. Hashimoto, *Appl. Catal. A* **397**, 121–126 (2011).

158. E. Kowalska, S. Rau, B. Ohtani, *J. Nanotech.* **361853**, 361811 (2012).

159. Q. Jia, D. Zhao, B. Tang, N. Zhao, H. Li, Y. Sang, N. Bao, X. Zhang, X. Xu, H. Liu, *J. Mater. Chem. A* **2**, 16292–16298 (2014).

160. T. Ohno, T. Higo, N. Murakami, H. Saito, Q. Zhang, Y. Yang, T. Tsubota, *Appl. Catal. B* **152–153**, 309–316 (2014).

161. M. C. Hidalgo, J. J. Murcia, J. A. Navío, G. Colón, *Appl. Catal. A* **397**, 112–120 (2011).
162. A. A. Ismail, A. Hakki, D. W. Bahnemann, *J. Mol. Catal. A* **358**, 145–151 (2012).
163. V. G. Pol, A. Gedanken, J. Calderon-Moreno, *Chem. Mater.* **15**, 1111–1118 (2003).
164. N. Perkas, V. G. Pol, S. V. Pol, A. Gedanken, *Cryst. Growth Design* **6**, 293–296 (2006).
165. N. Perkas, Z. Zhong, J. Grinblat, A. Gedanken, *Catal. Lett.* **120**, 19–24 (2008).
166. H. Bunazawa, Y. Yamazaki, *J. Power Sources* **190**, 210–215 (2009).
167. S. A. C. Carabineiro, B. F. Machado, R. R. Bacsa, P. Serp, G. Drazic, J. L. Faria, J. L. Figueiredo, *J. Catal.* **273**, 191–198 (2010).
168. S. A. C. Carabineiro, N. Bogdanchikova, M. Avalos-Borja, A. Pestryakov, P. B. Tavares, J. L. Figueiredo, *Nano Res.* **4**, 180–193 (2011).
169. Y. Mizukoshi, Y. Tsuru, A. Tominaga, S. Seino, N. Masahashi, S. Tanabe, A. Yamamoto, *Ultrason. Sonochem.* **15**, 875–880 (2008).
170. Z. Zhong, J. Lin, S.-P. Teh, J. Teo, F. M. Dautzenberg, *Adv. Funct. Mater.* **17**, 1402–1408 (2007).
171. X. Liu, A. Wang, X. Yang, T. Zhang, C.-Y. Mou, D.-S. Su, J. Li, *Chem. Mater.* **21**, 410–418 (2009).
172. X. Liu, A. Wang, T. Zhang, D.-S. Su, C.-Y. Mou, *Catal. Today* **160**, 103–108 (2011).
173. J. C. Bauer, D. Mullins, M. Li, Z. Wu, E. A. Payzant, S. H. Overbury, S. Dai, *Phys. Chem. Chem. Phys.* **13**, 2571–2581 (2011).
174. S. Yan, L. Gao, S. Zhang, L. Gao, W. Zhang, Y. Li, *Int. J. Hydr. Ener.* **38**, 18283–18284 (2013).
175. P. Qiao, S. Zou, S. Xu, J. Liu, Y. Li, G. M. L. Xiao, H. Loua, J. Fan, *J. Mater. Chem. A* **2**, 17321–17328 (2014).
176. X. Chen, Z. Cai, X. Chen, M. Oyama, *J. Mater. Chem. A* **2**, 5668–5674 (2014).
177. L. Prati, A. Villa, *Acc. Chem. Res.* **47**, 855–863 (2014).
178. B. Neppolian, C. Wang, M. Ashokkumar, *Ultrason. Sonochem.* **21**, 1948–1953 (2014).
179. A. Tanaka, S. Sakaguchi, K. Hashimoto, H. Kominami, *ACS Catal.* **3**, 79–85 (2013).
180. H. Zhu, Z. Guo, X. Zhang, K. Han, Y. Guo, F. Wang, Z. Wang, Y. Wei, *Int. J. Hydr. Ener.* **37**, 873–876 (2012).
181. A. Villa, M. Schiavoni, L. Prati, *Catal. Sci. Technol.* **2**, 673–682 (2012).
182. A. Horvath, A. Beck, G. Stefler, T. Benko, G. Safran, Z. Varga, J. Gubicza, L. Guczi, *J. Phys. Chem. C* **115**, 20388–20398 (2011).
183. M. C. Hidalgo, M. Maicu, J. A. Navío, G. Colon, *J. Phys. Chem. C* **113**, 12840–12847 (2009).
184. C. Caballero, J. Valencia, M. Barrera, A. Gil, *Powder Tech.* **203**, 412–414 (2010).
185. D. Caruntu, B. L. Cushing, G. Caruntu, C. J. O'Connor, *Chem. Mater.* **17**, 3398–3402 (2005).
186. A. Beck, G. Magesh, B. Kuppan, Z. Schay, O. Geszti, T. Benkó, R. P. Viswanath, P. Selvam, B. Viswanathan, L. Guczi, *Catal. Today* **164**, 325–331 (2011).
187. Z. Wang, E. V. Beletskiy, S. Lee, X. Hou, Y. Wu, T. Li, M. C. Kung, H. H. Kung, *J. Mater. Chem. A* **3**, 1743–1751 (2015).
188. N. Hickey, P. Arneodo Larochette, C. Gentilini, L. Sordelli, L. Olivi, S. Polizzi, T. Montini, P. Fornasiero, L. Pasquato, M. Graziani, *Chem. Mater.* **19**, 650–651 (2007).

189. H. Yin, Z. Ma, M. Chi, S. Dai, *Catal. Lett.* **136**, 209–221 (2010).
190. N. Zheng, G. D. Stucky, *J. Am. Chem. Soc.* **128**, 14278–14280 (2006).
191. T. Herranz, X. Deng, A. Cabot, Z. Liu, G. Soler-Illia, M. Salmeron, *Catal. Today* **143**, 158–166 (2009).
192. A. Villa, D. Wang, G. M. Veith, F. Vindigni, L. Prati, *Catal. Sci. Technol.* **3**, 3036–3041 (2013).
193. A. Quintanilla, V. C. L. Butselaar-Orthlieb, C. Kwakernaak, W. G. Sloof, M. T. Kreutzer, F. Kapteijn, *J. Catal.* **271**, 104–114 (2010).
194. J. D. Grunwaldt, M. Maciejewski, O. S. Becker, P. Fabrizioli, A. Baiker, *J. Catal.* **186**, 458–469 (1999).
195. J. Chou, E. W. McFarland, *Chem. Commun.* 1648–1649 (2004).
196. M. Tominaga, T. Shimazoe, M. Nagashima, I. Taniguchi, *J. Electroanal. Chem.* **615**, 51–61 (2008).
197. G. Martra, L. Prati, C. Manfredotti, S. Biella, M. Rossi, S. Coluccia, *J. Phys. Chem. B* **107**, 5453 (2003).
198. M. Comotti, W.-C. Li, B. Spliethoff, F. Schüth, *J. Am. Chem. Soc.* **126**, 917–924 (2006).
199. C. G. Long, J. D. Gilbertson, G. Vijayaraghavan, K. J. Stevenson, C. J. Pursell, B. D. Chandler, *J. Am. Chem. Soc.* **130**, 10103–10115 (2008).
200. L. D. Menard, F. Xu, R. G. Nuzzo, J. C. Yang, *J. Catal.* **243**, 64–73 (2006).
201. J. Llorca, A. Casanovas, M. Domınguez, I. Casanova, I. Angurell, M. Seco, O. Rossell, *J. Nano. Res.* **10**, 537–542 (2008).
202. R.-Y. Zhong, X.-H. Yan, Z.-K. Gao, R.-J. Zhang, B.-Q. Xu, *Catal. Sci. Technol.* **3**, 3013–3019 (2013).
203. J. A. Lopez-Sanchez, N. Dimitratos, C. Hammond, G. L. Brett, L. Kesavan, S. White, P. Miedziak, R. Tiruvalam, R. L. Jenkins, A. F. Carley, D. Knight, C. J. Kiely, G. J. Hutchings, *Nature Chem.* **3**, 551–556 (2011).
204. I. Gandarias, P. J. Miedziak, E. Nowicka, M. Douthwaite, D. J. Morgan, G. J. Hutchings, S. H. Taylor, *ChemSusChem* **8**, 473–480 (2015).
205. M. Luo, Y. Hong, W. Yao, C. Huang, Q. Xu, Q. Wu, *J. Mater. Chem. A* **3**, 2770–2775 (2015).
206. N. Naresh, F. G. S. Wasim, B. P. Ladewig, M. Neergat, *J. Mater. Chem. A* **1**, 8553–8559 (2013).
207. S. M. Ansar, F. S. Ameer, W. Hu, S. Zou, J. Charles U. Pittman, D. Zhang, *Nano Lett.* **13**, 1226–1229 (2013).
208. Y. Zhao, L. Jia, J. A. Medrano, J. R. H. Ross, L. Lefferts, *ACS Catal.* **3**, 2341–2352 (2013).
209. J. Monzo, M. T. M. Koper, P. Rodriguez, *ChemPhysChem* **13**, 709–715 (2012).
210. J. Chou, N. R. Franklin, S.-H. Baeck, T. F. Jaramillo, E. W. McFarland, *Catal. Lett.* **95**, 107–111 (2004).
211. Q. Yang, D.-Y. Chen, *Chin. J. Chem. Phys.* **25**, 352–358 (2012).
212. R. W. J. Scott, O. M. Wilson, R. M. Crooks, *J. Phys. Chem. B* **109**, 692–704 (2005).
213. B. D. Chandler, J. D. Gilbertson, in *Nanoparticles and Catalysis*, D. Astruc (ed.), Wiley-VCH (2007).

214. E. A. Kyriakidou, K. Khivantsev, T. M. Gostanian, O. S. Alexeev, M. D. Amiridis, *Appl. Catal. A* **504**, 482–492 (2015).

215. R. W. J. Scott, O. M. Wilson, R. M. Crooks, *Chem. Mater.* **16**, 5682–5688 (2004).

216. M. Nemanashi, R. Meijboom, *Catal. Lett.* **143**, 324–332 (2013).

217. S. Aquino Neto, T. S. Almeida, D. M. Belnap, S. D. Minteer, A. R. De Andrade, *J. Power Sources* **273**, 1065–1072 (2015).

218. C. L. Bianchi, P. Canton, N. Dimitratos, F. Porta, L. Prati, *Catal. Today*, **102–103**, 203–212 (2005).

219. D. Wang, A. Villa, F. Porta, D. Su, L. Prati, *Chem. Commun.* 1956–1958 (2006).

220. D. Wang, A. Villa, L. Prati, F. Porta, D. Su, *J. Phys. Chem. C* **112**, 8617–8622 (2008).

221. A. Villa, D. Wang, D. Su, G. M. Veith, L. Prati, *Phys. Chem. Chem. Phys.* **12**, 2183–2189 (2010).

222. Z. Konuspayeva, P. Afanasiev, T.-S. Nguyen, L. D. Felice, F. Morfin, N.-T. Nguyen, J. Nelayah, C. Ricolleau, Z. Y. Li, J. Yuan, G. Berhault, L. Piccolo, *Phys. Chem. Chem. Phys.* **17**, 28112–28120 (2015).

223. T. Benkó, A. Beck, K. Frey, D. F. Srankó, O. Geszti, G. Sáfrán, B. Maróti, Z. Schay, *Appl. Catal. A* **479**, 103–111 (2014).

224. E. G. Rodrigues, M. F. R. Pereira, X. Chen, J. J. Delgado, J. J. M. Órfão, *Ind. Eng. Chem. Res.* **52**, 17390–17398 (2013).

225. A. M. Venezia, L. F. Liotta, G. Pantaleo, V. L. Parola, G. Deganello, A. Beck, Z. Koppány, K. Frey, D. Horváth, L. Guczi, *Appl. Catal. A* **251**, 359–368 (2003).

226. S. Marx, F. Krumeich, A. Baiker, *J. Phys. Chem. C* **115**, 8195–8205 (2011).

227. L. Guczi, A. Becka, A. Horváth, Z. Koppány, G. Stefler, K. Frey, I. Sajó, O. Geszti, D. Bazin, J. Lynch, *J. Mol. Catal. A* **204–205**, 545–552 (2003).

228. A. Alshammari, A. Kockritz, V. N. Kalevaru, A. Bagabas, A. Martin, *Top. Catal.* **58**, 1069–1076 (2015).

229. S. W. Verbruggen, M. Keulemans, M. Filippousi, D. Flahaut, G. V. Tendeloo, S. Lacombe, J. A. Martens, S. Lenaerts, *Appl. Catal. B* **156–157**, 116–121 (2014).

230. J. Llorca, M. Domínguez, C. Ledesma, R. J. Chimentão, F. Medina, J. S. Sueiras, I. Angurell, M. Seco, O. Rossell, *J. Catal.* **258**, 187–198 (2008).

231. P. He, Y. Wang, X. Wang, F. Pei, H. Wang, L. Liu, L. Yi, *J. Power Sources* **196**, 1042–1047 (2011).

232. P. He, X. Wang, Y. Liu, X. Liu, L. Yi, *Int. J. Hydr. Ener.* **37**, 11984–11993 (2012).

233. T. Szumełda, A. Drelinkiewicz, R. Kosydar, J. Gurgul, *Appl. Catal. A* **487**, 1–15 (2014).

234. W. G. Menezes, V. Zielasek, K. Thiel, A. Hartwig, M. Bäumer, *J. Catal.* **299**, 222–231 (2013).

235. L.-C. Lee, C. Xiao, W. Huang, Y. Zhao, *New J. Chem.* **39**, 2459–2466 (2015).

236. H. Lang, S. Maldonado, K. J. Stevenson, B. D. Chandler, *J. Am. Chem. Soc.* **126**, 12949–12956 (2004).

237. B. J. Auten, H. Lang, B. D. Chandler, *Appl. Catal. B* **81**, 225–235 (2008).

238. B. D. Chandler, C. G. Long, J. D. Gilbertson, C. J. Pursell, G. Vijayaraghavan, K. J. Stevens, *J. Phys. Chem. C* **114**, 11498–11508 (2010).

239. Y.-J. Song, Y. M. L. P.-D. Jesús, P. T. Fanson, C. T. Williams, *J. Phys. Chem. C* **117**, 10999–11007 (2013).

240. T. Redjala, H. Remita, G. Apostolescu, M. Mostafavi, C. Thomazeau, D. Uzio, *Oil Gas Sci. Tech.* **61**, 789–797 (2006).

241. R. P. Doherty, J.-M. Krafft, C. Méthivier, S. Casale, H. Remita, C. Louis, C. Thomas, *J. Catal.* **287**, 102–113 (2012).

242. T. A. Yamamoto, T. Nakagawa, S. Seino, H. Nitani, *Appl. Catal. A* **387**, 195–202 (2010).

243. T. Pasini, M. Piccinini, M. Blosi, R. Bonelli, S. Albonetti, N. Dimitratos, J. A. Lopez-Sanchez, M. Sankar, Q. He, C. J. Kiely, G. J. Hutchings, F. Cavania, *Green Chem.* **13**, 2091–2099 (2011).

244. M. Yin, Y. Huang, Q. Lv, L. Liang, J. Liao, C. Liu, W. Xing, *Electrochimica Acta* **58**, 6–11 (2011).

245. B. D. Chandler, A. B. Schabel, C. F. Blanford, L. H. Pignolet, *J. Catal.* **187**, 367–384 (1999).

246. S. Albonetti, R. Bonelli, R. Delaigle, C. Femoni, E. M. Gaigneaux, V. Morandi, L. Ortolani, C. Tiozzo, S. Zacchini, F. Trifiro, *Appl. Catal. A* **372**, 138–146 (2010).

247. J. Barbier, in *Handbook of Heterogeneous Catalysis*, G. Ertl, H. Knözinger, J. Weitkamp (eds.), Wiley VCH, Weinheim, Vol. 1, p. 257 (1997).

248. F. Epron, C. Especel, G. Lafaye, P. Marécot, in *Nanoparticles and Catalysis*, D. Astruc (ed.), Wiley VCH, Weinheim, pp. 281–304 (2007).

249. C. Especel, D. Duprez, F. Epron, *C. R. Chimie* **17**, 790–800 (2014).

250. K. D. Beard, M. T. Schaal, J. W. V. Zee, J. R. Monnier, *Appl. Catal. B* **72**, 262–271 (2007).

251. J. Rebelli, A. A. Rodriguez, S. Ma, C. T. Williams, J. R. Monnier, *Catal. Today* **160**, 170–178 (2011).

252. Q. An, M. Yu, Y. Zhang, W. Ma, J. Guo, C. Wang, *J. Phys. Chem. C* **116**, 22432–22440 (2012).

253. Y. Guan, N. Zhao, B. Tang, Q. Jia, X. Xu, H. Liu, R. I. Boughton, *Chem. Commun.* **49**, 1524–11526 (2013).

254. A. Sarkany, À. Horvath, A. Beck, *Appl. Catal. A* **229**, 117–125 (2002).

255. T. V. Choudhary, C. Sivadinarayana, A. K. Datye, D. Kumar, D. W. Goodman, *Catal. Lett.* **86**, 1–8 (2003).

256. J. Barbier, P. Marécot, G. D. Angel, P. Bosch, J. P. Boitiaux, B. Didillon, J. M. Dominguez, I. Schifter, G. Espinosa, *Appl. Catal. A* **143**, 283 (1994).

257. E. P. Maris, W. C. Ketchie, M. Murayama, R. J. Davis, *J. Catal.* **251**, 281–294 (2007).

258. J. Rebelli, M. Detwiler, S. Ma, C. T. Williams, J. R. Monnier, *J. Catal.* **270**, 224–233 (2010).

259. M. B. Griffin, A. A. Rodriguez, M. M. Montemore, J. R. Monnier, C. T. Williams, J. W. Medlin, *J. Catal.* **307**, 111–120 (2013).

260. M. Haruta, H. Kageyama, N. Kamijo, T. Kobayashi, F. Delannay, *Stud. Surf. Sci. Catal.* **44**, 33–42 (1988).

261. G. J. Hutchings, M. R. H. Siddiqui, A. Burrows, C. J. Kiely, R. Whyman, *J. Chem. Soc. Faraday Trans.* **93**, 187–188 (1997).

262. S.-J. Lee, A. Gavriilidis, Q. A. Pankhurst, A. Kyek, F. E. Wagner, P. C. L. Wong, K. L. Yeung, *J. Catal.* **200**, 298–308 (2001).

263. B. E. Solsona, T. Garcia, C. Jones, S. H. Taylor, A. F. Carley, G. J. Hutchings, *Appl. Catal. A* **312**, 67–76 (2006).

264. B. Qiao, J. Zhang, L. Liu, Y. Deng, *Appl. Catal. A* **340**, 220–228 (2008).

265. S. Kudo, T. Maki, M. Yamada, K. Mae, *Chem. Eng. Sci.* **65**, 214–219 (2010).

266. T. Shodiya, O. Schmidt, W. Peng, N. Hotz, *J. Catal.* **300**, 63–69 (2013).

267. H. Wang, W. Fan, Y. He, J. Wang, J. N. Kondo, T. Tatsumi, *J. Catal.* **299**, 10–19 (2013).

268. C. Zhang, X. Cui, H. Yang, L. Zheng, Y. Deng, F. Shi, *Appl. Catal. A* **473**, 7–12 (2014).

269. H. Kozuka, S. Sakka, *Chem. Mater.* **5**, 222–228 (1993).

270. F. B. Li, X. Z. Li, *Appl. Catal. A* **228**, 15–27 (2002).

271. J. Fang, S.-W. Cao, Z. Wang, M. M. Shahjamali, S. C. J. Loo, J. Barber, C. Xue, *Int. J. Hydr. Ener.* **37**, 17853–17861 (2012).

272. H. Zhao, P. Zhang, U. Wang, W. Huang, S. Zhang, *J. Sol-Gel Sci. Technol.* **71**, 406–412 (2014).

273. E. Seker, E. Gulari, *Appl. Catal. A* **232**, 203–217 (2002).

274. M. A. Al-Daous, A. A. Manda, H. Hattorib, *J. Mol. Catal. A* **363–364**, 512–520 (2012).

275. J. Yu, L. Yue, S. Liu, B. Huang, X. Zhang, *J. Coll. Interf. Sci.* **334**, 58–64 (2009).

276. J. J. Pietron, R. M. Stroud, D. R. Rolison, *Nano Lett.* **2**, 545–549 (2002).

277. R. W. J. Scott, O. M. Wilson, R. M. Crooks, *Chem. Mater.* **16**, 5682–5688 (2004).

278. G. Budroni, A. Corma, *Ang. Chem. Int. Ed.* **45**, 3328–3331 (2006).

279. R. Nafria, P. R. D. L. Piscina, N. Homs, J. R. Morante, A. Cabot, U. Diaze, A. Corma, *J. Mater. Chem. A* **1**, 14170–14176 (2013).

280. S. Monyanon, S. Pongstabodee, A. Luengnaruemitchai, *J. Chin. Inst. Chem. Eng.* **38**, 435–441 (2007).

281. S. Pramanik, M. K. Mishra, G. De, *Cryst. Eng. Comm.* **16**, 56–63 (2014).

282. P. Dash, T. Bond, C. Fowler, W. Hou, N. Coombs, R. W. J. Scott, *J. Phys. Chem. C* **113**, 12719–12730 (2009).

283. H.-L. Jiang, T. Umegaki, T. Akita, X.-B. Zhang, M. Haruta, Q. Xu, *Chem. Eur. J.* **16**, 3132–3137 (2010).

284. R. W. J. Scott, C. Sivadinarayana, O. M. Wilson, Z. Yan, D. W. Goodman, R. M. Crooks, *J. Am. Chem. Soc.* **127**, 1380–1381 (2005).

285. Y.-S. Chi, H.-P. Lin, C.-N. Lin, C.-Y. Mou, B.-Z. Wan, *Stud. Surf. Sci. Catal.* **141**, 329–336 (2002).

286. S. Cheng, Y. Wei, Q. Feng, K.-Y. Qiu, J.-B. Pang, S. A. Jansen, R. Yin, K. Ong, *Chem. Mater.* **15**, 1560–1566 (2003).

287. Z. Konya, V. F. Funtes, I. Kiricsi, J. Zhu, J. W. Ager, M. K. Ko, H. Frei, A. P. Alivisatos, G. A. Somorjai, *Chem. Mater.* **15**, 1242–1248 (2003).

288. C. Aprile, A. Abad, H. Garcia, A. Corma, *J. Mater. Chem.* **15**, 4408–4413 (2005).

289. G. Lu, R. Zhao, G. Qian, Y. Qi, X. Wang, J. Suo, *Catal. Lett.* **97**, 115–118 (2004).

290. A. A. Ismail, D. W. Bahnemann, I. Bannat, M. Wark, *J. Phys. Chem. C* **113**, 7429–7435 (2009).

291. H. Zhu, B. Lee, S. Dai, S. H. Overbury, *Langmuir* **19**, 3974–3980 (2003).

292. J. Gu, L. Xiong, J. Shi, Z. Hua, L. Zhang, L. Li, *J. Solid State Chem.* **179**, 1060–1066 (2006).

293. H. Li, Z. Zheng, M. Cao, R. Cao, *Microp. Mesop. Mater.* **136**, 42–49 (2010).

294. B. Lee, Z. Maa, Z. Zhang, C. Park, S. Dai, *Microp. Mesop. Mater.* **122**, 160–167 (2009).

295. Z. Zheng, H. Li, T. Liu, R. Cao, *J. Catal.* **270**, 268–274 (2010).

296. C. Feral-Martin, M. Birot, H. Deleuze, A. Desforges, R. Backov, *React. Funct. Polym.* **67**, 1072–1082 (2007).

297. M. Schrinner, F. Polzer, Y. Mei, Y. Lu, B. Haupt, M. Ballauff, A. Goldel, M. Drechsler, J. Preussner, U. Glatzel, *Macromol. Chem. Phys.* **208**, 1542–1547 (2007).

298. B. Corain, C. Burato, P. Centomo, S. Lorac, W. Meyer-Zaikad, G. Schmidd, *J. Mol. Catal. A* **225**, 189–195 (2005).

299. C. Burato, P. Centomo, G. Pace, M. Favaro, L. Prati, B. Corain, *J. Mol. Catal. A* **238**, 26–34 (2005).

300. H. Miyamura, R. Matsubara, Y. Miyazaki, S. Kobayashi, *Angew. Chem. Int. Ed.* **46**, 4151–4154 (2007).

301. F. Klasovsky, M. Steffan, J. Arras, J. Radnik, P. Claus, *Open Phys. Chem. J.* **1**, 1–4 (2007).

302. S. Panigrahi, S. Basu, S. Praharaj, S. Pande, S. Jana, A. Pal, S. K. Ghosh, T. Pal, *J. Phys. Chem. C* **111**, 4596–4605 (2007).

303. Y.-C. Cao, Z. Wang, X. Jin, X.-F. Hua, M.-X.Liu, Y.-D. Zhao, *Coll. Surf. A* **334**, 53–58 (2009).

304. L. Balan, V. Melinte, T. Buruiana, R. Schneider, L. Vidal, *Nanotech.* **23**, 415705 (2012).

305. M. Cao, L. Zhou, X. Xu, S. Cheng, J.-L. Yao, L.-J. Fan, *J. Mater. Chem. A* **1**, 8942–8949 (2013).

306. F. Jiang, R. Li, I. Cai, W. Xu, A. Cao, D. Chen, X. Zhang, C. Wang, C. Shu, *J. Mater. Chem. A* **3**, 19433–19438 (2015).

307. V. K. Rao, T. P. Radhakrishnan, *ACS Appl. Mater. Interf.* **7**, 12767–12773 (2015).

308. S. Hermes, M.-K. Schroter, R. Schmid, L. Khodeir, M. Muhler, A. Tissler, R. W. Fischer, R. A. Fischer, *Angew. Chem. Int. Ed.* **44**, 6237–6241 (2005).

309. H.-L. Jiang, B. Liu, T. Akita, M. Haruta, H. Sakurai, Q. Xu, *J. Am. Chem. Soc.* **131**, 11302–11303 (2009).

310. L. Lili, Z. Xin, G. Jinsen, X. Chunming, *Green Chem.* **14**, 1710–1720 (2012).

311. B. Gole, U. Sanyal, P. S. Mukherjee, *Chem. Commun.* **5**, 4872–4875 (2015).

312. Z. Sun, G. Li, L. Liu, H.-o. Liu, *Catal. Commun.* **27**, 200–205 (2012).

313. J. Long, H. Liu, S. Wu, S. Liao, Y. Li, *ACS Catal.* **3**, 667–654 (2013).

314. J. Li, Q.-L. Zhua, Q. Xu, *Chem. Commun.* **50**, 5899–5901 (2014).

315. Y. Liu, S.-Y. Jia, S.-H. Wu, P.-L. Li, C.-J. Liu, Y.-M. Xu, F.-X. Qin, *Catal. Commun.* **70**, 44–48 (2015).

316. T. Miyazaki, R. Hasegawa, H. Yamaguchi, H. Oh-oka, H. Nagato, I. Amemiya, S. Uchikoga, *J. Phys. Chem. C* **113**, 8484–8490 (2009).

317. J.-S. Lee, J. Cho, C. Lee, I. Kim, J. Park, Y.-M. Kim, H. Shin, J. Lee, F. Caruso, *Nature Nanotech.* **2**, 790–795 (2007).

318. R. Banerjee, J. Novák, C. Frank, M. Girleanu, O. Ersen, M. Brinkmann, F. Anger, C. Lorch, J. Dieterle, A. Gerlach, J. Drnec, S. Yu, F. Schreiber, *J. Phys. Chem. C* **119**, 5225–5237 (2015).

319. S. Kumar, S. Zou, *J. Phys. Chem. B* **109**, 15707–15713 (2005).

320. P. Diao, D. F. Zhang, M. Guo, Q. Zhang, *J. Catal.* **250**, 247–253 (2007).

321. H. R. Cho, J. H. Lee, *Eur. J. Inorg. Chem.* **2015**, 640–645 (2015).

322. G. Milczarek, A. Ciszewski, *Coll. Surf. B* **90**, 53–57 (2012).

323. M. R. H. Nezhad, M. Aizawa, L. A. Porter Jr., A. E. Ribbe, J. M. Buriak, *Small* **1**, 1076–1081 (2005).

324. J. P. Spatz, S. Mossmer, C. Hartmann, M. Moller, T. Herzog, M. Krieger, H.-G. Boyen, P. Ziemann, B. Kabius, *Langmuir* **16**, 407–415 (2000).

325. B. R. Cuenya, S.-H. Baeck, T. F. Jaramillo, E. W. McFarland, *J. Am. Chem. Soc.* **125**, 12928–12934 (2003).

326. S. Kumar, S. Zou, *Langmuir* **25**, 574–581 (2009).

327. S. Kielbassa, A. Habich, J. Schnaidt, J. Bansmann, F. Weigl, H.-G. Boyen, P. Ziemann, R. J. Behm, *Langmuir* **22**, 7873–7880 (2006).

328. J. Bansmann, S. Kielbassa, H. Hoster, F. Weigl, H. G. Boyen, U. Wiedwald, P. Ziemann, R. J. Behm, *Langmuir* **23**, 10150–10155 (2007).

329. F. Behafarid, B. R. Cuenya, *Nano Lett.* **11**, 5290–5296 (2011).

330. F. Behafarid, J. Matos, S. Hong, L. Zhang, T. S. Rahman, B. R. Cuenya, *ACS Nano* **8**, 6671–6681 (2014).

331. M. Aizawa, J. M. Buriak, *Chem. Mater.* **19**, 5090–5101 (2007).

332. I. R. Laskar, S. Watanabe, M. Hada, H. Yoshida, J. Li, T. Iyoda, *Surf. Sci.* **603**, 625–631 (2009).

Chapter 9
Catalytic Properties of Gold Nanoparticles

Evgeny (Eugene) Beletskiy, Mayfair C. Kung and
Harold H. Kung

*Department of Chemical and Biological Engineering, Northwestern
University, Evanston, USA*

9.1 Introduction

The discovery that oxide-supported Au nanoparticles have unusual catalytic properties, such as in CO oxidation at the exceptionally low temperature of 203 K as reported by the group of Haruta,[1] has spurred years of intensive research. Now, more than thirty years later, new discovery about different aspects of CO oxidation over Au catalysts are continually forthcoming. In view of the vast amount of published literature, this review on Au catalysis cannot be exhaustive but can merely touch upon some of the interesting progresses that have taken place since the last publication of this book. We would also like to point out that although the scope of this chapter is limited to monometallic Au, another interesting area of Au catalysis is bimetallic Au-based catalysts, especially AuPd catalysts.[2–4] There are some extensive reviews on this rapidly evolving area of research.[5–7]

This chapter will cover various aspects of oxidation and hydrogenation reactions. Discussion about CO oxidation has to be linked to support effect as the activity of Au nanoparticles for this reaction is highly dependent on the nature of the support, especially with respect to the reducibility of the support. In fact, the different oxidation reactions share sufficient common aspects that a substantial fraction of this chapter will be devoted to oxidation reactions and the dependence of their catalytic properties on the nature of the support. We will start with the discussion of CO oxidation and models to explain the dependence of the activity on the size of the Au particles, the nature of the support and the presence of water. This will be followed

"

by other oxidation reactions using molecular oxygen: propane and alcohol oxidation, as well as epoxidation. A section will be devoted to a detailed discussion of reactions in which sacrificial reductants were used to generate hydroperoxide, which is the proximal oxidant. The chapter will conclude with discussion of hydrogenation reactions, which includes the role that Au plays in CO_2 utilisation.

9.2 CO Oxidation

Various hypotheses have been advanced for the unusual catalytic CO oxidation activity of supported Au nanoparticles. The earlier models primarily focus on electronic effect of small particles (quantisation of energy levels, electron transfer between support and metal, oxidation state) and geometric effects (coordination of surface metal atoms). Recent discussion of these include work by Visikovskiy *et al.*, which reported that *d*-band parameters changed markedly below a critical cluster size of Au ($\sim$150 atoms/cluster), a size that coincided with the sharp reduction of the average Au coordination number, contraction of Au–Au bond distance and onset of catalytic activity.[8] The group of Goodman showed that bilayer structures of Au on support were the most catalytically active for CO oxidation because they possessed electronic structure and chemical properties, distinct from bulk Au, that were attributed to the quantum size effect.[9,10] Herzing *et al.*, using aberration corrected microscope, correlated the highest CO oxidation activity with the presence of bilayer clusters over iron oxide-supported Au catalyst.[11] Another work ascribed the unique catalytic properties to geometric effect (Au atoms at corners and edges of a crystallite have lower coordination than atoms on close-packed surfaces as shown in Figure 2.5 of Chapter 2[12,13]).

Haruta, in his initial observation of low-temperature CO oxidation over supported Au catalysts, had already proposed that the active sites reside at the Au-support perimeter.[14] This concept was later expanded by Bond and Thompson.[15] Their model was derived from the reported literature, which indicated that (1) the catalytic activity depended strongly on the oxide support, (2) only samples containing small ($<$5 nm with the highest activity between 2 and 3 nm) Au particles were highly active and (3) catalytic activity

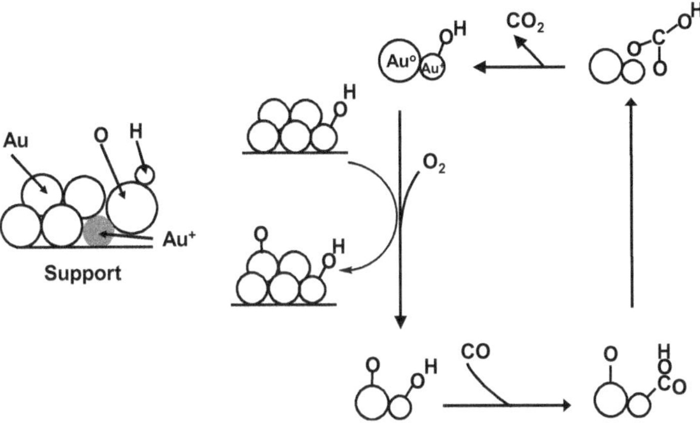

Figure 9.1   Model of active site for CO oxidation and the corresponding scheme for the reaction mechanism. Reprinted from Ref. 18 with permission from Elsevier Science.

could be generated upon deposition of titania on Au powders.[16] Their model suggested two possible roles of the support at the metal particle perimeter: one is to stabilise a positively charged Au cation at the perimeter, and the other is to act as a source and sink of protons and/or hydroxyls in the formation of a hydroxycarbonyl intermediate. Whereas the Bond and Thompson model was proposed for Au supported on transition metal oxide, the group of Kung extended it to explain their data for Au/alumina.[17,18] Their proposed active site (Figure 9.1) is an ensemble composed of metallic Au and Au–OH where the latter is situated at the interfacial perimeter site and stabilised by the support. The hydroxyl is associated with Au because thermal treatment at 373 K deactivated the catalyst whereas hydroxyls associated with alumina are expected to be stable to higher temperatures. Metallic Au was deemed necessary because an uncalcined catalyst, where Au was present predominantly in the ionic form, was inactive.

In the Kung's model, the reaction proceeds by insertion of an adsorbed CO into the Au^+–OH bond to form a hydroxycarbonyl. The hydroxycarbonyl is oxidised to a bicarbonate, which is then decarboxylated to regenerate the active site. Since hydroxycarbonyl reacts by oxidation, its surface density would be higher in the absence of oxygen. Using FTIR to monitor the catalyst, the Kung group passed a stream of 1% CO over Au/TiO_2 at 213 K, a low temperature chosen to minimise possible reaction of the hydroxycarbonyl with oxygen impurity, and observed a peak that was consistent with a hydroxycarbonyl.[19] They later collected additional evidence to support the catalytic role of perimeter sites by bromide poisoning of Au/TiO_2 catalyst.[20] They correlated the amount of adsorbed halide determined with X-ray absorption spectroscopy to the reduction in catalytic activity and the amount of adsorbed CO, and determined that the activity could be suppressed totally when the adsorbed halide was sufficient to block the Au-support perimeter, while CO adsorption was suppressed by about 60%; the remaining adsorption capacity was equivalent to the fraction of surface Au atoms not at the perimeter. A later report by Chandler *et al.*[21] on NaBr poisoning of Au/TiO_2 concluded that the poisoned sites correlated well with the edge and corner sites of the Au particles.

Work with model systems also provided support for the interfacial perimeter sites. The group of Flytzani–Stephanopoulos constructed a model system of nanotowers of alternating layers of Au and reducible CeO_2 on a silicon wafer in which only the sides of the nanotower were exposed to the reaction mixture.[22] They observed that the CO oxidation rates scaled linearly with the number of $Au–CeO_2$ interface. Insertion of an inert non-reducible SiO_2 layer between an Au and a CeO_2 layer to eliminate the $Au–CeO_2$ interface suppressed the activity. Thus, the active structure was proposed to reside in the $Au–CeO_2$ interface.

Using the inverse catalyst approach whereby oxide is deposited on the metal, Mullin and co-workers[23] decorated a Au(111) surface with Fe_2O_3 clusters. They observed that both a clean surface and one with complete coverage of a layer of Fe_2O_3 were inactive for CO oxidation. However, a partially covered surface was active, with the maximum activity observed at roughly 0.5 monolayer equivalent coverage. The data were consistent with the active sites being associated with atoms at the metal perimeter–support interface as exposed $Fe_2O_3–Au$ interfaces were only present on samples with partial surface coverage of Fe_2O_3.

Extensive research in recent years on supported Au powder catalysts has yielded other interesting and insightful clarification on the nature of the interfacial perimeter catalytic active sites and the reaction mechanism of carbon monoxide oxidation. The following studies captured the significant progress made. Although they were conducted at different temperatures, the data generated all supported the concept that oxygen activation originated at the interfacial active site.

The group of Behm,[24,25] based on quantitative temporal analysis of products (TAP) reactor measurements at 353 K, concluded that atomic oxygen, formed by dissociative adsorption of O_2 at the perimeter site, can react with CO. Based on IR spectroscopic evidence collected between 110 and 130 K and periodic density functional theory (DFT) calculations, Yates and co-workers suggested that for Au/TiO_2, O_2 is activated through di-σ bonding to the interfacial Au site and the Ti_{5c} site (five-fold-coordinated Ti) of the support in the form of Au–O–O–Ti. CO on TiO_2 diffuses rapidly to the Au–TiO_2 interface, where it reacts to form CO_2 via a CO–O_2 complex (Figure 9.2).[26] The intriguing dual site mechanism has attracted much attention, especially with regard to the activation of oxygen and the importance of the Au–O–O–Ti species at higher temperatures than 130 K, since molecular

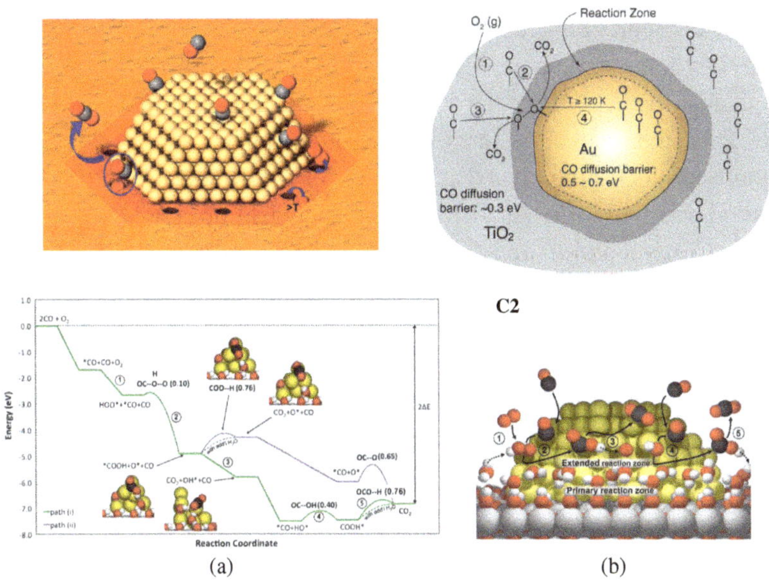

Figure 9.2 Different models of active sites and reaction pathways for CO oxidation. (a) O_2 dissociation into atomic O at perimeter site, which subsequently reacts with CO on adjacent Au. Reprinted from Ref. 25 with permission from the American Chemical Society. (b) O_2 forms Au–O–O–Ti that reacts with CO on TiO_2. Reprinted from Ref. 26 with permission from American Association for the Advancement of Science. C1 and C2: Proton transfer at the interface to adsorbed O_2 results in Au–OOH intermediate that reacts with CO. Reprinted from Ref. 28 with permission from American Association for the Advancement of Science.

O_2 was observed to desorbed at 170 K.[26] Koga *et al.*, based on their DFT calculation, proposed that even if Ti–OO–Au is formed, it is a transient species that would rapidly transform to more stable ones.[27] Thus, they attributed the importance of an Au perimeter site to its ability to activate CO, which readily reacts with both dissociated or undissociated O_2 on the neighbouring Ti_{5c} sites.

These proposals, however, did not adequately address the promotional effect of water on CO oxidation rates.[18,29,30] CO oxidation rates were observed to increase by orders of magnitude when low level of water (ppm range) was co-fed with the reactants over Au supported on metal oxides. Despite the large change in the reaction rates, the apparent activation energies appeared to be independent of the moisture content of the feed, suggesting that the mechanism of reaction remained unaltered with the addition of water.[29,30] Chandler and co-workers,[28] with a combination of experimental results on effects of water partial pressure, deuterium kinetic isotope effect and DFT computational results, proposed a water-mediated reaction mechanism for room temperature CO oxidation over Au/TiO_2. They concluded that an adsorbed *OOH in the form of Au–OOH is an important intermediate. This intermediate is formed by proton transfer at the interface that facilitates both O_2 binding and activation. Au–OOH reacts readily with CO adsorbed on Au to form adsorbed-(CO)OH (hydroxycarbonyl) that decomposes to CO_2 via proton transfer to water.[28] An adsorbed *OOH was shown to be consistent with the detailed kinetics and isotope effect data for CO oxidation over Au/Al_2O_3 catalyst.[31] The presence of an adsorbed *OOH intermediate is also consistent with the fact that significant amount of H_2O_2 formation was detected in oxidation catalysis involving Au,[32–35] an aspect that would be discussed in greater details later.

9.3 Hydrocarbon Oxidation in the Presence of H_2 or Other Sacrificial Reductants

Besides CO oxidation, Au on support oxides also catalyses a broad range of reactions using only molecular O_2, or O_2 together with sacrificial reductants such as H_2 or CO. The group of Haruta first discovered that, in the presence of both H_2 and O_2, selective epoxidation of propylene occurred over Au/TiO$_2$ and Au/Ti-MCM-41[36,37] and selective oxidation of propane to acetone occurred over Au/Ti-MCM-41.[37] Their work attracted much interest and significant efforts were expanded to develop a commercial process as well as research to generate mechanistic information. A popular mechanistic model is that Au catalyses the formation of hydrogen peroxide from H$_2$ and O$_2$, and the peroxide diffuses to the Ti site to effect oxidation of the hydrocarbons.

Recently, a different concept was being pursued by the Kung group that the interfacial perimeter site of Au and TiO$_x$ plays a significant role in such oxidation reactions. In order to test the concept, they designed controlled metal-support interface by decorating the Au particles on Au/SiO$_2$ with patches of Ti–SiO$_2$ where the TiO$_x$ moieties were isolated or of low nuclearity (Figure 9.3).

When tested for the oxidation of propane in the presence of H$_2$ and O$_2$, both Au/SiO$_2$ and Au/SiO$_2$ decorated with SiO$_2$ patches were inactive (Table 9.1, catalyst **I**). However, acetone was formed with high selectivity (60–80%) when the Au particles supported on silica were decorated with patches of Ti–SiO$_2$. The activity increased with Ti–SiO$_2$ coverage (Table 9.1; catalysts **II** and **III**).[38] Interestingly, if the Au crystallite surface was first decorated with SiO$_2$ patches and then patches of Ti–SiO$_2$ were added, the activity declined, but the selectivity for acetone remained.[39] In a separate series of experiments, the authors started with Au/TiO$_2$, which generated propene (no acetone) as the selective product in the same reaction (Table 9.1, catalyst **IV**). Decorating the Au particles of Au/TiO$_2$ with SiO$_2$ patches (Table 9.1, catalyst **V**) suppressed the activity without changing the selectivity. However, as with Au/SiO$_2$, decorating the Au particles on Au/TiO$_2$ with Ti–SiO$_2$ patches changed the selectivity to form acetone (Table 9.1, catalyst **VI**).[40] Taken together, these results were interpreted to support the interfacial perimeter site model, that Au–TiO$_x$ interface is necessary for acetone formation because (a) there was no acetone formation with Ti-free patches or no patches, (b) the activity for acetone formation declined when there were fewer Ti-containing patches, (c) the activity declined by increasing the average distance of Ti–SiO$_2$ from Au and (d) when TiO$_x$ moieties were of high nuclearity such as in TiO$_2$, the primary product was propene but the selectivity changed to acetone when the Au particles were decorated with Ti–SiO$_2$.

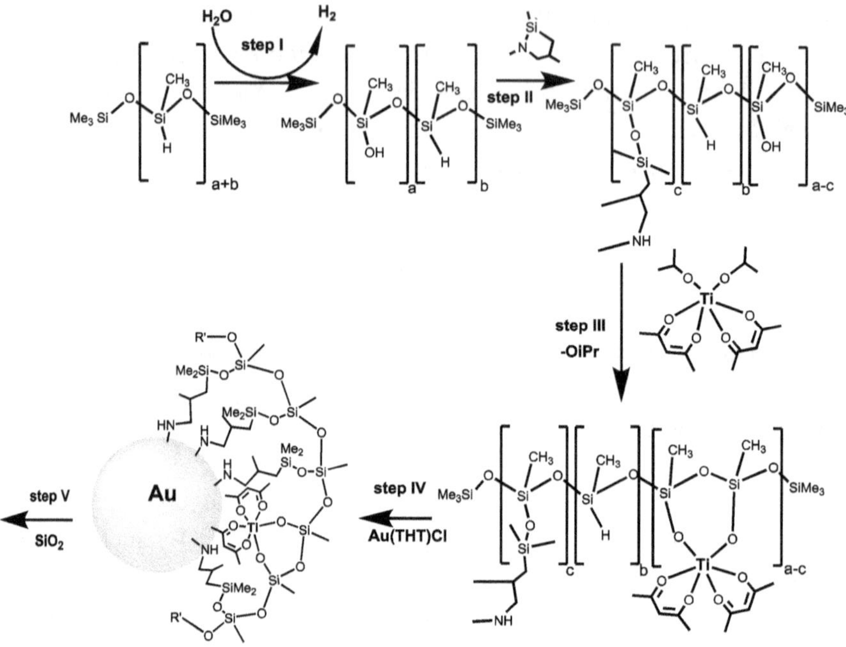

Figure 9.3   Scheme adapted from Ref. 38 showing the synthesis of TiO_x decorated Au/SiO_2. Step I: Oxidation of Si–H groups of polymethylhydrosilane to Si–OH groups over $Pd(OH)_2/C$ catalyst. Step II: Reaction of predetermined numbers of Si–OH groups with *N*-methyl-aza-2,2,4-trimethylsilacyclopentane to yield secondary amine functional groups. Step III: Reaction of remaining Si–OH groups with titanium diisopropoxide (bis-2,4-pentanedionate). Step IV: Binding of Au to amine functional groups and reduction of Au(I) by silane groups of the siloxane polymer. Step V: add solution to silica.

Mechanistic investigations by various groups on Au supported on titanosilicates pointed to the presence of a hydroperoxyl intermediate (˙OOH) in propylene epoxidation in the presence of a H_2–O_2 mixture. Based on the results of a combination of kinetics[41] and computational studies,[42] Delgass and co-workers concluded that Ti–OOH, formed at the Au–support interface, is an intermediate in the propene epoxidation reaction. One of the few direct observations of Ti–OOH was from the Au–Ba/Ti–SiO_2 (Ti-TUD)-catalysed epoxidation of propene [43], where a peak in the UV-visible spectrum was identified to be due to Ti–OOH. By following the transient behaviour of this absorption peak and with corroboration from XANES and FTIR data, the Ti–OOH species was determined to be a kinetically relevant species. The results of DFT calculation of hydrogen oxidation over

Table 9.1 Selective oxidation of propane on Ti–SiO$_2$-decorated gold catalysts.[38,40a]

| | Conv. | | Selectivity (%) | | |
Catalysts[b]	C$_3$H$_8$ (%)	D_p (nm)[c]	Acetone	Propene	CO$_2$
I. (SiO$_2$)/[Au/SiO$_2$][b]	0.09	4.2±1.5	67[d]	0	28
II. (Ti$_{0.7}$–SiO$_2$)/[Au/SiO$_2$][b]	0.56	4.4±1.8	68[d]	0	21
III. (Ti$_{1.5}$–SiO$_2$)/[Au/SiO$_2$][b]	0.71	4.4±1.7	71[d]	0	21
IV. Au/TiO$_2$[e]	2.2	2.9±1.1	2	68	29
V. (SiO$_2$)/[Au/TiO$_2$][e]	1.1	3.1±1.2	3	73	22
VI. (Ti–SiO$_2$)/[Au/TiO$_2$][e]	1.6	3.4±1.2	30[f]	5	50

[a]Reaction conditions: Catalyst weight = 350 mg except 5.5 mg Au/TiO$_2$, C$_3$H$_8$:O$_2$:H$_2$:He = 5:5:5:85, total flow rate = 30 mL/min, data shown were collected after 4.5 h time-on-stream at 220°C. The numbers in the subscript are the loadings of Ti (wt.%) determined by ICP-AES. [b](MO$_x$)/[Au/SiO$_2$] denotes MO$_x$ patches decorating a Au/SiO$_2$ catalyst. Au loading 0.5 wt.%, Ti loadings for the three samples were 0 for , 0.7, and 1.5 wt.%.[38] [c]Average Au particle diameter after catalytic testing estimated from STEM images. [d]Not listed are small amounts (4–6%) each of isopropanol and acetaldehyde also detected. [e] From Ref. 40. [f]11% acetaldehyde also detected.

Au catalysts suggested Ti–OOH at the interfacial perimeter site as an intermediate.[44]

Instead of H$_2$, CO can be used also as a sacrificial reductant. In the presence of a Au catalyst in a methanol–water mixture, a mixture of CO and O$_2$ generated a peroxide species (likely CH$_3$OOH), which effected the epoxidation of propene in the presence of a titanium silicalite-1 (TS-1) catalyst.[35] Interestingly, ^{18}O isotope labelling showed that the O in the epoxide originated entirely from O$_2$ and not from H$_2$O or methanol.

Even in the absence of hydrogen, the formation of hydroproxyl intermediate mediated by water or hydroxyl is possible. The group of Haruta showed that Ti–OOH species was formed as an intermediate in propylene epoxidation over a potassium promoted Au/TS-1 catalyst in a O$_2$/H$_2$O stream.[45]

9.4 Oxidation Using Molecular O$_2$

In addition to CO, supported Au are also active in other oxidation reactions using only molecular O$_2$. Two important classes of reactions, alcohol oxidation (Equation (9.1)) and epoxidation (Equation (9.2)) are discussed here.

$$RCH_2OH + O_2 \longrightarrow RCHO + H_2O \tag{9.1}$$

$$R_2C=CR_2 + 0.5O_2 \longrightarrow R_2C\underset{O}{\overset{\diagdown\diagup}{\text{---}}}CR_2 \tag{9.2}$$

R= aliphatic or aromatic group

Selective oxidation of alcohol is an important chemical transformation and is a promising route for the production of aldehydes, which are used in the production of resins, polyurethanes, perfumes and other fine chemicals. Au catalysts exhibit high selectivities to the corresponding aldehydes, higher than Pd or Au/Pd catalysts in organic solvents or under solvent-free conditions.[46] This is a rich area for research as significant controversy remains with respect to the effect of the Au particle size, the nature of the active site, as well as the support effect. In the discussion here, the focus will be on the liquid phase oxidation reactions. In general, selectivity in the oxidation of alcohol to aldehyde or ketone in organic solvents or under solvent-free conditions is high. The group of Hutchings[47] investigated the underlying reason for this high selectivity in the reaction of benzyl alcohol oxidation. They noted that although benzaldehyde (C_6H_5CHO) under ambient condition was prone to autoxidation to benzoic acid (C_6H_5COOH), the presence of even low concentrations of benzyl alcohol ($C_6H_5CH_2OH$) inhibited this process by hydrogen atom transfer to the acylperoxy radical that played a role in the autoxidation process. They observed that 1-octanol ($CH_3(CH_2)_7OH$) also had an inhibiting effect on the autoxidation of octanal ($CH_3(CH_2)_6CHO$), and the group is continuing their effort to understand the structural requirements needed in an inhibitor to quench the aldehyde autoxidation process.

For different Au catalysts, although the selectivity to aldehyde in general was high, the activity varied over quite a wide range. Abad *et al.*[46] examined the oxidation of cinnamyl alcohol ($C_6H_5CH=CHCH_2OH$) to cinnamaldehyde ($C_6H_5CH=CHCHO$) over various Au/TiO$_2$ catalysts and observed that the oxidation activity scaled linearly with the number of surface Au atoms. This suggested that the intrinsic activity per external gold atom was the same for Au of different particle sizes. Other studies, however, came to different conclusions. Chen *et al.*[48] observed that there was a mild dependence of the activity on the Au particle size for the oxidative dehydrogenation of benzyl alcohol over Au supported on hydrotalcite; the moles of alcohol reacted per mole of surface Au (turn over frequency [TOF]) increased by a factor of 1.7 when the Au particle size decreased from 12 to 2.1 nm. Panthi *et al.*[49] sintered the Au/Al_2O_3 catalysts containing Au particles too small to be detected by TEM and observed that, after sintering, the total activity and the activity per active site increased slightly, suggesting that larger Au particles were more active. Haider *et al.*[50,51] investigated benzyl alcohol oxidation over Au/TiO_2 and Au/CeO_2 and observed that medium size Au particles ($\sim$6.9 nm) were the most active. Zhu *et al.*[52] found the rates of benzyl alcohol oxidation on Au/activated carbon increased with increasing amounts of oxygen functional groups on the carbon support, indicating that the metal site alone was insufficient to describe the catalytic system and a possible role of the support for oxygen activation was proposed.

Recently, there is increasing evidence to indicate that the active sites for this reaction reside at the perimeter sites between Au and the support. Zhao *et al.*[53] decorated large Au particles on Ti-fibre with nanoparticles of metal oxides (CoO, NiO and Mn_3O_4) to create a high density of interfacial perimeter sites for the selective oxidation of benzyl alcohol to benzaldehyde. All the inverse catalysts were significantly more active than the parent Au/Ti-fibre catalyst. Similarly, Wang *et al.* attributed the higher activity in alcohol oxidation over CeO_2-modified Au@SBA-15 compared with Au@SBA-15 and CeO_2 to the creation of a large number of interfacial perimeter sites by confining the Au and CeO_2 in the channels of SBA-15.[54] Panthi *et al.*[49] investigated substituted benzyl alcohol oxidation over Au/Al_2O_3, Au/TiO_2 and Au/SiO_2 catalysts that contained Au nanoparticles of similar sizes. They used kinetic poisoning strategy with phenylethyl mercaptan as the poison and determined that $\sim$12% of the total Au was active for the selective 4-methoxybenzyl alcohol oxidation. This number agreed well with the number of Au atoms at or near the perimeter site. They further postulated that the Au atoms function as the hydride acceptor of the benzylic hydrogen. The activity and product selectivity of methanol oxidation differed for Au/ZnO and Au/TiO_2. But for both types of catalysts, the activity could be correlated to the number of Au atoms at the perimeter sites.[55]

Thus there appears to be a lack of consensus with respect to the particle size requirement of the Au for alcohol oxidation. The confusion may arise because correlations were drawn from the average particle size of Au in the sample and sometimes the size distributions of the Au particles were quite wide. The heterogeneity of the sample deterred quantitation of the surface exposed Au and that is why preparation of oxide supported Au with narrow distribution of Au particle size is paramount to Au catalysis. The different techniques in the preparation of Au catalysts, especially with respect to small Au nanoparticles supported on oxides, was discussed in detail in Chapter 8.

There is also intensive effort expanded to understand the reaction mechanism. Abad *et al.*[46] observed a primary isotope effect when they compared the rates of *p*-methylbenzyl and α-deutero-*p*-methylbenzyl alcohols and proposed the mechanism as depicted in Figure 9.4 for the aerobic oxidation of alcohol. The first step of the reaction is the formation of Au–alcoholate with positive Au ions. The Au–alkoxide complex would subsequently undergo a hydride shift, resulting in the formation of carbonyl

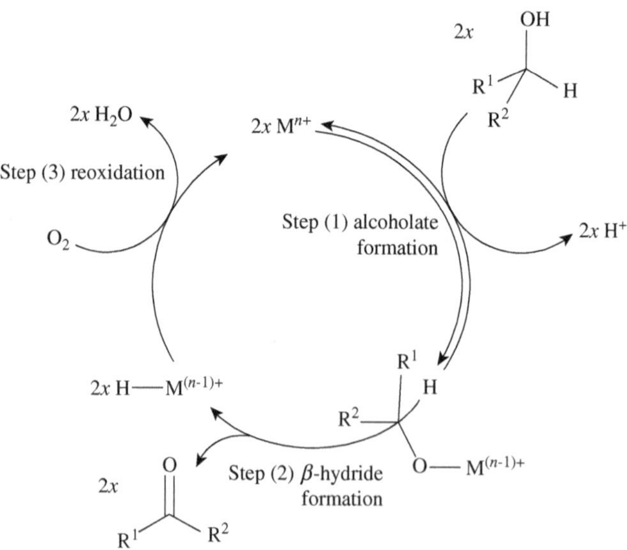

Figure 9.4   Reaction mechanism for the aerobic liquid phase oxidation of alcohol over supported Au catalysts in organic solvents. Reprinted from Ref. 46 with permission from Wiley-VCH Verlag GmbH and Company.

product and a Au–hydride intermediate. The observed primary isotope effect indicated that the second step is the rate-limiting step. Oxidation of the hydride by oxygen to form water returned the catalyst to its original metallic state and this is proposed to be a rapid step. The presence Au–H in an Au-catalysed alcohol oxidation reaction was confirmed by electron paramagnetic resonance and spin trapping.[56]

The above proposal is for the oxidative pathway. However, alcohol decomposition (dehydrogenation) can also proceed via a non-oxidative pathway. Based on DFT calculations, one proposed mechanism for ethanol dehydrogenation over metallic Au via the non-oxidative pathway involves dehydrogenation of the hydroxyl group of ethanol (C_2H_5OH) to form an adsorbed ethoxy and hydrogen as the rate-limiting step.[57] However, Shylesh *et al.* observed only a small kinetic isotope effect (1–1.2) for Au/hydrotalcite when deuterium labelling was at the hydroxyl (C_2H_5OH versus C_2H_5OD), whereas a larger kinetic isotope effect of 2.2 was observed when the deuterium label was at the carbon (C_2H_5OH versus C_2D_5OH).[58] Small or no kinetic isotope effects are observed when deuterium substitution for hydrogen are in bonds that are not being broken in the rate-determining step. Chen *et al.* also measured the kinetic isotope effect for benzyl alcohol ($PhCH_2OH$ versus $PhCD_2OH$) over Au/hydrotalcite, and observed a value of 2.7 for the oxidative reaction and 2.2 for the non-oxidative reaction.[48] These results suggested that for both oxidative and non-oxidative reactions of alcohol, the rate-determining step is the cleavage of the α-C–H bond. The reaction was significantly slower under non-oxidative than oxidative conditions for a given alcohol over the same catalyst.[48] Since both pathways shared the same rate-limiting step, then the presence of oxygen must have facilitated the C–H bond cleavage step. Using DFT calculations, the adsorbed O_2 was found to be activated to form OOH by abstraction of hydrogen from the OH group of water or alcohol.[59] The hydroperoxyl can subsequently decompose into atomic oxygen and hydroxyl. All three species (hydroperoxide, atomic oxygen and hydroxyl) are capable of abstracting the β-H from the alkoxide intermediate, thus forming the aldehyde. Thus the oxidative pathway can proceed much faster than in the absence of oxygen.[48] The proposed involvement of the hydroperoxyl intermediate in the abstraction of the β-H from the Au–alkoxide differed significantly from the proposal of Abad *et al.*[46] and from the observation of Au–H in aerobic oxidation.[56]

Clearly, more work is needed to resolve the conflicting results and proposals.

In contrast to the oxidative pathway, most of the reports in the literature agreed that the non-oxidative dehydrogenation of alcohol on Au catalysts is a structure-sensitive reaction (reaction characteristics such as rate and product distribution change with metal particle size), and the TOF increases with decreasing Au particle size.[48,57,58,60] The proposed mechanisms generally pointed to β-H elimination by low-coordination-number Au atoms as the major pathway for the cleavage of the alcohol C–H bond,[48,57,58] which explained the structure sensitivity of this reaction. In the presence of oxygen, since the formation of the hydroperoxyl intermediate could also occur on terrace sites and this hydroperoxyl intermediate is involved in the rate-limiting step, the dependence of the oxidative reaction rate on the Au particle structure would be less. Furthermore, there may be other factors that contribute significantly to the oxidation rate. Thus, there is the difference on the particle size effect on the rate of reaction between the oxidative and non-oxidative reactions.

The scheme in Figure 9.4 did not implicitly show a role for the support. However, it is known that for both oxidative and non-oxidative reaction of alcohol over supported Au catalyst, there is a strong dependence of the rate on the support.[46,48,49,58,60] Abad *et al.* proposed that the support stabilises the positive Au ions through interfacial Au-support interaction and facilitates oxygen activation for removal of the hydride.[46] Their conclusion regarding facilitating oxygen activation was based on the observation that the number of moles of cinnamyl alcohol reacted per mole of Au on Au/ceria catalysts depended strongly on the Au loading, with the catalyst of the lowest Au content being the most active. For two Au catalysts of different metal loadings but prepared in the same manner and having similar particle size distribution, the rate normalised to total Au varied by over a factor of three. Since the authors regarded ceria as an oxygen pump that facilitated the oxidation of the Au–H, they attributed the sample with a lower activity to be due to a reduction of free ceria surface area consequent of its higher Au coverage. However, there is one problem with their reasoning. The area covered by Au for a 4.4% weight Au loading sample would be less than $1 \, \text{m}^2\text{g}^{-1}$, which was inconsequential compared with $180 \, \text{m}^2\text{g}^{-1}$ of the nanoceria. Also, if removal of Au–H by oxidation was rapid, then oxygen activation should not be critical in determining the activity of the catalyst.

Shylesh *et al.* examined the role of basic sites in *n*-butanol oxidation.[58] They co-fed butanoic acid (C_3H_7COOH) and *n*-butanol (C_4H_9OH) over a Au/hydrotalcite catalyst. The inclusion of the acid was to neutralise the basic sites of hydrotalcite. The complete suppression of catalytic activity prompted them to propose that formation of the alkoxide intermediate ($C_4H_9O^-$) occurs on the surface of the basic support. This would imply that the hydrogen released in the formation of the alkoxide resides on Au (probably a low-coordination Au atom), which further imply that the reaction proceeds at the Au-support interfacial perimeter sites.

Oxidation of biomass-derived alcohol in water with O_2 is an area of interest because it is an environmentally friendly (green) process for valorisation of renewable feedstock. The rate of selective oxidation of alcohols in an aqueous phase over supported Au was accelerated at high-pH and the end product was the acid.[61] Under such a condition and using Au/C or Au/TiO$_2$ catalysts, oxidation of alcohol in water using $^{18}O_2 + H_2^{16}O$ did not produce any ^{18}O-containing products. But when a mixture of $^{16}O_2 + H_2^{18}O$ was used, ^{18}O isotope incorporation in the products was observed.[61] Control experiments confirmed that this incorporation was not due to isotopic exchange of the products with water. Assisted by DFT computational results, the reaction scheme shown in Figure 9.5 was proposed that explained the OH^- enhancement of the reaction rate and the isotopic distribution in the products. In this scheme, the OH^- in solution as well as those adsorbed on Au facilitate the deprotonation of the alcohol. In addition, adsorbed OH^- also lowers the barrier for C–H bond cleavage to form the aldehyde, which reacts with OH^- to form a germinal diol/alkoxy intermediate. This intermediate can rapidly exchange with OH^- in solution. In this manner, the ^{18}O label in water could be incorporated into the product. The activation barrier for the reaction of the aldehyde with the OH^- to form the $RCHOOH^-$ intermediate on Au was only 5 kJ mole^{-1}, whereas the barrier for the subsequent C–H activation of this intermediate to form acetic acid was higher at 20 kJ mole^{-1}, although the latter still proceeded readily.

Epoxides are valuable chemical intermediates but their production often requires the use of expensive oxidants such as H_2O_2 or alkylperoxide as shown in Equation (9.3).[62,63]

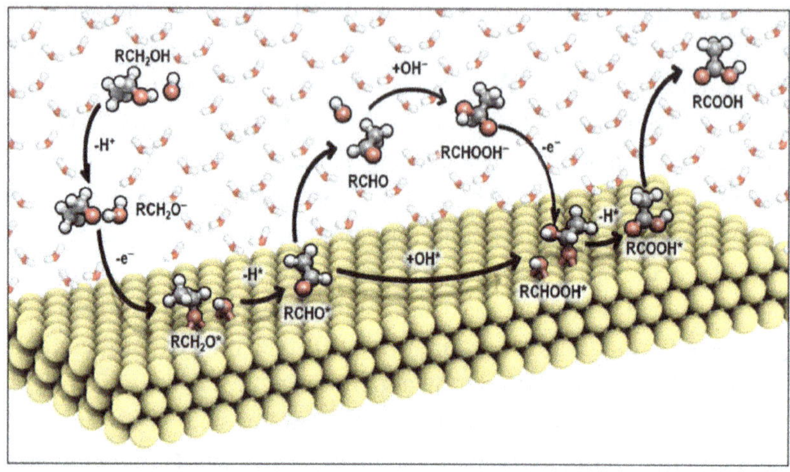

Figure 9.5   Reaction scheme for the oxidation of alcohols to acids over a Au surface in aqueous solution at high pH. Reprinted from Ref. 61 with permission from American Association for the Advancement of Science.

$$R_2C{=}CR_2 + R'OOH \longrightarrow R_2C\underset{O}{\overset{}{\diagdown\diagup}}CR_2 \quad +R'OH \tag{9.3}$$

R = aliphatic or aromatic group
R′ = alkyl or H

The only industrial processes that utilise molecular oxygen as an oxidant are the silver-catalysed epoxidation of ethylene and oxidation of butadiene to butene oxide. For both economic and environmental impact reasons, there is intense interest to use O_2 as the oxidant. However, due to the strong O–O bond in O_2, conditions necessary to activate this molecule also result in low product selectivity. One strategy that has been explored to overcome this limitation is to use the expensive peroxides as initiators while using O_2 as the terminal oxidant.[64,65] This strategy involves the addition of a catalytic amount of an initiator such as alkyl hydroperoxide or hydrogen peroxide to start the reaction and using molecular O_2 to continue and sustain the reaction. The role of the initiator is still under debate and will be discussed in more details later. Using dioxygen only, Turner *et al.* observed a sharp threshold in the size of the Au nanoparticles on an inert non-reducible support in the selective oxidation of styrene to styrene oxide without an initiators.[66] Whereas particles larger than 2 nm were inactive, smaller particles were able to oxidise styrene primarily to benzaldehyde with the minor product being styrene oxide. The selectivity for epoxide increased with use, being 12%, 23.7% and 27% for the fresh, first, and second reuse of the catalyst, respectively, but these changes were accompanied by a decrease in styrene conversions of 25.8%, 21.4% and 15.9%, respectively. Zhu *et al.* compared the reactivities of structurally well-defined and atomically monodisperse, thiolate-capped Au nanoclusters supported on SiO_2 for styrene oxidation with O_2. They tested Au of different cluster sizes and observed in all cases, benzaldehyde was the major product and styrene epoxide was the minor product, and that O_2 activation was more facile for smaller Au clusters.[67] In these two examples, there were no information on the progress of the reaction as a function of reaction time. Donoeva *et al.* followed the progress of the aerobic, solvent-free, stabiliser-free oxidation of cyclohexene (Equation (9.4)) at 65°C in a batch reactor over an Au/SiO_2 catalyst.[68] They reported an induction period during which time the Au particle size increased. The onset of the activity corresponded to the time when the mean particle size of the Au became >2 nm. Thus, they advocated that Au particles >2 nm were the active catalysts. The four major products of cyclohexene oxidation over Au/SiO_2 were cyclohexene oxide, 2-cyclohexen-1-ol, 2-cyclohexen-1-one, and cyclohexenyl hydroperoxide, with the latter being the main product. The group of Golovko attempted to tune the selectivity of the catalysts using Au on different supports,[69] and reported that for Au on WO_3 with the same loading of Au as that on silica, the cyclohexene epoxide selectivity improved from 7% to 35%.

Cyclohexene cyclohexene oxide 2-cyclohexen-1-one 2- cyclohexen-1-ol cyclohexenyl hydroperoxide

$$(9.4)$$

Other researches have also observed olefin oxidation activity using O_2 and Au catalysts with large Au particles. Alshammari *et al.* reported the use of Au/graphite in the catalysis of stabiliser-free cyclic alkene (C_6, C_7 and C_8) oxidation by molecular O_2 to form epoxides. The average Au particle size of this catalyst was 4.7 nm and the size distribution was from 1 to 12 nm.[70] In the aerobic oxidation of styrene, Wang *et al.* reported styrene epoxide formation using amino functionalised, porous polydivinylbenzene supported, positively charged bulk Au particles with particle sizes of 20–150 nm.[71]

Two relevant questions to this area of research are whether the reaction proceeds via homogeneous (using leached Au as the catalyst) or heterogeneous pathway and what is the role of the initiator used in the reaction. Although some of the epoxidation studies showed good recyclability of the catalysts,[68,72] the recycled catalysts were obtained by separation from the reaction mixture via centrifugation. Thus, if Au was leached into the solution during reactions, generally between 60°C and 80°C, to form soluble catalysts of small gold clusters, they could re-precipitate/re-adsorb onto the large area support during low-temperature centrifugation step used to separate the catalysts from the reaction solution. They would then be carried forth to subsequent reaction mixtures used to test recyclability of the catalyst. Ovoshchnikov *et al.*[69] tried to rule out the importance of dissolved Au with a series of experiments. They determined the Au content of a fresh catalyst and one that had been used with atomic absorption spectroscopy and noted that, although significant leaching from the fresh catalyst was observed, the Au content of the catalysts from subsequent reused catalysts appeared fairly constant within experimental error. However, the results do not exclude the possibility that dissolved Au, although very small in quantity, could be very active. For example, Oliver-Meseguer *et al.* observed that small clusters of gold in solution could catalyse various organic reactions at room temperature, even when their concentrations were only parts per billion.[73] Another control experiment conducted by Ovoshchnikov *et al.* involved replacing O_2 with argon.[69] After heating the mixture as in a reaction run, they removed the solid catalyst by hot filtration and found no activity with the filtrate. However, this result does not exclude the possibility that leaching of the active form of Au requires the presence of oxygen, i.e. the leached Au clusters have to be stabilised, possibly by the products of oxidation. The authors also reported that after 6 h of reaction, the cyclohexene oxidation reaction did not slow down upon removal of the Au/WO_3 catalyst from the reaction mixture. They attributed this to the presence of cyclohexenyl hydroperoxide formed in the reaction, which can catalyse autoxidation of cyclohexene. Since cyclohexenyl hydroperoxide was present in the reaction mixture before and after Au/WO$_3$ removal and the reaction rate did not slow down after the removal of Au/WO$_3$, then hydroperoxide appeared to be the species responsible for sustaining the reaction, whereas the role of Au appeared to be just the initial generation of the hydroperoxide. These authors tried to probe the role of cyclohexenyl hydroperoxide further by forming a solution with very little hydroperoxide via high dilution of the cylcohexene substrate. After 6 h in typical oxidative reaction conditions, the liquid was separated from the solid Au catalysts by hot filtration. Then fresh cyclohexene substrate was added to both the solid catalyst and the filtrate and catalytic activity was detected only with the solid catalyst and not with the filtrate. This experiment showed that solid Au catalyst was capable of initiating the reaction, but it did not clarify whether its role is to solely generate cyclohexenyl hydroperoxide, which, after reaching a critical concentration, effects a continuous reaction via an autocatalytic chain reaction pathway.

Related to the above discussion is the role of the initiator. There are a number of examples where the Au catalyst was deemed inactive using O$_2$ alone but significant activity was observed if an initiator such as hydroperoxide was added in catalytic amounts. The role of the initiator is still under intense discussion. Some researchers proposed a role for the initiator in the catalytic cycle,[74,75] while others proposed that the addition of the initiator was to counter the effect of the stabiliser, such as 2,6-bis(t-butyl)-4-methylphenol (BHT) or 3-octadecyl-(3,5-di-t-butyl-4-hydroxyphenyl)-propanoate (irganox 1076), sometimes added to commercial alkene to prevent their auto-oxidation.[70] There are yet unanswered questions for this model. It is not clear whether after the removal of the alkene stabiliser how fast would hydroperoxeride be formed from the olefin and what concentration of hydroperoxide is needed to initiate and maintain the free radical reaction. Donovea *et al.* noted that cyclohexenyl hydroperoxide (CyOOH) was present in trace amounts ($\sim$0.015%) in the stabiliser-free cyclohexene. In the presence of an initiator, Au of different sizes were all reported to be active in olefin epoxidation with O$_2$.[76–79] There appeared to be no abrupt cut-off in reactivity related to Au particle size, albeit higher activities were generally observed for smaller Au clusters[78] or particle sizes.[79]

Thus, questions remain as to what constitute the active catalyst for oxidation reactions using molecular oxygen. Although Au catalysts are needed, their role in the reaction is unclear. Similarly, the importance of the role of hydroperoxide needs further clarification. These are interesting research questions that should be investigated in order to advance Au-catalysed alkene oxidation using O_2 as the proximal oxidant.

9.5 Hydrogenation

Over 40 years ago, Bond *et al.* demonstrated for the first time that gold nanoparticles supported on silica, alumina or boehmite could be catalytically active in hydrogenation of unsaturated hydrocarbons at a relatively low temperatures.[80] Moreover, 1,3-butadiene and 2-butyne reacted with an unusual exclusive selectivity for monoreduction. Subsequently, discoveries of other highly selective hydrogenation reactions on gold nanoparticles involving alkynes, unsaturated carbonyls and nitro compounds were reported.[81,82]

Hydrogenation of alkynes is a powerful approach to synthesise Z-alkenes. In order to avoid the undesirable overreduction to alkanes, hydrogenation catalysts with intentionally suppressed activity were used, such as the Lindlar catalyst (Pd on $CaCO_3$ poisoned with $Pb(OAc)_2$).[83] Often, the suppression of activity is achieved using toxic additives (e.g. Pb), and the accompanying reduction of selectivity at higher conversions, *E–Z* isomerisation, as well as occurrence of oligomer formation are undesirable in practical processes, such as acetylene removal from the polymer-grade ethylene[84] or flavour and fine chemical synthesis.[85]

Choudhary *et al.* reported that 4.7 nm Au nanoparticles, prepared by temperature-programmed reduction–oxidation of $Au_6(PPh_3)_6(BF_4)_2$ on TiO_2, catalysed the reduction of acetylene to ethylene very selectively (>90%) even at high conversions.[86] However, the activity and stability of the catalyst were poorer than the Pd-based catalyst. Shao *et al.* demonstrated that the activity and stability of Au nanoparticles in alkyne hydrogenation could be improved by supporting them on graphene oxide.[87]

Phenylacetylene was hydrogenated at 333 K at 99% conversion with 99% selectivity to styrene. The unusually high activity and stability was ascribed to high concentrations of functional groups on graphene oxide that were able to stabilise highly dispersed rounded Au nanoparticles with high density of low-coordination sites. Analogous Pd-containing reference catalysts were somewhat more reactive, but yielded over-hydrogenated ethylbenzene product. The higher selectivity observed for Au-based relative to Pd-based catalysts was examined using DFT calculations by Pérez-Ramírez *et al.*[88] Their results indicated that the activation barriers for hydrogenation of $C\equiv C$ and $C=C$ bonds were similar over gold nanoparticles. However, alkyne adsorbed to the edges of the nanoparticles much more strongly than alkene, leading to its much higher rate of hydrogenation. On the other hand, both alkenes and alkynes adsorbed strongly on Pd (111). Thus, they competed equally for the same active sites resulting in poorer selectivity.

Selective hydrogenation of a functional group in the presence of a different functional group on the same molecule is a challenging task. Over some supported Au catalysts, the reduction of the nitro group was preferred over other functional groups such as olefinic and carbonyl groups, whereas over other Au catalysts, a reversal in selectivity was observed. The ensuing paragraphs would discuss the underlying reasons for these observed differences.

Reduction of nitro compounds into amines is a key step in the preparation of many dyes, pigments, pharmaceuticals and agrochemicals. The group of Corma found that Au/TiO_2 and Au/Fe_2O_3 catalysed the hydrogenation of substituted aromatic nitro compounds ($XC_6H_4NO_2$ where X was a reducible functional group) with high selectivities, yielding substituted anilines ($XC_6H_4NH_2$).[89] The dissociation of H_2 was the rate-limiting step and it was proposed to preferentially occur at the Au low coordination sites that were neutral and not directly bonded to the support.[90] Over Au/TiO_2, the intrinsic rate of hydrogenation (number of molecules transformed per h per Au atom) of nitrobenzene ($NO_2C_6H_5$) was only 2.2 times higher than styrene ($C_2H_3C_6H_5$). DFT calculations indicated that the interactions of nitrostyrene ($C_2H_3C_6H_5NO_2$) with Au(111) and Au(001) surfaces are weak and the optimised N–O and $C=C$ distances are almost identical to that of the isolated gas-phase molecule. On the other hand, calculations also show that the interaction of nitrostyrene with low-coordinated Au is much stronger but there are no preferential activation towards either one of the functional groups. Thus, additional factors have to be present to be responsible for the chemoselective hydrogenation of the nitro group.[91] Competitive adsorption study using FTIR indicated that although both the nitro and the olefinic groups adsorbed on Au/TiO$_2$, a strong preferential adsorption of the nitro group from nitrobenzene versus the olefinic in the styrene was observed. A distinct band at $1532\,\mathrm{cm}^{-1}$ in the IR spectrum was ascribed to the $\upsilon_{as}(NO_2)$ vibration frequency of a weakly adsorbed and catalytically relevant nitrostyrene adsorbed selectively via the nitro group on TiO$_2$ and Au/TiO$_2$. The above results together with DFT calculations led the authors to propose that the selective adsorption and activation of the nitro group in a configuration where the oxygen atoms of the nitro group interacted with two low-coordinated Au atoms located at the interface of Au–TiO$_2$ and 2.6–$2.7\,\mathrm{\AA}$ from the support oxygen contributed to the observed high selectivity in nitro group reduction of nitrostyrene over Au/TiO$_2$.

Contrary to the work of the group of Corma who demonstrated very high selectivity in the hydrogenation of nitro groups in aromatic compounds in the presence of carbonyl groups,[89] the group of Jin found a reversal in selectivity when hydrogenating nitroaromatics containing aldehyde groups.[92] In their work, aldehyde CHO group was reduced with an exclusive selectivity over NO$_2$ as nitrobenzaldehyde was hydrogenated into nitrobenzyl alcohol over Au$_n$(SG)$_m$ nanocluster catalysts (H-SG $=$ glutathione) of different sizes, including Au$_{15}$(SG)$_{13}$, Au$_{18}$(SG)$_{14}$, Au$_{25}$(SG)$_{18}$, Au$_{38}$(SG)$_{24}$ and captopril-capped Au$_{25}$(Capt)$_{18}$. The same trend was observed using Au$_{99}$(SPh)$_m$ clusters (SPh $=$ thiolate ligands). When these clusters were deposited on support, the exclusive selectivity was maintained but the activity was increased.[93] The improvement in activity was support-dependent, and the order of enhancement being CeO$_2$ $>$ TiO$_2$ $>$ SiO$_2$. The exclusive selectivity for aldehyde versus nitro group was also true for unsupported Au nanoparticles as well as Au nanorods.[94] In all the above systems, pyridine additive was required for the reaction to proceed and the role of pyridine was proposed to be involved in hydrogen activation. The authors proposed that pyridine promoted the heterolytic splitting of hydrogen by forming pyridinium (PyH$^+$), leaving the H$^-$ on Au.[94]

$$C_5H_5N + Au + H_2 \longrightarrow C_5H_5NH^+ + Au\text{–}H^- \qquad (9.5)$$

Additional insight into the chemoselectivity of the reaction might be gained from an analogous system of Au nanoparticles ligated by secondary

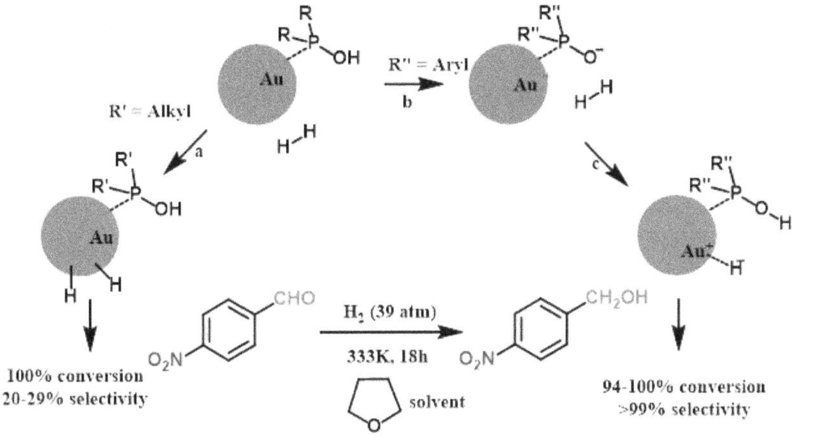

Figure 9.6 Scheme showing heterolytic versus homolytic dissociative adsorption of H_2 on phosphine oxide-stabilised gold nanoparticles. Adapted from Ref. 95.

phosphine oxides (R_2PHO)[95] published shortly after the Au nanorods work. Figure 9.6 is a schematic showing the importance of the nature of the phosphine oxide ligands in effecting the chemoselective reduction of nitrobenzaldehyde to nitrobenzyl alcohol. When the R group of phosphine oxide was aromatic (Aryl), the hydroxide form of the isomer binds to Au in a dissociative form as $R_2P–O^-$ with the formation of an induced surface Au^+ (Figure 9.6, step b). The anionic form of the bound phosphine oxide was confirmed spectroscopically. The phosphine oxide–Au acted as a Lewis acid–base pair that binds H_2 by heterolytic dissociation (Figure 9.6, step c). The resulting hydride led to highly selective CHO reduction over NO_2. On the other hand, if R was an alkyl group, the ligand is less acidic and binds to Au without dissociating the OH bond. In this case, H_2 binds by homolytic dissociation on Au and the selectivity for nitro group hydrogenation was low (20–29%). In this model, chemoselectivity in hydrogenation was dictated by the manner in which hydrogen is activated.

In addition to selectivity discrimination between nitro and carbonyl groups, it is also desirable to suppress the reactivity of olefinic groups in the presence of carbonyl functionality. Au nanoparticles ligated by aromatic secondary phosphine oxides were completely selective to the carbonyl functionality in α, β-unsaturated aldehydes,[95] despite the fact that selective C=O hydrogenation is thermodynamically (by ca. 35 kJ/mol) and kinetically unfavourable compared to the C=C bond reduction.[96] Zhu *et al.*[97] also achieved exclusive selectivity for allyl alcohols in the hydrogenation of unsaturated carbonyl compounds by using atomically precise ultrasmall $Au_{25}(SCH_2CH_2Ph)_{18}$ particles (0.97 nm metal–core diameter) under very mild conditions (1 atm, 273 K). DFT calculations indicated that the reaction proceeded via a heterolytic cleavage of H_2 to an exterior electron-deficient Au and the carbonyl O of the physically adsorbed substrates (Figure 9.7(b)). Polar solvents such as ethanol stabilised the partially hydrogenated intermediate of C=O hydrogenation via hydrogen bonding and lowered the energy of H_2 cleavage (0.90 eV). Hydride transfer from Au to the partially hydrogenated radical led to the alcohol product (Figure 9.7, transformation from b to c).[98]

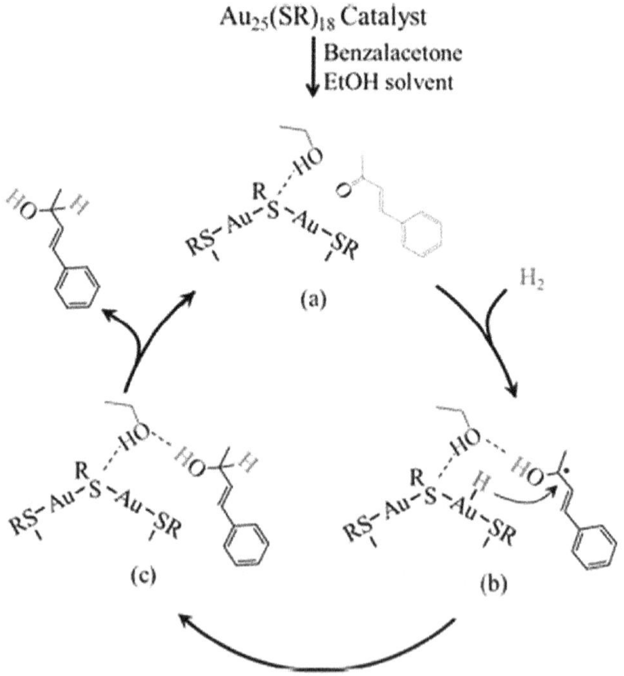

Figure 9.7 Scheme showing selective hydrogenation of benzalacetone to the corresponding unsaturated alcohol over the $Au_{25}(SR)_{18}$ catalysts. The $Au_{25}(SR)_{18}$ is represented by the RS–Au–SR–Au–SR motif of the open facets. Reprinted from Ref. 98 with the permission of American Chemical Society.

CO_2 hydrogenation appears to be a rapidly evolving and promising new arena for catalytic Au. CO_2 is a greenhouse gas and the largest contributor to global warming. Burning fossil fuels yields carbon dioxide and increases the atmospheric concentration of CO_2, which has become $>40\%$ higher since modern industrialisation. There has been significant effort to use CO_2 as a feedstock to produce valuable chemicals, and to capture it from the atmosphere. Since CO_2, a highly oxidised product, sits in a thermodynamic well, its conversion to other chemicals generally requires reduction reactions. CO_2 can be reduced sequentially to formic acid, formaldehyde, methanol, and eventually to methane. However, in order to make higher hydrocarbons, it is necessary to form C–C bond by various coupling reactions. Among these potential products, methanol is particularly attractive as processes have been developed to convert it into value-added chemicals, such as fuels and plastics.[99] Efficient activation of CO_2 is an essential step. Since Au nanoparticles are known to catalyse the hydrogenation of C=O bonds in aldehydes and ketones using H_2,[100] it is logical to pursue hydrogenation of CO_2 over Au catalysts.

Formation of formic acid and its derivative over supported Au catalyst has been investigated heavily. Hydrogenation of CO_2 to formic acid is thermodynamically uphill by 33 KJ/mol, but Fachinetti *et al.*[101] showed that in the presence of a stoichiometric amount of trimethylamine, the resulting formic acid could be trapped as a $HCOOH/NEt_3$ adduct. Since the Au/TiO_2 catalyst was in the form of extrudate sticks and immobilised in metal net cages in contact with the reactions solution, a continuous process for the production of catalyst-free and solvent-free adducts was possible. Pure formic acid can be recovered by exchanging with high boiling point amine. This seminal work has sparked considerable interest of research. Filonenko *et al.*[102] investigated the liquid phase hydrogenation of CO_2 and found that Au/Al_2O_3 was more active than Au/TiO_2 and DMF was the best solvent. They established that metallic Au was the active phase of the catalyst as the material after cyanide leaching, which selectively extract zero valent Au, showed no catalytic activity. Han *et al.*[103] have utilised methanol as a quenching reagent for HCOOH to form methyl formate. Au/ZrO_2 (2–3 nm) were utilised, and it was proposed that HCOOH intermediate was formed directly from CO_2 without involving water-gas shift reaction.

The reaction of three equivalents of H_2 with CO_2 to form methanol is another important reaction. The group of Haruta[104] was the first to demonstrate that gold nanoparticles, when properly dispersed on ZnO, could have higher activity than the conventional Cu/ZnO in the hydrogenation of CO_2 to form methanol. Yang *et al.*[105] used a model catalyst of Au and CeO_2 deposited on TiO_2 (110) and found improved selectivity to methanol in the low pressure (700 mTorr) hydrogenation of CO_2. They found that gold nanoparticles were nucleated preferentially on the surface defects of TiO_2, and a high density of $Au–CeO_x$ interface was obtained due to the high dispersity of ceria. XPS analysis revealed that under the reaction conditions, ceria was mostly in the form of Ce^{3+}, and that its presence increased the stability of carbonate intermediates. The $Au^0–Ce^{3+}$ interface polarisation, as well as the enhanced Ce^{3+}-carbonate interaction were proposed to contribute to increased adsorption and activation of CO_2 compared to Ti^{4+} based on DFT calculations. It was suggested that in this process CO_2 was first reduced to CO via the water-gas shift reaction, and CO was further hydrogenated to methoxide.

An innovative approach of CO_2 utilisation was recently introduced by Kondratenko *et al.*[106] for converting CO_2 with C_2H_4 and H_2 into propanol. The best oxo-selectivity (propanol plus propanal) was observed with the smallest nanoparticles of Au on the TiO_2 support. When C_2H_4 was cofed with CO_2 and H_2, propanol was detected. Without C_2H_4, the only products were CO and H_2O, indicating that propanol was formed from CO_2 and C_2H_4 and not through CO_2 hydrogenation.

9.6 Conclusions

Significant advances have been made in all fronts for a diverse range of Au-catalysed reactions. Even for a simple reaction like CO oxidation, investigations in the past five years have led to unforeseen insight into the active site and reaction mechanism. Currently for many reactions, there appears to be strong indication for the importance of the Au-support perimeter sites. Numerous conflicting proposals and explanations exist for the active site and reaction mechanisms of many of the reactions mentioned in this chapter. One of the reasons for the variety of conclusions about active sites is due to uncontrolled Au particle size distributions that differ from one preparation to another. There is clearly a need to develop easily scalable methods to synthesise catalysts with uniform active sites to facilitate fundamental understanding. For liquid phase oxidation reaction, the role of the Au in the catalytic cycle is not yet determined. The mechanism for Au leaching, and the significance of the leached Au and the hydroperoxide product remain open questions. In spite the impressive advances made in recent years, the field of Au catalysis remains as intriguing as before, and there is room for much more work before arriving at the level of understanding that permits making reliable predictions on catalytic activity and selectivity. The impetus to continue and expand the research on Au catalysis also stems from the fact that gold often surpasses other metal catalysts by being more active and more selective as seen in the numerous examples given in this chapter.

Acknowledgement

Evgeny V. Beletskiy acknowledges support by the U.S. Department of Energy, Office of Science, Office of Basic Energy Sciences under Award Number DOE DE-FG02-03-ER154757.

References

1. M. Haruta, N. Yamada, T. Kobayashi, S. Iijima, Gold catalysts prepared by coprecipitation for low-temperature oxidation of hydrogen and of carbon monoxide, *J. Catal.* **115**, 301–309 (1989).
2. A. Villa, S. J. Freakley, M. Schiavoni, J. K. Edwards, C. Hammond, G. M. Veith, W. Wang, D. Wang, L. Prati, N. Dimitratos, G. J. Hutchings, Depressing the hydrogenation and decomposition reaction in H_2O_2 synthesis by supporting AuPd on oxygen functionalized carbon nanofibers, *Catal. Sci. Technol.* (2015) Ahead of Print.
3. J. K. Edwards, S. F. Parker, J. Pritchard, M. Piccinini, S. J. Freakley, Q. He, A. F. Carley, C. J. Kiely, G. J. Hutchings, Effect of acid pre-treatment on AuPd/SiO2 catalysts for the direct synthesis of hydrogen peroxide, *Catal. Sci. Technol.* **3**, 812–818 (2013).
4. J. K. Edwards, S. J. Freakley, A. F. Carley, C. J. Kiely, G. J. Hutchings, Strategies for designing supported gold–palladium bimetallic atalysts for the direct synthesis of hydrogen peroxide, *Acc. Chem. Res.* **47**, 845–854 (2014).
5. A. Villa, D. Wang, D. S. Su, L. Prati, New challenges in gold catalysis: bimetallic systems, *Catal. Sci. Technol.* **5**, 55–68 (2015).

6. A. Wang, X. Y. Liu, C.-Y. Mou, T. Zhang, Understanding the synergistic effects of gold bimetallic catalysts, *J. Catal.* **308**, 258–271 (2013).

7. F. Gao, D. W. Goodman, Pd–Au bimetallic catalysts: Understanding alloy effects from planar models and (supported) nanoparticles, *Chem. Soc. Rev.* **41**, 8009–8020 (2012).

8. A. Visikovskiy, H. Matsumoto, K. Mitsuhara, T. Nakada, T. Akita, Y. Kido, Electronic d-band properties of gold nanoclusters grown on amorphous carbon, *Phys. Rev. B* **83**, 165428 (2011).

9. M. Valden, X. Lai, D. W. Goodman, Onset of catalytic activity of gold clusters on titania with the appearance of nonmetallic properties, *Science* **281**, 1647–1650 (1998).

10. M. S. Chen, D. W. Goodman, The structure of catalytically active gold on titania, *Science* **306**, 252–255 (2004).

11. A. A. Herzing, C. J. Kiely, A. F. Carley, P. Landon, G. J. Hutchings, Identification of active gold nanoclusters on iron oxide supports for CO oxidation, *Science* **321**, 1331–1335 (2008).

12. N. Lopez, T. V. W. Janssens, B. S. Clausen, Y. Xu, M. Mavrikakis, T. Bligaard, J. K. Nørskov, On the origin of the catalytic activity of gold nanoparticles for low-temperature CO oxidation, *J. Catal.* **223**, 232–235 (2004).

13. B. K. Min, C. M. Friend, Heterogeneous gold-based catalysis for green chemistry: Low-temperature co oxidation and propene oxidation, *Chem. Rev.* **107**, 2709–2724 (2007).

14. M. Haruta, S. Tsubota, T. Kobayashi, H. Kageyama, M. J. Genet, B. Delmon, Low-temperature oxidation of carbon monoxide over gold supported on titanium dioxide, α-ferric oxide, and cobalt tetraoxide, *J. Catal.* **144**, 175–192 (1993).

15. G. Bond, D. Thompson, Gold-catalysed oxidation of carbon monoxide, *Gold Bull.* **33**, 41–50 (2000).

16. Z. M. Liu, M. A. Vannice, CO and O2 adsorption on model Au–TiO2 systems, *Catal. Lett.* **43**, 51–54 (1997).

17. C. K. Costello, M. C. Kung, H. S. Oh, Y. Wang, H. H. Kung, Nature of the active site for CO oxidation on highly active Au/γ-Al$_2$O$_3$, *Appl. Catal. A* **232**, 159–168 (2002).

18. C. K. Costello, J. H. Yang, H. Y. Law, Y. Wang, J. N. Lin, L. D. Marks, M. C. Kung, H. H. Kung, On the potential role of hydroxyl groups in CO oxidation over Au/Al$_2$O$_3$, *Appl. Catal. a–General* **243**, 15–24 (2003).

19. J. D. Henao, T. Caputo, J. H. Yang, M. C. Kung, H. H. Kung, In Situ Transient FTIR and XANES Studies of the Evolution of Surface Species in CO Oxidation on Au/TiO$_2$, *J. Phys. Chem. B* **110**, 8689–8700 (2006).

20. S. M. Oxford, J. D. Henao, J. H. Yang, M. C. Kung, H. H. Kung, Understanding the effect of halide poisoning in CO oxidation over Au/TiO2, *Appl. Catal. A* **339**, 180–186 (2008).

21. B. D. Chandler, S. Kendell, H. Doan, R. Korkosz, L. C. Grabow, C. J. Pursell, NaBr poisoning of Au/TiO$_2$ catalysts: Effects on kinetics, poisoning mechanism, and estimation of the number of catalytic active sites, *ACS Catal.* **2**, 684–694 (2012).

22. Z. Zhou, S. Kooi, M. Flytzani-Stephanopoulos, H. Saltsburg, The role of the interface in CO oxidation on Au/CeO$_2$ multilayer nanotowers, *Adv. Funct. Mater.* **18**, 2801–2807 (2008).

23. T. Yan, D. W. Redman, W.-Y. Yu, D. W. Flaherty, J. A. Rodriguez, C. B. Mullins, CO oxidation on inverse Fe_2O_3/Au(111) model catalysts, *J. Catal.* **294**, 216–222 (2012).
24. M. Kotobuki, R. Leppelt, D. A. Hansgen, D. Widmann, R. J. Behm, Reactive oxygen on a Au/TiO_2 supported catalyst, *J. Catal.* **264**, 67–76 (2009).
25. D. Widmann, R. J. Behm, Activation of molecular oxygen and the nature of the active oxygen species for CO oxidation on oxide supported Au catalysts, *Acc. Chem. Res.* **47**, 740–749 (2014).
26. I. X. Green, W. Tang, M. Neurock, J. T. Yates, Jr., Spectroscopic observation of dual catalytic sites during oxidation of CO on a Au/TiO_2 catalyst, *Science* **333**, 736–739 (2011).
27. H. Koga, K. Tada, M. Okumura, Density functional theory study of active oxygen at the perimeter of Au/TiO_2 catalysts, *J. Phys. Chem. C*, (2015) Ahead of Print.
28. J. Saavedra, H. A. Doan, C. J. Pursell, L. C. Grabow, B. D. Chandler, The critical role of water at the gold–titania interface in catalytic CO oxidation, *Science* **345**, 1599–1602 (2014).
29. M. Date, M. Okumura, S. Tsubota, M. Haruta, *Angew. Chem. Int. Ed.* **43**, 2129 (2004).
30. M. Date, M. Haruta, Moisture effect on CO oxidation over Au/TiO_2 catalyst, *J. Catal.* **201**, 221–224 (2001).
31. M. Ojeda, B.-Z. Zhan, E. Iglesia, Mechanistic interpretation of CO oxidation turnover rates on supported Au clusters, *J. Catal.* **285**, 92–102 (2012).
32. S. E. Davis, B. N. Zope, R. J. Davis, On the mechanism of selective oxidation of 5-hydroxymethylfurfural to 2,5-furandicarboxylic acid over supported Pt and Au catalysts, *Green Chem.* **14**, 143–147 (2012).
33. W. C. Ketchie, M. Murayama, R. J. Davis, Promotional effect of hydroxyl on the aqueous phase oxidation of carbon monoxide and glycerol over supported Au catalysts, *Topics Catal.* **44**, 307–317 (2007).
34. D. Wang, A. Villa, D. Su, L. Prati, R. Schloegl, Carbon-supported gold nanocatalysts: Shape effect in the selective glycerol oxidation, *ChemCatChem* **5**, 2717–2723 (2013).
35. J. Jiang, S. M. Oxford, B. Fu, M. C. Kung, H. H. Kung, J. Ma, Isotope labelling study of CO oxidation-assisted epoxidation of propene. Implications for oxygen activation on Au catalysts, *Chem. Commun.* **46**, 3791–3793 (2010).
36. T. Hayashi, K. Tanaka, M. Haruta, Selective vapor-phase epoxidation of propylene over Au/TiO_2 catalysts in the presence of oxygen and hydrogen, *J. Catal.* **178**, 566–575 (1998).
37. Y. A. Kalvachev, T. Hayashi, S. Tsubota, M. Haruta, Vapor-phase selective oxidation of aliphatic hydrocarbons over gold deposited on mesoporous titanium silicates in the co-presence of oxygen and hydrogen, *J. Catal.* **186**, 228–233 (1999).
38. N. A. Mashayekhi, Y. Y. Wu, M. C. Kung, H. H. Kung, Metal nanoparticle catalysts decorated with metal oxide clusters, *Chem. Commun.* (Cambridge, United Kingdom), **48**, 10096–10098 (2012).
39. Y. Y. Wu, N. A. Mashayekhi, H. H. Kung, Au–metal oxide support interface as catalytic active sites, *Catal. Sci. Technol.* **3**, 2881–2891 (2013).
40. N. A. Mashayekhi, M. C. Kung, H. H. Kung, Selective oxidation of hydrocarbons on supported Au catalysts, *Catal. Today* **238**, 74–79 (2014).

41. B. Taylor, J. Lauterbach, G. E. Blau, W. N. Delgass, Reaction kinetic analysis of the gas-phase epoxidation of propylene over Au/TS-1, *J. Catal.* **242**, 142–152 (2006).

42. A. M. Joshi, W. N. Delgass, K. T. Thomson, Mechanistic implications of Au_n/Ti-lattice proximity for propylene epoxidation, *J. Phys. Chem. C* **111**, 7841–7844 (2007).

43. J. J. Bravo-Suarez, K. K. Bando, J. Lu, M. Haruta, T. Fujitani, S. T. Oyama, Transient technique for identification of true reaction intermediates: Hydroperoxide species in propylene epoxidation on gold/titanosilicate catalysts by X-ray absorption fine structure spectroscopy, *J. Phys. Chem. C* **112**, 1115–1123 (2008).

44. I. X. Green, W. Tang, M. Neurock, J. T. Yates, Low-temperature catalytic H_2 oxidation over Au nanoparticle/TiO_2 dual perimeter sites, *Angew. Chem. Int. Ed.* **50**, 10186–10189 (2011), S10186/10181–S10186/10115.

45. J. Huang, T. Akita, J. Faye, T. Fujitani, T. Takei, M. Haruta, Propene epoxidation with dioxygen catalyzed by gold clusters, *Angew. Chem. Int. Ed.* **48**, 7862–7866 (2009), S7862/7861–S7862/7813.

46. A. Abad, A. Corma, H. Garcia, Catalyst parameters determining activity and selectivity of supported gold nanoparticles for the aerobic oxidation of alcohols: the molecular reaction mechanism, *Chemistry* **14**, 212–222 (2008).

47. M. Sankar, E. Nowicka, E. Carter, D. M. Murphy, D. W. Knight, D. Bethell, G. J. Hutchings, The benzaldehyde oxidation paradox explained by the interception of peroxy radical by benzyl alcohol, *Nat. Commun.* **5**, 4332/4331–4332/4336 (2014).

48. J. Chen, W. Fang, Q. Zhang, W. Deng, Y. Wang, A comparative study of size effects in the Au-catalyzed oxidative and non-oxidative dehydrogenation of benzyl alcohol, *Chemistry* **9**, 2187–2196 (2014).

49. B. Panthi, A. Mukhopadhyay, L. Tibbitts, J. Saavedra, C. J. Pursell, R. M. Rioux, B. D. Chandler, Using thiol adsorption on supported au nanoparticle catalysts to evaluate au dispersion and the number of active sites for benzyl alcohol oxidation, *ACS Catal.* **5**, 2232–2241 (2015).

50. P. Haider, A. Urakawa, E. Schmidt, A. Baiker, Selective blocking of active sites on supported gold catalysts by adsorbed thiols and its effect on the catalytic behavior: A combined experimental and theoretical study, *J. Molec. Catal. A: Chem.* **305**, 161–169 (2009).

51. P. Haider, B. Kimmerle, F. Krumeich, W. Kleist, J.-D. Grunwaldt, A. Baiker, Gold-catalyzed aerobic oxidation of benzyl alcohol: effect of gold particle size on activity and selectivity in different solvents, *Catal. Lett.* **125**, 169–176 (2008).

52. J. Zhu, S. A. C. Carabineiro, D. Shan, J. L. Faria, Y. Zhu, J. L. Figueiredo, Oxygen activation sites in gold and iron catalysts supported on carbon nitride and activated carbon, *J. Catal.* **274**, 207–214 (2010).

53. G. Zhao, X.-P. Wu, R. Chai, Q. Zhang, X.-Q. Gong, J. Huang, Y. Lu, Tailoring nano-catalysts: Turning gold nanoparticles on bulk metal oxides to inverse nano-metal oxides on large gold particles, *Chem. Commun.* **51**, 5975–5978 (2015).

54. T. Wang, X. Yuan, S. Li, L. Zeng, J. Gong, CeO_2-modified Au@SBA-15 nanocatalysts for liquid-phase selective oxidation of benzyl alcohol, *Nanoscale* **7**, 7593–7602 (2015).

55. K. Kaehler, M. C. Holz, M. Rohe, A. C. van Veen, M. Muhler, Methanol oxidation as probe reaction for active sites in Au/ZnO and Au/TiO_2 catalysts, *J. Catal.* **299**, 162–170 (2013).

56. M. Conte, H. Miyamura, S. Kobayashi, V. Chechik, Spin trapping of Au–H intermediate in the alcohol oxidation by supported and unsupported gold catalysts, *J. Am. Chem. Soc.* **131**, 7189–7196 (2009).

57. M. Boronat, A. Corma, F. Illas, J. Radilla, T. Rodenas, M. J. Sabater, Mechanism of selective alcohol oxidation to aldehydes on gold catalysts: Influence of surface roughness on reactivity, *J. Catal.* **278**, 50–58 (2011).

58. S. Shylesh, D. Kim, C. R. Ho, G. R. Johnson, J. Wu, A. T. Bell, Non-oxidative dehydrogenation pathways for the conversion of C2–C4 alcohols to carbonyl compounds, *ChemSusChem*, (2015) Ahead of Print.

59. C.-R. Chang, X.-F. Yang, B. Long, J. Li, A water-promoted mechanism of alcohol oxidation on a Au(111) surface: Understanding the catalytic behavior of bulk gold, *ACS Catal.* **3**, 1693–1699 (2013).

60. W. Fang, J. Chen, Q. Zhang, W. Deng, Y. Wang, Hydrotalcite-supported gold catalyst for the oxidant-free dehydrogenation of benzyl alcohol: Studies on support and gold size effects, *Chemistry* **17**, 1247–1256 (2011), S1247/1241–S1247/1243.

61. B. N. Zope, D. D. Hibbitts, M. Neurock, R. J. Davis, Reactivity of the gold/water interface during selective oxidation catalysis, *Science* **330**, 74–78 (2010).

62. D. Yin, L. Qin, J. Liu, C. Li, Y. Jin, Gold nanoparticles deposited on mesoporous alumina for epoxidation of styrene: Effects of the surface basicity of the supports, *J. Molec. Catal. A: Chem.* **240**, 40–48 (2005).

63. M. G. Clerici, The activity of titanium silicalite-1 (TS-1): Some considerations on its origin, *Kinetics Catal.* (Translation of Kinetika i Kataliz), **56**, 450–455 (2015).

64. S. Bawaked, N. F. Dummer, N. Dimitratos, D. Bethell, Q. He, C. J. Kiely, G. J. Hutchings, Solvent-free selective epoxidation of cyclooctene using supported gold catalysts, *Green Chem.* **11**, 1037–1044 (2009).

65. M. S. Hamdy, Au-TUD-1: A new catalyst for aerobic oxidation of cyclohexene, *Microporous Mesoporous Mater.* **220**, 81–87 (2016).

66. M. Turner, V. B. Golovko, O. P. H. Vaughan, P. Abdulkin, A. Berenguer-Murcia, M. S. Tikhov, B. F. G. Johnson, R. M. Lambert, Selective oxidation with dioxygen by gold nanoparticle catalysts derived from 55-atom clusters, *Nature* **454**, 981–983 (2008).

67. Y. Zhu, H. Qian, M. Zhu, R. Jin, Thiolate-protected au nanoclusters as catalysts for selective oxidation and hydrogenation processes, *Adv. Mater.* **22**, 1915–1920 (2010).

68. B. G. Donoeva, D. S. Ovoshchnikov, V. B. Golovko, Establishing a Au nanoparticle size effect in the oxidation of cyclohexene using gradually changing Au catalysts, *ACS Catal.* **3**, 2986–2991 (2013).

69. D. S. Ovoshchnikov, B. G. Donoeva, B. E. Williamson, V. B. Golovko, Tuning the selectivity of a supported gold catalyst in solvent- and radical initiator-free aerobic oxidation of cyclohexene, *Catal. Sci. Technol.* **4**, 752–757 (2014).

70. H. Alshammari, P. J. Miedziak, T. E. Davies, D. J. Willock, D. W. Knight, G. J. Hutchings, Initiator-free hydrocarbon oxidation using supported gold nanoparticles, *Catal. Sci. Technol.* **4**, 908–911 (2014).

71. L. Wang, B. Zhang, W. Zhang, J. Zhang, X. Gao, X. Meng, D. S. Su, F.-S. Xiao, Positively charged bulk Au particles as an efficient catalyst for oxidation of styrene with molecular oxygen, *Chem. Commun.* **49**, 3449–3451 (2013).

72. S. Bawaked, N. F. Dummer, D. Bethell, D. W. Knight, G. J. Hutchings, Solvent-free selective epoxidation of cyclooctene using supported gold catalysts: an investigation of catalyst re-use, *Green Chem.* **13**, 127–134 (2011).

73. J. Oliver-Meseguer, J. R. Cabrero-Antonino, I. Dominguez, A. Leyva-Perez, A. Corma, Small gold clusters formed in solution give reaction turnover numbers of 107 at room temperature, *Science* **338**, 1452–1455 (2012).

74. M. Alvaro, C. Aprile, A. Corma, B. Ferrer, H. Garcia, Influence of radical initiators in gold catalysis: Evidence supporting trapping of radicals derived from azo-bis(isobutyronitrile) by gold halides, *J. Catal.* **245**, 249–252 (2006).

75. P. Lignier, F. Morfin, L. Piccolo, J.-L. Rousset, V. Caps, Insight into the free-radical chain mechanism of gold-catalyzed hydrocarbon oxidation reactions in the liquid phase, *Catal. Today* **122**, 284–291 (2007).

76. E. Tebandeke, C. Coman, K. Guillois, G. Canning, E. Ataman, J. Knudsen, L. R. Wallenberg, H. Ssekaalo, J. Schnadt, O. F. Wendt, Epoxidation of olefins with molecular oxygen as the oxidant using gold catalysts supported on polyoxometalates, *Green Chem.* **16**, 1586–1593 (2014).

77. M. D. Hughes, Y.-J. Xu, P. Jenkins, P. McMorn, P. Landon, D. I. Enache, A. F. Carley, G. A. Attard, G. J. Hutchings, F. King, E. H. Stitt, P. Johnston, K. Griffin, C. J. Kiely, Tunable gold catalysts for selective hydrocarbon oxidation under mild conditions, *Nature* **437**, 1132–1135 (2005).

78. Y. Zhu, H. Qian, R. Jin, An atomic-level strategy for unraveling gold nanocatalysis from the perspective of Au(n)(SR)m nanoclusters, *Chemistry* **16**, 11455–11462 (2010).

79. H. Alshammari, P. J. Miedziak, D. W. Knight, D. J. Willock, G. J. Hutchings, The effect of ring size on the selective oxidation of cycloalkenes using supported metal catalysts, *Catal. Sci. Technol.* **3**, 1531–1539 (2013).

80. G. C. Bond, P. A. Sermon, G. Webb, D. A. Buchanan, P. B. Wells, Hydrogenation over supported gold catalysts, *J. Chem. Soc. Chem. Commun.* 444–445 (1973).

81. M. Pan, A. J. Brush, Z. D. Pozun, H. C. Ham, W.-Y. Yu, G. Henkelman, G. S. Hwang, C. B. Mullins, Model studies of heterogeneous catalytic hydrogenation reactions with gold, *Chem. Soc. Rev.* **42**, 5002–5013 (2013).

82. M. Stratakis, H. Garcia, Catalysis by supported gold nanoparticles: Beyond aerobic oxidative processes, *Chem. Rev.* **112**, 4469–4506 (2012).

83. H. Lindlar, A new catalyst for selective hydrogenations, *Hel. Chim. Acta* **35**, 446–450 (1952).

84. Y. Azizi, C. Petit, V. Pitchon, Formation of polymer-grade ethylene by selective hydrogenation of acetylene over Au/CeO_2 catalyst, *J. Catal.* **256**, 338–344 (2008).

85. T. Mitsudome, K. Kaneda, Gold nanoparticle catalysts for selective hydrogenations, *Green Chem.* **15**, 2636–2654 (2013).

86. T. V. Choudhary, C. Sivadinarayana, A. K. Datye, D. Kumar, D. W. Goodman, Acetylene hydrogenation on Au-based catalysts, *Catal. Lett.* **86**, 1–8 (2003).

87. L. Shao, X. Huang, D. Teschner, W. Zhang, Gold supported on graphene oxide: an active and selective catalyst for phenylacetylene hydrogenations at low temperatures, *ACS Catal.* **4**, 2369–2373 (2014).

88. Y. Segura, N. Lopez, J. Perez-Ramirez, Origin of the superior hydrogenation selectivity of gold nanoparticles in alkyne + alkene mixtures: Triple-versus double-bond activation, *J. Catal.* **247**, 383–386 (2007).

89. A. Corma, P. Serna, Chemoselective hydrogenation of nitro compounds with supported gold catalysts, *Science* **313**, 332–334 (2006).

90. M. Boronat, F. Illas, A. Corma, Active sites for H_2 adsorption and activation in Au/TiO_2 and the role of the support, *J. Phys. Chem. A* **113**, 3750–3757 (2009).

91. M. Boronat, P. Concepcion, A. Corma, S. Gonzalez, F. Illas, P. Serna, A molecular mechanism for the chemoselective hydrogenation of substituted nitroaromatics with nanoparticles of gold on TiO_2 catalysts: A cooperative effect between gold and the support, *J. Am. Chem. Soc.* **129**, 16230–16237 (2007).

92. G. Li, D.-E. Jiang, S. Kumar, Y. Chen, R. Jin, Size Dependence of atomically precise gold nanoclusters in chemoselective hydrogenation and active site structure, *ACS Catal.* **4**, 2463–2469 (2014).

93. G. Li, C. Zeng, R. Jin, Thermally robust $Au_{99}(SPh)_{42}$ nanoclusters for chemoselective hydrogenation of nitrobenzaldehyde derivatives in water, *J. Am. Chem. Soc.* **136**, 3673–3679 (2014).

94. G. Li, C. Zeng, R. Jin, Chemoselective hydrogenation of nitrobenzaldehyde to nitrobenzyl alcohol with unsupported Au nanorod catalysts in water, *J. Phys. Chem. C* **119**, 11143–11147 (2015).

95. I. Cano, M. A. Huertos, A. M. Chapman, G. Buntkowsky, T. Gutmann, P. B. Groszewicz, P. W. N. M. van Leeuwen, Air-stable gold nanoparticles ligated by secondary phosphine oxides as catalyst for the chemoselective hydrogenation of substituted aldehydes: A remarkable ligand effect, *J. Am. Chem. Soc.* **137**, 7718–7727 (2015).

96. P. Claus, Heterogeneously catalyzed hydrogenation using gold catalysts, *Appl. Catal. A* **291**, 222–229 (2005).

97. Y. Zhu, H. Qian, A. Drake Bethany, R. Jin, Atomically precise Au25(SR)18 nanoparticles as catalysts for the selective hydrogenation of alpha, beta-unsaturated ketones and aldehydes, *Ang. Chem. Int. Ed.* **49**, 1295–1298 (2010).

98. R. Ouyang, D.-E. Jiang, Understanding selective hydrogenation of α, β-unsaturated ketones to unsaturated alcohols on the $Au_{25}(SR)_{18}$ cluster, *ACS Catal.* **5**, 6624–6629 (2015).

99. Y. Li, S. H. Chan, Q. Sun, Heterogeneous catalytic conversion of CO_2: A comprehensive theoretical review, *Nanoscale* **7**, 8663–8683 (2015).

100. Y. Zhu, H. Qian, B. A. Drake, R. Jin, Atomically Precise $Au_{25}(SR)_{18}$ Nanoparticles as catalysts for the selective hydrogenation of α, β-unsaturated ketones and aldehydes, *Angew. Chem. Int. Ed.* **49**, 1295–1298 (2010), S1295/1291–S1295/1295.

101. D. Preti, C. Resta, S. Squarcialupi, G. Fachinetti, Carbon dioxide hydrogenation to formic acid by using a heterogeneous gold catalyst, *Angew. Chem. Int. Ed.* **50**, 12551–12554 (2011).

102. G. A. Filonenko, W. L. Vrijburg, E. J. M. Hensen, E. A. Pidko, On the activity of supported Au catalysts in the liquid phase hydrogenation of CO_2 to formates, *J. Catal.* (2015) Ahead of Print.

103. C. Wu, Z. Zhang, Q. Zhu, H. Han, Y. Yang, B. Han, Highly efficient hydrogenation of carbon dioxide to methyl formate over supported gold catalysts, *Green Chem.* **17**, 1467–1472 (2015).

104. H. Sakurai, S. Tsubota, M. Haruta, Hydrogenation of CO_2 over gold supported on metal oxides, *Appl. Catal. A* **102**, 125–136 (1993).

105. X. Yang, S. Kattel, S. D. Senanayake, J. A. Boscoboinik, X. Nie, J. Graciani, J. A. Rodriguez, P. Liu, D. J. Stacchiola, J. G. Chen, Low pressure CO_2 hydrogenation to methanol over gold nanoparticles activated on a CeO_x/TiO_2 interface, *J. Am. Chem. Soc.* **137**, 10104–10107 (2015).

106. S. J. Ahlers, U. Bentrup, D. Linke, E. V. Kondratenko, An innovative approach for highly selective direct conversion of CO_2 into propanol using C_2H_4 and H_2, *ChemSusChem* **7**, 2631–2639 (2014).

Chapter 10
Plasmonic Photocatalysis

Ewa Kowalska

Institute for Catalysis, Hokkaido University, Sapporo, Japan

10.1 Introduction

Research on semiconductor photocatalysis has been extensively studied for more than 40 years for environmental purification of water and air,[1–3] self-cleaning and anti-fogging surfaces,[4] water splitting[5] and solar energy conversion.[6] It is expected that semiconductor photocatalysis driven by the sun, i.e. solar photocatalysis, could help in solving at least three of 'top ten problems of humanity' (proposed by Prof. Smalley),[7] i.e. (1) energy, (2) water and (3) environment, where abundant and clean energy would enable the resolution of all other problems. Solar energy conversion into electricity and fuel (H_2, CH_3OH) by novel photocatalytic systems has recently been extensively studied, e.g. inspired by nature, two-step photoexcitation systems, the so-called 'artificial photosynthesis'.[8]

The mechanism of semiconductor photocatalysis consists in (a) excitation of semiconductor with light energy equal or stronger than its band-gap energy ($E \geq E_g$), resulting in the formation of pairs of charge carriers: electrons and holes (e^-/h^+) (Figures 10.1(a)) and 10.1(b) charge carriers migration to the surface of semiconductor (or their undesired recombination in the bulk, 'bulk recombination') (Figures 10.1(b)) and 10.1(c) finally the reaction of charge carriers with various species adsorbed on the semiconductor surface (or their undesired recombination on the surface, 'surface recombination') (Figure 10.1(c)). In the presence of water and oxygen in the system, reactive oxygen species (ROS), such as superoxide ($^\bullet O_2^-$), hydrogen peroxide (H_2O_2) and hydroxyl radicals ($^\bullet OH$), are formed and easily react with all organic and inorganic compounds. Hydroxyl radicals are known as the strongest oxidant, which could be used safely in water treatment processes,

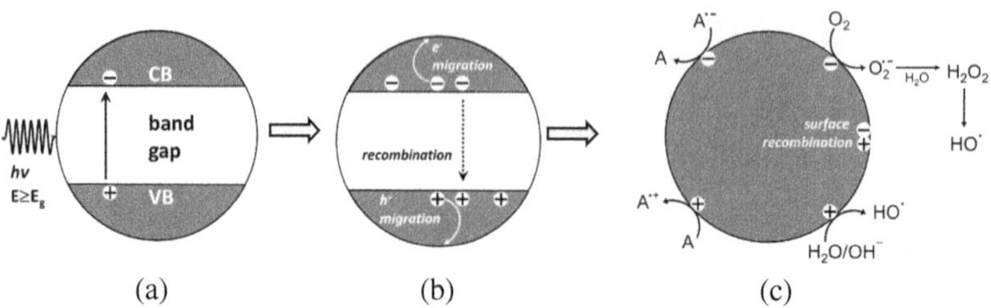

Figure 10.1 Schematic mechanism of semiconductor photocatalysis: (a) semiconductor excitation, (b) charge carriers migration to the surface and/or recombination, (c) reaction on semiconductor surface: CB, conduction band; VB, valence band; E_g, band-gap energy, e$^-$, -electron, h$^+$, hole, A, adsorbate.

and the processes/technologies, which use $^\bullet$OH are known as advanced oxidation processes/technologies (AOPs/AOTs). It should be pointed that ROS are generated by both oxidation and reduction pathways when oxygen is present in the system, i.e. formation of $^\bullet O_2^-$ by scavenging of electrons by oxygen (as shown in Figure 10.1(c)).

Titanium(IV) oxide is one of the most often used semiconducting photocatalysts because of its high photoreactivity, chemical stability, non-toxicity and high availability.[9] However, two shortcomings for applications should be overcome, recombination of charge carriers (e^-/h^+) and inactivity in the visible range of the solar spectrum (due to wide band-gap). Therefore, plenty of studies have been performed to improve performance of titania photocatalysts by morphology arrangements,[10] surface modification,[11] doping[12] and coupling with other semiconductors.[13] Gold and other noble metals (platinum, silver, iridium, palladium) in the form of adsorbed complexes[14,15] and metallic deposits[16–20] have been extensively investigated for improvement of photocatalytic activity of titania under UV irradiation.[21,22] Enhancement of photocatalytic activity mainly originates from prolongation of lifetime of charge carriers,[19] since noble metals serve as electron sinks, and thus accelerate the transfer of electrons from titania to substrates (e.g. protons to evolve hydrogen).[23,24]

Over the last decade, another property of noble metals (Au, Ag), i.e. absorption of visible light due to plasmon resonance, has been used to activate titania[25,26] and other wide band-gap semiconductors (CeO_2,[27,28] ZnO)[29] towards visible light. Although plasmonic properties of noble metals were observed 100 years ago, explained more than 30 years ago and commercially used in many fields as discussed in this book (e.g. biosensors,[30] chemical sensors,[31,32] nanolithography, nanophotonics,[33,34] surface-enhanced Raman spectroscopy (SERS)[30,32] and medicine (drug delivery, cancer therapy)[32] and optical data storage[35]) the examination of their usage for photocatalysis was started a few years ago. Despite the novelty of plasmonic photocatalysis, a large number of studies have already been performed to improve photoactivity and stability as well as to clarify the mechanism under irradiation with visible light; a few reviews on plasmonic photocatalysis have also been published.[36–40]

This chapter provides an introduction to plasmonic photocatalysis (mainly heterogeneous photocatalysis on gold–semiconductor nanostructures). Although under UV irradiation the main mechanism is due to semiconductor photoexcitation but not plasmonic activation, it will be shortly discussed to clearly show the difference between UV and visible light activation of gold-modified semiconductors. Then, exemplary applications of plasmonic photocatalysts for environmental purification and solar energy conversion will be presented. Finally, the recent reports on improvement of photocatalytic performance and perspectives of plasmonic photocatalysis will be discussed.

10.2 Function of Gold and Mechanism of Plasmon-assisted Reactions

The properties of both gold NPs and their support strongly influence the properties of resultant photocatalysts such as specific surface area, crystallinity, crystallite size, photoabsorption properties, morphology and thus resultant photocatalytic activities. Therefore, morphology, compositions and properties of photocatalysts will be shortly presented at the beginning of this section. Examples of structures are shown in Figure 10.2.

Gold NPs are usually deposited on titania, but other supports are also used, e.g. (i) other wide band-gap semiconductors ($ZrO_2^{[41]}$ and ZnO),[42] (ii) narrow band-gap semiconductors possessing activity under visible irradiation (CdS,[13] reduced graphene oxide (RGO),[44] CeO_2 and $Fe_2O_3^{[41]}$) and (iii) insulators ($SiO_2^{[41]}$, 'gold-assisted catalysis', due to plasmonic heating). Titania is used in the crystalline (anatase,[45] rutile,[46] brookite)[47]

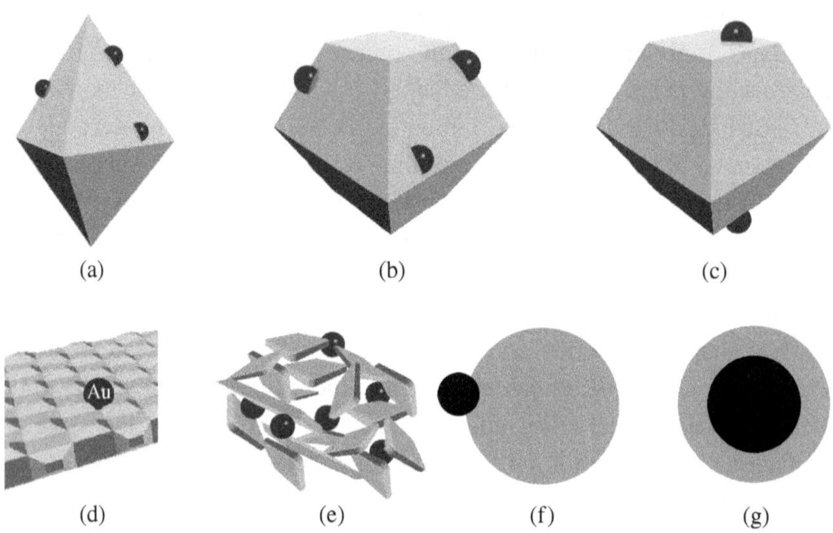

Figure 10.2 Exemplary nanostructure of plasmonic photocatalysts; (a–f) gold NPs deposited on (a) {101} facets of octahedral anatase titania, (b) {101} and (c) {001} facets of decahedral anatase titania, (d) titania mesocrystals, (e) titania flakes, (f) titania particle and (g) gold (core)–silica (shell).

and amorphous[48] forms or as their mixtures, e.g. P25 (anatase, rutile and amorphous titania)[45] and P90 (anatase and rutile).[49] Either first gold nanostructures are synthesised and then deposited on titania,[50] or (i) titania is synthesised on gold nanostructures,[51–53] (ii) gold cations are directly reduced on titania by chemical,[54] photochemical,[55] thermal[41,56] and sonochemical methods,[57] (iii) gold is deposited on titania by other methods (e.g. electrophoretic and sintering),[58] (iv) gold and titania are simultaneously synthesised during hydrothermal process,[59] laser pyrolysis[60] and hydrolysis of titanium tetrabutoxide[61]/isopropoxide[62] followed by titania crystallisation with simultaneous formation of gold nanoparticles (NPs).

Various shapes/structures of gold deposits are prepared such as clusters,[63] NPs,[64] hexagonal and triangular nanoplates,[65] nanospheres,[66] nanocages,[67] nanopillars, nanorods (NRs),[68,69] nanofilms,[70] nanoprisms/ nanotriangles[64] and irregular plate-like NPs.[71] Besides simple gold deposits on titania, other complex nanostructures are also synthesised such as three-dimensional (3D) network of aerogels of titania with incorporated gold NPs,[52] titania mesocrystals with gold NPs deposited on basal or lateral surfaces,[72] core–shells and Janus nanostructures[51,73] and reverse nanostructures where gold is covered with titania.[74] Several support structures are also used (e.g. titania film,[58,70] titania/silica film,[75] mesoporous titania,[53,59] 3D titania aerogels,[52] titania nanotubes,[54] titania mesocrystals,[72] nanofibrous titania mats[76] and ZnO nanorods).[42] Titania can be itself supported on other materials (i) to enhance specific surface area (alumina,[77] mesoporous silica,[55,62] protonated zeolite),[61] (ii) for easier recovery of photocatalysts after reaction in liquid phase (silica: glass slides, glass helix and Raschig rings),[63] (iii) to control the size and shape of gold NPs (mesoporous silica)[55] and (iv) to enhance photocatalytic activity by separation of charge carriers (fluorine-doped tin oxide (FTO) film,[58] FTO with CuO/SnO_2,[78] and zeolite).[61]

The kind and structure of the support as well as the method of gold deposition highly influence the properties of the resultant gold deposits, for example (i) larger gold NPs are formed on larger titania NPs[46] and on pure anatase than on anatase–rutile mixture,[70] (ii) pores in titania film nanostructure limit the size of deposited gold NPs,[58] (iii) the smallest gold NPs are formed by deposition–precipitation, then sol-immobilisation and finally by photodeposition method.[50] Thermal post-treatment is used for various purposes such as (1) titania crystallisation, (2) gold reduction, (3) removal of adsorbed impurities like Cl^- (from gold precursor), capping[79] and stabilising[58] agents of gold colloids, (4) enhancement of interaction between gold and support[58,71] and (5) gold aggregation.[58,80] The conditions of photocatalytic reaction can also change the structure and properties of photocatalysts, for example, due to gold reshaping as a result of temperature increase[81] or irradiation.[82]

The variety of photocatalyst structures and of reaction conditions are responsible for different results, which are explained by different mechanisms of plasmonic photocatalysis. It is reasonable that different mechanisms are proposed for different nanostructures, yet similar nanostructures are explained by different mechanisms, which makes discussion on plasmonic photocatalysis more complex. Furthermore, as developed in Section 10.3, the photocatalytic properties are tested for various applications including redox reactions such as purification of air and water, reduction of organic compounds (e.g. dehydrogenation of alcohols), self-cleaning surfaces, water splitting and photocurrent generation. Dyes are often selected as models for organic molecules because their photodegradation is easy to detect by UV-visible spectroscopy. However, it must be pointed that dyes must not be used for activity testing under visible light irradiation (although are often used), due to sensitisation of photocatalyst by them.[83,84] Although a large number of model compounds and tested reactions are reported, the level of photocatalytic activity (e.g. quantum yield), which depends on intrinsic properties of photocatalyst (e.g. amount and kind of electron traps), but not particular reaction, is the most crucial since under aerobic conditions, heterogeneous photocatalytic reaction involves generation of non-selective hydroxyl radicals (reacting with all compounds leading to their mineralisation). Therefore, chemical reactions will not be discussed here in detail to keep this chapter clear and simple for understanding (see cited references for detailed experimental procedures).

10.2.1 *Under UV Irradiation: Activation of Semiconducting Support*

Although under UV irradiation gold/semiconductor (mainly titania) photocatalysis cannot be considered as 'plasmonic photocatalysis', it should be introduced since future applications of those materials are focused on solar energy utilisation and thus both UV and visible range of sunlight must be taken into consideration.

As mentioned earlier, semiconductor is activated when light of energy stronger than its band-gap energy reaches its surface. Electrons are excited from valence band (VB) to conduction band (CB), resulting in the formation of charge carriers in VB (holes) and CB (electrons), as shown in Figure 10.1. Then the generated charge carriers migrate to the surface, where they react with adsorbed molecules. For aerobic conditions, reactive oxygen species (ROS) are efficiently formed during photocatalytic reactions. In the case of titania, energy of ca. 3 eV depending on its crystalline form must be absorbed. This means that only the UV range of the solar spectrum can be used, i.e. only ca. 4% of the solar energy. Additionally, as in the case of all semiconducting materials, recombination of charge carriers results in quantum efficiency much lower than 100%.

Therefore, semiconductors are modified or doped with various dopants/modifiers (Figure 10.3). For example, surface modification of

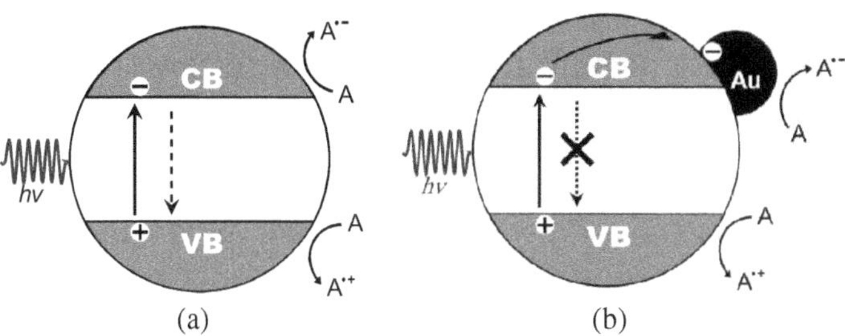

Figure 10.3 Schematic mechanisms of (a) semiconductor activation with energy larger than its band gap, and (b) electron transfer from semiconductor to gold NPs, where: ↑ — electron excitation from valence band (VB) to conduction band (CB), ↓ — electron recombination, A — adsorbate.

titania with noble metal NPs results in electron transfer from titania to the modifiers forming the Schottky barrier, which inhibits recombination.[85] The role of electron traps played by noble metal NPs is confirmed by various studies, for instance, through a decrease in photoluminescence (due to electron de-excitation across band-gap to recombine with holes),[53,86–89] increase of photocurrent quantum yield,[90] reduced lifetime of mobile electrons,[91] enhancement of photocatalytic activity[92] and generation of various ROS.[49,50]

However, scavenging of electrons by gold NPs could also lead to a decrease in photocatalytic activity, especially for reduction reactions occurring on titania surface, as a result of decreasing amount of available electrons, due to their transfer from titania to gold-deposited NPs. For example, Au/Ag bimetallic NPs embedded in titania film causes enhanced activity for photo-oxidation of methylene blue (model dye), owing to electron sinking inside Au/Ag, but reduced activity for photo-reduction of resazurin dye adsorbed on titania surface.[93] Similarly, a decrease in photocurrent generation is observed for composite electrode in which gold is deposited on the titania surface of a TiO_2/FTO (fluorine doped tin oxide) electrode (Au/TiO_2/FTO); this is due to the back electron transfer from titania to gold instead of to the FTO electrode.[58]

Light scattering on gold NPs is also proposed as another reason for better performance of Au/TiO_2 than bare TiO_2, especially in the case of semiconductor thin films for which efficient light harvesting is decisive for overall performance.[94]

Usually, introduction of gold causes high enhancement of photocatalytic performance, but for some particular structures and/or reactions a decrease in photocatalytic activity is also observed. This can be due to

(i) electron transfer to gold, as discussed above,

(ii) a decrease in free surface of titania, due to partial covering of titania surface by gold deposits, which can result in: (1) drop in photon flux reaching the titania surface (gold as a filter, the so-called inner filter effect),[95] (2) reduction of the titania/aqueous phase interface, leading to smaller interface between titania and reagents (e.g. reduced titania–electrolyte interface resulting in lower photocurrent level),[95]

(iii) competition of adsorption on titania surface between hydroxyl groups and gold deposits, especially in the case of reactions performed in organic media. For example, for cyclohexane oxidation, deposition of gold NPs on titania surface results in decreased amount of available hydroxyl groups, which, in consequence, decreases the amount of generated hydroxyl radicals during irradiation and thus efficiency of oxidation.[96]

In spite of the fact that UV light cannot excite the gold plasmon, two new mechanisms of gold NPs excitation by UV irradiation of Au/semiconductor nanostructures have been recently proposed:

(1) Plasmonic excitation of gold NPs deposited on ZnO nanorods (NRs); after ZnO NRs modification with gold, excitation with UV light at 325 nm results in an increase in photoluminescence at 385 nm and a decrease in the photoluminescence in the visible range (500–700 nm).[42] This suggests that electrons from ZnO are transferred to Au NPs, which reduces the green emission (less e^-/h^+ recombination as typical for gold-modified semiconductors under UV irradiation), but simultaneously this residual visible light energy (green emission) can excite the plasmon resonance of gold NPs. As a consequence 'hot' electrons are transferred from gold to CB of ZnO, which results in an increase in photoluminescence at the UV range.

(2) Direct excitation of gold NPs deposited on titania; excitation of electrons from the 5d band to the 6sp band of gold results in their interaction with air and in the formation of superoxide anion radicals (O_2^-),

which further react with hydroxyl groups at the titania surface to give hydroperoxy radicals (HO_2^-).[49]

In summary, under UV irradiation, wide band-gap semiconductor is photoexcited and gold NPs deposited on its surface prevent charge carrier recombination by scavenging of photo-excited electrons, which gives rise to enhanced level of photocatalytic activity. Moreover, gold NPs can increase UV absorption ability of photocatalyst due to enhanced scattering effect (light harvesting), but can also decrease it because of 'inner-filter effect'. Decrease in photocatalytic activity can also happen, due to the reduced free surface of semiconductor for adsorption of water/hydroxyl groups and reagents. Two new mechanisms of gold activation under UV irradiation have recently been proposed: (i) direct gold excitation resulting in electron transfer from the 5d to the 6sp band and (ii) indirect excitation of gold NPs (plasmonic activation) by visible light emitted during recombination of charge carriers in semiconductor.

10.2.2 *Under Visible Irradiation: Activation of Plasmon Resonance*

Since wide band-gap semiconductors are only active under UV irradiation $(E \geq E_g)$, when electrons are excited from VB to CB (semiconductor photo-excitation), other mechanisms must be responsible for semiconductor activation under visible light irradiation. Interestingly, even under visible light irradiation, gold has been considered as an electron sink in the degradation of dye on Au/RGO (reduced graphene oxide): Upon visible light irradiation, the excited electrons from dye are transferred via RGO to gold NPs on which ROS are formed to oxidize dye and cause its degradation.[44] A mechanism of dye degradation by self-oxidation during RGO sensitisation should be also considered.

Applications of localised surface plasmon resonance (LSPR) to photo-catalysis started at the beginning of this century but plasmonic features were only used for characterisation of gold NPs deposited on titania, i.e. their formation, properties (size and shape) and stability (under UV irradiation).[97] The first proof that plasmon resonance was responsible for the activity under visible light irradiation was presented by a comparison of photoabsorption

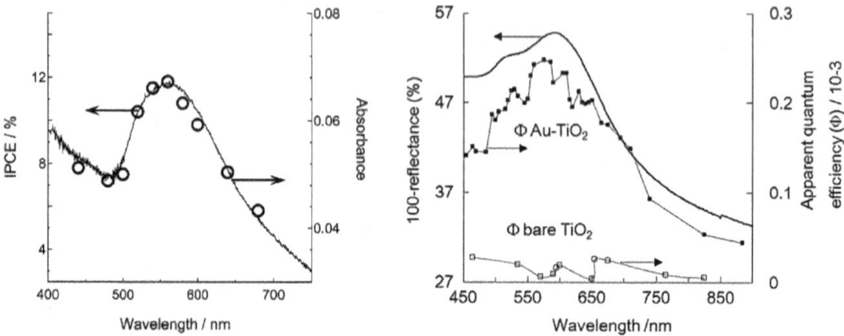

Figure 10.4   Comparison of absorption spectra of Au/TiO$_2$ with action spectra of (left) short-circuit photocurrent for Au-TiO$_2$ photoanode and (right) 2-propanol oxidation on bare (□) and gold (■) modified rutile titania; Reprinted with permission from Ref. 25. Copyright (2005) American Chemical Society, and adapted from Ref. 26.

spectra with respective action spectra for photocurrent generation[25] and for photooxidation of 2-propanol,[26] as shown in Figure 10.4.

Three main mechanisms are proposed for plasmonic photocatalysis under visible light: (1) charge transfer (mainly electron transfer), (2) energy transfer and (3) plasmonic heating (thermal activation). The two first mechanisms can be classified as 'plasmon-assisted photocatalysis' since energy/electron transfer happens directly under visible irradiation, while the last one (3) can be classified as 'plasmon-assisted catalysis' since LSPR activates catalytic (thermal) reaction, as shortly discussed in subsequent parts.

It is important to point that all mechanism studies must be performed under the sole visible light irradiation, i.e. after UV cutting off, to be sure that photocatalytic reaction is not initiated by UV irradiation. Therefore, because of the UV content in the solar spectrum, experiments performed under solar radiation cannot be used for mechanistic studies, but only to show the working ability of plasmonic photocatalysts under sunlight.

10.2.2.1 *Charge transfer (plasmon-assisted photocatalysis)*

In the case of plasmonic photocatalysis, charge transfer usually means electron transfer (i.e. transfer of 'hot' electrons after excitation of plasmon resonance). However, transfer of 'hot' holes has also been reported and will be shortly discussed further. Electron transfer from Au NPs to CB of titania was proposed for the first time by Tian and Tatsuma in 2005 for photocurrent generation on Au/TiO$_2$.[25] In 2007, Orlov *et al.* suggested the decomposition of organic compounds (OCs) by electron transfer from gold NPs to titania, but for gold NPs smaller than 3 nm with semiconducting rather than metallic character, based on semiconductor–semiconductor contact.[98]

In general, the mechanism of decomposition of organic compounds (OCs) on Au/TiO$_2$ under irradiation with visible light is proposed to be similar to that of activation of sensitisers, such as metal complexes or dyes (Figure 10.5),[26,99–101] and thus gold is sometimes called 'plasmonic photosensitiser'.[102] At first, incident photons are absorbed by gold nanoparticles through their LSPR excitation, then, an electron ('hot electron') is transferred from gold NPs into the CB of titania (Figure 10.5(a)). Then, the electron reduces molecular oxygen adsorbed on titania surface and the resultant electron-deficient gold NPs can oxidise OCs to recover its original metallic state (Figure 10.5(b)).[26,101]

Many proofs of electron transfer from gold to titania have been given in the literature, which involve different types of experiments: (i) femtosecond transient absorption spectroscopy with an IR probe of interband absorption of electrons injected from gold nanodots into the CB of

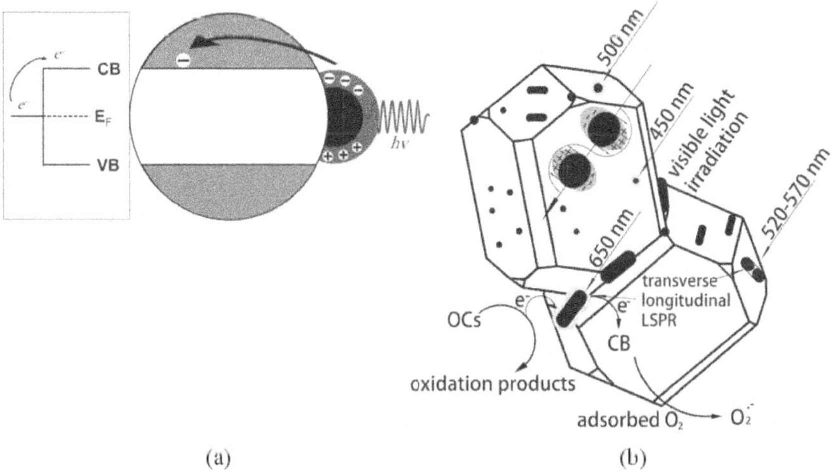

Figure 10.5   Schematic mechanism of electron transfer from gold NP to titania (a), and then to adsorbed oxygen with simultaneous electron transfer from OCs to gold (b). Adapted from Refs. 100 and 112.

titania,[103,104] (ii) comparison of the photocatalytic activity of Au/TiO$_2$ and Au/ZrO$_2$ for the oxidation of OCs under visible light irradiation,[105,106] (iii) shift of electrode potential (negative or positive photopotential) and generation of anodic or cathodic photocurrent depending on electrode configuration: ITO/TiO$_2$/Au or ITO/Au/TiO$_2$, respectively,[107,108] (iv) different mechanisms of OCs oxidation under UV and visible light irradiation (i.e. generation of electron–hole pairs and only electrons, respectively),[52] (v) inhibition of activity in the presence of other molecules on the surface of gold or between gold and titania (dodecanethiol,[58] 3-mercaptopropyltrimethoxysilane),[101] (vi) dissolution of metal due to its partial oxidation[74] and (viii) EPR experiments on Au/TiO$_2$ photocatalyst resulting in the detection of different species under irradiation with UV and visible light.[49,50,109]

Besides 'hot' electrons, participation of 'hot' holes is also proposed in the mechanism of photocurrent generation for gold NPs deposited into porous titania film.[58] However, the origin of 'hot' holes is not due to plasmonic excitation of gold NPs, but interband *d-sp* transition within the gold NPs. Therefore, 'hot' electrons and 'hot' holes are responsible for photocurrent generation at different excitation wavelengths, i.e. at plasmonic resonance range (ca. 550–650 nm) and at shorter wavelengths of ca. 410–440 nm, respectively. Similar observation has been reported for action spectrum of the oxidation of 2-propanol on gold-modified anatase, i.e. two main activity peaks at 450 and 560 nm.[101]

Modified mechanism of electron transfer without participation of the semiconductor (or with its negligible contribution) is also reported. For example, 'hot' electrons are responsible for the dissociation of H$_2$ under visible light irradiation of gold NPs deposited on titania.[110] However, titania practically does not participate in the mechanism of electron transfer (only 'a small, additional negative charge' transfer from titania to gold, due to oxygen vacancies). It is proposed that 'hot' electrons with energy between the vacuum level and work function of metal could be transferred into resonance of hydrogen molecule adsorbed on the surface of gold NPs, and causes its dissociation (Figure 10.6). Similarly, the electron transfer is proposed for unsupported gold NPs in the presence of hydrogen peroxide as an electron acceptor for selective oxidation of organic alcohols.[111]

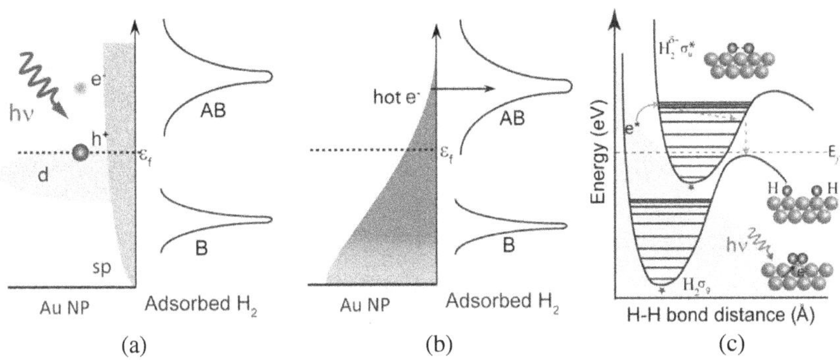

Figure 10.6   Schemes of (a) hot electron excitation in gold NP, where d: band electron–hole pair excited above the Fermi level upon plasmon decay, B: narrow bonding and AB: broad antibonding states of adsorbed H_2, (b) Fermi–Dirac type distribution of hot electrons permitting hot electron transfer into the anti-bonding state of H_2; (c) mechanism of hot electron-induced dissociation of H_2 on Au NP surface; Reprinted with permission from Ref. 110. Copyright (2013) American Chemical Society.

The recombination of charge carriers (i.e. back electron transfer from titania to gold; as typical under UV irradiation and discussed in Section 10.2.1) is responsible for low quantum yield of plasmonic photocatalysts under visible light irradiation. It is proposed that recombination depends on titania particle sizes. For titania particles of different sizes (9, 20, 30 and 50 nm) modified with 10-nm gold NPs, the best activity is achieved for larger ones, which delays the time of charge recombination because of the longer diffusion path in those particles.[104]

10.2.2.2 *Energy transfer (plasmon-assisted photocatalysis)*

Energy transfer between two different compounds can take place when they have closely matched energy levels. This is not expected for Au/TiO_2 since gold NPs plasmon energy (LSPR of ca. 2.2 eV) is lower than the band-gap of titania (ca. 3 eV). The first reports suggesting energy transfer from gold NPs to titania have been mentioned for pre-modified titania in order to make it able to absorb visible light and thus being active under visible light irradiation.

One of the first studies reporting energy transfer, also called plasmon energy transfer (PRET),[94] was shown by Liu *et al.* for photocurrent generation on 5-nm gold NPs on a titania film pre-modified with nitrogen and fluorine.[95] Five-fold and 66-fold photocurrent enhancement have been obtained under irradiation with 532 and 633 nm, respectively. The enhancement is attributed to the broader absorption spectrum in the visible range of pre-modified titania, and energy transfer between two matching energy levels of F-/N-titania and Au is proposed.

Other pre-modified titanias, e.g. with nitrogen,[60] and narrow band-gap semiconductors possessing visible light activity like $CuWO_4$ (2–2.5 eV)[94] show enhanced photocatalytic activity after addition of gold NPs, which is also attributed to PRET mechanism. Interestingly, titania possessing crystalline defects,[112] and amorphous titania with disorders resulting in localised electronic states inside the band gap[51] (i.e. electron traps [ETs]) are also good materials for energy transfer due to energy matching of these states and LSPR of gold.

Cushing *et al.* deepened the concept of energy transfer and proposed two types of mechanisms for gold-modified semiconductors, i.e. local electromagnetic field (LEMF) and resonant energy transfer (RET), as shown in Figure 10.7.[113] LEMF applies only to the case of energy matching between gold NPs and semiconductor (as discussed above), whereas RET can also create electron–hole pairs in the semiconductors with an energy lower than the band-gap energy (novel concept), due to non-radiative dipole–dipole energy transfer.

10.2.2.3 *Plasmonic heating (plasmon-assisted catalysis)*

The first report showing plasmon-assisted catalysis was published by Chen *et al.* in 2008. It showed that heated gold NPs could activate

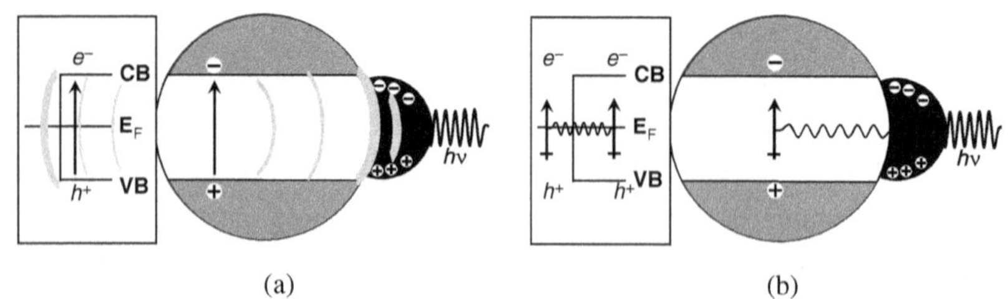

Figure 10.7   Schemes of energy transfer mechanisms from gold NPs (black deposits) to semiconductor of: (a) the band-gap energy matching the plasmon energy of gold NPs, i.e. local electromagnetic field transfer (LEMF), and (b) the band-gap energy lower than plasmon energy of gold NPs, i.e. resonant energy transfer (RET) transfer. Adapted from Ref. 113.

organic molecules to induce their oxidation by visible light at ambient temperature.[41] Gold was deposited on different supports, i.e. mainly on insulating oxides, ZrO_2 and SiO_2 with band-gaps of ca. 5 and 9 eV, respectively, for which neither visible light excitation of electrons from valence band nor energy/electron transfer could be expected. The experiments were performed in gas phase for oxidation of formaldehyde, methanol and cyclohexane. It was proposed that gold NPs could be heated quickly (2–3 min) to 100°C by light absorption, which led to formaldehyde oxidation. The reaction was attributed to activation of polar molecules (inactivity for cyclohexane) by electromagnetic field resulting in reaction with oxygen. However, the experiments were not performed under visible light of LSPR of gold NPs (500–600 nm), but at shorter and longer wavelengths of 400–500 and 600–700 nm, respectively. In addition, deposition of gold on semiconducting supports (Fe_2O_3 and CeO_2) resulted in a significant enhancement of photocatalytic activity suggesting that participation of other mechanisms (energy/electron transfer) could not be excluded for Au/Fe_2O_3 and Au/CeO_2 nanomaterials.

Plasmonic heating has been proposed as the main mechanism for catalytic reduction of CO_2 on Au/ZnO photocatalyst under irradiation with visible light.[29] The inactivity of bare ZnO and Au/SiO_2 under via irradiation, activity of bare ZnO under heating in the dark and the same distribution of reaction products (CO and CH_4) for heated ZnO and irradiated Au/ZnO prove that plasmonic heating by LSPR of gold NPs catalytically activates ZnO.

It is suggested that mechanism of plasmonic photocatalysis depends on semiconductor support. For example, both Au/TiO_2 and Au/ZnO are active under visible light irradiation, but Au/TiO_2 heated in the dark is inactive[114] and Au/ZnO is active.[29] It is deduced that the mechanism of photocatalysis is plasmon-assisted photocatalysis for Au/TiO_2 and plasmon-assisted catalysis (plasmonic heating) for Au/ZnO. It must be stressed that plasmonic heating requires highly energetic irradiation (usually laser intensity $> 10^5$ W/m^2)[14,29] to cause catalytic response. Therefore, it is expected that irradiation sources commonly used in photocatalysis laboratories (mercury lamps, xenon lamps and light emitting diodes (LED)) do not have sufficient energy to cause plasmon-assisted catalysis.

Plasmonic heating is also proposed as the main mechanism responsible for the photoactivity of gold NPs deposited on titania incorporated in mesoporous silica for CO oxidation[115] and capped with homogeneous catalysts for hydrolysis of methyl parathion ($Cu^{II}(bpy)$).[116] However, in the second case, two other mechanisms of plasmon-assisted photocatalysis are also possible, i.e. (i) energy transfer sensitising the Cu(II) metal centre, through the creation of $d-d$ excited state and (ii) electron transfer resulting in formation of more active Cu(I) catalyst (reduction of Cu(II) by 'hot electrons').

On the contrary, there are plenty of reports rejecting the mechanism of plasmonic heating, due to inactivity of gold-modified insulators or unsupported gold NPs in comparison to highly active gold-modified semiconductors.[101,112,113,117] Plasmonic heating is not considered as the main mechanism in the following cases:

(i) water splitting, which requires much higher energy (1.32 eV) than thermal energy generated by plasmonic heating, as confirmed by lack of activity for control experiments of gold deposited on glass and quartz,[95]

(ii) photocurrent generation, where identical temperature is obtained for bare titania and gold-modified titania under visible light irradiation (photocurrent is generated by 'hot' electrons transfer)[118] and

(iii) hydrogen dissociation, where heating of gold NPs in the dark results in negligible reaction, and direct photodissociation of hydrogen requires 10 times higher laser intensities than used for plasmon-assisted photocatalysis.[110]

Studies on activation energy also exclude plasmonic heating in the case of decomposition of organic compounds[28] and photocurrent generation.[119] For example, the apparent activation energy (2.4 kJ mol^{-1}) for mineralisation of formic acid on Au/CeO_2 is much lower than that for the thermal activation process (24–36 kJ mol^{-1} on Pt/Al_2O_3, Pt/C, $Pt/Ce_xZr_{1-x}O_2$). This indicates that the rate-determining step of mineralisation is different from that of oxygen activation over supported metals, suggesting that a photoinduction step (by plasmonic activation of Au) is involved in the mechanism.[28]

10.2.3 *Mechanism Dependence on Properties of Photocatalysts*

Various structures of plasmonic photocatalysts are proposed (shortly presented in Section 10.2 and in Figure 10.2), and they differ significantly in physicochemical properties (adsorption properties of the reagents, light absorption properties and resultant activities). It is therefore not surprising that different photocatalytic mechanisms are proposed. Moreover, several mechanisms may be involved in the (photo)catalytic reaction (e.g. plasmon-assisted photocatalysis and catalysis (in the dark) or plasmon-assisted photocatalysis and plasmon-assisted catalysis). In this section, a few examples of mechanism dependence on the properties of gold, support and their interface are shortly discussed for plasmonic photocatalysis (under visible light irradiation).

10.2.3.1 *Gold properties*

The dependence of the mechanism of plasmonic photocatalysis on the properties of gold NPs (size and shape) can be illustrated by several examples. For instance, a mechanism of plasmon-assisted photocatalysis is proposed for titania modified with gold NPs larger than 7 nm, while for smaller gold NPs ($Au^{\delta+}$, without detected LSPR), their electrodeficiency inhibits charge carriers recombination (CB electron sinking into Au NPs, as under UV excitation of titania, Figure 10.3(b)).[77] Unfortunately, the mechanism of generation of charge carriers on titania under visible light irradiation of Au NPs ($<$7 nm)/TiO_2 is not discussed in this chapter.

The interesting article by Tsukamoto *et al.* reports aerobic oxidation of alcohols by two simultaneous mechanisms, i.e. catalysis and plasmon-assisted photocatalysis, depending on gold properties. Only gold NPs with size of 3.4–3.7 nm, deposited at the interface between two titania crystalline phases (anatase and rutile), result in efficient plasmon-assisted electron transfer, four times higher than when the reaction is performed in the dark (catalysis).[109] This is attributed to an electron transfer from gold NPs to anatase via rutile and then to adsorbed oxygen resulting in the formation of oxygen anion, which can interact with a residual positive charge on gold NPs producing the peroxo-type anion. The lower photocatalytic activity for larger gold NPs (4.5–8 nm) results from reduced joint active sites (rutile/Au/anatase), and thus catalytic oxidation (in the dark) becomes the predominant mechanism of oxidation. An optimal size of gold NPs has also been suggested for hydrogen dissociation, but for almost twice as large gold NPs (7.4 nm)[110] as those for alcohol oxidation (as discussed above). The increase in gold size results in a decrease in the available Au surface area, and hence chemisorption probability of hydrogen. However, the mechanism of hydrogen dissociation, which involves two simultaneous mechanisms, plasmonic heating and plasmon-assisted photocatalysis (electron transfer), does not depend on gold particle size, but on reaction temperature: an increase in temperature from 297 to 475 K results in increase in contribution of plasmonic heating in total activity.

On the other hand, several papers report an increase in photocatalytic activity with increase in gold NPs size.[26,51,101,120] Increase of activity in oxidative degradation of phenol was found almost proportional to an increase in gold NPs size for Au/TiO$_2$ prepared by a water-in-oil microemulsion method, i.e. two times for gold NPs from 50 to 100 nm on anatase titania, and seven times for gold NPs from 10 to 70 nm on rutile titania.[120] Similarly, increase in crystallite size from 8 to 60 nm for gold NPs photodeposited on various commercial titanias results in linear increase in the photooxidation rate of OCs.[26] The highest level of activity was obtained on rutile titania with gold deposits of various sizes and shapes (sphere-like and nanorods). The polydispersity of gold NPs leads to broader LSPR, improves the photoabsorption properties, and accordingly is proposed to enhance photocatalytic activity via plasmon-assisted electron transfer from gold NPs to titania (proved by inactivity of reference samples: Au/SiO$_2$ and Au/-thiol-/TiO$_2$). Although fine gold NPs were almost inactive,[101] some Au/TiO$_2$ samples prepared in special conditions to form much smaller gold NPs (a few nanometres and narrow LSPR), deposited next to each other ('hot spots'), exhibited very high photocatalytic activity, similar to the most active Au/TiO$_2$ samples possessing very broad LSPR. Therefore, a different mechanism for those photocatalysts possessing gold 'hot spots' is expected (i.e. energy transfer).[121]

Activity enhancement with an increase in the size of gold NPs is also observed for hydrogen generation on TiO$_2$/Au Janus-type nanostructures of 30, 50 and 70 nm. On the contrary to above-discussed reports (for degradation of organic compounds),[101,120] energy-transfer mechanism (but not electron transfer) has been proposed for hydrogen generation, due to more-intense plasmonic near-field for larger gold NPs according to discrete dipole approximation (DDA) simulations.[51]

Furthermore, it is expected that large gold NPs should improve photocatalytic performance because of enhanced light scattering on them (light harvesting).[94] Therefore, different mechanisms are suggested for different configuration of photoactive films. For example, besides light scattering on gold NPs, energy transfer is proposed as the main mechanism of photocurrent generation when gold NPs are deposited at the interface of semiconductor ($CuWO_4$) film electrolyte. In contrast, enhanced light scattering and lack of energy transfer (so no photocurrent generation) is suggested for the film in which gold NPs are deposited at the back contact of the $CuWO_4$ film with FTO (fluorine-doped tin oxide).

10.2.3.2 *Support properties*

The mechanism of plasmonic photocatalysis is also dependent on the properties of the support, for example, on its crystallinity. Energy transfer has been suggested for amorphous titania, i.e. from excited gold NPs (30–70 nm) to localised electronic states inside the band-gap of titania, as confirmed by enhancement of plasmonic near-fields (DDA simulations) at the interface of titania–gold for Janus nanostructures only in the case of amorphous titania.[51] However, no experimental proof has been presented for the same Janus nanostructure of crystalline titania (except smaller NPs of gold (5 nm) deposited on titania P25).

An interesting report has recently been published on the influence of thickness of the titania layer on the reaction mechanism in the case of tri-layered gold(core)/silver/titania nanorods (as shown in Figure 10.8).[74] Electron transfer is the predominant mechanism for thin titania layers (<10 nm) since silver is released in solution as Ag^+ because of its oxidation after electron transfer to CB of titania. In contrast, energy transfer is predominant mechanism in the case of thick titania layers (≥ 10 nm). The optimal thickness of titania shell (10 nm) is not consistent with the electron transfer mechanism, where increase in thickness of titania shell should result in a decrease in photocatalytic activity as a result of increase in probability of

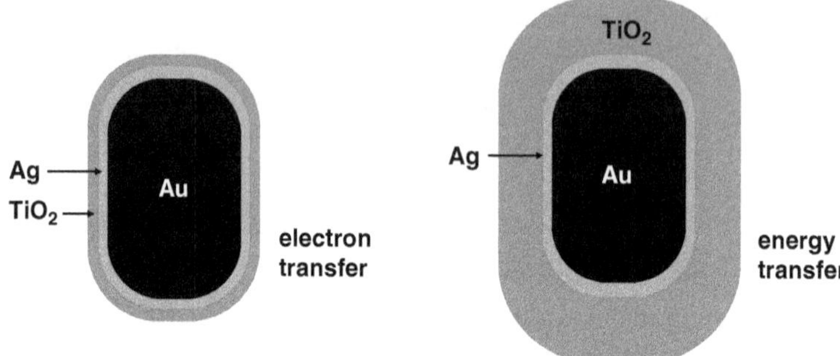

Figure 10.8   Schemes showing core(gold) shell(Ag) structures covered with titania layer of different thickness for which mechanism of electron (left) and energy (right) transfers are proposed under visible light irradiation. Adapted from Ref. 74.

trapping electrons before their arrival at the shell surface. Therefore, the optimal thickness of titania shell should be explained by energy transfer mechanism (i.e. with increase in titania shell thickness larger volume of titania shell can receive the plasmonic energy up to optimal shell thickness, over which energy transfer should decrease, since electric field decreases exponentially with the distance from the metal surface).

In contrast, for 10-nm gold NPs deposited on titania, increase in size of titania NPs from 9 to 50 nm leads to an increase in photocatalytic performance and has been explained as the mechanism of electron transfer.[122] An increase in the size of titania NPs results in delay of back electron transfer to gold NPs (see Section 10.2.2.1). Similarly, an increase in the photocatalytic activity with an increase in titania size has been reported for 15 commercial titania samples of different sizes, but it has been explained by the influence of gold NP size (the larger crystallite sizes of titania are, the larger are gold crystallites deposited on them,[101] and increase in gold amount leads to the formation of larger gold NPs possessing higher activity).[123]

10.2.3.3 *Interaction interface between gold and support*

Under UV irradiation, nanostructures of gold and semiconductor exhibit much higher photocatalytic activity than physical mixture of two components, proving that electron transfer from semiconductor to gold is crucial and the interface is very important.[43] It is expected that strong electronic interaction between gold and support should be observed for photocatalysts prepared by direct reduction of gold cations by electrons coming from support (e.g. photogenerated electrons from irradiated semiconductor (photodeposition), reaction with support anions (Se^{-2} from CdSe)). Indeed, higher photocatalytic activities under visible light irradiation are observed for Au/TiO_2 photocatalysts prepared by photodeposition than impregnation method with the same titania samples possessing the same amount of gold NPs of similar sizes, which indicates electron transfer as predominant mechanism.[101]

The importance of gold–titania interface is also attested for a three-dimensional (3D) network of aerogels with two kinds of nanostructures: one in which 5-nm gold NPs are either incorporated in the network (3D Au–TiO$_2$) or deposited in the porosity (DP Au/TiO$_2$), as shown in Figure 10.9.[52,124] 3D Au–TiO$_2$ exhibits much higher photocatalytic activity than DP Au/TiO$_2$. Although two mechanisms have been proposed, i.e. electron transfer and resonant energy transfer (RET), it is likely that enhanced contact between gold and titania in 3D nanostructures should result in higher probability of electron transfer.

The crucial role of interface between gold and titania is also pointed in other reports:

(i) The mechanism of electron transfer; in the case of gold NPs deposited at the interface between anatase and rutile, which allows efficient electron

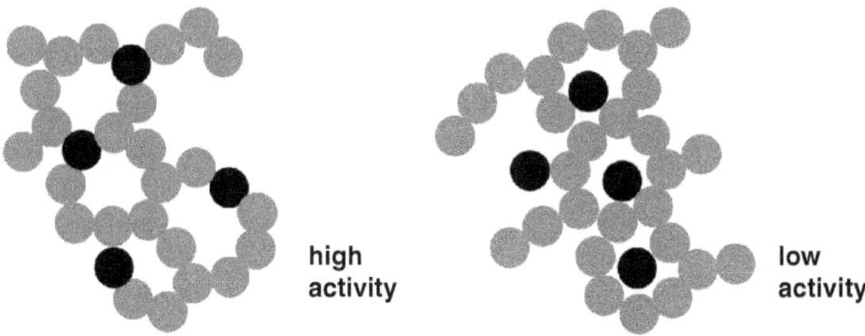

Figure 10.9   Schemes showing titania aerogels with gold NPs (black dots): (left) incorporated inside the network of titania (3D Au–TiO$_2$), and (right) in porosity of titania (DP Au–TiO$_2$). Adapted from Refs. 52 and 124.

transfer from gold via rutile to anatase and then to adsorbed oxygen (see Section 10.2.2.1)[109]

(ii) Change of predominant mechanism from electron to energy transfer; when an insulator layer is inserted between the semiconductor and gold (i.e. in the case of Au@SiO_2@Cu_2O composite), the energy transfer mechanism of RET-type (see Section 10.2.2.2) is predominant.[113]

10.3 Application

Although research on plasmonic photocatalysis started only a few years ago, various possible applications have already been proposed, such as environmental purification (wastewater treatment, water purification, gas treatment, surface purification), solar energy conversion (solar electricity and solar fuel) and organic synthesis. Examples of applications of plasmonic photocatalysis (under visible light) are discussed in the present section. Photocatalytic activities under UV irradiation (semiconductor activation) and solar radiation (simultaneous excitation of semiconductor and gold) are also shortly presented to show a wide range of possible application of gold-modified semiconductor photocatalysts.

10.3.1 *Environmental Purification*

10.3.1.1 *Water and wastewater treatment*

Gold-modified titania is used under UV irradiation for activity enhancement (inhibition of charge carriers recombination as discussed in Section 10.2.1 and Figure 10.3(b)), for example, during oxidative decomposition of pollutants in liquid phase such as formic acid,[60] stearic acid,[125] salicylic acid,[126] antibiotics (tetracycline hydrochloride,[90] trimethoprim)[127] and dyes (Methylene Blue (MB),[76,88] and Methyl Orange (MO)).[61] The enhancement of activity is also observed during reductive decomposition of some pollutants such as nitrobenzene and MB.[43] Usually, significant enhancement of activity is observed after gold deposition, i.e. (i) two times higher activity for degradation of oxalic acid on Au NPs/P25,[128] Congo Red dye on Au NPs/TNTs (TiO_2 nanotubes)[54] and MB dye on

Au NPs/anatase,[88] (ii) four times higher for reduction of 2,2′-dipyridyl disulphide (RSSR) dye on Au NPs/TiO$_2$,[129] oxidation of MB dye on Au NPs/TiO$_2$-SiO$_2$ and Au NRs/TiO$_2$-SiO$_2$[55] and oxidation of R6G dye on Au NPs/ZnO NRs[42] and (iii) more than 20 times faster oxidation rate of benzene on Au NPs/TNTs.[130] However, in some cases, only a slight increase in degradation efficiency is observed after gold deposition, for example, for the degradation of Plasmocorinth B[131] and MB,[76] or even inhibition of photocatalytic activity, for example, for *Bacillus subtilis* bacteria degradation.[131]

One of the first studies under natural solar radiation (India, 2006) showed twice higher activity of Au/TiO$_2$ than that of bare titania.[63] Similarly, twice higher photocatalytic activities have been obtained after deposition of gold NPs on titania nanotubes pre-modified with 12-phosphotungstic acid for degradation of Congo Red dye under simulated solar radiation,[54] and titania/polyethene beads for degradation of Rhodamine-6-G dye under natural solar radiation (in Egypt).[132]

The first report under visible light irradiation did not show activity of gold–titania photocatalyst.[59] Probably, it is caused by too small gold NPs (nanosized), the tested reaction (acetone oxidation) or too slow reaction (usually reaction rate under visible light is even two orders of magnitude lower than that under UV).[101] Similarly, lack of activity under visible light irradiation for titania modified with nanosized gold has been observed for formic acid degradation.[60] The significant improvement of photocatalytic activity has been shown for dye degradation (e.g. Congo Red,[54] MB,[72] Rhodamine B (RhB),[72] MO,[61,113] Acid Orange 7 (AO7)),[133] but it cannot attest for a plasmonic effect due to sensitisation mechanism for dyes.

Plasmonic photocatalysis under irradiation with visible light has been proved for oxidative degradation of colourless compounds, such as (i) methyl *tert*-butyl ether (MTBE) to carbon dioxide (mineralisation),[77] (ii) 2-propanol to acetone[67,123,134] and carbon dioxide,[134] (iii) cinnamyl alcohol to cinnamaldehyde,[78] (iv) methyl parathion to 4-nitrophenolate,[116] (v) phenol[133] to benzoquinone, hydroquinone, muconic acid, muconic aldehyde and oxalic acid.[120] Complete decomposition (mineralisation) of organic molecules does not always proceed, due to stabilisation of reaction intermediates.[77]

10.3.1.2 *Gas phase purification (and artificial photosynthesis)*

Similar to purification of liquid phase, at first, gold NPs were used for the separation of charge carriers generated on UV-irradiated semiconductor, and thus enhanced photocatalytic activities were obtained for degradation of acetone,[59,62,70] benzene,[130] acetaldehyde,[135] formaldehyde,[53] benzene,[136] 2-propanol,[74] nitrogen monoxide[137] and carbon monoxide.[138] It has been proposed that gold, besides accepting electrons, could also activate oxygen molecule forming Au–O• species.[135] Further activity enhancement can be obtained by (i) increase of surface area for more efficient adsorption of pollutants (e.g. by deposition of Au–TiO$_2$ on materials with large surface area (mesoporous silica, MCM-41),[62] and formation of mesoporous Au–TiO$_2$ structures (nanocomposite microspheres))[53] and (ii) inhibition of charge carriers recombination (e.g. by efficient electron transfer in semiconductor nanowires (ZnO)).[136]

Recently photocatalytic activity of gold–semiconductor photocatalysts has been reported for gas-phase purification also under visible light, for example, for degradation of acetaldehyde, benzene, 2-propanol, carbon monoxide and carbon dioxide.[78] Gold deposition on inactive ZnO nanowires results in efficient degradation of benzene (only two times slower than under UV irradiation).[136] Oxidation of 2-propanol to acetone occurs in gas phase on gold NRs covered with silver shell and then with amorphous titania (Figure 10.8). Activity strongly depends on the morphology, and reaches an optimal value for Au/Ag atomic ratio = 1:5 and 10-nm titania layer.[74]

Nowadays, one of the most crucial reaction involving solar energy is artificial photosynthesis for two main reasons to (1) limit CO$_2$ content in the environment and (2) convert solar energy to chemical energy. Plasmonic photocatalysts have already been proposed for both purposes (the latter will be discussed in Section 10.3.2). Example of CO$_2$ and H$_2$ conversion to CH$_4$ and CO has been shown by Wang *et al.* for ZnO modified with gold NPs.[29] Visible light irradiation can thermally activate ZnO through gold plasmonic heating similarly to common catalytic reactions, and methane is mainly formed at lower temperature, but carbon monoxide at higher one.

10.3.1.3 *Self-cleaning of surfaces*

Titania coating on various surfaces such glass, concrete, metal or wood has two main purposes: self-cleaning by decomposition of adsorbed pollutants and anti-fogging due to superhydrophilicity. Therefore, it is highly important to keep the superhydrophilic properties of titania after its modification. Fortunately, deposition of gold NPs on titania surface does not influence significantly hydrophilic properties of titania film (increase in contact angle from 53.7° to 80° under dark and from 6.4° to 9.1° under UV irradiation), while photocatalytic activity (e.g. mineralisation of stearic acid) is enhanced.[125]

Support for the photocatalysts can influence their hydrophilic properties (e.g. the largest contact angle is observed for Au/Ag-TiO_2 photocatalyst supported on titanium and steel (ca. 60°), while the smallest one on glass surface (ca. 20°)).[93] In addition, support influences the change rate of hydrophilic properties during UV irradiation, i.e. shorter time of irradiation is sufficient to achieve superhydrophilicity on glass (only 20 min) than on titanium and steel (60 min). However, the most efficient decomposition of MB dye under UV irradiation is achieved on titanium support.

Recently, photocatalytic activity of gold-modified titania films has been reported also under visible light irradiation. Methanol dissociatively adsorbs on bare and gold (5 wt.%, 2–3 nm) modified titania (12-nm anatase) 3D aerogel under dark conditions.[52] Under UV irradiation, methoxyl groups act as hole-traps and extend lifetimes of mobile electrons (longer in the case of gold-modified than bare titania), whereas, under visible light irradiation, only gold-modified nanostructures are active, due to plasmonic excitation of gold NPs.

Photocatalytic activity for decomposition of microbiological pollutants have been reported in the dark and under irradiation with UV and/or visible light.[139] Although the inhibition of fungal growth on Au/TiO_2 under visible light irradiation is usually low (only ca. 30% larger than that for bare titania, depending on properties of titania and gold NPs), the inhibition of sporulation, and thus mycotoxins generation, is significant, as shown in Figure 10.10 (the lack of formed droplets containing mycotoxins in the presence of Au/TiO_2 support).

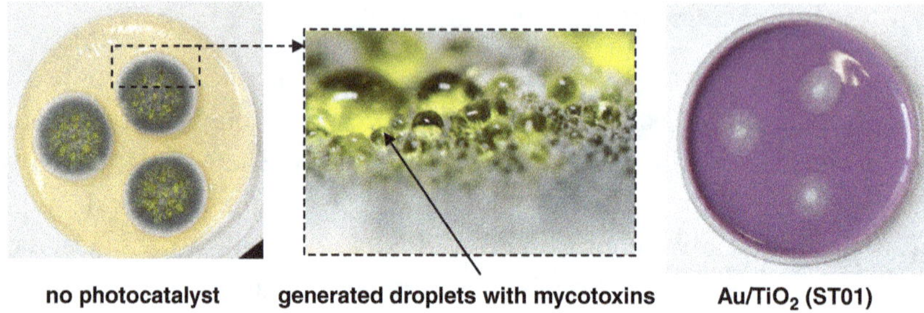

Figure 10.10   Images of 8-day grown of *Penicillium chrysogenum* under visible light irradiation (62 h, λ >400 nm) in the absence (left) and in the presence (right) of Au/TiO$_2$ photocatalysts suspended in the fungal media (MEA); (centre) cross-section of mycelium with produced mycotoxins; adapted from Ref. 139.

10.3.2 *Solar Energy Conversion*

Recently, gold-modified semiconductors have been used for the conversion of solar energy to electricity (solar cells) and fuels (chemical energy). This is shortly presented in this section.

10.3.2.1 *Photocurrent generation*

Contradictory results have been presented after semiconductors modification with gold (i.e. increase and decrease of IPCE (incident photon-to-current efficiency) depending on irradiation ranges). For example, photoanode of tungsten trioxide bearing plasmonic gold–polyoxometalate shows enhancement of IPCE mainly under UV (up to 470 nm) and inactivity at LSPR range, which indicates only catalytic effect of Au–PMo$_{12}$O$_{40}^{3-}$.[140] Similarly, modification of CuWO$_4$ (E_g of 2–2.5 eV) with gold NPs covered with 2 nm layer of titania results in photocurrent enhancement only at wavelengths shorter than LSPR of gold (<470 nm) with a maximum at 390 nm.[94] On the contrary, an increase in photocurrent under visible light and decrease under UV irradiation is observed for porous titania film after modification with gold NPs.[58] The transfer of charge carriers is responsible for activity under visible light, and at the same, for reduced activity under UV, where photo-generated electrons are scavenged by gold NPs instead of being transfer to an FTO electrode. A five-fold increase in photocurrent under visible light irradiation is also observed after C-TiO$_2$ (titania doped with carbon) modification with gold nanocages.[67]

Recently, enhanced activities for overall UV-visible range of irradiation have been reported for Au–TiO$_2$ aerogels[124] and hematite nanoflake arrays decorated with gold NPs.[102] The interface between gold and semiconductor, as well as uniform distribution of gold NPs, is crucial for efficient charge transfer. For example, 55- and 23-fold increase in IPCE is achieved after titania modification with 8 wt.% of gold incorporated inside titania network and in the porosity of the network (shown in Figure 10.9), respectively.[124]

Moreover, gold NPs are used to improve the performance of dye-sensitised solar cells (DSSCs), due to enhanced scattering of light.[141] The first use of gold in DSSCs was reported by McFarland and Tang in 2003 for DSSCs composed of photoreceptor layer (merbomin dye) deposited on gold/titania/titanium film (dye/Au/TiO$_2$/Ti: 10–50 nm Au film/200 nm TiO$_2$ film/Ti (an ohmic metal)).[142] Four-step photon-to-electron conversion was proposed: (i) light absorption by dye, (ii) transfer of excited electrons from dye to gold and their traverse the metal, (iii) electron transfer above Schottky barrier to CB of titania ('internal electron emission'), (iv) electron collecting at back ohmic contact (Ti). Although weak visible response in the absence of photoreceptor (dye) was explained as a result of existence of defect levels inside titania, plasmonic activation of titania could not be excluded.

Enhancement of IPCE and power conversion efficiency (PCE) has been also observed for DSSCs modified with gold NRs, but only at wavelengths shorter than LSPR (300–500 nm) and explained as enhanced light scattering and a decrease in back reaction probability of electrons with I$_3^-$ ions in the electrolyte.[141]

10.3.2.2 *Fuel generation*

Research on fuel generation on gold-modified semiconductors mainly focuses on half-reaction of water splitting (water reduction) in the absence or presence of sacrificial hole scavengers. At first, the studies were performed under UV irradiation. For example, it has been shown that the properties of the semiconductor are crucial for efficient hydrogen evolution in the presence of methanol as hole scavenger, for example, the highest reaction rates are obtained for fine particles of anatase (ca. 10 nm) and large ones of rutile (ca. 550 nm) after deposition of 2 wt.% of gold. These could result from their high specific surface area and crystallinity of titania, respectively.[101] The polymorphic form of titania has an influence on the activity, which varies as follows: P25 > anatase > brookite > rutile.[86] For low amount of deposited gold (0.05 and 0.1 wt.%), the specific surface area of titania is decisive for efficient hydrogen evolution, while for higher gold loading (2 wt.%), similar activities are obtained independently on the type of titania.[123] The nature of sacrificial hole scavengers also influences the resultant activity. It varies as follows: glycerol > ethylene glycol > methanol > ethanol for all tested Au/TiO$_2$ photocatalysts.[86] Activity also depends on polarity, the number of hydroxyl groups and standard oxidation potential of the scavenger. For instance, the efficiency of hydrogen evolution for gold NPs deposited on titania nanorods and on P25 decreases in the order: triol (glycerol) > diol (1,2-ethanediol, 1,2-propanediol) > ethanol > 1-propanol.[87]

Studies on hydrogen evolution indicate that the optimal size of gold deposits and optimal content of gold depend on a hole scavenger and on the properties of titania support, i.e. crystalline form, crystallinity and specific surface area.[143] For example, the highest hydrogen evolution is found for 2.7 nm Au NPs in 0.5 wt.% gold on titania P25 with methanol as a hole scavenger,[144] 3.6 nm NPs in 0.5 wt.% Au on titania nanorods and 5.5 nm NPs in 1.5 wt.% Au on titania P25 with ethanol as hole scavenger,[87] and 6 nm NPs in 1 wt.% Au on titania (independently on the type of titania) with oxalic acid as a hole scavenger.[145] Generally, the existence of optimal gold amount and gold NP size is explained as the result of compromise between electron-scavenging capacity (to avoid charge carriers recombination) and 'inner-filter' effect (preventing transmission of light to titania through gold deposits on its surface). Priebet *et al.* has proposed that in addition to the polymorphic form of titania (anatase, rutile, brookite) and the size of gold NPs, interaction between gold and titania also strongly influences the rate of hydrogen evolution from methanol–water mixture.[50] EPR measurements indicate that abundant surface vacancies and surface hydroxyl groups enhance the stabilisation of separated charge carriers, whereas Ti^{+3} in the support lattice hampers an efficient electron transport.

An interesting report on biomass conversion under conditions similar to solar radiation (Ne lamp with mainly visible spectrum) shows that wastewater from sugar industry can be an efficient electron donor for hydrogen evolution, and that four-fold increase in activity is achieved after 0.5 wt.% of gold is deposited on titania supported on silica.[146]

Although many reports show high photocatalytic activity of Au/TiO$_2$ for hydrogen evolution under UV irradiation, the large majority of them report no activity under visible light irradiation. This can be explained in the terms of electron transfer mechanism: since hydrogen evolution occurs on the surface of gold NPs, but under visible light irradiation, electrons are transferred from gold to titania (opposite direction than under UV) inducing formation of electron-deficient gold. To overcome this problem, the use of additional co-catalyst has been proposed, for example of platinum NPs.[72] In the case of gold deposition on basal surface of titania mesocrystals, platinum should be deposited on lateral surfaces, which leads to better charge separation and transfer of electrons from gold to platinum NPs through the titania network, as shown in Figure 10.11.

Recently, hydrogen evolution under visible light irradiation in the absence of second co-catalyst (metal) has been reported in two works. The first one shows efficient hydrogen evolution for Janus nanostructures of amorphous titania and 50 nm gold NPs under irradiation with wavelengths longer than 400 nm.[51] In comparison, core(Au)–shell(TiO$_2$) nanostructure, bare titania and bare gold NPs show much lower hydrogen evolution of ca. 58%, 1% and ≪1%, respectively. Janus structure is proposed to be the preferential one because of strong localisation of LSPR close to the Au–TiO$_2$ interface, as presented in Figure 10.12 (white region around Au NP). The second article reports the evolution of hydrogen by trapping of 'hot' electrons (from gold NPs) by lattice Ti^{4+}, and subsequent reduction of protons by

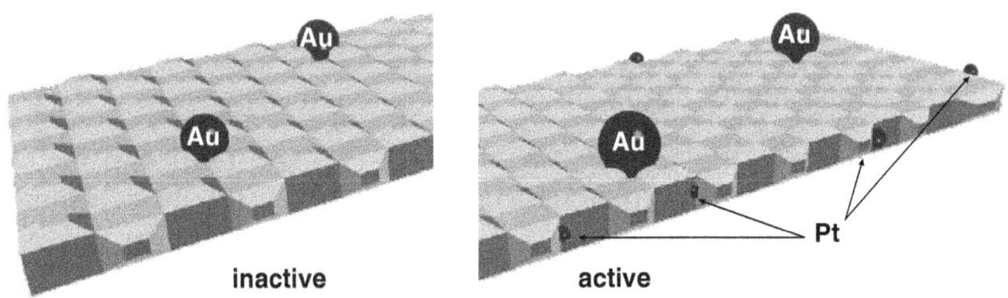

Figure 10.11 Schemes showing titania mesocrystals: (left) single modified with gold NPs, and (right) co-modified with gold NPs and platinum NPs. Adapted from Ref. 72.

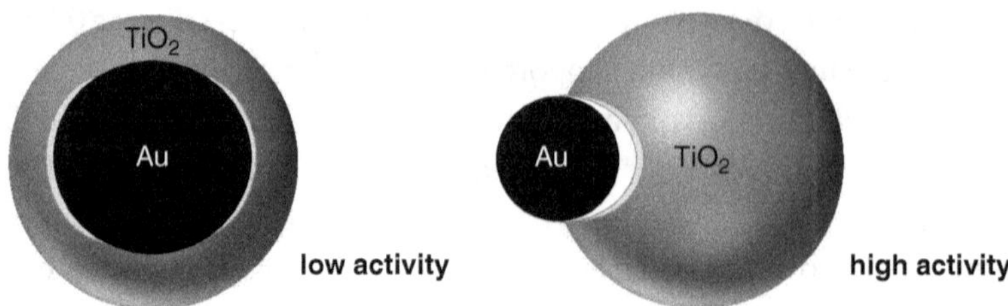

Figure 10.12   Schemes showing (left) core(Au)-shell(TiO$_2$) and (right) Janus nanostructures used for hydrogen evolution;white areas indicate plasmonic near-field distribution; adapted from Ref. 51.

formed Ti^{3+}.[50] Hydrogen is mainly evolved from rutile/anatase mixed samples, whereas pure rutile shows low activity and pure anatase and brookite samples are inactive. Therefore, the 'synergetic effect' enabling efficient charge separation, ascribed to additional electron transfer between two titania polymorphs is suggested, i.e. 'hot' electron transfer from gold to rutile and then to anatase.

10.3.3 *Synthesis of Organic Compounds*

Several synthesis reactions have been tested on gold-modified semiconductor photocatalysts under UV and visible light irradiation. For example, selective oxidation of methanol into methyl formate is achieved on titania modified with Ag/Au alloy under UV irradiation.[85] The high selectivity and conversion result from the structure of the metallic alloy, allowing the efficient separation of charge carriers (*spd* hybridisation resulting in interband electron transitions). Negative charges on the alloy surface are responsible for dissociation of oxygen, which is the rate-determining step in the reaction.

Usually gold deposition on semiconductor results in high enhancement of activity under UV irradiation and plasmon-activation of photocatalyst under visible irradiation.[109] However, in some cases, a decrease of activity is reported, for example, gold deposition on titania results in decrease in UV photooxidation of cyclohexane.[96] It is correlated with the decrease in the amount of available hydroxyl groups on titania surface after gold deposition, which reduces the amount of formed hydroxyl radicals that can react with cyclohexane to generate cyclohexyl radicals. Therefore, in the case of reactions performed in organic media (lack of water in reaction environment), gold deposition could result in a decrease in photocatalytic activity under UV irradiation.

Under visible light irradiation, the reduction of nitrobenzene (in methanol solution) to nitrosobenzene and aniline occurs on $Au/LaTiO_3$ photocatalyst for an optimal gold loading of 3 wt.%.[89] Selective oxidation of various alcohols (in toluene solution) to ketones and aldehydes by plasmonic excitation of gold NPs (<5 nm) deposited on titania P25 is also reported, for example, with 99% conversion of 1-phenylethanol to acetophenone.[109]

10.4 Strategies for Activity and Stability Enhancement

Various strategies have been proposed for the improvement of performance (photoactivity and stability) of plasmonic photocatalysts, such as (i) change of properties of gold, support and their interface and (ii) coupling of plasmonic photocatalysts with other materials (semiconductors, metals, insulators, homogeneous photocatalysts, dark catalysts), as shortly presented in this section.

10.4.1 *Nano-architecture Arrangement*

10.4.1.1 *Gold properties: Extension of action for overall solar spectrum*

The preparation of different sizes and shapes of gold deposits for the enhancement of photocatalytic activity has been extensively investigated and contradictory results have been published. For example, reduction in gold size results in an increase in photocatalytic activity for oxidation of alcohols,[109] acetone[59] and oxalic acid (e.g. a decrease in gold NPs size from 18 to 5 nm results in a two-fold increase in activity).[128] However, Plasmocorinth B dye decomposes faster after gold aggregation.[131] Similarly, an increase in gold size results in the enhancement of hydrogen evolution for dehydrogenation of isopropyl alcohol,[51] photocurrent generation,[58] phenol oxidation,[120] 2-propanol oxidation[26] and reduction of RSSR (2,2'-dipyridyl disulphide) and nitrobenzene.[129] The existence of an optimal size of gold crystallites depending on support properties is also reported, for example, 5.3 nm on titania P25 and 7.7 nm on anatase.[70] Furthermore, it has also been shown that the Fermi energy increases with growing in mean size of gold NPs in the range of 3–13 nm, resulting in more efficient electron transfer from gold NPs to titania under UV irradiation.[129]

The gold loading also significantly influences the resultant photocatalytic activity, and similarly the gold particle size, contradictory results can be found in the literature, i.e. an increase of photoactivity with an increase[123] and a decrease[59] in the amount of deposited gold. The existence of optimal gold amount (2–3 wt.%,[109] 0.1 wt.%)[123] or optimal thickness of gold film (15 nm)[70] is also proposed. It is important to notice that other properties can change with an increase in the amount of gold, such as particle size, interface between gold and support, light absorption properties ('inner filter' effect and/or light scattering), and thus all of these properties influence the resultant photocatalytic activity.

At present, improvement of photocatalytic activity mainly focuses on more efficient light harvesting, which means the ability of use of overall solar spectrum (a vast range of photoabsorption). To achieve this goal, modification of titania with gold deposits possessing various sizes and shapes is proposed.

The first report showing that higher activity was obtained for broader absorption range was presented for large particles of rutile titania modified with both gold NPs and NRs of various sizes.[101] In those samples, LSPR possesses two main absorption peaks at shorter (ca. 520 nm) and longer (ca. 600 nm) wavelengths, due to the presence of gold NRs (transverse and longitudinal LSPR, see Chapter 3). Since then, gold NRs have been extensively studied for plasmon-assisted photocatalysis, for example, gold NR as a core for trilayered $Au/Ag/TiO_2$,[74] and gold NRs with different aspect ratio.[134] NRs are more efficient nanostructures than NPs, mainly because of their ability of absorbing light in a broader wavelength range, i.e. 400–1000 nm for NRs against 400–700 nm for NPs.[55]

Finally, it is important to point that various chemical compounds used for gold nanoparticles synthesis (see Chapters 6 and 7) are not easy to decompose and completely eliminate.[65] Therefore, a part of them or their by-products can still be present in the final photocatalyst and have an influence on the observed activity. In addition, the properties of gold NPs can change during application, for example, because of aggregation of gold NPs during suspension stirring, subsequent detachment of gold deposits from the support or partial oxidation of gold by electron transfer to titania.[121] Therefore, long-term activity and recycling tests showing stability of plasmonic photocatalysts must be more extensively investigated.

10.4.1.2 *Support properties and interface between gold and support*

Surface properties of support significantly influence the electronic properties of the resultant material. For example, two opposite behaviours are proposed for non-stoichiometric titania interfaces (i.e. transfer of electrons): (i) from Ti atoms to Au atoms for the Ti-rich interface, when Au and Ti atoms are in contact and (ii) from Au atoms to O atoms for O-rich interface when Au atoms are in contact with O atoms.[147] The resultant electronic structure of Au/TiO_x is influenced by the stoichiometry of titania, i.e. for the Ti-rich interface, Au d-components are found above the Fermi energy, due to the hybridisation with the Ti $3d$ orbitals, while for O-rich interface, Au $5d$ bands are broaden due to strong hybridisation with the O $2p$ orbitals.[147]

To improve the photocatalytic performance of plasmonic photocatalysts (under visible light), various modifications of support properties are pro posed, mainly to hinder back electron transfer from CB of titania to gold NPs. For example, an improved structure of support is proposed by Z. Bian *et al.*, i.e. 'advanced superstructure system'.[72] This structure is based on titania mesocrystals in which electrons from gold NPs migrated through the titania nanocrystal networks from the basal surfaces to the edges of the plate-like of mesocrystals, where they are temporarily stored for further reactions, as shown in Figure 10.13(a). Due to this anisotropic electron flow, which hinders the recombination of electrons and holes in the gold NPs, this structure results in longer electron lifetime, and therefore enhanced photocatalytic activity by more than one order of magnitude, as compared to that of conventional particulate Au/TiO_2 system (Figure 10.13(c)), is obtained.

Partial covering of gold NPs deposited on large titania particles with fine titania crystallites ($TiO_2/Au/TiO_2$) has been also proposed to extend the interface between gold and titania.[121] The resultant $TiO_2/Au/TiO_2$ nanostructure shows strong plasmonic near-field enhancement (by finite

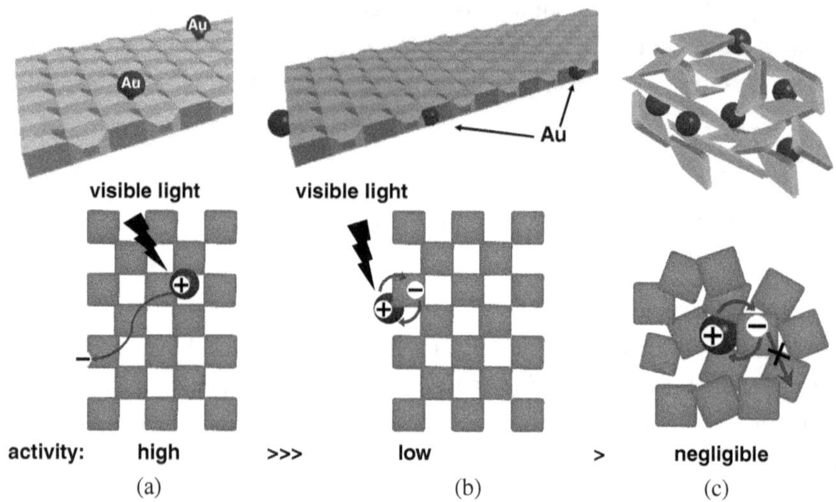

Figure 10.13   Schemes showing Au/TiO$_2$ photocatalysts used for degradation of organic compounds under visible light; gold deposited on (a) basal and (b) lateral surfaces of titania network, and (c) on titania NPs; adapted from Refs. 72.

difference time domain method (3D-FDTD)) and enhanced photocatalytic activity for OCs decomposition under visible irradiation.

As mentioned above in Section 10.2.3.3, the interface between gold and support plays a crucial role in photocatalytic properties, and accordingly removal of molecules present on the surface of gold or between gold and titania causes an increase in photocatalytic activity. For example, removal of stabilising agents from nanoporous titania film results in a significant enhancement of generated photocurrent.[58] A thermal treatment is also applied for enlargement of interface, for example, (i) partial embedding of gold NPs into the titania film structure, i.e. titania doping[58] and (ii) titania aggregation[91] (which can result in inter-particle electron transfer in the formed aggregates: between bare and modified titania). Similarly, gold NPs of small sizes are able to enter inside the titania pores, resulting in larger interface between gold and titania, and hence higher activity (under UV light).[126]

Localisation of gold NPs on the support surface is crucial, as well.[72] In the case of titania mesocrystals (described above), the deposition of gold on the basal surfaces results in efficient transfer of electrons from gold NPs to the lateral titania surfaces through nanocrystal networks under visible irradiation, as shown in Figure 10.13(a). This results in prolongation of electron lifetime, and therefore enhanced photocatalytic activity, when compared to the structure with gold NPs deposited on the lateral surfaces (Figure 10.13(b)).

10.4.2 *Hybrid Nanostructures*

Advanced nanostructures can improve the performance of plasmonic photocatalysts, by their coupling with other photocatalysts, catalysts and insulators for more efficient electron/energy transfer and improvement of photoabsorption and adsorption properties. In this section, a few examples of novel nanostructures will be presented.

10.4.2.1 *Heterogeneous nanostructures: Plasmonic photocatalysts and other solid materials*

The deposition of Au–TiO_2 on additional supports is proposed to (i) inhibit the electron–hole recombination (protonated zeolite,[61] CuO,[141] Cu/SnO_2[77]), (ii) enhance the photocatalytic activity by heterojunction between Au–TiO_2 and the support ($Zn_2Ti_3O_8$ and $ZnTiO_3$),[87] (iii) increase the plasmon-activation efficiency by efficient electron transfer from gold via titania to support (CuO,[141] Cu/SnO_2[77]), (iv) extend the surface area for efficient pollutants adsorption (zeolite,[61] CuO),[141] (v) improve the photocatalyst recovery (Fe_2O_3)[47] and (vi) promote the photocatalytic activity by better dispersion of gold NPs on support (CuO).[141]

For example, modification of titania with Zn^{2+} results in the formation of new $Zn_2Ti_3O_8$ and $ZnTiO_3$ phases, allowing heterojunctions between them and titania and consequently extending the lifetime of charge carriers, which gives rise to enhanced photocatalytic activity.[90]

A nanocomposite material with three components, a magnetic core (Fe_3O_4), TiO_2 bridging layer and gold deposits[48] is proposed for easier recovery of photocatalyst after use, and for other non-photocatalytic applications, for example, as magnetic resonance imaging (MRI) contrast and magnetic drug delivery agents.

Improved water splitting under visible light is obtained on tungsten trioxide photoanode modified with gold–polyoxometalate NPs. Enhanced visible activity of WO_3 is due to the presence of gold (dark catalytic effect and/or energy transfer), while polyoxometalates inhibit recombination of charge carriers.[142]

Hybrid nanostructure (bi-overlayer plasmonic photocatalyst) consisting of two photocatalysts (Au/TiO_2 and Cu/SnO_2) deposited on opposite sides of FTO shows promoted visible-light activities. Degradation of organic compounds occurs on Au/TiO_2, while simultaneous oxygen reduction reaction (ORR) takes place on Cu/SnO_2.[78] The high activity of this photocatalyst (four times higher than Au/TiO_2) is caused by an efficient separation of charge carriers, attributed to electron transfer from Au to CuO ($Au \rightarrow TiO_2 \rightarrow FTO \rightarrow SnO_2 \rightarrow CuO$).

Another example of hybrid nanostructure is proposed for CO oxidation on gold-loaded hedgehog-shaped titania nanospheres additionally decorated with CuO of narrow band-gap.[138] This system is advantageous because (i) CuO promotes the dispersion of Au NPs on the support, (ii) CuO reinforces the electron interaction between Au NPs and titania (efficient charge mediator, which can be excited with visible light), (iii) CuO promotes an efficient electron transfer from Au NPs to titania and (iv) CuO allows efficient oxygen adsorption and activation.

10.4.2.2 *Heterogeneous–homogeneous photocatalysts (plasmonic photocatalysts–metal complexes)*

Coupling of gold with metal complexes for photocatalysis is efficient in DSSCs, due to enhanced light scattering on gold NPs,[108] enhanced electron transfer[143] and inhibition of back electron reaction,[108] as already discussed in Section 10.3.2.1.

At present, coupling of gold/titania with dye complex is proposed for more efficient light harvesting since gold and dye can absorb visible light at different ranges of solar radiation and both can activate titania.[148,149] Although titania co-modification with gold and ruthenium complex ($[(bpy)_2Ru(bpdpH_4)]Cl_2$) results in a broadening of photoabsorption (200–800 nm: band-gap absorption by titania below 400 nm, metal to ligand charge transfer (MLCT) of metal complex at ca. 400–500 nm and LSPR of gold at ca. 520–600 nm), the activity under visible light irradiation (2-propanol oxidation) is only improved in comparison with gold-modified titania. However, the highest activity is obtained for titania modified with ruthenium complex. Under UV irradiation (acetic acid oxidation and dehydrogenation of methanol), the best activity is achieved for hybrid nanostructures, and the enhancement factor depends on the type of titania support.[148] Furthermore, small changes in ruthenium complex structure result in a significant change of photocatalytic performance, for example, the direct bonding of bipyridine motif with phosphonate groups allows better electronic contact and electron transfer than in the case of their bridging by methylene group.[149] Finally, the sequence of titania modification (gold deposition then ruthenium adsorption or *vice versa*) significantly influences the resultant properties of gold deposits (e.g. reshaping), and hence the photocatalytic activities under both UV and visible irradiation.

10.4.2.3 *Bimetallic plasmonic photocatalysts*

Titania nanostructures modified with two noble metals (gold, silver, platinum, palladium and copper) in the form of alloys, core–shells and monometallic NPs are proposed to enhance photocatalytic activities. Bimetallic nanostructures seem advantageous, due to broadening and intensification of photoabsorption properties, for example, although LSPR band of silver NPs is located at shorter wavelengths than that of gold (ca. 390 nm), it is almost four times more intense.[74] Therefore, Au–Ag nanostructures show much more intense and broader LSPR than monometallic counterparts.[74,150] Besides experimental studies, a lot of theoretical studies focus on the prediction of absorption properties of bimetallic NPs, for example, by DDA for Au(core)–Au/Ag(shell)[151] and finite difference time domain method (3D-FDTD) for Au(core)–Ag(shell).[150]

It is proposed that in the case of bimetallic NPs, more electrons are photo-generated as a result of interband transitions in the alloy (due to *spd* hybridisation) under UV irradiation.[85] Indeed, significant enhancement of activity under UV irradiation is observed for gold-based bimetallic (Au/Ag, Au/Cu, Au/Ni) nanostructures tested in various reactions such as phenol oxidation, methanol dehydrogenation, acetic acid oxidation and methyl orange oxidation.[120,125,150,152–154]

Similarly, to hybrid nanostructures composed of gold/titania/dye, bimetallic photocatalysts exhibit higher activity under UV irradiation,

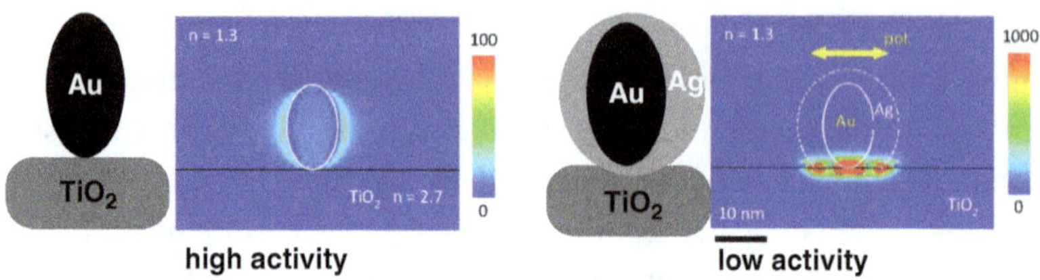

Figure 10.14   Schemes showing (left) Au/TiO$_2$ and (right) Ag/Au/TiO$_2$ with respective near-field simulations (3D-FDTD). Adapted from Ref. 150.

but lower activity under visible light irradiation than monometallic photocatalysts.[153,155] It is proposed that 'hot' electron from one metal could sink in the other one, instead of being transfer to CB of titania. Enhancement of activity is only observed when titania is modified with two metals in the form of single monometallic deposits.[150] Interestingly, the decrease in activity for Au(core)–Ag(shell)/TiO$_2$ nanostructures indirectly supports the hypothesis of a charge transfer mechanism (instead of energy transfer mechanism), since for those nanostructures of poor activity, a strong enhancement of the plasmonic near-field is observed, as shown in Figure 10.14.

Enhanced activity under visible light irradiation has only been observed for reverse nanostructure when titania is a covering layer instead of being a support, that is, gold NRs are covered with shell of silver and finally with layer of amorphous titania, as shown in Figure 10.8.[74]

10.5 Conclusions

A new field of heterogeneous photocatalysis with plasmon-assisted (photo)catalysis on gold-modified solid materials (mainly semiconductors) activated by LSPR has emerged in less than 10 years. Although research on plasmonic photocatalysis is recent, a lot of articles have already been published on their preparation, mechanism investigations and possible applications, as well as on nanoarchitecture arrangement for activity and stability improvement. Nanoarchitecture of plasmonic photocatalysts (properties of gold NPs, properties of support and their interactions) governs the reaction mechanism and the application, and hence an inactive structure for a given reaction or application may show excellent performance for another one.

It is thought that in a near future those photocatalysts (pure or coupled with other nanomaterials) will be commercially used for environmental purification and solar energy conversion. It is expected that fundamental studies and theoretical calculations will help to explain the reaction mechanism and to design more efficient and stable plasmonic photocatalysts.

Acknowledgements

I would like to thank Prof. C. Louis for valuable comments, Mr. Kunlei Wang for schematic drawings, Dr. Shuaizhi Zheng and Dr. Zhishun Wei for their invaluable advice.

Appendix

Action spectrum (AS) — the efficiency with which irradiation results in a photochemical reaction plotted (often in the form of quantum yield/efficiency) as a function of the irradiation wavelength (e.g. Figure 10.4).

Artificial photosynthesis — chemical process that replicates the photosynthesis; a process that converts sunlight, water and carbon dioxide into carbohydrates (or other compounds, i.e. solar fuel) and oxygen.

CB — conduction band.

DSSCs — dye-sensitised solar cells (kind of thin-film solar cells); a photoelectrochemical system based on a semiconductor designed such that titania NPs are coated with light-sensitive dye and surrounded by electrolyte, which is sandwiched between another electrolyte and a cathode.

Electron trap (ET) — a defect or chemical impurity in a semiconductor or insulator, which captures mobile electrons. There are two kinds of ETs: shallow-level ETs and deep-level ETs in the sense of energy required to remove an electron or hole from the trap to the valence or conduction band. Therefore, prolonged or shortened lifetime of charge carriers is observed for shallow- or deep-level ETs, respectively.

FTO — fluorine-doped tin oxide; used as a transparent conducting film (TCF).

Inner-filter effect (IFE) — decrease in light absorption due to deposits (modifiers, e.g. gold NPs, impurity) adsorbed on the surface of photocatalysts.

Janus particle/nanostructure — particles/nanostructures whose surfaces of both hemispheres are different from a chemical point of view [e.g. gold–titania (Figure 10.12(right))].

Light harvesting — systems or compounds use to collect more of the incoming light, e.g. by (i) light scattering on nanoparticles, (ii) light trapping inside nanostructures (inverse-opal) and (iii) extension of absorption towards overall solar spectrum (by antenna).

LEMF — local electromagnetic field transfer (kind of PRET); for energy matching between plasmonic NPs (LSPR) and band-gap of semiconductor (Figure 10.7(a)).

LSPR — localised surface plasmon resonance; the result of the confinement of a surface plasmon in a NP of size comparable to or smaller than the wavelength of light used to excite the plasmon.

MB — methylene blue (dye); often used as a model pollutant for photocatalytic activity tests.

NPs — nanoparticles; particles between 1 and 100 nanometres in size.

NRs — nanorods; one morphology of nanoscale objects (each of their dimensions range from 1–100 nm) with standard aspect ratios (length divided by width) of 3–5.

OCs — organic compounds.

Quantum efficiency (yield) (Φ) — a radiation-induced process in which the number of times a specific chemical reaction occurs per photon absorbed by the system; since for heterogeneous photocatalysis amount of absorbed photons is difficult to be evaluated (due to light scattering on NPs), the apparent quantum efficiency ((Φ_{app}) is used, in which amount of all photons entering the system is measured (absorbed and scattered).

Photocurrent — the electric current, as the result of exposure of a photosensitive device to irradiation.

P25 — commercial titania (AEROXIDE®TiO$_2$ P 25 from Evonik (before Degussa P25)) possessing one of the highest photocatalytic activities.

PRET — plasmon energy transfer; semiconductor excitation by plasmonic near-field energy of noble metal NPs.

RET — resonant energy transfer (kind of PRET); for energy mismatching between plasmonic NPs (LSPR) and band-gap of semiconductor (Figure 10.7(b)).

ROS — reactive oxygen species.

Superhydrophilicity — water dropped onto titania forming no contact angle (almost 0 degrees) under light irradiation.

VB — valence band.

References

1. M. R. Hoffmann, S. T. Martin, W. Y. Choi, D. W. Bahnemann, *Chem. Rev.* **95**, 69 (1995).
2. N. Serpone, A. V. Emeline, S. Horikoshi, V. N. Kuznetsov, V. K. Ryabchuk, *Photochem. Photobiol. Sci.* **11**, 1121 (2012).
3. J. C. D'Oliveira, G. Al-Sayyed, P. Pichat, *Environ. Sci. Technol.* **24**, 990 (1990).
4. A. Fujishima, T. N. Rao, *Proc. Indian Acad. Sci. Chem. Sci.* **109**, 471 (1997).
5. A. Fujishima, K. Honda, *Nature (London, U. K.)* **238**, 37 (1972).
6. N. J. Cherepy, G. P. Smestad, M. Grätzel, J. Z. Zhang, *J. Phys. Chem. B* **101**, 9342 (1997).
7. R. E. Smalley, in *MIT Forum*, River Oaks, USA, 2003.
8. R. Abe, *J. Photoch. Photobiol. C* **11**, (2010).
9. A. Mills, R. H. Davies, D. Worsley, *Chem. Soc. Rev.* **22**, 417 (1993).
10. Z. Wei, E. Kowalska, J. Verrett, C. Colbeau-Justin, H. Remita, B. Ohtani, *Nanoscale* **7**, 12392 (2015).
11. D. Mitoraj, H. Kisch, *Angew Chem. Int. Ed.* **47**, 9975 (2008).
12. R. Asahi, T. Morikawa, T. Ohwaki, K. Aoki, Y. Taga, *Science* **293**, 269 (2001).
13. K. Lalitha, G. Sadanandam, V. D. Kumari, M. Subrahmanyam, B. Sreedhar, N. Y. Hebalkar, *J. Phys. Chem. C* **114**, 22181 (2010).
14. L. Zang, C. Lange, I. Abraham, S. Storck, W. F. Maier, H. Kisch, *J. Phys. Chem. B* **102**, 10765 (1998).
15. B. Ohtani, S.-I. Nishimoto, *J. Phys. Chem. B* **97**, 920 (1993).
16. A. Sclafani, M.-N. Mozzanega, J. M. Herrmann, *J. Catal.* **168**, 117 (1997).
17. G. R. Banmwenda, S. Tsubota, T. Nakamura, M. Haruta, *J. Photochem. Photobiol. A-Chem.* **89**, 177 (1995).
18. B. Ohtani, M. Kakimoto, H. Miyadzu, S. Nishimoto, T. Kagiya, *J. Phys. Chem.* **92**, 5773 (1988).
19. E. Kowalska, H. Remita, C. Colbeau-Justin, J. Hupka, J. Belloni, *J. Phys. Chem. C* **112**, 1124 (2008).
20. G. A. Hope, A. J. Bard, *J. Phys. Chem.* **87**, 1979 (1983).
21. N. Jaffrezic-Renault, P. Pichat, A. Foissy, R. Mercier, *J. Phys. Chem.* **90**, 2733 (1986).
22. S. Nishimoto, B. Ohtani, H. Kajiwara, T. Kagiya, *J. Chem. Soc.* **79**, 2685 (1983).
23. P. Pichat, J. M. Herrmann, J. Disdier, H. Courbon, M. N. Mozzanega, *Nouv. J. Chim.* **5**, 627 (1981).

24. B. Ohtani, Y. Ogawa, S.-I. Nishimoto, *J. Phys. Chem. B* **101**, 3746 (1997).
25. Y. Tian, T. Tatsuma, *J. Am. Chem. Soc.* **127**, 7632 (2005).
26. E. Kowalska, R. Abe, B. Ohtani, *Chem. Commun.* 241 (2009).
27. H. Kominami, A. Tanaka, K. Hashimoto, *Chem. Commun.* **46**, 1287 (2010).
28. H. Kominami, A. Tanaka, K. Hashimoto, *Appl. Catal. A: Gen.* **397**, 121 (2011).
29. C. J. Wang, O. Ranasingha, S. Natesakhawat, P. R. Ohodnicki, M. Andio, J. P. Lewis, C. Matranga, *Nanoscale* **5**, 6968 (2013).
30. A. J. Haes, C. L. Haynes, A. D. McFarland, S. G. C. Van Duyne, R. P. S. Zou, *MRS Bull.* **30**, 368 (2005).
31. F. Stietz, F. Trager, *Philos. Mag.* **79**, 1281 (1999).
32. N. Halas, *MRS Bull.* **30**, 362 (2005).
33. H. A. Atwater, S. Maier, A. Polman, J. A. Dionne, L. Sweatlock, *MRS Bull.* **30**, 385 (2005).
34. K. Okamoto, I. Niki, A. Shvarster, G. Maltezos, Y. Narukawa, T. Mukai, Y. Kawakami, A. Scherer, *Phys. Status Solidi A* **204**, 2103 (2007).
35. P. Zijlstra, J. W. M. Chon, M. Gu, *Nature* **459**, 410 (2009).
36. K. Ueno, H. Misawa, *J. Photoch. Photobiol. C* **15**, 31 (2013).
37. S. W. Verbruggen, *J. Photoch. Photobiol. C* **24**, 64 (2015).
38. S. Sarina, E. R. Waclawik, H. Y. Zhu, *Green Chem.* **15**, 1814 (2013).
39. Z. Z. Lou, Z. Y. Wang, B. B. Huang, Y. Dai, *Chemcatchem* **6**, 2456 (2014).
40. W. B. Hou, S. B. Cronin, *Adv. Funct. Mater.* **23**, 1612 (2013).
41. X. Chen, H.-Y. Zhu, J.-C. Zhao, Z.-F. Zheng, X.-P. Gao, *Angew. Chem. Int. Ed.* **47**, 5353 (2008).
42. Z. Yi, Y. Chen, J. Luo, Y. Y. X. Kang, P. Ye, X. Gao, Y. Tang, Y. Yi, *Plasmonics* **10**, 1373 (2015).
43. R. Costi, A. E. Saunders, E. Elmalem, A. Salant, U. Banin, *Nano Lett.* **8**, 637 (2008).
44. Z. G. Xiong, L. L. Zhang, J. Z. Ma, X. S. Zhao, *Chem. Commun.* **46**, 6099 (2010).
45. F. Moreau, G. C. Bond, *Appl. Catal. A—Gen.* **302**, 110 (2006).
46. X. Bokhimi, R. Zanella, A. Morales, *J. Phys. Chem. C* **111**, 15210 (2007).
47. W. F. Yan, B. Chen, S. M. Mahurin, V. Schwartz, D. R. Mullins, A. R. Lupini, S. J. Pennycook, Dai, S. S. H. Overbury, *J. Phys. Chem. B* **109**, 10676 (2005).
48. B. L. Oliva, A. Pradhan, D. Caruntu, C. J. O'Connor, M. A. Tarr, *J. Mater. Res.* **21**, 1312 (2006).
49. I. Caretti, M. Keulemans, S. W. Verbruggen, S. Lenaerts, S. Van Doorslaer, *Top. Catal.* **58**, 776 (2015).
50. J. B. Priebe, J. Radnik, A. J. J. Lennox, M. M. Pohl, M. Karnahl, D. Hollmann, K. Grabow, U. Bentrup, H. Junge, M. Beller, A. Bruckner, *Acs Catal* **5**, 2137 (2015).
51. Z. W. Seh, S. W. Liu, M. Low, S.-Y. Zhang, Z. Liu, A. Mlayah, M.-Y. Han, *Adv. Mater.* **24**, 2310 (2012).
52. D. A. Panayotov, P. A. DeSario, J. J. Pietron, T. H. Brintlinger, L. C. Szymczak, D. R. Rolison, J. R. Morris, *J. Phys. Chem. C* **117**, 15035 (2013).
53. J. G. Yue, L. Yu, S. W. Liu, B. B. Huang, X. Y. Zhang, *J. Colloid Interface Sci.* **334**, 58 (2009).
54. A. Pearson, H. D. Zheng, Kalantar-zadeh, K. S. K. Bhargava, V. Bansal, *Langmuir* **28**, 14470 (2012).

55. T. Okuno, G. Kawamura, H. Muto, A. Matsuda, *J. Sol-Gel Sci. Technol.* **74**, 748 (2015).
56. M. V. Dozzi, L. Prati, P. Canton, E. Selli, *Phys. Chem. Chem. Phys.* **11**, 7171 (2009).
57. P. D. Cozzoli, M. L. Curri, C. Giannini, A. Agostiano, *Small* **2**, 413 (2006).
58. L. J. Brennan, F. Purcell-Milton, A. S. Salmeron, H. Zhang, A. O. Govorov, A. V. Fedorov, Y. K. Gun'ko, *Nanoscale Res. Lett.* **10**, 1 (2015).
59. S. V. Awate, A. A. Belhekar, S. V. Bhagwat, R. Kumar, N. M. Gupta, *Int. J. Photoenergy* 789149 (2008).
60. S. Bouhadoun, C. Guillard, F. Dapozze, S. Singh, D. Amans, J. Boucle, N. Herlin-Boime, *Appl. Catal. B-Environ.* **174**, 367 (2015).
61. J. W. Sun, N. Liu, S. R. Zhai, Z. Y. Xiao, Q. D. An, D. Z. Huang, *Mater. Sci. Semicond. Process.* **25**, 286 (2014).
62. N. P. Tangale, A. A. Belhekar, K. B. Kale, S. V. Awate, *Water Air Soil Poll.* **225**, 1 (2014).
63. R. S. Sonawane, M. K. Dongare, *J. Mol. Catal. A—Chem.* **243**, 68 (2006).
64. A. Ahmad, S. Senapati, M. I. Khan, R. Kumar, M. Sastry, *J. Biomed. Nanotechnol.* **1**, 47 (2005).
65. J. Sharma, K. P. Vijayamohanan, *J. Colloid Interface Sci.* **298**, 679 (2006).
66. A. M. Schwartzberg, T. Y. Olson, C. E. Talley, J. Z. Zhang, *J. Phys. Chem. B* **110**, 19935 (2006).
67. L. Q. Liu, T. D. Dao, R. Kodiyath, Q. Kang, H. Abe, T. Nagao, J. H. Ye, *Adv. Funct. Mater.* **24**, 7754 (2014).
68. J. Zhu, L. Q. Huang, J. W. Zhao, Y. C. Wang, Y. R. Zhao, L. M. Hao, Y. Lu, M. *Mat. Sci. Eng. B-Solid.* **121**, 199 (2005).
69. N. R. Jana, L. Gearheart, C. J. Murphy, *Adv. Mater.* **13**, 1389 (2001).
70. B. Cojocaru, S. Neatu, Sacaliuc-Parvulescu, E. F. Levy, V. I. Parvulescu, H. Garcia, *Appl. catal. B-Environ.* **107**, 140 (2011).
71. F. Menegazzo, M. Signoretto, D. Marchese, F. Pinna, M. Manzoli, *J. Catal.* **326**, 1 (2015).
72. Z. F. Bian, T. Tachikawa, P. Zhang, M. Fujitsuka, T. Majima, *J. Am. Chem. Soc.* **136**, 458 (2014).
73. J. Li, H. C. Zeng, *Angew. Chem. Int. Ed.* **44**, 4342 (2005).
74. Y. Horiguchi, T. Kanda, K. Torigoe, H. Sakai, M. Abe, *Langmuir* **30**, 922 (2014).
75. H. Ko, Y. Y. Mizuhata, M. Kajinami, A. S. Deki, *Thin Solid Films* **491**, 86 (2005).
76. X. D. Wang, J. Choi, D. R. G. Mitchell, Y. B. Truong, I. L. Kyratzis, R. A. Caruso, *Chemcatchem* **5**, 2646 (2013).
77. V. Rodriguez-Gonzalez, R. Zanella, G. del Angel, R. Gomez, *J. Mol. Catal. A: Chem.* **281**, 93 (2008).
78. S.-I. Naya, T. Kume, T. Okumura, H. Tada, *Phys. Chem. Chem. Phys.* **17**, 18004 (2015).
79. M. Bowker, C. Morton, J. Kennedy, H. Bahruji, J. Greves, W. Jones, P. R. Davies, C. Brookes, P. P. Wells, N. Dimitratos, *J. Catal.* **310**, 10 (2014).
80. M. Manzoli, F. Menegazzo, M. Signoretto, G. Cruciani, F. Pinna, *J. Catal.* **330**, 465 (2015).
81. Y. Liu, E. N. Mills, R. J. Composto, *J. Mater. Chem.* **19**, 2704 (2009).

82. E.-S. A. M. Al-Sherbini, *Mater. Chem. Phys.* **121**, 349 (2010).

83. X. Yan, T. Ohno, K. Nishijima, R. Abe, B. Ohtani, *Chem. Phys. Lett.* **429**, 606 (2006).

84. B. Ohtani, *Phys. Chem. Chem. Phys.* **16**, 1788 (2014).

85. C. H. Han, X. Z. Yang, G. J. Gao, J. Wang, H. L. Lu, J. Liu, M. Tong, X. Liang, Y. *Green Chem.* **16**, 3603 (2014).

86. W. T. Chen, A. Chan, Z. H. N. Al-Azri, A. G. Dosado, M. A. Nadeem, Sun-Waterhouse, D. H. Idriss, Waterhouse, G. I. N. *J. Catal.* **329**, 499 (2015).

87. A. G. Dosado, W. T. Chen, A. Chan, D. Sun-Waterhouse, G. I. N. Waterhouse, *J. Catal.* **330**, 238 (2015).

88. F. B. Li, X. Z. Li, *Appl. Catal. A.—Gen.* **228**, 15 (2002).

89. I. A. Mkhalid, *J. Alloy Compd.* **631**, 298 (2015).

90. N. Smirnova, V. Vorobets, O. Linnik, E. Manuilov, G. Kolbasov, A. Eremenko, *Surf. Interface Anal.* **42**, 1205 (2010).

91. J. T. Carneiro, T. J. Savenije, G. Mul, *Phys. Chem. Chem. Phys.* **11**, 2708 (2009).

92. B. Mei, C. Wiktor, S. Turner, A. Pougin, G. van Tendeloo, R. A. Fischer, M. Muhler, J. Strunk, *ACS Catal.* **3**, 3041 (2013).

93. S. Kundu, A. Kafizas, G. Hyett, A. Mills, J. A. Darr, I. P. Parkin, *J. Mater. Chem.* **21**, 6854 (2011).

94. M. Valenti, D. Dolat, G. Biskos, A. Schmidt-Ott, W. A. Smith, *J. Phys. Chem. C* **119**, 2096 (2015).

95. Z. Liu, W. Hou, P. Pavaskar, M. Aykol, S. B. Cronin, *Nano Lett.* **11**, 1111 (2011).

96. J. T. Carneiro, C. C. Yang, J. A. Moma, J. A. Moulijn, G. Mul, *Catal. Lett.* **129**, 12 (2009).

97. V. Subramanian, E. Wolf, P. V. Kamat, *J. Phys. Chem. B* **105**, 11439 (2001).

98. A. Orlov, D. A. Jefferson, M. Tikhov, R. M. Lambert, *Catal. Commun.* **8**, 821 (2007).

99. W. Macyk, G. Burgeth, H. Kisch, *Photochem. Photobiol. Sci.* **2**, 322 (2003).

100. G. Burgeth, H. Kisch, *Coord. Chem. Rev.* **230**, 41 (2002).

101. E. Kowalska, O. O. Mahaney, P. R. Abe, B. Ohtani, *Phys. Chem. Chem. Phys.* **12**, 2344 (2010).

102. L. Wang, X. M. N. Zhou, T. Nguyen, P. Schmuki, *Chemsuschem* **8**, 618 (2015).

103. A. Furube, L. Du, K. Hara, R. Katoh, M. Tachiya, *J. Am. Chem. Soc.* **129**, 14852 (2007).

104. L. Du, A. Furube, K. Hara, R. Katoh, M. Tachiya, *Thin Solid Films* **158**, 861 (2009).

105. S.-I. Naya, M. Teranishi, T. Isobe, H. Tada, *Chem. Commun.* **46**, 815 (2010).

106. S.-I. Naya, A. Inoue, H. Tada, *J. Am. Chem. Soc.* **132**, 6292 (2010).

107. N. Sakai, Y. Fujiwara, M. Arai, K. Yu, T. Tatsuma, *J. Electroanal. Chem.* **628**, 7 (2009).

108. N. Sakai, Y. Fujiwara, Y. Takahashi, T. Tatsuma, *Chem. Phys. Chem.* **10**, 766 (2009).

109. D. Tsukamoto, Y. Shiraishi, Y. Sugano, S. Ichikawa, S. Tanaka, T. Hirai, *J. Am. Chem. Soc.* **134**, 6309 (2012).

110. S. Mukherjee, F. Libisch, N. Large, O. Neumann, L. V. Brown, J. Cheng, J. B. Lassiter, E. A. Carter, P. Nordlander, N. J. Halas, *Nano Lett.* **13**, 240 (2013).

111. G. L. Hallett-Tapley, M. J. Silvero, M. Gonzalez-Bejar, M. Grenier, J. C. Netto-Fereira, J. C. Scaiano, *J. Phys. Chem. C* **115**, 10784 (2011).

112. W. Hou, Z. Liu, P. avaskar, P Hsuan W. Hung, S. B. Cronin, *J. Catal.* **277**, 149 (2011).

113. S. K. Cushing, J. T. Li, F. K. Meng, T. R. Senty, S. Suri, M. J. Zhi, M. Li, A. D. Bristow, N. Q. Wu, *J. Am. Chem. Soc.* **134**, 15033 (2012).

114. W. B. Hou, W. H. Hung, P. Pavaskar, A. Goeppert, M. Aykol, S. B. Cronin, *Acs Catal* **1**, (2011).

115. R. M. Mohamed, E. S. Aazam, *Int. J. Photoenergy* 137328 (2011).

116. S. A. Trammell, R. Nita, M. Moore, D. Zabetakis, E. Chang, D. A. Knight, *Chem. Commun.* **48**, 4121 (2012).

117. C. G. Silva, R. Juarez, T. Marino, R. Molinari, H. Garcia, *J. Am. Chem. Soc.* **133**, 595 (2011).

118. M. S. Son, J. E. Im, K. K. Wang, S. L. Oh, Y. R. Kim, K. H. Yoo, *Appl. Phys. Lett.* **96**, 023115 (2010).

119. Y. Nishijima, K. Ueno, Y. Yokata, K. Murakoshi, H. Misawa, *J. Phys. Chem. Lett.* **1**, 2031 (2010).

120. A. E. Zielińska-Jurek, J. Kowalska, W. Sobczak, W. Lisowski, B. Ohtani, A. Zaleska, *Appl. Catal. B — Environ.* **101**, 504 (2011).

121. E. Kowalska, L. Rosa, S. Juodkazis, B. Ohtani, (2016) under preparation.

122. L. C. Du, A. Furube, K. Yamamoto, K. Hara, R. Katoh, R. M. Tachiya, *J. Phys. Chem. C* **113**, (2009).

123. E. Kowalska, S. Rau, B. Ohtani, *J. Nanotechnol.* 361853 (2012).

124. P. A. DeSario, J. J. Pietron, D. E. DeVantier, T. H. Brintlinger, R. M. Stroud, D. R. Rolison, *Nanoscale* **5**, 8073 (2013).

125. A. Kafizas, S. Kellici, J. A. Darr, I. P. Parkin, *J. Photoch. Photobiol. A* **204**, 183 (2009).

126. Z. Pap, A. Radu, I. J. Hidi, G. Melinte, L. Diamandescu, T. Popescu, L. Baia, V. Danciu, M. Baia, *Chinese J. Catal.* **34**, 734 (2013).

127. S. Oros-Ruiz, R. Zanella, B. Prado, *J. Hazard. Mater.* **263**, 28 (2013).

128. V. Iliev, D. Tomova, L. Bilyarska, G. Tyuliev, *J. Mol. Catal. A-Chem.* **263**, 32 (2007).

129. T. Kiyonaga, M. Fujii, T. Akita, H. Kobayashi, H. Tada, *Phys. Chem. Chem. Phys.* **10**, 6553 (2008).

130. S. V. Awate, R. K. Sahu, M. D. Kadgaonkar, R. Kumar, N. M. Gupta, *Catal. Today* **141**, 144 (2009).

131. L. Armelao, D. Barreca, G. Bottaro, A. Gasparotto, C. Maccato, Maragno, C. E. Tondello, U. L. Stangar, M. Bergant, D. Mahne, *Nanotechnology* **18**, 375709 (2007).

132. S. A. Elfeky, A. S. A. Al-Sherbini, *J. Nanomater.* 570438 (2011).

133. G. N. Wang, X. F. Wang, J. F. Liu, X. M. Sun, *Chem.-Eur. J.* **18**, 5361 (2012).

134. L. Q. Liu, S. X. Ouyang, J. H. Ye, *Angew. Chem.* **52**, 6689 (2013).

135. S. S. Malwadkar, R. S. Gholap, S. V. Awate, P. V. Korake, M. G. Chaskar, N. M. Gupta, *J. Photoch. Photobiol. A* **203**, 24 (2009).

136. H. Yu, H. Ming, H. Zhang, H. Li, K. Pan, Y. Liu, F. Wang, J. Gong, Z. Kang, *Mater. Chem. Phys.* **137**, 113 (2012).

137. J. Hernandez-Fernandez, A. Aguilar-Elguezabal, S. Castillo, B. Ceron-Ceron, R. D. Arizabalo, M. Moran-Pineda, *Catal. Today* **148**, 115 (2009).

138. K. Yang, K. Huang, L. L. Lin, X. Chen, W. X. X. Dai, Z. Fu, *J. Power Sources* **284**, 194 (2015).

139. E. Kowalska, Z. Wei, B. Karabiyik, M. Janczarek, M. Endo, K. Wang, P. Rokicka, A. Markowska-Szczupak, B. Ohtani, *Adv. Sci. Tech.* **93**, 174 (2014).

140. R. Solarska, K. Bienkowski, S. Zoladek, A. Majcher, T. Stefaniuk, P. J. Kulesza, J. Augustynski, *Angew. Chem. Int. Ed.* **53**, 14196 (2014).

141. C. H. Fang, H. L. Jia, Chang, S. Q. F. Ruan, P. Wang, T. Chen, J. F. Wang, *Energ. Environ. Sci.* **7**, 3431 (2014).

142. E. W. McFarland, J. Tang, *Nature (London, U. K.)* **421**, 616 (2003).

143. J. A. Ortega-Mendez, C. R. Lopez, E. Pulido Melian, O. Gonzales Diaz, J. M. Dona Rodriguez, D. Fernandez Hevia, M. Macias, *Appl. catal. B-Environ.* **147**, 439 (2014).

144. S. Oros-Ruiz, R. Zanella, R. Lopez, A. Hernandez-Gordillo, R. Gomez, *J. Hazard. Mater.* **263**, 2 (2013).

145. A. Kmetyko, K. Mogyorosi, P. Pusztai, T. Radu, Z. Konya, A. Dombi, K. Hernadi, *Materials* **7**, 7615 (2014).

146. M. Ilie, B. Cojocaru, V. I. Parvulescu, H. Garcia, *Int. J. Hydrogen Energy* **36**, 15509 (2011).

147. K. Okazaki, Y. Morikawa, S. Tanaka, K. Tanaka, M. Kohyama, *Phys. Rev. B* **69**, 235404 (2004).

148. E. Kowalska, K. Yoshiiri, Z. Wei, S. Zheng, E. Kastl, H. Remita, S. Rau, B. Ohtani, *Appl. Catal. B–Environ.* **178**, 133 (2015).

149. S. Zheng, Z. Wei, K. Yoshiiri, M. Braumuller, B. Ohtani, S. Rau, E. Kowalska, *Photochem. Photobiol. Sci.* **15**, 69 (2016).

150. E. Kowalska, M. Janczarek, L. Rosa, S. Juodkazi, B. Ohtani, *Catal. Today* **230**, 131 (2014).

151. X. B. Xu, Z. Yi, X. B. Li, Y. Y. Wang, X. Geng, J. S. Luo, B. C. Luo, Y. G. Yi, Y. J. Tang, *J. Phys. Chem. C* **116**, 24046 (2012).

152. Z. B. Hai, N. El Kolli, D. B. Uribe, P. Beaunier, M. J. Yacaman, M. Vigneron, J. A. Etcheberry, S. Sorgues, C. Colbeau-Justin, J. F. Chena, H. Remita, *J. Mater. Chem. A* **1**, 10829 (2013).

153. Z. B. Hai, N. El Kolli, J. F. Chen, H. S. Remita, *New J. Chem.* **38**, 5279 (2014).

154. A. L. Luna, E. Novoseltceva, E. Louarn, P. Beaunier, E. Kowalska, B. Ohtani, M. A. Valenzuela, H. Remita, C. Colbeau-Justin, *Appl. Catal. B-Environ.* **191**, 18 (2016).

Chapter 11

Electrical Generation of Light from Plasmonic Gold Nanoparticles

Eric Le Moal, Gérald Dujardin and
Elizabeth Boer-Duchemin

*Institut des Sciences Moléculaires d'Orsay,
Université Paris-Sud — CNRS, Orsay, France*

11.1 Introduction

Gold nanoparticles are used extensively in many aspects of science. In particular, their optical properties are exploited in diverse areas such as biosensing,[1] molecular rulers,[2] organic solar cells[3] and photocatalysis,[4] to name a few. In all these applications, light is used to excite the surface plasmon resonance and/or intra and interband transitions in the gold nanoparticles.[5,6] However, the ability to use electrons rather than photons to induce light from gold nanoparticles opens up horizons which would otherwise be unattainable.

Two main methods may be considered practically for obtaining light from the electrical excitation of gold nanoparticles: one can use low energy ($\sim$2 eV) tunnel electrons from a scanning tunnelling microscope (STM)[7,8] or high-energy ($\sim$30 keV) electrons from a scanning electron microscope.[9,10] In both cases, the main advantage of this method is the small size of the excitation when using electrons as compared to photons. Whereas the size of a photonic probe is limited by diffraction to a few hundred nanometres, the size of an electronic probe is truly nanoscale.[11,12] As will be shown in the following, such extremely high spatial selectivity enables completely new insights into the optical properties of gold nanoparticles. Indeed, excitation with a spatial selectivity on the order of 10 nm may not only be used to investigate single gold nanoparticles but moreover to provoke the emission of light by exciting specific locations inside a single gold nano-object. Low-energy electrical excitation of a gold nanoparticle is of further interest for applications such as electron/photon transduction at the nanoscale. Using this technique, an electrical signal may be converted into a photonic one. This will lead to new applications in future optoelectronic devices.

From a historical point of view, electron-to-photon energy conversion has played an emblematic role in physics over the last 150 years. The invention of the Crook's tube in 1879, where the energy of an electrical discharge is converted into light, has led to a number of discoveries; X-rays by W. Röntgen in 1895, the fluorescent tube by T. Edison in 1895 and the discovery of the electron by J.J. Thomson in 1897.

More recently, the conversion of energy from high-energy (keV) electrons to light via metallic films was first predicted by Ferrell[13] in 1958 and experimentally observed by Steinmann[14] and Brown *et al.*[15] in 1960. Concerning low-energy ($\sim$2 eV) tunnel electrons, the first observation of energy conversion into light dates back to 1976 when Lambe and McCarthy[16] discovered this new method for the generation of light. This light emission method was extended to tunnel electrons from the STM in 1988 by Gimzewski *et al.*,[17,18] a few years after the invention of the STM by Binnig and Rohrer.[19]

In this chapter, we review recent studies of electron-to-photon energy conversion in gold nanoparticles. In Section 11.2, we describe the fundamental mechanisms for the electrical generation of light from gold nanoparticles when using low-energy ($\sim$2 eV) tunnel electrons from a scanning tunnelling microscope or high-energy ($\sim$30 keV) electrons from a scanning electron microscope. In Section 11.3, we report on examples of such electron–photon transduction experiments in various gold nanostructures.

11.2 Light from Electrons via Gold Nanoparticles: Mechanisms and Experimental Set-ups

11.2.1 *Light from the Low-energy Electrical Excitation of Gold: Biased Tunnel Junctions*

As their colleagues were working on increasing gas mileage to over 3 miles per gallon,[20] Lambe and McCarthy of the Ford Motor Company discovered a new low-voltage method for generating light from electrons.[16,21] Their device was made of a slightly oxidised aluminum electrode criss-crossed with a roughened gold counter-electrode. This formed a metal–insulator–metal structure with an insulating layer whose thickness was on the order of 3 nm. When a voltage difference of 2–4 V was applied between the two electrodes, visible light, whose colour changed 'from deep red at low voltage to orange to blue white as the voltage was increased'[16] could be seen by eye. Similarly, as mentioned above, a bit more than 10 years later, light emitted from the nanometre-sized junction formed between the tip of a scanning tunnelling microscope and a metallic sample was observed.[17] Where did this light come from?

Both of the above experiments involve a tunnel junction. When a potential difference is applied between two conductors separated by a few nanometres or less, current may flow, despite the lack of any physical, electrical connection. The resulting quantum mechanical tunnel current depends exponentially on the distance between the two electrodes and is explained in more detail in Figure 11.1(a). Discovered in the late 1920s, this current is central in the scanning tunnelling microscope.[19]

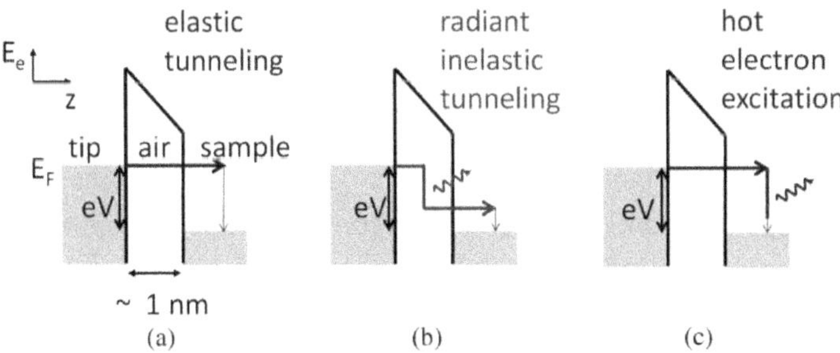

Figure 11.1   Electron tunnelling and the emission of light. (a) Elastic tunnelling: when a potential difference V is applied between two conducting materials (e.g. between an STM tip and sample) that are separated by $\approx$1 nm, a current may flow despite the lack of a direct electrical contact. In quantum mechanics, when an electron encounters a potential barrier (here represented by the air gap), there is a non-zero probability that the electron may 'leak' through this barrier. This probability decays exponentially with increasing distance between the two electrodes. This tunnelling phenomenon is elastic since the electron has the same energy on either side of the barrier. Once in the sample, the electron most often loses its excess energy via non-radiative processes. (b) Radiative inelastic electron tunnelling[22]: a small percentage of the tunnel electrons undergoes an inelastic transition where the electron releases all or part of its energy, which may be converted to light. (c) Hot electron excitation or elastic tunnelling followed by the emission of light. In this scenario, the electron first tunnels elastically and then loses part or all of its excess energy radiatively.

But how does tunnel current lead to the emission of light? In their seminal paper,[16] Lambe and McCarthy divided the phenomenon into two parts: a source term and an emission term.[7] We will begin here by discussing the source of the excitation: inelastic electron tunnelling.

11.2.1.1 *Excitation*

When an electron tunnels from one side of a potential barrier to the other, most of the time the electron does so elastically, i.e. it does not lose any energy (see Figure 11.1(a)). However, every 1 in a 100 (Ref. 23) to 1 in 10 (Ref. 24) tunnel electrons crosses the barrier inelastically, or in other words, it loses part or all of its excess energy (see Figure 11.1(b)). It is this released energy that can lead to the emission of a photon.

While inelastic electron tunnelling (IET) is most often considered the source of the light from biased tunnel junctions, hot electron excitation has also been proposed as a possible mechanism (see Figure 11.1(c)), especially in light of results that are poorly explained in the IET picture.[25] In the hot electron scenario, the electrons tunnel elastically then lose their energy in the electrode. Several arguments have been put forward against the hot electron explanation: first of all, since the electrons lose their energy in the electrode, the receiving electrode material should have more influence on the result; however, in asymmetric or 'bimetal' junctions (i.e. with different materials on either side of the tunnel barrier), clear differences have not been seen as a function of junction polarity.[16,26] Also, since the major loss mechanisms include phonon excitation and electron–electron interactions and the generation of excited electrons which decay non-radiatively, it seems reasonable that these mechanisms would dominate when a hot electron has entered the material.[17] Finally, it has been estimated by calculation that the photon emission rate due to inelastic tunnelling should dominate over that of hot electron excitation by about three orders of magnitude.[27]

Here, we will determine how the intensity of the IET photon source varies with emission frequency (i.e. the energy difference between the initial and final states) independent of its surroundings. To do this we can first use Fermi's golden rule to estimate the elastic tunnelling rate, w_{el}, by taking into account the fact that both the initial and final states are part of a continuum[27]:

$$w_{\text{el}} = \frac{2\pi}{\hbar} |H'_{\text{el}}|^2 \int_{E_F - eV}^{E_F} \rho_{\text{tip}}(E)\, \rho_{\text{sample}}(E)\, \mathrm{d}E,$$

$$w_{\text{el}} \approx \frac{2\pi}{\hbar} |H'_{\text{el}}|^2 \rho_{\text{tip}}(E_F)\, \rho_{\text{sample}}(E_F)\, eV$$

with H'_{el} is the perturbation Hamiltonian, which represents the tunnelling, ρ the localised electronic density of states or e-LDOS of the tip and sample respectively, V the applied voltage, E_F the Fermi energy of the tip, e the charge on an electron and $\hbar$ the reduced Planck's constant. Here it is also assumed that the perturbation Hamiltonian and the densities of states vary slowly with energy in the interval $E_F - eV < E < E_F$.

The expression for the inelastic tunnelling rate is quite similar, but the perturbing Hamiltonian and integration boundaries are different:

$$w_{\text{inel}} = \frac{2\pi}{\hbar} |H'_{\text{inel}}|^2 \int_{E_F - eV + \hbar\omega}^{E_F} \rho_{\text{tip}}(E)\, \rho_{\text{sample}}(E - \hbar\omega)\, \mathrm{d}E,$$

$$w_{\text{inel}} \approx \frac{2\pi}{\hbar} |H'_{\text{inel}}|^2 \rho_{\text{tip}}(E_F)\, \rho_{\text{sample}}(E_F)\, (eV - \hbar\omega).$$

Thus, the expression for the probability of an electron to tunnel inelastically is:

$$P_{\text{inel}} = \frac{w_{\text{inel}}}{w_{\text{inel}} + w_{\text{el}}} \approx \frac{w_{\text{inel}}}{w_{\text{el}}} \propto \left(1 - \frac{\hbar\omega}{eV}\right).$$

Three remarks must be made with regard to this expression. The first is that we see that $\hbar\omega$ must be smaller or equal to eV, which is consistent with energy conservation considerations. The second is the lower the energy of the generated excitation, the more probable it is. Finally, it must not be forgotten that this section refers to the characteristics of the source, independent of its surroundings, which will be taken into account in Section 11.2.1.2. This IET source is often approximated as a vertical oscillating electromagnetic dipole.[23,28]

Already in the early theoretical papers about light from tunnel junctions,[28–32] the IET source term was attributed to the 'fluctuations in time of the tunnel current'[32] and it was described that these fluctuations had 'the character of shot noise'.[30] More recently, this connection has been investigated in more detail both experimentally and theoretically. Schneider *et al.*[33] have recently demonstrated experimentally that the emission of light as a function of junction conductance reaches a minimum at the quantum conductance $G_0 = 2e^2/h$ (where e is the charge of an electron and h is the Planck's constant). Such behaviour is also predicted by quantum shot noise reduction theory and has been demonstrated in a completely different system.[34] The fact that the light emission behaves the same way as the shot noise strongly suggests that these two are intimately related. Recent publications have also theoretically underlined the link between quantum shot noise and the emission of light as described by inelastic electron tunnelling.[35,36]

11.2.1.2 *Emission*

Let us recall the geometry of a tunnel junction experiment: in the case of an STM, for example, a tip with a $\sim$50 nm radius is located about a nanometre from the sample. (Note that it is an atomic-sized 'microtip' on this larger tip that can lead to atomic resolution in STM.) The light from our IET source is thus 'trapped' in this very confined space (hundreds of times smaller than the wavelength) unless the local surroundings of the tunnel junction act as an antenna[29,37] and radiates this excitation to the far field. The antenna action of the surroundings depends strongly on the shape and materials of the tunnel junction and in particular on the existing localised surface plasmon resonances (see Chapter 3). The effective IET dipole source can thus radiate to the far-field thanks to the localised 'gap plasmon' mode of the junction. Here is where we see that it is important that the junction be made of 'plasmonic' materials (i.e. supporting plasmons in the frequency range of interest) such as gold. While the initial spectrum is inherently broadband as seen above since the inelastic electrons need not lose a specific amount of energy, it is the local geometry and material properties that will strongly influence the final spectrum and intensity of the emitted light.

While the inelastic tunnel rate has been theoretically estimated to be relatively high (see Section 11.2.1.1) experimentally about 10,000 tunnel electrons are required for one detected photon.[26] This difference between experiment and theory is attributed to losses (e.g. instead of exciting a surface plasmon that decays radiatively, the excitation may decay non-radiatively). A solution to this problem is to optimise the junction geometry by designing an optical antenna, which will reduce losses and enhance the photon emission (see Section 11.4).

11.2.1.3 *Probe size and experimental apparatus for the local electrical excitation of gold nanoparticles with low energy electrons*

As will be discussed in more detail in Section 11.3.2 the major advantage of the electrical excitation of gold nanoparticles over *optical* excitation is that the excitation probe itself is nanoscale in size. Even though the tunnel current is restricted to an atomic-sized area for a sharp tip, the effective excitation area is larger. If we approximate the STM tip by a sphere of about 100 nm diameter and consider the distance between the tip and sample to be on the order of 1 nm, the lateral extent of the localised surface plasmon mode[28,31] that may exist between the tip and sample is on the order of 10 nm.[37] This 'gap plasmon' mode between the tip and sample may be considered as the exciting probe.

Two main types of experimental set-ups have been used for the low-energy excitation of gold nanoparticles with electrons. In the first case, the sample is (most often) in an ultrahigh vacuum (UHV) chamber, and a series of optics couples the light emitted from above the sample to a detector and/or spectrometer (see Figure 11.2(a)).[33,40–48] The advantages of this setup is that the risk of sample contamination is strongly reduced since the experiment is carried out in UHV, both transparent and opaque samples may be used, and in certain systems low-temperature measurements are also possible. However, only light emitted into a reduced solid angle is collected.

In the second type of experimental set-up, an STM head operating in air is mounted on an inverted optical microscope (Figure 11.2(b)).[39,49] Here all the light emitted through the transparent sample is collected by the microscope objective and focused on a detector and/or spectrometer. In this set-up, a major advantage is that the light-emitting areas on the sample may be identified from real-space images. Information on the direction (angle) of the emitted light may also be obtained (Fourier-space imaging). In this set-up, however, the sample will not be as free from contamination as in a UHV system, and the substrates must be transparent.

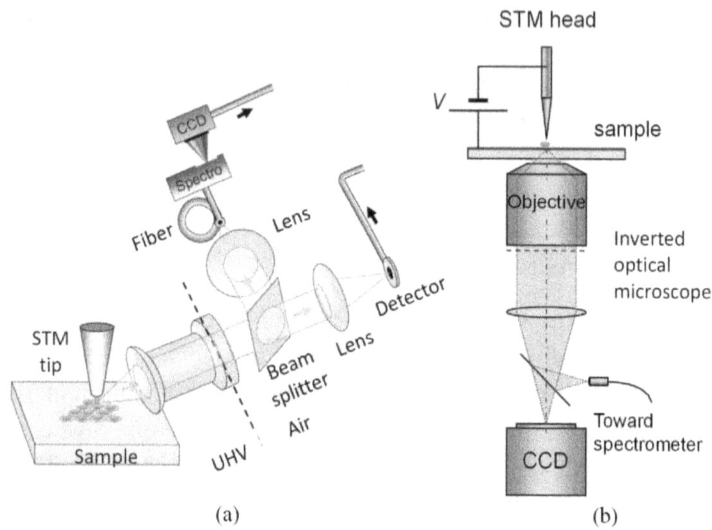

Figure 11.2   STM electrical excitation of gold nanoparticles. (a) Ultrahigh vacuum system, with the emitted light detected above the sample. Adapted from Ref. 38. (b) Ambient system: STM head coupled to an inverted optical microscope. The emitted light is detected below the transparent substrate.[39]

11.2.2 *Light from the High-energy Electrical Excitation of Gold: Cathodoluminescence*

The emission of light from the interaction of high-energy ($\sim$10 keV) electrons with matter, discovered in the mid-nineteenth century was omnipresent in our modern lives in the form of television screens and computer monitors until the recent advancement of other technologies. Cathode ray tubes emit light of different colours when a high-energy electron beam excites luminescent coatings of different constitutions. The emission of light from metals such as gold upon high-energy electron bombardment was first investigated theoretically and experimentally in the 1950s and 1960s. While there exist many different mechanisms for the emission of light from matter irradiated by high-energy electrons,[9] two processes dominate in metals such as gold[50]; they are direct transition radiation and the decay of surface plasmons.

11.2.2.1 *Excitation and emission*

Predicted by Ginsburg and Frank in 1946,[51] transition radiation (TR) may be described intuitively in the following way[9,50]: the electric field of an

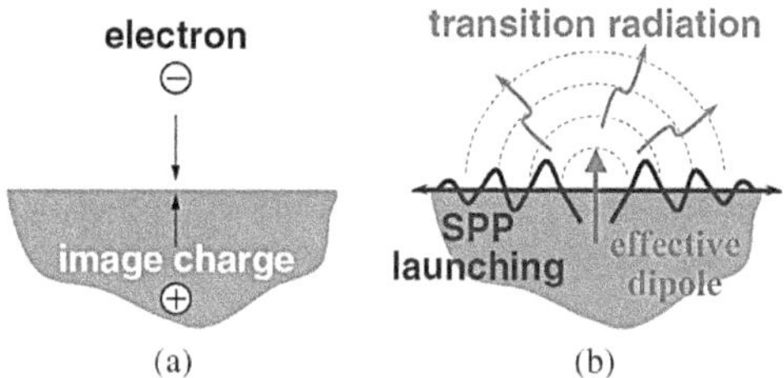

Figure 11.3   The emission of light from the high-energy electron excitation of gold. (a) The incoming electron induces an image charge in the sample. (b) When the electron crosses the interface the image charge vanishes, leading to an effective dipole that may emit light (transition radiation) or excite surface plasmons. Reprinted with permission from Ref. 9. Copyright (2010) by the American Physical Society.

electron approaching a metallic surface will interact with the free charges in the metal, leading to the formation of an image charge of opposite sign (see Figure 11.3(a)). When the electron reaches the interface, the image charge suddenly vanishes — this event may be approximated as a vertical oscillating dipole. Radiation from this effective dipole may either decay as radiation in the far-field (TR), or excite surface plasmons.[12]

Note also that if a fast electron passes close to (instead of incident on), a gold nanoparticle (i.e. within a distance of about 0.2λ, where λ is the detected wavelength and the electron velocity is roughly half that of the speed of light), photons may be excited in a similar fashion. When the electron is either incident upon or passes close to a gold nanoparticle, in both cases, the evanescent electric field of the electron creates time-varying currents in the metal particle. This leads to the electromagnetic field that is detected in the far field.[9,50,52]

The effective oscillating dipole source can either directly emit photons to the far field or excite the surface plasmon modes of the gold nanoparticle, which then may decay radiatively. Due to its localised nature and short ($<$fs) interaction time, fast electrons lead to broadband emission,[53] which may again be tailored by the electromagnetic environment.

11.2.2.2 *Cathodoluminescence and the radiative local electromagnetic density of states*

If an excited molecule is placed before a reflecting interface, its fluorescence lifetime will vary periodically with the molecule's distance from the surface. This phenomenon was studied by Drexhage[54] and is explained classically by considering the light emitted by the molecule and that which is reflected by the interface — when the two interfere constructively at the molecule's position the emission is enhanced, but if the interference is destructive, the emission is suppressed. This behaviour may be described by the electromagnetic local density of states (EM-LDOS). The full EM-LDOS may be defined as the time-averaged power transferred from a point dipole to its environment. On the other hand, the radiative EM-LDOS is related to the power transferred from the same point dipole to the far-field as light.[55] It is this latter quantity that is of interest for us for the high-energy electrical excitation of light from gold nanoparticles.

If the excitation mechanism for fast electrons may be approximated by an effective dipole, then the decay of this dipole into transition radiation and surface plasmons should be governed by the radiative EM-LDOS (projected along the direction of the electron beam).[12] Experimental results showing the expected 'donut' angular distribution of the emitted transition radiation (see Figure 11.4)[50] and good agreement between the calculated and measured emission probability for surface plasmon radiation have been produced.[12] Errors due to this approximation exist since the fast electron excites radiation along its trajectory while the effective oscillating dipole is fixed in space.[12] As will be shown in Section 11.3.2, cathodoluminescence may be used to map the radiative EM-LDOS of gold nanoparticles.

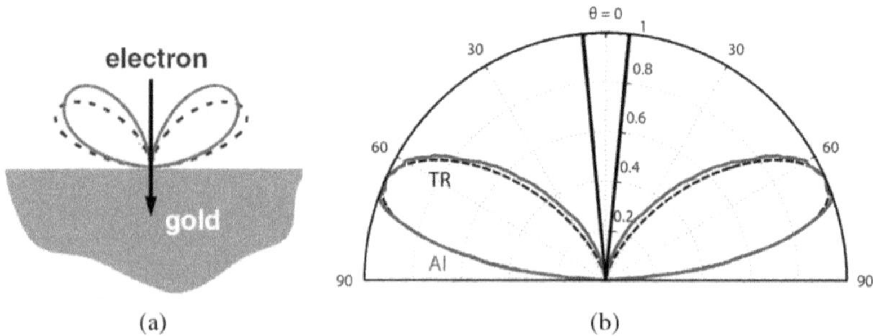

Figure 11.4　Angular dispersion of the light emitted from a metal upon irradiation with a high energy electron beam. (a) Rigorous calculation of the emission pattern (solid curve) for transition radiation and comparison with the emission pattern for a vertical dipole (dashed curve). Reprinted with permission from Ref. 9. Copyright (2010) by the American Physical Society. (b) The same 'donut' shape is found experimentally (solid curve: data from an Al film, dashed curve: transition radiation calculation). Adapted from Ref. 50.

Note that electron energy loss spectroscopy may be used to map the full EM-LDOS,[56] at least in 2D systems.[9,57]

11.2.2.3 *Probe size and experimental apparatus for the local electrical excitation of gold nanoparticles with high-energy electrons*

Just as in the case of the STM, the size of the current spot does not reflect the effective size of the excitation area. As an example, we consider a 30 keV electron beam from a field-emission source that is focused to a $\sim$5 nm spot on the sample. An upper bound to the effective source size may be determined by the extension of the evanescent field of the incoming electron, which may be approximated by $v/2\omega$, where v is the electron velocity and ω is the frequency of light. This gives an upper bound of about 16 nm for the effective size of this 30 keV electron beam with a 5 nm spot size ($\lambda = 600$ nm).[12] (Note that state-of-the-art systems can now achieve a focused beam size of less than an angstrom![58])

Cathodoluminescence may be carried out in a scanning electron microscope (SEM)[59] or incorporated with electron energy loss spectroscopy (EELS) in a scanning transmission electron microscope (STEM)[10] (see Figure 11.5). The key element in such set-ups is a parabolic mirror that collects and directs the light emitted above the sample. As in the case of STM excitation, spectra and images showing the spatial or angular distribution of the emitted light may be obtained. These experiments are carried out in ultrahigh vacuum.

11.3 Recent Achievements in the Electrical Generation of Light from Gold Nanoparticles

The ability of metallic nanostructures to emit light upon excitation with electrons has found a number of uses, both for fundamental studies and technological applications. In most cases, the emission of light occurs through the excitation of surface plasmons (see Chapter 3). Metallic nanostructures of very different sizes have been investigated, from atomic chains of a few atoms[60] to nanoparticles of a few nanometres in diameter,[45,61] as well as regular arrays of such nanoparticles,[40] and larger nanocrystals

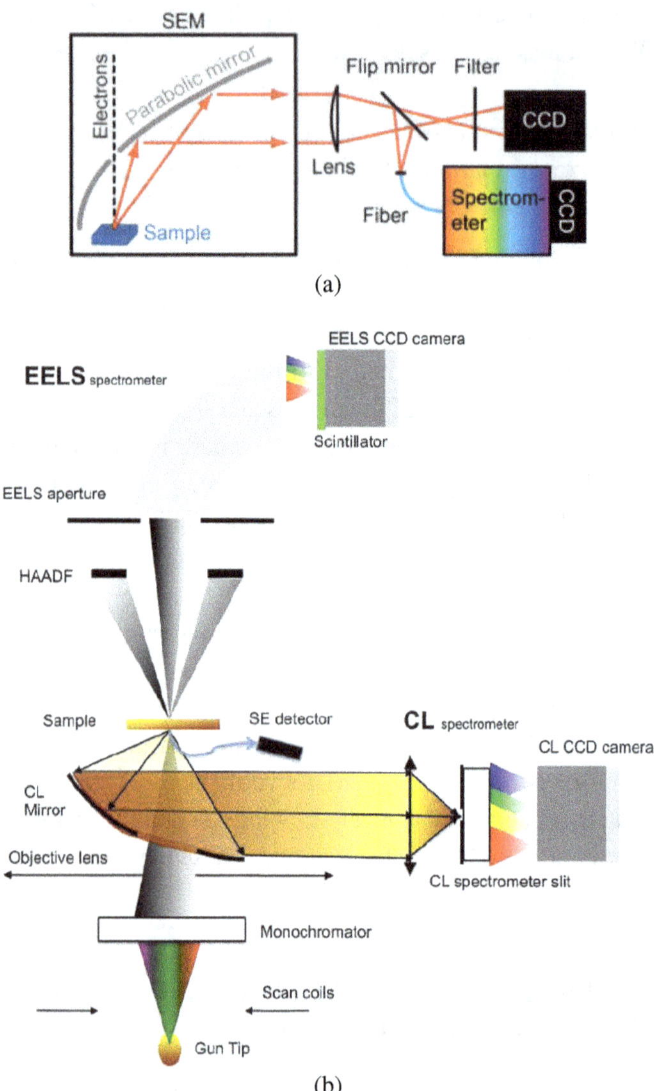

Figure 11.5　Cathodoluminescence systems. (a) In a scanning electron microscope: high-energy ($\approx$30 keV) electrons reach the sample through a hole in the parabolic mirror, which directs the light emitted from the sample to the detectors. Adapted from Ref. 59. (b) In a scanning transmission electron microscope, a similar system may be incorporated with EELS. Reproduced from Ref. 10 with permission of The Royal Chemistry Society.

of tens to hundreds of nanometres in size.[62–64] Past and present research has essentially concentrated on silver and gold nanoparticles since surface plasmons have comparatively low losses in these materials in the visible frequency range.[65] Gold nanoparticles are at the centre of the most recent breakthroughs, presumably due to the major progress achieved in the chemical synthesis of gold colloids,[66] as well as the fact that gold is chemically inert. In particular, electrically driven light emission from gold nanoparticles has recently led to the development of light nanosources for future device applications.[67,68]

11.3.1 *A Probe of Nanoscale Electronic Phenomena*

From the earliest studies, the emission of light resulting from the electrical excitation of metallic nanostructures with tunnel electrons has been envisaged as a spectroscopic tool for the exploration of their optical and electronic properties.[69–72] Why does this emitted light carry information, not only about the optical but also the electronic properties of the sample? As explained in Section 11.2.1, the phenomenon that leads to the electrical generation of light from a tunnel junction may actually be divided into a source term and an emission term.[7] The source term is related to the tunnel current and defines the amplitude of the source dipole in the tunnel junction, while the emission term describes the capacity of this source dipole to radiatively emit to the far field. Thus, the emitted light depends on the electronic properties of the sample and on the electron transport process via the source term. As a result, not only information about the optical properties of the sample may be obtained but also important aspects of its electronic structure that are otherwise inaccessible without resorting to electron spectroscopy.

For instance, STM-induced light emission has been used to investigate electron confinement in a gold nanostructure (see Figure 11.6), i.e. the spatial variations in the probability of finding the electrons on the gold surface. This experiment exemplifies the role of the local density of electronic states (*e*-LDOS) underneath the STM tip in the light excitation process. The STM topography image (Figure 11.6(a)), obtained at constant tunnel current, does

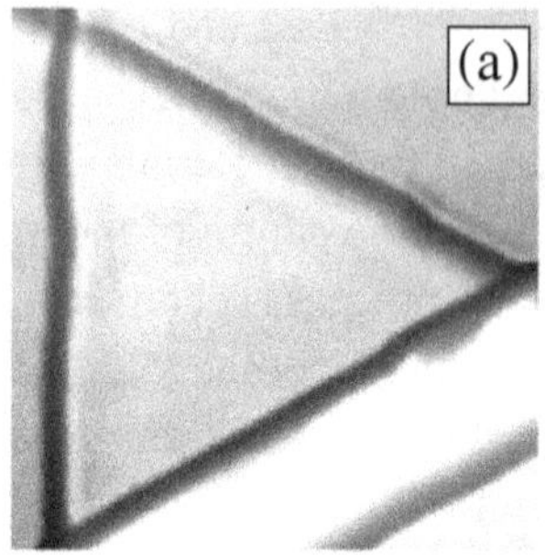

Figure 11.6   Electron confinement in a gold nanostructure revealed by the emission of light upon electrical excitation with tunnel electrons in an STM. (a) $11 \times 11\,\mathrm{nm}^2$ constant-current STM image of a triangular island on Au(111). (b) Energy-resolved ($1.65 < h\nu < 1.7\,\mathrm{eV}$) map of the detected photon intensity ('photon map') acquired simultaneously with the STM image in (a): the image contrast is due to the variations of the local density of electronic states (e-LDOS). Data acquired with a sample bias of 1.73 V and a tunnelling current of 300 nA. Adapted with permission from Ref. 72. Copyrighted by the American Physical Society.

not show the internal electronic structure of the gold nanostructure. Conversely, the 'photon map' (Figure 11.6(b)), which is obtained by detecting the emitted photons within a narrow energy range during the measurement of the STM topography, clearly reveals the standing waves of the e-LDOS inside the gold nanostructure. This may be understood by realising that a particular photonic wavelength corresponds to a particular electron state with a specific electron wavelength. These results confirm that the generation of light in STM can depend on the electronic properties of the nanostructure underneath the STM tip.

11.3.2   *Selective Electrical Excitation and Imaging of the Plasmonic Modes of Gold Nanoparticles*

The overwhelming advantage of the electrical excitation as opposed to the optical excitation of gold nanoparticles is the nanometric spatial extent, and resulting unmatched modal selectivity of the excitation. Due to the diffraction limit, a light beam may be focused at best to a spot whose width is about half the wavelength. This may be large compared to typical nanoparticle sizes. Thus, upon illumination with visible light, the incident excitation field may be almost uniform on the scale of the nanoparticle. Conversely, the electronic current tunnelling between a sharp electrode (or STM tip) and a metal nanoparticle is by nature extremely localised due to the fact that the tunnel current decays exponentially with distance between the electrode and particle. As well, a beam of high-energy electrons (in a scanning or transmission electron microscope) may be focused to a spot of truly nanometric size. This brings about a supplementary degree of control over the excitation of the plasmonic modes of a nanoparticle. Indeed, the rate of emitted light upon excitation with electrons is dependent on the local density of electromagnetic modes (EM-LDOS) at the excitation site, i.e. at the location of the tunnel junction or at the impact point of the electron beam,[55] as discussed in Section 11.2.2.2.

Each plasmonic mode of a metallic nanoparticle has a specific EM-LDOS spatial distribution, which mirrors the electric field induced around the particle at the resonant energy (wavelength) of the mode. As an example, the selective excitation of the plasmonic modes of a gold nanoprism with high-energy (100 keV) electrons is shown in Figure 11.7. The dipolar and higher-order multipolar modes of the particle have their highest EM-LDOS at the vertex and near the edge of the nanoprism, respectively; therefore, when the electron beam is focused on the vertex, it selectively excites

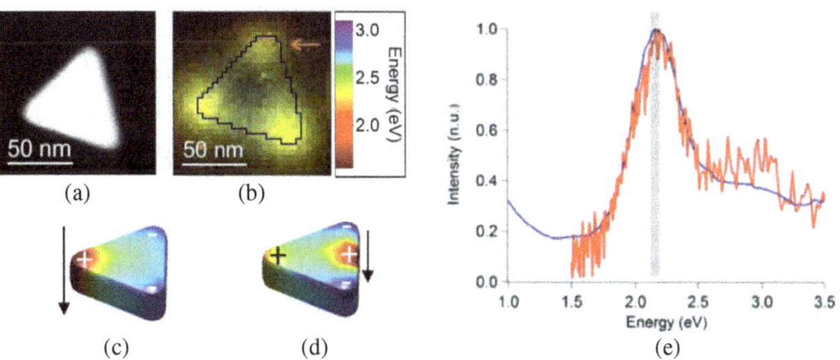

Figure 11.7   Selective mode excitation of a gold nanoparticle with high-energy electrons. (a) A high-angle annular dark field image of a 60 nm long, 30 nm thick gold nanoprism deposited on graphene (realised with a scanning transmission electron microscope). (b) Cathodoluminescence (CL) image of the same nanoparticle, in false colour, showing at what energy and from which locations the CL light is emitted. (c,d) Simulated surface charges induced by the electron beam (represented as arrows) focused on the vertex in (c) and on the side of the nanoprism in (d). In (c), we see that mainly the dipolar mode is excited as opposed to (d) where higher-order multipolar modes are present. (e) CL spectra recorded with the electron beam focused on a nanoprism vertex (the blue line is an electron energy loss measurement). Adapted with permission from Ref. 63. Copyright (2015) American Chemical Society.

the dipolar mode. The cathodoluminescence spectrum shown in panel e of Figure 11.7 has been recorded while focusing the electron beam on a nanoparticle vertex. The spectrum exhibits a narrow peak at about 2.2 eV, which indeed corresponds to the dipolar mode resonance as determined from calculations. The preferential excitation of this dipolar mode when the electron beam is at a nanoprism vertex is also confirmed by the cathodoluminescence image in panel b where bright spots are seen at all three vertices. This demonstrates that single modes or a combination of modes may be selectively excited by choosing the adequate excitation site, where the local density of these modes is high (and comparatively low for the other modes). Gold nanoparticles are particularly interesting platforms for this selective excitation because they often exhibit intense, spectrally narrow, plasmon resonances, mainly because of the low losses in gold. This, in turn, leads to well-defined mode patterns. In particular, areas of high local EM-density of states (hot spots) may be observed in crystalline gold nanoparticles with sharp geometrical features or, even more so, in assemblies of interacting gold nanoparticles.

Since electrical excitation is extremely local and the resulting light emission varies as a function of the EM-LDOS, scanning the excitation over a gold nanoparticle provides a 'map' of its plasmonic modes with nanoscale spatial resolution. This approach has been used on a variety of single gold nanoparticles (platelets,[63,73] truncated tetrahedra,[74] decahedra,[75,76] stars and shells[77]), mostly using high-energy electron beams. Features in the spectra and emission patterns may be assigned not only to the dipolar modes, but also to higher-order multipolar modes, tip modes or wedge modes using this technique. Similar measurements have been reported for groups of interacting gold nanoparticles designed to form optical antennas[78] or resonators,[79] as will be discussed further in Section 11.3.4 In such particle assemblies, the plasmonic modes of the individual particles coherently couple to yield collective modes, whose resonances are often associated with a strongly anisotropic optical response. Moreover, these collective modes may exhibit highly localised peaks in the EM-LDOS, especially in gaps between gold nanoparticles. As in the case of single gold nanoparticles, these hot spots provide activation sites for efficient and selective excitation of these collective modes with electrons. As discussed below (see Section 11.3.3) this has been used recently to generate nanosources of light with controlled directionality.

11.3.3 *A Single Gold Nanoparticle as an Electrically Driven Nanosource of Light*

The emission of light from the electrical excitation of a gold nanoparticle is essentially due to the radiative decay of the excited plasmonic modes. When electrons excite a combination of the particle modes, the coherent superposition of their radiation (at a given frequency) may yield an asymmetric emission pattern with, in some cases, high directionality. Since the combination of excited modes is determined by the excitation site, electrical excitation provides an efficient means to control the emission of light from gold nanoparticles, well beyond the possibilities of excitation with light. This has been reported for single gold nanoparticles on a transparent electrode (indium-tin-oxide coated glass) upon excitation with tunnel electrons from the tip of an STM.[62] In this case, the gold nanoparticles have a truncated bitetrahedral shape with an edge length of about 130 nm. As shown in Figure 11.8, changing the excitation site (by moving the STM tip) along a line running from one edge of the particle to the opposite vertex changes the maximum emission direction in a continuous fashion. This effect is ascribed to the fact that the respective contributions of the vertical and horizontal dipolar modes of the particle to the total emission pattern vary as a function of the excitation site. When the STM is located on the centre of the upper face of the particle, only the vertical dipolar mode is excited, yielding an azimuthally isotropic emission pattern. Off-centring the STM tip leads to anisotropic emission patterns, with the strongest directionality when the excitation is located on the vertices of the particle. This is because of the increasing contribution from its horizontal dipolar mode. Emission directionality may thus be reversed, from one side of the particle to the other. In this way, a single gold nanoparticle plays the role of an electrically driven nanosource of light with a controllable emission pattern. Similar observations have been reported for single gold nanodisks on silicon upon excitation with high-energy (30 keV) electrons in a scanning electron microscope.[80] The resulting emission of light is directional with a clear dependence on the location of the electron excitation on the gold nanoparticles. This has been explained in terms of a coherent superposition of electric dipole, magnetic dipole and electric quadrupole contributions.

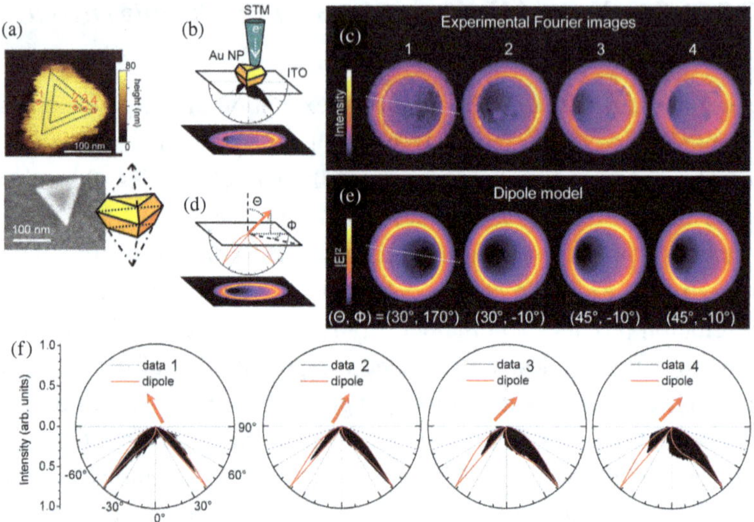

Figure 11.8   A single gold nanoparticle as an electrically driven nanosource of light with a controllable emission pattern. (a) STM topography image and scanning electron micrograph of a gold nanoparticle of truncated bitetrahedral shape, deposited on an indium-tin oxide (ITO) layer on glass. (b) Schematics of the local, electrical excitation of the particle with low-energy ($\sim$3 eV) electrons tunnelling from the tungsten tip of an STM and the resulting light emission pattern in the glass substrate. (c) Emission patterns recorded for the four different excitation sites indicated in (a). (d)–(f) Comparison with the theoretical emission patterns of a tilted electric dipole on an air–glass interface, calculated using a simple analytical model. Panel (f) shows intensity profiles taken from (c) and (e). Adapted with permission from Ref. 62. Copyright (2013) American Chemical Society.

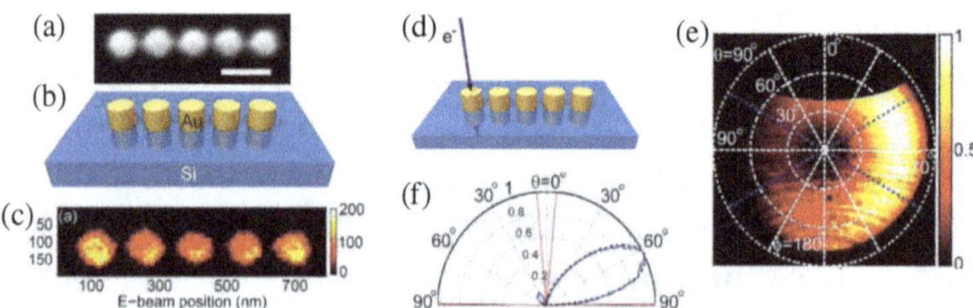

Figure 11.9   A gold nanoparticle array as an electrically driven optical antenna. (a) Electron micrograph of a plasmonic Yagi-Uda antenna of gold particles (diameter 98 nm, height 70 nm, on Si). The scale bar represents 200 nm. (b) Schematic of the antenna. (c) Spatially resolved excitation maps of the nano-antenna showing cathodoluminescence (CL) intensity for $\lambda_0 = 500$ nm as function of the high-energy (30 keV) electron beam position. (d) Schematic of the excitation geometry: the electron beam is focused on the first particle of the array. (e) CL emission intensity for $\lambda_0 = 500$ nm (integrated over a 40 nm bandwidth) as a function of angle. (f) Cross-cut of the image shown in (e). Highly directional emission of light is observed in the forward direction. Adapted with permission from Ref. 78. Copyright (2011) American Chemical Society.

11.3.4 *A Gold Nanoparticle Array as an Electrically Driven Optical Antenna or Resonator*

Besides single gold nanoparticles, specially designed assemblies of gold nanoparticles have also been considered for the fabrication of electrically driven nanosources of light with controllable emission properties. For instance, it has been reported that an array of five gold nanodots, aligned in a row, may behave as an optical antenna (often referred to as Yagi-Uda nanoantenna in reference to its radiofrequency counterpart). Such a nanoantenna can convert incident high-energy electrons into highly directional propagating light.[78] As shown in Figure 11.9, the directionality is obtained when focusing the electron beam onto one end of the row. The emission direction reverses when the excitation spot is moved to the opposite end of the row. This effect results from the coherent sum of the radiation from each induced dipole in the chain of gold nanodots. The electrical excitation of a nanodot at one end of the array and the interaction between the nanodots induce a plasmon oscillation in the array that has a phase profile which matches the emission of light in the forward direction. This leads to light beaming away from the excitation point. Because the electrical excitation is local, changing the excitation site changes the phase relation between the plasmon oscillations in the nanodots, and thus modifies the interference pattern of their radiation. In other words, the array of gold nanodots has collective modes whose radiative decay yields highly anisotropic emission patterns. The EM-LDOS of these collective modes is high at the end of the row and low at its centre, as visible in the excitation map of Figure 11.9(c).

Another application of the electrical excitation of gold nanoparticle arrays has been proposed where the coherent interaction of the plasmonic particles provides control over the emission spectrum.[79] In this application (see Figure 11.10), three gold nanorods are arranged to look like a *dolmen* (a prehistoric burial tomb). Two of the nanorods mimic the dolmen walls and a larger one plays the role of the dolmen roof. With this configuration, one can take advantage of a very interesting effect in ensembles of closely interacting gold nanoparticles. This effect is called plasmon hybridisation.[81] Using the analogy of molecular orbitals, two neighbouring metallic nanostructures may see their native plasmonic modes merging into higher-energy (anti-bonding) and lower-energy (bonding) modes. Here, plasmon

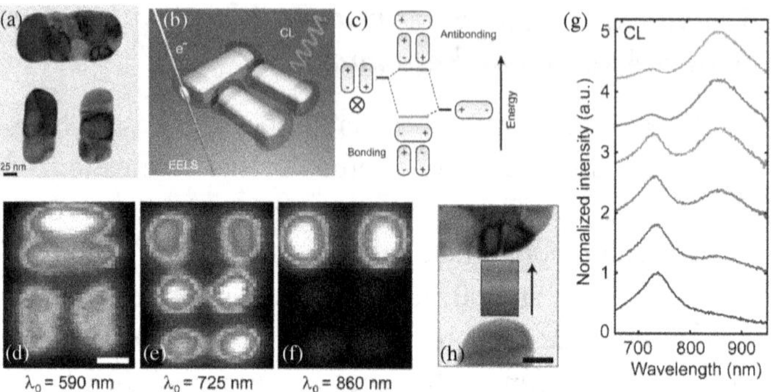

Figure 11.10 A set of three gold nanoparticles as an electrically driven optical nanoresonator. (a) Bright-field transmission electron micrograph and (b) schematic of a plasmonic dolmen structure composed of a resonantly coupled monomer and dimer excited using high-energy electrons. (c) Hybridisation scheme of the plasmonic modes for the dolmen. (d–f) 2D cathodoluminescence images for $\lambda_0 = 590$, 725 and 860 nm, i.e. at different mode energies respectively, integrated over a 30 nm bandwidth. Scale bar: 50 nm. (g) CL spectra recorded with the electron beam focused at different places along a vertical line segment within the gap between the right-hand part of the dimer and the monomer, as indicated in the zoom of image (a) shown in (h). The excitation position may be identified by comparing the colour of each spectrum in (g) and the colour within the framed rectangular area in (h). Adapted with permission from Ref. 79. Copyright (2015) American Chemical Society.

hybridisation involves the native modes of the dolmen walls and those of the dolmen roof. The resulting anti-bonding and bonding modes have very different mode density distributions over the dolmen (see Figures 11.10(d)–11.10(f)); therefore, they may be electrically excited with high selectivity by choosing the proper excitation site. Emission of light at different energies may thus be activated on request (see Figure 11.10(g)). In this way, an electrically driven nanosource of light with a controllable emission spectrum is built out of a set of three gold nanoparticles.

11.4 Towards On-chip Applications of Electron-to-photon Energy Conversion Using Gold Nanoparticles

As shown in the above-mentioned examples, gold nanoparticles and nanoparticle arrays are nanoscale electron-light converters with highly adaptable emission properties. An important aspect of these electron-light converters is their potential integration with nanoelectronics. These nanodevices may find attractive applications in the areas of on-chip information transfer and optical interconnects in response to the increasing demand for

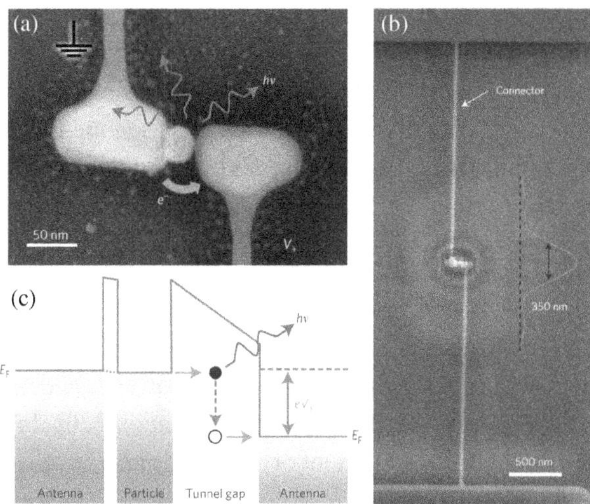

Figure 11.11   Integrated nanosource of light based on a single gold nanoparticle inserted in the gap of an electrically driven optical antenna. (a) Electron micrograph of the lateral tunnel structure. (b) Electron micrograph of the structure superimposed with the emission spot showing an Airy pattern with a full width at half maximum of 350 nm. (c) Tunnel gap with applied voltage. The electron tunnels inelastically and the released energy is used to excite a photon. Reprinted by permission from Macmillan Publishers Ltd: *Nature Photonics* Ref. 67, copyright (2015).

higher information capacity technologies. In this context, electrical excitation with low-energy electrons is preferable over its high-energy counterpart because it is compatible with state-of-the-art electronics and can be operated under ambient conditions. Tunnel junctions may be fabricated on-chip with stacked or electromigrated electrodes.

Most recently, an integrated nanosource of light based on a single gold nanoparticle within the gap between two electrodes has been developed.[67] In this device, the gold nanoparticle plays a dual role: it 'completes' the electrical nanocircuitry by forming a tunnel junction with the electrodes (see Figure 11.11) and it spectrally filters the broadband optical emission of the tunnel junction. Indeed, the localised surface plasmon of a gold nanoparticle may have a spectrally narrow resonance, whereas the emission of light from a tunnel junction is spectrally broad (see Section 11.2.1.1). This may be detrimental to specific applications, especially when wavelength selection multiplexing is demanded. Just as a gold nanoparticle illuminated with white light may scatter colored light, the gold nanoparticle in the tunnel junction 'colours' the emitted light because it most efficiently scatters the driving energy at the resonance frequency of its localised surface plasmon resonance. Remarkably, this effect is extremely sensitive to the position of the gold nanoparticle within the electrode gap. This means that the emission spectrum of this integrated light nanosource may be tuned by changing the position of the nanoparticle in the gap. Note also that the gold electrodes play the role of an optical antenna and enhance the emission of radiation.

11.5 Conclusion

The electrical generation of light from gold nanoparticles and nanostructures has a long history, which dates back to the 1960s and 1970s. In recent years, it has benefited from the development of new particle synthesis methods and advances in instrumentation, for both low-energy (STM) and high-energy (electron microscope) excitation. Our present understanding of how tunnel electrons and high-energy electrons interact with gold nanoparticles for producing light has been discussed in this chapter. However, fundamental questions still remain such as the role of tunnel current fluctuations in the emission of light from a biased tunnel junction.

In the case of both an STM and an electron microscope, the effective size of the electrical probe is of the order of 10 nm. This is much smaller than the probe size of a light source (e.g. a laser), which is of the order of several hundreds of nanometres in size. We have shown in this chapter that this $\sim$10 nm probe size is an overwhelming advantage for exploring the optical properties of individual gold nanoparticles or nanometre-scale arrangements of gold nanoparticles. As a result, the directionality and/or the spectral range of the emitted light may be controlled by site-specific electron excitation of the gold nanoparticles.

We emphasise that, in most cases, electrical generation of light from gold nanoparticles occurs through the excitation of surface plasmons, which, in turn, decay radiatively. Therefore, these nanosources of electrons are primarily nanosources of surface plasmons. Of particular interest is the case of a tunnel junction, which can be miniaturised at the nano-scale and coupled to a specially designed assembly of plasmonic gold nanoparticles. In this way, fully integrated plasmonic nanoscale devices may be envisaged.

References

1. J. N. Anker, *et al.* Biosensing with plasmonic nanosensors. *Nat. Mater.* **7**, 442–453 (2008).
2. B. M. Reinhard, M. Siu, H. Agarwal, A. P. Alivisatos, and J. Liphardt, Calibration of dynamic molecular rulers based on plasmon coupling between gold nanoparticles. *Nano Lett.* **5**, 2246–2252 (2005).
3. M. Notarianni, *et al.* Plasmonic effect of gold nanoparticles in organic solar cells. *Sol. Energy* **106**, 23–37 (2014).
4. G. L. Hallett-Tapley, *et al.* Supported gold nanoparticles as efficient catalysts in the solventless plasmon mediated oxidation of *sec*-phenethyl and benzyl alcohol. *J. Phys. Chem. C* **117**, 12279–12288 (2013).
5. R. A. Farrer, F. L. Butterfield, V. W. Chen, and J. T. Fourkas, Highly efficient multiphoton-absorption-induced luminescence from gold nanoparticles. *Nano Lett.* **5**, 1139–1142 (2005).
6. J. Zheng, C. Zhou, M. Yu, and J. Liu, Different sized luminescent gold nanoparticles. *Nanoscale* **4**, 4073–4083 (2012).
7. L. Douillard, and F. Charra, High-resolution mapping of plasmonic modes: photoemission and scanning tunnelling luminescence microscopies. *J. Phys. Appl. Phys.* **44**, 464002 (2011).
8. S. Kano, T. Tada, and Y. Majima, Nanoparticle characterization based on STM and STS. *Chem Soc Rev* **44**, 970–987 (2015).
9. F. J. García de Abajo, Optical excitations in electron microscopy. *Rev. Mod. Phys.* **82**, 209–275 (2010).
10. M. Kociak, and O. Stéphan, Mapping plasmons at the nanometer scale in an electron microscope. *Chem. Soc. Rev.* **43**, 3865 (2014).
11. E. Boer-Duchemin, T. Wang, Le E. Moal, and G. Dujardin, Electrically driven surface plasmon nanosources. *Proc. of SPIE* **9361**, 93610R–93610R–11 (2015).
12. M. Kuttge, *et al.* Local density of states, spectrum, and far-field interference of surface plasmon polaritons probed by cathodoluminescence. *Phys. Rev. B* **79**, 113405 (2009).
13. R. A. Ferrell, Predicted radiation of plasma oscillations in metal films. *Phys. Rev.* **111**, 1214–1222 (1958).
14. W. Steinmann, Experimental verification of radiation of plasma oscillations in thin silver films. *Phys. Rev. Lett.* **5**, 470–472 (1960).
15. R. W. Brown, P. Wessel, and E. P. Trounson, Plasmon reradiation from silver films. *Phys. Rev. Lett.* **5**, 472–473 (1960).
16. J. Lambe, and S. L. McCarthy, Light emission from inelastic electron tunneling. *Phys. Rev. Lett.* **37**, 923 (1976).
17. J. K. Gimzewski, J. K. Sass, R. R. Schlitter, and J. Schott, Enhanced photon-emission in scanning tunnelling microscopy. *Europhys. Lett.* **8**, 435–440 (1989).
18. J. K. Gimzewski, B. Reihl, J. H. Coombs, and R. R. Schlittler, Photon emission with the scanning tunneling microscope. *Z. Für Phys. B Condens. Matter* **72**, 497–501 (1988).
19. G. Binnig, H. Rohrer, C. Gerber, and E. Weibel, Surface studies by scanning tunneling microscopy. *Phys. Rev. Lett.* **49**, 57–61 (1982).
20. The Morning Record. Meriden, Connecticut; Feb. 16 1976, p. A11.

21. Two strips make light work. *New Scientist* 216 (1976).

22. G. Dujardin, E. Boer-Duchemin, E. Le Moal, A. J. Mayne, and Riedel, D. DIET at the nanoscale. *Surf. Sci.* **643**, 13–17 (2016).

23. P. Johansson, Light emission from a scanning tunneling microscope: Fully retarded calculation. *Phys. Rev. B* **58**, 10823 (1998).

24. L. C. Davis, Theory of surface-plasmon excitation in metal–insulator–metal tunnel junctions. *Phys. Rev. B* **16**, 2482–2490 (1977).

25. J. R. Kirtley, T. N. Theis, J. C. Tsang, and D. J. DiMaria, Hot-electron picture of light emission from tunnel junctions. *Phys. Rev. B* **27**, 4601–4611 (1983).

26. R. Berndt, J. K. Gimzewski, and P. Johansson, Inelastic tunneling excitation of tip-induced plasmon modes on noble-metal surfaces. *Phys. Rev. Lett.* **67**, 3796 (1991).

27. B. N. J. Persson, and A. Baratoff, Theory of photon emission in electron tunneling to metallic particles. *Phys. Rev. Lett.* **68**, 3224 (1992).

28. Y. Uehara, Y. Kimura, S. Ushioda, and K. Takeuchi, Theory of visible-light emission from scanning tunneling microscope. *Jpn. J. Appl. Phys. Part 1-Regul. Pap. Short Notes Rev. Pap.* **31**, 2465–2469 (1992).

29. D. Hone, Mühlschlegel, B. and D. J. Scalapino, Theory of light emission from small particle tunnel junctions. *Appl. Phys. Lett.* **33**, 203 (1978).

30. R. W. Rendell, and D. J. Scalapino, Surface plasmons confined by microstructures on tunnel junctions. *Phys. Rev. B* **24**, 3276 (1981).

31. P. Johansson, R. Monreal, and P. Apell, Theory for light emission from a scanning tunneling microscope. *Phys. Rev. B* **42**, 9210 (1990).

32. B. Laks, and D. L. Mills, Photon emission from slightly roughened tunnel junctions. *Phys. Rev. B* **20**, 4962–4980 (1979).

33. N. L. Schneider, G. Schull, and R. Berndt, Optical probe of quantum shot-noise reduction at a single-atom contact. *Phys. Rev. Lett.* **105**, 026601 (2010).

34. A. Kumar, L. Saminadayar, D. C. Glattli, Y. Jin, and B. Etienne, Experimental test of the quantum shot noise reduction theory. *Phys. Rev. Lett.* **76**, 2778–2781 (1996).

35. J.-T. Lü, R. B. Christensen, and M. Brandbyge, Light emission and finite-frequency shot noise in molecular junctions: From tunneling to contact. *Phys. Rev. B* **88**, 045413 (2013).

36. K. Kaasbjerg, and A. Nitzan, Theory of light emission from quantum noise in plasmonic contacts: Above-threshold emission from higher-order electron-plasmon scattering. *Phys. Rev. Lett.* **114**, 126803 (2015).

37. R. W. Rendell, D. J. Scalapino, and B. Mühlschlegel, Role of local plasmon modes in light emission from small-particle tunnel junctions. *Phys. Rev. Lett.* **41**, 1746–1750 (1978).

38. G. Schull, E. Boer-Duchemin, G. Comtet, and G. Dujardin, Emission de la lumière sous la pointe d'un microscope à effet tunnel. *Reflets Phys.* **38**, 4 (2014).

39. T. Wang, E. Boer-Duchemin, Y. Zhang, G. Comtet, and G. Dujardin, Excitation of propagating surface plasmons with a scanning tunnelling microscope. *Nanotechnology* **22**, 175201 (2011).

40. F. Silly, A. O. Gusev, A. Taleb, F. Charra, and M.-P. Pileni, Coupled plasmon modes in an ordered hexagonal monolayer of metal nanoparticles: A direct observation. *Phys. Rev. Lett.* **84**, 5840 (2000).

41. T. Lutz, *et al.* Molecular orbital gates for plasmon excitation. *Nano Lett.* **13**, 2846–2850 (2013).

42. X. H. Qiu, Vibrationally resolved fluorescence excited with submolecular precision. *Science* **299**, 542–546 (2003).

43. Z. C. Dong, *et al.* Generation of molecular hot electroluminescence by resonant nanocavity plasmons. *Nat. Photonics* **4**, 50–54 (2009).

44. T. Frederiksen, G. Foti, F. Scheurer, V. Speisser, and G. Schull, Chemical control of electrical contact to sp2 carbon atoms. *Nat. Commun.* **5**, 3659 (2014).

45. N. Nilius, N. Ernst, and H. J. Freund, Photon emission spectroscopy of individual oxide-supported silver clusters in a scanning tunneling microscope. *Phys. Rev. Lett.* **84**, 3994–3997 (2000).

46. P. Dawson, and M. G. Boyle, Light emission from scanning tunnelling microscope on polycrystalline Au films — what is happening at the single-grain level? *J. Opt. Pure Appl. Opt.* **8**, S219–S226 (2006).

47. Y. Uehara, T. Fujita, and S. Ushioda, Scanning tunneling microscope light emission spectra of Au (110)–(2 × 1) with atomic spatial resolution. *Phys. Rev. Lett.* **83**, 2445 (1999).

48. V. Sivel, R. Coratger, F. Ajustron, and J. Beauvillain, Photon emission stimulated by scanning tunneling microscopy in air. *Phys. Rev. B* **45**, 8634–8637 (1992).

49. P. Bharadwaj, A. Bouhelier, and L. Novotny, Electrical excitation of surface plasmons. *Phys. Rev. Lett.* **106**, 226802 (2011).

50. B. J. M. Brenny, T. Coenen, and A. Polman, Quantifying coherent and incoherent cathodoluminescence in semiconductors and metals. *J. Appl. Phys.* **115**, 244307 (2014).

51. V. L. Ginsburg, and I. M. Frank, Radiation of a uniformly moving electron due to its transition from one medium to another. *Sh. Eksp. Teor. Fiz.* **16**, 1528 (1946).

52. A. P. Potylitsyn, Transition radiation and diffraction radiation. Similarities and differences. *Nucl. Instrum. Methods Phys. Res. Sect. B Beam Interact. Mater. At.* **145**, 169–179 (1998).

53. M. Kociak, *et al.* Seeing and measuring in colours: Electron microscopy and spectroscopies applied to nano-optics. *Comptes Rendus Phys.* **15**, 158–175 (2014).

54. K. H. Drexhage, in *Progress in Optics*, E. Wolf (ed.), Elsevier, Vol. 12, pp. 163–232 (1974).

55. A. Losquin, and M. Kociak, Link between cathodoluminescence and electron Energy loss spectroscopy and the radiative and full electromagnetic local density of states. *ACS Photonics* **2**, 1619–1627 (2015).

56. F. J. García de Abajo, and M. Kociak, Probing the photonic local density of states with electron energy loss spectroscopy. *Phys. Rev. Lett.* **100**, 106804 (2008).

57. U. Hohenester, H. Ditlbacher, and J. R. Krenn, Electron-energy-loss spectra of plasmonic nanoparticles. *Phys. Rev. Lett.* **103**, 106801 (2009).

58. www.nion.com | Nion Products. (2015), at http://www.nion.com/products.html.

59. T. Coenen, B. J. M. Brenny, E. J. Vesseur, and A. Polman, Cathodoluminescence microscopy: Optical imaging and spectroscopy with deep-subwavelength resolution. *MRS Bull.* **40**, 359–365 (2015).

60. C. Chen, C. A. Bobisch, and W. Ho, Visualization of Fermi's golden rule through imaging of light emission from atomic silver chains. *Science* **325**, 981–985 (2009).

61. P. Myrach, N. Nilius, and Freund, H.-J. Photon mapping of individual Ag particles on MgO/Mo(001). *Phys. Rev. B* **83**, 035416 (2011).

62. E. Le Moal, *et al.* An electrically excited nanoscale light source with active angular control of the emitted light. *Nano Lett.* **13**, 4198–4205 (2013).

63. A. Losquin, *et al.* Unveiling nanometer scale extinction and scattering phenomena through combined electron energy loss spectroscopy and cathodoluminescence measurements. *Nano Lett.* **15**, 1229–1237 (2015).

64. E. Le Moal, *et al.* Engineering the emission of light from a scanning tunneling microscope using the plasmonic modes of a nanoparticle. *Phys. Rev. B* **93**, 035418 (2016).

65. P. R. West, *et al.* Searching for better plasmonic materials. *Laser Photonics Rev.* **4**, 795–808 (2010).

66. M. R. Gonçalves, Plasmonic nanoparticles: fabrication, simulation and experiments. *J. Phys. Appl. Phys.* **47**, 213001 (2014).

67. J. Kern, *et al.* Electrically driven optical antennas. *Nat. Photonics* **9**, 582–586 (2015).

68. Y. Vardi, E. Cohen-Hoshen, G. Shalem, and I. Bar-Joseph, Fano resonance in an electrically driven plasmonic device. *Nano Lett.* **16**, 748–752 (2016).

69. A. Downes, M. E. Taylor, and M. E. Welland, Two-sphere model of photon emission from the scanning tunneling microscope. *Phys. Rev. B* **57**, 6706 (1998).

70. R. Nishitani, T. Umeno, K. Suga, A. Kasuya, and Y. Nishina, Measurements of photon intensity map for metal particles by scanning tunneling microscopy. *Mater. Sci. Eng. A* **217–218**, 99–102 (1996).

71. Y. Uehara, T. Iida, K. J. Ito, M. Iwami, and S. Ushioda, Optical observation of single-electron charging effect in metallic particles. *Phys. Rev. B* **65**, 155408 (2002).

72. G. Schull, M. Becker, and R. Berndt, Imaging confined electrons with plasmonic light. *Phys. Rev. Lett.* **101**, 136801 (2008).

73. L. Gu, *et al.* Resonant wedge-plasmon modes in single-crystalline gold nanoplatelets. *Phys. Rev. B* **83**, 195433 (2011).

74. P. Das, T. K. Chini, and J. Pond, Probing higher order surface plasmon modes on individual truncated tetrahedral gold nanoparticle using cathodoluminescence imaging and spectroscopy combined with FDTD simulations. *J. Phys. Chem. C* **116**, 15610–15619 (2012).

75. P. Das, and T. K. Chini, Spectroscopy and imaging of plasmonic modes over a single decahedron gold nanoparticle: A combined experimental and numerical study. *J. Phys. Chem. C* **116**, 25969–25976 (2012).

76. V. Myroshnychenko, *et al.* Plasmon spectroscopy and imaging of individual gold nanodecahedra: a combined optical microscopy, cathodoluminescence, and electron energy-loss spectroscopy study. *Nano Lett.* **12**, 4172–4180 (2012).

77. A. C. Atre, *et al.* Nanoscale optical tomography with cathodoluminescence spectroscopy. *Nat. Nanotechnol.* **10**, 429–436 (2015).

78. T. Coenen, E. J. R. Vesseur, A. Polman, and A. F. Koenderink, Directional emission from plasmonic yagi–uda antennas probed by angle-resolved cathodoluminescence spectroscopy. *Nano Lett.* **11**, 3779–3784 (2011).
79. T. Coenen, *et al.* Nanoscale spatial coherent control over the modal excitation of a coupled plasmonic resonator system. *Nano Lett.* **15**, 7666–7670 (2015).
80. T. Coenen, Bernal F. Arango, Femius A. Koenderink, and A. Polman, Directional emission from a single plasmonic scatterer. *Nat. Commun.* **5**, 3250 (2014).
81. E. Prodan, C. Radloff, N. J. Halas, and P. Nordlander, A Hybridization model for the plasmon response of complex nanostructures. *Science* **302**, 419–422 (2003).

Chapter 12
Surface Structures of Gold and Gold-based Bimetallic Nanoparticles

Shamil Shaikhutdinov

Department of Chemical Physics, Fritz Haber Institute of the Max Planck Society, Berlin, Germany

12.1 Introduction

Since Roman times, it has been recognised that gold dispersed in glasses gives rise to fascinating optical phenomena, which were (much later, of course) rationalised on the basis of physical properties of gold in highly dispersed state (see Chapter 2). Regarding its chemical properties, gold has long been considered as an inert material since bulk gold does not react easily with many molecules typically present in the ambient atmosphere, the property that renders gold the most noble metal. Chemistry of gold has only recently received much attention after superior catalytic proper-ties observed for highly dispersed gold in a number of industrially impor-tant reactions such as low-temperature CO oxidation, selective oxidation of propene, water gas shift reaction, selective hydrogenation of acetylene, to name a few[1–5] (see also Chapter 9). Obviously, the performance of gold in heterogeneous catalysts is primarily determined by geometrical and elec-tronic structures of gold species, which are surface in nature. The surface structures become very complex in the case of nanosized gold particles, which expose facets of different orientations as well as substantial amounts of undercoordinated atoms. The issue becomes even more complicated in the presence of a support (typically, an oxide with the high surface area), which may not necessarily behave as an inert 'spectator', but rather as a key player, which controls the reactivity of gold, for example, through stabilisation of particular structures of gold species and gold/support interface.

The atomic-level understanding of structure, composition and electronic state of surfaces can only be obtained by surface-sensitive techniques, which are commonly employed in surface science. The recent surge in study of gold catalysts[1–5] has been accompanied by a corresponding swell of interest in the surface science of gold. In spite of the enhanced activities worldwide, the surface chemistry of the small gold particles is still far from being well understood at the atomic scale. Partially, this is due to the fact that the real catalytic systems, consisting of Au nanoparticles deposited on oxide supports with poorly defined surface structures, are too complex for making precise and unambiguous structural characterisation of such systems, which may, in turn, be linked to their functional properties. This renders careful model studies a necessity. Within the so-called surface science approach,[6–16] the structural complexity of a real catalyst is reduced to a well-defined model. The latter, however, should retain the main features of the real system such as nature of a support, high dispersion of a supported metal, etc.

Numerous surface science studies are currently being performed in order to understand the fundamental aspects of catalysis by gold, in particular addressing the key question: why gold behaves as a catalyst when its dimensions are reduced to the nanometre scale (see, for instance, Refs. 16–23).

This chapter will focus on surface structures of gold nanoparticles deposited on well-defined planar oxide supports. A number of model systems will be discussed here in order to illustrate the approach to surface chemistry of gold and also to demonstrate the complexity of the gold–support interaction. After description of model systems in Section 12.2, we will briefly discuss surface structures of gold single crystals (Section 12.3). Basic structural motifs of Au nanoparticles are discussed in Section 12.4. The preparation of thin film supports will be addressed in Section 12.5. Then in Section 12.6, we show basic principles of gold deposition onto planar supports. Section 12.7 presents case studies of size, support end environmental effects observed on Au/oxide model systems. Then we address supported gold species in the form of single-layer islands, which represent gold in a two-dimensional state causing certain differences as compared to a three-dimensional counterpart (Section 12.7). Finally, the effects of gold alloying with other metals will be discussed in Section 12.8.

Certainly, this chapter is not aimed at providing a comprehensive review of the current studies on surface science of gold. It rather demonstrates opportunities and limitations of the approach towards a deeper understanding of surface chemistry of gold in highly dispersed state.

12.2 Background

In planar model systems, metal particles are deposited onto an electrically conducting oxide single crystal or a thin metal-oxide film grown on a metal single crystal substrate as schematically shown in Figure 12.1. The system allows the facile application of a large variety of surface-sensitive techniques for precise system characterisation such as the following:

- Electron spectroscopy (e.g. ultraviolet and X-ray photoelectron spectroscopy (UPS/XPS), in particular using the synchrotron light source, and Auger electron spectroscopy (AES)). The methods allow to determine the elemental surface composition, the oxidation states of the constituting elements and the valence band structure.
- Ion spectroscopy, e.g. low-energy ion surface scattering (LEISS) and secondary ion mass spectrometry (SIMS). The methods provide information on the composition of the topmost surface layers in combination with an elemental depth-profile analysis upon surface sputtering.

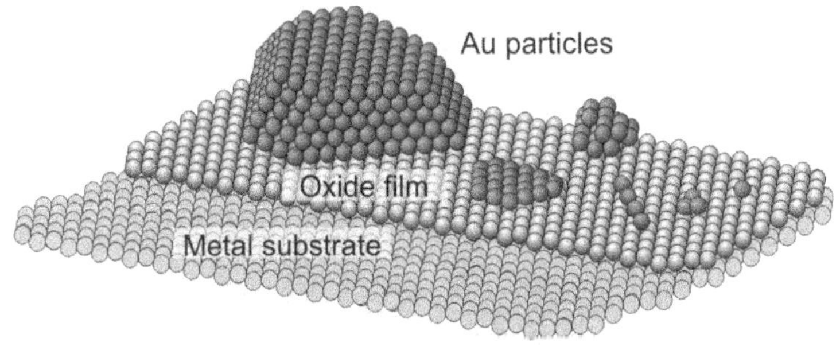

Figure 12.1   Model planar systems for studying surface structures and reactivity of gold nanoparticles supported on well-defined thin oxide films.

- Vibrational spectroscopy, e.g. high-resolution electron energy loss spectroscopy (HREELS) and infrared reflection-absorption spectroscopy (IRAS). The methods monitor the lattice vibrations (phonons) and also the vibrations of adsorbed species.
- Low-energy electron diffraction (LEED), surface X-ray diffraction (SXRD) and grazing incident small-angle X-ray scattering (GISAXS). The methods are usually employed for ordered structures and interfaces.
- Scanning probe microscopy, e.g. scanning tunnelling microscopy (STM) and atomic force microscopy (AFM). These seem to be the only methods that allow to determine morphology (topography) of the surface nanostructures with atomic resolution.

In addition to the above techniques, adsorption and reaction of molecules on these surfaces can readily be studied by temperature programmed desorption (TPD) as the planar systems do not suffer from diffusion limitations existing for powdered and microporous materials. Kinetics and elementary steps of chemical reactions at surfaces can be monitored by molecular beam technique, particularly in combination with IRAS (see, for example, Ref. 24).

Although most of the surface sensitive techniques can only be applied in high and ultra-high vacuum (UHV), while catalytic processes occur at ambient pressures, *in situ* methods have been developed to bridge the so-called pressure gap. Sum frequency generation (SFG), polarisation-modulation IRAS (PMIRAS), ambient-pressure XPS (APXPS) and high-pressure STM techniques allow one to carry out studies at more realistic pressure conditions.[25–30] Furthermore, *in situ* structural characterisation can be performed simultaneously with reactivity measurements using gas chromatography, in essence, in the same way as it is routinely used in catalytic experiments.

Basically, a research strategy of model 'surface science' studies includes the following 'elementary' steps:

— Preparation of a planar support and its characterisation with respect to morphology, surface composition and structural defects;
— Deposition of metal particles onto the support and monitoring the nucleation and growth process;

— Structural characterisation of the supported metal particles, in particular their size, shape, electronic state and thermal stability;

— Adsorption studies of molecules participating in a target catalytic reaction as well as of the 'probe' molecules, adsorption of which is sensitive to the surface structure;

— Reactivity studies under nearly realistic conditions to establish reaction kinetics as a function of the gas composition and temperature.

Structure–reactivity relationships, such as size, support and environmental effects, observed from such experiments provide important, if not crucial information for understanding surface chemistry of metal nanoparticles and elucidating reaction mechanisms.

12.3 Surface Structures of Gold Single Crystals

Although the catalytic nature of gold is most likely linked to size and structure issues, it is instructive here to address first the atomic structures of gold single crystal surfaces. In addition, the extended surfaces are well suited for studying surface reconstructions, if any, which may occur under reaction conditions.

Gold shares a face centred cubic (*fcc*) lattice. The low Miller indexed surfaces of gold, such as (111), (110) and (100), shown in Figure 12.2, are all known to undergo surface reconstruction in UHV. Perhaps, the most famous and still intriguing is the reconstruction of the Au(111) surface, unique for pure *fcc* metals, which is referred to as a 'herringbone' reconstruction after first atomically resolved STM images[31] (see Figure 12.3(a)). The reconstruction can be described by a complex stacking-fault-domain model of the topmost layer having higher surface density of atoms than in the bulk layers[31–33] and rationalised in terms of surface states that arise due to the interaction of *sp* and *d* states, a consequence of its relativistic nature[34] (see Chapter 2).

The Au(110) surface reconstructs into the (1×2) surface, which is formed by 'missing rows' along[1 10] direction as shown in inset in Figure 12.3(b).[35] This reconstruction gives rise to three different types of surface atoms: on top of row, side of row and trench atoms. The sides of

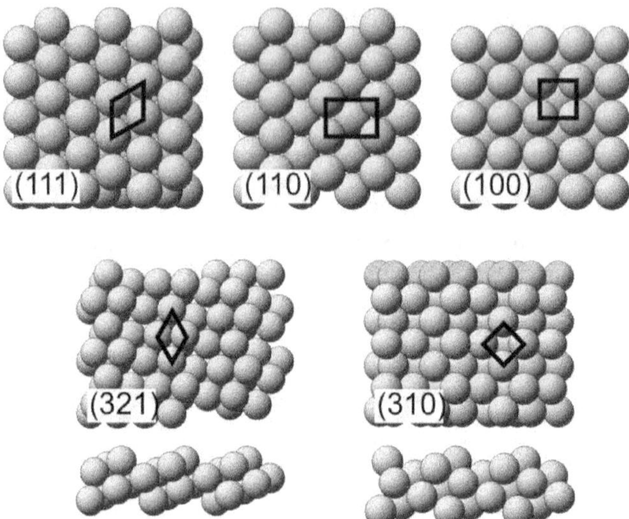

Figure 12.2   (Top panel) Top views of unreconstructed (111), (110) and (100) surfaces of fcc metals. The unit cells are indicated. (Bottom panel) Top and cross views of the stepped Au(321) and Au(310) surfaces. The (111) and (100) unit cells of the micro-facets are indicated.

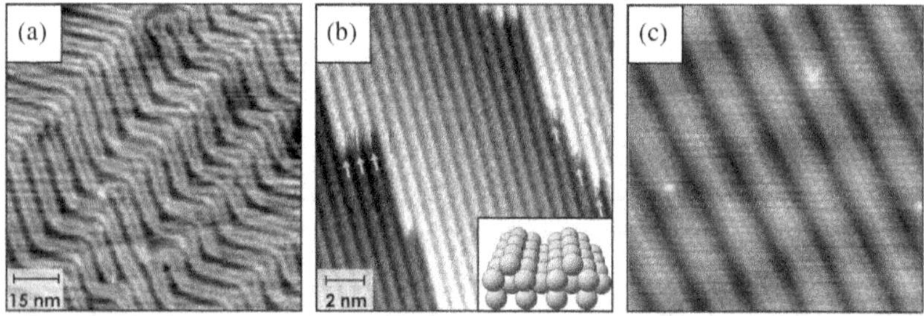

Figure 12.3   STM images of (a) the 'herringbone' reconstruction of Au(111); (b) the (1×2) 'missing row' reconstruction of Au(110) (adapted from Ref. 35. Copyright (1992) by the American Physical Society); (c) the 'hexagonal' reconstruction of Au(100) (reproduced from Ref. 36).

row atoms are arranged in the same manner as the (111) surface and often referred to as (111) microfacets.

The reconstruction of the Au(100) surface apparently depends on preparation and remains controversial. It has originally been assigned to (1×5) 'adding row' reconstruction.[37] Later, more complex surface structures have been observed (see Figure 12.3(c)), such as (5×20),[38] a (5×20) with rotation,[39] a hexagonal (28×5) R0.6°,[40] to name a few.

It has been proposed that the reconstructed surfaces may experience different combinations of these structures depending upon step density and surface temperature.[41] The reconstruction is generally believed to be limited to the first layer, thus indicating that the more compact surface arrangement is favoured to a degree where the energy cost due to a lack of commensuration with the layer underneath can be overcome.[42]

These reconstructions of the gold surfaces are quite stable under vacuum conditions. For example, the herringbone reconstruction of Au(111) was seen at $\sim$850 K,[43] but can be lifted upon adsorption of certain gases such as CO. SXRD studies[44] revealed lattice expansion of the Au(111) surface exposed to CO at 300 K at pressures between 0.1 and 530 mbar, although the herringbone reconstruction was preserved. Even more extensive surface transformations of Au(111) were observed upon exposure to 110 mbar CO at 600 K. Similar behaviour was found for Au(110) at CO pressures above 0.1 mbar.[45] STM results showed significant surface roughening and a lifting of terrace anisotropy. Since the morphological changes observed for both Au(111) and Au(110) surfaces remained after evacuation, CO dissociation might have occurred at high pressures accompanied by carbon deposition.

A limited number of studies were reported on gold single crystal surfaces with higher Miller indexes, e.g. Au (991), Au(430), Au(221), Au(332), Au(310) and Au(321).[46–48] The latter two surfaces are shown in Figure 12.2. The (321) surface forms (111) microfacets, whereas the (310) surface exhibits (100)-like facets, with the step atoms on both surfaces having a coordination number 6. Comparison of adsorption properties of CO (which is very often used for the characterisation of the gold-based catalysts) on the two surfaces revealed that CO binding to gold is not only dependent of the coordination number but the exact geometrical surface structure.[48] The studies on the stepped Au surfaces are helpful for understanding the reactivity of the low-coordinated gold atoms, which become dominating the surfaces of gold nanoparticles.

12.4 Morphology of Gold Nanoparticles: General Considerations

In principle, small metal particles expose different facets resulting from a truncation of single crystals. In principle, the surface energies of (*hkl*)

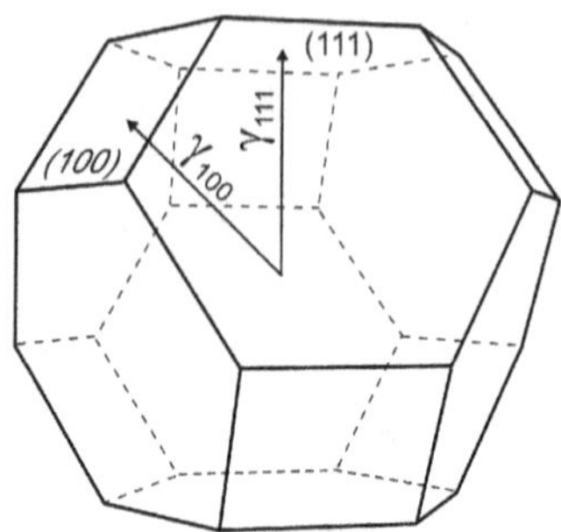

Figure 12.4   The Wulff construction of (fcc) metal polyhedral particles comprising (111) and (100) faces.

planes, described as the Gibbs free energy per unit area, determine the morphology of particles. Having obtained the free energies of the single crystal surfaces (γ_{hkl}), one can predict an equilibrium crystal shape of an unsupported particle (in the same environment as single crystal surfaces) for a given volume using the so-called Wulff construction.[49–51] Representing the surface free energy as a function of direction by vectors of length γ_{hkl}, the minimum energy shape at a constant volume is the inner envelope of the normals to this surface (see Figure 12.4).

Beyond the truncated octahedron, shown in Figure 12.4, and their twinned variants, the possible structures include also decahedral and icosahedral motifs,[51–53] as schematically shown in Figure 12.5. The latter structures have a relatively high strain energy.[54]

Obviously, the above described thermodynamic approach is valid only for a particle that reached an equilibrium shape. Note also that this construction neglects any contribution of 'edge' and 'corner' sites and can, therefore, be applied only to relatively large particles, where the facet dimensions are much larger than the surface unit cells. Basically, metal particles at moderate sizes (several tens of nanometres) are generally consistent with the Wulff construction.[51] The situation becomes more complicated upon reducing the size down to several nanometres. Surface stress, size-dependent lattice parameter changes, twin boundaries, and ambient conditions, all these factors may influence the morphology of metal particles.[51] For supported particles, the role of substrate may be included in the construction by replacing the free surface energy of the contact plane with an effective surface energy (γ^*), which is the difference between the interface energy and the

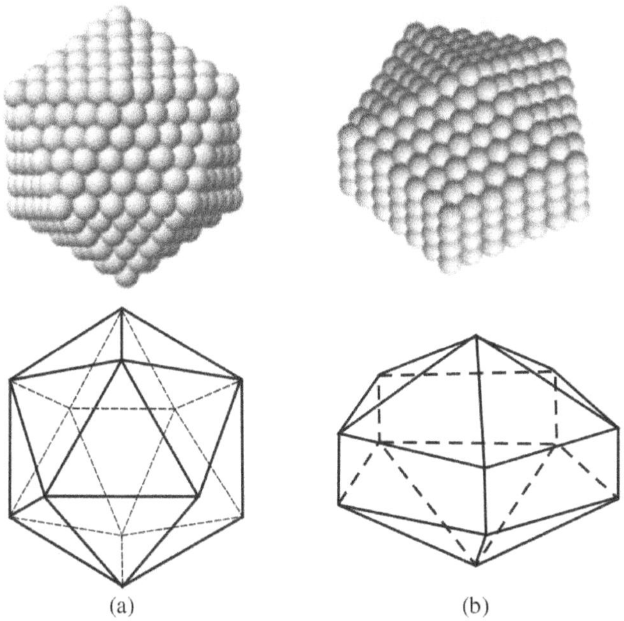

(a)   (b)

Figure 12.5   Icosahedral (a) and decahedral (b) structures of gold nanoparticles.

surface energy of the substrate[50]:

$$\gamma^{*} = \gamma_{\text{interface}} - \gamma_{\text{support}}. \tag{12.1}$$

This gives the equilibrium shape truncated at the interface (see Figure 12.6)[11]:

$$\Delta h / h_i = W_{\text{adh}} / \gamma_{\text{metal}(i)}, \tag{12.2}$$

where W_{adh} is the adhesion energy, which is defined as the energy per unit area to pull the system apart to its constituents, and relates to the surface energies through the equation:

$$W_{\text{adh}} = \gamma_{\text{metal}(i)} + \gamma_{\text{support}} - \gamma_{\text{interface}} = \gamma_{\text{metal}(i)} - \gamma^{*}. \tag{12.3}$$

It is evident from Equations (12.2) and (12.3) that, for negative values of γ^{*}, the particle height becomes smaller than the corresponding radii of the unsupported particle. Clearly, the higher the adhesion energy, the more flattened the particle is. Therefore, flattening of metal particles may be considered as the physical manifestation of the strong metal–support interaction.

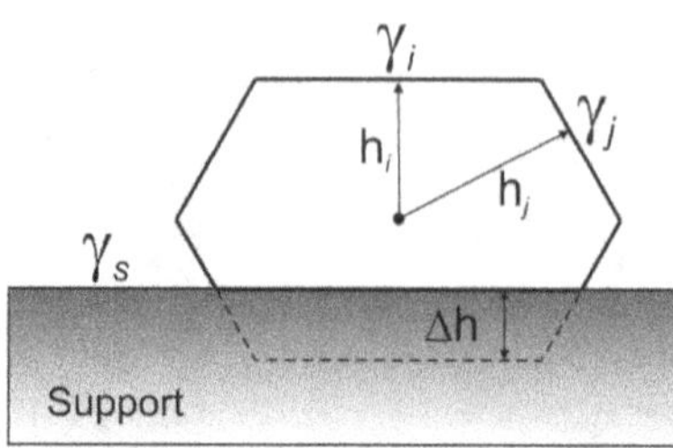

Figure 12.6   Schematic representation of the Wulff construction for supported particles (see text).

High-resolution transmission electron microscopy (HRTEM) studies revealed that the structure of very small Au nanoparticles ($d \sim 1$ nm) fluctuated significantly even at room temperature (the phenomenon was later referred to as 'quasi-melting'[51,55,56]), as the energy difference between conformations calculated for gas-phase clusters is comparable with kT.[55] Although several effects may contribute to this phenomenon including an electron beam induced overheating, it appears that the substrate plays a critical role in the stabilisation of the particle in particular states. Indeed, the lack of such structural transitions on Au deposited onto vacuum-cleaved MgO microcubes was explained by the high interfacial energy, which must be overcome when the particles are epitaxially oriented to the support.[57] Also *in situ* HRTEM monitoring[58] of a gold single particle, growing on a MgO(100) step while continuous gold deposition in vacuum, revealed that the particle rearranged continuously, apparently to maintain the lowest energy structure. Shape modulation from tetragonal pyramid to truncated tetragonal pyramid or the inverse was observed repeatedly. Examining the surface area ratios of Au(100) and Au(111) facets during the particle growth and repetitions in the truncation pattern were seen to emerge as the cluster grew until reaching a size of over 1400 atoms. It should be mentioned, however, that the kinetics of particle growth may also affect the resulting morphology of deposited Au nanoparticles (see below).

Atomic-scale imaging of metal particles has been substantially improved by using aberration-corrected scanning transmission electron microscopy (ac-STEM).[59] In particular, in the case of Au/MgO system discussed above, STEM studies revealed semi-coherent interfacial epitaxy and coordinate-dependent surface contraction for the fcc (001) oriented Au nanoparticles (2–3 nm in diameter), suggesting that their interaction with the substrate is relatively weak. In addition, a significant change in interfacial separation distance from 2.47 ± 0.12 Å for the fcc (001) oriented Au nanoparticles to 3.07 ± 0.11 Å for the fcc (111) oriented Au nanoparticles has been observed.[60]

12.5 Planar Supports

Following the research strategy presented in Section 12.2, the first step in the preparation of supported gold model systems is the fabrication of a planar support with a well-defined structure. Among those, oxide single crystals, such as MgO, TiO_2, sapphire (Al_2O_3) and quartz (SiO_2), were of first choice in early studies.[11,16] However, the lack of electrical conductivity (most oxides are wide gap semiconductors or insulators) renders their use for surface-science studies very difficult, except the partially reduced forms of rutile $TiO_2(110)$ crystal, which seems to be the most widely studied support used for gold model catalysts. In contrast, thin oxide films grown on metal single crystals provide good electrical and thermal conductivity for a facile application of surface science techniques. In addition, the surface selection rules[61] (i.e. only vibrations with dipole moment changes normal to the metal surface can be detected by IRAS) often ease assignment of infrared bands.

Various methods for preparation of thin oxide films are reported in the literature (see Refs. 8, 9, 13, 62 and 63). The first one is the direct oxidation of a metal single crystal as schematically shown in Figure 12.7. However, this preparation primarily results in amorphous or polycrystalline overlayers (e.g. SiO_2/Si, Al_2O_3/Al)[64,65] or crystalline films with a high density of defects (e.g. NiO(100)/Ni(100).[66] Formation of a crystalline $Cr_2O_3(111)$ film on Cr(110) appears as the only successful example for such preparation.[67]

Another route is the oxidation of bimetallic or alloyed surfaces where the element having higher affinity for oxygen segregates to the surface

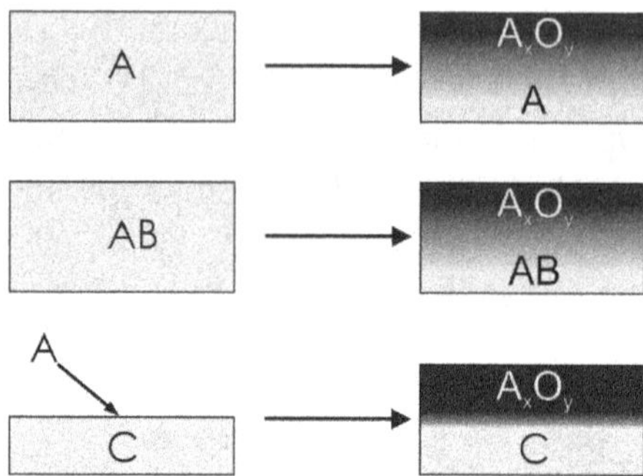

Figure 12.7   Preparations of well-ordered thin oxide films by oxidation of a metal single crystal (top), by oxidation of bimetallic surfaces (middle), and by deposition/oxidation of metal overlayer on the second metal single crystal (bottom).

and becomes oxidised, ultimately forming thin oxide film at elevated temperatures. Perhaps, the best examples of the crystalline films fabricated using this approach are the alumina ultrathin films grown on NiAl(110), NiAl(111) and $Ni_3Al(111)$ single crystals.[13,68–70]

In a great majority of cases, however, oxide films are prepared by vapour deposition of one metal (to be oxidised) onto the second metal (usually, noble metal) single crystal either in oxygen ambient or in vacuum followed by post-annealing in vacuum or in oxygen, respectively. In addition, a metal substrate can be pre-covered with atomic oxygen to ease oxidation of incoming metal atoms and to prevent their migration into the substrate bulk. Numerous oxide films were grown in this way, such as $Fe_3O_4(111)$ and $Fe_2O_3(0001)$ films on Pt(111),[63] MgO(001)/Ag(001),[71] $CeO_2(111)$/Ru(0001),[72,73] $V_2O_3(0001)$/Pd(111),[74] $V_2O_5(001)$/Au(111),[75] SiO_2/Ru(0001),[76] to name a few.

It is noteworthy that highly oriented pyrolytic graphite (HOPG) has also been used as a planar substrate for gold nanoparticles due to its flatness and good electrical conductivity. However, graphite only shows a weak interaction with gold, thus resulting in relatively large Au particles even at room temperature.[77] To some extent, the gold particle density and size can be varied by ion sputtering and/or oxidation, which result in etched pits on the HOPG surface, which, in turn, serve as nucleation centres for diffusing gold ad-atoms.[23,77]

12.6 Gold Deposition on Planar Supports

12.6.1 *Physical Vapour Deposition*

Most commonly gold particles are deposited onto planar oxide supports in vacuum using physical vapour deposition (PVD), which provides the best cleanliness of the systems under study. There are various realisations of PVD depending on how a metal vapour is produced, e.g. by thermal evaporation, by sputtering of metal target using high-energy ions, by laser ablation, etc. (see below).

For the thermal evaporation, gold is placed into a crucible (usually made of high melting point materials such as W and Mo) that is heated resistively or using electron beam (see Figure 12.8). Since gold has a relatively high vapour pressure, the reasonable flux can be obtained at the temperatures close to or just above the melting point of gold (1337 K). The most simple, 'home-built' evaporators can be made of a gold wire wrapped around tungsten wire, which is heated by passing the electric current to first form a small droplet of gold. The deposition flux of gold atoms from these sources is typically in the range of 10^{14} atoms $cm^{-2}s^{-1}$, which can be measured

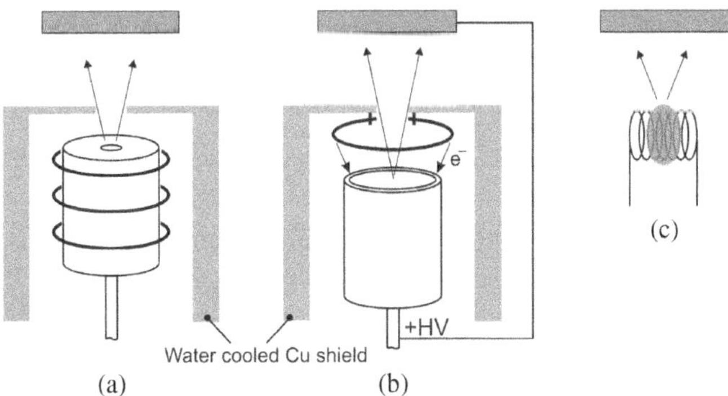

Figure 12.8 PVD sources commonly used for gold deposition on planar supports. A crucible filled with metallic gold is heated resistively (a) or by electron bombardment (b). In (b), the substrate is biased at the same high-voltage potential as the crucible (typically 800–1000 V) to prevent accelerating of charged gold species towards the substrate. Evaporators have normally a water-cooled Cu shield to maintain the UHV conditions during the deposition. (c) The 'home-built' evaporator consisting of tungsten filament with a small Au droplet formed upon resistive heating of the Au wire wrapped around the tungsten filament.

either by a quartz microbalance or directly by STM using the support on which Au atoms adsorb with a high sticking coefficient.

Note that other recently developed PVD methods include a glow plasma discharge or a cathodic arc to vapourise the target material. In the pulsed laser deposition method, a high power laser ablates material from the target into a vapour. Although the experimental set-up may be fairly simple, the ablation process itself is extremely complex involving the interaction between the laser and a solid target material, plasma formation and the transport of material across the vacuum to the substrate. All these methods are rarely used for PVD of pure metals like gold, but rather for multi-elemental compounds and materials with low vapour pressures.

From any deposition source, the incoming Au atoms first adsorb onto the surface, then diffuse and ultimately aggregate into clusters and nanoparticles as schematically shown in Figure 12.9. In principle, two types of nucleation modes may occur at surfaces: homogeneous and heterogeneous. For the homogeneous nucleation, an immobile nucleus is formed on regular surface sites by aggregation of several atoms. The critical island size is defined as the size above which the islands are stable. In other words, by addition of further atoms, the nuclei will grow, whereas islands up to this size can dissolve again. For the simplest situation, when a dimer is already a stable species (the critical size is one atom), theoretical considerations show[9,78] that the saturation density of islands N depends on the diffusion coefficient D and the deposition flux F as $N \sim (F/D)^{1/3}$. Then no further nuclei form, and all diffusing atoms stick to existing islands. (For more sophisticated analysis the readers refer to Ref. 9 and references therein.)

However, if defects are present on the surface, the ad-atoms may be trapped at these sites, thus forming nucleation centres for subsequent

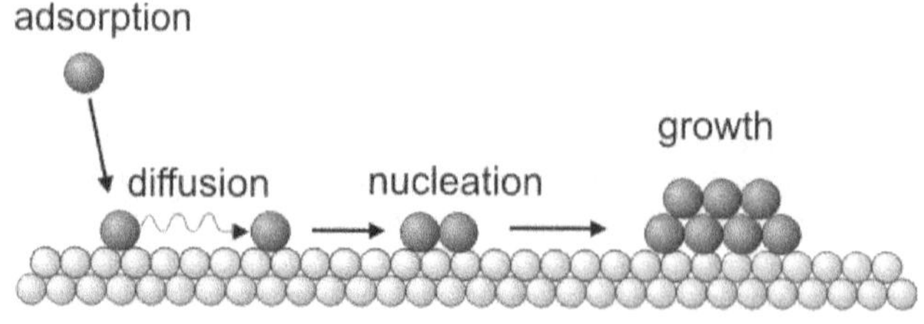

Figure 12.9   Elementary steps of the formation of Au particles by physical vapour deposition.

growth, which is called heterogeneous nucleation. If the interaction of ad-atoms with defects is strong and the defect density is relatively high, the metal particle density will be independent of the deposition flux. (It is important to note here that, using electron beam assisted evaporators and other methods resulting in high-energy charged species (typically 500–1000 eV), one needs to take precautions against accelerating of the ionised metal atoms towards a substrate, which may create additional defects upon collision as nucleation centres for subsequent gold particles. Although there are no yet systematic studies on this effect, it is strongly recommended to bias the support, at least, at the same potential as the metal source, see Figure 12.8(b)).

The nucleation and growth of gold deposited by PVD on oxide supports is often governed by the defect structure of the support (i.e. point defects [for example, oxygen vacancies], step edges, impurities) as a result of relatively weak interaction of gold with oxides and hence high surface diffusivity of gold ad-atoms. Such effect may even be used for decorating surface defects to count them.[73]

Certainly, the substrate temperature can alter the growth processes. Indeed, the Au ad-atoms may escape from defect sites at elevated temperatures and keep on diffusing across the surface. In addition, the particles usually sinter and/or coalesce at high temperatures. Very often, the size and the shape of the gold aggregates formed at low temperatures is kinetically limited as illustrated in Figure 12.10 for Au deposited on the $Fe_3O_4(111)$ films. When deposited at $\sim 100\,K$, the Au particles exhibit a broad size distribution and poorly defined (irregular) shape, whereas the annealed at 500 K system exposes well-faceted particles with a more uniform particle size distribution.

In principle, a range of particle sizes between 1 and 10 nm can readily be prepared by PVD. A degree of size control can be achieved by optimising the coverage, substrate temperature and deposition flux.

12.6.2 *Cluster Deposition*

In the range of very small sizes, a 'soft-landing' of mass-selected clusters pre-formed in the gas phase was suggested as highly controllable and flexible method[79,80] albeit experimentally very sophisticated and much more

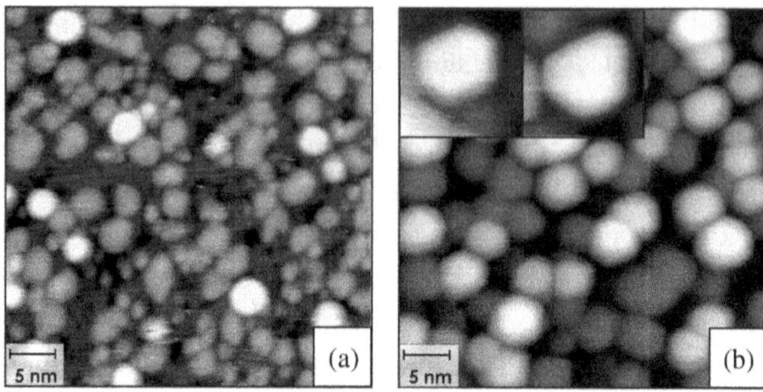

Figure 12.10   Typical STM images of Au particles deposited by PVD on a $Fe_3O_4(111)$ film at 100 K (a) and after annealing to 500 K (b). The characteristic shapes of the annealed Au particles are shown in the inset.

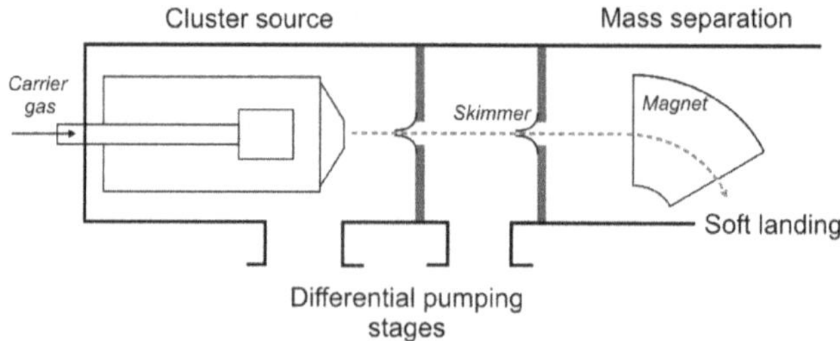

Figure 12.11   Schematic representation of a mass-selected cluster deposition. Mass separation employs deflection of charged gold clusters in an electric or magnetic field. Here, segmented magnet is only shown, for simplicity.

'expensive' than PVD. The method allows, in principle, precise control over the cluster size. The latter is particularly important for studies in the so-called non-scalable regime of a gold particle size, where adding or removing only one gold atom to the cluster may considerably change its functional properties, e.g. catalytic performance.[81] Note that using this cluster deposition one has to take precautions against coalescence of deposited clusters, which implies very low metal coverages.

Basically, the apparatus includes a cluster beam source, a mass filter and facilities for soft landing deposition (Figure 12.11). In the beam source, a metal vapour is ejected into a flow of a cooled inert gas, where the condensation of a supersaturated vapour into clusters occurs. The metal vapour can be generated by laser ablation, sputtering, a pulsed or continuous arc, or thermally.[79,82] Depending on the type of source, the gas pressure inside the condensation zone ranges from a few millibar to a few bar. The cluster size distribution depends on the time till the mixture of carrier gas and clusters exits the aperture into the vacuum region, which can be controlled, for example, by the gas flow rate. The expansion accelerates the clusters, which, in the limit of large pressure drop across the aperture ('free-jet expansion'), are thermally equilibrated with the bath gas and have the same velocity distribution irrespective of their size.

Mass separation and selection of the clusters is based on the deflection of the charged species either in an electric or magnetic field. The Wien filter applies orthogonal electric and magnetic fields to charged particles. Therefore, the equipment additionally requires a cluster ionizer (except sputtering-based beam sources, which contain substantial amounts of charged species). Some cluster sources even employ two mass filters: A high-resolution instrument such as time-of-flight analyser to measure the mass spectrum in the free beam, and a lower resolution high-throughput device to narrow the native mass distribution prior to deposition. Mass separation of neutral clusters is still challenging (for more details, see Refs. 82 and 83).

Soft-landing of clusters means that the clusters do not fragment upon collision with a substrate. The kinetic energy of clusters to fulfill this regime is typically set to 0.1 eV per atom. Following the well-known Maxwell–Boltzmann velocity (v) distribution function, i.e. $f(v) \sim (m/2\pi kT)^{3/2} v^2 \exp(-mv^2/2kT)$, the soft-landing regime may be achieved by proper choosing the mass m and temperature T of a carrier gas. For example, for Ar at 100 K, the velocity at the peak of the distribution is about 200 m s^{-1}, which corresponds to the impact energy of an Au cluster about 0.04 eV/atom. However, in the case of He at 300 K as the carrier gas, this would yield 1.5 eV/atom that falls into the range of medium impact energies (1–10 eV/atom).[83] In addition, soft landing can be achieved by decelerating the charged clusters by electric field while approaching a support (see for example, Refs. 84 and 85).

Several experiments demonstrated that soft landing of metal clusters is possible, indeed.[81,86,87] STM characterisation of Au_n^+ ($n = 1-8$) clusters deposited on $TiO_2(110)$ showed that, starting from dimers, Au_2^+ the clusters could be deposited intact, and no cluster agglomeration occurred,

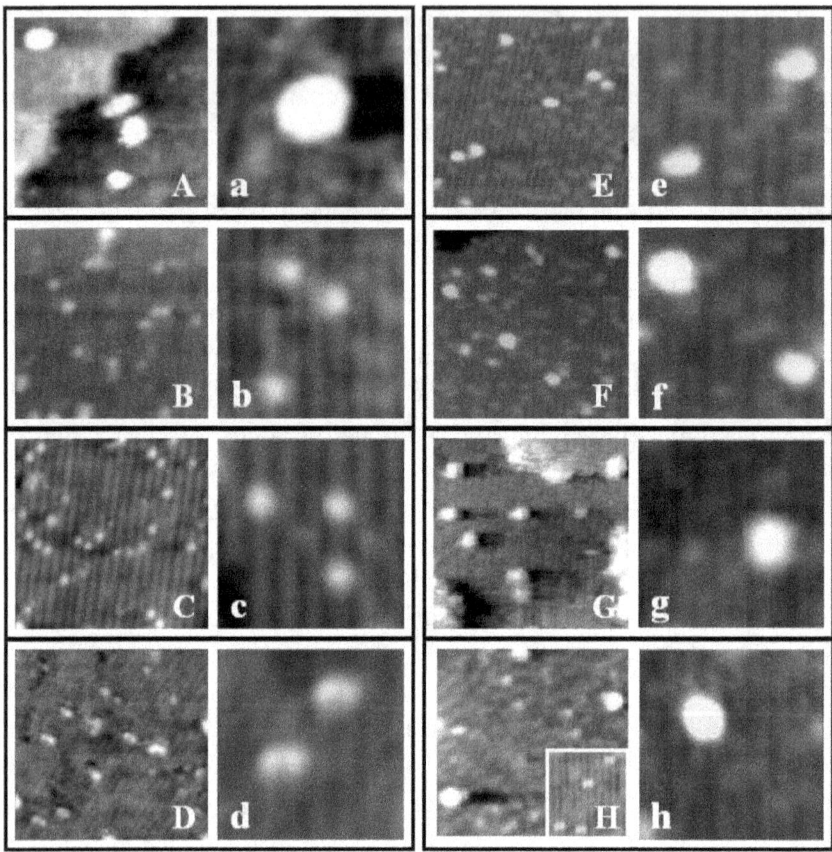

Figure 12.12   STM images 14 nm × 14 nm (uppercase letters) and 5 nm × 5 nm (lowercase letters) of Au_n^+ clusters deposited on $TiO_2(110)$ at 300 K ($n = 1 - 8$ for (a–h), respectively). The bright protruding spots are the Au clusters, the dim spots between the rows of 6.5 Å apart are bridging oxygen vacancies of $TiO_2(110)$. (Reprinted with permission from Ref. 87 Copyright (2005). American Chemical Society.)

at least, at room temperature,[87] as shown in Figure 12.12. Interestingly, single-atom deposition ($n = 1$) revealed no small Au clusters on surface (see Figure 12.12(a)). These observations indicate that Au monomers are highly mobile on the $TiO_2(110)$ surface, leading to aggregation into larger clusters, containing on the orders of tens of atoms, on average.

Finally, the deposition onto a thin buffer layer of rare gas[88] (typically Xe) formed on a substrate cooled down to cryogenic temperatures (ca. 40 K) may also lead to the non-fragmented deposition of metal clusters. Subsequent heating to desorb the weakly bound buffer layer is accompanied by adsorption of the clusters onto a substrate.

In principle, mass-selected cluster deposition is not limited to the very small clusters. In fact, modern deposition sources can produce clusters up to $\sim$70,000 atoms, depending on the target material. For example, $Au_{(923\pm23)}$ nanoclusters were produced using a magnetron-sputtering gas-condensation cluster beam source.[89] Characterisation of deposited 'giant' Au clusters by ac-STEM showed the formation of icosahedral, decahedral and face-centred cubic isomers (see Section 12.4) within a set of populations, with each population corresponding to a specific set of formation conditions. Tuning these conditions, one can eliminate completely all icosahedral clusters, which are commonly found under other conditions, thus providing a tool to the preparation of nanoclusters containing a dominant or single isomer.[90]

12.6.3 *Reactive Deposition Methods*

The buffer layer assisted deposition mentioned above was further extended to PVD of metals on 'amorphous solid water' films or, simply saying, on ice formed by water adsorption on a planar substrate kept at temperatures below 130 K.[91] After metal deposition, the sample is heated to the room temperature to desorb weakly bound water molecules. The formation and aggregation of particles on ice films is a complex process that is driven by the dewetting, islanding and sublimation of ice on heating and is yet not well understood.[92] Nonetheless, water-assisted deposition of gold on silica films showed that one can independently control particle size (in the range of several nanometres as observed) and only change particle density by repeating the whole deposition procedure.[93]

Chemical vapour deposition (CVD) is based on use of volatile complexes as precursors, which decompose on a target substrate kept at elevated temperatures. The CVD can be performed by direct sublimation of solid precursor in vacuum or in the carrier gas such as nitrogen or hydrogen, the latter is additionally used as the reducing agent. Some preparations include passing the carrier gas through the precursor solution ('bubbling').

Usually, the gold(I) and gold(III) precursors are organometallic compounds such as alkyl(phosphine)gold(I) and dimethylgold(III) β-diketonates and their derivatives.[94–97] In general, synthesis of the gold precursors is often very complex and shows a low yield. In addition, many precursors are light and/or air-sensitive, and as such are difficult to handle. Dimethylgold(III) carboxylates were recently suggested as viable precursors due to their sufficient volatility and thermal stability.[98]

The principal advantage of CVD over PVD is the possibility to uniformly cover rough and/or porous surfaces without having shade zones, which are inevitable in PVD using a directional beam of the Au atoms.[95] However, the CVD method suffers from the lack of control of cleanliness arising from the stripping of the ligands and decomposition at surface. It is noteworthy that this method is characterised by high deposition rates, thus resulting in granular films rather than isolated nanoparticles, and is, therefore, primarily aimed at deposition of gold thin films in microelectronics, metal coatings, etc. There are only few surface science studies on CVD-prepared gold nanoparticles. In particular, AFM measurements revealed significant sintering of Au particles on $TiO_2(110)$ under an ambient atmosphere at temperatures as low as 363 K.[99] However, pre-treatment of the surface with ultraviolet radiation before CVD prevented particle agglomeration. It was suggested that such behaviour resulted from the presence of hydroxyl groups, which formed as a result of photo-induced dissociation of adventitious water adsorbed on titania surface from residual gases in vacuum used ($\sim 10^{-6}$ mbar), since the titania surface is known to be active in water adsorption. It was proposed that the hydroxyl groups react strongly with the gold (I)-phosphine precursor used and thereby brought about the formation of highly stable small gold particles, which showed limited agglomeration even at 493 K in air.

12.6.4 *Deposition from Solution*

Certainly, the planar supports in the form of thin films or single crystals are suited for gold deposition from solution as well, since such supports do not suffer from diffusion limitations associated with a porosity of real catalytic supports. For such deposition, one can, in principle, use the same procedures and treatments as described in Chapter 8 of this book. Again, gold nanoparticles can already be formed in solution and only need to be brought onto the support (reduction–deposition method), or the solution contains gold precursor to be thermally reduced at inorganic support (deposition–reduction method). To date, majority of such studies on planar supports utilises the first approach. The surface science studies aimed at fundamental understanding of deposition–precipitation processes are, in essence, in the premature state (see Ref. 100).

Putting a few droplets of gold-containing solution onto the support and subsequent drying in ambient typically results in a non-uniform material distribution across the surface. To have it spatially more uniform, a spin coating technique is employed, i.e. a few droplets of solution is put on the centre of the substrate, which is then rotated at certain speed in order to spread the material by centrifugal force. Another approach is dipping the substrate into the solution and pulling it slowly out, which is usually referred to as a dip-coating method.

12.6.5 *Deposition of Ordered Particles*

Recently, the deposition methods have been developed, which aim at a simultaneous control over particle size as well as distance between the Au particles. The latter parameter may be important for gold application in photonics and sensors, and in catalysis of diffusion-controlled reactions. One such approach includes deposition of a monolayer of Au filled micelles (first prepared in solution, see also Chapters 6 and 7) with a different length of the polymeric ligand.[101] The preparation allows deposition of Au particles of the mean particle size ranged between 1 and 15 nm with very narrow particle size distribution. The micelles can then be brought to the planar substrate by dip-coating.[101,102] The resulting systems revealed hexagonally arranged pattern of Au nano-dots, with the average distance between Au nanoparticles being controlled by the total length of the diblock copolymers (see Figure 12.13). The nature of substrate seems not to play a considerable role on particle spatial distribution.[101] Removing of the polymeric shell can be achieved by annealing in UHV or by treatment in oxygen plasma. However, this issue remains the most uncertain in these experiments.

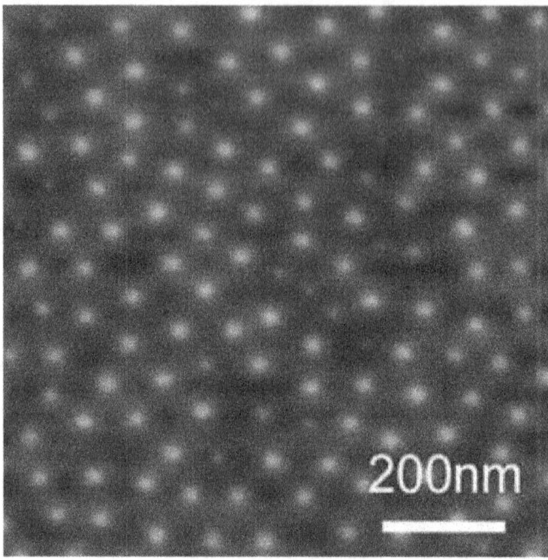

Figure 12.13 AFM image of gold nanoparticles deposited by dip-coating of inverse micelles on ultrathin TiC films (adapted from Ref. 103). The Au particles were synthesised using polystyrene-*block*-poly(2-vynilpyridine) as an encapsulating agent. The mean particle height is around 2 nm. (Note that the lateral size of the particles appears much larger than 2 nm due to the tip-sample deconvolution effects.)

Another approach for ordering metal particles on surfaces employs electron beam lithography (EBL) for the production of a patterned template followed by metal deposition using PVD, as nicely demonstrated first for Pd and Pt particles on various supports.[104–107] Briefly, a highly collimated electron beam is exposed to a thin layer of polymeric resist (such as poly(methylmethacrylate)) spin-coated on a flat substrate (e.g. a Si(100) wafer covered by the native oxide film or coated with thin film of alumina). The electron irradiation decomposes the polymer backbone, making it possible to dissolve the exposed polymer in a developing solution. Then a metal film is deposited on the surface by PVD, and the remaining polymer is removed with acetone. In the first experiments with Pt, the resulting systems showed highly ordered arrays of metallic particles about 30 nm in lateral size with 100 nm periodicity.[104]

Recently, a combined EBL and PVD preparation has been applied to gold, in particular for surface-enhanced Raman scattering (SERS) applications.[108] Using two different preparations (i.e. lift-off and plasma etching), a variety of nanostructures (disks, holes, gratings and other complex nanostructures) were fabricated. Large SERS enhancement was observed, which was dependent on the precise geometry of the nanostructures.

Note also that the EBL was effectively used for substrate patterning and subsequent deposition of ligand protected gold particles, which selectively bound to the patterned substrate.[109] Another possibility is patterning of Langmuir–Blodgett films of colloidal gold.[110,111] More about nanolithography methods in application to gold can be found in the review.[112]

To conclude this section, it is fair to say that the preferential choice in using one or another deposition method depends on many factors and is primarily determined by objectives of a study. On the one hand, all methods based on non-vacuum depositions have an inherent tendency to introduce contaminations into the system. On the other hand, the most clean, physical vapour and cluster depositions are hardly suited for preparations of technically relevant systems on large scale.

12.7 Surface Science Studies of Gold Nanoparticles

12.7.1 *Nucleation and Growth*

Gold on a $TiO_2(110)$ single-crystal support is one of the most widely studied planar model system involving gold.[16–18] The (110) surface of TiO_2 rutile is the most stable surface consisting of alternating rows of titanium and oxygen atoms with half of the titanium atoms covered by so-called bridging oxygen. These oxygens are relatively weakly bound and can be removed upon high-temperature annealing (in order to form an electrically conducting support), ultimately creating oxygen vacancies, which can strongly influence the support's chemistry[113] (see also Figure 12.12).

On the basis of electron and ion spectroscopy results, it was proposed that PVD-deposited gold at room temperature first grows on $TiO_2(110)$ two-dimensionally (2D) at very low coverage, which then subsequently changes to three-dimensional (3D) growth with increasing coverage.[114,115] At higher deposition temperatures, the growth mode is more 3D from the onset, indicating that the 2D growth at low temperatures is a kinetically limited mode (see also below).[114–117] A very similar behaviour was later observed for gold deposited on thin titania films grown on Ru(0001)[118] further supporting the concept that thin oxide films are, indeed, suitable supports for studying highly dispersed metal particles.

Early high-resolution STM studies, corroborated by density functional theory (DFT) calculations, showed a direct relationship between gold particle nucleation and surface oxygen vacancies on $TiO_2(110)$.[119] Further studies,[120] however, have revealed that the $TiO_2(110)$ surface is very sensitive to the traces of water in the UHV background leading to the formation of surface hydroxyl (OH) groups, which likely have influenced previous studies on gold deposition on this support. Indeed, hydroxyls were found to promote gold sintering and strongly affect the particle size distribution.[121]

When gold was deposited on $TiO_2(110)$ at 300 K and then annealed to 770 K the $(111)_{Au}//(110)_{TiO_2}$ epitaxial relationship was observed, whereas deposition at 770 K preferentially gave rise to the $(112)_{Au}//(110)_{TiO_2}$ epitaxy.[16,122] Interestingly, the Au lattice does not appear to undergo any deformation in spite of the minimal strain that must be overcome to match the TiO_2 epitaxy, indicating that the interaction between gold and titania is rather weak. On the other hand, the epitaxial relationship between titania and gold is strong enough, such that particles can grow in an Au(111) epitaxy up to rather large sizes.[16]

On $CeO_2(111)$ thin films, at the lowest coverage, gold preferentially nucleated on terraces, presumably on point defects, e.g. oxygen vacancies, typical for ceria, which is famous for its facile oxygen uptake and release behaviour.[123] At increasing gold coverage, particles are also formed at the step edges exposing a large variety of low-coordinated sites (Figure 12.14(a)). Defects density on terraces could be increased considerably by high-temperature annealing in UHV. Gold deposition on this 'reduced' surface resulted in the particles that were much smaller when compared to deposition of the same amount of gold on the fully oxidised ceria surface (see Figure 12.14). Only at high coverage, the Au particles grew homogeneously on the flat $CeO_2(111)$ terraces.[124] Gold exhibited a 3D growth mode from the onset resulting in nanoparticles, which were fairly stable towards sintering in vacuum at elevated temperatures. Increasing amounts of deposited gold essentially did not change the aspect (height/width) ratio of the particles, i.e. in agreement with thermodynamic considerations discussed in Section 12.3.

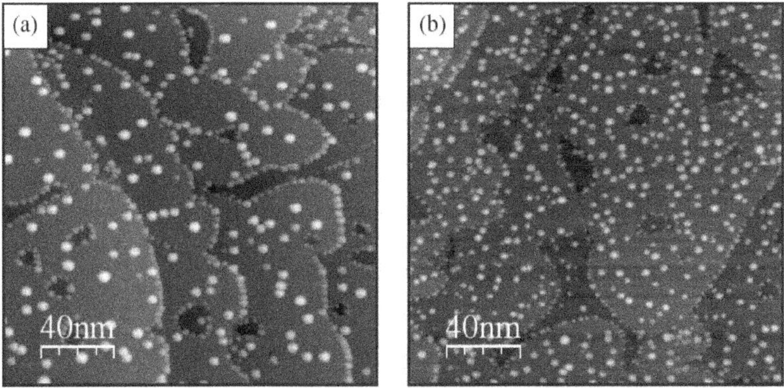

Figure 12.14   STM images of Au particles deposited on fully oxidised (a) and partially reduced (b) $CeO_2(111)$ films grown on Ru(0001). For the latter, the fully oxidised film was annealed at 1000 K in UHV prior to the deposition of the same amount of gold as for image (a).

Surprisingly, gold deposited onto crystalline alumina thin films did not show preferential nucleation as previously observed for Pd and Ag on the same support. The hemi-spherical particles were found to be fairly randomly distributed on the surface.[125,126] Such a behaviour has been attributed to the particular structure of the thin alumina film grown on NiAl(110), which results in a strong interaction with the gold single atoms.[127]

In principle, one can derive the adhesion energy between a metal and a support, if the precise geometry of supported particles is known. For the hemi-spherical metal particles, mainly exposing (100) and (111) facets, and contacting a substrate via the (111) plane (see Figure 12.15), an analysis leads to the following expression for the adhesion energy[128]:

$$W_{adh} = \gamma_{111}\left(2 - \frac{3}{\sqrt{2}}\frac{h}{w}\frac{s+1}{2s+1}\right),\tag{12.4}$$

where γ_{111} is the surface energy of the (111) surface, h (w) is height (width) of particles; s is the ratio of top-facet side lengths (s_{100}/s_{111}).

For example, STM images of annealed gold particles on a $Fe_3O_4(111)$ thin film showed[125] well-faceted particles, exhibiting mostly hexagonal shape of top facets as shown in Figure 12.10(b). In addition, the histogram analysis of STM images revealed that height of the particles was multiple of $\sim$2.5 Å (which is close to the height of a monolayer of gold in the (111)

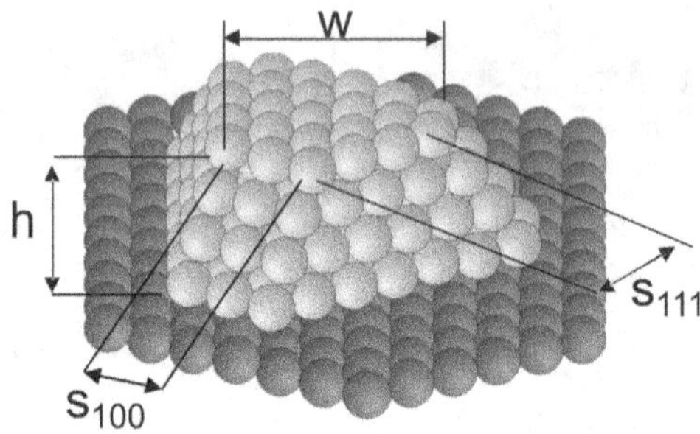

Figure 12.15   Structural characteristics used for deriving the adhesion energy between a metal particle and a support.

orientation). All these findings suggest that particles grow by increasing the number of the atomic layers parallel to the surface. Although atomic resolution of the top facets was not achieved, it seems plausible that the top facets show up the (111) surface owing to small ($\sim$3%) lattice mismatch between Au(111) and Fe_3O_4(111) surfaces. Based on the structural parameters derived from STM and using the theoretical value of 1.28 J/m^2 for Au(111) as calculated by DFT,[129] Equation (12.4) yields 2.3 J/m^2 for the adhesion energy between Au and the iron oxide. For comparison, Pd particles and Pt particles deposited on the same support showed the energies of 3.1 and 3.8 J/m^2, respectively,[130] using the values of 1.92 J/m^2 for Pd(111) and 2.3 J/m^2 for Pt(111) from the same DFT calculations, for consistency.[129] It is evident that Au interacts with the iron oxide surface more weakly than do Pt and Pd.

In the above examples, gold was deposited using PVD, which typically resulted in the mean particle size ca. 2–5 nm, except deposition at cryogenic temperatures where limited surface diffusion leads to the stabilisation of single Au ad-atoms or the formation of very small aggregates.[21,131,132] For the preparation of monodispersed Au particles below 1 nm in size, deposition of mass-selected clusters seems to be the only method (see Section 12.6.2). As mentioned above, the Au$_n^+$ ($n = 2-8$) clusters, adsorbed on TiO_2(110), showed no cluster agglomeration at room temperature[87] (Figure 12.12). In contrast, gold single atoms sintered rapidly and formed larger aggregates similarly to the samples prepared by PVD. In addition, the STM study revealed that supported Au$_5$–Au$_8$ clusters all exhibited 3D structures, albeit theory and gas-phase experiments indicated their planar configuration. These findings suggest that support, indeed, has certain stabilisation effect on particular structures of gold clusters. Interestingly, no direct evidence was found in these experiments for oxygen vacancies on TiO_2 to be required for binding the mass-selected clusters. Basically, the same conclusion has been drawn after revision of previous STM results of PVD-deposited Au particles.[121]

The role of oxygen vacancies on the structure of gold clusters have been addressed using MgO(100) thin films.[81,133] Experimental and theoretical studies of mass-selected Au_8^+ clusters deposited on perfect and defect-rich MgO films revealed that the so-called colour (or F–) centres play a crucial role in the reactivity of gold clusters in CO oxidation. DFT analysis showed that, although the Au_8 cluster adsorbed on F-centre is only slightly distorted as compared to the gas phase neutral cluster, it binds much more stronger to the defect, and the charge transfer from the oxygen vacancy to the Au_8 cluster occurs, as observed by IR spectroscopy of CO as a probe molecule.[81] It should be pointed out noted that final charge state of (initially positively charged) mass-selected Au clusters is case-sensitive.

12.7.2 *Particle Size Effects*

The ability of the cluster deposition technique to vary a cluster size atom-by-atom allowed monitoring size-dependent reactivity of the gold clusters, in particular in the CO oxidation. TPD and pulsed molecular beam studies of supported Au_n clusters revealed non-scalable activity towards CO_2 formation in such that clusters below Au_8 on MgO[81,133] (Figure 12.16) and below Au_7 on TiO_2(110)[134,135] were essentially inert. At increasing cluster size, some sort of oscillatory behaviour was obtained for reactivity (Figure 12.16(b)). Another important observation, which came out from these studies, is structural flexibility ('fluxionality'[81]) of the gold-based systems, which seem to adopt the optimum structure in their response to the gas adsorption and the most favourable reaction pathway.

Mass-selected Au clusters consisting of less than 10 atoms were deposited on HOPG surfaces pre-sputtered by Ar^+ ions in order to increase density of the defects acting as nucleation centres.[23] STM inspection of Au_7 clusters suggested the absence of sintering upon deposition.[136] Only Au_8 can be significantly oxidised using atomic oxygen source and subsequently

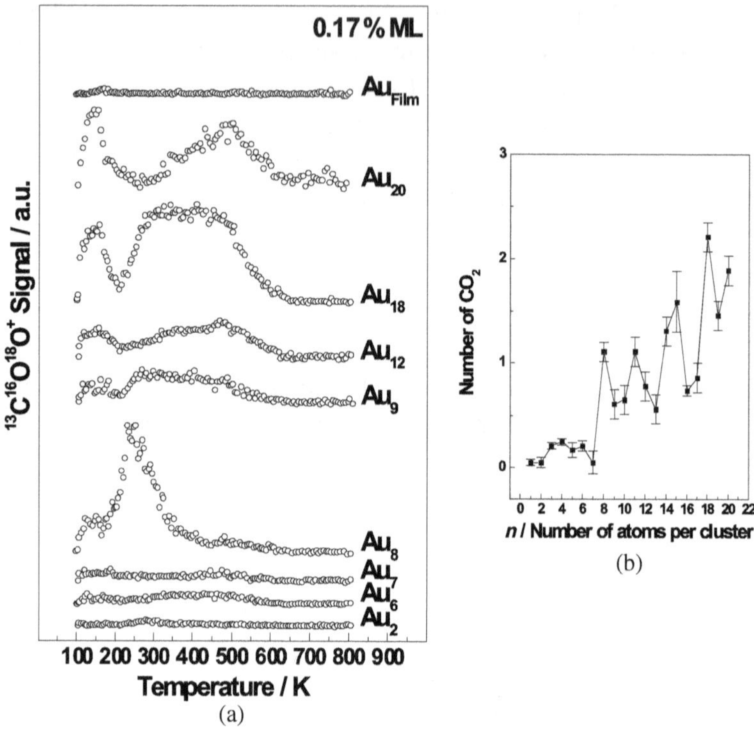

Figure 12.16   (a) TPD signal of $^{13}C^{16}O^{18}O$ production upon sequential adsorption of $^{18}O_2$ and ^{13}CO at 90 K on Au$_n$ clusters deposited onto defect-rich MgO thin films. The reactivity of Au$_n$ is expressed in (b) as the number of formed CO_2 molecules per cluster. (Reproduced with permission from Arenz *et al.*[81] Copyright Wiley-VCH Verlag GmbH & Co. KGaA.)

reduced by CO, as revealed by XPS. The results were somewhat different for Au clusters deposited on thermally grown silica films on Si(111), thus indicating importance of metal–support interactions for reactivity of gold.

TPD and IRAS studies of CO adsorption on gold nanoparticles deposited onto various oxide supports clearly showed a particle size effect in that small particles adsorb CO more strongly.[125,137] At low Au coverages and low deposition temperatures, i.e. at the conditions providing the formation of small Au aggregates, CO was found desorbing at temperatures as high as 300 K. Such relatively strongly bound states have never been observed on Au single crystal surfaces[46,48,138,139] and, therefore, have been associated with highly uncoordinated gold atoms present on the surface of the smallest particles.[137] However, these states disappear upon annealing to 400–500 K, which is accompanied by strong reduction of the CO uptake due to gold sintering, in full agreement with STM results pointing out low thermal stability of small gold clusters. Indeed, the spectra for annealed samples become very similar to those obtained on the stepped gold surfaces. Comparison of CO TPD spectra of gold deposited on various oxide supports showed that, for a given nanoparticle size, the interaction of CO with Au particles is essentially identical.[125] These findings suggest that the support effects frequently reported in the literature for real catalytic systems, particularly in oxidation reactions, could be associated with the oxide-dependent size distribution of gold nanoparticles and atomic structure of the gold–oxide interface, which are, in turn, controlled by defect structures of oxide supports. In addition, as mentioned above for the titania support, the presence of OH surface species either in the course of the catalyst preparation or under reaction conditions may also play a critical role in the support effects reported.

12.7.3 *Environmental Effects*

A majority of studies on the Au model catalysts was conducted under well-controlled, but vacuum-compatible conditions. Relatively unexplored are potential modifications of these systems in a reactive environment at nearly atmospheric pressures. In this respect, many fundamental questions still remain, including the possibility of reaction-induced morphological changes in the system. From general bond conservation considerations, a weakening of the Au–Au bonds within the cluster may occur when an Au cluster strongly interacts with the reactive gas, ultimately leading to a disruption of the structure of the metal particle. Also, adsorption-induced reshaping of gold particles is predicted by DFT calculations of an Au_{79} cluster ($\sim$1 nm in diameter) reacting with CO.[140] The strength of interaction between the gold nanoparticle and the support may play a significant role in determining the stability of these systems under realistic pressures and temperatures. Although Au particle reshaping was experimentally proven on Au/MgO catalysts with the help of high-resolution electron microscopy,[141] surface studies of gold nanoparticles under realistic reaction conditions remain scarce. To some extent, this is due to the lesser number of techniques available in the field of surface science to carry out experiments in ambient other than vacuum. Below, we show few examples of such studies in order to illustrate environmental effects on gold.

We first address to results obtained by STM, which may operate, in essence, in any atmosphere if the sample is electrically conducting. For example, no morphological changes of the Au particles deposited on thin FeO(111) films were observed in oxygen and hydrogen environments at pressures up to 2 mbar at room temperature.[142] However, in CO and CO + O_2 (2:1) atmospheres, the destabilisation of Au atoms located at the step edges was observed even at $\sim 10^{-3}$ mbar leading to the formation of mobile Au species, which diffused across the surface. The results were rationalised in terms of a stronger interaction of gold with CO, as compared to O_2 and H_2, which significantly weakens the Au/support interaction for the smallest particles.

STM studies[143,144] of Au/TiO$_2$(110) revealed a form of Ostwald ripening (the growth of large particles at the expense of small particles) by exposing to 14 mbar of O_2 at 300 K. This process appeared to be even stronger in the stoichiometric mixture of O_2 and CO. Further *in situ* STM studies[144,145] revealed that an initial uniformly sized group of Au particles underwent severe Ostwald ripening at 450 K in 0.7 mbar of O_2. The presence of oxygen served to weaken Au–Au bonds, thereby promoting sintering. Note, however, that STM images taken under reaction conditions showed that the behaviour of particles that were initially of the same size could be quite different, as some particles decreased in size or even disappeared, while others seemed to remain stable.

The morphology of the supported Au particles on CeO$_2$(111) films was studied by STM *in situ* and *ex situ* in CO, O_2 and CO + O_2 environment at room temperature.[124] No visible changes were observed after exposing Au particles to pure O_2 up to ~ 10 mbar. In the pure CO ambient, Ostwald ripening emerged above ~ 1 mbar. Meanwhile, sintering of the Au particles was observed in CO + O_2 (1:1) mixture at much lower pressure ($\sim 10^{-3}$ mbar) that mainly occurred along the step edges. The results indicate that the structural stability of the Au/CeO$_2$ surfaces is intimately connected with its reactivity in the CO oxidation reaction. The results revealed both similarities and differences with the Au/TiO$_2$(110)[145] and Au/FeO(111)[142] systems, suggesting that the oxide support is deeply involved in the stabilisation of the supported Au nanoparticles.

To address the structure of adsorbed species on gold under reaction conditions, SFG technique seems to be well-suited, as it was nicely demonstrated for CO adsorption on Pd nanoparticles in the wide range of pressures.[146] Comparative studies of CO adsorption on pure and ion-sputtered gold surfaces by SFG[147] revealed the important role of the defect sites in the CO adsorption. In addition, a promoter such as iron oxide was found to enhance CO adsorption, which is believed to take place at the interface between gold and iron oxide. In addition, SFG was successfully applied to surfactant-coated Au nanoparticles ($\sim$15 nm) deposited onto Si wafer.[148] Here, polar orientation and degree of conformational order of adsorbed surfactants could be addressed. Nonetheless, it is fair to say that the SFG studies on gold in catalytic reactions are scarce, indeed.

The instrumentation of synchrotron-based, ambient pressure XPS is much more sophisticated and hence costly, that in fact prevents this technique to be present in many laboratories. Recent *in situ* AP XPS studies[149,150] were particularly focused on interaction of gold with oxygen, which appears to be the most critical step in oxidation reactions on gold. The results showed that molecular oxygen does not oxidise Au at room temperature, either in the form of supported particles on $TiO_2(110)$ or in bulk (foil) form at pressures of up to 1 Torr.[149] In addition, the experiments demonstrated that X-rays play a critical dual role during *in situ* measurements and that care must be taken to carry out experiments and interpret spectra, especially when using intense synchrotron radiation.

By using ozone as a more strongly oxidising agent, the formation of a surface oxide on gold foil was monitored by AP XPS.[150] However, this phase was unstable and decomposed under vacuum and even in the presence of ozone, but at higher temperatures. It was found that the surface oxidation led to structural modifications of the gold surface, which is accompanied by the formation of low-coordinated Au atoms.

12.8 Two-dimensional Gold

In the course of preparation of gold model catalysts on planar oxide supports, gold was frequently observed in form of islands consisting of just a single layer of atoms. In addition, STM studies combined with reactivity measurements of Au deposited onto a $TiO_2(110)$ single crystal and a titania thin film in the CO oxidation reaction in the mbar pressure range[151–153] showed that maximum catalytic activity for these clusters coincides with the metal-to-semiconductor transition (determined by tunnelling spectroscopy), which in turn coincides with a transition from 2D islands into 3D nanoparticles. Behind the general interest to the physics of low-dimensional materials, such observations resulted in a closer look on the atomic structure of two-dimensional gold as catalytically active species.

On ultrathin MgO(001) films grown on Ag(001), gold first forms flat, single-layer islands, which develop into a nearly complete wetting layer with increasing Au coverage (Figure 12.17(a)).[154] Whereas the single-layer Au islands dominate the surface of oxide films of 2–3 monolayers (ML) in nominal thickness, 3D nanoparticles are only formed on 8 ML-thick MgO(001) films. Combined experimental and DFT studies showed that the formation of 2D gold in this system is primarily controlled by the charge transfer from the support through an oxide film. Gold tends to increase the contact area with the support, as this maximises the charge transfer into the gold affinity levels. Interestingly, this charge is localised at the island's rim. The respective

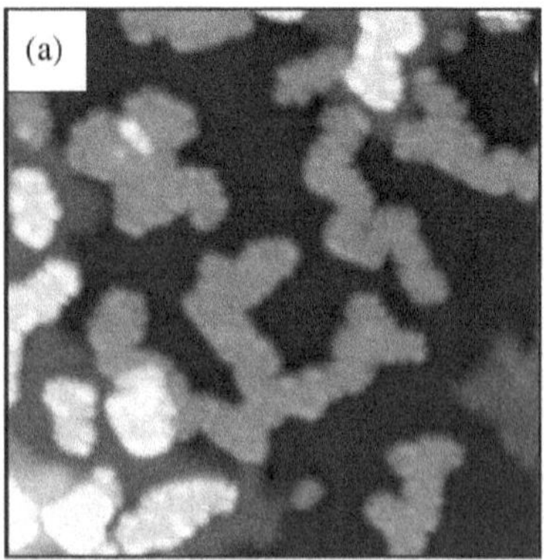
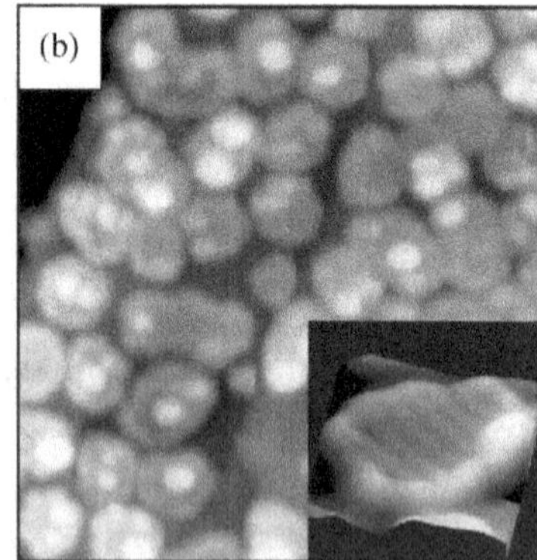

Figure 12.17   STM images of single-layer Au islands formed upon deposition of gold on 3 ML-thick MgO(001) film grown on Ag(001) (adapted from Ref. 154) (a) and Mo-doped, 10 ML-thick CaO(001) film grown on Mo(001) (adapted from Ref. 155) (b). Image sizes are 30 nm×30 nm (a), and 50 nm×50 nm (b). The inset in (b) shows a high-resolution image in perspective view that shows a stripe-like pattern on the Au surface, which is assigned to a Moiré structure formed between the square CaO(001) and the hexagonal Au(111) lattices.

electronic states are able to store one extra electron per low-coordinated edge atom, and become filled up with transfer electrons although the island interior remains neutral. The charge localisation in the low-coordinated, edge atoms suggests the 2D Au islands to be potentially active for adsorption and chemical reactions involving electron-accepting molecules.

Although the above presented complex metal-oxide-metal (Au–MgO–Ag) structures can hardly be implemented in the catalysis design right away, the concept of charge-mediated control of the gold particle shape can, in principle, be transferred to bulky oxide supports as well, providing a suitable charge source in the oxide lattice in a near-surface region. The latter can be achieved by doping of oxide by other cations, or it may even occur as a result of 'self-doping' by a small amount of impurities usually present in oxide support materials. Following this idea, high-valence dopants may serve as charge donors and provide extra electrons in the same way as a metal-supported ultrathin oxide film. Accordingly, low-valence dopants will show acceptor character and hence accommodate electrons from suitable adsorbates. Therefore, donors in an oxide lattice will have a similar effect on the particle shape as a metal support underneath an ultrathin oxide film. This concept has recently been validated for CaO(001) films grown on Mo(001). It was found by STM that, on the doped film, the gold spreads out into extended monolayer islands (Figure 12.17(b)), while the 3D growth is observed on the non-doped surface.[155]

12.9 Au-based Bimetallic Nanoparticles

Bimetallic nanoparticles often show synergy effects in that they exhibit properties distinct from those of mixture of monometallic particles.[156] In principle, bimetallic particles may form fully mixed alloys or a core–shell structure if, for example, composed of immiscible metals. Note again that surface composition and hence atomic structures may deviate from those present in the bulk due to surface segregation, the degree of which depends on many parameters such as lattice mismatch, particle size, temperature and ambient conditions. In this respect, Au-containing bimetallic systems received much attention on a theoretical ground by using *ab initio* calculations and Monte Carlo simulations.

As in the case of pure gold (Section 12.3), we first address general considerations applied to surface structures of bulky systems. Gold has much lower surface energies when compared to other noble metals (Pt, Pd) and close to that of Ag, only.[157] Therefore, it is not totally surprising that even small concentration of Au in Pt results in a surface layer composed almost exclusively of Au, as judged by AES.[158] STM, LEED and LEISS measurements of a clean $Cu_3Au(100)$ surface, representing a classical ordering alloy, revealed an Au-rich terminated layer out of two possible surface terminations.[159] For the $Au_3Pd(100)$ surface, STM with a so-called chemical contrast revealed the surface enriched by Au.[160] Figure 12.18 displays the high-resolution STM image where brighter protrusions are assigned to the surface Pd atoms, which are clearly at a much lower concentration when compared to the equilibrium distribution in the crystal bulk. In addition, ion scattering and tensor LEED analysis suggested that segregated topmost layer may have lattice constants different (in this case, smaller) from their respective bulk value.

Basically, similar trend (Au segregation at surface) holds true also for bimetallic nanoparticles. Theoretical studies of both, 40 atoms[161] and

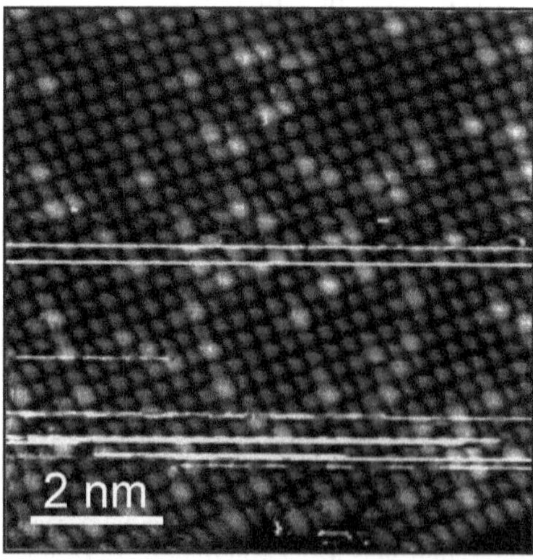

Figure 12.18   STM image of the clean $Au_3Pd(100)$ surface prepared by argon sputtering at 625 K. 'White' atoms are palladium. (Adapted from Ref. 160.) The streaks are due to the STM tip instability.

1654 atoms[157] Au–Pt clusters, revealed an Au-shell and Pt-core structure as thermodynamically the most favourable one. A strong surface Au enrichment was observed in all of the Au–Pt nanoparticles studied[162] and the surface segregation of Au was more pronounced at increasing particle sizes.

On the other hand, Monte Carlo simulations of Au–Ag nanoparticles showed Ag rather than Au segregating at the surface.[163] (Note that the surface energies of Ag and Au crystals are very close.[157]) The surface segregation is composition, size and temperature dependent. The surface Ag fractions are higher in the Ag-richer or larger-sized particles at low temperature. The resulting structures are formed upon the competition and balance between surface segregation and alloy formation. In addition, the calculated distribution of Au ensembles (e.g. monomers, dimers, trimers) on the particle surface depends on the Au:Ag compositional ratio and temperature. It is believed that metal surface distribution affects the reactivity and selectivity of bimetallic catalysts in structure-sensitive reactions. As an example, acetoxylation of ethylene to vinyl acetate (VA) was found to be strongly promoted by adding gold into a Pd catalyst. Comparative studies of the Au(100) and Au(111) surfaces as a function of Pd coverage showed the enhanced rates of VA formation for low Pd coverages relative to high Pd coverages, which were assigned to the critical reaction site for VA synthesis consisting of two non-contiguous, suitably spaced, Pd monomers.[164] The role of Au is to isolate single Pd sites that facilitate the coupling of critical surface species to product, while inhibiting the formation of undesirable reaction by-products.

Certainly, the surface segregation may substantially be changed when the Au-based nanoparticles are exposed to the ambient containing oxidising or reducing agent. In this case, the other metal usually having higher affinity for oxygen than gold will be prone to segregate to the surface. Such effect was theoretically predicted for the $Ag_3Pd(111)$ surface in an oxygen atmosphere.[165] Whereas a minimal segregation energy stabilises Ag-terminated surface structures in UHV, the much stronger oxygen bonding favours increasingly Pd-rich terminations in atmospheres with higher oxygen content.

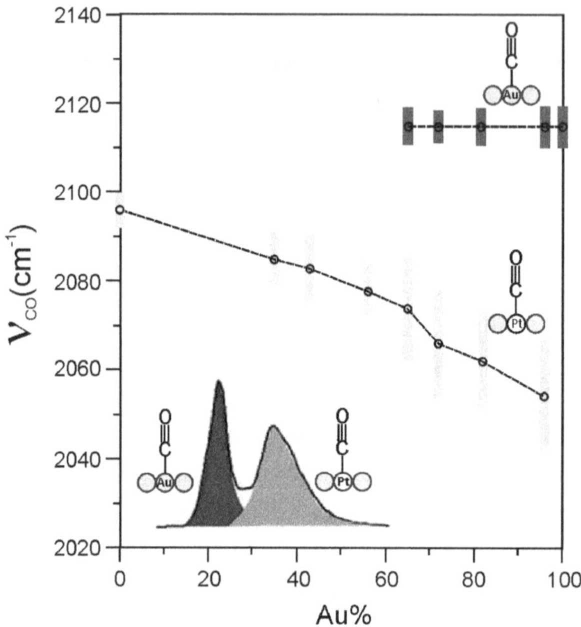

Figure 12.19 CO stretching frequencies of the two bands, assigned to terminal CO on Au and Pt atop sites (see inset), as a function of Au concentration in the Au–Pt nanoparticles. (Adapted from Ref. 167.) The bars represent the peak width at half maximum.

The experimental surface science study of bimetallic nanoparticles is not trivial, as most of the electron spectroscopy methods have surface sensitivity comparable with the particle size. More sensitive is low-energy ion scattering spectroscopy (LEISS) that, for instance, allowed to determine surface composition of the Au–Pd bimetallic particles vapour-deposited onto silica films.[166] The results showed Au surface enrichment, indeed, as predicted by theory, although to the lower extent as compared to extended Au–Pd surfaces. More straightforward are using vibrational spectroscopy of probe molecules like CO, albeit such studies also need complementary theoretical input. As an example, inset in Figure 12.19 depicts the ν(CO) stretching region in IR spectrum of CO adsorbed on silica supported Au–Pt particles.[167] The observed two bands are assigned to CO adsorbed in Au and Pt atop sites, respectively. In contrast to CO on Au, atop CO on Pt showed a substantial shift as a function of Au concentration, assigned to the electronic effect.

12.10 Concluding Remarks

In this chapter, we addressed surface structures of gold. Surface science studies, using state-of-the-art surface-sensitive techniques, applied to well-defined model planar systems allow one to rationalise chemical and catalytic properties of gold nanoparticles, which are very distinct from their bulky counterparts. In particular, it appears that the (oxide) support plays the critical role in stabilising atomic configuration, charge states and electronic structure of the ultra-small Au aggregates. The approach also allows monitoring the particle size effects as well as the effects caused by ambient conditions. The *in situ* experiments remain highly demanding for elucidating the reaction mechanisms.

Acknowledgements

I would like to thank all my co-workers, whose names appear in the references below, for their tremendous work in the laboratories of the group Structure and Reactivity in the Department of Chemical Physics of the Fritz-Haber Institute (Berlin) headed by Prof. Hans-Joachim Freund.

References

1. M. Haruta, *The Chemical Record* **3**, 75 (2003).
2. M. Haruta, *Gold Bulletin* **37**, 27 (2004).
3. A. S. K. Hashmi, G. J. Hutchings, *Angewandte Chemie International Edition* **45**, 7896 (2006).
4. M. C. Kung, R. J. Davis, H. H. Kung, *The Journal of Physical Chemistry C* **111**, 11767 (2007).
5. G. C. Bond, C. Louis, D. T. Thompson, *Catalysis by Gold* (Imperial College Press, London, 2006).
6. D. W. Goodman, *Surface Review and Letters* **2**, 9 (1995).
7. D. W. Goodman, *The Journal of Physical Chemistry* **100**, 13090 (1996).
8. H.-J. Freund, *Angewandte Chemie International Edition in English* **36**, 452 (1997).
9. C. T. Campbell, *Surface Science Reports* **27**, 1 (1997).
10. P. L. J. Gunter, J. W. Niemantsverdriet, F. H. Ribeiro, G. A. Somorjai, *Catalysis Reviews* **39**, 77 (1997).
11. C. R. Henry, *Surface Science Reports* **31**, 231 (1998).
12. D. R. Rainer, D. W. Goodman, *Journal of Molecular Catalysis A: Chemical* **131**, 259 (1998).

13. M. Bäumer, H.-J. Freund, *Progress in Surface Science* **61**, 127 (1999).

14. H.-J. Freund, *Surface Science* **500**, 271 (2002).

15. H.-J. Freund, G. Pacchioni, *Chemical Society Review* **37**, 2224 (2008).

16. F. Cosandey, T. E. Madey, *Surface Review and Letters* **08**, 73 (2001).

17. R. Meyer, C. Lemire, S. K. Shaikhutdinov, H. J. Freund, *Gold Bulletin* **37**, 72 (2004).

18. M. Chen, D. W. Goodman, *Accounts of Chemical Research* **39**, 739 (2006).

19. B. K. Min, C. M. Friend, *Chemical Reviews* **107**, 2709 (2007).

20. J. A. Rodriguez, S. Ma, P. Liu, J. Hrbek, J. Evans, M. Pérez, *Science* **318**, 1757 (2007).

21. T. Risse, S. Shaikhutdinov, N. Nilius, M. Sterrer, H.-J. Freund, *Accounts of Chemical Research* **41**, 949 (2008).

22. J. Gong, C. B. Mullins, *Accounts of Chemical Research* **42**, 1063 (2009).

23. D.-C. Lim, C.-C. Hwang, G. Gantefor, Y. D. Kim, *Physical Chemistry Chemical Physics* **12**, 15172 (2010).

24. J. Libuda, H. J. Freund, *Surface Science Reports* **57**, 157 (2005).

25. G. A. Somorjai, *Applied Surface Science* **121–122**, 1 (1997).

26. G. Rupprechter, *Catalysis Today* **126**, 3 (2007).

27. D. C. Meier, D. W. Goodman, *Journal of the American Chemical Society* **126**, 1892 (2004).

28. B. M. Hendriksen, S. Bobaru, J. M. Frenken, *Topics in Catalysis* **36**, 43 (2005).

29. D. C. Tang, K. S. Hwang, M. Salmeron, G. A. Somorjai, *The Journal of Physical Chemistry B* **108**, 13300 (2004).

30. M. Salmeron, R. Schlögl, *Surface Science Reports* **63**, 169 (2008).

31. J. V. Barth, H. Brune, G. Ertl, R. J. Behm, *Physical Review B* **42**, 9307 (1990).

32. M. A. Van Hove, R. J. Koestner, P. C. Stair, J. P. Bibérian, L. L. Kesmodel, I. Bartoš, G. A. Somorjai, *Surface Science* **103**, 189 (1981).

33. U. Harten, A. M. Lahee, J. P. Toennies, C. Wöll, *Physical Review Letters* **54**, 2619 (1985).

34. V. Heine, L. D. Marks, *Surface Science* **165**, 65 (1986).

35. J. K. Gimzewski, R. Berndt, R. R. Schlittler, *Physical Review B* **45**, 6844 (1992).

36. https://en.wikipedia.org/wiki/File:Atomic_resolution_Au100.JPG

37. D. G. Fedak, N. A. Gjostein, *Acta Metallurgica* **15**, 827 (1967).

38. D. G. Fedak, N. A. Gjostein, *Surface Science* **8**, 77 (1967).

39. H. Melle, E. Menzel, Superstructures on Spherical Gold Crystals, *Zeitschrift für Naturforschung A*, 1978, p. 282.

40. K. Yamazaki, K. Takayanagi, Y. Tanishiro, K. Yagi, *Surface Science* **199**, 595 (1988).

41. N. Wang, Y. Uchida, G. Lehmpfuhl, *Surface Science* **284**, L419 (1993).

42. T. Kunio, T. Yasumasa, K. Kunio, A. Kazuhiro, Y. Katsumichi, *Japanese Journal of Applied Physics* **26**, L957 (1987).

43. A. R. Sandy, S. G. J. Mochrie, D. M. Zehner, K. G. Huang, D. Gibbs, *Physical Review B* **43**, 4667 (1991).

44. K. F. Peters, P. Steadman, H. Isern, J. Alvarez, S. Ferrer, *Surface Science* **467**, 10 (2000).

45. Y. Jugnet, F. J. Cadete Santos Aires, C. Deranlot, L. Piccolo, J. C. Bertolini, *Surface Science* **521**, L639 (2002).

46. C. Ruggiero, P. Hollins, *Surface Science* **377–379**, 583 (1997).

47. M. Borbonus, R. Koch, O. Haase, K. H. Rieder, *Surface Science* **249**, L317 (1991).

48. C. J. Weststrate, E. Lundgren, J. N. Andersen, E. D. L. Rienks, A. C. Gluhoi, J. W. Bakker, I. M. N. Groot, B. E. Nieuwenhuys, *Surface Science* **603**, 2152 (2009).

49. G. Wulff, Zur Frage der Geschwindigkeit des Wachsthums und der Auflösung der Krystallflächen, *Zeitschrift für Kristallographie*, 1901, p. 449.

50. W. L. Winterbottom, *Acta Metallurgica* **15**, 303 (1967).

51. L. D. Marks, *Reports on Progress in Physics* **57**, 603 (1994).

52. C. L. Cleveland, U. ndman, T. G. Schaaff, M. N. Shafigullin, P. W. Stephens, R. L. Whetten, *Physical Review Letters* **79**, 1873 (1997).

53. J. A. Ascencio, M. Pérez, M. José-Yacamán, *Surface Science* **447**, 73 (2000).

54. C. L. Cleveland, U. Landman, *The Journal of Chemical Physics* **94**, 7376 (1991).

55. P. M. Ajayan, L. D. Marks, *Physical Review Letters* **63**, 279 (1989).

56. M. Mitome, Y. Tanishiro, K. Takayanagi, *Z Phys D–Atoms, Molecules and Clusters* **12**, 45 (1989).

57. S. Giorgio, C. Chapon, C. R. Henry, G. Nihoul, J. M. Penisson, *Philosophical Magazine A* **64**, 87 (1991).

58. T. Kizuka, N. Tanaka, *Physical Review B* **56**, R10079 (1997).

59. J. Liu, *ChemCatChem* **3**, 934 (2011).

60. Y. Han, R. Ferrando, Z. Y. Li, *The Journal of Physical Chemistry Letters* **5**, 131 (2014).

61. P. Hollins, *Encyclopedia of Analytical Chemistry*, John Wiley & Sons, Ltd, 2006.

62. S. A. Chambers, *Surface Science Reports* **39**, 105 (2000).

63. W. Weiss, W. Ranke, *Progress in Surface Science* **70**, 1 (2002).

64. J. G. Chen, J. E. Crowell, J. T. Yates Jr, *Surface Science* **185**, 373 (1987).

65. F. Rochet, S. Rigo, M. Froment, C. d'Anterroches, C. Maillot, H. Roulet, G. Dufour, *Advances in Physics* **35**, 237 (1986).

66. M. Bäumer, D. Cappus, H. Kuhlenbeck, H. J. Freund, G. Wilhelmi, A. Brodde, H. Neddermeye, *Surface Science* **253**, 116 (1991).

67. R. Rohr, M. Bäumer, H. J. Freund, J. A. Mejias, V. Staemmler, S. Muller, L. Hammer, K. Heinz, *Surface Science* **389**, 391 (1997).

68. R. M. Jaeger, H. Kuhlenbeck, H. J. Freund, M. Wuttig, W. Hoffmann, R. Franchy, H. Ibach, *Surface Science* **259**, 235 (1991).

69. R. Franchy, J. Masuch, P. Gassmann, *Applied Surface Science* **93**, 317 (1996).

70. C. Becker, J. Kandler, H. Raaf, R. Linke, T. Pelster, M. Dräger, M. Tanemura, K. Wandelt, *Journal of Vacuum Science & Technology A* **16**, 1000 (1998).

71. J. Wollschläger, J. Viernow, C. Tegenkamp, D. Erdös, K. M. Schröder, H. Pfnür, *Applied Surface Science* **142**, 129 (1999).

72. D. R. Mullins, P. V. Radulovic, S. H. Overbury, *Surface Science* **429**, 186 (1999).

73. J. L. Lu, H. J. Gao, S. Shaikhutdinov, H. J. Freund, *Surface Science* **600**, 5004 (2006).

74. S. Surnev, M. G. Ramsey, F. P. Netzer, *Progress in Surface Science* **73**, 117 (2003).

75. S. Guimond, J. M. Sturm, D. Göbke, Y. Romanyshyn, M. Naschitzki, H. Kuhlenbeck, H.-J. Freund, *The Journal of Physical Chemistry C* **112**, 11835 (2008).

76. D. Löffler, J. J. Uhlrich, M. Baron, B. Yang, X. Yu, L. Lichtenstein, L. Heinke, C. Büchner, M. Heyde, S. Shaikhutdinov, H. J. Freund, R. Włodarczyk, M. Sierka, J. Sauer, *Physical Review Letters* **105**, 146104 (2010).

77. H. Hövel, I. Barke, *Progress in Surface Science* **81**, 53 (2006).

78. D. R. Frankl, J. A. Venables, *Advances in Physics* **19**, 409 (1970).

79. P. Milani, S. Iannotta, *Cluster Beam Synthesis of Nanostructured Materials* (Springer-Verlag Berlin Heidelberg, 1999).

80. K. Wegner, P. Piseri, H. V. Tafreshi, P. Milani, *Journal of Physics D: Applied Physics* **39**, R439 (2006).

81. M. Arenz, U. Landman, U. Heiz, *ChemPhysChem* **7**, 1871 (2006).

82. C. Binns, *Surface Science Reports* **44**, 1 (2001).

83. C. Binns, In *Handbook of Metal Physics*, J. A. Blackman (ed.), Elsevier, p. 49 (2008).

84. H. Haberland, M. Mall, M. Moseler, Y. Qiang, T. Reiners, Y. Thurner, *Journal of Vacuum Science & Technology A* **12**, 2925 (1994).

85. D. A. Eastham, B. Hamilton, P. M. Denby, *Nanotechnology* **13**, 51 (2002).

86. K. Bromann, H. Brune, C. Félix, W. Harbich, R. Monot, J. Buttet, K. Kern, *Surface Science* **377–379**, 1051 (1997).

87. X. Tong, L. Benz, P. Kemper, H. Metiu, M. T. Bowers, S. K. Buratto, *Journal of the American Chemical Society* **127**, 13516 (2005).

88. L. Huang, S. J. Chey, J. H. Weaver, *Physical Review Letters* **80**, 4095 (1998).

89. S. Pratontep, S. J. Carroll, C. Xirouchaki, M. Streun, R. E. Palmer, *Review of Scientific Instruments* **76**, 045103 (2005).

90. S. R. Plant, L. Cao, R. E. Palmer, *Journal of the American Chemical Society* **136**, 7559 (2014).

91. E. Gross, Y. Horowitz, M. Asscher, *Langmuir* **21**, 8892 (2005).

92. J. S. Palmer, S. Sivaramakrishnan, P. S. Waggoner, J. H. Weaver, *Surface Science* **602**, 2278 (2008).

93. E. Gross, M. Asscher, M. Lundwall, D. W. Goodman, *The Journal of Physical Chemistry C* **111**, 16197 (2007).

94. T. H. Baum, C. R. Jones, *Journal of Vacuum Science & Technology B* **4**, 1187 (1986).

95. E. Feurer, H. Suhr, *Applied Physics A* **44**, 171 (1987).

96. E. Szłyk, P. Piszczek, I. Łakomska, A. Grodzicki, J. Szatkowski, T. Błaszczyk, *Chemical Vapor Deposition* **6**, 105 (2000).

97. P. D. Tran, P. Doppelt, *Journal of the Electrochemical Society* **154** D520 (2007).

98. A. A. Bessonov, N. B. Morozova, N. V. Gelfond, P. P. Semyannikov, S. V. Trubin, Y. V. Shevtsov, Y. V. Shubin, I. K. Igumenov, *Surface and Coatings Technology* **201**, 9099 (2007).

99. K.-i. Fukui, S. Sugiyama, Y. Iwasawa, *Physical Chemistry Chemical Physics* **3**, 3871 (2001).

100. M. Sterrer, H.-J. Freund, *Catalysis Letters* **143**, 375 (2013).

101. G. Kästle, H. G. Boyen, F. Weigl, G. Lengl, T. Herzog, P. Ziemann, S. Riethmüller, O. Mayer, C. Hartmann, J. P. Spatz, M. Möller, M. Ozawa, F. Banhart, M. G. Garnier, P. Oelhafen, *Advanced Functional Materials* **13**, 853 (2003).

102. B. R. Cuenya, S.-H. Baeck, T. F. Jaramillo, E. W. McFarland, *Journal of the American Chemical Society* **125**, 12928 (2003).

103. L. Ono, B. Roldán-Cuenya, *Catalysis Letters* **113**, 86 (2007).

104. A. Eppler, J. Zhu, E. Anderson, G. Somorjai, *Topics in Catalysis* **13**, 33 (2000).

105. J. Grunes, J. Zhu, M. Yang, G. Somorjai, *Catalysis Letters* **86**, 157 (2003).

106. S. Johansson, E. Fridell, B. Kasemo, *Journal of Catalysis* **200**, 370 (2001).

107. K. Wong, S. Johansson, B. Kasemo, *Faraday Discussions* **105**, 237 (1996).
108. Y. Weisheng, W. Zhihong, Y. Yang, C. Longqing, S. Ahad, W. Kimchong, W. Xianbin, *Journal of Micromechanics and Microengineering* **22**, 125007 (2012).
109. P. M. Mendes, S. Jacke, K. Critchley, J. Plaza, Y. Chen, K. Nikitin, R. E. Palmer, J. A. Preece, S. D. Evans, D. Fitzmaurice, *Langmuir* **20**, 3766 (2004).
110. M. H. V. Werts, M. Lambert, J.-P. Bourgoin, M. Brust, *Nano Letters* **2**, 43 (2002).
111. M.-V. Meli, R. B. Lennox, *Langmuir* **19**, 9097 (2003).
112. N. Stokes, A. McDonagh, M. Cortie, *Gold Bulletin* **40**, 310 (2007).
113. U. Diebold, *Surface Science Reports* **48**, 53 (2003).
114. L. Zhang, R. Persaud, T. E. Madey, *Physical Review B* **56**, 10549 (1997).
115. S. C. Parker, A. W. Grant, V. A. Bondzie, C. T. Campbell, *Surface Science* **441**, 10 (1999).
116. K. S. Ashok, K. Andrei, Y. Fan, D. W. Goodman, *Japanese Journal of Applied Physics* **42**, 4795 (2003).
117. N. Spiridis, J. Haber, J. Korecki, *Vacuum* **63**, 99 (2001).
118. T. Diemant, H. Hartmann, J. Bansmann, R. J. Behm, *Journal of Catalysis* **252**, 171 (2007).
119. E. Wahlström, N. Lopez, R. Schaub, P. Thostrup, A. Rønnau, C. Africh, E. Lægsgaard, J. K. Nørskov, F. Besenbacher, *Physical Review Letters* **90**, 026101 (2003).
120. O. Bikondoa, C. L. Pang, R. Ithnin, C. A. Muryn, H. Onishi, G. Thornton, *Natural Materials* **5**, 189 (2006).
121. D. Matthey, J. G. Wang, S. Wendt, J. Matthiesen, R. Schaub, E. Lægsgaard, B. Hammer, F. Besenbacher, *Science* **315**, 1692 (2007).
122. F. Cosandey, L. Zhang, T. E. Madey, *Surface Science* **474**, 1 (2001).
123. M. Baron, O. Bondarchuk, D. Stacchiola, S. Shaikhutdinov, H. J. Freund, *The Journal of Physical Chemistry C* **113**, 6042 (2009).
124. J. L. Lu, H. J. Gao, S. Shaikhutdinov, H. J. Freund, *Catalysis Letters* **114**, 8 (2007).
125. S. K. Shaikhutdinov, R. Meyer, M. Naschitzki, M. Bäumer, H. J. Freund, *Catalysis Letters* **86**, 211 (2003).
126. C. Winkler, A. Carew, R. Raval, J. Ledieu, R. McGrath, *Surface Review and Letters* **08**, 693 (2001).
127. N. Nilius, M. V. Ganduglia-Pirovano, V. Brázdová, M. Kulawik, J. Sauer, H. J. Freund, *Physical Review Letters* **100**, 096802 (2008).
128. K. H. Hansen, T. Worren, S. Stempel, E. Lægsgaard, M. Bäumer, H. J. Freund, F. Besenbacher, I. Stensgaard, *Physical Review Letters* **83**, 4120 (1999).
129. L. Vitos, A. V. Ruban, H. L. Skriver, J. Kollár, *Surface Science* **411**, 186 (1998).
130. Z. H. Qin, M. Lewandowski, Y. N. Sun, S. Shaikhutdinov, H. J. Freund, *The Journal of Physical Chemistry C* **112**, 10209 (2008).
131. M. Yulikov, M. Sterrer, M. Heyde, H.-P. Rust, T. Risse, H.-J. Freund, G. Pacchioni, A. Scagnelli, *Physical Review Letters* **96**, 146804 (2006).
132. H. M. Benia, X. Lin, H. J. Gao, N. Nilius, H. J. Freund, *The Journal of Physical Chemistry C* **111**, 10528 (2007).
133. A. Sanchez, S. Abbet, U. Heiz, W. D. Schneider, H. Häkkinen, R. N. Barnett, U. Landman, *The Journal of Physical Chemistry A* **103**, 9573 (1999).

134. S. Lee, C. Fan, T. Wu, S. L. Anderson, *The Journal of Physical Chemistry B* **109**, 11340 (2005).

135. S. Lee, C. Fan, T. Wu, S. L. Anderson, *Journal of the American Chemical Society* **126**, 5682 (2004).

136. D. C. Lim, R. Dietsche, M. Bubek, T. Ketterer, G. Ganteför, Y. D. Kim, *Chemical Physics Letters* **439**, 364 (2007).

137. C. Lemire, R. Meyer, S. K. Shaikhutdinov, H. J. Freund, *Surface Science* **552**, 27 (2004).

138. D. A. Outka, R. J. Madix, *Surface Science* **179**, 351 (1987).

139. J. M. Gottfried, K. J. Schmidt, S. L. M. Schroeder, K. Christmann, *Surface Science* **536**, 206 (2003).

140. K. P. McKenna, A. L. Shluger, *The Journal of Physical Chemistry C* **111**, 18848 (2007).

141. S. Giorgio, M. Cabié, C. R. Henry, *Gold Bulletin* **41**, 167 (2008).

142. D. Starr, S. Shaikhutdinov, H.-J. Freund, *Topics in Catalysis* **36**, 33 (2005).

143. X. Lai, D. W. Goodman, *Journal of Molecular Catalysis A: Chemical* **162**, 33 (2000).

144. A. Kolmakov, D. W. Goodman, *Catalysis Letters* **70**, 93 (2000).

145. A. Kolmakov, D. W. Goodman, *Surface Science* **490**, L597 (2001).

146. T. Dellwig, G. Rupprechter, H. Unterhalt, H. J. Freund, *Physical Review Letters* **85**, 776 (2000).

147. O. Hakkel, Z. Pászti, A. Berkó, K. Frey, L. Guczi, *Catalysis Today* **158**, 63 (2010).

148. T. Kawai, D. J. Neivandt, P. B. Davies, *Journal of the American Chemical Society* **122**, 12031 (2000).

149. P. Jiang, S. Porsgaard, F. Borondics, M. Köber, A. Caballero, H. Bluhm, F. Besenbacher, M. Salmeron, *Journal of the American Chemical Society* **132**, 2858 (2010).

150. A. Y. Klyushin, T. C. R. Rocha, M. Havecker, A. Knop-Gericke, R. Schlogl, *Physical Chemistry Chemical Physics* **16**, 7881 (2014).

151. X. Lai, T. P. S. Clair, M. Valden, D. W. Goodman, *Progress in Surface Science* **59**, 25 (1998).

152. M. Valden, X. Lai, D. W. Goodman, *Science* **281**, 1647 (1998).

153. M. Valden, S. Pak, X. Lai, D. W. Goodman, *Catalysis Letters* **56**, 7 (1998).

154. M. Sterrer, T. Risse, M. Heyde, H.-P. Rust, H.-J. Freund, *Physical Review Letters* **98**, 206103 (2007).

155. X. Shao, S. Prada, L. Giordano, G. Pacchioni, N. Nilius, H.-J. Freund, *Angewandte Chemie International Edition* **50**, 11525 (2011).

156. R. Ferrando, J. Jellinek, R. L. Johnston, *Chemical Reviews* **108**, 845 (2008).

157. K. Yun, Y.-H. Cho, P.-R. Cha, J. Lee, H.-S. Nam, J. S. Oh, J.-H. Choi, S.-C. Lee, *Acta Materialia* **60**, 4908 (2012).

158. S. E. Hörnström, L. I. Johansson, A. Flodström, *Applied Surface Science* **26**, 27 (1986).

159. H. Niehus, C. Achete, *Surface Science* **289**, 19 (1993).

160. M. Aschoff, S. Speller, J. Kuntze, W. Heiland, E. Platzgummer, M. Schmid, P. Varga, B. Baretzky, *Surface Science* **415**, L1051 (1998).

161. D. T. Tran, R. L. Johnston, *Proceedings of the Royal Society of London A: Mathematical, Physical and Engineering Sciences* **467**, 2004 (2011).

162. L. Deng, W. Hu, H. Deng, S. Xiao, *The Journal of Physical Chemistry C* **114**, 11026 (2010).

163. L. Deng, W. Hu, H. Deng, S. Xiao, J. Tang, *The Journal of Physical Chemistry C* **115**, 11355 (2011).

164. M. Chen, D. Kumar, C.-W. Yi, D. W. Goodman, *Science* **310**, 291 (2005).

165. J. R. Kitchin, K. Reuter, M. Scheffler, *Physical Review B* **77**, 075437 (2008).

166. K. Luo, T. Wei, C. W. Yi, S. Axnanda, D. W. Goodman, *The Journal of Physical Chemistry B* **109**, 23517 (2005).

167. D. Mott, J. Luo, A. Smith, P. N. Njoki, L. Wang, C.-J. Zhong, *Nanoscale Research Letters* **2**, 12 (2007).

Chapter 13

Theoretical Studies of Gold Nanoclusters in Various Chemical Environments: When the Size Matters

Hannu Häkkinen

University of Jyväskylä, Jyväskylä, Finland

13.1 Introduction

Gold nanoparticles (AuNPs) exhibit a rich array of interesting and important electronic, optical, chemical and catalytic properties, which has sparked a huge interest in gold-based systems in several interdisciplinary areas, leading to an explosive growth in the volume of both experimental and theoretical research.[1,2] (For reviews on gold cluster and nanoparticles, see Ref. 3). A large variety of AuNPs differing by their size (1–100 nm), shape and surrounding can been synthesised (see Chapters 6 and 7). Among them the nanoparticles termed as clusters are made of a countable number of atoms (less than 150 atoms), which corresponds to particles smaller than 2 nm, and they offer a unique playground where the most elaborate *ab initio* calculation can accurately reproduce and interpret the experimental situations. This chapter contains an overview of the developments of structural determination of gold clusters in gas phase and as stabilised chemically by various ligands. Many gold particles with precise structure and properties can now be synthesised by wet chemistry and they can be handled as normal chemicals: stored, modified and functionalised for applications in medical therapy, biolabelling, sensing, nanoelectronics and catalysis. In recent years, understanding of the stability, surface chemistry and functionalisation of these interesting building blocks of nanomatter has taken a quantum leap. This is facilitated by simultaneous breakthroughs in experimental and theoretical fronts concerning accurate structural determinations of thiolate-stabilised gold clusters of 1–3 nm in diameter, in conjunction with computational studies.

Computational studies on these systems are challenging for many reasons. The chemical and physical properties of gold can be understood only if relativistic effects are taken properly into account (Chapter 2). This makes gold a very peculiar metal, which gives rise to specific properties.[2] The size of many systems imposes a numerical burden for codes based on the density functional theory. The systems are complex, since the ligand-passivated nanocluster exhibits properties that call for an understanding of both the surface (covalent) chemistry between metal atoms and molecules, and the origins of the 'metallic' properties of the gold core.

In dealing with the electronic properties, bulk gold is a good (6s) free-electron metal. These properties result from electrons shared between an infinite number of gold atoms and it is natural to ask up to what extent this free-electron behaviour shows up in the electronic structure when the number of atoms is reduced until forming the nanocluster. (For an authoritative review on cluster production techniques and analysis of electronic properties of simple metal clusters, see Ref. 4). Nanoparticles larger than 2 nm are known to be 'metallic', i.e. they have an appreciable density of electron states at the Fermi level (no HOMO–LUMO gap) and exhibit a typical surface plasmon at about 520 nm. Smaller particles will have a finite HOMO–LUMO gap due to constriction of the delocalised Au(6s) derived states in the finite volume. A rough but instructive estimate follows an elementary discussion of a 3D free-electron metal.[5] For such a system, the density of electron levels depends on the energy as $D(\varepsilon) \propto \sqrt{\varepsilon}$. At the Fermi energy ε_F (energy of the highest occupied electron state), the mean spacing of electron levels is $\delta(\varepsilon_F) = 1/D(\varepsilon_F) = 2\varepsilon_F/3Nz)$ where N is the number of atoms and z is the valence. Thus, decreasing the size of the particle (N) one increases the energy gap at the Fermi energy. It is relevant to compare this energy gap to thermal excitations that are of the order of kT (0.025 eV at room temperature). When the gap exceeds kT, quantum size effects due to discreteness of the energy levels become dominant. For this rough estimate, we can treat gold as monovalent, i.e. $z = 1$ and take a free electron value of $\varepsilon_F = 5.5$ eV, which gives a limiting particle size of $N = 150$ atoms corresponding to a critical diameter $d = 1.7$ nm. Below that size, gold clusters are expected to turn from 'metallic' to 'semiconducting'.

This contribution deals with theoretical studies of gold clusters with a countable number of atoms (below 150). This regime is distinct from the larger nanoparticles that are discussed in most of the other chapters of this book. In the nanocluster regime, the atomic and electronic structures are intimately related, thus the computational results presented herein provide an overview over the achievements in understanding thermodynamic stability of gold clusters of increasing sizes: starting from a gold triangle with 3 atoms up to clusters made of 150 atoms. The relationship between structure and chemical properties, the transition from 'semiconducting' to 'metallic' electronic properties, the reactivity, the ligand-bond formation and the catalytic activity of these clusters are also tackled.

Given the huge body of published work in the area, this discussion is not meant to be an exhaustive review, rather it reflects personal views of the author who has had the privilege of being heavily involved in an extensive theoretical and experimental collaborative network since the late 1990s. At the same time, it provides a certain glimpse on the timeline of development of ideas, which is hoped to be beneficial particularly for a novice entering this exciting field.

Revised in 2015, this contribution now includes also a brief discussion of central computational methods and a brief overview of gold-based bimetallic nanoclusters, where significant advances have been recently made particularly in the field of nanoclusters stabilised by organic ligands, made by wet chemistry.

13.2 Computational Methods

Finding 'globally' optimised structures of nanoclusters is generally an extremely challenging problem due to a very rapid increase in the dimension of the configurational space as the system size increases. Independent of the level of theory that is used to calculate the interactions (forces) between the atoms, the optimisation algorithms need to be able to sample the configurational space in a representative way, exploring a large number of possible energy minima. Standard methods include genetic algorithms, simulated annealing and basin hopping strategies. A fairly recent review of various optimisation methods has been given by Ferrando.[6]

Once a representative set of candidate structures is found, several criteria can be used for validation. A comparison of total energies of the candidates gives an idea about their thermodynamic stability. Isomeric structures have different shapes, which can be used as a criterion when comparing to mobility experiments in the gas phase. Computationally obtained structural coordinates can also be used to calculate, e.g. electron diffraction patterns or infrared frequencies and intensities to be compared to available experimental data. Clusters supported on solid substrates can be imaged by transmission electron microscopy (TEM) and/or scanning tunnelling microscopy (STM), and in the most favourable cases atomic-scale or near atomic-scale structural information can be obtained, which is then directly comparable to simulations. It can be expected that in the coming years, with the improved resolution of TEM and STM and rapidly rising computational power to handle ever larger systems, the combined approach of simulation and microscopy will be a powerful combination to look at these systems.

In recent years, one particular sub-field of cluster science, namely that of ligand-stabilised noble metal clusters, has grown very rapidly and is presently in an active discovery phase. Tens of cluster compounds of a general form $M_x L_y$ (M = metal, L = ligand) have been synthesised and many have been successfully crystallised. The current status of this field is summarised in a recent book.[7] The availability of a large database of crystal structures gives an excellent starting point for theory to understand the properties of these novel building blocks of nanomaterials by using density functional theory (DFT) for the ground-state properties or time-dependent density functional theory (TDDFT) for optical properties (linear absorption and circular dichroism). Atomistic DFT and TDDFT methods can currently handle noble metal clusters up to about 500 metal atoms.

The structural database for the ligand-stabilised clusters facilitates also parametrisation of the atomic interactions into force fields that are useful for classical molecular dynamics (MD) simulations where the clusters are interacting with the environment. By using effective MD algorithms such as GROMACS, multi-million atom simulations have been already performed looking at the interaction between thiolated gold clusters and model lipid membranes.[8]

13.3 Clusters in Gas Phase

A few experimental methods have been developed to produce clusters in gas phase. They can be considered as model systems since they are made of pure gold but the challenge relies in the control of the actual size and the dispersity of the population being produced (see Chapters 6 and 7). This section deals with clusters made of 3–50 gold atoms and focuses on their thermodynamic stability and the evolution of their density of states as the number of atoms increases. Determination of absolute structures of metal nanoclusters in gas phase is experimentally a very challenging task. Cluster beams are controlled by electrostatic fields and most experimental techniques for structure analysis work conveniently only for beams of charged clusters.[4]

13.3.1 *Cationic Clusters Au_N^+*

The first systematic structure determination for a range of cluster cations came from ion mobility experiments by the Karlsruhe group for $4 \leq N \leq 20$.[9] In the ion mobility experiment, an electrostatic field accelerates size-selected, charged clusters through a drift tube, filled by inert carrier gas (such as He and Ar). Collisions to the molecules of the carrier gas provide a drag force affecting the flight time through the tube. The flight time is measured and gives access to the collision cross-sections. This latter is also obtained from theoretical calculations (density functional theory, DFT) of a set of structure candidates for a given cluster size. The comparison between measured and computed cross-sections is crucial to conclude on the actual geometry of gold clusters.

The measured cross-sections up to $N = 20$ are shown in Figure 13.1 together with the calculated ones for the energetically most favourable cluster isomers up to $N = 13$. A close-up of the ground-state structures as well as of some close-lying isomers for $N = 6, 8, 10$ is shown in Figure 13.2. Comparison to the theory shows that up to $N = 7$, the measured data points essentially coincide with the calculated values for the ground-state structures, all planar: a triangle, a rhombus, an 'hour-glass shape', a triangle and a centred hexagon for $N = 3$–7, respectively. All of these geometries are simply fragments of a close-packed hexagonal plane. This preference

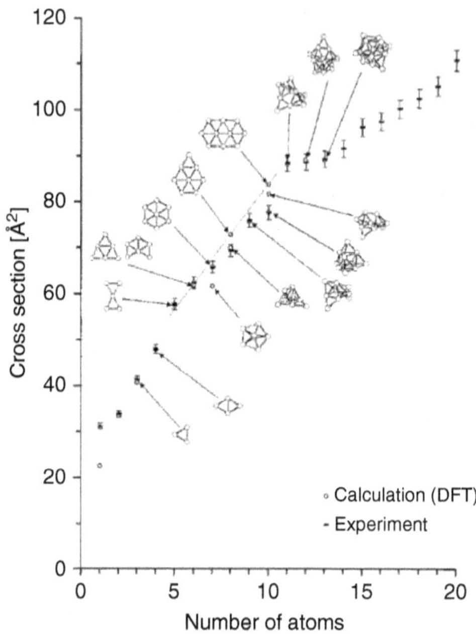

Figure 13.1   Measured and predicted ion mobility cross-sections for gold cluster cations. Reproduced from Ref. 9 by permission. Copyright 2002 American Institute of Physics.

of gold for arranging into planar geometry is a good illustration of the achievements yielded by the combination between *ab initio* calculations and experimental observations. For $8 \leq N \leq 13$, an equally good match is observed for three-dimensional structures, which in many cases can be described as slightly relaxed fragments of fcc bulk structure, note e.g. the tetrahedral structure for $N = 10$ (isomer II in Figure 13.2). For $N = 8$, 10 a close-lying isomer (8-II and 10-II) gave the best correspondence with the measured cross-section. Here, the isomers were within 0.1 eV from the calculated ground state structure. Planar structures 8-I and 10-III, while predicted being energetically as good as the 3D structures 8-II and 10-II, were not observed in the experiment. The comparison shows that while theory is indispensable for interpretation of the experimental data, care should be exercised for taking theoretical structure prediction for granted, since little details in the ways the electron–electron interaction is treated in DFT may affect the sensitive energy balance between structure isomers of different dimensionality. The structural database of gas-phase gold clusters (both

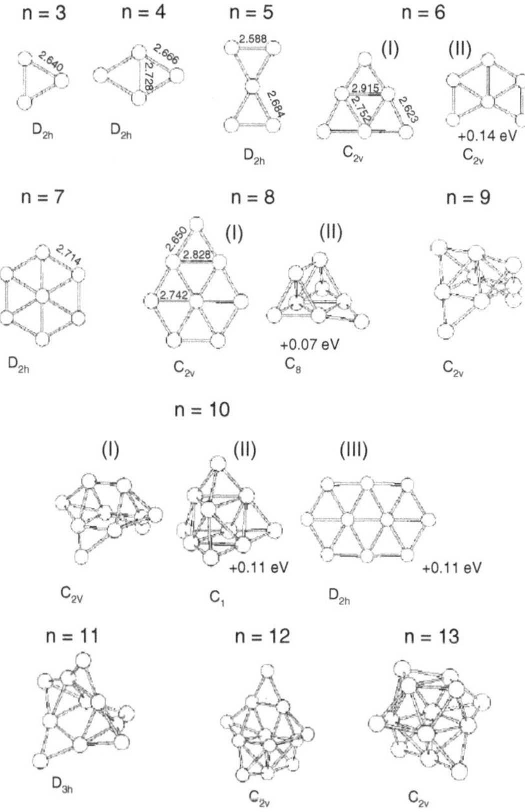

Figure 13.2   Calculated of the most stable of gold cluster cations (low-energy isomers). Up to seven atoms, the clusters arrange into planar geometry. Reproduced from Ref. 9 by permission. Copyright 2002 American Institute of Physics.

cations and anions, to be discussed next) has served as a valuable benchmark for developing and testing reliable DFT methods for gold clusters for the past 10 years.

13.3.2 *Anionic Clusters* Au_N^-

Anionic clusters are amenable to structure determination by the ion mobility method as explained before. Besides cross-sections, other experimental approaches such as photoelectron spectroscopy (PES) or trapped-ion electron diffraction (TIED) are used to confirm the structures. PES uses a UV or visible laser to produce photoelectrons whose detection provides information on the detachment energies of electrons from negatively charged clusters (analogous to the well-known photoelectric effect on bulk metal surfaces). The vertical detachment energy (VDE) is the energy required to remove the extra electron from the anion. This information can be compared to the structure of electron states obtained in a DFT calculation. In the TIED method, a 'cloud' of size-selected, charged clusters is trapped by a set of electrostatic fields and is irradiated by a collimated electron beam. The scattering function of electrons can be measured, and the method can be considered to be analogous to powder X-ray diffraction. The experimental effort on anionic gold clusters intensified around 2001–2003 with several reports from ion mobility and PES experiments supported by DFT calculations.[9–11] A remarkable tendency of gold cluster anions to favour planar structures up to fairly large sizes was quickly discovered. Therefore, an intriguing question is to determine at what size the 2D to 3D transition occurs when the number of atoms increases in gold clusters.

The Karlsruhe ion mobility experiment reported in 2002[10] convincingly set the 2D/3D transition size to $N = 12$. For $N = 12$, a bimodal arrival time distribution was observed and two distinct cross-sections measured, allowing for a conclusion that for this cross-over size, both planar and three-dimensional structure isomers were present in the beam. It was concluded that for all the other sizes only one isomer was observed. For most cases, the ground state structure predicted by DFT gave also the best fit to the experimental cross-section, exceptions included 4-, 10- and 13-atom clusters where the first isomer above ground-state gave the best fit.

A complementary follow-up work involved high-resolution photoelectron spectroscopy (HR-PES) with the data interpreted via DFT calculations for the density of single-electron states (DOS) in the cluster anion for an extensive set of cluster isomers in the size range $4 \leq N \leq 14$.[11] In the single-particle interpretation, the measured PES data contains information about the distribution of binding energies of valence electrons detached from the cluster by a photon. These energies are also accessible with DFT calculations, including the energy of the weakest bound electron (vertical detachment energy (VDE), whose counterpart is the ionisation potential for a finite neutral cluster and the work function for bulk metal surface). Comparing these 'fingerprints' may thus provide a more sensitive measure to

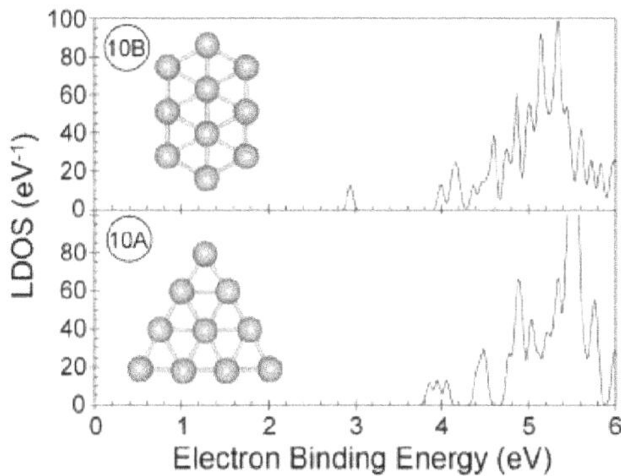

Figure 13.3   The theoretical DOS of the two lowest-energy structures of Au_{10}^-. Adapted from Ref. 11 by permission. Copyright 2002 American Chemical Society.

judge the presence of a given isomer or isomers in the beam, as compared to making structure assignments based on the collision cross-section, which is a single number for a given cluster isomer.

The theoretical work reported in Ref. 11 confirmed the energetic stability of the earlier reported planar structures. In addition, comparison to the measured PES data (VDE values and spectral details) gave evidence of isomers that were present in the cluster beam for $N = 4, 8, 12$, and 13. As a prominent example, the theory predicts two close-lying planar structures for $N = 10$: a triangular and an elongated close-packed flake, see 10A and 10B in Figure 13.3. The two independent DFT calculations[10,11] gave a consistent result by predicting the triangular structure 10A to be the ground state and the elongated isomer 10B to be 0.12 eV (Ref. 11) or 0.15 eV (Ref. 10) higher in energy. The geometrical cross-section of the isomer 10B fits better with the measured collision cross-section. The two structures deviate significantly from each other regarding the VDE value: for the ground state, it is calculated to be 3.86 eV (Ref. 11) or 4.02 eV (Ref. 10) and for the higher-energy isomer, it is 2.94 eV (Ref. 11) or 3.08 eV (Ref. 10). Early PES studies[12] assigned an experimental VDE of Au_{10}^- to be around 3 eV, which would be consistent with the higher energy isomeric structure. However, the high-resolution experiment revealed that the low-energy feature in the PES is due to a minor isomer present in the beam and the experimental VDE for the dominant Au_{10}^- structure in the beam was determined to be

3.91 eV, which is in an excellent agreement with the theoretical VDE values calculated for the ground state.[11] It has been shown recently that these isomers have different reactivities with molecular oxygen and their relative abundance can be controlled by source conditions.[13]

Not only does thermodynamic stability rule the cluster formation but kinetics may play even a more drastic role in enabling the co-existence of cluster isomers of different dimensionality in the beam; the 2D/3D crossover size $N = 12$ is one example. DFT calculations give a consistent large energy separation of about 0.5 eV for the most favourable 2D and 3D structures. This energy difference is far too large to be explained by thermal population and interconversion of structures under room temperature conditions in the beam. The most likely explanation is in the kinetics of formation of two structural 'families', planar and three-dimensional, and their fast cooling in the beam. This was simulated via DFT-based tight-binding molecular dynamics simulations of the cooling process that showed that in the 2D/3D crossover region, supercooling to 'wrong dimensionality' is possible.[14] It has to be noted that the standard DFT calculations probing energetics of isomers are strictly valid for $TH = 0$ conditions only (no entropy effects included) and for that reason real molecular dynamics studies like the one reported in Ref. 14, using forces from semi-empirical theory such as tight-binding (or in some cases from DFT is the computer resources allow) should prove useful for probing thermal/entropy effects present in cluster beams.

Why does gold favour such large planar atomic structures? A systematic theoretical study of noble metal clusters Cu_7^-, Ag_7^- and Au_7^-, carried out in 2002, gave at least partial answers.[15] The stability of the planar structures was traced back to strong relativistic effects in bonding in gold,[2] which induce s–d hybridisation, contraction of the Au–Au bond length and a significant overlap of d-orbitals.[15,16]

It is interesting to consider the effects of a particular feature of the electronic structure of $N = 12$ clusters that has been determined to be the 2D/3D crossover size. As mentioned in the Introduction, a subset of the full electronic structure of gold clusters consists of Au(6s)-derived delocalised electron states.[17] Their global symmetries can be analysed and related to a planar quantum-dot model (Figure 13.4).[18] In that model, an electron number of 12 constitutes a shell closing. After the shell closure at 12 electrons,

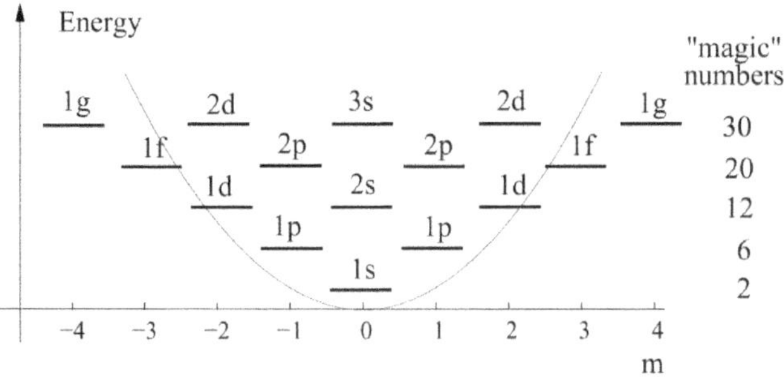

Figure 13.4   A schematics of electron states in a planar harmonic quantum dot. Gold clusters containing a 'magic number' of electrons have closed shell orbitals and exhibit a higher stability.

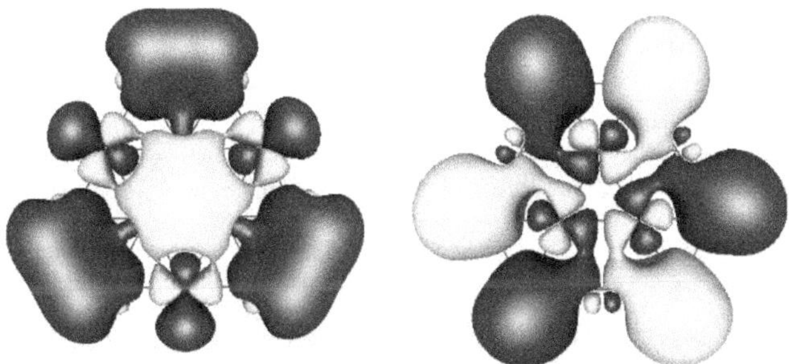

Figure 13.5   Left: HOMO-1 state of D_{3h} Au_{12}^- cluster. Right: the HOMO state. 2S and 1F symmetries are seen.

a new 8-electron shell opens with 2P and 1F symmetries, leading to the next 'magic number' in a plane, 20. Figure 13.5 shows the two highest occupied orbitals for D_{3h} Au_{12}^-, for which the effective 6s-electron count is 13. Indeed the highest state, the singly occupied molecular orbital (SOMO) shows a clear 1F symmetry and the state below that has a clear 2S symmetry of the 6-electron 2S1D shell. This shell closing is responsible for the large HOMO–LUMO gap of neutral Au_{12}, visible in the experimental photoelectron spectrum of the anion.[12]

13.3.3 *From Flakes to Cages to Tubes: Anionic Clusters with N = 13–24*

Two recent experimental investigations for larger anionic clusters, a PES study[17] and a trapped ion electron diffraction study,[19] confirmed the early

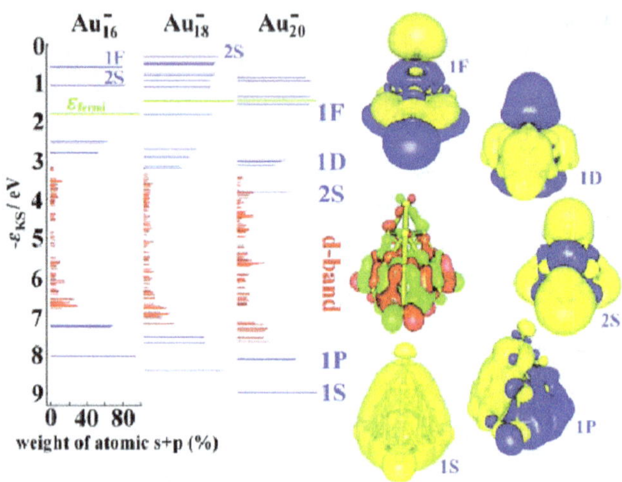

Figure 13.6   Left: Angular-momentum analysis of Kohn–Sham eigenstates for Au_{16}^-, Au_{18}^- and Au_{20}^-. The states marked in blue have high local sp weight on atoms; globally they are free-electron-states over the cluster. A visualisation of such states for Au_{20}^- is on the right. Reproduced from Ref. 17 by permission. Copyright 2007 Wiley.

conclusions from mobility experiments regarding the 2D/3D structural cross-over at the size Au_{12}^-. The experimental data were compared to the corresponding theoretical quantities using the same isomer database.[17] These studies confirmed the earlier photoelectron results for tetrahedral structures of Au_{16}^{-20} and Au_{20}^{-21} Furthermore, these investigations showed the gradual transformation of the optimal structure from a near-planar, flat 'cage' ($N = 13, 14$) to a tetrahedral cage, evolving finally to tubular structures for $N > 20$.

Figure 13.6 shows the computed energy-level diagram of electrons for cluster anions with $N = 16, 18, 20$, and visualisation of selected orbitals for Au_{20}^- (the special Au_{16}^- case is discussed in the next section). Angular momentum analysis of the orbitals of Au_{20}^- shows that they can be generally divided into two major classes, the Au(5d) derived 'band' of states in the energy range –3 to –8 eV, and to delocalised Au(6s) derived states above and below the 5d-band. A very important conclusion is obtained from these calculations: qualitatively, this medium-size cluster already shows all the characteristics of the bulk band of gold, where the free-electron-like 6s-derived band crosses the 5d-derived band and extends all the way from the gamma-point to the Fermi surface.[21] The upper edge of the 5d-band is about $2\,\mathrm{eV}$ below the Fermi surface, which also qualitatively matches the behaviour of the states in the Au_{20}^{-} cluster. The Au_{20}^{-} then shows the expected free-electron-like shell filling pattern of 21 electrons: $1S^2 1P^6 2S^2 1D^{10} 1F^1$ with $1F^1$ as SOMO, confirming that the large HOMO–LUMO gap of the neutral Au_{20} is the one between 20 and 21 electrons in the free-electron model. Many more examples of the free-electron behaviour of medium-size gold clusters can be analyzed, and some are given below.

13.3.4 Au_{16}^{-} : *The Smallest Golden Cage and the Manifestation of Shell Closing of 18 Delocalised Electrons*

In 2006, a combined photoelectron spectroscopy and density functional study[20] concluded that the experimentally observed isomer of the Au_{16}^{-} anion has tetrahedral symmetry and the geometry can be derived from the previously discovered[21] T_{d}-symmetric Au_{20}^{-} by removing the four vertex atoms and allowing for an outward relaxation of the four face-centred atoms, which yields a structure that was coined as the 'smallest golden cage' (see Figure 13.7).

Since the early 1990s, it had been known that Au_{16}^{-} exhibits an anomalously high VDE.[12] In a follow-up study it was shown that the high VDE of Au_{16}^{-} is due to a tendency to complete a shell of 18 delocalised electrons leading to a stability of the dianionic Au_{16}^{2-} cluster with a predicted HOMO–LUMO gap of $1.5\,\mathrm{eV}$.[23] In the shell-model notation, the relevant shells are $1s^2 1p^6 1d^{10}$ with the 1d shell split by the symmetry. In terms of the T_{d}-symmetric crystal-field split molecular orbitals, the relevant configuration is $a_1^2 t_2^6 e^4 t_2^6$. The 2s shell is at high energies (and thus unoccupied) due to the fact that a radial node is not supported by the hollow cage. The cage maintains its robust geometry, with a minor Jahn–Teller deformation, over several charge states ($q = -1, 0, +1$), forming spin doublet, triplet and quadruplet states according to the Hund's rules. The cage is roomy enough so that it can be doped endohedrally by guest atoms.

Despite the extensive experimental and theoretical work, one still cannot be fully conclusive about all the factors that drive the structural transitions

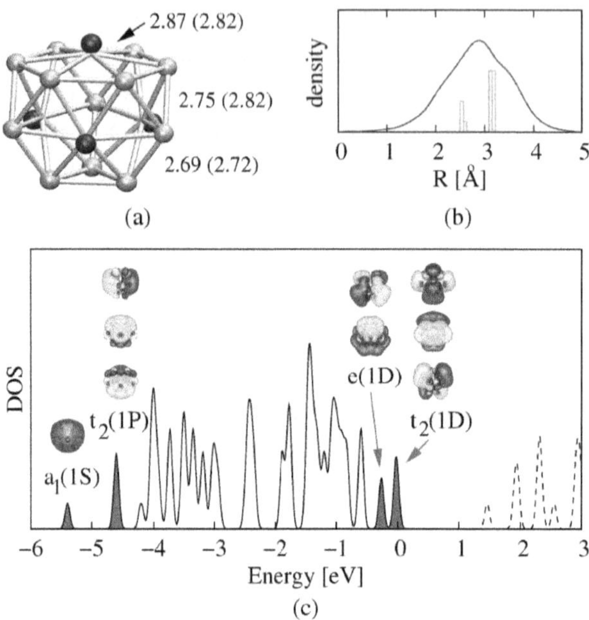

Figure 13.7   Structure of Au$_{16}^q$. The numbers are values of the indicated Au–Au distances in the dianion, $q = -2$ (in parentheses for the cation, $q = +1$). The 'stick' framework indicates the structure of the cation, and the 'balls' are drawn from the coordinates of the dianion. Only the dark-coloured face atoms move significantly during relaxation between different charge states. (b) Radial distribution of atoms (bars) and electrons (line) for the dianion. The radius of the cage is ∼2.5 Å. (c) Density of Kohn–Sham molecular orbitals (DOS, folded by 0.05 eV Gaussians) of Au$_{16}^{2-}$. The HOMO state is at zero energy, and the empty states are denoted by a dashed line. The shaded and labelled peaks denote the delocalised, T_d crystal-field split states that are derived from a spherical cage confinement (angular momentum labelling in parentheses). These MOs are also visualised. t_2 is the six-fold degenerate HOMO orbital (3 spatial orbitals ×2 for spin). Reproduced from Ref. 23 by permission of the PCCP Owner Societies 2007.

as the size of gold cluster anions increases up to about $N = 20$. As discussed above, relativistic bonding effects are indicated from theory to be one factor in stabilising the planar clusters up to $N = 12$. The evolution of 'flat cages' or biplanar structures joint by edges for slightly larger sizes can also be thought as caused by the tendency for quasi-planarity. As discussed above, Au$_{16}^-$ is a 3D hollow cage and Au$_{20}^-$ is a piece of fcc bulk structure. However, clusters that are slightly larger that $N = 20$ develop cage- or tube-like structures again and do not follow the motif of simply adding atoms to the fcc crystallite of $N = 20$.

13.3.5 *Anionic Clusters with N > 30*

Since the early 1990s, the Au_{34}^- anion has been known to have a very large gap in the photoelectron spectrum,[12] reflecting a large HOMO–LUMO gap for the corresponding neutral cluster at 34 s-electrons in a configuration $1s^2 1p^6 1d^{10} 2s^2 1f$.[14] A recent combination of TIED and UV-VIS PES methods in conjunction to density functional theory calculations shed light onto its most likely atomic structure.[24] The best fit to the measured TIED data was found for a C_3 structure that can be constructed from a more symmetric C_{3v} geometry via a twist (see Figure 13.8), which increases the surface packing density, similar to what has been found for helical gold nanowires.[25]

In 2004, HR-PES investigation of cold mass-selected Cu_N^-, Ag_N^- and Au_N^- with $N = 53-58$ revealed interesting systematic behaviour.[26] In principle PES spectra are direct images of the electronic density of states. In the bulk form, Cu, Ag and Au have an electronic structure characterised by a half-filled and rather free-electron-like band, formed from the atomic *s*-orbitals. This band overlaps with the *d*-band, which lies some eV below the Fermi energy/level and is formed by the rather localised atomic *d* orbitals. It

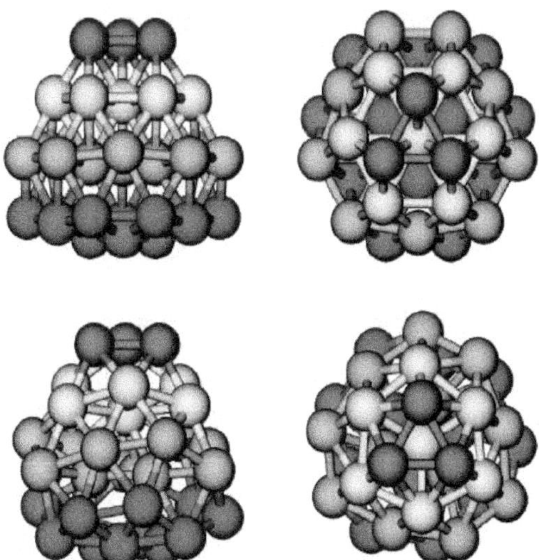

Figure 13.8 Two views of predicted low-energy C_{3v} (top) and C_3 (bottom) structures of Au_{34}^-. The C_3 structure gives the best match with experimental electron diffraction and photoelectron data. Reproduced from Ref. 24 by permission. Copyright 2007 Wiley.

was observed that while the spectra of copper and silver clusters were practically identical in the upper valence band region, gold clusters exhibited a totally different spectral structure with only one band, although this band was heavily detailed. The series of highly degenerate peaks for Cu and Ag is a signature of high symmetry in the atomic structure. Indeed, by comparing the experimental PES to theoretical DOS calculated for several candidate structures, it was unambiguously concluded that copper and silver clusters in that size range have 55-icosahedron based ground-state structures (Figure 13.9). On the other hand, several low-lying, low-symmetry isomers were found for Au_{55}^-. None of these structures produced DOS that would match satisfactorily the observed photoelectron spectral features. It is possible that the true ground-state structure was not found or that in reality there are many low-energy structures present even at low temperatures (around 200 K). The preference for Au_{55}^- to have low-symmetry structures as opposed to symmetric ones was traced to relativistic bonding effects whose most dramatic effect is the shortening of the interatomic bond length and the increase of bulk modulus. Finally, one can remark that the highly symmetric Au_{55}^- metastable isomers, like an icosahedron, support degenerate 2P and 1G shell states (1G split by the I_h symmetry as for Ag_{55}^- and Cu_{55}^-). The large energy gap at around 5 eV electron binding energy in Figure 13.9 for I_h Au_{55}^- can be identified as the '34 electron gap'. This fact has relevance for discussion of symmetric cores of ligand-protected gold clusters discussed in the next section.

13.4 Ligand-protected Nanocluster

13.4.1 *Synthesis of Ligand-protected Gold Nanoparticles*

Several solution-phase preparation methods to grow and stabilise gold nanocluster with various ligands exist, starting from the classic work of Faraday in the mid-1800s (for a more detailed discussion, see Chapters 1 and 6).[27–33] At the beginning of this section, two methods including stabilisation by phosphines (Schmid) or thiols (Brust–Schiffrin) are briefly discussed as they relate to the systems where extensive computational work

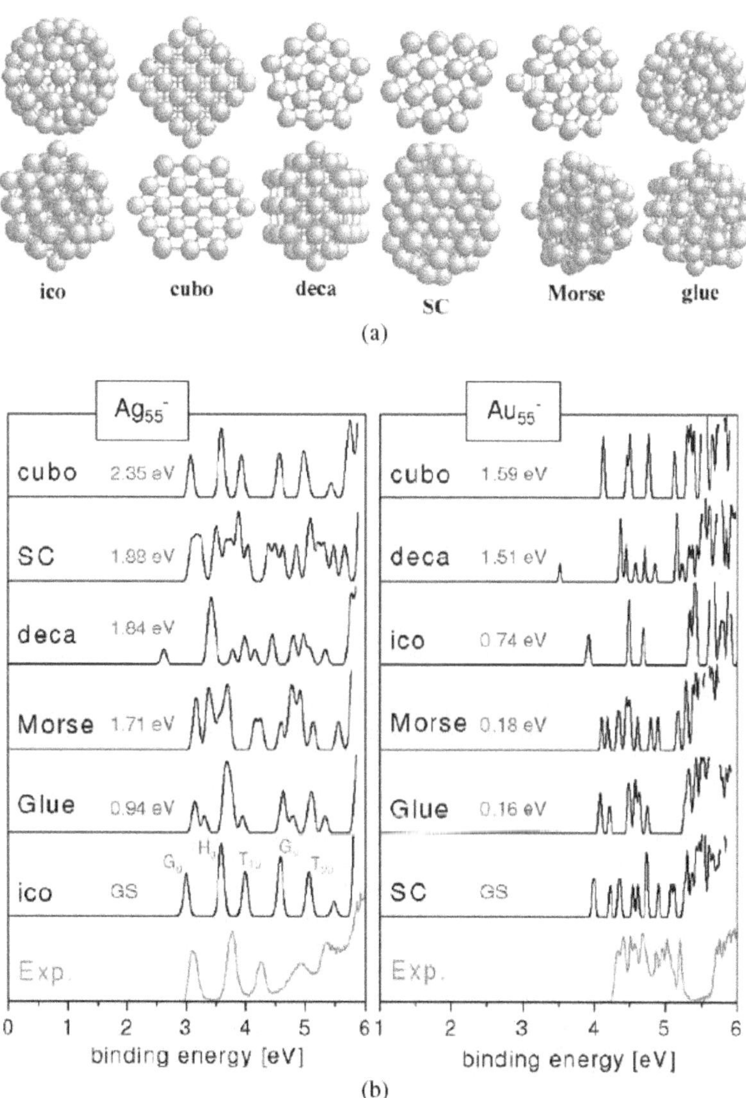

Figure 13.9   Theoretical DOS for a number of structure isomers for Ag_{55}^- and Au_{55}^-. Comparison to the experimental PES shows that the observed Ag_{55}^- isomer is the icosahedral one. For gold, none of the structure candidates gives a perfect match. Reproduced from Ref. 26 by permission. Copyright 2004 American Physical Society.

is available. Ligand-protected clusters are also often termed as monolayer-protected clusters (MPC).

In 1981, Schmid *et al.*[29] presented a method to stabilise 1–2 nm gold clusters with a narrow size-distribution by triphenylphosphines (TPP). The method involves reduction of the gold salt by a stream of diborane gas through benzene solution. This synthesis was a significant milestone that has spurred enormous research activities in characterisation and utilisation of thus-formed Au–TPP nanoparticles. A representative cluster of this nanoparticle material was labelled as $Au_{55}TPP_{12}Cl_6$ where the Au_{55} core was thought to have an fcc-type close-packed geometry. However, a later careful experimental analysis concluded that the cluster material produced by Schmid *et al.*'s method is heterogeneous in size (diameters ranging between 1 and 3 nm with a maximum at 1.4 nm) and structure and degrades upon storage in air.[30] In 1997, Hutchison *et al.* presented a refinement of the method where TOA+ (tetra octyl ammonium cation) is used to transfer $AuCl_4^-$ (aq) ions to TPP–toluene solution.[28] Similarly, the reducing agent $NaBH_4$ (aq) is transferred to the toluene solution by TOA+. This synthesis produces 1.4 nm (average diameter) Au–TPP particles. The particles can be stored in powder form and used for ligand-exchange reactions with various thiols, which further stabilises them and enhances their functionality.[32] It is to be noted, however, that the detailed atomic compositions and structures of the dominant Au–TPP clusters formed by these methods remain unknown to date and the above assignments cannot be taken literally. In fact, a DFT study suggested alternative compositions, slightly larger than the generic 'Au_{55}', which produce a good stability both from the points of view of the electronic structure and steric protection.[31]

Brust *et al.* reported the preparation of thiol-stabilised gold clusters in 2–8 nm range in 1994 in a two-phase synthesis where $AuCl_4^-$ (aq) ions were transferred to organic phase by TOA+ and reduced by $NaBH_4$ in the presence of dodecanethiol ($HSC_{12}H_{25}$).[33] This method has proven to be the most useful and versatile one to yield air-stable thiolate-protected clusters that can be handled much like ordinary chemicals. The method is amenable for use with a wide range of different thiols, including water-soluble ones. Later refinements to this method involved a one-phase modification and an improved control of thiol-to-gold ratio to yield truly monodispersed clusters in 1–2 nm range. Several clusters are now atomically resolved via X-ray crystallography.[7]

Although inert in the bulk phase, gold has rich complexation chemistry and is able to form chemical bonds in many oxidation states from $-I$ to $+V$.[1] Many complexes display Au^I–Au^I bonds whose length is in the 3.0–3.3 Å range; this interaction is generally termed aurophilic and the weak attraction between the filled $Au(d^{10})$ electron shells of two neighbouring atoms is of the order of a strong hydrogen bond. Gold is the metal where this effect is the strongest and its origin is to be found in the relativistic effects already mentioned (Chapter 2).[34] It is worth noting that even well before the breakthrough of total structure determination of thiolate-stabilised clusters, many sub-nanometre cluster complexes stabilised by a combination of phosphines and halides were crystallographically characterised.[35] Notable among them are various undecagold Au_{11} and tredecagold Au_{13} clusters that are inter-related via disproportionation reactions, which is a special redox reaction where some gold atoms are oxidised and others are reduced. The largest crystallographically characterised phosphine–halide stabilised cluster to date is the complex $Au_{39}TPP_{14}Cl_6^q$.[36,37] These early achievements of synthesis and structure characterisation are valuable from theoretical point of view; recent work[38] has found unifying principles to understand the stability of both phosphine–halide and thiolate-protected Au clusters on the basis of common features in the electronic structure, as discussed in Section 13.3.7.

13.4.2 *The Noble Metal–Thiolate Bond*

The affinity of organothiolate ligands (SR: sulphur bound to an organic radical R) to noble metal centres is a key interaction in diverse fields such as metal extraction from ores,[39] formation of metal–thiolate complexes that can act as therapeutic agents,[40] stabilisation and functionalisation of metal surfaces by self-assembled monolayers (SAMs)[41,42] and stabilising nanoparticle growth by forming a 'curved SAM' onto the metal core of the nanoparticle. The metal–SR bond is rather strong and competes with the metal–metal bond. For instance, DFT calculations with a generalised gradient correction (GGA) typically give a value 2.5 eV for the Au–SR bond, which is slightly larger than the experimentally known dissociation

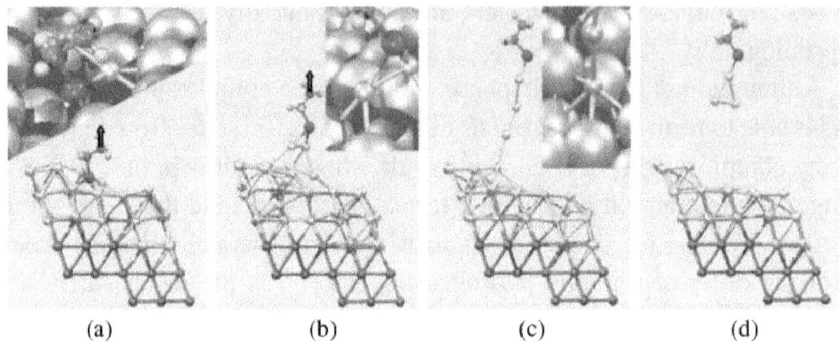

(a)   (b)   (c)   (d)

Figure 13.10 Snapshots of different stages of pulling an ethylthiolate molecule from a gold surface. The strong Au–S bond acts as a clamp and an atomic gold chain is formed. Reproduced from Ref. 43 by permission. Copyright 2002 American Physical Society.

energy of 2.3 eV for the gas phase gold dimer Au_2. Using the simplest bond-counting argument, creating an atom vacancy on the principal crystal surfaces of gold costs energy in relation to the lost nearest neighbour Au–Au 'bonds', scaled to the bulk cohesive energy (3.8 eV), as follows: 2.9 eV for Au(111), 2.5 eV for Au(100) and 2.2 eV for Au(110). These numbers are comparable to the Au–SR bond strength. In fact, *ab initio* molecular dynamics simulations have shown[43] that it is possible to pull an atomic Au chain out of gold surface by using the thiolate–Au bond as a 'clamp'. The formed chain finally breaks at the Au–Au bond (Figure 13.10).

Based on the above discussion, a strong reconstruction of thiolate-SAM-covered noble metal surfaces can thus be expected. Due to the buried interface, it is only recently that a consistent picture of the RS adsorption configuration has emerged. It is now established that thiolates drive pronounced reconstruction of all noble metal surfaces.[42] Scanning tunnelling microscopy (STM) measurements of a monolayer of methyl thiolates (SMe) on Cu(111) show that the adsorbates occupy four-fold hollow positions on a pseudo-(100) reconstructed surface.[44] Low-energy electron diffraction (LEED) measurements of SR adsorption on Ag(111) have revealed a $(\sqrt{7} \times \sqrt{7}R19°)$ surface cell.[45] Several models for the structure of SR on Au(111) have been proposed based on experiments during the past few years: (i) adsorption atop Au ad-atoms forming AuSR units,[46] (ii) adsorption as RSAuSR complexes[47] and (iii) a combined model that comprises $(AuSR)_x$ polymers and SR adsorbed at surface point defects.[48] Among these

456

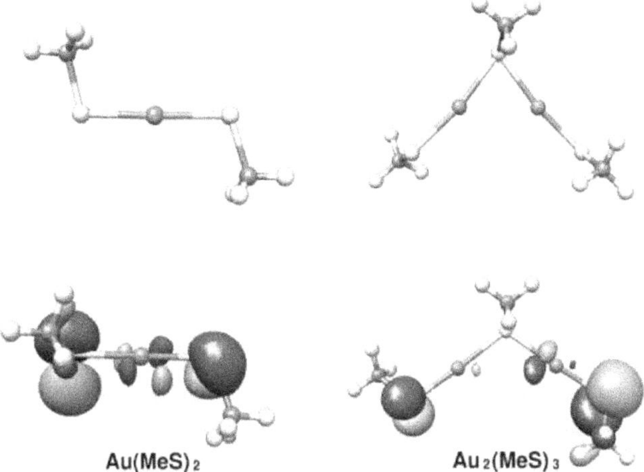

Figure 13.11 Structures for neutral (top) and anionic (bottom) Au(SMe)$_2$ and Au$_2$SMe$_3$. The Kohn–Sham orbitals for the HOMO levels are shown for the charged systems. Reproduced from Ref. 50 by permission. Copyright 2010 American Chemical Society.

three models, DFT calculations are in favour of the formation of RSAuSR complexes, which have been shown to yield the observed c(4 × 2) super-structure on Au(111), energetically superior to earlier models.[49]

It is extremely interesting to note that similar RSAuSR complexes also protect gold cores when the curved SAM forms in thiolate-protected clusters. These complexes are polymeric and can be written generally as SR(AuSR)$_x$; the structures with $x = 1$ and $x = 2$ are shown for R=Methyl in Figure 13.11 in neutral and anionic forms.[50] The Au–S bond in these complexes is of covalent type, with a very small polarisation character (electron loss of 0.1 e to sulphur).[50,51] The capacity of MeSAuSMe complex to localise an electron is indicted by a large electron affinity (EA). The adiabatic EA is calculated to be 2.6 eV. The highest weights of the HOMO level in the anionic complex is found at the sulphur atoms. The properties of MeS(AuSMe)$_2$ are similar to that of MeSAuSMe.

13.4.3 *Early Theoretical Models*

Concentrating here on thiolate-protected clusters, the early theoretical models around the mid-1990s employed pre-parametrised potential functions for interatomic interactions and structures were explored via classical molecular dynamics methods, a considerable computational challenge at the time. A prevailing structural concept was the one with an atomically 'smooth' Au/S interface and compact gold core. This paralleled the understanding of a similar smooth interface in the planar SAMs on the Au(111) surface. Much of the work of that era is highlighted in Refs. 52 and 53.

Then, around 1998, the first electronic structure calculations of the protected clusters concentrated on the $Au_{38}(SR)_{24}$ cluster, where the structure motif for the Au_{38} core was drawn from classical simulations for medium-sized gold clusters.[53–55]

But the most significant improvement occurred when realising that the gold atoms within the core do not have the same status. However, the most significant breakthrough occurred when one realised that the gold atoms of the core were different from those of the external shell. This gave rise to the 'divide and protect' concept.

13.4.4 *The 'Divide and Protect' Concept*

In 2006, a novel 'divide and protect' structural concept was introduced.[56] The new structural model emerged from density functional calculations with improved exchange-correlation functionals. In short, this approach revealed a peculiar composition of the gold cluster with a gold core protected by gold-thiolate tetra-units $(AuSR)_4$. It occurs through the 'etching' of Au atoms from the Au_{38} core by sulfur, as shown by refined DFT calculations and it leads to the formation of six square-like $(AuSR)_4$ units (Figure 13.12). Consequently, the composition $Au_{38}(SR)_{24}$ could be written as $Au_{14}[(AuSR)_4]_6$. The gold atoms in the cluster were found to be in two distinct chemical states, the ones in Au_{14} core essentially neutral ('metallic') and those inside the 'rings' oxidised (formally Au^I). The binding energy of one $(AuSR)_4$ units to the Au_{14} core was found to be quite weak, about 1 eV. A follow-up systematic study on the $(AuSR)_x$ units showed that they are polymeric for $x \geq 4$, i.e. the binding energy per one AuSR unit saturates.[51]

Comparison to the earlier alternative structure model with a disordered gold core[55] showed that the calculations modified also that structure and

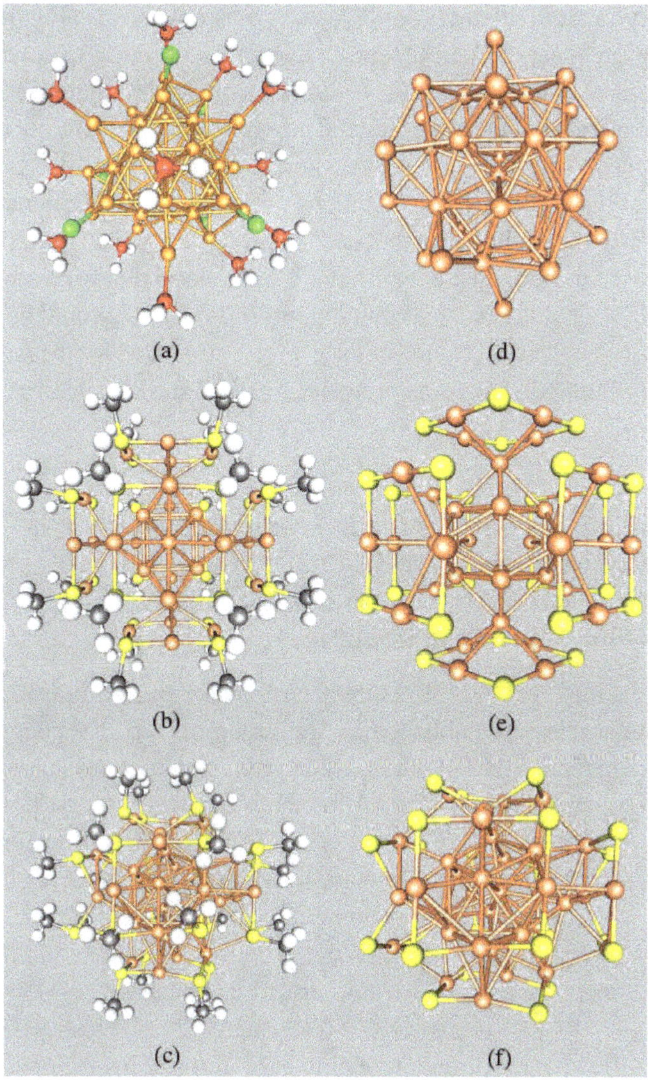

Figure 13.12 (a) A phosphine/chloride passivated $Au_{39}(PH_3)_{14}Cl_6^-$ cluster; (b), (c) two isomers of $Au_{38}(SR)_{24}$; (d) Au_{39} core, shown by a 90° rotation about the horizontal axis on the left; (e) Au–S framework of B, with 45° rotation about the vertical axis on the left; and (f) Au–S framework of C. Au: orange-brown, S: yellow, P: red, Cl: green, C: dark grey, H: white. Reproduced from Ref. 56 by permission. Copyright 2006 American Chemical Society.

a tendency to form distinct $(AuSR)_4$ units was observable (Figure 13.12), signalling an energetic competition between two driving factors: to optimise the number of metallic Au–Au interactions and to optimise the number of covalent Au–S interactions. The energetic competition is due to the improved description of the Au–Au interaction strength (it has been known that the earlier DFT calculations overestimated that interaction significantly). Structurally, the new concept provided also an attractive model for optimised packing in the ligand shell, since steric repulsion among long (such as C_{12}) or bulky (such as glutathione GSH) ligands could be avoided. In 2007, a related model ('core-in-cage') was introduced for the $Au_{25}(SR)_{18}$ cluster consisting of a Au_7 core protected by two $(AuSR)_3$ and one $(AuSR)_{12}$ units.[57]

13.4.5 *The Experimental Breakthroughs*: *X-Ray Crystallography for All-thiolate Protected Au_{102} and Au_{25} Clusters and the Success of the Superatom Model*

In 2007, the first ever total structure determination of an all-thiol-protected gold nanoparticle was published by the group of R.D. Kornberg[58] based on X-ray diffraction at 1.1 Å resolution from single crystals containing a distinct compound with 21 kDa Au core mass, protected by p-MBA ligands (p-MBA $=$ *para* mercapto benzoic acid, SC_6H_4COOH). Specifically, the composition was determined as $Au_{102}(p\text{-MBA})_{44}$ and the crystal unit cell was observed to contain an enantiomeric pair of these clusters.

A subsequent thorough analysis of the atomic structure and full density functional treatment of the electronic structure of the $Au_{102}(p\text{-MBA})_{44}$ cluster (with all of its 762 atoms and 3366 valence electrons) resulted in a clear picture of the identity of the protecting gold–thiolate ligands and the gold core, and the underlying reasons for the thermodynamic stability of this compound.[38] It was found that the atomic structure of the $Au_{102}(p\text{-MBA})_{44}$ compound (Figure 13.13) consists of an approximately D_{5h}-symmetric Au_{79} metallic core with a protective gold–thiolate layer of composition $Au_{23}(p\text{-MBA})_{44}$. The $Au_{23}(p\text{-MBA})_{44}$ layer can further be decomposed into $RS(AuSR)_x$ units with 19 units for $x = 1$ and 2 units for $x = 2$, which are anchored to the core via sulphur in atop positions. The Au_{core}–S–Au_{ligand}

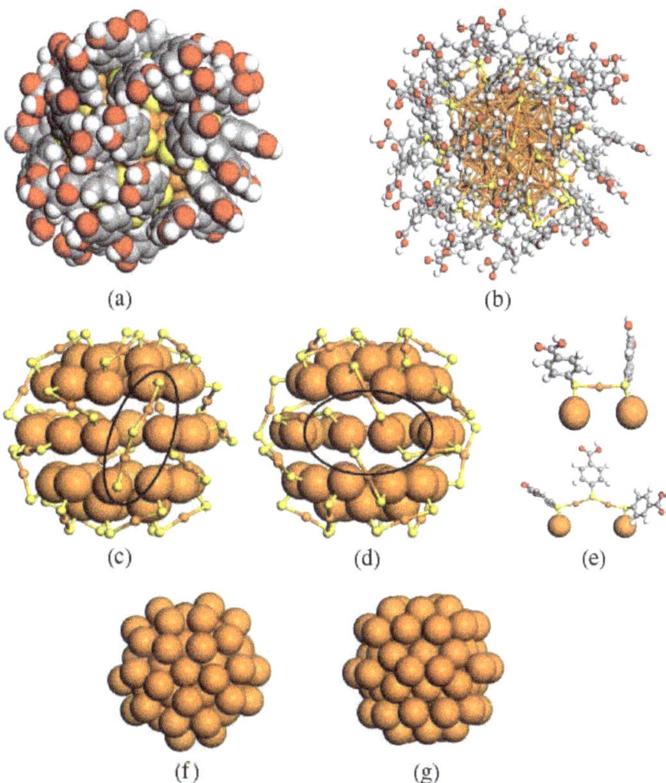

Figure 13.13   Core–shell structure of the $Au_{102}(p\text{-MBA})_{44}$ cluster. (a) Space-filling and (b) ball-stick representations of the $Au_{102}(p\text{-MBA})_{44}$ nanoparticle. Au: orange, S: yellow, C: grey, O: red, H: white. (c, d) Two views of the 40-atom surface of the Au_{79} core, together with the passivating $Au_{23}(p-\text{MBA})_{44}$ ligand shell. The Au atoms in the ligand shell are depicted by the smaller orange spheres. The 'structure defects' at the core–mantle interface (two Au atoms with two Au–S bonds, and a long $RS\text{-}(AuSR)_2$ unit) are highlighted. (e) Close-up of the protecting $RS\text{-}(AuSR)_x$ unit with $x = 1,2$. (f, g) Two views of the Au_{79} core, which has a symmetry of D_{5h} (within 0.4 Å tolerance). To associate the atoms to grey scale, please see the subfigure e. Reproduced from Ref. 38 by permission. Copyright 2008 National Academy of Sciences.

angle is close to 90 degrees and the Au_{ligand} atoms are linearly coordinated with two sulphurs. Hence $Au_{102}(p\text{-MBA})_{44}$ is more accurately described in the formulation $Au_{79}[p\text{-MBA}(Au\ p\text{-MBA})]_{19}\ [p\text{-MBA}(Au\ p\text{-MBA})_2]_2$. The gold atoms in the cluster are in two distinct chemical states: the 79 core Au atoms (Au_{core}) are in a metallic (Au^0) state whereas the 23 Au atoms (Au_{ligand}) that belong to protecting $RS\text{-}(AuSR)_x$ units are oxidised. Consequently, the composition evokes the predicted 'divide and protect' structural motif.[56] The total number of RS(AuSR)$_x$ units, 21, is intimately related to the electronic stability of the particle.

A confirmation of the metallic character of the Au$_{79}$ core came through analysis of radial difference in the cumulative induced charge when the Au$_{102}$(p-MBA)$_{44}$ compound was made either cationic or anionic (that is, remove or add one electron, re-calculate the electron density, and analyse the radial density difference, see Figure 13.14). In both cases, the

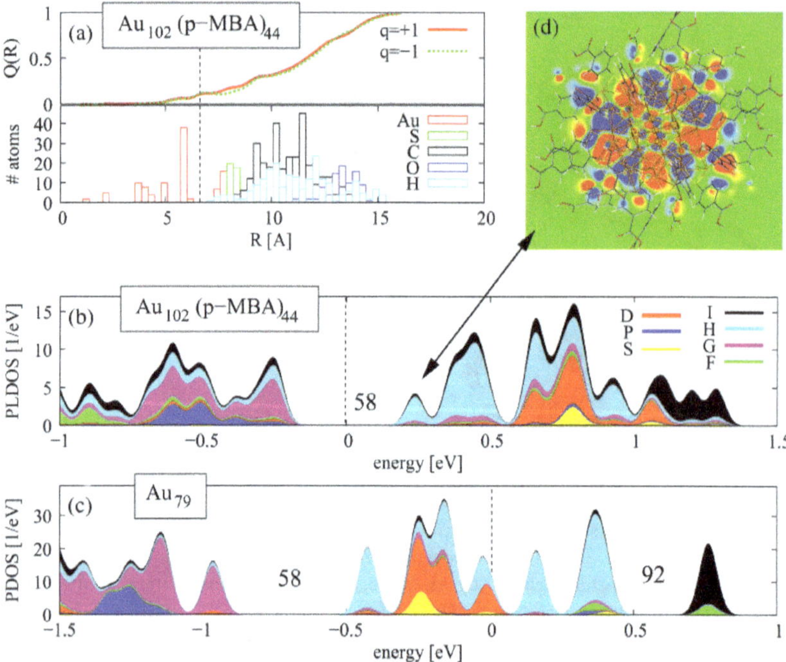

Figure 13.14   Electronic structure analysis of the Au$_{102}$(p-MBA)$_{44}$ cluster. (a) The radial dependence of the integrated induced charge Q(R) upon removing (red curve) and adding (green curve) one electron to the neutral Au$_{102}$(p-MBA)$_{44}$ cluster (top panel), and the radial distribution of atoms (lower panel). The dashed line indicates a midpoint between the surface of Au$_{79}$ core and the Au-thiolate layer. $Q(R) = 4\pi \int^R \Delta\rho(r)r^2 \, dr$ where $\Delta\rho(r) = \rho^0(r) - \rho^q(r)$ is the induced charge-difference from two DFT calculations for the neutral and charge particle. (b) The angular-momentum-projected local electron density of states (PLDOS) (projection up to the I-symmetry, i.e. $l = 6$) for the Au$_{79}$ core in Au$_{102}$(p-MBA)$_{44}$. (c) The same for the bare Au$_{79}$ without the Au–thiolate layer. (d) A cut-plane visualisation of the LUMO-state of the Au$_{102}$(p-MBA)$_{44}$ cluster. Note the H-symmetry (10 angular nodes) at the interface between the Au$_{79}$ core and the gold–thiolate layer. In (b), the zero energy corresponds to the middle of the HOMO–LUMO gap, while in (c), the zero energy is at the HOMO level (dashed lines). Shell-closing electron numbers are indicated in (b, c). Reproduced from Ref. 38 by permission. Copyright 2008 National Academy of Sciences.

major portion (90%) of the induced charge was found in the $Au_{23}(p\text{-}MBA)_{44}$ shell. Virtually no change was observed inside a radius of 5 Å and only 10% of the induced charge resides at the interface between the Au_{79} core and the $Au_{23}(p\text{-}MBA)_{44}$ protective layer ($5\text{Å} < R < 7\text{Å}$). Since a metallic cluster accepts charge only at its surface, it could be directly concluded that the electronic structure of the Au_{79} should feature delocalised-electron shell structure, just as the smaller bare metallic Au clusters.

The calculated HOMO–LUMO gap of $Au_{102}(p\text{-}MBA)_{44}$ cluster was found to be appreciable, about 0.5 eV, indicating a major electron shell closing. The analysis of angular momentum character of the electron states of the bare Au_{79} core and the full compound revealed the exact mechanism how this shell closing is obtained in the protected cluster (Figure 13.14). The stability of the $Au_{102}(p\text{-}MBA)_{44}$ particle is due to several co-existing factors: (i) formation of a compact, symmetric-enough metal core that can support clear electron–shell structure, (ii) complete chemical protection (passivation) of the core surface by $RS\text{-}(AuSR)_x$ units, the number of which has to be 'just right' so that (iii) a major gap is exposed in the electron shell structure. The resulting superatom is a thermodynamically stable species at ambient conditions just as the ordinary atoms in the Periodic Table. Very recently, a combined experimental-theoretical investigation on the NIR absorption by $Au_{102}(p\text{-}MBA)_{44}$ in solution and solid phases documented a full spectroscopic characterisation of this cluster in a wide mid-IR–NIR–VIS–UV range.[59]

The existence of two different protective units in $Au_{102}(p\text{-}MBA)_{44}$, RSAuSR and $RS(AuSR)_2$ may seem surprising. However, polymeric 'zig-zag' chains or rings of such units are known (see Refs. 37, 55 and references therein). The $RS(AuSR)_2$ unit may exist (at least) in two conformations, a sharp-angle or wide-angle 'V' with the central Au–S–Au angle around 100 or 124 degrees, respectively. While the latter one was found in $Au_{102}(p\text{-}MBA)_{44}$, the 'sharp-angle V' unit was found to offer an ideal building block to protect a much smaller particle $Au_{25}(SR)_{18}$ where the composition could be written as $Au_{13}[RS(AuSR)_2]_6$.[57] The central Au_{13} core is a slightly distorted icosahedron with the six $RS(AuSR)_2$ ligands octahedrally arranged around the core (Figure 13.15). Taking the cluster to be anionic ($q = -1$) renders the system as an eight-electron 'superatom' (the

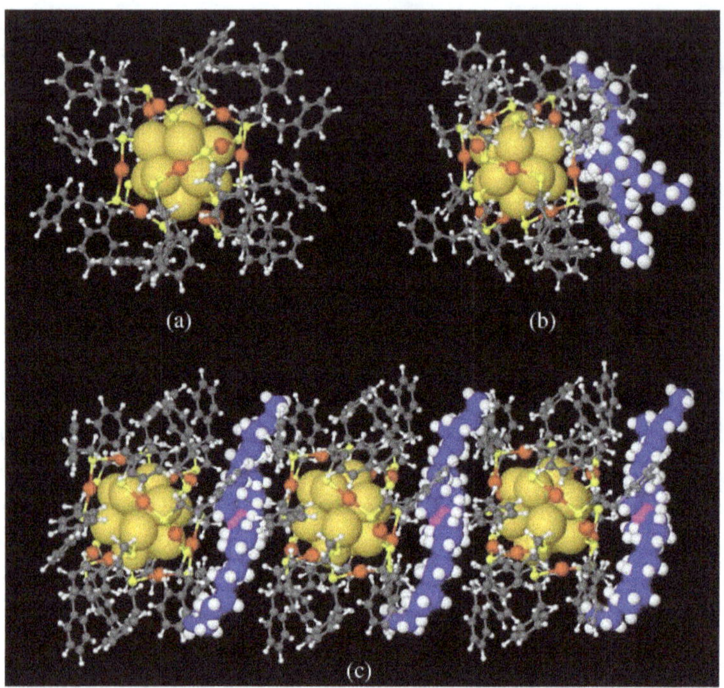

Figure 13.15   (a) Geometry of the $Au_{25}(SR)_{18}^{-1}$ in the gas phase and (b) with the TOA+ counter-ion in the crystal. SR = SEtPh. (c) shows the unit cell in (b), replicated three times along the c-axis of the crystal. Reproduced from Ref. 64 by permission. Copyright 2010 American Chemical Society.

six SR(AuSR)$_2$ units localise in a total of six electrons from the 14-electron (6s) shell structure of Au_{13}^{-} and in spherical harmonics notation, the 6s-derived shells are $1S^2 1P^6$).

Remarkably, two simultaneous and independent experiments[61,62] confirmed the structural prediction for $Au_{25}(SR)_{18}^{-}$. The cluster was passivated by using the SCH_2CH_2Ph ligand. The unit cell of the crystal was found to contain also a TOA^+ counter-ion (tetra octyl ammonium, a phase-transfer agent) which confirmed the anionic state of the gold cluster (Figure 13.15). The theoretical analysis revealed that the close-to-icosahedral Au_{13} is quite rigid also with respect to charge $q = 0, 1$, confirming in part the earlier experimental results that indicated robustness of optical spectra with respect to charges $q = -1, 0, 1$.[63] The robustness can be understood straightforwardly from the superatom model: it was observed that the Au_{13} core remains rigid for different charge states, hence the major optical transitions of $q = 0,1$ clusters are still over the 'HOMO–LUMO gap' of the $q = -1$ cluster, since transitions inside the same angular-symmetric shell (1P → 1P) are forbidden by the dipole selection rule.[60,64] Soon afterwards, a more detailed analysis of the optical excitations of a model cluster $Au_{25}(SR)_{18}^-$ followed.[62,65] The observation of $Au_{25}(SR)_{18}^-$ added yet another member to the 'family' of ligand-protected clusters with eight delocalised electrons (see below), a family that was born by early predictions[66] and synthesis[67] of $[Au_{13}(PMe_2Ph)_{10}Cl_2]^{3+}[(PF_6)_3]^{3-}$ complex.

13.4.6 *Phosphine-stabilised Au$_{11}$ and Au$_{39}$ Clusters: Superatoms with 8 and 34 Electrons*

Not only thiol moiety can play a stabilising role for gold clusters, but also phosphine provides interesting bonding capability with gold. Various Au_{11} and Au_{13}-based phosphine–halide passivated clusters have been char-acterised in the solid state by X-ray diffraction since the late 1970s (for review, see Ref. 35). The undecagold compounds generally have the for-mula $Au_{11}(PR_3)_7X_3$ where X = halide or thiolate, and the gold skeleton often has an approximate C_{3v} symmetry. A recent investigation dealt with the electronic structure of clusters $Au_{11}(PH_3)_7(SMe)_3$ and $Au_{11}(PH_3)_7Cl_3$, which are homologous models for a recently reported thiolate-stabilised cluster $Au_{11}(S\text{-}4\text{-}NC_5H_4)_3(PPh_3)_7$.[68]

The calculated HOMO–LUMO gaps of these compounds are 1.5 eV for X = SMe and 2.1 eV for X = Cl.[38] The dominant angular momentum char-acter of the states around the gap was found to change from P-symmetry to D-symmetry. In the delocalised electron model, this corresponds to clos-ing of the eight-electron (in configuration 1S21P6) gap. This gap expo-sure is due to the fact that the three halide or thiolate ligands localise one electron each out of the 11 conduction electrons from the gold core. It is interesting to note that a halide and a thiolate ligand act here in analogous roles, although the character of the Au–Cl bond is more 'iono-covalent' than that of the Au–SR bond. The seven phosphine ligands act as weak surfac-tants in both systems, without modifying the electron shell structure of the gold core.

A tredecagold compound $[Au_{13}(PMe_2Ph)_{10}Cl_2][PF_6]_3$ was experimentally characterised in 1981,[67] confirming earlier theoretical predictions of stable ligand-protected icosahedral gold clusters.[66] The three hexafluorophosphate anions stabilise the triple-cationic gold compound in the crystal structure. The calculated HOMO–LUMO gap for the homologous relaxed $Au_{13}(PH_3)_{10}Cl_2^{3+}$ compound is 1.8 eV, very similar to the undecagold compounds.

In 1992, the $Au_{39}(PPh_3)_{14}Cl_6^q$ (q is the total charge) compound was isolated and crystallised, and for 15 years remained the largest 'soluble' cluster with an unambiguously determined structure.[36] The geometrical arrangement of the Au_{39} core of this cluster is close to D_3 symmetry and can be also described as two hexagonal close-packed (hcp) crystallites, joined together by 30° twist (Figure 13.12). There is only one fully coordinated gold atom in the centre of a hexagonal anti-prismatic cage. The calculated HOMO–LUMO gap was found to be large, 0.8 eV, for the anionic compound ($q = -1$).[38] The angular momentum analysis of the electron states around the gap showed that the gap closes a band of states that have dominantly F-character while the states above the gap have a major G-character. The F-shell closing indicates an effective conduction electron count of 34 in the gold core. This is consistent with the fact that there are six ionocovalent AuCl bonds at the surface, thereby reducing the effective count of delocalised electrons from 40 to 34.

13.4.7 *The Unifying Superatom Concept*

The above analysis of precisely known compositions and structures of all-(mono)thiolate, phosphine–halide or phospine-(mono)thiolate protected gold clusters suggests that all these compounds can be expressed by a formula[38]

$$(L_s \cdot Au_N X_M)^q, \tag{13.1}$$

where the gold cluster (core size N) is protected by M electron-withdrawing ligands X and s 'weak' ligands that do not affect the effective free-electron-count of the gold core but they complete the sterical protection (typically

L = phosphine). The compound may have an overall charge q. It is essential to recognise the identity of the electron-withdrawing ligands X in the 'divide and protect' concept. For instance, in the $Au_{25}(SR)_{18} = Au_{13}[RS(AuSR)_2]_6$ cluster $X = RS(AuSR)_2$ ($N = 13$ and $M = 6$) and $Au_{102}(SR)_{44}$ has two types of X ligands, 19 of RSAuSR type and two of RS(AuSR)2 type as discussed in Section 13.3.5 (thus $N = 79$ and $M = 21$). In phosphine–halide protected clusters, $X =$ halide. All the shell closing numbers (hence the electronic stability) n_e can be evaluated with an 'effective gold valence' $(v_A) = 1$ from

$$n_e = Nv_A - M - q. \tag{13.2}$$

The result of this analysis is summarised in Table 13.1. A compound having a closed electron shell and a complete chemical protection of the metal core can be called a 'noble-gas superatom'.

Table 13.1 Experimentally determined band gaps for free gas-phase gold cluster anions from photoelectron spectroscopy vs. theoretical DFT values (PBE functional) for HOMO–LUMO gaps of passivated gold cluster compounds that correspond to $n_e = 8, 34$, and 58 conduction-electron shell closings. N_{core} is the number of gold atoms in the cluster core. For details see Ref. 38.

Shell closing, n_e	Experiment		Theory (Ref. 38)		
	Cluster	Gap (eV)	Cluster compound	N_{core}	Gap (eV)
8e ($1S^2 1P^6$)			$Au_{11}(PH_3)_7(SMe)_3$	11	1.5
8e			$Au_{11}(PH_3)_7Cl_3$	11	2.1
8e			$Au_{13}(PH_3)_{10}Cl_2^{3+}$	13	1.8
8e			$Au_{25}(SMe)_{18}^-$	13	1.2
34e ($8e + 1D^{10}2S^2 1F^{14}$)	Au_{34}^- (Refs. 9, 21)	1.0	$Au_{39}Cl_6(PH_3)_{14}^-$	39	0.8
58e ($34e + 2P^6 1G^{18}$)	Au_{58}^- (Refs. 21, 23)	0.6	$Au_{102}(p\text{-}MBA)_{44}$	79	0.5
58e			$Au_{102}(SMe)_{44}$	79	0.5

13.4.8 *Use of the Superatom Concept to Understand the Reactivity of Gold Clusters: Dioxygen Activation and CO Oxidation*

Surfaces of bulk gold are chemically inert, but finely dispersed gold particles with size below a few nanometres are catalytically active species for O–O, C–C and C–H bond activation.[69–74] Notably, when gold particles are catalytically active for oxidation reactions (essentially CO oxidation), they function well at ambient temperature and pressure (see Chapter 9). This propensity makes gold-based nanoscale catalysts interesting systems to exploit for green chemistry. Intensive research over two decades has concluded that several factors are contributing to this behaviour. Many active gold catalysts are prepared on reducible oxides and strong interactions between the support and the particle may create active sites at the periphery of the particle/support interface. These interactions may also induce charge transfer to or from the particle. From purely geometric arguments, small particles have always a high ratio of low-coordinated edge and corner atoms to well-coordinated atoms in the middle of facets, which may increase the reactivity provided that the low-coordinated gold atoms bind the reactants more effectively. Room-temperature vibrations of atoms in very small clusters can be soft, leading to fluctuations of the shape and structure, which may further lower reaction barriers.[71,75] Less attention has been paid to quantum size effects that are known to dominate the physical and chemical properties of small isolated metal clusters in vacuum ($N \leq 150$ atoms, $d \leq 1.7$ nm). Metal clusters with such few atoms have molecule-like distinct electronic states, and all the chemistry then arises from the interactions between fairly few frontier orbitals of the cluster and the reactant molecules.

First experimental and theoretical indications on the catalytic activity of the smallest substrate-supported gold clusters, just a few atoms in size, were observed over a decade ago.[70] Temperature-programmed desorption experiments performed on model catalysts prepared by soft-landing mass-selected gold clusters on a magnesia surface concluded that an eight-atom cluster is the smallest one that catalyses CO oxidation. Accompanying density functional calculations showed that a model cluster Au_8 with two atomic layers can indeed bind both reactants, CO and O_2, and activate

O_2 by transferring electron charge to the $2\pi^*$ orbital of di-oxygen.[70,71] The charging mechanism was later verified by Fourier-transform infrared absorption study of CO stretch frequencies at the Au_8–MgO reaction centre.[72] Recently, two independent experiments gave strong indications on the catalytic activity of ~ 1 nm or even sub-nanometre supported Au clusters. Gold catalysts prepared on inert supports from ligand-protected clusters containing initially about 50 Au atoms were shown to be effective for partial oxidation of styrene.[86] The catalytic particles had a size distribution that was peaked around 1.5 nm. On the other hand, an aberration-corrected scanning transmission electron microscopy study concluded that the most active species catalysing CO oxidation on iron-oxide support are bi-layer 0.5 nm clusters containing only about 10 Au atoms.[73] Although this claim has recently been challenged,[74] these observations clearly call for theoretical understanding on the size-dependence of binding and activation of oxygen at gold nanoclusters and the possible role of the quantum-size effects.

Ligand-protected and chemically passivated Au clusters provide an ideal testing ground for these studies, since the composition and structure of several clusters are by now known precisely and their electronic structure is well understood as discussed above. A recent study considered clusters with the overall size (including the ligand shell) extending up to 2.4 nm and demonstrated by density functional calculations that quantum size effects are instrumental for binding of dioxygen to the nanocatalyst.[76] Partial removal of the protective phosphine–halide or gold–thiolate layer activates the cluster with an occupied electron state over an energy gap that originates from the HOMO–LUMO gap of the fully protected cluster. Transfer of this electron to dioxygen $2\pi^*$ orbital stabilises dioxygen adsorption in an activated superoxo O_2^- form. A clear correlation between the energy gap and the binding energy of the O_2^- species is found. Surprisingly, only the smallest clusters with an overall size of 1.2–1.8 nm (Au core size 0.9 to 1.5 nm) and energy gaps larger than 0.5 eV are able to bind O_2^* appreciably (Table 13.2). Comparison to O_2 adsorption on bulk Au(111) surface shows that the favourable binding of O_2^* to the smallest clusters requires proper alignment of $2\pi^*$ orbitals in the energy gap of the cluster to facilitate electron transfer from the cluster to O_2. Hybridisation of $2\pi^*$ orbitals with the gold 5d derived band weakens the bonding, and this mechanism is responsible for the weaker binding of O_2 to larger clusters. On the smallest clusters the oxidation reaction $2CO + O_2 \rightarrow {<}2CO_2$ proceeds effectively via the Langmuir–Hinshelwood mechanism with reaction barriers that are below $0.7\,eV$, indicating low-temperature activity.

The effect of the electron counting rule (Equation (13.2)) for understanding the reactivity of the partially protected gold cluster can be demonstrated here in case of cluster **1** in Table 13.2. Removal of a single halogen ligand increases the electron count to n^*+1, thus a new electron state in the gold core is occupied. This state is located over an energy gap, which originates from the HOMO–LUMO gap of the fully protected parent cluster. This mechanism turns the partially protected cluster electropositive and thus reactive towards adsorption of electronegative O_2. The fully protected cluster **1** has 8 electrons in the gold core in a configuration $1S^2 1P^6$ of centre-of-mass spherical harmonics and removing one Cl from the ligand shell modifies the electronic structure to $1S^2 1P^6 1D^1$, i.e. the gold core has nine itinerant electrons. The ninth electron occupies the D-symmetric state. Upon O_2 adsorption, this state depletes completely and the electron is transferred to one of the $2\pi^*$ orbitals of O_2, initially empty in the gas-phase triplet O_2. Occupation of this orbital and stretch of the O–O bond length ($1.31\,\mathring{A}$ vs. $1.24\,\mathring{A}$ for the calculated O–O bond in the neutral triplet O_2) verifies that the dioxygen is activated to the superoxo O_2^- state. When the size of the gold cluster increases, the HOMO–LUMO gap of the fully protected cluster decreases (Table 13.2), and thus the energy gain of transferring charge to the $2\pi^*$ orbital of O_2 decreases. Concomitantly, the degree of hybridisation of the molecular O_2 states with the Au states increases. These two factors play a role in the weakening of the Au–O bond. Cluster **6**, which has a very small energy gap ($< 0.1\,eV$), functions already at the bulk limit of metallic Au(111) surface **7**, displaying a 'zero-order' Au–O bond and a metastable O_2 molecular adsorption.

13.5 Gold-based Bimetallic Clusters

The field of ligand-stabilised gold and gold-based nanoclusters is in an active discovery phase. New crystal structures are reported almost monthly.

Table 13.2   Dioxygen adsorption on various partially protected gold clusters.

Fully protected cluster	Total diameter (nm)	Core diameter (nm)	n_e	HOMO–LUMO gap (eV)	Ligand removed	n_e*	BE(O_2) (eV)	d(Au–O) (Å)	d(O–O) (Å)
1. $Au_{11}(PH_3)_7Cl_3$	1.2	0.9	8	2.03	Cl	9	0.95	2.17	1.31
2. $Au_{25}(SR)_{18}^{-1}$	1.6	0.9	8	1.25	$Au_2(SR)_3$	9	0.72	2.24	1.31
3. $Au_{25}(SR)_{18}^{-1}$	1.6	0.9	8	1.25	$2xAu_2(SR)_3$	10	0.62	2.36	1.30
								2.20	1.32
4. $Au_{39}(PH_3)_{14}Cl_6^{-1}$	1.8	1.5	34	0.85	Cl	35	0.59	2.25	1.32
5. $Au_{102}(SR)_{44}$	2.2	1.5	58	0.53	$Au(SR)_2$	59	0.08	2.19	1.30
6. $Au_{144}(SR)_{60}$	2.4	1.7	84	0.08	$Au(SR)_2$	85	−0.15	2.22	1.29
7. Au(111) surface				0			−0.54	2.26	1.24

Note: n_e is the electron count from Equation (13.2). Removal of the ligand(s) changes the electron count to n_e*. BE(O_2) is the binding energy of O_2 to the activated cluster. d(Au–O) is the distance between the adsorbed molecule and the closest Au atom. d(O–O) is the intramolecular bond length. R = methyl. From Ref. 76.

A recent status of the field is summarised in a recent book,[7] with two chapters dedicated to controlled synthesis of bimetallic nanoclusters.[77,78] Several groups have worked on the doping of the $Au_{25}(SR)_{18}$ cluster by silver, copper and palladium.[79–81] Theoretical calculations have predicted the most stable site as the centre of the icosahedral metal$_{13}$ core, which was recently confirmed by single-crystal X-ray studies.[82,83] From the electronic structure point of view, all the metal-doped $Au_{25}(SR)_{18}$ clusters confirm the eight-electron superatom configuration as cluster anions. Interestingly, another eight-electron superatom system was recently discovered, composed of the icosahedral Au_{13} cluster with mixed ligand layer bearing thiolate and pyridyl-phosphine groups.[84]

In 2013, two groups (Zheng and Bigioni) reported simultaneously a breakthrough in crystallographic characterisation of nanometre-scale thiolate protected silver clusters of the form $Ag_{44}(SR)_{30}^{4-}$ where the aromatic thiolate can be both organo-soluble[85] or water-soluble.[86] The metal core of this cluster is a two-shell Keplerate Ag_{32} construction of a hollow 12-atom icosahedron inside a 20-atom dodecahedron. The core is capped by six three-dimensional $Ag_2(SR)_5$ units in octahedral positions. The synthesis of this cluster is easily scalable to macroscopic quantities. An interesting detail is the fairly high native charge (4−), which means that the cluster is stabilised by four positive counter-cations. Density functional analysis of the electronic structure and optical properties revealed that the cluster is a very good example of a 18-electron 'superatom', with the occupation configuration of $1s^2\,1p^6\,1s^{18}$. This configuration is stabilised by the hollow core geometry, which pushes the 2s (LUMO) orbital up in energy.[85] The optical absorption spectrum has several peaks in the UV-VIS range and the aromaticity of the ligand layer contributes to most of the excitations. At the same time, Zheng group reported a bimetallic $Au_{12}Ag_{32}(SR)_{30}^{4-}$ cluster where the gold is found in the first shell of the core.[85] Later, the Zheng group reported a series of copper–gold clusters of the same geometry, namely $(Au_{12+n}Cu_{32}(SR)_{30+n})^{4-}$ ($n = 0,2,4,6$), where copper is found also in the ligand layer,[87] and another 18-electron, metal$_{44}$ cluster of the form $Au_{24}Ag_{20}(2\text{-}SPy)_4(PA)_{20}Cl_2$ that has a mixed ligand layer of pyridylthiolates (2-SPy) and phenylacetylenes (PA) featuring a direct metal–carbon bond.[88]

The group of Dass has done recent interesting work where gold-based thiolated clusters with about 1.5 nm metal core diameter have been doped by various amounts of silver,[89] palladium[90] or copper atoms.[91] Mass spectrometry yields a general stable compound as $Au_{144-x}M_x(SR)_{60}$ (M = Pd, Cu, Ag) where the maximum dopant concentration has been about 15%. Addition of the dopant metal has been shown to affect the electronic structure and optical properties of this particle in a dramatic way. Extensive theoretical calculations based on the structural model of Lopez-Acevedo *et al.* from 2008 (Ref. 92) showed that the all-gold $Au_{144}(SR)_{60}$ cluster is 'on the verge' of becoming plasmonic, i.e. it can support a surface plasmon resonance, which becomes more dominant for slightly larger clusters such as a model cluster $Au_{314}(SR)_{96}$.[93] The calculations predict that doping by silver or copper in the inner part of the metal core enhances the surface plasmon resonance for the mixed $Au_{144-x}M_x(SR)_{60}$ as shown in Figures 13.16 and 13.17 in the case of copper. This phenomenon is currently understood in terms of internal charge-transfer effects and d-band shifts between the metals.[94,95] Thiolate-stabilised bimetallic gold-based clusters thus seem to offer an ideal material system where plasmonic properties are sensitive and can be tuned by metal doping, and the final cluster material is stable in ambient conditions. The lower limit of gold-based plasmonic nanoparticles is thus pushed down to a remarkably small value of about 1.5 nm diameter of the metal core.

13.6 Outlook

Traditionally, the 'phosphine chemistry' and the 'thiolate chemistry' have been regarded as separate branches to prepare ligand-protected gold nanoparticles; no general, unifying theoretical concepts have been available to understand and classify the wealth of experimental information on the well-defined, discrete compounds. The recent experimental and theoretical advances discussed in this contribution provide now certain guiding principles for molecular-precision synthesis and functionalisation of these exciting building blocks of nanomaterials that are finding applications in diverse fields of biolabelling, photonics, sensing and nanocatalysis. The early suggestions to use 'magic' metal clusters as 'superatoms' to build novel materials may perhaps now be realised by ligand-protected gold clusters. This is obviously a huge open field for high-level theory and

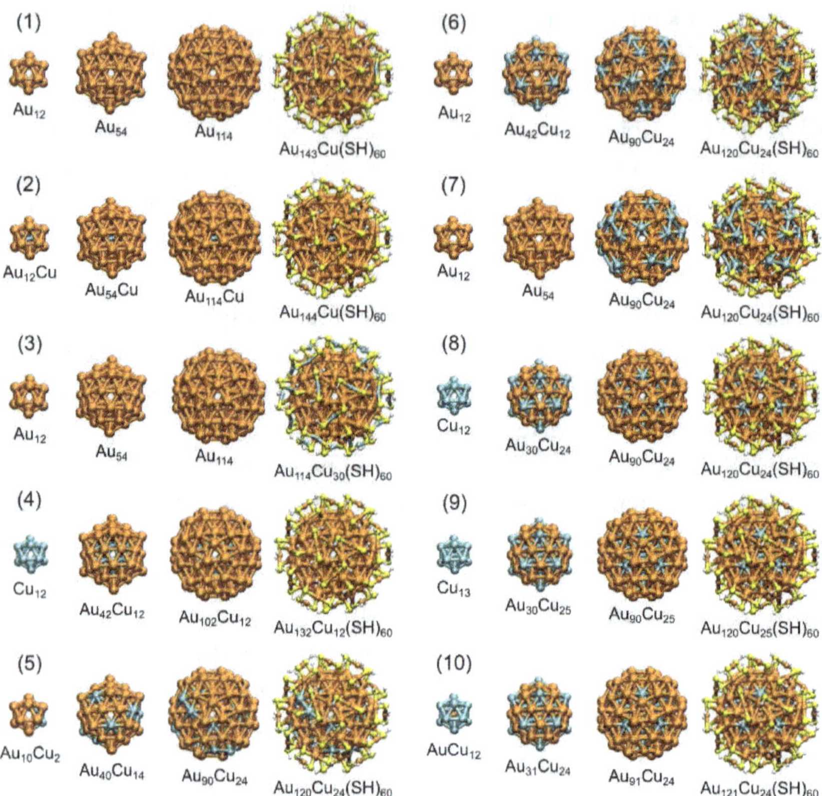

Figure 13.16   Model structures **1–10** used in calculations for a bimetallic thiolate protected cluster $Au_{144-x}Cu_x(SH)_{60}$ ($x \leq 23$). Au: orange, Cu: blue, S: yellow and H: white. Reproduced from Ref. 95 by permission. Copyright 2015 American Chemical Society.

computations. However, theory always needs concrete contact points to ongoing experiments. To close this contribution, a few 'burning questions' are briefly mentioned.

Yet unknown cluster compositions and structures. The definite total-structure determinations discussed here have indicated spherical electron shell closings of 8 ($Au_{11}PR_3X_3$, $Au_{25}(SR)_{18}^{-1}$), 34 ($Au_{39}PR_{14}Cl_6^{-1}$) and 58 ($Au_{102}(SR)_{44}$) electrons, as well as the 14-electron shell closing in a nano-rod-shaped metal core ($Au_{38}(SR)_{24}$). Many other shell closings and the corresponding compositions are waiting to be discovered. On the small side, possibilities include 2, 18 and 20 electrons for roughly spherical metal cores and 16 electrons for a strongly oblate core. It is interesting

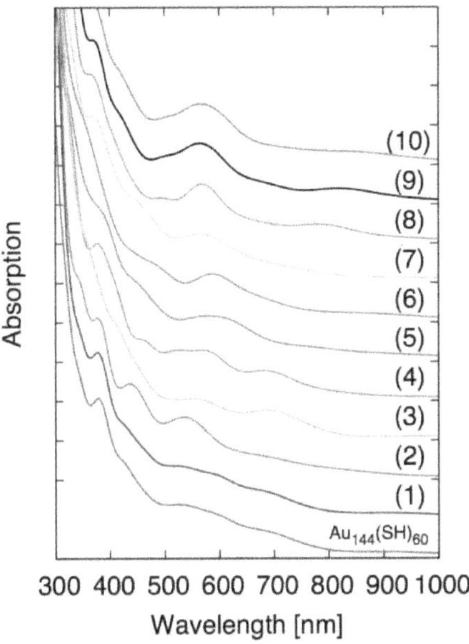

Figure 13.17   Computed optical absorption spectra of structures **1–10** shown in Figure 13.16 as compared to the all-gold Au$_{144}$(SH)$_{60}$. Reproduced from Ref. 95 by permission. Copyright 2015 American Chemical Society.

to note that compositions Au$_{39}$(SR)$_{23}^{96}$ and Au$_{40}$(SR)$_{24}^{97}$ have been determined from high-resolution mass spectrometry; those would correspond to 16-electron systems. On the larger side, a composition of Au$_{68}$(SR)$_{34}$ that would correspond to the 34-electron shell closing (if the cluster is neutral) has been reported.[98] One can also ponder the upper limit of the cluster size where the electron shell-closing effects would still be significant for stabilisation of the cluster via opening a HOMO–LUMO gap. An interesting case is that of the so-called 29 kDa Au cluster (140–150 gold atoms and 50–60 thiolates): recent high-resolution mass-spectrometry investigations have reported compositions of Au$_{144}$(SR)$_{59}$,[99] Au$_{144}$(SR)$_{60}$ and Au$_{146}$(SR)$_{59}^{100}$ and Au$_{144}$(SR)$_{60}$.[101] A theoretical model, fitting very well with the reported powder X-ray data, has been presented for Au$_{144}$(SR)$_{60}^{92}$ and the stability of this cluster is assigned to the high geometrical symmetry of the atomic arrangement (the itinerant electron count of the cluster is 84, nowhere near any electronic shell closing).

Ligand-exchange reactions. The detailed structural knowledge of several clusters facilitates now computational studies of the mechanisms of place-exchange reactions of thiolates, routinely used for functionalisation of the nanoparticle surface. This is a sub-field of great interest where no theoretical insight is currently available. An interesting question is that whether the obtained cluster stabilities and geometries can be retained by functionalising ligands such that, e.g. a toxic phosphine-stabilised gold cluster could be made non-toxic via exchange to amine ligands that are more amenable to biological applications.

Ligand-protected clusters as nanocatalysts. As discussed in Section 13.4.8, activation of ligand-protected gold clusters by partial removal of the protective layer can make them robust electropositive species that can bind and reduce dioxygen and catalyse ambient CO oxidation reactions. Quantum size effects, characterised by the magnitude of the HOMO–LUMO gap of the parent cluster, determine the binding energy of dioxygen to the activated sub-nanometre gold cluster. By 'ligand-engineering', it seems possible to tune the number of occupied electron states in the metal core and thus control the catalytic activity. Several routes for achieving this could be explored. Synthesis of ligand-protected gold clusters has now reached a stage where exploration of techniques to fabricate robust material systems from these building blocks for catalysis has started.[102,103] Concerning thiolate-protected clusters, the recent study[102] indicated that although sulphur has traditionally been considered as poison for a catalyst, controlled extent of thiolate-protection can be beneficial since it aids achieving near-monodispersity and can be used to limit the particle size to the active sub-2nm region. Partial removal of the thiolate layer could be achieved by suitable thermal treatment.[102] In the solution phase, dynamical ligand-exchange reactions may be the key to understand the recently reported activity of $Au_{25}(SR)_{18}$ cluster for selective hydrogenation of unsaturated ketones and aldehydes[104] and the electrochemical reduction of oxygen.[105] Gold has potential as an effective catalyst for low-temperature activation of O=O, C=O, C=C and C–H bonds, and a better understanding of ligand-protected gold nanocluster could pave a way to explore catalytic systems made out of these particles for practical applications in green chemistry.

Acknowledgements

I wish to thank all my collaborators from 1990s to date. Each collaboration has taught me some new aspects about the physics and chemistry of bare, supported and passivated gold clusters and nanoparticles. The author's work is supported by the Academy of Finland. The work would not have been possible without significant and sustained support for computational resources from the Finnish IT Center for Science (CSC) and the European PRACE-organisation.

References

1. H. Schmidbaur, *Gold: Progress in Chemistry, Biochemistry and Technology*, Wiley, Chichester, 1999.
2. P. Pyykkö, *Chem. Rev.* **88**, 563 (1988).
3. G. Schmid, *Chem. Rev.* **92**, 1709 (1992); M.-C. Daniel and D. Astruc, *Chem. Rev.* **104**, 293 (2004); P. Pyykkö, *Angew. Chem. Int. Ed.* **43**, 4412 (2004); P. Pyykkö, *Inorg. Chim. Acta* **358**, 4113 (2005); R. W. Murray, *Chem. Rev.* **108**, 2688 (2008); H. Häkkinen, *Chem. Soc. Rev.* **37**, 1847 (2008); C. M. Aikens, *J. Phys. Chem. Lett.* **2**, 99 (2011); R. Sardar, A. M. Funston, P. Mulvaney and R. W. Murray, *Langmuir* **25**, 13840 (2009); J. F. Parker, C. A. Fields-Zinna and R. W. Murray, *Acc. Chem. Res.* **43**, 1289 (2010).
4. W. A. de Heer, *Rev. Mod. Phys.* **65**, 611 (1993).
5. C. Kittel, *Introduction to Solid State Physics*, Wiley & Sons, Hoboken, NJ, 1996, pp. 146–151.
6. R. Ferrando, 'Global optimization of free and supported clusters', in *Metal Clusters and Nanoalloys: From Modelling to Applications*, edited by M. Marisal, A. Oviedo and E. P. Marcos Leiva, Springer, Berlin, 2013.
7. H. Häkkinen and T. Tsukuda, *Protected Metal Clusters: From Fundamentals to Applications*, Elsevier, Amsterdam, 2015.
8. E. Heikkilä, A. G. Gurtovenko, H. Martinez-Seara, H. Häkkinen, I. Vattulainen and J. Akola, *J. Phys. Chem. C* **116**, 9805 (2012); R. C. van Lehn and A. Alexander-Katz, *J. Phys. Chem. C* **117**, 20104 (2012); E. Heikkilä, H. Martinez-Seara, A. A. Gurtovenko, M. Javanainen, H. Häkkinen, I. Vattulainen and J. Akola, *J. Phys. Chem. C* **118**, 11131 (2014).
9. S. Gilb, P. Weis, F. Furche, R. Ahlrichs and M. M. Kappes, *J. Chem. Phys.* **116**, 4094 (2002).
10. F. Furche, R. Ahlrichs, P. Weis, C. Jacob, S. Gilb, T. Bierweiler and M. M. Kappes, *J. Chem. Phys.* **117**, 6982 (2002).
11. H. Häkkinen, B. Yoon, U. Landman, X. Li, H.-J. Zhai and L.-S Wang, *J. Phys. Chem. A.* **107**, 6168 (2003).
12. K. J. Taylor, C. L. Pettiette-Hall, O. Cheshnovsky and R. E. Smalley, *J. Chem. Phys.* **96**, 3319 (1992).

13. W. Huang and L. S. Wang, *Phys. Chem. Chem. Phys.* **11**, 2663 (2009).

14. P. Koskinen, H. Häkkinen, B. Huber, B. von Issendorff and M. Moseler, *Phys. Rev. Lett.* **98**, 15701 (2007).

15. H. Häkkinen, M. Moseler and U. Landman, *Phys. Rev. Lett.* **89**, 33401 (2002).

16. H. Grönbeck and P. Broqvist, *Phys. Rev. B* **71**, 73408 (2005).

17. B. Yoon, P. Koskinen, B. Huber, O. Kostko, B. von Issendorff, H. Häkkinen, M. Moseler and U. Landman, *ChemPhysChem* **8**, 157 (2007).

18. E. Janssens, H. Tanaka, S. Neukermans, R. E. Silverans and P. Lievens, *New J. Phys.* **5**, 46 (2003).

19. X. Xing, B. Yoon, U. Landman and J. H. Parks, *Phys. Rev. B*, **74**, 165423 (2006).

20. S. Bulusu, X. Li, L.-S. Wang and X. C. Zeng, *Proc. Nat. Acad. Sci. USA*, **103**, 8326 (2006).

21. J. Li, X. Li, H.-J. Zhai and L.-S. Wang, *Science* **299**, 864 (2003).

22. I. Opahle, PhD Thesis, Technical University of Dresden, 2001.

23. M. Walter and H. Häkkinen, *Phys. Chem. Chem. Phys.* **8**, 5407 (2006).

24. A. Lechtken, D. Schooss, J. R. Stairs, M. N. Blom, F. Furche, N. Morgner, O. Kostko, B. von Issendorff and M. M. Kappes, *Angew. Chem. Int. Ed.* **46**, 2944 (2007).

25. Y. Kondo and K. Takayanagi, *Science* **289**, 606 (2000).

26. H. Häkkinen, M. Moseler, O. Kostko, N. Morgner, M. A. Hoffmann and B. von Issendorff, *Phys. Rev. Lett.* **93**, 093401 (2004).

27. M. Faraday, *Philos. Trans. R. Soc. London* **147**, 145 (1857); D. Thompson, *Gold Bull.* **40**, 4 (2007); J. Turkevich, *Gold Bull.* **18**, 86 (1985); J. Turkevich, P. C. Stevenson and J. Hillier, *Discuss. Faraday Soc.* **11**, 55 (1951); J. Turkevich, P. C. Stevenson and J. Hillier, *J. Phys. Chem.* **57**, 670 (1953).

28. J. A. Dahl, B. L. S. Maddux and J. E. Hutchison, *Chem. Rev.* **107**, 2228 (2007).

29. G. Schmid, R. Pfeil, R. Boese, F. Bandermann, S. Meyer, G. H. M. Calis and J. W. A. van der Velden, *Chem. Ber.* **114**, 3634 (1981).

30. D. H. Rapoport, W. Vogel, H. Cölfen and R. Schlögl, *J. Phys. Chem.* **101**, 4175 (1997).

31. M. Walter, M. Moseler, R. L. Whetten and H. Häkkinen, *Chem. Sci.* **2**, 1583 (2011).

32. L. O. Brown and J. E. Hutchison, *J. Am. Chem. Soc.* **119**, 12384 (1997).

33. M. Brust, M. Walker, D. Bethell, D. J. Schiffrin and R. Whyman, *J. Chem. Soc. — Chem. Commun.* 801 (1994).

34. P. Pyykkö, *Chem. Rev.* **97**, 597 (1997).

35. H. Schmidbaur, *Gold: Progress in Chemistry, Biochemistry and Technology*, Wiley, Chichester, 1999, p. 511.

36. B. K. Teo, X. Shi and H. Zhang, *J. Am. Chem. Soc.* **114**, 2743 (1992).

37. B. K. Teo and H. Zhang, *Coord. Chem. Rev.* **143**, 611 (1995).

38. M. Walter, J. Akola, O. Lopez-Acevedo, P. D. Jadzinsky, G. Calero, C. J. Ackerson, R. L. Whetten, H. Grönbeck and H. Häkkinen, *Proc. Acad. Natl. Sci. USA*, **105**, 9157 (2008).

39. D. E. Rawlings, *Ann. Rev. Microbiol.* **41**, 279 (2002).

40. C. F. Shaw, *Chem. Rev.* **99**, 2589 (1999).

41. A. Ulman, *Chem. Rev.* **96**, 1533 (1996); J. C. Love, L. A. Estroff, J. K. Kriebel, R. G. Nuzzo and G. M. Whitesides, *Chem. Rev.* **105**, 1103 (2005).

42. D. P. Woodruff, *Phys. Chem. Chem. Phys.* **10**, 7211 (2008).

43. D. Kruger, H. Fuchs, R. Rousseau, D. Marx and M. Parrinello, *Phys. Rev. Lett.* **89**, 186402 (2002).

44. S. M. Driver and D. P. Woodruff, *Surf. Sci.* **457**, 11 (2000).

45. A. L. Harris, L. Rothberg, L. H. Dubois, N. J. Levinos and L. Dhar, *Phys. Rev. Lett.* **64**, 2086 (1990).

46. M. G. Roper, M. P. Skegg, C. J. Fisher, J. J. Lee, D. P. Woodruff and R. G. Jones, *Chem. Phys. Lett.* **389**, 87 (2004).

47. P. Maksymovych, D. C. Sorescu and J. T. Yates, Jr., *Phys. Rev. Lett.* **97**, 146103 (2006).

48. A. Cossaro, R. Mazzarello, R. Rousseau, L. Casalis, A. Verdini, A. Kohlmeyer, L. Floreano, S. Scandolo, A. Morgante, M. L. Klein and G. Scoles, *Science* **321**, 943 (2008).

49. H. Grönbeck, H. Häkkinen and R. L. Whetten, *J. Phys. Chem. C*, **112**, 15940 (2008).

50. K. A. Kacprzak, O. Lopez-Acevedo, H. Häkkinen and H. Grönbeck, *J. Phys. Chem. C* **114**, 13571 (2010).

51. H. Grönbeck, M. Walter and H. Häkkinen, *J. Am. Chem. Soc.* **128**, (2006) 10268.

52. R. L. Whetten, J. T. Khoury, M. M. Alvarez, S. Murthy, I. Vezmar, Z. L. Wang, P. W. Stephens, C. L. Cleveland, W. D. Luedtke and U. Landman, *Adv. Mater.* **8**, 428 (1996).

53. C. L. Cleveland, U. Landman, T. G. Schaaff, M. N. Shafigullin, P. W. Stephens and R. L. Whetten, *Phys. Rev. Lett.* **79**, 1873 (1997).

54. H. Häkkinen, R. N. Barnett and U. Landman, *Phys. Rev. Lett.* **82**, 3264 (1999); R. N. Barnett, C. L. Cleveland, H. Häkkinen, W. D. Luedtke, C. Yannouleas and U. Landman, *Eur. Phys. J D* **9**, 95 (1999); H. Grönbeck and W. Reoni, *Int. J. Quantum Chem.* **80**, 598 (2000); I. L. Garzón, K. Michaelian, M. R. Beltrán, A. Posada-Amarillas, P. Ordejón, E. Artacho, D. Sánchez-Portal and J. M. Soler, *Phys. Rev. Lett.* **81**, 1600 (1998).

55. I. L. Garzon, C. Rovira, K. Michaelian, M. R. Beltrán, P. Ordejón, J. Junquera, D. Sánchez-Portal, E. Artacho and J. M. Soler, *Phys. Rev. Lett.* **85**, 5250 (2000).

56. H. Häkkinen, M. Walter and H. Grönbeck, *J. Phys. Chem. B*, **110**, 9927 (2006).

57. T. Iwasa and K. Nobusada, *J. Phys. Chem. C* **111**, 45 (2007).

58. P. D. Jadzinsky, G. Calero, C. J. Ackerson, D. A. Bushnell and R. D. Kornberg, *Science* **318**, 430 (2007).

59. E. Hulkko, O. Lopez-Acevedo, J. Koivisto, Y. Levi-Kalisman, R. D. Kornberg, M. Pettersson and H. Häkkinen, *J. Am. Chem. Soc.* **133**, 3752 (2011).

60. J. Akola, M. Walter, R. L. Whetten, H. Häkkinen and H. Grönbeck, *J. Am. Chem. Soc.* **130**, 3756 (2008).

61. M. W. Heaven, A. Dass, P. S. White, K. M. Holt and R. W. Murray, *J. Am. Chem. Soc.* **130**, 3754 (2008).

62. M. Zhu, C. M. Aikens, F. J. Hollander, G. C. Schatz and R. Jin, *J. Am. Chem. Soc.* **130**, 5883 (2008).

63. Y. Negishi, N. K. Chaki, Y. Shichibu, R. L. Whetten and T. Tsukuda, *J. Am. Chem. Soc.* **129**, 11322 (2007).

64. J. Akola, K. A. Kacprzak, O. Lopez-Acevedo, M. Walter, H. Gronbeck and H. Hakkinen, *J. Phys. Chem. C* **114**, 15986 (2010).

65. C. M. Aikens, *J. Phys. Chem. C* **112**, 19797 (2008).

66. D. M. P. Mingos, *J. Chem. Soc. Dalton Trans.* **13**, 1163–1169 (1976).

67. C. E. Briant, B. R. C. Theobald, J. W. White, L. K. Bell, M. P. Mingos and A. J. Welch, *J. Chem. Soc. Chem. Commun.* **5**, 201–202 (1981).

68. K. Nunokawa, S. Onaka, M. Ito, M. Horibe, T. Yonezawa, H. Nishihara, T. Ozeki, H. Chiba, S. Watase and M. Nakamoto, *Organomet. Chem.* **691**, 638 (2006).

69. G. J. Hutchings and R. Joffe, *Appl. Catal.* **20** 215 (1986); M. Haruta, T. Kobayashi, H. Sano and N. Yamada, *Chem. Lett.* **2**, 405 (1987); M. Haruta, *Catal. Today* **36**, 153 (1997); G. C. Bond and D. Thompson, *Cat. Rev. — Sci. Eng.* **41**, 319 (1999); M. Valden, X. Lai and W. Goodman, *Science* **281**, 1647 (1998); M. D. Hughes, Y. J. Xu, P. Jenkins, P. McMorn, P. landon, D. I. Enache, A. F. Carley, G. A. Attard, G. J. Hutchings, F. King, E. H. Stitt, P. Johnston, K. Griffin and C. J. Kelly, *Nature* **437**, 1132 (2005); M. Turner, V. B. Golovko, O. P. H. Vaughan, P. Abdulkin, A. Berenguer-Murcia, M. S. Tikhov, B. F. G. Johnson and R. M. Lambert, *Nature* **454**, 981 (2008); T. Ishida and M. Haruta, *Angew. Chem. Int. Ed.* **46**, 7154 (2008); R. Meyer, C. Lemire, S. Shaikhutdinov and H. J. Freund, *Gold Bull.* **37**, 72 (2004); N. Lopez, T. V. W. Janssens, B. S. Clausen, Y. Xu, M. Mavrikakis, T. Bligaard and J. K. Nørskov, *J. Catal.* **223**, 232 (2004).

70. A. Sanchez, S. Abbet, U. Heiz, W.-D. Schneider, H. Häkkinen, R. N. Barnett and U. Landman, *J. Phys. Chem. A* **103**, 9573 (1999).

71. H. Häkkinen, S. Abbet, A. Sanchez, U. Heiz and U. Landman, *Angew. Chem. Int. Ed.* **42**, 1297 (2003).

72. B. Yoon, H. Häkkinen, U. Landman, A. S. Wörz, J.-M. Antonietti, S. Abbet, K. Judai and U. Heiz, *Science* **307**, 403 (2005).

73. A. A. Herzing, C. J. Kiely, A. F. Carley, P. landon and G. J. Hutchings, *Science* **321**, 1331 (2008).

74. Y. Liu, C.-J. Jia, J. Yamasaki, O. Terasaki and F. Schüth, *Angew. Chem. Int. Ed.* **49**, 5771 (2010).

75. K. A. Kacprzak, J. Akola and H. Häkkinen, *Phys. Chem. Chem. Phys.* **11**, 6359 (2009).

76. O. Lopez-Acevedo, K. A. Kacprzak, J. Akola and H. Häkkinen, *Nature Chem.* **2**, 329 (2010).

77. Y. Negishi, Y. Niihori and W. Kurashige, 'Controlled synthesis: Composition and interface control', Chapter 3 in Ref. 7.

78. Y. Wang, H. Yang and N. Zheng, 'Structural engineering of heterometallic nanoclusters', Chapter 4 in Ref. 7.

79. Y. Negishi, T. Iwai and M. Ide, *Chem. Commun.* **46**, 4713–4715 (2010).

80. Y. Negishi, W. Kurashige, Y. Niihori, T Iwasa and K. Nobusada, *Phys. Chem. Chem. Phys.* **12**, 6219 (2010).

81. Negishi, K. Munakata, W. Ohgake and K. Nobusada, *J. Phys. Chem. Lett.* **3**, 2209 (2012).

82. C. Kumara, C. M. Aikens and Dass, *J. Phys. Chem. Lett.* **5**, 461 (2014).

83. M. A. Tofanelli, T. W. Ni, B. D. Phillips and C. J. Ackerson, *Inorg. Chem.* **55**, 999–1001 (2016).

84. H. Yang, Y. Wang, J. Lei, L. Shi, X, Wu, V. Mäkinen, S. Lin, Z. Tang, J. He, H. Häkkinen, L. Zheng and N. Zheng, *J. Am. Chem. Soc.* **135**, 9568 (2013).

85. H. Yang, Y. Wang, H. Huang, L. Gell, L. Lehtovaara, S. Malola, H. Häkkinen and N. Zheng, *Nat. Commun.* **4**, 2422 (2013).

86. A. Desireddy, B. E. Conn, J. Guo, B. Yoon, R. N. Barnett, B. M. Monahan, K. Kirschbaum, W. P. Griffith, R. L. Whetten, U. Landman and T. P. Bigioni, *Nature* **501**, 399 (2013).

87. H. Yang, Y. Wang, J. Yan, X. Chen, X. Zhang, H. Häkkinen and N. Zheng, *J. Am. Chem. Soc.* **136**, (2014) 7197.

88. Y. Wang, H. Su, C. Xu, G. Li, L. Gell, S. Lin, Z. Tang, H. Häkkinen and N. F. Zheng, *J. Am. Chem. Soc.* **137**, (2015) 4324.

89. C. Kumara and A. Dass, *Nanoscale* **3**, 3064–3067 (2011).

90. N. Kothalawala, C. Kumara, R. Ferrando and A. Dass, *Chem. Commun.* **49**, 10850–10852 (2013).

91. A. C. Dharmaratne and A. Dass, *Chem. Commun.* **50**, 1722 (2014).

92. O. Lopez-Acevedo, J. Akola, R. L. Whetten, H. Grönbeck and H. Häkkinen, *J. Phys. Chem. C* **113**, 5035 (2009).

93. S. Malola, L. Lehtovaara, J. Enkovaara and H. Häkkinen, *ACS Nano* **7**, 10263 (2013).

94. S. Malola and H. Häkkinen, *J. Phys. Chem. Lett.* **2**, 2316 (2011).

95. S. Malola, M. Hartmann and H. Häkkinen, *J. Phys. Chem. Lett.* **5**, 515 (2015).

96. Y. Negishi, K. Nobusada and T. Tsukuda, *J. Am. Chem. Soc.* **127**, 5261 (2005).

97. H. Qian, Y. Zhu and R. C. Jin, *J. Am. Chem.* **132**, 4583 (2010).

98. A. Dass, *J. Am. Chem. Soc.* **131**, 11666 (2009).

99. N. K. Chaki, Y. Negishi, H. Tsunoyama, Y. Shichubu and T. Tsukuda, *J. Am. Chem. Soc.* **130**, 8608 (2008).

100. C. A. Fields-Zinna, R. Sardar, C. A. Beasley and R. W. Murray, *J. Am. Chem. Soc.* **131**, 16266 (2009).

101. H. Qian and R. C. Jin, *Nano Lett.* **9**, 4083 (2009).

102. N. Zheng and G. D. Stucky, *J. Am. Chem. Soc.* **128**, 14278 (2006).

103. Y. Liu, H. Tsunoyama, T. Akita and T. Tsukuda, *Chem. Commun.* **46**, 550 (2010); H. Tsunoyama, N. Ichikuni, H. Sakurai and T. Tsukuda, *J. Am. Chem. Soc.* **131**, 7086 (2009).

104. Y. Zhu, H. Qian, B. A. Drake and R. C. Jin, *Angew. Chem. Int. Ed.* **49**, 1295 (2010).

105. W. Chen and S. W. Chen, *Angew. Chemie Int. Ed.* **48**, 4386 (2009).

Chapter 14
Optical and Thermal Properties of Gold Nanoparticles for Biology and Medicine

Romain Quidant

Plasmon Nanooptics Group, ICFO — The Institute of Photonic Sciences, Barcelona, Spain

14.1 Introduction

Light plays an increasing role in life science, especially with the recent developments of new optical techniques that enable both imaging biological processes down to the molecular level and treating patients. In parallel, recent advances in nanotechnologies have brought new tools to medicine, both to diagnose and cure diseases. In this chapter, the aim is to discuss recent research that sits at the convergence of optics, nanotechnology and health. This research is based on the extraordinary optical properties of gold nanoparticles (AuNPs) supporting localised surface plasmon (LSP) resonances. We discuss how plasmonic properties can turn AuNPs into efficient nanosources of either light or heat for biological and medical applications.

LSP resonances of NPs are associated with a dramatic increase of both their absorption and scattering cross-sections. For gold, such resonances occur in the visible range of the spectrum and their features (central wavelength and position) are determined by the particle geometry and environment.[1] Upon suitable illumination matching the resonance conditions, the light is efficiently coupled to the nanoparticle. Part of the coupled light is efficiently scattered: (i) in the near field, leading to an enhanced field at the particle surface and (ii) to the far field, making the NP acting as an efficient optical antenna. The remaining part of the energy is absorbed and dissipated into heat, creating an increase of the metal temperature. While the optical and thermal response of AuNPs are intrinsically connected, it has been shown that AuNPs can be engineered to act more specifically as efficient point-like sources of either light or heat[2,3] (see Chapter 3). In this chapter, we discuss how these properties open an extraordinary potential in life sciences. The chapter is organised into three sections. The first section introduces the use of plasmonic nanostuctures as efficient biosensors for biomolecule detection. In particular, we discuss what governs their sensitivity and show that lithographically prepared gold nanostructures can be designed to achieve enhanced performances. We will see in the second section that AuNPs offer great opportunities as contrast agents in bioimaging towards efficient optical diagnosis. Finally, we review in the last section the use of AuNPs as point-like source of heat for photothermal cancer therapy.

14.2 Gold Nanoparticles for Biomolecule Sensing

14.2.1 *LSP Sensing: Concept and Motivation*

Since surface plasmon modes are bound to the metal, their resonance features are naturally strongly sensitive to a local change of the refractive index. This dependence to the surrounding dielectric function that explicitly appears in the expression of the dispersion relations of both extended (SPR) and localised surface plasmon resonances (LSPR) (see Chapter 3) is the foundation of the use of surface plasmons for optical sensing. Plasmon sensors based on SPR supported by extended flat gold films have been widely studied and have led to several commercial devices that are broadly used as table-top systems by chemists and biologists in the detection of biomolecules and study of their specific binding. SPR sensors usually monitor changes in the resonance condition (via the incident angle) associated to the modification of the surrounding refractive index. They have been applied to many different contexts including unravelling the biological mechanism, drug design, virology, etc. Like SPR at extended flat metal films, LSPR supported by AuNPs can be used for sensing. In this case, a typical sensing experiment consists in monitoring the frequency shift in the LSPR resonance.

The use of 3D AuNPs instead of homogeneous gold films is motivated by at least three advantages: (i) First, unlike SPR, LSPR supported by particles can be directly coupled to free space light without needing any bulk prism or grating, which limits integration. (ii) Along the same lines, the use of nanosized particles offers possibilities of integrating a large number of sensing sites on a chip for parallel assays. (iii) Finally, while the SPR configuration is expected to be more sensitive to bulk refractive index changes, i.e. a homogeneous change of refraction index over the whole dielectric superstrate in contact with the metal film (as for liquid and gas sensing), LSPR sensors have the potential to be more sensitive to shallow changes of refractive index as those induced by the binding of small molecules at the metal surface.[4] This can be easily understood when considering that the magnitude of the resonance shift is directly related to the spatial overlap between the plasmonic mode and the volume occupied by the target molecules when binding to the metal (see details in next subsection). Indeed, NPs offer much more flexibility than films to tailor the spatial distribution of the electromagnetic mode and therefore of the sensing volume. In the next subsection, we will discuss how exploiting the electromagnetic coupling between adjacent AuNPs can enhance the sensing sensitivity.

14.2.2 *Sensitivity of LSPR Sensors*

The maximum resonance shift of LSPR sensors upon binding of molecules is determined by various parameters related to the sensor itself, its surrounding medium and the target molecule. It has been established that such dependence can be formalised into the following equation, adopted from exact theory for planar SPR sensing, and accepted to some extent as a valid approximation for LSPR:

$$\Delta\lambda \approx m(n_{\text{analyte}} - n_{\text{medium}})(1 - e^{-2d/L_d}), \tag{14.1}$$

where m is the bulk sensitivity in refractive index unit (RIU) change per nm, n_{analyte} and n_{medium} are the refractive indexes of the adsorbing molecule and medium surrounding the system, d is the thickness of the adsorbed layer and L_d is the electromagnetic field decay length. Equation (14.1) suggests that there are quite a number of parameters that influence the change in the resonance. In the case in which the system is adopted for particular analyte in a given medium (n_{analyte}, n_{medium} and d are then fixed), to increase resonance shift, one can tune m and L_d. m and L_d are strongly correlated, accounting for the near-field enhancement and modal distribution, respectively. Equation (14.1) shows that the resonance shift and thus the sensor sensitivity can be increased by concentrating the plasmonic mode into a nanoscale volume and maximising its spatial overlap with the volume occupied by the analyte.

While simple and intuitive, the 2D model of Equation (14.1) fails in accounting for the inhomogeneous spatial 3D distribution of the near field in LSPR sensing as well as the non-uniform assembly of molecules due to steric hindrance, etc. It supposes indeed a uniform molecular coverage, and consequently does not apply to single or few molecular binding events.

One of the more rigorous approaches, based on perturbation theory, yields a more complicated expression, attempting to address the important differences between LSP and SPP. Derived initially from dielectric non-lossy resonators, Equation (14.2) gives more insight into the contributions of spatial near-field distribution around 3D nanoparticles and the location and size of the perturbation:

$$\Delta\lambda \approx \lambda \frac{(n_{\text{analyte}} - n_{\text{medium}}) \int_{V_{\text{analyte}}} \varepsilon_{\text{medium}} |\mathrm{E}|^2 \mathrm{d}V}{n_{\text{medium}} \int_{V_{\text{nearfield}}} \varepsilon_{\text{metal}} |\mathrm{E}|^2 \mathrm{d}V} = \lambda \frac{\Delta n}{n_{\text{medium}}} C \quad (14.2)$$

in which V_{analyte} and $V_{\text{nearfield}}$ are the volumes occupied by the analyte and the plasmonic mode, respectively. Equation (14.2) clearly highlights that the resonance wavelength shift is directly proportional to the mode-analyte overlap, C.

14.2.3 *State of the Art in LSP Sensing: From Single Particle to Engineered Architectures*

The simplest implementation of LSPR sensing is based on large random ensembles of AuNPs.[5] The particles are conjugated with a receptor that has specific binding affinity with the target molecule to be detected. By monitoring the LSPR resonance shift induced by the antigen binding, one gets direct information about its concentration and binding coefficient.[6] Motivated by the need to get rid of size dispersion and agglomeration that tend to broaden the resonance bandwidth and thus limit the shift read-out, Raschke and co-workers demonstrated LSPR sensing on a single gold nanosphere immobilised on glass.[7] In their experiment, scattering spectroscopy using dark-field microscopy was used to monitor the specific binding between biotin and streptavidin in the micromolar range. While single-particle sensing is *a priori* very attractive to integrate a large number of sensors for parallel assays, it is, in practice, incompatible with simple detection schemes. However, a fair signal-to-noise ratio can be achieved with a moderate number of particles, maintaining an overall smaller size as compared to typical SPR pads.

As the magnitude of the resonance shift to a given change of the dielectric environment is strongly dependent on the optical near-field distribution associated with the LSPR, the geometry of the AuNP has a strong influence on the sensor sensitivity. Recent studies have demonstrated that particle shapes with sharp edges, such as nanorices,[8] nanostars[9] and bipyramids[10] that lead to intense hot spots feature up to several folds higher sensitivities as compared to spherical AuNPs. Alternatively, other geometries as, for instance, gold nanorings,[11] prepared by projection lithography (also known as colloidal lithography), show remarkable sensing properties.[12]

Beyond isolated particles and random ensembles, further control of the sensitivity can be achieved by accurately engineering the electromagnetic coupling between adjacent AuNPs. The concept of plasmon-mode engineering for improving plasmonic sensing was first suggested in 2004, in a theoretical proposal by Enoch and co-workers.[13] The proposed configuration consists of a periodic ensemble of plasmonic dimers formed by two adjacent gold cylinders separated by a nanometre-sized air gap (Figure 14.1). Upon illumination, linearly polarised along the particle alignment, a strong gradient of surface charges is created across the gap, leading to a concentration of light within the gap region.[14] The results showed that an array of gold dimers is about five times more sensitive as compared to an array of isolated particles. Similar calculations performed more recently by using another numerical method have confirmed the enhanced sensitivity of gold dimers.[15]

Based on these numerical predictions, arrays of dimers with different gap sizes were fabricated by e-beam lithography on a glass substrate.[16] In order to maximise the level of reproducibility in the gap size all over each of the arrays, the conventional positive resist process combined with

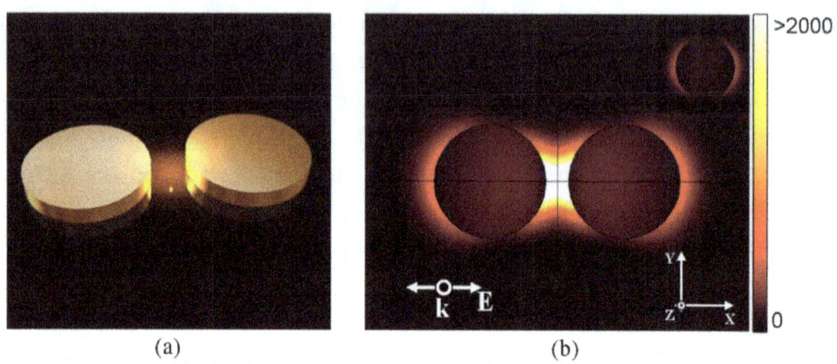

Figure 14.1   (a) Plasmonic dimer formed by two gold cylinders. (b) FEM calculations of the norma-lised electric field intensity distribution at resonance (659 nm). (Inset) Single gold disk at resonance.

lift-off, usually used in plasmonics, was substituted by an alternative process based on negative resist combined with reactive ion etching. Using this process, arrays of gold cylinder dimers (40 nm in height, 100 nm in diameter) and with gaps as small as 10 nm were successfully fabricated. The sensing properties of the fabricated structures were tested by measuring the extinction resonance shifts after binding bovine serum albumin (BSA) to a self-assembled monolayer of mercaptoundecanoic acid (MUA). The experimental data are summarised in Figure 14.2 in which the resonance shift induced by BSA binding is plotted as a function of the gap size. Two different regimes can be identified. For gap sizes greater than 60 nm, the resonance shift is small and nearly constant. Under these conditions, the weak near-field coupling between the adjacent particles forming the dimers behave similar to isolated particles. A drastically different regime is observed while decreasing the gap size from 60 nm to contact. The shift increases exponentially until reaching a maximum for a gap of about 30 nm. For this gap size, the sensitivity to the BSA binding is about five times larger than with isolated particles, in good agreement with the predictions of Ref. 13. Further decrease of the gap leads to a dramatic drop of the shift followed by a second maximum. In order to understand this discrete evolution, one needs to consider the geometry and the binding properties of the BSA molecule. BSA is an elongated 14 nm molecule that tends to bind perpendicularly to the MUA layer. The maximum resonance shift thus corresponds to a gap size for which two BSA molecules bounded across the gap fill in the

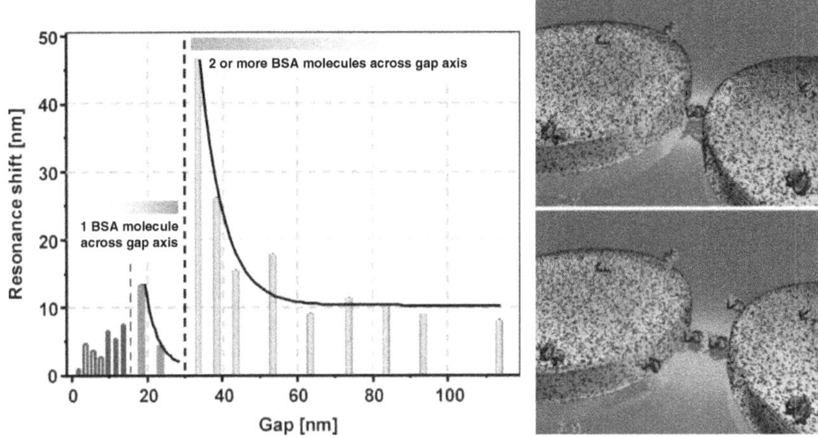

Figure 14.2 Enhanced detection of biomolecules in an array of gold dimers: (Left) Evolution with the dimer gap size of the resonance shift induced by BSA binding. (Right) Artistic view illustrating both binding regimes.

whole sensing volume. For slightly shorter gaps (of about 25 nm), only one molecule can fit across the gap decreasing the spatial overlap with the dimer mode and subsequently, leading to a large drop of the resonance shift. Further decreasing the gap size leads to a second maximum corresponding to a single molecule filling in the gap.

Exploiting this same concept of plasmon mode engineering, several other approaches have been considered to increase the sensitivity of LSP-based plasmonic sensors.[17,18] Liu *et al.*[17] proposes the use of coupled plasmonic geometries with low symmetry to exploit the optical equivalent to electromagnetically induced transparency (EIT) in atom physics. This coherent phenomenon leads to a sharp transparent window in the LSPR extinction peak that can be used to improve the resonance shift read-out. Along the same strategy, Liu and co-workers have recently used EIT in planar metamaterials to achieve narrow resonance line widths featuring enhanced sensitivity to the surrounding refractive index.[21]

14.2.4 *Towards Integrated Biosensing Platforms*

While recent studies have mostly focused on investigating new configurations with enhanced sensitivity, and testing them with model molecular systems, the use of LSPR sensing to address concrete biological problems is only at its premises. The increasing interactions between physicists, chemists and medical doctors render possible testing LSPR sensing with pertinent biological molecules. For instance, lots of effort is currently focused on detecting very low concentrations of cancer markers in blood patients, as a novel strategy to achieve earlier diagnosis and treatment monitoring of cancer.

Reliable tracking of biomolecules will though require the integration of LSPR sensors into a microfluidic platform,[19] which could enable multiple assays in a short time from a very tiny amount of sample but also ensure a high level of reproducibility.

The current extensive research in this direction starts giving very promising results, which should soon provide compact and turnkey analytical devices to be used in the biomedical discipline. A first important step towards this goal is the work by Endo and collaborators that demonstrated for the first time the feasibility of achieving multiplexed detection of antigens using a core–shell structured nanoparticle layer.[20] More recently, we combined for the first time LSPR sensing with gold nanoantennas and advanced two-layer microfluidics to achieve detection of clinical levels of cancer markers from a single drop of human serum as shown in Figure 14.3.[21] LSPR sensing in arrays of plasmonic nanoholes has also been combined with lens-free

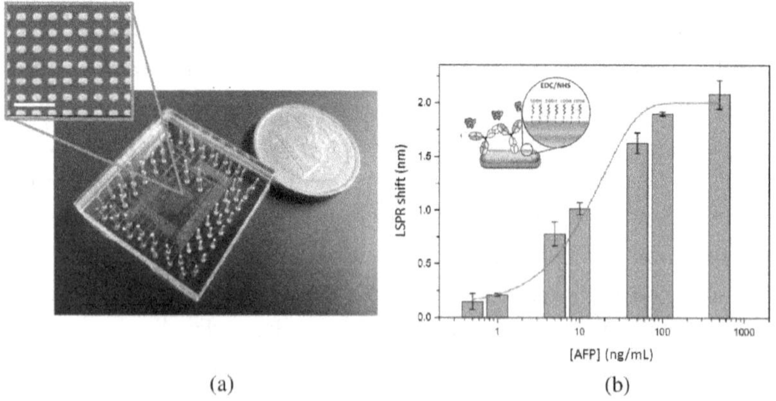

(a)   (b)

Figure 14.3   LSPR chip for the detection of cancer markers in human serum: (a) Picture of the chip next to a one-euro coin. The scale bar in the SEM picture of the array of gold nanosensors is 500 nm. (b) Calibration curve obtained on a single chip for AFP. From this data, we determine that the minimum concentration detected in 0.35 ng/mL (from Ref. 21).

microscopy to lead to a handheld LSPR sensing device.[22] Beyond the detection of antigen, LSPR sensing appears to be very well suited for the detection of exosomes that show potential for cancer diagnostics.[23]

14.3 Gold Nanoparticles as Contrast Agents for Bio-imaging: Application to Cancer Diagnosis

Bio-imaging is an important discipline that has the potential to advance biology and help in early detection and treatment monitoring of disease. Imaging of biological samples is usually based on the use of contrast agents designed to target to specific biomarkers and monitor biological processes down to the molecular level.

While quantum dots and fluorescent molecules have been widely used as contrast agents, AuNPs offer several key advantages:

 (i) The light absorbed and scattered by AuNPs can be strong at their LSPR resonance, despite their small size.
 (ii) AuNPs do not exhibit limitations related to the power of the incident laser such as photobleaching and blinking, making imaging over a long period of time possible.
(iii) The dimensions and geometry of AuNPs can be changed to tune their LSPR resonance into the near-infrared region of the spectrum, in which optical penetration into biological samples is longer.
(iv) Unlike semiconductors, quantum dots or most organic dyes, AuNPs are fully biocompatible and non-toxic.

The main optical imaging techniques using AuNPs as a contrast agent can be classified into three main families: linear microscopy, nonlinear microscopy and photo-acoustic imaging. In this section, we discuss the concept and specificities of each approach and review some of their respective recent advances.

14.3.1 *Linear Imaging Techniques*

Among the different techniques that can strongly benefit from the use of AuNPs as a contrast agent, we start by discussing the first family of imaging modalities that are based on linear optical processes.

14.3.1.1 *Reflectance microscopy*

Advanced reflectance-based optical methods for *in vivo* imaging of superficial tissue (OCT, RCM, etc.) have recently been proposed as non-invasive clinical-imaging techniques, in particular for mapping structural changes associated with the development of cancer. Such techniques are based on the change in the reflective properties of affected tissues. Despite their promise to perform real-time optical biopsies, the modest contrast of regions of early malignancy is usually too low to be of significant clinical value. Here AuNPs show great potential to enhance the reflection contrast. In that case, each nanoparticle is used as a nanoreflector. By proper functionalisation and conjugation of the metal surface (see Chapter 7), the particle can selectively target the region of interest, making it more reflective.

Combining confocal reflectance microscopy with AuNP labelling, cancer cervical cells, both isolated and 3D-agglomerated into a phantom mimicking epithelial tissues, were successfully imaged by Sokolov and co-workers.[24] Exploiting the over-expression of epithelial growth factor receptor (EGFR), cancer epithelial cells were targeted with 12 nm gold nanoparticles bioconjugated with monoclonal anti-EGFR antibodies (see Figure 14.4). First, data show that the gold nanospheres preferentially bind to the surface of the cells. Moreover, the actual distribution of EGFR markers within the phantom can be assessed showing a huge potential for *in vivo* cancer detection and treatment monitoring. More recently, other studies have applied AuNP-enhanced reflectance to imaging of other cancer markers like prostate-specific membrane antigen (PSMA) using AuNPs conjugated with aptamers.[25]

14.3.1.2 *Dark-field microscopy*

Beyond the high reflectivity of AuNPs, one can exploit their extraordinary optical properties associated to their LSPR resonances. A simple way to use AuNPs as a contrast agent is based on their ability to efficiently scatter light with frequencies within their plasmon band. El Sayed and co-workers first suggested fast-screening cancer cells targeted with AuNPs using dark-field microscopy.[26] Unlike conventional (bright) transmission microscopy, in a dark-field microscope, the numerical aperture (NA) of the illumination condenser does not overlap with the NA of the collection objective lens.

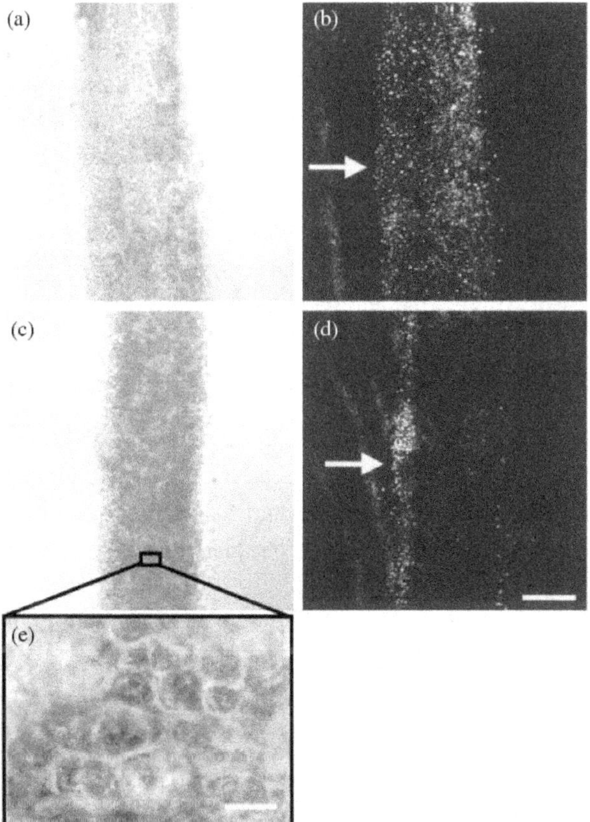

Figure 14.4   Transmittance (a, c, and e) and reflectance (b and d) images of engineered tissue constructs labelled with anti-EGFR/gold conjugates. The tissue constructs consist of densely packed, multiple layers of cervical cancer (SiHa) cells. The contrast agents were added on top of the tissue phantoms in 10% PVP solution in phosphate-buffered saline (PBS) (a and b) or in pure PBS (c and d). After incubation for 30 min at room temperature, the phantoms were transversely sectioned with a Krumdieck tissue slicer, and the sections were imaged using the Zeiss Leica inverted laser scanning confocal microscope with X10 (a–d) objective. A small spot on a tissue construct was imaged using X40 oil immersion objective to show high density of the epithelial cells in the phantom (e). Reflectance images were obtained with 647 nm excitation. Arrows show the surfaces exposed to the contrast agents. The scale bars are (a–d) 200 μm and (e) 20 μm (from Ref. 24).

In other words, in absence of any scattering centres, the incident light rays are not collected and lead to a black (dark) background. This technique thus enables us to image tiny objects with a large signal over noise. For instance, a dark field was used to perform the first scattering spectroscopy of individual plasmonic nanoparticles.[27]

The data of El Sayed *et al.* on epithelial cancer cell lines targeted with gold nanospheres conjugated to anti-EGFR show that this non-scanning, fast and simple imaging technique enables us to discriminate between cancer cells and healthy cells after incubation within a solution of the conjugated AuNPs. Additionally, when combined with linear scattering spectroscopy, the level of agglomeration of the AuNPs can be assessed. However, the significant direct scattering from the cell can be a major drawback when dealing with low concentrations of markers, as it strongly limits the minimum number of AuNPs that can be detected.

14.3.1.3 *Enhanced-fluorescence microscopy*

Another approach relies on exploiting the ability of AuNPs to enhance the fluorescence yield of fluorophores located at their vicinity. Through the adjustment of the particle resonance and the fluorophore–metal distance, the excitation and decay channels of the molecule can be substantially modified.[28] For very short distances, typically shorter than 10 nm, the fluorescence gets quenched as the non-radiative decay channels prevail over the radiative ones. For larger distances, the balance is inverted and fluorescence is enhanced as the result of a combination of processes including enhanced light absorption, enhanced radiative decay and enhanced reemission of fluorescence to the far field (antenna effect).

While near-infrared fluorophore-like indocyanine green (ICG) has been extensively used for molecular imaging, it usually suffers from low quantum yields in aqueous media (typically of about 1%) that limit the imaging sensitivity. Recently, it was shown that the emission of weak NIR fluorophores, such as ICG, could be dramatically enhanced to about 80% by gold nanoshells consisting of spherical dielectric core coated with a thin gold layer.[29] ICG molecules were positioned at an optimum distance of about 10 nm by growing a silica layer around the nanoparticle surface. Interestingly, the spacer layer can also incorporate some Fe_3O_4 magnetic nanoparticles that enables using the same complex for both fluorescence imaging and magnetic resonance imaging (MRI). Despite its lower spatial resolution, MRI is complementary to fluorescence since it penetrates tissues to depth of several centimetres.

Bardhan and co-workers first demonstrated the suitability of their fluorescent–magnetic contrast agent conjugated to anti-HER2 (human epidermal growth factor 2) for *in vitro* imaging breast cancer cells that over-express HER2.[31] Very recently, the same authors extended their approach to a mice model to track the circulation of the complex through the mice body over several days.[30] *In vivo* fluorescent imaging was used to monitor the concentration of complexes after their injection in the blood flow (Figure 14.5). Their data nicely show that due to their bioconjugation, the complexes remain within the tumour region for longer than in

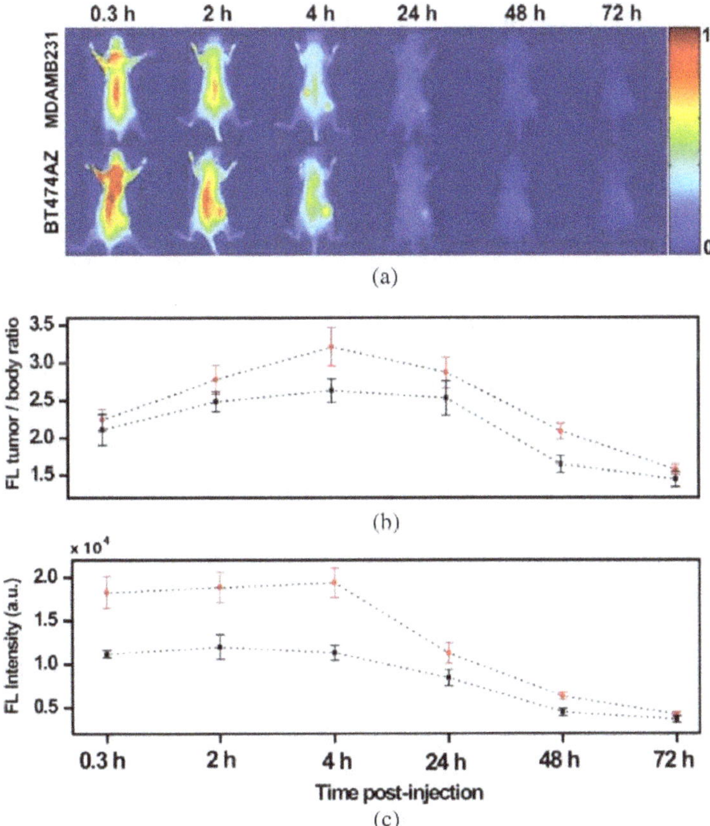

Figure 14.5   *In vivo* fluorescence monitoring of gold nanoshell/ICG complexes in mice: (a) NIR images of mice with HER2 low expressing MDAMB231 xenografts (top) and HER2 overexpressing BT474AZ xenografts (bottom) at 0.3, 2, 4, 24, 48 and 72 h after injection of nanocomplexes. (b) Fluorescence (FL) intensity of tumour-to-body ratio at different time points of mice with BT474AZ xenografts and MDAMB231 xenografts and showing maximum fluorescence at 4 h. (c) Fluorescence intensity comparison of tumours only between BT474AZ and MDAMB231 showing 71.5% increase in signal at 4 h in BT474AZ (from Ref. 30).

healthy tissues. There is thus a time window of several hours during which the fluorescent map enables detection and/or treatment (see Figure 14.5) of cancer tissues. It was also observed that the complexes were fully eliminated from the mice organism after 72 h.

14.3.2 *Nonlinear Imaging Techniques*

Another attractive aspect of plasmonic AuNPs is their ability to dramatically enhance, through their intense local fields, weak nonlinear optical processes occurring either in the particle itself or in its vicinity.

14.3.2.1 *Multiphoton imaging*

Among the nonlinear imaging modalities that can benefit from enhanced plasmonic fields, let's first mention second harmonic generation (SHG) microscopy. SHG (also known as frequency doubling) is a nonlinear optical process, in which photons from a pulsed laser source interacting with a non-linear material are effectively 'combined' to form new photons with twice the energy, and therefore half the wavelength of the incident photons.[32] A unique feature of SHG is that it requires the presence of an asymmetric distribution of the second harmonic sources (non-centrosymmetry). Thus cell membranes, their proteins and their crucial contribution to cellular physiology are ideally suited to interrogation with such technique. However, SHG is a process with poor efficiency that, in practice, requires advanced detection schemes. It was shown that complexes formed by nonlinear dyes attached to AuNPs could be used to strongly enhance SHG at the membrane of cells.[33]

Similarly to SHG, third harmonic generation (THG) microscopy, in which incident photons at frequency ω lead to photons at 3ω, can be boosted by plasmon nanooptics. The absence of asymmetry requirements, combined with the high third-order susceptibility $\chi^{(3)}$ of gold,[34] makes AuNPs excellent examples of efficient sources of THG for cell imaging, without any need for additional nonlinear molecules. In the experiment by Yelin and co-workers, AuNPs were grown into the cells from tiny gold seeds to a size at which they were resonant with the near infrared light from a femto-second titanium–sapphire laser source.[35]

14.3.2.2 *SERS imaging*

Another imaging modality that can strongly benefit from AuNPs is Raman imaging. Raman spectroscopy is a spectroscopic technique used to study vibrational, rotational and other low-frequency modes in a system. It relies on inelastic scattering, or Raman scattering, of monochromatic light, usually from a laser in the visible, near infrared or near ultraviolet range. The laser light interacts with molecular vibrations, phonons or other excitations in the system, resulting in the energy of the laser photons being shifted up or down. Consequently, the shift in energy gives information about the phonon modes in the system and hence is very powerful to identify the presence of given specie via its structural fingerprint. However, Raman signal is very weak (typically 1 out of 10^7 photons) and the integration times needed render imaging very tedious in practice. Nevertheless, it was shown that metallic nanostructures could dramatically enhance the Raman cross-section by more than 10 orders of magnitude[36,37] to reach levels of the signal comparable to fluorescence and enable single-molecule measurements. The exact mechanism of the so-called surface enhanced Raman scattering (SERS) has, however, been a matter of debate for many years between two main theories, based on electromagnetic and chemical mechanisms, respectively. In the electromagnetic theory, the local field enhancement near resonant AuNPs augments both the excitation field experienced by the molecules and their emitted Raman signal. While the electromagnetic theory of enhancement can be applied regardless of the molecule being studied, it does not fully explain the magnitude of the enhancement observed in many systems. The chemical mechanism involves charge transfer between the chemisorbed species and the metal surface. It only applies to specific cases though, where the molecules have formed a chemical bond with the surface. After many years of debate, nowadays, it is pretty well accepted that in many SERS experiments, both mechanisms may coexist.[37]

While SERS was discovered in the 1970s, it has only recently been exploited for bio-imaging.[38] The approach considered by several groups is based on preparing a complex consisting of a gold nanoparticle combined to an efficient Raman-active molecule. After a proper bioconjugation, the complex can be used to target cancer cells to map the distribution of cancer markers.[39] Lately, the technique has been extended to *in vivo* imaging on a mouse model.[40] The main advantage of SERS imaging over other approaches is the possibility to simultaneously detect multiplex analytes by exploiting the sharper bandwidth of Raman peaks as compared to fluorescence peaks.

14.3.2.3 *Two-photon induced luminescence*

In addition to frequency generation and Raman scattering, there has also been a growing interest in exploiting a $\chi^{(3)}$-based phenomenon known as two-photon absorption. Such a process is based on the simultaneous absorption of two photons with the same energy that excites the molecules into higher energy state. In this context, it was shown that the strong local field resulting from the agglomeration of AuNPs at the surface of cells could be used to dramatically enhance the two-photon absorption and thus the subsequent fluorescence of adjacent molecules of the cell membrane.[35] Alternatively, one can directly exploit the two-photon-induced luminescence (TPL) from AuNPs. Such a process, first reported by Mooradian in the 1960s,[41] is slightly different from two-photon absorption in molecules, which requires simultaneous absorption of two coherent photons. TPL-based imaging has recently received lot of attention in the plasmon nanooptics community as a powerful technique to probe the near-field optical response of plasmonic nanostructures.[42] Recently, Durr and co-workers used TPL microscopy to perform targeted imaging of cancer cells in three-dimensional tissue phantoms.[43] In their experiment, gold nanorods designed to be resonant at 760 nm were bioconjugated with EGFR to target skin cancer cells (Figure 14.6). Their data show the distribution of the nanorods at the cells membrane. Discrete bright spots within the cytoplasm are indicative of endosomal uptake of EGFR receptors with nanorods inside the cells. More recently, the same approach has been combined with 3D imaging *in vivo* to characterise the intestinal blood vessels of a mouse.[44]

14.3.3 *Photo-acoustic Imaging*

To close this section on the use of AuNPs as a contrast agent in bio-imaging, we review recent advances in photo-acoustic microscopy for *in vivo* imaging. Photo-acoustic imaging is based on the photo-acoustic effect: upon illumination with laser pulses at optical frequencies, biological tissues

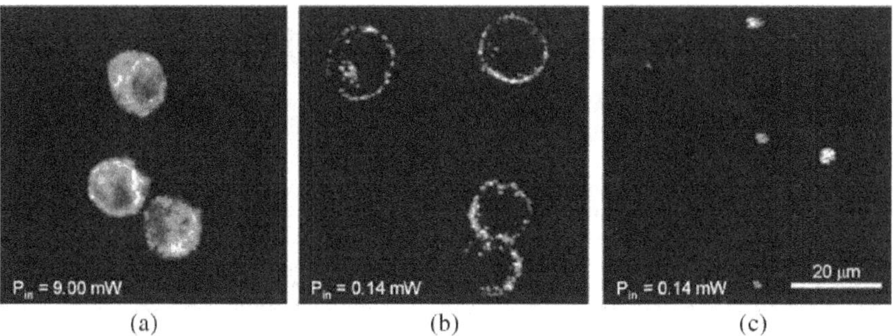

Figure 14.6 Two-photon luminescence (TPL) images of cancer cells placed on a coverslip from a cell suspension. (a) TPL image of unlabelled cells. (b) TPL image of nanorod-labelled cells. Imaging required 9 mW of excitation power in unlabelled cells to get same signal level obtained with only 140 μW for nanorod labelled cells, indicating that TPL from nanorods can be more than 4000 times brighter than TPAF from intrinsic fluorophores. (c) TPL image of non-specifically labelled cells (from Ref. 43).

adsorb part of the delivered energy and convert it into heat, leading to transient thermo-elastic expansion and thus wideband (e.g. MHz) ultrasonic emission. The generated ultrasonic waves are then detected by ultrasonic transducers to form an image with a spatial resolution down to tens of micrometres. Upon near-infrared illumination, the penetration depth is on the order of centimetres, making this modality very well suited for *in vivo* imaging. The optical absorption in biological tissues can be due to endogenous molecules such as hemoglobin or melanin. Since blood usually has much larger absorption levels than the surrounding tissues, there is sufficient endogenous contrast for photo-acoustic imaging to visualise blood vessels. Recent studies have shown that photo-acoustic imaging can be used *in vivo* for tumour angiogenesis monitoring, blood oxygenation mapping, functional brain imaging and skin melanoma detection. However, when aiming at imaging regions with low absorption, the use of contrast agents is required to provide sufficient signal over noise. In this context, the photo-thermal properties of gold nanoparticles make them very attractive candidates to boost the ultrasonic signal.[45] As a practical example, Wang and co-workers have recently demonstrated efficient intravascular imaging of macrophages using agglomeration of gold nanospheres as a novel strategy towards monitoring of cardiovascular diseases.[46] Interestingly, multi-wavelength imaging (changing the wavelength of the optical illumination) is an efficient way to

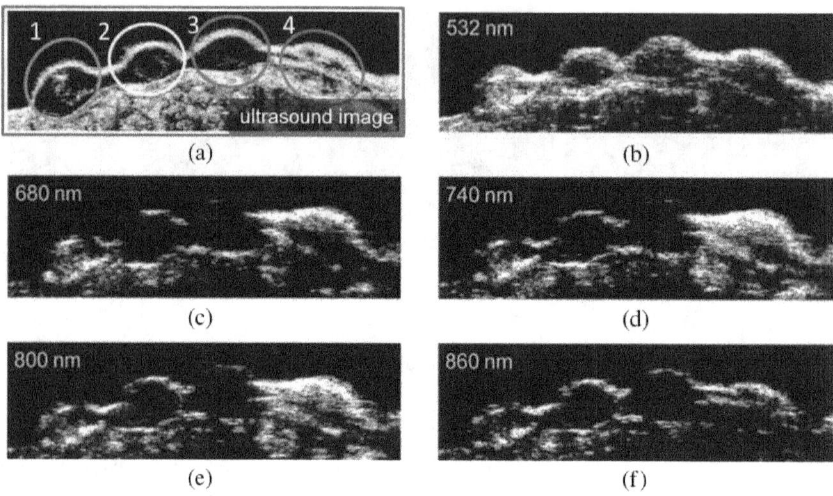

Figure 14.7   Ultrasound (a) and photoacoustic (b-f) images of gelatin implants in mouse tissue *ex vivo* at laser illumination wavelength 532, 680, 740, 800 and 860 nm, respectively. The gelatin implants containing the cells with targeted AuNPs (1), control A431 cells (2), the A431 cells mixed with mPEG-SH coated AuNPs (3) and NIR dye (4) are indicated on the ultrasound image. Tuning the incident wavelength enables discriminating agglomerated AuNPs targeted to cancer cells from the non-targeted AuNPs. The images measure 44 mm laterally and 11 mm axially (from Ref. 47).

discriminate the signal from the agglomerated AuNPs targeted to cancer cells from the non-targeted AuNPs (Figure 14.7).[47]

In this section, we have reviewed some of the main bio-imaging modalities in which AuNPs can behave as an efficient contrast agent. Lately there has been a clear tendency towards multimodal plasmonic imaging in which a single nanoprobe could be used by multiple imaging methods; for instance, to confirm the development of a disease.

Following this trend, in the following section, we will discuss how the same complexes used for imaging can lead to promising novel cancer therapy based on local photo-heating.

14.4   Photothermal Properties of Gold Nanoparticles and their Application to Photothermal Cancer Therapy

In vivo local delivery of heat has raised a growing interest in particular for local tissue ablation. Conventional methods include laser-induced therapy, microwave and radio frequency ablation, and magnetic and focused ultrasound ablation. However, all of these approaches suffer from a common limitation, which arises from the fact that heating is non-specific, hence it leads to the damage of healthy tissues. Alternatively, the use of magnetic fields to heat magnetic particles targeted to tissues was suggested. Similarly, in 2003 Hirsch and co-workers proposed to use AuNPs as local heat sources controllable by an external laser illumination.[48] As well as ablation, it has also been suggested to exploit a more moderate temperature increase in gold complexes for local drug delivery. Interestingly, photothermal therapy is fully compatible with molecular targeting used for diagnosis and the same gold complex could thus be used to detect the disease and treat it. In this last section, we first discuss the optimisation of heat generation in AuNPs before reviewing the latest advances in photothermal cancer therapy and thermal-induced drug delivery.

14.4.1 *Optimising Heat Generation in Gold Nanoparticles*

When a metal nanoparticle is illuminated, part of the intercepted light is scattered in the surroundings while the other part gets absorbed and ultimately dissipated into heat.[49] The efficiency of each of these processes can be characterised by σ_{sca} and σ_{abs}, the elastic scattering and the absorption cross-sections, respectively. The sum of these two processes leads to light attenuation characterised by the extinction cross-section σ_{ext}:

$$\sigma_{\text{ext}} = \sigma_{\text{sca}} + \sigma_{\text{abs}}. \tag{14.3}$$

Depending on the size and the shape of the nanoparticle, the balance between scattering and absorption can vary substantially.[49,50] For instance, while small gold spheres (<10 nm in diameter) mainly act as invisible nano-sources of heat,[51] scattering processes dominate for diameters larger than ~ 50 nm.[49] Here, we focus on the absorption processes and the subsequent heat generation. The general expression of the absorption cross-section for a nanoparticle illuminated by a plane wave is (in MKS units):

$$\sigma_{\text{abs}} = \frac{k}{\varepsilon_0 |E_0|^2} \int_V \Im(\varepsilon_\omega) |E(r)|^2 dr, \tag{14.4}$$

where $k = 2\pi n / \lambda_0 = n\omega / c$ is the wave vector, n is the refractive index of the surrounding medium, ε_ω is the permittivity of the nanoparticle material at frequency ω, E_0 the electric field amplitude of the incoming light considered as a plane wave and $E(r)$ the total electric field amplitude. $\Im(\varepsilon_\omega)$ denotes the imaginary part of the dielectric function. The integral is calculated over the nanoparticle volume V. The power of heat generation Q inside the nanostructure is directly proportional to σ_{abs}:

$$Q = \sigma l = \sigma c \varepsilon_0 |E_0|^2, \tag{14.5}$$

where $l = nc\varepsilon_0 |E_0|^2$ is the intensity of the incoming light and using Equation (14.4) we obtain:

$$Q = \frac{n^2}{2}\Im(\varepsilon_\omega)\int_V |E(r)|^2 dr = \int_V q(r)dr, \tag{14.6}$$

where $q(r) = (n^2\omega/2)\Im(\varepsilon_\omega)|E(r)|^2$ is the volumetric power density of heat generation. This expression shows that the quantity of generated heat is governed by the electric field intensity within the metal. Consequently, the drastic influence of the particle geometry on the plasmon mode distribution offer some degree of control for designing efficient nanoheat sources, remotely controllable by laser illumination.

Recently the green dyadic method (GDM)[10,11] has been used to quantify the influence of the geometry of a gold AuNPs on its heating efficiency. The GDM makes it possible to map the spatial distribution of the heat power density inside the nanoparticles, providing further insight into the influence of the particle shape and the illumination conditions on the origin of heat. Figure 14.8 displays calculations of heat power spectra $Q(\omega)$ for different geometries of gold nanoparticles surrounded by water and illuminated by a plane wave. We fix the intensity of the incoming light at $1\,\text{mW}\mu\text{m}^{-2} = 10^5\,\text{W}\cdot\text{cm}^{-2}$.

To illustrate the influence of the particle geometry, the heat generation of a sphere progressively elongating into a rod-like structure at a constant volume ($4\pi r_{\text{eff}}/3$, where $r_{\text{eff}} = 25\,\text{nm}$) is shown in Figure 14.8 (left panel). The successive nanorods aspect ratios are 1:1 (sphere), 1.4:1, 2:1 and 3:1. Two major features arise from the calculations. First, the LSP resonance markedly depends on the nanoparticle shape. A redshift is indeed expected for nanorods compared with spheres.

Beyond the resonance redshift, a substantial increase of the heating efficiency is observed, by a factor 5 from the sphere to the 3:1 nanorod. The GDM can be efficiently employed to understand this feature. Figure 14.8

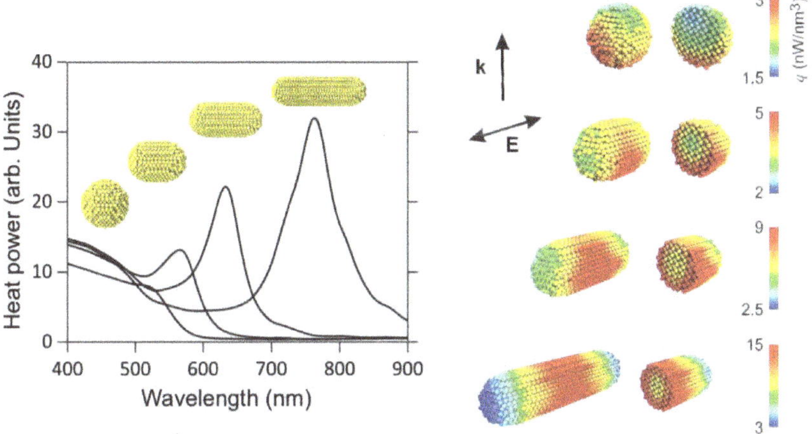

Figure 14.8   Heat generation in gold nanoparticles: (Left) Evolution of the heat power spectrum with the particle aspect ratio (at constant gold volume) (Right) 3D mapping of the heat power density computed for the four nanoparticles shape at their respective plasmon resonance.

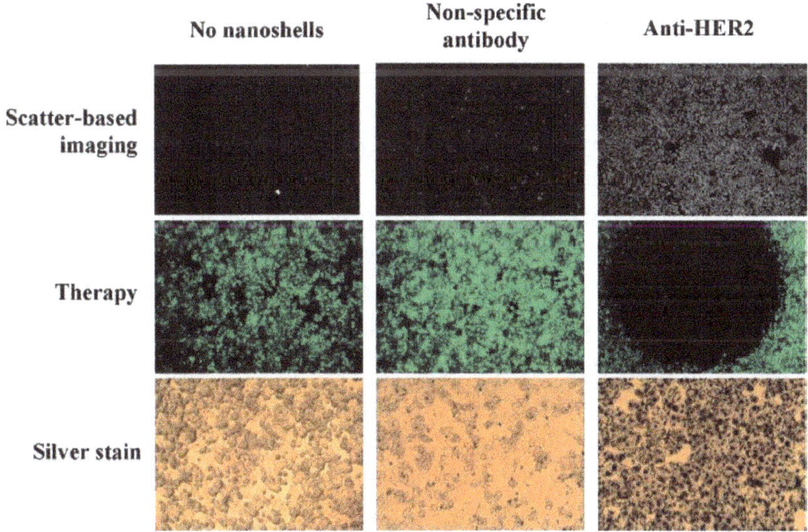

Figure 14.9   Combined imaging and therapy of SKBr3 breast cancer cells using HER2-targeted nanoshells. Scatter-based dark-field imaging of HER2 expression (top row), cell viability assessed via calcein staining (middle row) and silver stain assessment of nanoshell binding (bottom row). Cytotoxicity was observed in cells treated with a NIR-emitting laser following exposure and imaging of cells targeted with anti-HER2 nanoshells only. Note increased contrast (top row, right column) and cytotoxicity (dark spot) in cells treated with a NIR- emitting laser following nanoshell exposure (middle row, right column) compared to controls (left and middle columns) (from Ref. 56).

(right panel) represents the distribution of the heat power density $q(r)$ around and across each of the geometries. Interestingly, for a sphere excited at the LSP resonance, the heat generation arises mainly from the outer part of the particles facing the incoming light. Consequently, the major part of the nanoparticle sees weak electric field intensity and thus does not contribute to heating. However, for elongated nanorods, the field can further penetrate the inner part of the particle, thus making the whole metal volume more efficiently involved in the heating process. It should be underlined that the heat generation mainly arises from the central part of the nanorods because the extremities undergo charge accumulation, which leads to a weaker electric field inside the structure. Using another numerical method, systematic temperature calculations have been performed on other particles geometries.[52]

The temperature increase induced by gold nanorods randomly agglomerated on a glass surface has recently been measured experimentally by using fluorescence polarisation anisotropy (FPA).[2,53]

14.4.2 *Photothermal Therapy (Thermal Ablative Therapy)*

In the original work by Hirsch and co-workers,[48] gold nanoshells, formed by a dielectric core surrounded by a thin gold layer, were tuned to be resonant in the near-infrared region where the transmission through tissues is optimal. Human breast carcinoma cells incubated with nanoshells *in vitro* were found to have undergone photothermally induced death upon exposure to NIR light (820 nm, 35 W/cm^2). Conversely, cells without nanoshells displayed no loss in viability after the same periods and conditions of NIR illumination. Moreover, exposure to low doses of NIR light (820 nm, 4 W/cm^2) in solid tumours treated with metal nanoshells reached average maximum temperatures capable of inducing irreversible tissue damage ($\Delta T = 37.4 \pm 6.6°\,$C) within 4–6 min (Figure 14.9). Importantly, controls treated without nanoshells demonstrated a significantly lower average temperature increase on exposure to NIR light ($\Delta T < 10°$ C). Shortly afterwards, the feasibility of this approach was successfully tested *in vivo* on a mice model.[54] Gold nanoshells coated with polyethylene glycol (PEG) were intravenously injected into the mice. The tumour was then illuminated with a diode laser during three minute sessions. After 10 days of treatment, complete resorption of the tumour was observed. More than 90 days after the treatment, all treated mice remained healthy and free of tumours.

Since then, numerous studies have been carried out to push this initial proposal towards clinical trials. Special attention has been given to investigating new particle geometries and their specific targeting to cancer cells (see Section 14.2). In recent years, the company Nanospectra.[55] has started clinical tests on head and neck cancer using gold nanoshells.

14.5 Drug Delivery

Beyond tissue ablation, moderate temperature increases below the damage threshold can be used for the control delivery of molecules, which is in high demand, especially for drug delivery and gene therapy. A first step towards this goal is to exploit photothermal heating of AuNPs to control dehybridisation and release of DNA in cells. In the recent experiment of Lee and co-workers,[57] thiol-modified sense oligonucleotides were attached to gold nanorods before hybridising anti-sense oligonucleotides. When the minimum temperature on the complex reaches the melting temperature of the short duplex, the short double-stranded oligonucleotides denature. The anti-sense oligonucleotides are released and are allowed bind to a portion of the corresponding mRNA into the cell. Lately this approach has been extended to deliver other types of molecules. In the approach by Huschka and co-workers.[58] DAPI (4′,6-diamidino-2-phenylindole), a water-soluble blue fluorescent dye, is intercalated into the dsDNA of a gold nanoshell–daDNA complex. The nanoshell–dsDNA–DAPI complexes were incubated with H1299 lung cancer cells. Upon illumination with an 800 nm CW laser, corresponding to the peak resonant wavelength of the nanoshell complexes, the DAPI molecules were released from the nanoshell complexes. Subsequent to release, the DAPI diffused through the cytoplasm and into the cell nucleus, where it preferentially bound and stained the nuclear DNA. *A priori*, this strategy could be extended to a multitude of other guest molecules that associate with the host dsDNA carrier including small organic fluorophores, steroid hormones and therapeutic molecules. For example, the quest to find dsDNA intercalators that inhibit the uncontrollable replication of tumour cells comprises an entire field of cancer research.

14.6 Conclusion

Throughout this chapter, we have seen that the extraordinary optical properties of AuNPs offer a set of new opportunities to improve diagnosis and therapy of diseases such as cancer. Enhanced optical fields near lithographically prepared gold nanostructures can be exploited for sensing purposes to detect low concentrations of biomolecules. Their integration into microfluidic chip may soon lead to novel compact analytical platforms able to track cancer markers in the blood of patients. Also, conjugated colloidal AuNPs act as bright, stable and biocompatible contrast agents for *in vivo* biomolecule tracking in biological tissues, opening new horizons in optical diagnosis. Remarkably, one can exploit the ability of these same conjugated AuNPs to heat up upon suitable illumination to either destroy cancer cells or control intra-cellular drug delivery.

Despite their great potential, most of the above-mentioned approaches based on AuNPs are still being developed in research laboratories. Prior to becoming new tools for physicians with positive repercussion on health care, further optimisation is required and some issues need to be addressed. For instance, for applications in which AuNPs need to be injected into the patient's body, extensive toxicity tests still need to be performed to assess potential accumulation in the patient organs.

References

1. U. Kreibig and M. Vollmer, *Optical Properties of Metal Clusters* (Springer Verlag, Berlin, 1995).
2. G. Baffou, C. Girard and R. Quidant, *Phys. Rev. Lett.* **104**, 136805 (2010).
3. G. Baffou, R. Quidant and C. Girard, *Appl. Phys. Lett.* **94**, 153109–153111 (2009).
4. M. Svedendahl, S. Chen, A. Dmitriev *et al.*, *Nano Lett.* **9**, 4428–33 (2009).
5. T. Okamoto, I. Yamaguchi and T. Kobayashi, *Opt. Lett.* **25**, 372–74 (2000).
6. M. P. Kreuzer, R. Quidant, G. Badenes *et al.*, *Biosens. Bioelectron.* **21**, 1345–49 (2006); P. Englebienne, A. Van Hoonacker and M. Verhas, *Analyst* **126**, 1645–51 (2001); P. Englebienne, *Analyst* **123**, 1599–603 (1998).
7. G. Raschke, S. Kowarik, T. Franzl *et al.*, *Nano Lett.* **3**, 935–38 (2003).
8. H. Wang, D. W. Brandl, F. Le *et al.*, *Nano Lett.* **6**, 827–32 (2006).
9. C. L. Nehl, H. W. Liao and J. H. Hafner, *Nano Lett.* **6**, 683–88 (2006).
10. K. M. Mayer, F. Hao, S. Lee *et al.*, *Nanotechnology* **21**, 255503 (2010).
11. J. Aizpurua, P. Hanarp, D. S. Sutherland, M. Kall, G. W. Bryant and F. J. G. de Abajo, *Phys. Rev. Lett.* **90**, 057401 (2003).

12. E. M. Larsson, J. Alegret, M. Kall *et al.*, *Nano Lett.* **7**, 1256–63 (2007).
13. S. Enoch, R. Quidant and G. Badenes, *Opt. Express* **12**, 3422–27 (2004).
14. J. P. Kottmann and O. J. F. Martin, *Opt. Express* **8**, 655–63 (2001).
15. P. K. Jain and M. A. El-Sayed, *Nano Lett.* **8**, 4347–52 (2008).
16. S. S. Acimovic, M. P. Kreuzer, M. U. Gonzalez *et al.*, *ACS Nano* **3**, 1231–37 (2009).
17. N. Liu, T. Weiss, M. Mesch *et al.*, *Nano Lett.* **10**, 1103–07 (2010); N. Verellen, Y. Sonnefraud, H. Sobhani *et al.*, *Nano Lett.* **9**, 1663–67 (2009).
18. A. B. Evlyukhin, S. I. Bozhevolnyi, A. Pors *et al.*, *Nano Lett.* **10**, 4571–77 (2010).
19. T. Thorsen, S. J. Maerkl and S. R. Quake, *Science* **298**, 580–84 (2002).
20. T. Endo, K. Kerman, N. Nagatani *et al.*, *Anal. Chem.* **78**, 6465–75 (2006).
21. S. S. Acimovic, M. A. Ortega, V. Sanz *et al.*, *Nano Lett.* **14**, 2636–41 (2014).
22. A. E. Cetin, A. F. Coskun, B. C. Galarreta *et al.*, *Light Sci. Appl.* **3**, e122 (2014).
23. H. Im, H. Shao, Y. I. Park *et al.*, *Nat. Biotechnol.* **32**, 490–95 (2014).
24. K. Sokolov, M. Follen, J. Aaron *et al.*, *Cancer Res.* **63**, 1999–2004 (2003).
25. D. J. Javier, N. Nitin, M. Levy *et al.*, *Bioconjug. Chem.* **19**, 1309–12 (2008).
26. I. H. El-Sayed, X. H. Huang and M. A. El-Sayed, *Nano Lett.* **5**, 829–34 (2005).
27. T. Klar, M. Perner, S. Grosse *et al.*, *Phys. Rev. Lett.* **80**, 4249–52 (1998).
28. P. Anger, P. Bharadwaj and L. Novotny, *Phys. Rev. Lett.* **96**, 113002–113005 (2006); E. Dulkeith, A. C. Morteani, T. Niedereichholz, T. A. Klar, J. Feldmann, S. A. Levi, F. van Veggel, D. N. Reinhoudt, M. Moller and D. I. Gittins, *Phys. Rev. Lett.* **89**, 203002–203005 (2002); S. Kuehn, U. Hakanson, L. Rogobete and V. Sandoghdar, *Phys. Rev. Lett.* **97**, 017402–017405 (2006).
29. R. Bardhan, N. K. Grady, J. R. Cole *et al.*, *ACS Nano* **3**, 744–52 (2009); R. Bardhan, N. K. Grady and N. J. Halas, *Small* **4**, 1716–22 (2008).
30. R. Bardhan, W. Chen, M. Bartels *et al.*, *Nano Lett.* **10**, 4920–28 (2010).
31. R. Bardhan, W. Chen, C. Perez-Torres *et al.*, *Adv. Functional Mater.* **19**, 3901–09 (2009).
32. R. W. Boyd, *Nonlinear Optics*, 2nd ed. (Academic Press, New York, 2003).
33. G. Peleg, A. Lewis, O. Bouevitch *et al.*, *Bioimaging* **4**, 215–24 (1996).
34. J. Renger, R. Quidant, N. van Hulst *et al.*, *Phys. Rev. Lett.* **104**, 046803 (2010).
35. D. Yelin, D. Oron, S. Thiberge *et al.*, *Opt. Express* **11**, 1385–91 (2003).
36. S. Nie and S. R. Emory, *Science* **275**, 1102–6 (1997).
37. E. C. Le Ru, E. Blackie, M. Meyer *et al.*, *J. Phys. Chem. C* **111**, 13794–803 (2007).
38. J. Kneipp, H. Kneipp, W. L. Rice *et al.*, *Anal. Chem.* **77**, 2381–85 (2005).
39. S. Lee, H. Chon, M. Lee *et al.*, *Biosens. Bioelectron.* **24**, 2260–63 (2009); H. Park, S. Lee, L. Chen *et al.*, *Phys. Chem. Chem. Phys.* **11**, 7444–49 (2009); S. Lee, S. Kim, J. Choo *et al.*, *Anal. Chem.* **79**, 916–22 (2007).
40. M. V. Yigit, L. Zhu, M. A. Ifediba *et al.*, *ACS Nano* **5**, 1056–66 (2011).
41. A. Mooradia, *Phys. Rev. Lett.* **22**, 185 (1969).
42. P. Ghenuche, S. Cherukulappurath, T. H. Taminiau, N. F. van Hulst and R. Quidant, *Phys. Rev. Lett.* **101**, 116805–116808 (2008); P. Muhlschlegel, H. J. Eisler, O. J. F. Martin *et al.*, *Science* **308**, 1607–09 (2005); P. J. Schuck, D. P. Fromm, A. Sundaramurthy, G. S. Kino and W. E. Moerner, *Phys. Rev. Lett.*, **94**, 017402–017405 (2005); M. R. Beversluis, A. Bouhelier and L. Novotny, *Phys. Rev. B* **68**, 115433–115432 (2003).

43. N. J. Durr, T. Larson, D. K. Smith *et al.*, *Nano Lett.* **7**, 941–45 (2007).

44. S.-C. Tang, Y.-Y. Fu, W.-F. Lo *et al.*, *ACS Nano* **4**, 6278–84 (2010).

45. X. Yang, S. E. Skrabalak, Z.-Y. Li *et al.*, *Nano Lett.* **7**, 3798–802 (2007); M. Eghtedari, A. Oraevsky, J. A. Copland *et al.*, *Nano Lett.* **7**, 1914–18 (2007); S. Mallidi, T. Larson, J. Aaron *et al.*, *Opt. Express* **15**, 6583–88 (2007).

46. B. Wang, E. Yantsen, T. Larson *et al.*, *Nano Lett.* **9**, 2212–17 (2009).

47. S. Mallidi, T. Larson, J. Tam *et al.*, *Nano Lett.* **9**, 2825–31 (2009).

48. L. R. Hirsch, R. J. Stafford, J. A. Bankson *et al.*, *PNAS* **100**, 13549–54 (2003).

49. P. K. Jain, K. S. Lee, I. H. El-Sayed *et al.*, *J. Phys. Chem. B* **110**, 7238–48 (2006).

50. N. Harris, M. J. Ford and M. B. Cortie, *J. Phys. Chem. B* **110**, 10701–07 (2006); K. S. Lee and M. A. El-Sayed, *J. Phys. Chem. B* **109**, 20331–38 (2005); K. L. Kelly, E. Coronado, L. L. Zhao *et al.*, *J. Phys. Chem. B* **107**, 668–77 (2003).

51. A. Gaiduk, M. Yorulmaz, P. V. Ruijgrok *et al.*, *Science* **330**, 353–56 (2010); S. Berciaud, D. Lasne, G. A. Blab *et al.*, *Phys. Rev. B* **73** (2006).

52. G. Baffou, R. Quidant and F. Javier Garcia de Abajo, *ACS Nano* **4**, 709–16 (2010).

53. G. Baffou, M. P. Kreuzer, F. Kulzer *et al.*, *Opt. Express* **17**, 3291–98 (2009).

54. D. P. O'Neal, L. R. Hirsch, N. J. Halas *et al.*, *Cancer Lett.* **209**, 171–6 (2004).

55. http://www. nanospectra.com/ (accessed 26 February 2016).

56. C. Loo, A. Lowery, N. J. Halas *et al.*, *Nano Lett.* **5**, 709–11 (2005).

57. S. E. Lee, G. L. Liu, F. Kim *et al.*, *Nano Lett.* **9**, 562–70 (2009).

58. R. Huschka, O. Neumann, A. Barhoumi *et al.*, *Nano Lett.* **10**, 4117–22 (2010).

Chapter 15
Physical and Chemical Processes for Gold Nanoparticles and Ionising Radiation in Medical Contexts

Fred Currell and Balder Villagomez-Bernabe

Centre for Plasma Physics, School of Mathematics and Physics, The Queen's University of Belfast, Belfast, UK

After a review of key concepts in radiobiology, radiation–matter interactions, radiotherapy and radiosensitisation, the physical and chemical mechanisms at work when gold nanoparticles are irradiated in aqueous media are discussed with reference to both experimental results and theoretical predictions. In this context, various mechanisms able to cause biological damage are considered within a unified framework. The importance of these mechanisms for gold nanoparticle-enhanced radiotherapy is emphasised along with their more general importance for the whole field of nanomedicine employing gold nanoparticles.

15.1 Introduction

As is illustrated in Figure 15.1, the medical application of gold nanoparticles is a rapidly growing area, and about 50% of all research involving gold nanoparticles also involves medicine. About 50% of these publications also refer to radiation. Although one has to be careful interpreting data based on word occurrence alone, it is clear that the medical applications of ionising radiation used alongside gold nanoparticles is a significant and

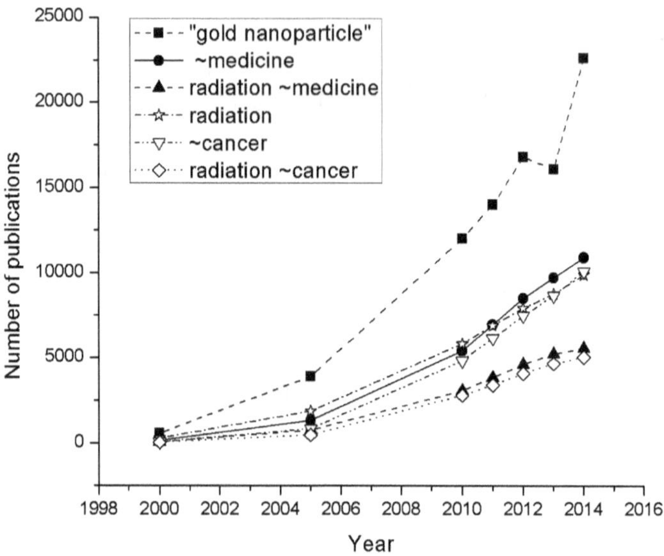

Figure 15.1   Trend in publications of papers concerning the medical applications of gold nanoparticles. The searches (ignoring patents and conference proceedings) were performed using Google scholar.[1] In all cases the occurrence of the string *gold nanoparticle* was required. Additionally, the string $\sim$ medicine or $\sim$ cancer and/or *radiation* was required as indicated in the legend. The $\sim$ character allows Google to include words carrying the same meaning to be included (e.g. medical).

still-growing area of research. As these search statistics indicate, the major medical application for which the combination of gold nanoparticle and ionising radiation is being investigated is cancer-related treatments. Other applications include medical imaging (again, predominantly associated with cancer care) and radioretinopathy.

This chapter will focus on the applications related to cancer care as this area is the most significant. Furthermore, it is the area where physical/ chemical processes play a key role. The basic idea is that by preferentially directing the nanoparticles to the tumour site followed by the application of radiotherapy should lead to an increase in cell killing for the same radiation dose. However, much that is discussed in this chapter is of relevance to the whole field of nanomedicine, even when radiation is not deliberately used in conjunction with the nanoparticles. We need to be cognisant of the interactions between ionising radiation and gold because X-rays provide the most common medical imaging modality. Hence, when gold nanoparticles are used in any nanomedicine-based intervention, consideration should be given to the interactions between the nanoparticles and radiation.

Cancer is widely acknowledged as one of the major societal challenges of the 21st century. It is amongst the leading causes of morbidity and mortality with about 14 million new cases and 8.2 million cancer-related deaths worldwide in 2012 alone. The number of new cases is predicted to rise by a further 70% in the next two decades.[2]

Clearly there is an urgent need for innovative new approaches to treating cancer. It is important to understand that cancer is an umbrella term for a range of diseases affecting various parts of the body and it is unlikely that a 'one-size fits all' approach will be found. However, gold nanoparticles selectively taken up by tumours and then exposed to ionising radiation represent an attractive approach that appears to be applicable to solid tumours in general. It offers promise to improve all treatment scenarios where external beam radiotherapy is used.

The modes of interaction appear to be predominantly physical and chemical, rather than biological when nanoparticles are subject to radiation in biological systems. One key consideration is that when gold nanoparticles are introduced to the body and exposed to radiation a whole new set of processes become possible, related to the very high atomic number atoms ($Z = 79$) present, compared to the body (predominantly $Z < 10$). As discussed in Section 15.1, this high atomic number facilitates a new set of physical interactions on the nanoscale. These considerations apply equally well to other nanoparticles with high atomic number introduced into the body.

Radiation is often considered a 'double-edged sword' — it can both be used to kill tumour cells, a desired outcome, but it can also induce mutagenesis in healthy cells, an undesired outcome that could eventually lead to new incidences of cancer. Ionising radiation is used both for cancer therapy and at lower doses for imaging diagnostics. Indeed, it provides the most common imaging modality used medically. The interactions described in this chapter act to amplify the effects of radiation so they are relevant to the whole field of nanomedicine where AuNPs are used (Chapters 14 and 16), since there is potential for unwanted side effects when they are used alongside diagnostic X-rays.

Figure 15.2 shows the classes of processes that can occur when aqueous media containing nanoparticles is irradiated. These interactions are key

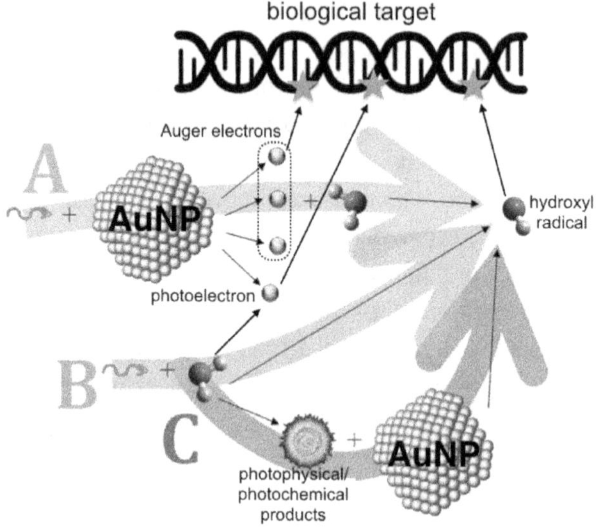

Figure 15.2   Schematic diagram showing the types of processes that can lead to biological damage when aqueous media containing gold nanoparticles (AuNP) is irradiated in biological systems. Based on figure from Ref. 33.

ones when considering the combination of radiation and AuNPs in medical contexts because we and all biological systems are made predominantly of water. DNA is considered the critical target for radiation damage in biological systems (see Section 15.1.1). This damage can occur in a number of ways: In the absence of nanoparticles an incoming photon can ionize the water. Either the photoelectron this produces or downstream chemical products produced can lead to biological damage. Here the hydroxyl radical is particularly important, it being responsible for between 50% and 70% of DNA damage.[3] The biological damage can either lead to desirable cell killing/inactivation or undesired mutagenesis, depending on the situation. The processes considered under pathways B in Figure 15.2 are the ones involved in traditional external beam radiotherapy.

The addition of the nanoparticle facilitates new interactions. The atoms in the nanoparticle can be photoionised (pathways A in Figure 15.2) to add to the yield of photoelectrons. The biological damage due to these photoelectrons is discussed in Section 15.2.2. However, the gold atoms in the nanoparticle have far more electron orbitals than the hydrogen or oxygen in the water. This difference means that inner-shell ionisation followed by production of an Auger cascade can occur. The vacancy created by inner-shell ionisation can be filled in one of two ways — radiative decay or Auger decay. In radiative decay, an electron can 'fall' from a higher orbital to fill the vacancy with the excess energy being carried away by a photon. In Auger decay, the electron gives its energy instead to another electron bound in the atom, the result being that the second electron now has enough energy to detach from the atom it was bound to. In an atom with many electron orbitals such as gold, the result is two vacancies, both possibly in core electron shells. These two vacancies can also possibly be filled by Auger decay with two more electrons escaping from the ion. In this fashion, several electrons can escape after one inner-shell ionisation event. This process is called an Auger cascade. The consequences of this type of process are discussed in Section 15.2.1.

Alternatively, some of the photophysical or photochemical products from the interaction of the photon and the water can be converted to the hydroxyl radical (pathways C in Figure 15.2). Evidence for these pathways is discussed in Section 15.3.

After a brief review of radiobiology (Section 15.1.1) and an introduction to the concept of radiosensitisers (Section 15.1.2), the basic interactions between ionising radiation and matter are discussed in Section 15.1.3. This discussion then provides the framework on which the mechanisms shown in Figure 15.2 are elucidated.

15.1.1 *Radiobiology*

Radiobiology is a well-established and complex field of study. Its goal is to understand and possibly direct the behaviour of biological systems when they are exposed to ionising radiation. One of the major considerations is the ability of cells to proliferate after a dose of radiation is applied. DNA is considered the critical target with induction of double-strand breaks correlating well with loss of a cell's ability to proliferate, often the result is fatality to the cells concerned. Experimentally the inability to proliferate is often characterised using the clonogenic assay whereby a known number of cells adhering to the base of a plastic well and supported by a nutrient-rich biological media are subjected to ionising radiation. The density of cells is chosen so that they have not formed a confluent layer in the bottom of the well. Indeed, none of the cells should be touching each other so they can be considered largely to be independent entities. After irradiation, the cells are placed into an incubator and allowed to form colonies, the number of colonies being counted after typically 8–12 days, depending on the cell line. The fraction of cells that go on to form colonies is referred to as the survival fraction (SF). The SF is a function of radiation dose (D), which is measured in Gray. Typically the SF reduces monotonically as the radiation dose increases (i.e. more radiation leads to less survival).

The SF is often described by the linear quadratic (LQ) model, which is governed by the equation

$$\text{SF} = \text{SF}_0\exp\left[-\alpha D - \beta D^2\right], \tag{15.1}$$

where SF_0 is a term representing the platting efficiency, i.e. it accounts for experimental artifacts, which cause cells to not go on to form colonies in the absence of radiation. α and β are cell-line and condition-dependent parameters. Although it is somewhat controversial, α is often connected with a sequence of events triggered by a single primary photon (hence, the linear dependence with dose), which go on to change the state of a cell so that it no longer forms a colony and β is connected with two separate primary photons each triggering a sequence of events (leading to a quadratic dose dependence), which in combination, lead to the cell being unable to go on to form a colony, i.e. cell killing/inactivation occurs.

The chains of events described above are (somewhat controversially) linked to induction of DNA strand-breaks in a cell, with induction of a double-strand break (i.e. breaks on both strands of the DNA typically 20 base pairs or less apart). The induction of double-strand breaks correlates well with cell death.[4] α can be associated with a cascade of events triggered by a single photon and producing two or more agents (e.g. a solvated electron and a hydroxyl radical), each of which goes on to cause a suitable DNA strand break. Alternatively, the primary photon can give rise to a single-strand break that happens to lie close to and on the opposite strand to a residual DNA strand break. Such residual strand breaks are common and our DNA repair machinery is repairing them all the time.

If one accepts this picture, and furthermore, that the numbers of these lethal events (i.e. events that stop cell proliferation, effectively rendering the cell to be dead) are random and uncorrelated, then it follows that they are described by the Poisson statistical distribution. Recalling that for a mean number of events a, the probability of observing n events is given by

$$P_n = a^n \exp\left(-a\right)/n!, \tag{15.2}$$

one can identify the probability of cells surviving with the induction of no lethal events, i.e. all cells that survive and proliferate do not have a lethal event. The probability of this happening for any given cell equates to the survival fraction (treating all the cells as identical), which reduces Equation (15.2) for the particular case $n = 0$ (i.e. no lethal events) to

$$P_0 = \exp\left(-a\right). \tag{15.3}$$

Comparing this result with Equation (15.1) and recalling that SF_0 is to account for experimental (non-radiation-related artifacts), one comes to a result that the mean number of lethal events is given by

$$a = \alpha D + \beta D^2. \tag{15.4}$$

15.1.2 *Radiotherapy and Radiosensitisers*

The goal of radiotherapy is to bring about sufficient killing of the malignant cells or at least to stop them proliferating while avoiding damage to the healthy cells as far as possible. The radiation is delivered in a series of fractions, typically given to a patient on alternate days. One of the particular challenges here is that the dose associated with cure (i.e. eradication of the tumour) is not very different from the dose associated with development of complications. This challenge means that the dose and how it conforms to the tumour are very carefully controlled. Each patient is imaged, the tumour is demarked and a fully personal treatment plan is produced — in this sense, radiotherapy is the first form of personal medicine.

The probability of a tumour being controlled is called the tumour control probability (TCP) and it is represented by a sigmoid function with respect to the dose. The probability of there being complications in the normal tissue is called the normal tissue complication probability (NTCP). It follows directly from these definitions that the probability of cure without complication (PCWC) is given by

$$\text{PCWC (d)} = \text{TCP (d)} \times (1 - \text{NTCP(d)}). \tag{15.5}$$

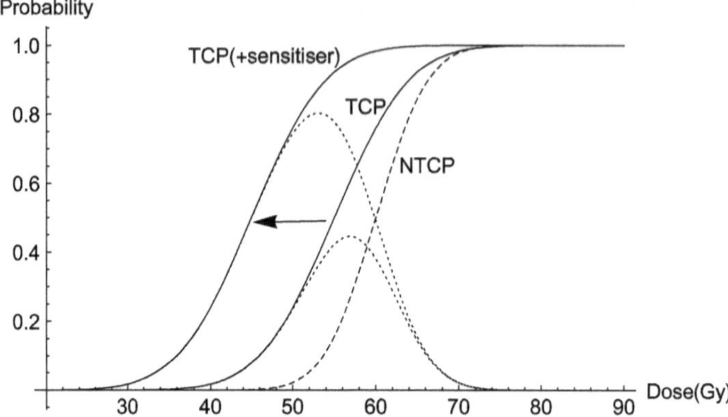

Figure 15.3   Schematic figure showing the TCP (solid line) and NTCP (dashed line) as a function of radiation dose. The TCP is illustrated for two scenarios, one with a TCD_{50} of 55 Gy and a second with a TCD_{50} of 45 Gy, to illustrate the action of a radiosensitiser targeted specifically to the tumour. In each case, the PCWC is shown by the dotted line.

The TCP is characterised by the radiation dose required to give a control probability of 50%, and is denoted by TCD_{50}, i.e. tumour control dose 50%. It is somewhat analogous to the lethal dose (LD_{50}), used in studying effects of drugs on biological systems. Figure 15.3 uses these concepts to illustrate in a very general way the potential benefit of a radiosensitising agent. Suppose the radiosensitising agent has the effect of increasing the cell killing, localised to the tumour alone. Such an agent would result in a decrease in the TCD_{50}. As can be seen in Figure 15.3, even a modest change in the TCD_{50} value can have a very significant effect on the PCWC, both increasing its maximum and also widening its distribution, thereby providing a bigger margin for any errors, which might occur in the treatment planning or the treatment delivery process.

Although these arguments are very simplistic and ignore many of the complications involved, they very clearly point to the potential benefits of targeted (to the tumour) radiosenstisers. Gold nanoparticles (or indeed any nanoparticle containing heavy atoms and specifically targeted to the tumour) represent a very good candidate for this approach to improving cancer therapy. Gold has received the majority of attention in this capacity, as it is considered to be biocompatible and easy to functionalise.

With gold nanoparticles, targeting occurs through their localised preferential uptake to the tumour mass. This preferential uptake can be brought about through passive means, with the leaky vasculature structure facilitating entry into the tumour mass. Alternatively, the nanoparticles can be deliberately functionalised with an antibody coating designed to provide preferential binding to certain cell receptors.[5] These nanoparticles often enter the tumour cells, although it is not absolutely clear that this is a prerequisite for them being effective radiosensitisers.

15.1.3 *Basic Principles of the Interactions of Radiation with Matter*

When ionising radiation interacts with matter, a number of processes occur. As the name implies, the radiation has sufficient energy to ionise the constituents of the matter it interacts with, through the photoelectric effect (i.e. the emission of electrons due to absorption of photons). The matter is ionised in this reaction, leading to subsequent chemical change. The electron thus liberated has an energy equal to the difference between the photon's energy and the energy of the atomic orbital from which it originated. Hence, the photoelectron can also potentially ionise the matter it travels through, being a type of ionising radiation itself.

Compton scattering (or incoherent scattering, the inelastic scattering of a photon by one of the bound electrons in the AuNP) can also produce ionisation and hence chemical change. Because this is a scattering process, the incident photon is not fully absorbed, and instead is scattered inelastically with the balance of energy being shared between the binding energy required to liberate the electron from the atom and the electron's final kinetic energy. The photon can also scatter elastically in a process called coherent (or Rayleigh) scattering.

Another relevant process is pair production. As first elucidated by Dirac, the vacuum can be thought of not as simply empty space but as a 'sea' of electrons with negative energy. If one of these negative electrons can receive enough energy from a photon, it can be plucked from the negative energy sea, to become a positive energy electron (i.e. the type of free electron we have already been discussing). When this happens, a 'hole' is left in the sea of electrons. This hole is the anti-particle of the electron (called the positron). When sufficient energy is provided by a photon, an electron–positron pair can be created with a combined kinetic energy equal to the difference between the initial photon energy and twice the rest of energy of the electron (1.022 MeV in total).

More details of these processes can be found in Ref. 6. A characteristic of these processes is that they generally result in an increased number of particles (i.e. photons, electrons and positrons), each having a lower energy. As a result, a high-energy photon beam creates a shower of lower energy photons, electrons and positrons as it traverses matter, gradually losing primary (high energy) photons as it penetrates into the matter. The likelihood of a given process occurring can be expressed in terms of the absorption cross-section. This quantity can be scaled with the density of the medium through which the high-energy photons are travelling (i.e. typically the patient) to give a characteristic length scale over which the process can be expected to happen. Figure 15.4 shows the contributions of the various processes to the total absorption cross-section for water across the entire range of energies of relevance to radiotherapy. This entire energy range accounts for both the primary photon beam (typically MeV energy) and all of the lower energy 'daughter' particles produced.

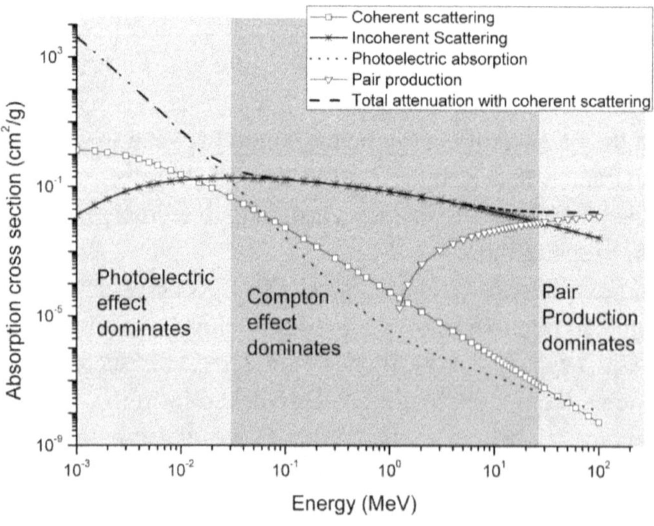

Figure 15.4   The total attenuation of photons in water and the processes contributing to this attenuation as a function of photon energy. This data was derived from the NIST X-ray Attenuation Databases.[7]

Because we (humans) are comprised almost exclusively of water and other low atomic number species, this figure provides a guide to the processes at work in a typical radiotherapy treatment. The electrons thus liberated or the chemical products resulting from ionisation are the species able to cause the DNA strand breaks (or other biological damage) described in Section 15.1.1. Note that due to the cascade nature of the processes involved, a single photon can lead to more than one species capable of breaking a DNA strand and hence, can be the sole primary cause of induction of a double strand break (DSB). Indeed, this type of process is one of the contributory factors to the parameter α in Equation (15.1).

When matter interacts with a high atomic number species like gold, the absorption cross-section is generally significantly larger. This is because most of the processes contributing to the absorption are interactions between the photon and the bound electrons, higher atomic number corresponding to higher numbers of electrons per atom. Furthermore, there are atomic orbitals with higher binding energies. For each of these binding energies, there is a threshold energy below which the electron cannot be liberated from the atom. When the photon energy increases and crosses one of these thresholds, there is a discontinuity in the cross-section for the photoelectric effect, which, in turn, leads to a discontinuity in the absorption cross-section. These discontinuities are called absorption edges and they are characteristic of the atomic species.

Atoms bind together due to the interaction of the valence shell electrons. Because the energies associated with the processes involved in Figure 15.5 are much higher than the energies associated with these electrons, the data presented in this figure is equally applicable to gold contained within nanoparticles, i.e. the orbitals concerned with the processes shown in Figure 15.5 are not effected by the bonding between atoms in the nanoparticle. In addition to the absorption being much greater, the photoelectric effect plays a much bigger role and over a wider energy range in AuNPs. Furthermore, the high-energy photons can remove electrons from very deep-lying orbitals, e.g. the 1s orbital (threshold associated with the edge at 78 keV) or from the 2s or 2p orbitals (thresholds 11.9–14.4 keV).

Figure 15.6 shows the mass energy absorption coefficients for gold and water over the energy range typically of interest in radiotherapy. This quantity provides a measure of how much matter a high-energy photon beam

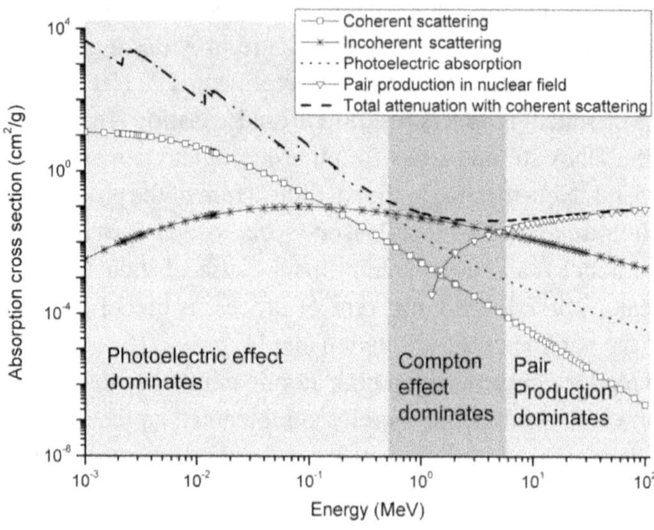

Figure 15.5   The total attenuation of photons in gold and the processes contributing to this attenuation as a function of photon energy. This data was derived from the NIST X-ray Attenuation Databases.[8]

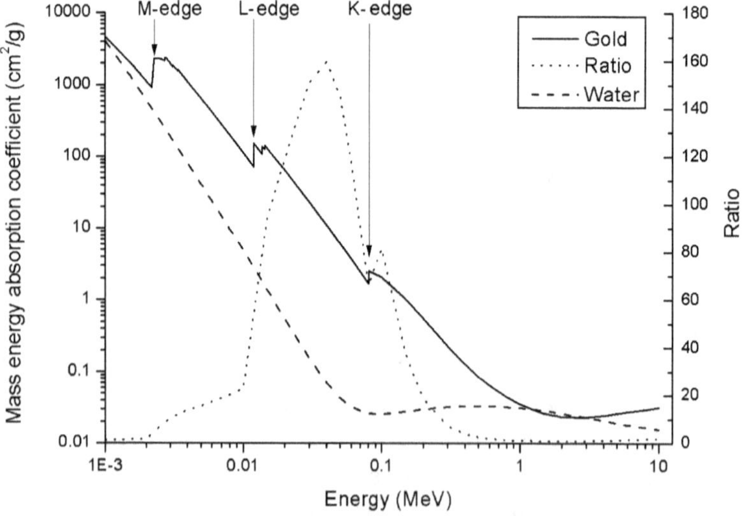

Figure 15.6   Comparison of the photon mass energy absorption coefficients for gold and water. From Refs. 9 and 10. The sharp discontinuities observed in the mass energy absorption coefficient for gold are called absorption edges and they are due to the photon energy crossing a threshold for ionisation. When this happens, electrons can be ionised out of the shell indicated, leading to an increase in the absorption.

must pass through to have its energy decreased by a factor of 1/e. This figure shows that gold can be over 100 times as effective at absorbing radiation as water. Hence, even a small amount of gold can lead to significant absorption of energy. However, the energy at which the ratio has a maximum is significantly lower than the primary energy associated with radiotherapy machines. A typical photon spectrum from one of these machines is shown in Figure 15.7. These photons are produced by impacting a high-energy (25 MeV in this case) electron beam onto a tungsten target.

Figure 15.7(a) also shows which photo-processes are dominant as a function of atomic number and photon energy while Figure 15.7(b) shows the spectrum of primary photons and secondary electrons occurring in a typical radiotherapy session. The photon spectrum[12] corresponds to that of a typical medical linac. The electron spectrum was derived by using this photon spectrum in a Geant4[13] simulation of radiation processes occurring in radiotherapy, where a cubic block of water 10 cm on each side was irradiated by using a 25 MeV linac.

It is important to reconsider the experience gained from traditional radiotherapy in the light of Figure 15.7. For the low atomic number species, Compton processes are dominant across almost the full energy range concerned. However, for high atomic number species such as gold, the Compton processes are dominant over a much smaller energy interval with the photoelectric effect and pair production dominating at the low- and high-energy ends of the energy range encountered in radiotherapy, when considering both primary photons and the daughter particles. It is for this reason that a detailed account is required of the physical mechanisms at work when gold nanoparticles are introduced to the system. This account is given in the next section.

15.2 Physical Processes

The first attempts at describing the physical processes simply considered the gold as a continuous, highly absorbing addition to the biological media.[14,15] However, while this approach provides a good approximation to the average increase in physical dose, it ignores the fact that the gold is present in the form of nanoparticles. Furthermore, it does not provide a good account of

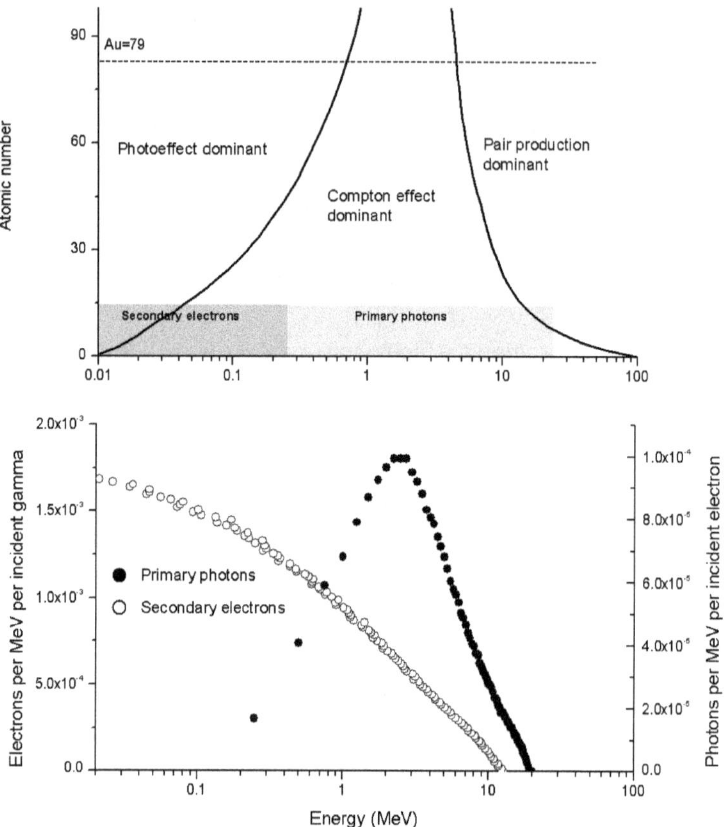

Figure 15.7 (a) Illustration of the domains over which the various photoprocesses are dominant as a function of atomic number and photon energy. Data taken from Ref. 11. (b) The spectra of electron and photon energies encountered within a typical radiotherapy scenario, where a linear accelerator (linac) with a 25 MeV electron beam is used to irradiate a cube made of water and 10 cm on each side. The linac spectrum (black circles) was taken from Ref. 12. The electron spectrum (white circles) was produced using Geant4.[13] Since linacs of this type (electron energies typ. 6–25 MeV) are routinely used for cancer therapy, this simulation illustrates a typical radiotherapy scenario.

the level of dose enhancement seen in cell biology experiments.[16,17] One needs a description that intrinsically accounts for the nanoscale effects.

15.2.1 *Nanoscale Local Effect Description*

When a photon of sufficiently high energy (above the M-shell edge shown in Figure 15.6) interacts with a gold atom inside a nanoparticle, it can cause photoionisation from the M-shell. If the energy is high enough the ionisation can be from the L-shell or even the K-shell. When this happens, a photoelectron is created. However, the resultant ionised atom is still unstable and it can stabilise either by Auger processes or X-ray fluorescence or a combination of both. An example of the type of process that can occur is shown in Equation (15.6).

$$\mathrm{Au} + h\upsilon \rightarrow \mathrm{Au}^+ \left[1s^{-1}\right] + e^-$$

$$\rightarrow \mathrm{Au}^+ \left[2l_1^{-1}\right] + h\upsilon' + e^-$$

$$\rightarrow \mathrm{Au}^{2+} \left[nl_2^{-1}, n'l_3^{-1}\right] + h\upsilon' + 2e^- \dots \tag{15.6}$$

The notation used here is derived from the standard way atomic transitions are described. n and l denote principle and orbital angular momentum numbers of the electrons concerned. The superscript -1 indicates an electron vacancy. In this particular set of reactions, an electron is photoionised from the K-shell to leave a 1s vacancy. This vacancy then rapidly stabilises by X-ray fluorescence (the dominant channel for a 1s vacancy) to leave a vacancy in the $n = 2$ shell and producing the photon $h\upsilon'$. Radiative stabilisation of this vacancy is slower than for the 1s vacancy and instead an Auger decay can happen. This process occurs because the electrons are sensitive to each other's electric fields. The coupling between the electrons means that one electron can fall from a higher orbital into the $n - 2$ vacancy. However, instead of a photon being liberated to maintain the energy balance, one of the other electrons is able to escape the ion, leaving it doubly charged. The system now has two vacancies, each of which also has the potential to decay by the Auger process. In practice, there is a complex mixture of Auger and fluorescence decay happening with up to 10 Auger electrons being liberated from a single photoionisation event.[18]

The photoionisation can happen on any atom of the nanoparticle. Hence, before interacting with the surrounding biological medium, each electron liberated from one of the atoms in the nanoparticle typically has to pass many other gold atoms. As an electron travels across the nanoparticle, it can undergo collisions with these atoms, resulting in it changing direction and losing energy.

Low-energy electrons are more susceptible to scattering effects having a lower range than higher energy electrons.[19] Furthermore, this effect increases with both atomic number and the density of the medium through which the electrons are travelling. As has been discussed in some detail by Wardlow *et al.*,[20] this observation implies that many of the electrons will not be able to escape from the nanoparticle and those which do will have their energies attenuated. However, the Auger electrons created near the surface of the nanoparticle can escape from it. They will subsequently travel a short distance in the surrounding biological medium (predominantly water). Therefore, they deposit their energy very close to the nanoparticle. This occurrence is shown in Figure 15.9, where a full set of electron paths following an inner-shell ionisation of a gold atom near the nanoparticle's surface is displayed.

In Figure 15.9, the incoming photon is illustrated by the green line. The solid line from left to right and going slightly upwards is the photoelectron, which in this case happens to traverse the whole nanoparticle with very little scattering. The line proceeding down and to the right is the highest energy Auger electron (from the 5–12 keV group of electrons shown in Figure 15.8). The small 'wiggly' structures leaving this electron's track are spurs created by ionisation of the water in which the nanoparticle is embedded. The complex, 'wiggly' structure seen predominantly to the top and right of the nanoparticle is where the rest of the Auger electrons deposit their energy in a highly localised zone where many water ionisations and other chemical reactions can occur. The highly localised energy deposition associated with this region results in a radial dose distribution, which is very large near the NP surface and which falls off rapidly as one moves away from the surface. An example of a calculated radial dose distribution is shown in Figure 15.10.

This radial dose distribution can be used in the local effect model (LEM)[21,22] to give an estimate of the biological effect due to this process. This approach has been used by a number of research groups[23–27] to give an account of the biological effect of heavy atom nanoparticles being irradiated. Equation (15.4) shows that the mean number of lethal events is a nonlinear function of dose. Integrating this dose distribution over the whole of a cell, the total number of lethal events is given by

$$N_{\text{tot}} = \frac{1}{V_{\text{cell}}} \int_{\text{cell}} \left(\alpha D\left(r\right) + \beta D\left(r\right)^2 \right) \mathrm{d}V, \tag{15.7}$$

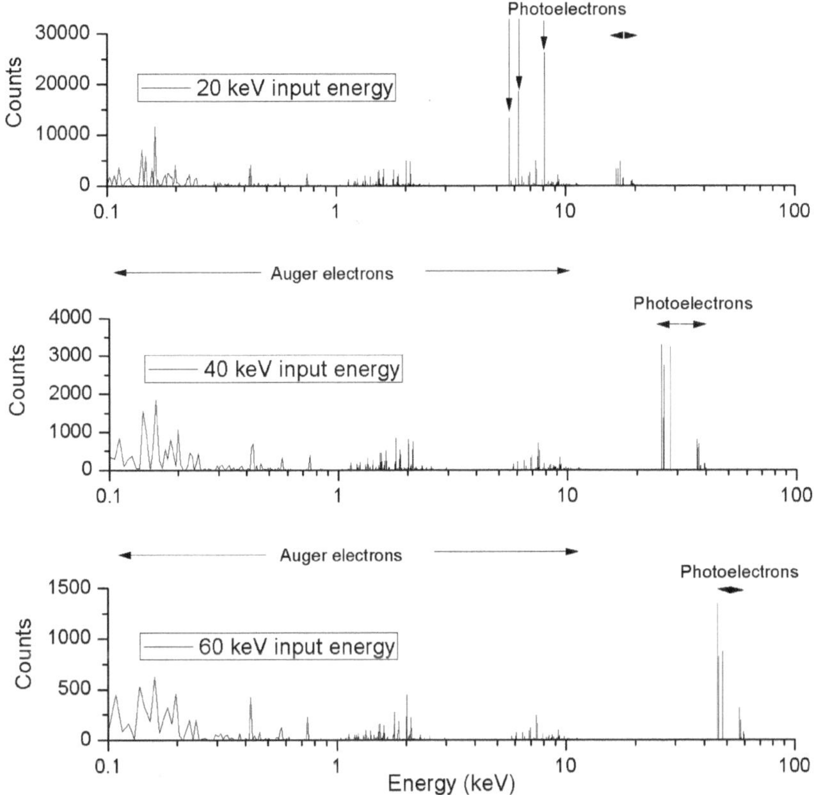

Figure 15.8　All-electron spectrum calculated using Geant4[13] for irradiation of gold atoms with monoenergetic radiation of various energies. The Auger electrons are found predominantly with energies below 3 keV, but there is a group between 5 and 12 keV (seen most clearly in the 40 and 60 keV spectra). Note the energies of the Auger electrons do not change as the photon energy is increased whereas, the photoelectrons energies do increase, exactly as expected.

where V_{cell} is the volume of the cell. The largest contributions to this integral come from the regions very close to the surface of the AuNPs. Here $D(r)$ can be many orders of magnitude greater than the dose far from the nanoparticles, which is essentially the dose that would be received without nanoparticles.

As is illustrated in Figure 15.9, these regions of high-radiation dose are small compared to the cell. However, according to Equation (15.7), they result in a very large contribution to N_{tot}, i.e. they make a large contribution to the probability that the cell is killed. Again assuming the probability

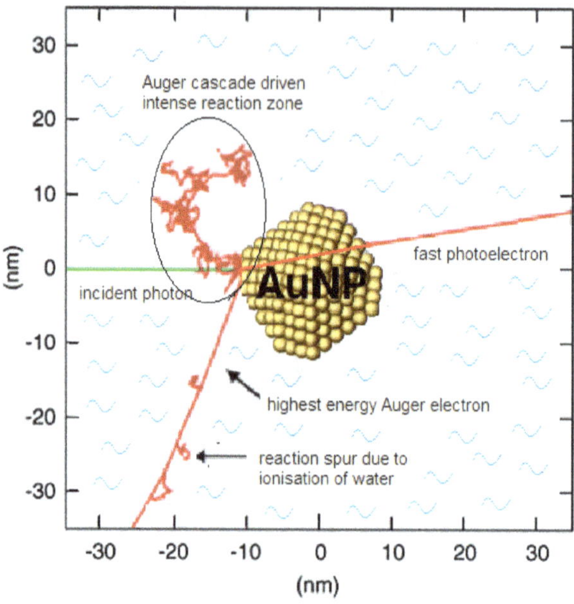

Figure 15.9   Schematic illustration of the typical electron tracks produced following inner-shell ionisation of a gold atom near the nanoparticle's surface.[18] Here a 20 nm diameter AuNP (gold atoms not to scale) is pictured in the centre of a 60 nm by 60 nm square, surrounded by water. An incident photon comes in left to right (green line) to cause an inner-shell ionisation on a gold atom near the NP surface. The result is the creation of a fast photoelectron, and several Auger electrons in a cascade process. Many of these electrons interact with the water in a very small region, giving rise to the intense reaction zone where there are many ionisation and excitation events.

distribution for N_{tot} to be subject to Poisson statistics, the total number of lethal events can be related to the survival fraction of the cells by

$$\text{SF} = \exp\left(-N_{\text{tot}}\right). \tag{15.8}$$

This equation follows directly from the same reasoning as was used to generate Equation (15.2). The framework for this type of calculation is exactly the same as the LEM used in planning heavy ion therapy, for example at the Heidelberg heavy ion therapy facility.[28] Accordingly, the model is well tested in a real clinical context. However, the interpretation of the results produced in a nanoparticle context has to be modified in the light of experimental results (discussed in Section 15.2.2). Notwithstanding considerations of the interpretation, predictions of the cell survival curve in the presence of nanoparticles can be made from this approach. Figure 15.11 shows an example of one such prediction.

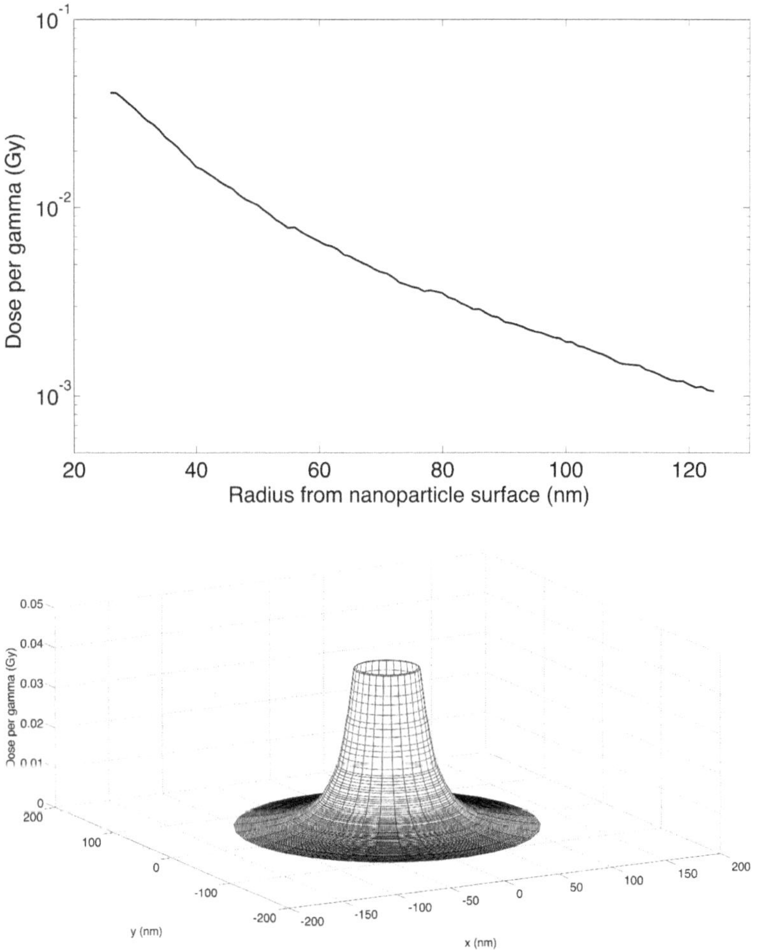

Figure 15.10   Radial dose distribution in the surrounding water due to irradiation of a 25 nm radius AuNP with a 50 keV monoenergetic photon beam. Left: radial dose distribution in 2D from NP radius (25 nm) out to 125 nm. Right: RDD in 3D from a radius of 25–125 nm.

Since its first implementation to describe the biological effect of irradiating MDA-231 (breast cancer cell line), i.e. cells that have been exposed to nanoparticles,[18] this approach has been widely used.[23–27] The X-ray energies used in this description are far lower than the typical primary photon energies used in radiotherapy. To properly consider the effects with higher energy photons in a therapy context, one must imbed the nanoparticles in

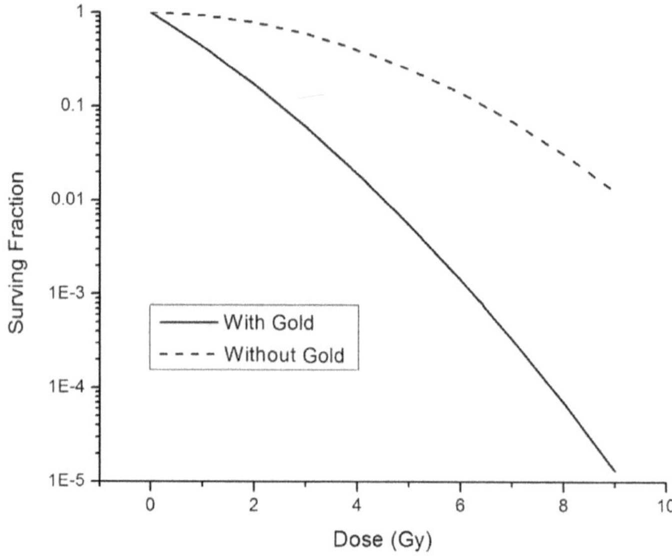

Figure 15.11   Survival fraction curve for breast tumour cells (MDA-231) when they are doped with AuNP of 25 nm radius at 500 μg/mL concentration and are irradiated with a 50 keV gamma beam.

a sea of both photons and electrons with distributions like those shown in Figure 15.6(b).[25] In this case, the photons and electrons with energies close to the maximum in the ratio shown in Figure 15.6 (i.e. those produced by earlier interactions with the biological medium or other parts of the patient) will give rise to the large enhancement to cell killing observed.

15.2.2 *Nanoparticle Imaging and the Role of the Photoelectrons*

The interpretation given in Section 15.2.1 was that the lethal events are associated with DNA damage. However, the same mathematical treatment would apply for damage to other sub-cellular biological targets. Indeed, the interpretation that it is only DNA damage that leads to cell death/inactivation is challenged by images showing the sub-cellular distribution of the gold.[29,30] Recall that the Auger electrons do not travel very far (typically <50 nm, see Figure 15.9) while they deposit their energy. However, the images of the gold nanoparticle subcellular distribution show the nanoparticles to be outside the nuclei of the cells with the nuclear membrane lying in the way.

McQuaid *et al.*[30] examined the energy dependence of both cell death and DNA damage to cells when monoenergetic (synchrotron) radiation is applied directly to thin samples containing cells. In this condition, there is not a shower of lower energy photons or electrons contributing to the kind of spectrum shown in Figure 15.6. This factor considerably simplifies the analysis. Good agreement was found with LEM predictions for the cell survival. However, good agreement was not found with the LEM for the DNA damage (see Figure 15.12). Since the cell damage was assessed 1 hour post-irradiation, this finding suggests that the kinds of physico-chemical processes illustrated in Figure 15.2 are not causing the DNA damage. Better agreement is found for the DNA damage with a model that only considers the role of the photoelectron.[30]

The agreement of the LEM with the cell survival data but its inability to predict the observed DNA damage calls into question the interpretation (at least in the context of nanoparticle enhanced radiation effects) of the lethal events being associated with DNA strand breaks (outlined in Section 15.1.1). Essentially the gold nanoparticles, through the Auger emission's localisation are acting as a nanoscale probe of radiation processes. The formalism used for the LEM calculations can remain unchanged but one does not associate the lethal events specifically with DNA damage. It is possible that the AuNPs do not even need to enter the cell to be effective, perhaps the cascade of Auger electrons could cause sufficient cell membrane damage to kill/inactivate cells.

As well as being of fundamental interest, these findings point to another potential benefit of using nanoparticles as dose-enhancing agents. Because the AuNPs are not inside the cell nucleus, the DNA damage observed is mediated by the photoelectrons. However, as is described by the LEM, the high levels of cell killing are due to the high levels of local dose near the nanoparticle surface, away from the cell nucleus. Therefore, it is reasonable to suppose that there will be less genomic alteration in cells which are subject to radiation and have taken up nanoparticles that survive, compared to cells that have not taken up nanoparticles. Although this is an experimentally unverified assertion, it suggests that there will be less mutagenic complications associated with nanoparticle-enhanced therapy. Put simply, the local Auger effects either kill the cells outright or they survive with little DNA damage.

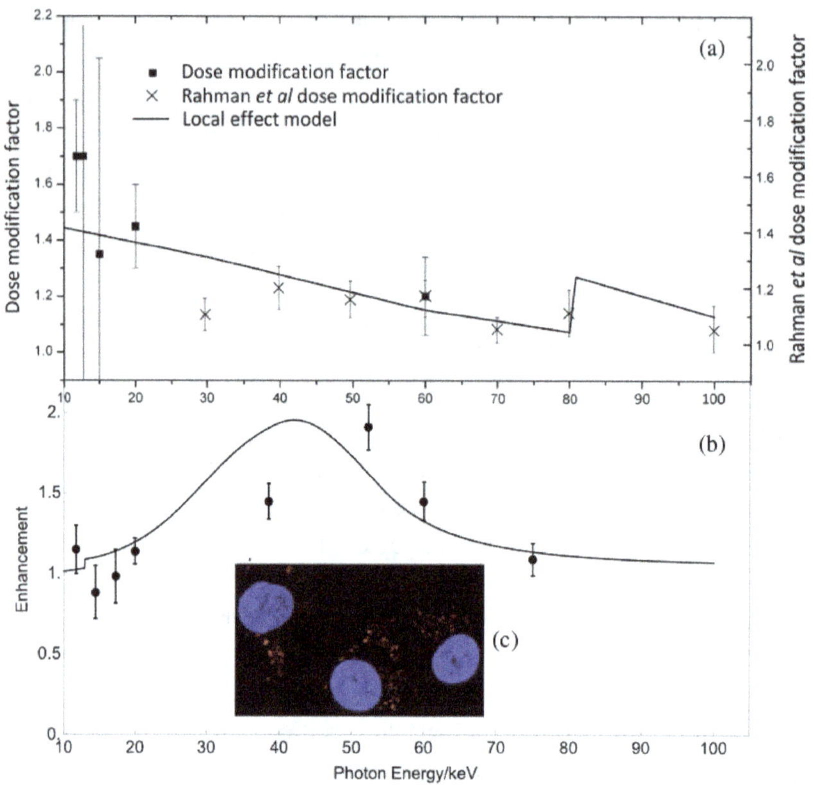

Figure 15.12   (a) Cell survival dose modification factor (i.e. the ratio of cell killing with to that without the addition of AuNPs) measured using monoenergetic radiation[30] and compared to the LEM. (b) DNA damage enhancement measured using monoenergetic radiation and compared to a model considering the photoelectron only. (c) A multiphoton image of the gold nanoparticle uptake, taken from Ref. 30. The cell nuclei are shown with the blue stain whilst the signal from the nanoparticles is shown as orange.

15.3 Chemical Processes

In describing the physical processes (Section 1.2), we have ignored the fact that the AuNP will usually have some kind of coating layer. The coating is usually thin and composed of a low atomic number material and therefore, will allow electrons to pass more easily than the gold. Hence, it will not play a significant role in the electron-driven physical processes described in Section 15.2. However, such a coating can have a role in chemical processes symbolised by pathway C in Figure 15.1.

Using fluorescent probes, Mizawa and Takahasi showed that gold nanoparticles can generate a range of reactive oxygen species (ROS).[31] This finding of itself would not be enough to infer new (chemical) processes, beyond those already discussed in Section 15.2, were taking place in the system. That is to say that pathways A of figure 15.2 can lead to the generation of ROS (predominantly but not exclusively the hydroxyl radical). Mizawa and Takahasi[32] found that there was an inverse relationship between the nanoparticle diameter and the yield of ROS, something they attributed to catalytic activity (pathways C) of Figure 15.2.

Cheng *et al.*[30] also reported enhancements to the yield of the hydroxyl radical due to a range of nanoparticles (including gold) under irradiation. They used the concept of turnover frequency (TOF) to examine the catalytic nature of the reactions. TOF is the number of chemical reactions catalysed per catalytic site (surface atom in this context) per unit time. They found the TOF decreased with available surface area, suggesting that some other reagent limits the reaction rather than the availability of nanoparticle surface sites. They also found that the catalytic process was coating dependent.

Sicard-Roselli *et al.*[33] used monochromatic synchrotron radiation to further examine the energetics of chemical production processes. The benefit of this approach is that one can readily calculate the total energy absorbed in the nanoparticles. The analysis of the results showed that the yield of the hydroxyl radical was greater than could be accounted for using pathways A of Figure 15.2. This yield is consistent with the view of a catalytic process as expressed by Mizawa and Takahashi[31] but it adds the extra insight that the initial interaction is in the water, not the nanoparticle. Wardlow *et al.*[20] further analysed this phenomenon, accounting for the loss of energy experienced by electrons travelling through the nanoparticles to better determine an upper bound on the yield of OH due to pathways A.

Although there is a significant body of evidence showing that catalysis of radiolytic products at the gold nanoparticle surface is possible, there is not (as yet) a fully quantitative model of this phenomenon. Comparing this evidence to the situation described in Section 15.2, the result is not surprising because the details of the chemistry are yet to be determined. This is clearly important, ripe for future research as it impacts on developing future therapeutic products significantly. Furthermore, it has

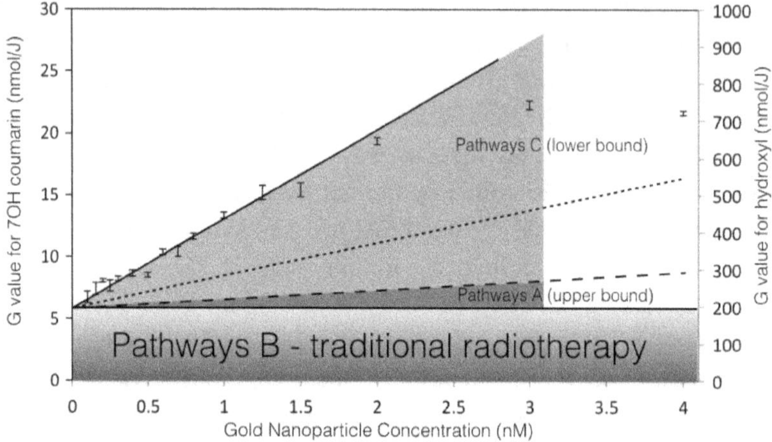

Figure 15.13　The G-value of OH as a function of gold nanoparticle concentration.[33] Three lines are shown, indicating the slope of the G-value in the low-concentration linear part of the curve, the maximum slope possible due to pathways A as dictated by simple energy considerations (dotted line) and the maximum slope possible due to pathways A when loss of energy of the electrons in the nanoparticle is taken into account (dashed line) 20. The remaining contribution is attributed to pathways C.

the potential to impact across the whole field of nanomedicine involving gold nanoparticles due to the prevalent use of diagnostic X-rays in medicine.

Recall the electron emission component (pathways A of Figure 15.2) is sufficient to account for the cell survival seen in experiments.[18] This statement implies that the reactions corresponding to pathways C of Figure 15.2 are not driving cell death under the conditions studied, possibly due to the cell's innate radical scavenging capabilities. It could be that if the right conditions were achieved, the extra 'chemical' enhancement to the OH yield shown in Figure 15.13 would be realised through utilising pathway C of Figure 15.2. This then implies that there is the possibility for even better enhancement of cancer therapy.

15.4　Conclusions and Future Outlook

The physical and chemical mechanisms at work when gold nanoparticles are irradiated in aqueous media have been discussed with both experimental results and theoretical predictions being presented. A range of mechanisms leading to biological damage has been considered. Although this article has focused on gold nanoparticles, the processes are very general. The mechanisms discussed in Section 15.2 all apply to any nanoparticle containing a significant fraction of heavy atoms and they point to excellent prospects to enhance radiotherapy in patients. The chemical mechanisms discussed in Section 15.3 are less well understood and appear to be highly dependent on the nanoparticle coating. Here there is clearly a need for much more research aiming to elucidate the chemical mechanisms at work. This research will then form a research-base upon which the construction of next-generation nano-therapeutics can be developed.

The interactions of radiation and the nanoparticles should be considered for any cancer-related nanoparticle, because radiotherapy is used in most cancer treatments, it could prove beneficial, providing extra efficacy. In contrast, because X-rays are such an ubiquitous imaging modality, the potential side-effects of nanoparticles interacting with ionising radiation should also be considered. Hence, consideration of radiation and the mechanisms described in this chapter should feature in the development of any gold nanoparticle being considered for medical applications.

Acknowledgements

We thank the many colleagues and collaborators with whom we have formed the ideas described in this chapter and acknowledge Diamond Light Source for time on beamlines B16 and I15, facilities integral to developing the concepts described.

References

1. Online: https://scholar.google.co.uk/ searches performed 28 January 2016.
2. Online: http://www.who.int/mediacentre/factsheets/fs297/en/ retrieved 2 January 2016.
3. C. Von-Sonntag, *The Chemical Basis of Radiation Biology* (Taylor & Francis, London). 1987.
4. J. Dahm-Daphi, C. Sass, and W. Alberti, Comparison of biological effects of DNA damage induced by ionizing radiation and hydrogen peroxide in CHO cells. *Int. J Radiat. Biol.* **76**, 67–75 (2000).

5. P. Ghosh, G. Han, M. De, C. K. Kim, and V. M. Rotello, Gold nanoparticles in delivery applications. *Adv. Drug. Deliv. Rev.* **60**, 1307–1315 (2008).

6. E. B. Podgorsak, *Radiation Oncology Physics: A Handbook for Teachers and Students*, E. B. Podgorsak (ed.), Chapter 1, 'Basic Radiation Physics' (International Atomic Energy Agency, Vienna) (2005).

7. Online: http://physics.nist.gov/cgi-bin/Xcom/xcom3_2 retrieved 14 January 2016.

8. Online: http://physics.nist.gov/cgi-bin/Xcom/xcom3_1 retrieved 14 January 2016.

9. Online: http://physics.nist.gov/PhysRefData/XrayMassCoef/ElemTab/z79.html retrieved 14 January 2016.

10. Online: http://physics.nist.gov/PhysRefData/XrayMassCoef/ComTab/water.html retrieved 14 January 2016.

11. Online: http://physics.nist.gov/cgi-bin/Xcom/xcom2 retrieved 14 January 2016.

12. D. Sheikh-Bagheri, and D. W. O. Rogers, Monte Carlo calculation of nine megavoltage photon beam spectra using the BEAM code. *Med. Phys.* **29**, 391–402 (2002).

13. S. Agostinelli, J. Allisonn, K. Amako, J. Apostolakis, H. Araujo, P. Arce, M. Asai, D. Axen, S. Banerjee, G. Barrand, F. Behner, L. Bellagamba, J. Boudreau, L. Broglia, A. Brunengo, H. Burkhardt, S. Chauvie, J. Chuma, R. Chytracek, G. Cooperman, G. Cosmo, P. Degtyarenko, A. Dell'Acqua, G. Depaola, D. Dietrich, R. Enami, A. Feliciello, C. Ferguson, H. Fesefeldt, G. Folger, F. Foppiano, A. Forti, S. Garelli, S. Giani, R. Giannitrapani, D. Gibin, J. J. Gomez-Cadenas, I. Gonzalez, G. Gracia-Abril, G. Greeniaus, W. Greiner, V. Grichine, A. Grossheim, S. Guatelli, P. Gumplinger, R. Hamatsu, K. Hashimoto, H. Hasui, A. Heikkinen, A. Howard, V. Ivanchenko, A. Johnson, F. W. Jones, J. Kallenbach, N. Kanaya[i], M. Kawabata, Y. Kawabata, M. Kawaguti, S. Kelner, P. Kent, A. Kimura, T. Kodama, R. Kokoulin, M. Kossov, H. Kurashige, E. Lamanna, T. Lampén, V. Lara, V. Lefebure, F. Lei, M. Liendl, W. Lockman, F. Longo, S. Magni, M. Maire, E. Medernach, K. Minamimoto, P. Mora-de-Freitas, Y. Morita, K. Murakami, M. Nagamatu, R. Nartallo, P. Nieminen, T. Nishimura, K. Ohtsubo, M. M. Okamura, S. O'Neale, Y. Oohata, K. Paech, J. Perl, A. Pfeiffer, M. G. Pia, F. Ranjard, A. Rybin, S. Sadilov, E. Di-Salvo, G. Santin, N. Sasaki[e], Y. Savvas[as], S. Sawada[ab], S. Scherer[af], V. Sei[aw], D. Sirotenko[i, al], N. Smith[g], H. Starkov[f], J. Stoecker[af], Sulkimo[ah], T. Takahata, S. Tanaka, E. Tcherniaev, E. Safai-Tehrani, M. Tropeano, P. Truscott, H. Uno, L. Urban, P. Urban, M. M. Verderi, A. Walkden, W. Wander, H. Weber, J. P. Wellisch, T. Wenaus, D. C. Williams, D. Wright, T. Yamada, H. Yoshida, and D. Zschiesche, GEANT4 — a simulation toolkit. *Nuclear Instruments and Methods.* **A 506**, 250–303 (2003).

14. S. H. Cho, Estimation of tumour dose enhancement due to gold nanoparticles during typical radiation treatments: a preliminary Monte Carlo study *Phys. Med. Biol.* **50** 163–173 (2005).

15. S. J. McMahon, M. H. Mendenhall, S. Jain, and F. Currell, Radiotherapy in the presence of contrast agents: A general figure of merit and its application to gold nanoparticles. *Phys. Med. Biol.* **53**, 5635–5651 (2008).

16. S. Jain, J. A. Coulter, A. R. Hounsell, K. T. Butterworth, S. J. McMahon, W. B. Hyland, F. M. Muir, G. R. Dickson, K. M. Prise, F. J. Currell, J. M. O'Sullivan, D. G. Hirst, J. A. Coulter, K. T. Butterworth, G. Schettino, G. Dikson, and A. R. Hounsell, Cell-specific

radiosensitization by gold nanoparticles at megavoltage radiation energies. *IJROBP (Red Journal)* **79** (2), 531–539 (2011).

17. J. A. Coulter, W. B. Hyland, J. Nicol, and F. J. Currell, Radiosensitising nanoparticles as novel cancer therapeutics — pipe dream or realistic prospect? *Clinical Oncology* **25** (10), 593–603 (2013).

18. S. J. McMahon, W. B. Hyland, F. M. Muir, S. Jain, J. M. O'Sullivan, K. M. Prise, D. G. Hirst, and F. J. Currell, Biological consequences of nanoscale energy deposition near irradiated heavy atom nanoparticles. *Sci. Rep.* **1**, 18 (2011).

19. J. Meesungnoen, J. P. Jay-Gerin, A. Filali-Mouhim, and S. Mankhetkorn, Low-energy electron penetration range in liquid water. *Radiat. Res.* **158**, 657–660 (2002).

20. N. Wardlow, C. Polin, B. Villagomez-Bernabe, and F. J. Currell, A simple model to quantify radiolytic production following electron emission from heavy-atom nanoparticles irradiated in liquid suspensions. *Rad. Res.* **184**, 518–532 (2015).

21. M. Scholz, and G. Kraft, Calculation of heavy ion inactivation probabilities based on track structure, x ray sensitivity and target size. *Radiat. Prot. Dosim.* **52**, 29–33 (1994).

22. T. Elsässer, and M. Scholz, Cluster effects within the local effect model. *Radiat. Res.* **167**, 319–329 (2007).

23. E. Lechtman, S. Mashouf, N. Chattopadhyay, B. M. Keller, P. Lai, Z. Cai, R. M. Reilly, and J. P. Pignol, A Monte Carlo-based model of gold nanoparticle radiosensitization accounting for increased radiobiological effectiveness *Phys. Med. Biol.* **58**, 3075–3087 (2013).

24. C. Wälzlein, E. Scifoni, M. Krämmer, and M. Durante, Simulations of dose enhancement for heavy atom nanoparticles irradiated by protons. *Phys. Med. Biol.* **59**, 1441 (2014).

25. S. J. McMahon, W. B. Hyland, F. M. Muir, J. A. Coulter, S. Jain, K. T. Butterworth, G. Schettino, G. Dikson, A. R. Hounsell, J. M. O'Sullivan, K. M. Prise, D. G. Hirst, and F. J. Currell, Nanodosimetric effect of gold nanoparticles in megavoltage radiation therapy. *Radiother. Oncol.* **100** (3), 412–416 (2011).

26. K. T. Butterworth, S. J. McMahon, F. J. Currell, and K. M. Prise, Physical basis and biological mechanisms of gold nanoparticle radiosensitization. *Nanoscale* **4**, 4830–4838 (2012).

27. Y. Linn, S. McMahon, M. Scarpelli, H. Paaganetti, and J. Schuemann, Comparing gold nano-particle enhanced radiotherapy with protons, megavoltage photons and kilovoltage photons: a Monte Carlo simulation. *Phys. Med. Biol.* **59**, 7675–7689 (2014).

28. M. Krämmer, E. Scifoni, F. Schmitz, O. Skol, and M. Durante, Overview of recent advances in treatment planning for ion beam radiotherapy. *Eur. Phys. J. D* **68**, 306–311 (2104).

29. S. W. Botchway, J. A. Coulter, and F. J. Currell, Imaging intracellular and systemic in vivo gold nanoparticles to enhance radiotherapy. *Br J Radiol* **88**, 20150170 (2015).

30. H. N. McQuaid, F. M. Muir, L. E. Taggart, S. J. McMahon, J. A. Coulter, W. B. Hyland, S. Jain, K. T. Butterworth, G. Schettino, K. M. Prise, D. G. Hirst, S. W. Botchway, and

F. J. Currell, Imaging and radiation effects of gold nanoparticle in tumour cells. *Sci. Rep.* **6**, 19442 (2016).

31. M. Mizawa, and J. Takahashi, Generation of reactive oxygen species induced by gold nanoparticles under x-ray and UV Irradiations. *Nanomed. Nanotechnol. Biol. Med.* **7** (5), 604–614 (2011).

32. N. N. Cheng, Z. Starkewolf, R. A. Davidson, A. Sharmah, C. Lee, J. Lien, and T. Guo, Chemical Enhancement by Nanomaterials Under X-ray Irradiation. *J. Am. Chem. Soc.* **134** (4), 1950–1953 (2012).

33. C. Sicard-Roselli, E. Brun, M. Gilles, G. Baldacchino, C. Kelsey, H. McQuaid, C. Polin, N. Wardlow, and F. J. Currell, A New Mechanism for Hydroxyl Radical Production in Irradiated Nanoparticle Solutions. *Small* **10** (16), 3338–3346 (2014).

Chapter 16
Gold Nanoparticles for Sensors and Drug Delivery

Christian Villiers

Institut Albert Bonniot, INSERM, Grenoble, France

16.1 Gold Nanoparticles for Health

16.1.1 *Overview and Societal Issues*

Because of the extraordinary development of nanotechnology over the past two decades, the term 'nanoparticles' appears increasingly in many fields of biology as well as in medicine. Gold nanoparticles has emerged as a new tool that opens the way to new diagnostics, new therapies so long not possible by traditional technologies. Their nanoscale size, the various capping agents they can accommodate as well as their tunable optical properties make them particularly attractive for the emerging field of nanomedicine. The aim of this chapter is to review important applications of these biological materials such as real-time diagnostics, label-free detection, cell tracking, etc. Nanoparticles may also be used for drug delivery and for the detection and/or treatment of tumours. When dealing with nanoparticles introduced in a living body, two aspects must also be addressed: are they biocompatible and are they stealth?

However, the repeated use of the term 'nanomaterial' results in arousing public attention and induces suspicion that must be taken into account. It is important to study in detail every aspect of the use of these materials, including their potential toxicity, since any changes whether in their shape, size or composition, induce different reaction of the body and therefore potential toxicity. Any new product or new modification must be very carefully studied for their biological applications; in parallel, toxicity tests should be performed even if, until now, the use of gold particles are considered to be particularly safe (see Chapter 17). The applications of these particles will certainly keep increasing; these new materials bring in a revolution into many aspects of medical treatment. It must be emphasised that they should allow a more accurate and more efficient treatment with less collateral damage.

16.1.2 *Surface Modification of Gold Nanoparticles*

Gold nanoparticles (AuNPs) are seldom used as bare nano-objects in biology. Indeed, surface functionalisation, also termed surface coating, is essential to give specific properties to the AuNPs (see Chapter 7 for details). Before discussing these properties in the next sections, it is crucial to pay attention to some potential side effects caused by surface functionalisation.

Aggregation: The surface coating of AuNPs can affect the stability of the colloidal suspension and lead to aggregation and sometimes precipitation of gold. This effect must be absolutely avoided when particles are injected into the body for diagnosis or treatment. From a different perspective, AuNP aggregation may be the desired effect when they are used in some bio-molecule detection schemes or protein quantification using *in vitro* techniques because these methods are based on particle–particle interactions. Such interaction modifies their optical response or affects other physical characteristics.

Elimination. After having fulfilled their mission, AuNPs need to be eliminated from the body and this elimination strongly depends on their surface coating. Requirements concerning their elimination rate from circulating fluids may vary according to the objective of the injection: it may be fast for diagnosis and slow for treatment (tumour imaging versus tumour treatment for example).

Stealth. It is very important that AuNPs remain invisible for the immune system, in order not to be treated as foreign elements, and to ensure that immune reaction will be triggered after the particle injection.

Targeting. Particles must specifically recognise their target, particularly in the case of treatment against tumours. This specific target recognition is ensured by proper surface functionalisation.

Requirements may differ according to the use of AuNPs and will be satisfied by modifying the surface, form and size of the gold nanoparticles. At this stage, it is worth noting that AuNPs interact naturally with proteins when they are incubated in the presence of serum.[1] Although the formation of such a protein shell improves the stability of the colloidal gold suspension, it has also several drawbacks: proteins do not bind uniformly and hence some of them can interact with the AuNPs in a proportion not related to their amount in the serum; furthermore, the affinity of these proteins for the gold surface is weak and some exchange may occur when the AuNPs move in the body, meaning that the shell surrounding the AuNPs will not have a stable composition, which is unsuitable in the case of medical treatment. Moreover, whereas the affinity of free serum proteins for cell receptors is most often very weak, the concentration of a few identical proteins on the same particle can induce sufficient avidity to allow their interaction with the receptors present at the surface of many circulating cells, and such interaction may trigger adverse effects like cell activation or particle internalisation. The chemical composition, size, shape and surface characteristics of nanoparticles affect their binding to proteins, which may impact on the interaction of AuNPs with cells and tissues.[2] It has also been shown that such protein–particle interactions can induce protein unfolding, which may modify their capacity to bind with receptors and therefore their internalisation and elimination by circulating cells.[3] In summary, to prevent their aggregation and to allow specific binding to target cells or receptors, it is very important to control the capacity of the nanoparticles to interact with their environment and the best way to achieve this is to cover gold nanoparticles with protective molecules, which prevent further interaction with irrelevant proteins.

Surface modifications are usually realised in two steps: the first step is the binding of a molecule (the linker) followed by the binding of the molecule of interest (spacer, targeting, coating, etc.); these surface modifications are developed in Chapter 7.

16.1.3 *Gold Nanoparticles and Biocompatibility*

As explained above, the purpose of surface modification is to form a protective layer that must help the AuNP fulfill its mission and be stable throughout the AuNP activity.

It has been shown that thiol (–SH) radicals bind to gold nanoparticles.[4,5] The multiplication of anchors using high sulphur content molecules such as dihydrolipoic acid[6] increases the efficiency of grafting, and seriously limits the desorption of the coating. This kind of bond is widely used to attach a large variety of molecules (biological components or linkers) to the AuNP surface. For example, gadolinium chelates were fixed to AuNP via dithiolated derivatives of diethylenetriaminepenta acetic acid.[7] Indeed, most of the molecules used to create a protective monolayer around the AuNP are modified to bear a thiol moiety.

These modifications may be used to control particle reactivity and to induce either hydrophobicity or hydrophylicity to direct them to lipidic or aqueous areas, respectively. By choosing the surfactant molecules, it is possible to adjust the reactivity of the surface of the particles to their environment. Many protocols have been developed for application in aqueous medium and in these conditions, SH-bearing molecules also possess carboxylic groups that give stability to the particles. Such molecules are particularly interesting because they cover the particles with stabilising negative charges; furthermore, these groups can be used to attach other molecules of biological interest. Poly(ethylene glycol) (PEG) is also often used to coat particles as this compound reduces non-specific adsorption of proteins and provides greater AuNPs stability by preventing interactions between particles by steric hindrance.[8]

Nanoparticle biocompatibility must be considered separately from their toxicity, which is addressed in Chapter 17. When the particles are injected, it is important to avoid (1) their recognition by immune cells as a foreign element, (2) their interaction with serum proteins, which may lead to complement activation or blood coagulation and (3) their detection by any cellular receptors and especially by phagocyte receptors, which may induce their internalisation by circulating cells like activated macrophages.

As far as biocompatibility is concerned, particles must be considered as a whole (core and shell). Even if AuNPs are often considered as non-cytotoxic[9] they may induce modification of the cells' functional activities.[10,11] As the shell is in direct contact with serum proteins and cells, this aspect is certainly the decisive factor for biocompatibility. By itself, most of metal core may induce inflammation if there is a partial solubilisation and release of ions or if an oxydo-reduction process occurs, but this is not the case for gold. Gold toxicity of AuNPs could be revealed if the shell is unstable, after its destruction by enzymes following endosomal internalisation, for example, or in the case of accumulation in cells leading to steric hindrance.

A widely used strategy for AuNP functionalisation relies on strongly interacting molecules with thiol moiety, which may substitute or complement the shell of stabilising molecules. Peptides and oligonucleotides have been attached with such an approach. The functionalised nanoparticles can then be arranged according to the number of bound molecules with the help of techniques like metal ion affinity chromatography, which allows the selection of homo-functionalised nanoparticles.[12] When direct binding of the molecule of interest on gold is not possible, an alternative approach relies on linking this molecule to the shell formed by the stabilising surfactant. The most common protocol consists of the formation of a covalent link between an amino group present on the biological molecule and a carboxylic group present on the shell, using a chemical reagent such as N-ethyl-N'-(3-dimethylaminopropyl)- carbodiimide (EDC) as shown Figure 16.1.[13]

In these conditions, virtually all molecules can be attached to gold nanoparticles. Even if this chemical reaction is relatively well understood, problems remain in adjusting the optimal conditions of the reaction and precisely characterising the final product. In fact, two aspects have to be controlled when this technique is used: first, bridging between particles must be avoided because it would lead to the formation of aggregates; and second, the homogeneity of the binding, as it is important to produce substituted particles as homogenous as possible. However, it is sometimes crucial to be able to assess the relative amount of bound molecules on each gold particle,[14] though to date a method for controlling the number of molecules bound to one AuNP has yet to be published. Even measuring this relative surface coverage of proteins is not straightforward.

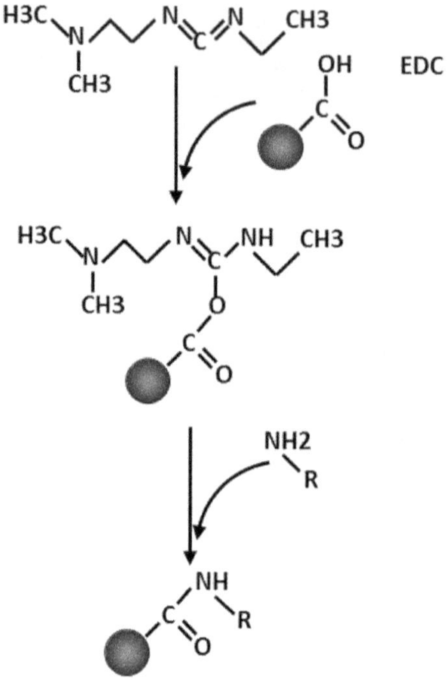

Figure 16.1   Schematic representation of EDC reaction with a carboxyl group on AuNPs. Proteins are attached to AuNPs taking advantage of the presence of a carboxylic moiety in the stabilising shell of the AuNP. N-ethyl-N'-(3-dimethylaminopropyl)-carbodiimide (EDC) reacts with the carboxyl group and forms an O-acylisourea intermediate. The intermediate reacts then with the amine on another molecule (R). The final reaction leads to the formation of an amine bond between the AuNPs and the molecule R.

16.2 Gold Nanoparticles for Diagnosis

In medicine, the diagnosis is the identification of disease by the analysis of symptoms or biological alterations evaluated by measurement of relevant elements: these elements can be biological molecules (proteins, hormones, mRNA), which are characterised or quantified with various techniques. Furthermore, intra-cellular or intra-corporal localisation of cells or of some of these molecules may be also of importance in diagnosis. The techniques implicated for these *in vivo* analysis must be non-destructive and non-toxic, which is possible thanks to the use of AuNPs.[15]

16.2.1 *Detection of Gold Nanoparticles Using Optical Techniques*

Metallic nanoparticles dispersed in aqueous or biological media exhibit important optical contrast, which makes them easily detectable. This effect is related to the localised surface plasmon resonance (LSPR) of gold nanoparticles described in Chapter 3. When applying the coating strategies described above for AuNPs, it is possible to target specific regions in the body and so localise a protein or a region of interest as accumulation of particles in this region induces a strong optical contrast. Observation using an optical microscope in specific modes such as phase contrast or differential interference contrast makes it possible to visualise AuNPs with diameters above 30 nm by direct observation. Furthermore, as the scattered wavelength (colour) and scattering cross-section (brilliance) varies with the shape and size of particles respectively, different kinds of AuNPs may be injected simultaneously as their binding may be differentiated by their colours or brilliances. According to the Mie theory, small particles (less than 20 nm), exhibit poor scattering intensity such that only their optical absorption is measured.[16]

When particles are in suspension, their mutual distances ($>1000\,$nm) make the dipolar interactions between them impossible, and their optical response is therefore identical to the one of isolated particles. But when this mutual distance becomes less than their diameter, the wavelength of maximum absorption is modified and the particle aggregation can be monitored. This effect was used for the first time by Leuvering[17] for the quantification of biomolecules. This technique named sol particle immunoassay (SPIA) is based on the modification of absorption wavelength due to particle agglutination: polyclonal antibodies are grafted onto 50 nm gold nanoparticles and the interaction of these antibodies with the different binding sites (epitopes) present on the corresponding antigen induces their aggregation and hence the bringing together of the particles. This method was applied to detect various molecules in urine or blood serum such as the selenoprotein[18] or cystatin[19] for example; further developments based on the sensitivity of the plasmon resonance to mutual coupling of AuNPs can be found in Chapter 3.

16.2.1.1 *SPR-based techniques*

Surface plasmon resonance (SPR) is observed when light excites electrons at the surface of metals, inducing a non-propagative evanescent wave. The first instruments using SPR for the detection of interactions between biological molecules were developed in the 1990s (this technique is presented in the Chapter 3). They generally use the surface of flat crystals covered with a thin layer of gold (50 nm) to detect biomolecular interactions, which perturb the evanescent wave and induce a change in the refractive index at the interfacial layer. However, as the field associated with the evanescent light decays exponentially with the distance normal to the surface, SPR can only detect biomolecular interactions occurring near the metal surface (i.e. closer than 100 nm). For a wide range of molecules, the modification of the refractive index is linearly correlated to their molecular weight, such that the SPR signal depends almost exclusively on the mass of the ligands bound to the gold surface. Nevertheless, as the detection limit was shown to be approximately 1 pg (1 picogram = 1×10^{-12} g) per square millimetre, it is extremely challenging to detect small molecules at low concentration: for example, detection of molecules of 1 kDa is very difficult at concentrations below 0.1–0.2 μM. The sensitivity of the detection can be improved by an artificial increase of the molecular weight of the molecule; this is achieved by adding a label that interacts with the molecule of interest already bound to the sensor and eventually damps the plasmon waves.[20] Therefore, the plasmon signal is switched off if the bio-recognition takes place (see Figure 16.2). He *et al.* have used 12 nm gold nanoparticles to improve the detection of oligonucleotides,[21] reaching a threshold of 10 pM for a 24-mer oligonucleotide (corresponding to a surface density less than 8×10^8 molecules per square centimetre), which represents a 1000-fold improvement of the sensitivity compared to standard detection of oligonucleotides. Wang *et al.*[22] have also used gold nanoparticles to increase the sensitivity of the detection by SPR: they use an indirect competition assay to measure very low nucleotide concentration (10 nM) with a complex anti-adenosine aptamers conjugated to gold nanoparticles. However, real-time measurement is the great advantage of SPR technology, which is lost with this approach since detection is realised only after the formation of multilayers made of the ligand fixed on the gold layer, the molecule of interest and the AuNPs. The same observation can be made in the case of the quartz-crystal balance[23] and micro-cantilevers,[24] for which the increased sensitivity obtained by the addition of markers bearing gold nanoparticles is detrimental to real-time measurement; nevertheless, this approach was used to detect heavy contamination through the presence of metallothionein for example.[25]

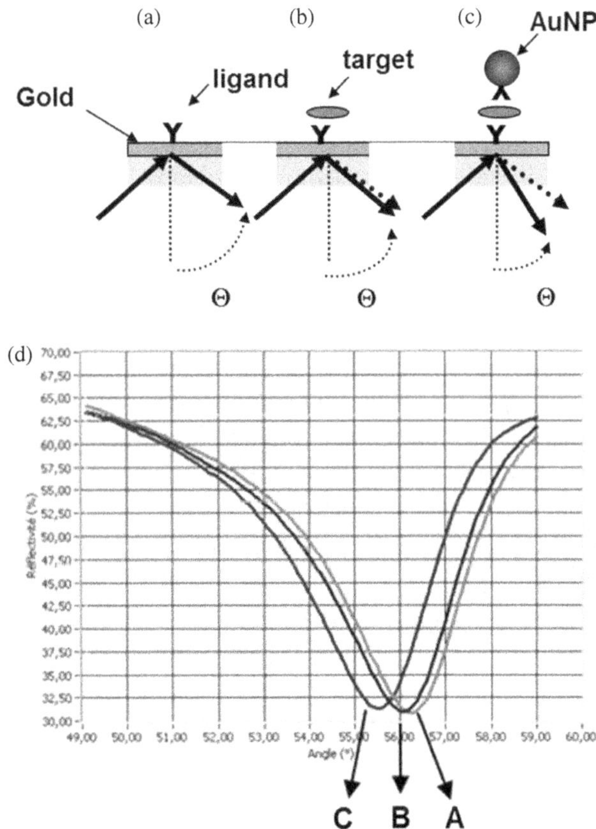

Figure 16.2   Conventional surface plasmon resonance using a gold film. (a) A thin gold film is located at the interface between a crystal and a ligand of the molecule of interest. (b) When a small molecule (target) binds to the ligand, the change in the deflection angle (θ) is low due to a too small modification of the refractive index. (c) Then, the binding of AuNPs to the target increases the deflection angle, which allows the detection of the molecule of interest. (d) Comparison of the deflection curves obtained in the three different conditions A, B and C.

Another approach is described by Matsui and colleagues[26] who used the molecular-imprinted polymers (MIP) technology. They succeeded in increasing the sensitivity of SPR with gold nanoparticles by neither losing the ability of real-time measurement nor needing any complex labelling. The molecular imprinted polymers are formed as follows: (1), the imprinted polymer is formed by polymerisation of an appropriate monomer in the presence of the target molecule, (2) the target molecule, which is embedded in the matrix, is then extracted, leaving a complementary cavity.

The imprinted polymer, which also contains AuNPs, is immobilised on a gold surface of a sensor chip and is ready to use.

Detection begins when the medium containing the target molecule to be quantified is added. The target molecule binds to the complementary cavity, and the local electromagnetic field between the nanoparticles and the gold film is expected to be enhanced by the binding of this molecule, making the sensor chip highly sensitive. This strategy was used to measure a very low concentration of a herbicide (atrazine, molecular weight: 215 Da), the authors being able to detect 5 pM of this very small molecule in acetonitrile.[26] More recently, using gold nanoparticles together with reduced graphene oxide, this technique was applied by Yao *et al.* for the detection of very low concentration of ractopamine (this molecule forbidden in Europe is used as additive for pig husbandry in America).[27]

The association of gold nanoparticles with SPR is doubly interesting: in addition to the damping of the evanescent wave, there may be coupling between the gold layer and the particles. These results prompted Englebienne to go further and replace the gold layer by gold nanoparticles: by functionalising them with antibodies, the binding of the corresponding antigen causes a change in the extinction spectrum, which can be monitored at 600 nm with a conventional spectrophotometer.[28,29] In this case, the main difficulty is the heterogeneity of the particle size, which reduces the sensitivity of the measurement. To circumvent this problem, gold nanorods were used as their shape minimises the impact of heterogeneity. Moreover, in addition to the conventional extinction peak at 530 nm, they absorb at greater wavelengths, which may be in the near infrared range provided the aspect ratio of the nanorod (ratio of length over diameter) is significant. The advantage of measurement in the near infrared is that the absorption due to serum proteins or blood components is highly reduced and the sensitivity of the particles to the local refractive index changes is greater.[30]

16.2.1.2 *Fluorescence*

Fluorescence is commonly used in biology for protein detection, cell or tissue characterisation, etc. As shown for cyanine dyes[31] or quantum dots,[32] most fluorescent compounds are quenched in the vicinity of gold nanoparticles, and this effect can be used for sensing strategies. Thus, detection and quantification of molecules can be realised using competition assays based on quenching suppression: the molecule to be detected is linked to a fluorescent probe. If it is attached to the AuNP, the distance between the nanoparticle and the fluorophore is reduced and fluorescence is quenched. Before the detection steps, AuNPs are first functionalised with a ligand or an antibody, which possesses a good affinity for the molecule. A molecule analogue to the target molecule and equipped with a different fluorophore is then used to specifically bind to the functionalised AuNP. Detection can now take place. As long as this analogue remains fixed near the NP, its associated fluorescence is quenched. When the target molecule is added to the solution, it competes with the analogue for the binding site, inducing the analogous release and consequently the suppression of the fluorescence quenching. The higher the target molecule concentration, the more important the competition, and consequently, the stronger the fluorescent signal.[33]

Fluorescence quenching can also be used for protease activity assay: Ray and colleagues[34] used a fluorescent polynucleotide fragment (Cy3-labelled nucleic acid) covalently linked to gold nanoparticles; initially, the fluorescence is quenched, but, after addition of both a complementary DNA fragment and a nuclease, the double-stranded DNA is cleaved, which releases the dye and suppresses the quenching. Nuclease activity can be monitored in real time through the augmentation of fluorescence; indeed, the rate of fluorescence increase is directly linked to the enzyme activity.

As the quenching of fluorescence by gold nanoparticles is strongly related to the distance between the probe and the particles,[34] this feature has been used to detect molecular interactions: if the binding of one molecule to another leads to a modification of the distance between a fluorescent dye and a gold particle, the interaction can be followed in real time by fluorescence recording. Dubertret and colleagues[35] described a hybrid material composed of (1) a single-stranded DNA molecule, (2) a 1.4 nm gold nanoparticle and (3) a fluorescent dye that is highly quenched by gold: the DNA molecule forms a hairpin, which brings together the nanoparticle and the dye inducing the quenching of the fluorescence. The addition of complementary DNA impairs hairpin formation, the consequence of which is an increase of the fluorescence by a factor of several thousands. The amount of complementary DNA is directly related to the quantity of measured fluorescence; the same technic was proposed for the detection of blood-circulating mRNA for the detection of specific tumour markers.[36]

In addition to the quenching of fluorescence by metal nanoparticles, there are also findings on metal-enhanced fluorescence[37] (see Chapter 14).

Martini *et al.*[38] have shown that AuNPs can be used to enhance fluorescence yield of nanoparticles: indeed, fluorescent probes embedded in silica particles are often used to study intracellular or intra-corporeal circulation. In order to increase the sensitivity, the amount of dye incorporated within one silica particle should be high. However, this method is limited because the increase of the concentration of fluorescent materials in the particles leads to self-quenching of the luminescence. Nevertheless, incorporation of AuNPs as a central core in the silica shell, together with the fluorescent dyes, suppresses almost entirely the phenomenon of fluorescence quenching: such an architecture exhibits a quantum yield as high as 80% compared to isolated fluorescein for an inter-dye distance of 3 nm in such particles whereas it is less than 15% in the absence of gold.

16.2.1.3 *Modification of absorbance*

Detection and quantification of proteins, hormones, drugs, and pesticides. Many devices using gold nanoparticles have been developed for the detection of proteins, hormones and pesticides. The most popular system is based on a membrane sheet (made of nitrocellulose or polyvinylidene difluoride (PVDF)) where the solution to be analysed undergoes lateral diffusion. The detection is associated with immune reaction with specific antibodies linked to nanoparticles (Figure 16.3). When the solution containing target molecules is deposited on the membrane sheet, antibodies migrate with the mobile phase and bind their antigen in solution, which prevents them from binding to a test line where the same antigen is covalently fixed to the membrane. If there is no antigen in the solution, the binding sites of the antibody remain free to interact with the test line, and the resulting accumulation of nanoparticles induces the development

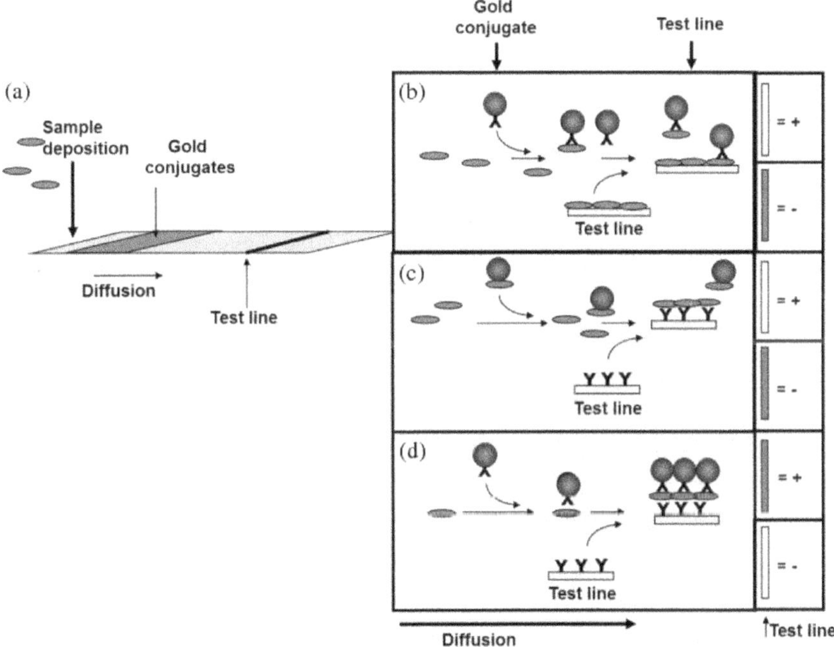

Figure 16.3   Detection of biological molecules using lateral diffusion of AuNPs. (a) After deposition of the sample, the lateral diffusion has two effects: first, the molecules of interest bind to the gold conjugates; then when reaching the test line, they bind to the ligand, which was deposited on this line. The diffusion is performed on a nitrocellulose membrane. (b) Example of antibodies bound to AuNPs: in this case, there is a competition between the molecules in the sample and the same molecule on the test line for the binding of the antibodies. (c) Example of molecules of interest bound to AuNPs: there is a competition between the molecules in the sample and the same molecules bearing AuNPs for the binding to antibodies fixed on the test line. (d) Example of antibodies bound to both the AuNPs and the test line: the molecules of interest in the sample react first with the antibodies bound to AuNPs and then with a second antibody fixed on the test line. These two antibodies bound to the molecules of interest on two different and non-competitive binding sites. As indicated in the right part of the drawing, the presence of the molecule to be analysed in the sample leads to the staining (c) or not (a,b) of the test line by gold nanoparticles.

of a coloured line: the increase of the amount of antigen in the medium induces a decrease in the amount of antibodies bonded to the test line and consequently a diminution of the coloration. These tests were used to characterise small molecules like hormones, pesticides or drugs.[39,40] In order to improve the sensitivity of the test, particles were directly linked to antigens instead of antibodies; in this case, there is a competition between the antigens fixed to the particles and the same molecules free in the medium (the one to be quantified), for the binding to the antibodies on the test line. By adjusting the number of molecules fixed to the nanoparticles, the authors were able to improve the sensitivity compared to the traditional device.[41]

For the detection of large proteins like prostate antigen against which at least two different monoclonal antibodies are commercially available, a direct assay is used: one antibody is linked to the nanoparticles and the other one is fixed on the test line: in this case, the target molecules bind to both antibodies bearing nanoparticles and those fixed to the membrane leading to the formation of a coloured line, whereas, in the absence of antigen, there is no binding and no visible line.[42]

The development of molecular biology prompted several laboratories to build devices based on the same principle as those used for protein quantification but suitable for the rapid detection of polynucleotides. It is possible to attach a small oligonucleotide (A) to the membrane (test line) and another one (B) to gold nanoparticles. In theory, a third oligonucleotide containing complementary sequences for both oligonucleotides (A and B) should lead to the accumulation of gold on the test line[43]; unfortunately, the technique used to make copies of the nucleotidic sequence to be detected is based on polymerase chain reaction (PCR); this method used to rapidly produce many copies of a fragment of DNA for diagnostic or research purposes, generates a double-stranded polynucleotide, which blocks the reaction, making it necessary to perform an asymmetric PCR much more difficult to optimise as shown in Figure 16.4.

For these reasons, lateral flow devices are not used for the measurement of oligonucleotides, all technical problems being not yet solved. Another possibility is to incorporate a flag in the amplified oligonucleotide, and to use anti-flag antibodies fixed to both the test line and the nanoparticles.[44] The interest of such a method is limited due to the small number of flags

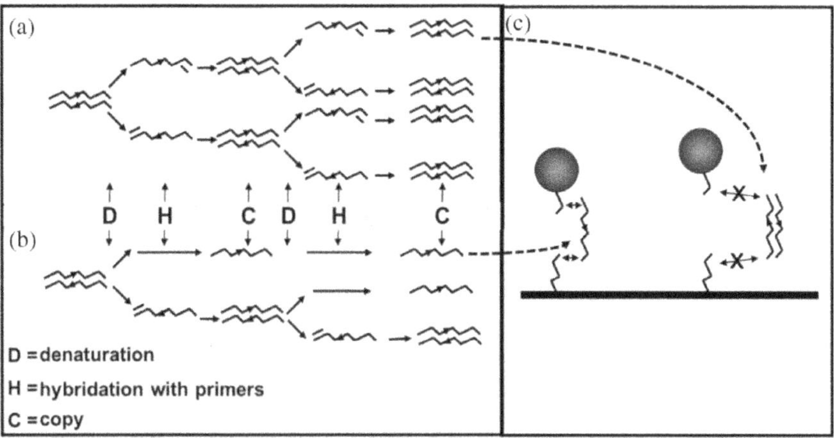

Figure 16.4   Use of AuNPs for the dosage of oligonucleotides. The amount of oligonucleotides is usually increased by polymerase chain reaction (PCR). Normal PCR leads to the formation of double-stranded polynucleotides (a) whereas asymmetric PCR leads to single-stranded polynucleotides (b). Detection of polynucleotides by two complementary sequences is possible only with single-stranded and not with double-stranded polynucleotides (c).

available, allowing only a few oligonucleotides to be detected in the same device.

The sensitivity of these methods may be increased by binding an enzyme to the gold nanoparticles: using HRP, He *et al.*[45] were able to lower the detection limit of nucleotides to a concentration of 0.01 pM, which is 1000 times lower than in prior works, without modification of the instrumentation for reading.

These devices have many applications: they are compact, portable, easy to use by non-specialists without special material and can be kept at room temperature as they are freeze-dried; furthermore, they are not expensive.

Gold nanoparticles and bio-barcodes. The goal of bio-barcodes is to identify, in one experiment, very low concentrations of different serum or cellular soluble proteins

The bio-barcodes use a cascade of reactions for (1) specific detection, (2) transcription and (3) amplification of the signal. The first step is the recognition of the target protein: this is realised by specific antibodies bound to magnetic beads, which are retained by a permanent magnet, while the unbound proteins are washed away. For the second step, gold nanoparticles bearing both specific antibodies and oligonucleotide sequences were added; the antibodies allow the specific binding of the gold nanoparticles to the proteins retained by the magnetic particles. After removal by washing of unbound material, bound gold nanoparticles are eluted and then further retained by an oligonucleotide fixed to the chip and which is complementary to the sequence fixed on the gold nanoparticles. This corresponds to the transcription stage of the detection: the binding, which is at the beginning specific for the protein, is transformed into a binding specific for an oligonucleotide. The last step is a silver amplification of the signal associated with the particles (Figure 16.5): Ag(I) reacts with the gold nanoparticle surface in the presence of reducing agents such as hydroquinone, resulting in nanoparticle-promoted deposition of silver on the particles.[46] This method increases gold nanoparticle size and induces the augmentation of

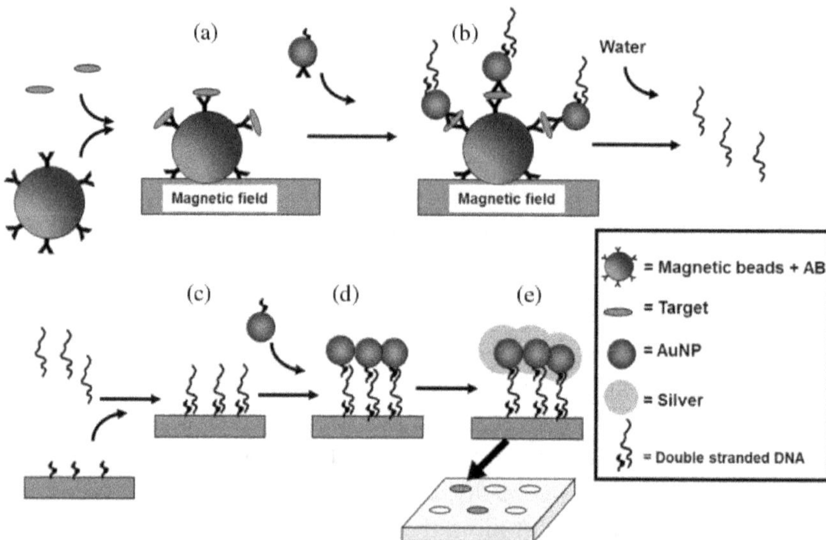

Figure 16.5   Use of AuNPs for a bio-barcode. (a) The sample containing the molecule of interest is mixed with magnetic beads bearing antibodies specific for this molecule, and the formed complex is retained on a magnet. (b) AuNPs are added to the mixture; they are covered by two types of molecules: a double-stranded polynucleotide and an antibody able to fix the molecule of interest but using a binding site different and non-competitive with the one used by the first antibody. (c) Then, the polynucleotide, which is released by addition of water, migrates and may interact with another polynucleotide with a complementary sequence and fixed to a chip. (d) Gold nanoparticles are added; they are covered with a small polynucleotide also complementary to the first one. (e) The detection is then amplified by silver deposition. Multi assays are possible using an array where various polynucleotides (one per molecule to be analysed) are spotted on a chip.

the signal by over five orders of magnitude. Simultaneous detection of different proteins is achieved thanks to a distribution on a two-dimensional array of oligonucleotides, which are complementary to those attached to the gold nanoparticles, with each oligonucleotide sequence corresponding to one specific antibody. Since the early approach in 2003,[47] different modifications of the protocol allow the proposal of disposable chips with a colorimetric reading[48] or in association to an evanescent wave fluorescent biosensor.[49] Nair *et al.*[50] have shown that in theory, the limit of such detection is at the sub-atto-molar level. Similar approach with gel assay was used for the detection of micro RNA.[51]

Gold nanoparticles and amperometric detection. Quantitative analysis of analytes by electric current is a very sensitive method. The detection is based on the modification of conductivity between two electrodes after the molecules of interest are bound. The use of gold nanoparticles as a conductance amplifier for the quantification of proteins was first published by Velev *et al.*[52] A full automatic platform was developed by Molecular Circuitry and the method was applied to detect nucleic acids.[53] The major problem with this technology is that the presence of nanoparticles between the electrodes (due to their binding to the target molecule) is not sufficient to amplify the electrical current; it requires several cycles of silver deposition for sufficient signal amplification.

More recently, Diessel[54] has shown that it is possible to reduce this coating by performing a real-time monitoring of the resistance; however, the gold nanoparticles were too far from each other on the chip and were surrounded by insulating biological molecules such that direct electrical measurement was not possible without silver amplification. Kim and his colleagues have introduced a conducting polymer (polyaniline) into the medium after immobilisation of the gold nanoparticle-conjugated antibodies on the target protein; in this case, the signal modification is 4.7 times faster when compared with plain gold and the maximum was 2.3-fold higher than that obtained using a photometric system under the same analytical conditions (Figure 16.6).[55,56]

Gold can also be used as an electron carrier, the idea being to transport the electrons generated from redox enzymatic reactions to electrodes via gold nanoparticles; measurement of the resulting current gives information about the presence and amount of the enzyme substrate. In these devices, the

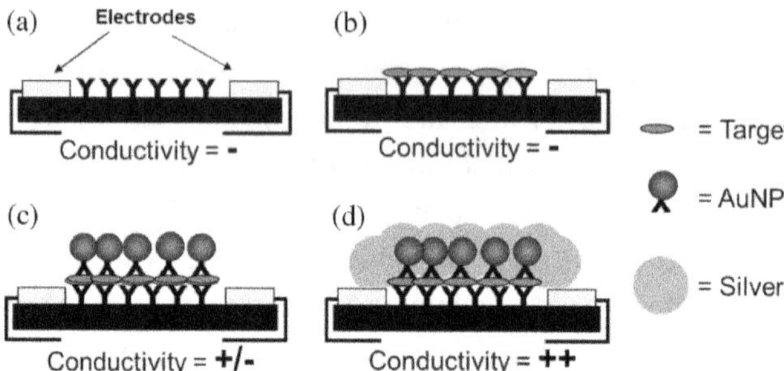

Figure 16.6   Amperometric biosensors based on the use of AuNPs. (a) Antibodies specific to the molecule of interest are deposited between gold electrodes. In these conditions, the conductivity is very low. (b) The target molecule contained in the sample medium binds to the antibodies without modification of the conductivity. (c) AuNPs bearing a second anti-target antibody are then added, and bound to the chip: the presence of AuNPs between the electrodes is generally not sufficient to increase the conductivity. (d) Deposition of silver on the gold nanoparticles increases the conductivity, leading to a signal, which is related to the presence of the target molecule.

enzyme is directly conjugated to the surface of the particles, which are fixed to the electrode of the chip. The enzyme binds preferentially to the particle and not to the electrode surface because the available surface is larger so that a greater amount of enzyme can be accommodated; furthermore, it seems that the shape of the small particles facilitates a close contact between the enzyme and the conductive surface and thus the electronic exchanges.

16.2.2 *Tomography and Gold Nanoparticles*

Visualisation of organs, cells or tumours is of paramount necessity for analysis and then treatment of many diseases. Among the many contrast agents that can be used, gold nanoparticles may have a particularly important role due to the variety of usable items in terms of shape and size: spherical nanoparticles, nanostars, -rods, -shells, -cage and -cubes as reviewed by Dreaden *et al.*[57] but optical imaging of gold nanoparticles is limited because of the low penetration of light into tissues (usually few millimetres). The penetration capacity of biological tissue by opto-acoustic-based imaging is of few centimetres (maximum 5 cm) due to the use of near infrared source (800–1200 nm) for monolayer-protected cluster (MPC) particles (2 nm),[58] but even in these conditions, this technique is too limited to be used in human applications.

Another consequence of light absorption by gold nanoparticles is the induction of fluid heating in their immediate environment, which can be measured by two methods: photothermal[59] and photoacoustic[60] imaging: in the first case, the conversion of the light into heat induces a punctual increase of the local temperature. The resulting pressure wave propagating into the surrounding tissue improved the spatial resolution of the tomographic images.[61] In the second case, the imaging results from the expansion of the liquid due to local heating around the AuNPs, creating a microphone-detectable wave.

Gold nanoparticles also interact with electrons, because of the high molecular weight of Au; in consequence, these particles have a very important contrasting effect for electron-based technologies and are thus widely used for transmission electron microscopy (TEM).[62]

Furthermore, the ability of AuNPs to induce X-ray scattering is also sufficient to provide significant X-ray imaging.[63]

Using positron emission tomography (PET) camera to monitor small tumours is particularly favourable because of the sensitivity of nuclear imaging; however, the radioactive molecules used for these processes have a very short biological half-life, which limits their use, while the nanoparticles remain much longer and enable tumour imaging. Indeed, radioactive gold (^{198}Au) may be used as a beta-emitter (99%, 0.96 MeV), but the radioactive element half-life of this compound is relatively short ($t_{1/2} = 2.69$ days), which limit its use.[64–66]

New development of nanoparticles is based on the association of two or three materials with interesting characteristics: as indicated previously, gold is a very good contrast agent for X-ray-based computed tomography whereas gadolinium, which has seven unpaired electrons and a large magnetic moment, influence drastically water proton relaxation and exhibits an excellent magnetic resonance imaging (MRI) contrast. Tian *et al.*[67] have associated Gd nanoparticles with gold nanoparticles for the preparation of a MRI/CT bimodal imaging agent. The results revealed high longitudinal relaxivity in MRI and excellent computerised tomography (CT) imaging performance. Similar association was used for following cells after transplantation of human neural stem cells.[68] Interestingly, very low false-positive cells were observed (1%) but 30% of false negative cells were recorded, indicating that the determination of the number of transplanted cells is strongly underestimated. The metal combination can also be used to simultaneously perform imaging and treatment of tumours as shown by Miladi *et al.*[69]: in this case, the presence of gold nanoparticles in the tumour reduces the radiation dose by a factor of two while keeping the same efficiency.

16.3 Gold Nanoparticles for Medical Treatment

16.3.1 *Gold Nanoparticles as Delivery Vehicles*

As indicated in the first part of this chapter, many molecules can be bound to AuNPs, and this feature is widely exploited for medical treatment. In this case, AuNPs can be considered as a vector for a precise delivery of molecules at the required place. If immunisation is the goal, antigens must be targeted to areas of high immunological activity; if tumour treatment is sought, active molecules must exclusively concentrate at the level of malignant cells in order first to facilitate their eradication and second to reduce adverse effects, which may result from the presence of these very harmful molecules near non-malignant cells. The delivery of such molecules can be realised using gold nanoparticles as carriers. For this purpose, they are fixed with a stable link to the particles that drive them into the area of interest, and at this stage, the molecule can be released by breaking the bond that links them to the particle either by proteolysis or by hydrolysis. These different possibilities are discussed in the following paragraphs.

16.3.1.1 *Problem for specific delivery*

The purpose of cell targeting is to precisely deliver a molecule to an organ or a group of cells and simultaneously avoid all aspecific interactions with non-targeted cells or tissues. Indeed, all cells can ingest nanoparticles but the amount of material found inside the cells may vary dramatically according to the cell type and the internalisation pathway involved: pinocytosis, receptor-mediated endocytosis, phagocytosis etc. In the case of particle targeting, the interaction with the cells must be as specific as possible. The amount of material internalised by pinocytosis and phagocytosis is directly related to two factors: the concentration of the particles and their stealth, i.e. their capacity to be invisible, in particular for immune cells. The biocompatibility of the particles has been previously documented in this chapter: when AuNPs are invisible to the immune system, phagocytosis by macrophages or dendritic cells is reduced, and particles remain in the circulating fluids, allowing their routing to and interaction with the targeted cells. In these conditions, the amount of specific interactions with targeted cells is much more important than aspecific ones with the other cells. Specific binding of AuNPs to membrane proteins expressed by the targeted cells ensures a great specificity but such interaction does not systematically lead to the internalisation of the particles. For example, some membrane proteins induce an intracellular signalisation after binding of their ligand but no internalisation as it is the case for toll-like receptors (the proteins in charge of the detection of foreign materials; viral ARN non-human glycol-proteins, etc.). Whereas other proteins are internalised upon binding of their ligand, as it is the case for transferrin receptors for example. There is no possibility of inducing the internalisation of a protein that normally does not have this capacity.

Concerning tumour treatment, internalisation of particles is not always required: whereas drug delivery may require internalisation, hyperthermia is effective even without internalisation. The challenge is to target AuNPs to receptors present only on tumour cells, but the main problem is that only a few proteins are expressed exclusively by these cells and most of these are not accessible from the outside of the cells without permeabilisation. The use of peptide for cancer therapy is an important field, which is difficult to summarise;[70] the success of these therapies is variable and require further studies. There is also the possibility to target proteins present on the endothelial cells forming the neovascularisation of the tumour as many of them have already been characterised: CD44,[71] VEGF,[72] $\alpha v \beta 3$ and $\alpha V \beta 5$[73,74] or VCAM[75] for example.

16.3.1.2 *Gold nanoparticles and drug transport*

The penetration of particles into a tumour is often facilitated by the so-called enhanced permeability and retention (EPR) effect observed in the vessels around tumours, but this permeability varies from tumour to tumour;[76] furthermore, binding particles to membrane proteins at the surface of tumour cells does not systematically induce their internalisation, but even in the case of endocytosis, the presence of particles in endosomes or lysosomes is not sufficient to induce cell death. Indeed, as documented previously, toxicity of gold nanoparticles is too low to induce cell death *per se.* In order to kill the tumour, particles have to be linked to active molecules. Two strategies are possible: either these molecules are able to correct the abnormality of the tumour by gene therapy, or they are cytotoxic and specifically eliminate tumour cells. In order to exercise their activity, particles must exit from the intracellular compartment to the cytosol or directly cross the extracellular membrane; for this, they can be coated by peptides like trans-activating transcriptional activator (TAT) that disrupt the membrane and allow their passage into the cytosol.[77]

For gene therapy, it has been shown that small-interfering RNA (siRNA) could be bound to particles. Shim *et al.*[78] first attached siRNA to a polymer (polyethyleimine) via a pH-sensitive link,[79] which is itself fixed to the gold nanoparticles. Once in the intracellular compartments, the link between siRNA and the polymer breaks, due to the environmental pH, which is slightly acid. Then, the released siRNA can act and block the expression of the corresponding protein. As pH is slightly acid inside the tumour, RNA is also released in their vicinity, and then siRNA enters the cells and plays its role.

Particles can also be used for carrying anti-tumour drugs. These are attached either at the AuNP surface or to structures connected to the particles. Molecules are then transported into or near the cell of interest together with the particles. Under these conditions, the accumulation is much greater and much faster than with molecules injected alone without the targeting element. Curnis *et al.*[80] have attached the tumour necrosis-factor (TNF) to the targeting peptide RGD on the AuNPs and injected the complex into animals, in combination with a chemotherapeutic treatment: this TNF strategy yields a very good anti-tumour effect. This cytokine is very toxic and usually difficult to use in systemic injection, but in this case, the targeting of the molecule with nanoparticles allows the use of very low concentration (sub-nanograms), avoiding adverse effects after injection. Such composition was tested for the treatment of human cancer.[81] More recently, Amreddy *et al.* described a more complicated construction made of gold nanorods targeted to transferrin receptor for the delivery of Doxorubicin.[82]

Gold nanoparticles were also used for the transport of platinum-based drugs that constitute the most widely used anti-cancer molecule.[83] The efficacy of these molecules is improved by their binding to carriers such as gold nanoparticles, which can transport large numbers of Pt complexes. The specificity of Pt delivery to cancer cells is obtained by vectors, which recognise specific targets.

In addition to the binding of drugs at their surface, the use of gold nanoparticles is interesting due to their optical properties that makes possible the simultaneous detection and localisation of AuNPs.[63] Patra *et al.*[84] indicate that gold nanoparticles are particularly suitable for pancreatic cancer because of their unique physico-chemical properties, such as ultra-small size, large surface area to mass ratio, high surface reactivity, presence of surface plasmon resonance (SPR) bands, biocompatibility and ease of surface functionalisation. The use of AuNPs targeted to the epidermal growth factor receptor (EGFR), which is over-expressed on these tumours, leads to increased efficacy of traditional chemotherapeutics.

Kang *et al.*[85] used gold nanoparticles to induce DNA damage in tumour cells: two peptides are fixed on the particles, one containing the RGD sequence for tumour targeting, the second corresponding to a nuclear localisation signal, which induces the traffic of the particle to the nucleus: this particle localisation leads first to the blockage of cell division and second to cell apoptosis. This work shows that nanoparticles localised in the cell nucleus can specifically affect cellular functions as reviewed by Dreaden *et al.*[86]

16.3.2 *Heat Reaction*

As indicated previously in this chapter, light at the wavelength corresponding to the plasmon resonance is absorbed by the AuNPs, leading to the excitation of free electrons in the gold structure; the damping of these electron oscillations within the nanoparticles is accompanied by a thermal dissipation of energy and leads to a temperature increase within the nanoparticles and their environment (see Chapter 4). This phenomenon can be used for photo-thermal imaging as well as tumour treatment because of the impact of small temperature changes on cell viability.

Cells are very sensitive to temperature variation and die above 42°C, the normal value for human cells being 37°C. This fact can be exploited to use localised hyperthermia to specifically kill tumour cells. To achieve this objective, there are two prerequisites: particle accumulation in the tumour cells and illumination of the nanoparticles with sufficient energy to induce heating (Figure 16.7). As discussed previously, in theory, AuNPs can target cancer cells by encapsulating the particles with molecules, which have a high affinity for the proteins expressed only by the tumour. This leads to particle accumulation in the tumour area due to their binding to the membrane with or without the ensuing internalisation; in both cases, the consequence of the light absorption will be the same. As indicated previously, the absorption resulting from plasmon resonance occurs at a wavelength that varies according to the size,[87] shape[88] and structure[89] of the particles: the optimal parameters were found to be around 40 nm diameter for spherical gold nanoparticles, between 20 and 70 nm length for nanorods and 50–100 nm total diameter for core–shell structures where the gold outer shell has a thickness of 7–10 nm. In these cases, the maximum absorption is measured at 800 nm. In other words, by adjusting the nanoparticles physical

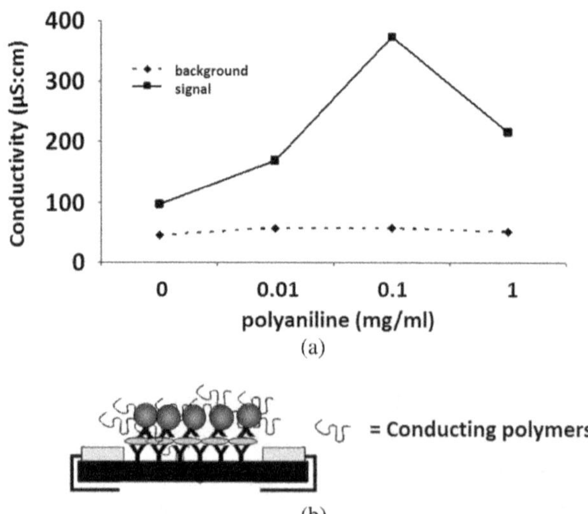

Figure 16.7   Enhanced electron transfer in the presence of conducting polymers. The polymers enhance the conduction by forming a bridge between adjacent AuNPs leading to increased signal. Adapted from Ref. 55.

characteristics, it is possible to modify the wavelength corresponding to the maximum light energy absorption (see Chapter 4). This property is very important because the main limitation of photo-thermal effects is the light absorption by animal tissues: a very small fraction of the light energy reaches the particles depending on the depth of the tumour. Tissue absorption varies according to the wavelength and is minimal in the infrared spectrum though, but in these conditions, the absorption remains fairly high. This means that such treatment can only be achieved for tumours close to the skin in the case of external illumination. For deep tumours, this technology is difficult to apply even using IR wavelengths; however, we may imagine that, in this case, it would be possible to use intra-body laser illumination.

The temperature increase induced by illumination of AuNPs can also be used to release drugs, proteins or polynucleotides transported to the tumour area (Figure 16.8). To date, this approach has not yet been used for tumour treatment, but it has been shown, for example, that small hairpin RNA (shRNA) delivered near a tumour may control its proliferation.[90] Stehr and colleagues have shown that AuNPs can be used to rapidly increase the temperature and liberate a polynucleotide,[91] aggregated proteins at the particle surface may be also solubilised by raising the temperature.[92] The photo induced heating can be used to liberate molecules from containers: the principle is to confine the molecule of interest in a container, which is itself fixed to the AuNPs; the hyperthermia induces the rupture of the wall of the container, and the release of the molecules.[93] It can be performed outside or inside the cell after internalisation:[94] a low number of laser pulses allows local heating of the particles and rupture of the container without any toxic effect on the cell, and cell apoptosis can be avoided: the thermal energy generated in these conditions is too low to have an impact on cell viability. The consequences of such a release range from specific inhibition of molecules or cell functions to cell labelling.[95] Recently, Kim *et al.*[96] have used a combination of nanoparticles made of methotrexate-loaded PLGA with gold/iron half-shell conjugated with RGD peptide for rheumatoid arthritis treatment. Upon near infrared irradiation, local heat induces the drug release: the combination of magnetic (iron) and heat (gold) provides an enhanced therapeutic effect using only 0.05% dose compared to free methotrexate therapy generally used for such pathology.

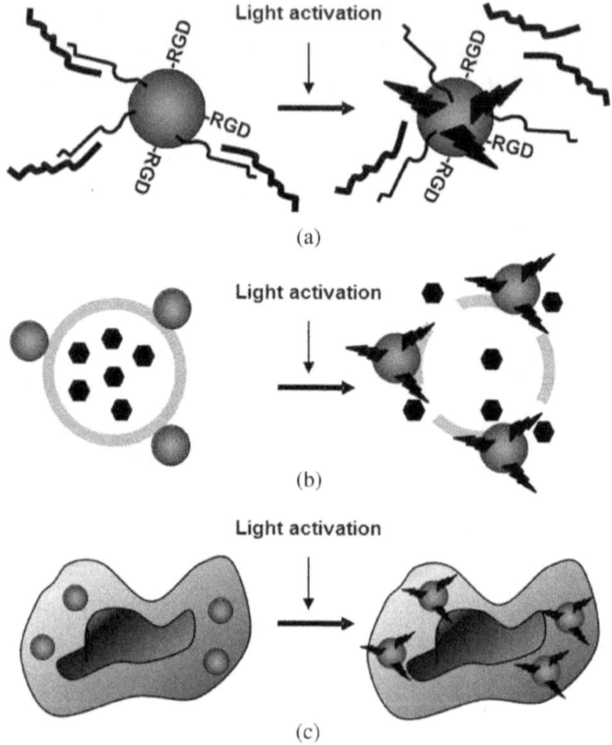

Figure 16.8   Clinical use of hyperthermia induced by AuNPs. (a) The polynucleotides are delivered at a specific place: the AuNPs bearing both a targeting motif (RGD sequence for example) and a double-stranded polynucleotide are injected. The particles, which accumulate in the region of interest, are illuminated by a laser. The hyperthermia in this case is sufficient to liberate the polynucleotide but too low to kill the cells. (b) Drug delivery: drugs are located in containers bound to AuNPs bearing targeting motifs at their surface. The illumination of AuNPs induces hyperthermia, which leads to the rupture of the container and the liberation of the drug. (c) Cell killing: the AuNP-bearing targeting motifs are injected. Their accumulation around the cells induced or not their endocytosis. In both cases, light illumination induces the production of hyperthermia; the number of laser pulses is determined to increase the temperature to a lethal level.

16.4   Other Biological Applications of Gold Nanoparticles

16.4.1   *Localisation of Proteins in Tissues*

16.4.1.1   *Electronic microscopy*

Identification of specific cells in a complex mixture or localisation of intracellular compartments such as endosomes, lysosomes or mitochondria in

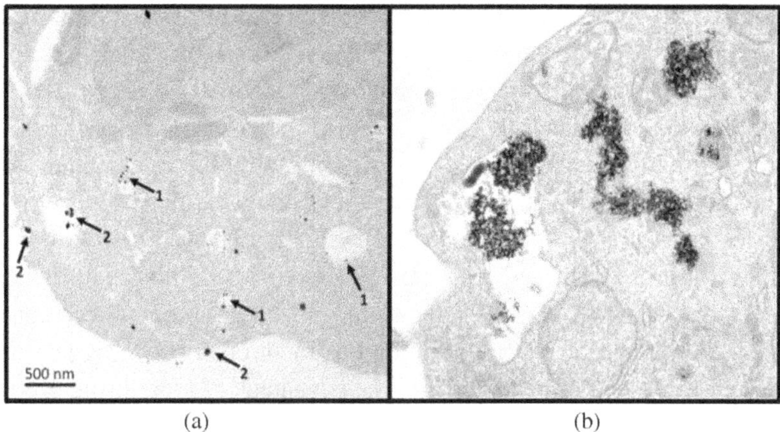

(a)   (b)

Figure 16.9   Cellular localisation of gold nanoparticles using electron microscopy. Cells were cut in very thin slices and analysed with an electron microscope. (a) Cellular proteins were labelled with antibodies bearing 5 nm (1) or 20 nm (2) gold nanoparticles, specific for lysosome-associated membrane protein (LAMP) and transferrin receptor respectively. (b) Cells were first incubated in the presence of gold nanoparticles (50 nm) before electron microscopy analysis; the nanoparticles are clearly visible in the peripheral cellular compartments (endosomes).

cells is a permanent challenge in many biological fields. As shown in Figure 16.9, the use of electron microscopy is probably one of the best known illustrations of intracellular protein localisation based on AuNPs. AuNPs are providing a powerful contrast enhancement for transmission electron microscopy (TEM). Antibodies are covalently attached to AuNPs so that they can be easily traced. The interaction of these antibodies with cell proteins is investigated: the antibody can target a protein localised in a cellular or intracellular structure and due to the presence of the AuNP, these events can be monitored by TEM. In the absence of such NP labelling, it is very difficult to characterise cellular compartments such as endosomes, lysosomes, Golgi apparatus, etc. by electron microscopy due to either a very low contrast or the absence of specific morphological characteristics. Cell or tissue slices are fixed on a support (cover slide or culture dish) and then permeabilised, allowing particles to pass through the cellular membrane in order to interact specifically with the antigen against which the antibody is directed.

Thus, gold nanoparticles provide an excellent contrast for observation using a transmission electron microscope, whereas it is much more difficult to observe the same nanoparticles with an optical microscopy due to the resolution limit which is theoretically a few hundreds of nanometres.

Klein *et al.* showed that detection of particles of 60 nm is reliable, succeeding in counting such nanoparticles with a confocal microscope. Furthermore, AuNPs observed in back-scattering mode induce a phase shift in the light signal, due to the cellular environment, which thereby facilitates their observation.[97] Of course, cell labelling is also possible without permeabilisation; in these conditions, nanoparticles do not enter the cells and can only interact with components localised on the outer face of the cellular membrane. The optical limitations mentioned above remain after permeabilisation but this method has the advantage of monitoring proteins or structures in real-time since the cellular integrity is not altered, allowing dynamic analysis. For example, if one AuNP is attached to an antibody that targets a membrane protein, it will be fixed with high affinity to this membrane and will follow its movements in the cell. Therefore, if the nanoparticle is found inside the cell after incubation, it means that the protein has been internalised by the cell and, in these conditions, a precise localisation may be performed using TEM as demonstrated by Cabezon *et al.*[98]: they have coupled AuNP to monoclonal antibodies directed against the transferrin receptors and demonstrate that the receptors may follow two intracellular routes: a classical one where the receptors enter intracellular processes of vesicular fusion accumulating in late endosomes, multivesicular bodies or lysosomes. A second traffic is possible for a small percentage of the vesicles which fuse with the abluminal membrane and open to the basal membrane. Such use of AuNP may be one of the most promising strategies to explain the transfer through the blood-brain barrier and to overcome it for drug delivery.

Dynamic analysis of the intracellular movements of the protein can be simply carried out by modifying the incubation time. In theory, these techniques, combined with the small size of AuNPs, make it possible to monitor all the endocytic processes: phagocytosis, pinocytosis, receptor-mediated endocytosis, etc. But attention must be paid to the necessary absence of impact of this tracker on endocytosis: nanoparticles should neither perturb the process nor bind to proteins, which could influence their internalisation. Indeed, it is worth noting that despite their small size, nanoparticles could interfere with the internalisation of receptors.[99] AuNPs may be a good alternative to fluorochromes for cellular imaging by confocal microscopy: fluorescent probes are often used because of their ease of use, but are very unstable due to photobleaching, whereas AuNPs are very stable.

Nanoparticle visualisation is also possible by photo-acoustic imaging: when the distance between particles decreases, the wavelength of the localised plasmon resonance shifts to a higher value. When the beads are equally distributed at the cell surface, the distance between the particles is too great to induce an interaction between the surface plasmons: in these conditions, when they are illuminated at a wavelength above the plasmon resonance frequency, no signal is recorded. Whereas, when antibody binding occurs on proteins that are aggregated at the cell surface, the gold nanoparticles are close to each other and their resulting interaction leads to light absorption and a photo-acoustic signal can be measured. In consequence, the gathering of receptors (capping) at the cell surface, which may be the result of the binding of their ligands, can be monitored by the variation of photo-acoustic signals.[100]

The development of multi-functional nanoplatform for both diagnostic and therapy is a great challenge. Li *et al.* reported an interesting construction of hyaluronic acid modified Fe3O4@Au core/shell nanostars for multi-detection (Figure 16.10): magnetic resonance, computed tomography and thermal imaging allowing photothermal therapy of tumours.[71] The authors show that laser treatment of the tumour after nanostars injection induces a strong reduction of the tumour volume together with a 100% survival of the animals after 60 days, which is significantly higher than without treatment.

16.4.1.2 *Reflection/fluorescence*

Fluorescent gold nanoparticles have been successfully synthesised and their use as probes for cellular targeting or imaging was largely developed but their use in the biological system is particularly difficult. The dark field microscopy imaging may be enhanced by utilising gold nanoparticles with high scattering properties. Straightforward image is most often used for cellular detection. Diffraction induced by the nanoparticles was used in various experiments: to demonstrate the binding of antibodies[101] or to localise the particles in the cell.[85] However, this phenomenon can be used only for *in vitro* experiments.

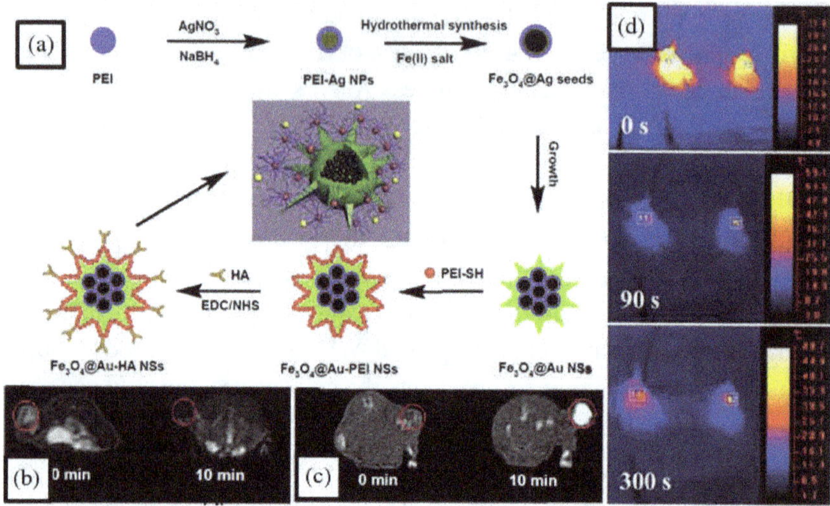

Figure 16.10   Hyaluronic acid modified Fe$_3$O$_4$@Au core/shell nanostars for multi-detection. (a) Schematic illustration of the synthesis of hyaluronic acid modified Fe$_3$O$_4$@Au nanostars. (b) T2-weighted magnetic images and (c) computed tomography of tumours before and 10 min after intratumoral injection of the nanostars. (d) Photothermal images of two tumours bearing mice injected with medium alone (left) and medium with nanostars following laser (915 nm) irradiation at the indicated time. Adapted from Li *et al.*[71]

It is also possible to visualise the presence of the nanoparticles by reflection (Figure 16.11); Klein *et al.* studied the relationship between the size of nanoparticles and their ability to be observed by confocal microscopy. In solution, the particles less than 60 nm cannot be detected while in the intracellular medium, particles of 15 nm showed a very high diffusion rate allowing a quantitative detection of the particles.[97]

16.4.2 *Immunisation Using Gene Gun*

The principle of the gene gun is to send gold nanoparticles with sufficient kinetic energy so that they cross the cell membrane, and carry into the cells the molecules fixed on their surface. The main interest of this technique is to deliver molecules into the cells without receptor limitation and whatever the nature of the cells. Particle acceleration is obtained using various devices such as bullets, gas cartridges or electric discharges.[102] The gene gun is used for the introduction of coding DNA into the cellular cytoplasm. One of the

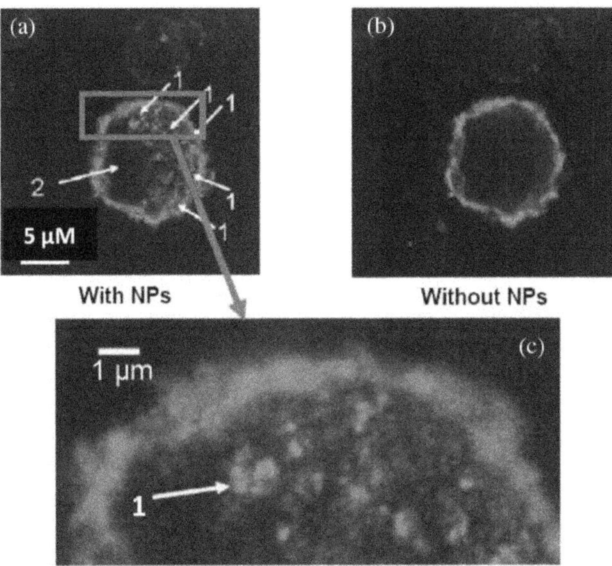

Figure 16.11   Intracellular localisation of gold nanoparticles by confocal microscopy. After their phagocytosis, gold nanoparticles were localised by confocal microscopy using reflection. Cells were also stained with FITC-labelled cholera toxin, which react with membrane gangliosides. (a) FITC fluorescence and reflection. (b) Fluorescence alone. (c) Magnification of (a) as indicated: 1 nanoparticles, 2 nucleus.

most important applications of the gene gun is the introduction of plasmid in plants, the use of such material being particularly suitable for crossing highly resistant cellulosic walls found around plant cells.[103]

The gene gun is also used to inject DNA into animal cells.[104] Indeed, skin is a highly immune-reactive tissue containing abundant antigen-presenting cells,[105] and consequently, it provides a favourable site for DNA immunisation. Such immunisation was obtained after injection of plasmide coding for the antigen alone or together with immune activators.[106]

16.4.3 *Gold Nanoparticles and Fingerprints*

Detection of fingerprints must be performed on various substrates and should lead to the clearest result with a good contrast between the ridge pattern and the background. Researchers have shown that one such strategy is to immerse the substrate in colloidal gold at low pH; in these operating conditions, particles are bound to the print. The signal may be further amplified by catalytic deposition of other metals as shown by Schnetz,[107] a method slightly improved later by Fairley.[108] Stauffer and colleagues have simplified the staining by using gold for both labelling and amplification,[109] whereas others have amplified the signal by multi-metal deposition[110] or luminescence.[111] These different techniques were reviewed by Becue.[112]

16.5 Conclusions

As shown, the number of techniques applied in biology and based on the use of gold nanoparticles is important and continues to increase. Of course, in some cases, gold could be replaced by other materials, but, contrary to many other metals, AuNPs are easy to manufacture, the particles are stable and not susceptible to oxidation.

Because of its optical properties, gold remains the metal with the largest potentialities; many techniques are based on the modifications of plasmon resonance upon interaction of gold particles with proteins, peptides or with other molecules alone or linked to other gold particles. For these experiments, the size, shape or structure of the gold particles may be changed, allowing the optimisation of the measurements. Because of the great surface of the particles compared to their volume, there is a great number of binding sites for molecules of interest (cellular targeting) and for molecules inducing the stealth of the particles. These technologies are moving towards the use of multi-modal platform for cellular targeting, specific delivery of molecules, hyperthermia, imagery, etc. Moreover, the structure of the particle may evolve to combine the advantages of various metals such as Si/Gd or Fe/Gd.

Most of the studies show that AuNPs are non-toxic, although analysis should continue because each change in the structure or composition of the nanoparticles may cause a modification of their processing by the cells and therefore potentially of their toxicity. We must keep in mind that particles can cross biological barriers and thus reach areas of the body normally protected like the brain,[113] sperm[114] or ovary.[115]

All these results indicate clearly that the use of AuNPs with respect to benefits of the optical and chemical characteristics, which are specific to this metal, and we can assume that for both diagnostic and treatment, new applications will be developed to increase the sensitivity, selectivity and efficiency of these tools; the only limitation of these developments are those of the human imagination.

References

1. M. A. Dobrovolskaia, B. W. Neun, S. Man, X. Ye, M. Hansen, A. K. Patri, R. M. Crist, S. E. McNeil, *Nanomedicine* **10**, 1453 (2014).
2. Z. J. Deng, M. Liang, M. Monteiro, I. Toth, R. F. Minchin, *Nat Nanotechnol* **6**, 39 (2011).
3. A. Selva Sharma, M. Ilanchelian, *J Phys Chem B* **119**, 9461 (2015).
4. R. G. Nuzzo, D. L. Allara, *J Am Chem Soc* **105**, 4481 (1983).
5. A. C. Templeton, W. P. Wuelfing, R. W. Murray, *Acc Chem Res* **33**, 27 (2000).
6. B. Garcia, M. Salome, L. Lemelle, J. L. Bridot, P. Gillet, P. Perriat, S. Roux, O. Tillement, *Chem Commun (Camb)* 369 (2005).
7. C. Alric, J. Taleb, G. Le Duc, C. Mandon, C. Billotey, A. Le Meur-Herland, T. Brochard, F. Vocanson, M. Janier, P. Perriat, S. Roux, O. Tillement, *J Am Chem Soc* **130**, 5908 (2008).
8. A. G. Kanaras, F. S. Kamounah, K. Schaumburg, C. J. Kiely, M. Brust, *Chem Commun (Camb)* 2294 (2002).
9. E. E. Connor, J. Mwamuka, A. Gole, C. J. Murphy, M. D. Wyatt, *Small* **1**, 325 (2005).
10. M. C. Senut, Y. Zhang, F. Liu, A. Sen, D. M. Ruden, G. Mao, *Small* **12**, 631 (2016).
11. C. L. Villiers, H. Fritas, R. Couderc, M.-B. Villiers, P. Marche, *J Nanopart Res* **12**, 55 (2010).
12. R. Levy, Z. Wang, L. Duchesne, R. C. Doty, A. I. Cooper, M. Brust, D. G. Fernig, *Chembiochem* **7**, 592 (2006).
13. G. A. Craig, P. J. Allen, M. D. Mason, *Methods Mol Biol* **624**, 177 (2010).
14. T. Pellegrino, R. A. Sperling, A. P. Alivisatos, W. J. Parak, *J Biomed Biotechnol* **2007**, 26796 (2007).
15. M. Rai, A. P. Ingle, S. Birla, A. Yadav, C. A. Santos, *Crit Rev Microbiol* 1 (2015).
16. M. A. van Dijk, A. L. Tchebotareva, M. Orrit, M. Lippitz, S. Berciaud, D. Lasne, L. Cognet, B. Lounis, *Phys Chem Chem Phys* **8**, 3486 (2006).
17. J. H. Leuvering, P. J. Thal, M. van der Waart, A. H. Schuurs, *J Immunoassay* **1**, 77 (1980).
18. M. Tanaka, Y. Saito, H. Misu, S. Kato, Y. Kita, Y. Takeshita, T. Kanamori, T. Nagano, M. Nakagen, T. Urabe, T. Takamura, S. Kaneko, K. Takahashi, N. Matsuyama, *J Clin Lab Anal* **30**, 114 (2016).
19. M. Tanaka, K. Matsuo, M. Enomoto, K. Mizuno, *Clin Biochem* **37**, 27 (2004).
20. E. E. Bedford, J. Spadavecchia, C. M. Pradier, F. X. Gu, *Macromol Biosci* **12**, 724 (2012).
21. I. He, D. Musick, S. R. Nicewarner, F. G. Salinas, S. J. Benkovic, M. J. Natan, C. Hkeating, *J Am Chem Soc* **122**, 9071 (2000).

22. J. Wang, A. Munir, Z. Li, H. S. Zhou, *Biosens Bioelectron* **25**, 124 (2009).

23. X. C. Zhou, S. J. O'Shea, S. F. Y. Li, *Chem Commun* 953 (2000).

24. M. Su, S. U. Li, P. D. V. *Appl Phys Lett* **82**, 3562 (2003).

25. N. Kim, S. H. Son, *J Nanosci Nanotechnol* **15**, 6188 (2015).

26. J. Matsui, M. Takayose, K. Akamatsu, H. Nawafune, K. Tamaki, N. Sugimoto, *Analyst* **134**, 80 (2009).

27. T. Yao, X. Gu, T. Li, J. Li, J. Li, Z. Zhao, J. Wang, Y. Qin, Y. She, *Biosens Bioelectron* **75**, 96 (2016).

28. P. Englebienne *Analyst* **123**, 1599 (1998).

29. P. Englebienne, A. V. Van Hoonacker, M. Verhas, *Analyst* **126**, 1645 (2001).

30. M. M. Miller, A. A. Lazarides, *j Phys Chem B* **109**, 21556 (2005).

31. R. Chhabra, J. Sharma, H. Wang, S. Zou, S. Lin, H. Yan, S. Lindsay, Y. Liu, *Nanotechnology* **20**, 485201 (2009).

32. K. W. Kuo, T. H. Chen, W. T. Kuo, H. Y. Huang, H. Y. Lo, Y. Y. Huang, *J Nanosci Nanotechnol* **10**, 4173 (2010).

33. E. Oh, M. Y. Hong, D. Lee, S. H. Nam, H. C. Yoon, H. S. Kim, *J Am Chem Soc* **127**, 3270 (2005).

34. P. C. Ray, A. Fortner, G. K. Darbha, *J Phys Chem B* **110**, 20745 (2006).

35. B. Dubertret, M. Calame, A. J. Libchaber, *Nat Biotechnol* **19**, 365 (2001).

36. P. S. Randeria, W. E. Briley, A. B. Chinen, C. M. Guan, S. H. Petrosko, C. A. Mirkin, *Cancer Treat Res* **166**, 1 (2015).

37. J. R. Lakowicz *Plasmonics* **1**, 5 (2006).

38. M. Martini, P. Perriat, M. Montania, R. Pansu, C. Julien, O. Tillement, S. Roux, *J Phys Chem C* **113**, 17669 (2009).

39. B. S. Delmulle, S. M. De Saeger, L. Sibanda, I. Barna-Vetro, C. H. Van Peteghem, *J Agric Food Chem* **53**, 3364 (2005).

40. G. P. Zhang, X. N. Wang, J. F. Yang, Y. Y. Yang, G. X. Xing, Q. M. Li, D. Zhao, S. J. Chai, J. Q. Guo, *J Immunol Methods* **312**, 27 (2006).

41. J. Aveyard, P. Nolan, R. Wilson, *Anal Chem* **80**, 6001 (2008).

42. C. Fernandez-Sanchez, C. J. McNeil, K. Rawson, O. Nilsson, H. Y. Leung, V. Gnanapragasam, *J Immunol Methods* **307**, 1 (2005).

43. J. Aveyard, M. Mehrabi, A. Cossins, H. Braven, R. Wilson, *Chem Commun (Cambridge)* 4251 (2007).

44. T. Suzuki, M. Tanaka, S. Otani, S. Matsuura, Y. Sakaguchi, T. Nishimura, A. Ishizaka, N. Hasegawa, *Diagn Microbiol Infect Dis* **56**, 275 (2006).

45. Y. He, S. Zhang, X. Zhang, M. Baloda, A. S. Gurung, H. Xu, X. Zhang, G. Liu, *Biosens Bioelectron* **26**, 2018 (2011).

46. T. A. Taton, C. A. Mirkin, R. L. Letsinger, *Science* **289**, 1757 (2000).

47. J. M. Nam, C. S. Thaxton, C. A. Mirkin, *Science* **301**, 1884 (2003).

48. E. D. Goluch, J. M. Nam, D. G. Georganopoulou, T. N. Chiesl, K. A. Shaikh, K. S. Ryu, A. E. Barron, C. A. Mirkin, C. Liu, *Lab Chip* **6**, 1293 (2006).

49. M. Trevisan, M. Schawaller, G. Quapil, E. Souteyrand, Y. Merieux, J. P. Cloarec, *Biosens Bioelectron* **26**, 1631 (2010).

50. P. R. Nair, M. A. Alam, *Analyst* **135**, 2798 (2010).

51. H. Lee, J. E. Park, J. M. Nam, *Nat Commun* **5**, 3367 (2014).

52. O. D. Velev, E. W. Kaler, *Langmuir* **15**, 3693 (1999).

53. S. J. Park, T. A. Taton, C. A. Mirkin, *Science* **295**, 1503 (2002).

54. E. Diessel, K. Grothe, H. M. Siebert, B. D. Warner, J. Burmeister, *Biosens Bioelectron* **19**, 1229 (2004).

55. J. H. Kim, J. H. Cho, G. S. Cha, C. W. Lee, H. B. Kim, S. H. Paek, *Biosens Bioelectron* **14**, 907 (2000).

56. G. Lai, H. Zhang, T. Tamanna, A. Yu, *Anal Chem* **86**, 1789 (2014).

57. E. C. Dreaden, L. A. Austin, M. A. Mackey, M. A. El-Sayed, *Ther Deliv* **3**, 457 (2012).

58. G. Wang, T. Huang, R. W. Murray, L. Menard, R. G. Nuzzo, *J Am Chem Soc* **127**, 812 (2005).

59. X. Liu, Y. Chen, H. Li, N. Huang, Q. Jin, K. Ren, J. Ji, *ACS Nano* **7**, 6244 (2013).

60. C. L. Bayer, S. Y. Nam, Y. S. Chen, S. Y. Emelianov, *J Biomed Opt* **18**, 16001 (2013).

61. R. Meir, M. Motiei, R. Popovtzer, *Nanomedicine (Lond)* **9**, 2059 (2014).

62. T. M. Mayhew, C. Muhlfeld, D. Vanhecke, M. Ochs, *Ann Anat* **191**, 153 (2009).

63. L. E. Cole, R. D. Ross, J. M. Tilley, T. Vargo-Gogola, R. K. Roeder, *Nanomedicine (Lond)* **10**, 321 (2015).

64. R. E. Gosselin *J Gen Physiol* **39**, 625 (1956).

65. M. K. Khan, L. D. Minc, S. S. Nigavekar, M. S. Kariapper, B. M. Nair, M. Schipper, A. C. Cook, W. G. Lesniak, L. P. Balogh, *Nanomedicine* **4**, 57 (2008).

66. K. C. Black, Y. Wang, H. P. Luehmann, X. Cai, W. Xing, B. Pang, Y. Zhao, C. S. Cutler, L. V. Wang, Y. Liu, Y. Xia, *ACS Nano* **8**, 4385 (2014).

67. C. Tian, L. Zhu, F. Lin, S. G. Boyes, *ACS Appl Mater Interfaces* **7**, 17765 (2015).

68. F. J. Nicholls, M. W. Rotz, H. Ghuman, K. W. MacRenaris, T. J. Meade, M. Modo, *Biomaterials* **77**, 291 (2016).

69. I. Miladi, C. Alric, S. Dufort, P. Mowat, A. Dutour, C. Mandon, G. Laurent, E. Brauer-Krisch, N. Herath, J. L. Coll, M. Dutreix, F. Lux, R. Bazzi, C. Billotey, M. Janier, P. Perriat, G. Le Duc, S. Roux, O. Tillement, *Small* **10**, 9 (2014).

70. Y. F. Xiao, M. M. Jie, B. S. Li, C. J. Hu, R. Xie, B. Tang, S. M. Yang, *J Immunol Res* **2015**, 761820 (2015).

71. J. Li, Y. Hu, J. Yang, P. Wei, W. Sun, M. Shen, G. Zhang, X. Shi, *Biomaterials* **38**, 10 (2015).

72. Y. Pan, H. Ding, L. Qin, X. Zhao, J. Cai, B. Du, *J Biomed Nanotechnol* **9**, 1746 (2013).

73. K. Li, Z. Zhang, L. Zheng, H. Liu, W. Wei, Z. Li, Z. He, A. C. Larson, G. Zhang, *Nanomedicine (Lond)* **10**, 2185 (2015).

74. H. S. Qhattal, X. Liu, *Mol Pharm* **8**, 1233 (2011).

75. S. Gosk, T. Moos, C. Gottstein, G. Bendas, *Biochim Biophys Acta* **1778**, 854 (2008).

76. J. M. de la Fuente, C. C. Berry, *Bioconjug Chem* **16**, 1176 (2005).

77. M. S. Shim, C. S. Kim, Y. C. Ahn, Z. Chen, Y. J. Kwon, *J Am Chem Soc* **132**, 8316 (2010).

78. M. S. Shim, Y. J. Kwon, *Bioconjug Chem* **20**, 488 (2009).

79. F. Curnis, A. Gasparri, A. Sacchi, R. Longhi, A. Corti, *Cancer Res* **64**, 565 (2004).

80. S. K. Libutti, G. F. Paciotti, A. A. Byrnes, H. R. Jr. Alexander, W. E. Gannon, M. Walker, G. D. Seidel, N. Yuldasheva, L. Tamarkin, *Clin Cancer Res* **16**, 6139 (2010).

81. N. Amreddy, R. Muralidharan, A. Babu, M. Mehta, E. V. Johnson, Y. D. Zhao, A. Munshi, R. Ramesh, *Int J Nanomedicine* **10**, 6773 (2015).

82. J. S. Butler, P. J. Sadler, *Curr Opin Chem Biol* **17**, 175 (2013).

83. C. R. Patra, R. Bhattacharya, D. Mukhopadhyay, P. Mukherjee, *Adv Drug Deliv Rev* **62**, 346 (2010).

84. B. Kang, M. A. Mackey, M. A. El-Sayed, *J Am Chem Soc* **132**, 1517 (2010).

85. E. C. Dreaden, M. A. El-Sayed, *Acc Chem Res* **45**, 1854 (2012).

86. H. Maeda, J. Wu, T. Sawa, Y. Matsumura, K. Hori, *J Control Release* **65**, 271 (2000).

87. S. Link, M. A. El-Sayed, *J Phys Chem B* (1999) 4212.

88. B. Nikoobakht, M. A. El-Sayed, *Chem Mater* **15**, 1957 (2003).

89. C. Loo, A. Lin, L. Hirsch, M. H. Lee, J. Barton, N. Halas, J. West, R. Drezek, *Technol Cancer Res Treat* **3**, 33 (2004).

90. S. M. Ryou, S. Kim, H. H. Jang, J. H. Kim, J. H. Yeom, M. S. Eom, J. Bae, M. S. Han, K. Lee, *Biochem Biophys Res Commun* **398**, 542 (2010).

91. J. Stehr, C. Hrelescu, R. A. Sperling, G. Raschke, M. Wunderlich, A. Nichtl, D. Heindl, K. Kurzinger, W. J. Parak, T. A. Klar, J. Feldmann, *Nano Lett* **8**, 619 (2008).

92. M. J. Kogan, N. G. Bastus, R. Amigo, D. Grillo-Bosch, E. Araya, A. Turiel, A. Labarta, E. Giralt, V. F. Puntes, *Nano Lett* **6**, 110 (2006).

93. A. G. Skirtach, C. Dejugnat, D. Braun, A. S. Susha, A. L. Rogach, W. J. Parak, H. Mohwald, G. B. Sukhorukov, *Nano Lett* **5**, 1371 (2005).

94. A. G. Skirtach, J. munoz, O. Kreft, K. Köhler, A. P. Alberola, H. Möhwald, W. J. Parak, G. B. Sukhorukov, *Angew Chem Int Ed* **45**, 4612 (2006).

95. C. M. Pitsillides, E. K. Joe, X. Wei, R. R. Anderson, C. P. Lin, *Biophys J* **84**, 4023 (2003).

96. H. J. Kim, S. M. Lee, K. H. Park, C. H. Mun, Y. B. Park, K. H. Yoo, *Biomaterials* **61**, 95 (2015).

97. S. Klein, S. Petersen, U. Taylor, D. Rath, S. Barcikowski, *J Biomed Opt* **15**, 036015 (2010).

98. I. Cabezon, G. Manich, R. Martin-Venegas, A. Camins, C. Pelegri, J. Vilaplana, *Mol Pharm* **12**, 4137 (2015).

99. S. Bhattacharyya, R. Bhattacharya, S. Curley, M. A. McNiven, P. Mukherjee, *Proc Natl Acad Sci U S A* **107**, 14541 (2010).

100. S. Mallidi, T. Larson, J. Aaron, K. Sokolov, S. Emelianov, *Opt Express* **15**, 6583 (2007).

101. M. P. Melancon, W. Lu, Z. Yang, R. Zhang, Z. Cheng, A. M. Elliot, J. Stafford, T. Olson, J. Z. Zhang, C. Li, *Mol Cancer Ther* **7**, 1730 (2008).

102. D.-R. Chen, C. H. wendt, Pui, D. Y. H. *J Nanopart Res* **2**, 133 (2000).

103. V. M. Ramesh, S. E. Bingham, A. N. Webber, *Methods Mol Biol* **274**, 301 (2004).

104. S. Kuriyama, A. Mitoro, H. Tsujinoue, T. Nakatani, H. Yoshiji, T. Tsujimoto, M. Yamazaki, H. Fukui, *Gene Ther* **7**, 1132 (2000).

105. P. W. Lee, S. H. Hsu, J. S. Tsai, F. R. Chen, P. J. Huang, C. J. Ke, Z. X. Liao, C. W. Hsiao, H. J. Lin, H. W. Sung, *Biomaterials* **31**, 2425 (2010).

106. R. Weiss, M. Gabler, T. Jacobs, T. W. Gilberger, J. Thalhamer, S. Scheiblhofer, *Vaccine* **28**, 4515 (2010).

107. B. Schnetz, P. Margot, *Forensic Sci Int* **118**, 21 (2001).

108. C. Fairley, S. M. Bleay, V. G. Sears, N. NicDaeid, *Forensic Sci Int* **217**, 5 (2012).

109. E. Stauffer, A. Becue, K. V. Singh, K. R. Thampi, C. Champod, P. Margot, *Forensic Sci Int* **168**, e5 (2007).
110. M. Zhang, H. H. Girault, *Analyst* **134**, 25 (2009).
111. A. Becue, A. Scoundrianos, C. Champod, P. Margot, *Forensic Sci Int* **179**, 39 (2008).
112. A. Becue, S. Moret, C. Champod, P. Margot, *Biotech Histochem* **86**, 140 (2011).
113. M. Shilo, A. Sharon, K. Baranes, M. Motiei, J. P. Lellouche, R. Popovtzer, *J Nanobiotechnol* **13**, 19 (2015).
114. U. Taylor, D. Tiedemann, C. Rehbock, W. A. Kues, S. Barcikowski, D. Rath, *Beilstein J Nanotechnol* **6**, 651 (2015).
115. J. K. Larson, M. J. 3rd, Carvan, J. G. Teeguarden, G. Watanabe, K. Taya, E. Krystofiak, R. J. Hutz, *Nanotoxicology* **8**, 856 (2014).

Chapter 17

What About Toxicity and Ecotoxicity of Gold Nanoparticles?

Marie Carrière

Laboratoire Lésions des Acides Nucléiques, Commissariat à l'Energie Atomique, Université Joseph Fourier — CNRS, Grenoble, France

17.1 Introduction

While production and use of nanoparticles in commercial products increase exponentially, the perception of risk also becomes more acute, as these new substances may generate new adverse effects, both on human health and on the environment. As mentioned previously in this volume, gold nanoparticles (AuNPs) are promising tools for diagnostic and therapeutic purposes (see Chapter 16). Before launching any medical protocol using AuNPs, their innocuousness has to be proven. Gold colloids have been used for years for therapeutic purposes (see Chapter 1) and this safe use suggests that AuNPs should also be safe. However, the properties of materials at the nanoscale, i.e. in the 1–100 nm size range, are so different from the properties of the bulk material that it is reasonable to revisit their toxicological and eco-toxicological impact. During the last decade, several research groups have published valuable data proving that AuNPs exert moderate toxic effects on eukaryotic cells, on animal models and on several organisms representing different levels of ecosystems. These toxic effects greatly depend on the AuNP size and surface coating or stabilising agent, the chemical agents used for these surface modifications often being themselves more harmful than AuNPs *per se*. Note also that most of the data collected to date have been obtained after exposure of the organisms to very high concentrations of AuNPs, which do not reflect a real exposure of humans or a real release in the environment. Still in recent years, some studies carried out at more realistic concentrations have emerged, especially on environmental models. The present chapter is a survey of these data and tries to assess the risks generated by AuNP known to date.

17.2 Impact of Gold Nanoparticles on Human Health

17.2.1 *The Toxicological Approach, Applied to Nanoparticles*

Risk is commonly recognised as the combination of hazard (H) and exposure (E). If there is no exposure, then even if the substance is hazardous, there will be no risk. If exposure occurs, but the substance is safe, then there will be no risk either. Risk assessment and management are thus only possible if data related to these two elements are available. Hazard is an inherent property of the considered substance, while the extent of exposure is dependent on multiple variables and scenarios. There is an increasing literature related to NP hazard assessment, while the extent of data related to exposure is still low. Consequently, even if many advances have been made in the field of NP toxicology over the last two decades, it is today not possible to precisely answer the question of risk related to NP exposure.

Evaluating exposure to NPs is a challenging task. Several exposure scenarios can be imagined: exposure of workers at their working place (occupational exposure), accidental exposure of populations and intentional exposure of patients for medical purpose since AuNPs are foreseen as future therapeutic and diagnostic agents. In this scenario of medical application, exposure is strictly controlled since the applied dose is known. However, depending on the route of application (intravenous injection, instillation, inhalation, skin deposition, etc.), AuNPs still have to cross physiological barriers to reach their final destination. The effective dose, reaching the target organ, will thus not strictly be the applied dose. For example, dermal penetration of NPs has been extensively studied, and it is now recognised that most NPs do not cross undamaged skin.[1] Conversely some NPs have been shown to reach the brain when instilled in an animal's nose. Moreover, when directly injected intravenously, NPs diffuse through the whole body and reach various target organs. Each organ receives a particular quantity of NP, which depends both on the organ morphology and physiology. The dose reaching each organ also depend on NP physico-chemistry, especially their size, shape and surface properties, i.e. surface charge, hydrophilicity or hydrophobicity, the latter often deriving from NP surface coating. Regarding the surface coating, exposure depends on the initial coating that result from the synthesis process, but also on the coating that forms upon contact with biological fluids and biomolecules. The so-called corona that forms on the surface of AuNPs will define the target organ, the affinity for cell membrane transporters, receptors or other biomolecules, potential accumulation and release from target organs and finally toxicological outcomes. A plethora of reports currently describes the influence of the protein corona that forms on the surface of NPs in biological media, but one must keep in mind that proteins are not the only biomolecules that adsorb on the surface of NPs and subsequently define the bio-nanointeractions.[2] It is important to note here that due to their interesting properties and particularly their chemical stability, i.e. absence of dissolution in biological media, their ease to detect in biological media and low background concentration in biological tissues, their very controlled and well-established synthesis processes, AuNPs have served as benchmark NPs for characterisation of bio-nanointeractions. A recent report, based on AuNPs, shows how nanoparticle–cell interactions can be predicted based on corona proteins and NP physico-chemical properties, using a quantitative structure activity relationship (QSAR) approach.[3]

Various degrees of toxic effects would then appear on the target organs. Depending on the required therapy, NPs then sometimes have to cross the cellular membrane to reach their final intracellular target. Several routes are available for NPs to reach the intracellular environment, from simple diffusion through cell membrane to cell uptake through specific transporters, if the NP is complexed to a specific ligand (see Chapter 16). Finally most of the reports published to date rather show that the most probable route of intracellular accumulation is endocytosis, which includes several pathways: macropinocytosis; clathrin- or caveolae-mediated endocytosis; and clathrin/caveolae-independent endocytosis (for review, see Ref. 4). To summarise, the extent of exposure thus varies with the route of application and the physico-chemistry of applied NP.

Today, the question of hazard of chemical substances is regarded intensely, through various legislations such as the REACH European Union regulation, which deals with Registration, Evaluation, Authorisation and restriction of Chemical substances. This regulation, implemented on 1 June 2007, aims at 'improving the protection of human health and the environment from the risks that can be posed by chemicals, while enhancing the competitiveness of the EU chemicals industry. It also promotes alternative methods for the hazard assessment of substances in order to reduce the number of tests on animals' (http://echa.europa.eu/regulations/reach). Practically, users, manufacturers and importers must register their chemical substances, in a volume-triggered system: all substances produced and imported at more than one ton per year have to be registered. For substances produced or imported at more than 10 tons per year, users, manufacturers and importers are also required to gather physico-chemical, toxicological and ecotoxicological information, that they have to provide together with a chemical safety report. All this information is centralised on a database, run by the European Chemical Agency (ECHA). Substances of 'very high concern' are considered separately. They have to be registered even if their production or importation is lower than one ton per year. But sufficient toxicological data are needed to classify a substance as of 'very high concern'. Progressive substitution of the most dangerous substances is demanded, when substitutes are available. This regulation, in its present format, is applicable to NPs since they are, *per se*, a chemical substance. In the present format of the regulation, AuNPs, if produced by more than one ton per year, have to be registered and evaluated as a chemical substance and not as a nanoscaled particle, i.e. they will be registered if produced by more than one ton per year. Given the toxicological data currently available, it is probable that AuNPs cannot be considered as substances of 'very high concern'. The limitations of REACH regulation appear as soon as one understands that the toxicity of AuNPs greatly differs from the toxicity of Au ions. Moreover, the methods currently used for regulatory toxicity testing have been validated for toxicity assessment of soluble substances, not for NPs. A working group was formed in July 2008, in order to study the possibility of adaptation of the present regulation, to improve it in terms of substances at the nanoscale. The European Commission envisages revisions of the REACH annexes, and for this purpose, a public consultation was launched on 13 June 2013, to

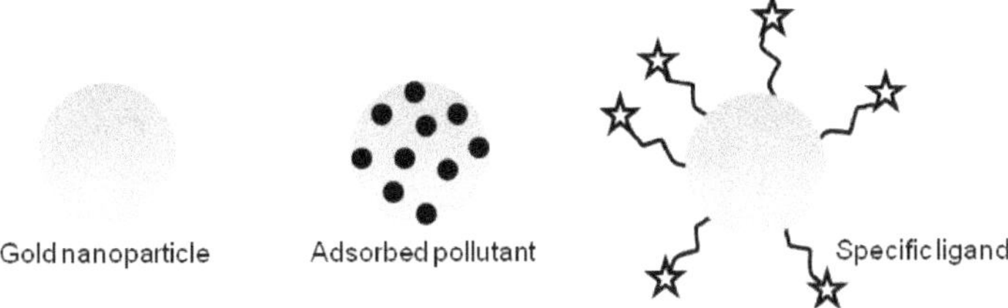

Figure 17.1   Surface modification of a AuNPs, either by adsorption of a pollutant, or by functionalisation with a ligand, specific for a chosen cell membrane transporter.

which several institutions responded by publishing position papers. However, the introduction of new information requirements for nanomaterials will not be published in REACH annnexes before the end of the 2015–2018 nanomaterials workplan.

Technically, studying the toxicity of materials at the nanoscale is a challenging task and requires innovative strategies and methodologies. As compared to chemical pollutants, the specific properties of NPs, such as their high specific surface area, render them highly reactive. Specific surface area increases when NP size decreases. Consequently, NP reactivity will depend on their size (see Chapter 16), and the smallest NPs will probably have a different toxicological profile than larger ones. Moreover, their high surface reactivity would trigger an interaction with various molecules and among them with pollutants (Figure 17.1). Indirect toxicity is thus likely to appear in addition to the inherent toxicity of the NP, if toxic pollutants such as synthesis residues are adsorbed on the surface of NPs get released in intracellular compartments.

Adsorbed molecules may modify NP fate/biodistribution in living organisms. NPs will improbably cross the physiological barriers freely; their mobility will rather be governed by their size and/or by the nature of functional molecules linked or adsorbed on their surface. In the specific case of AuNPs, their high affinity for thiol-bearing molecules would render them highly affine for some proteins or thiolated biomolecules. This affinity is interesting for the construction of nanocomposites for therapeutic purposes; gaining affinity for some specific cell surface transporters would enable cellular internalisation (Figure 17.2) (see Chapter 16).

High affinity of AuNP surface for thiol-bearing ligands may conversely cause problems if AuNPs reach physiological fluids or become internalised

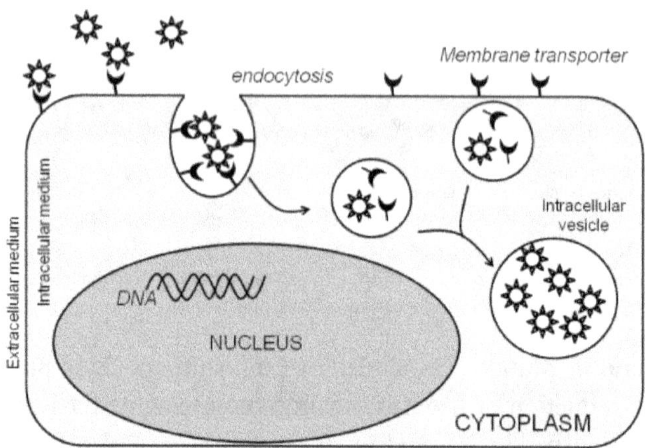

Figure 17.2   Cell internalisation of a AuNPs functionalised with a specific ligand, targeted to a cell membrane transporter.

in cells, lose their surface coating due to enzymatic processes or due to sequestration in the lysosomal compartment (pH5). These AuNPs would then possibly deplete fluids or cell cytoplasm from essential molecules, such as major the molecular antioxidant of cells, glutathione. For all these reasons, it is all the more important to deeply characterise AuNP's physico-chemistry (size, porosity, specific surface area, surface charge, presence of synthesis residues on nanoparticle's surface, etc.) before launching any toxicology experiment.

Another point is that due to their high surface reactivity, NPs may adsorb some components of toxicology assay kits. This phenomenon has been reported for carbon-based NPs such as carbon nanotubes[5] and may also be true for AuNPs. Moreover, the plasmon resonance band of AuNP sometimes interferes with toxicity assays, which are often based on the measurement of absorbance of colorimetric dyes. For instance, the MTT assay (3-(4,5-dimethylthiazol-2-yl)-2,5-diphenyl tetrazolium bromide assay), classically used for cytotoxicity evaluation by quantifying the activity of mitochondrial enzymes, ends up with the measurement of absorbance at 562 nm, which is in the range of some AuNP plasmon band. Interference between these assays and AuNPs would lead to misinterpretation of toxicological data.[6] It is thus strongly recommended to validate toxicity data with at least two independent assays in order to have a clear proof of NP toxicity or harmlessness.

Lastly, the guidelines for toxicity testing published by national or international organisations (OECD, ISO, DIN, etc.) have been validated and are considered valuable for chemical compounds, not for NPs. Several working groups are currently dedicated to the validation of these tests for assessment of NP toxicity.

Concerning the toxicity studies related to AuNPs, on *in vivo* (animal) or *in vitro* (cultured cells) models, inter-comparison of the published data is difficult, since different authors use different biological models, AuNPs with different shapes and diameters, coated with different ligands or still covered with different synthesis residues. However, general trends can be ruled, as described thereafter.

17.2.2 *Biokinetics and Target Organs of Gold Nanoparticles after Systemic Exposure*

Several studies describe the target organs and biokinetics of AuNPs when intravenously injected in rodents. Two major parameters have been shown to play an important role in AuNPs or nanorods biodistribution and biokinetics: their primary diameter and their surface coating.

Small AuNPs (diameter: 1.9 nm) have been tested by the Hainfeld team, whose first concern was to enhance radiotherapy treatments. After a single injection in mice of AuNPs, in suspension in phosphate buffer saline (PBS), NPs were rapidly cleared through the kidneys, without causing any harmful adverse effects to the animals.[7] Four years after, Semmler-Benke *et al.* showed that the biodistribution pattern of 1.4 nm AuNPs differed from that of 18 nm AuNPs, after intravenous injection in rats at $0.01 \mu g \cdot g^{-1}$ of body weight.[8] These NPs were coated with TPPMS ($Ph_2PC_6H_4SO_3Na$). The smaller NPs were preferentially eliminated via the urinary and hepatobiliary route, and thus recovered in urine and faeces, but 3.7% of 1.4 nm size NPs still circulated in the bloodstream after 24 h. Conversely, within 24 h, almost all 18 nm NPs had been removed from the blood and accumulated in the liver and spleen. More recently, the biodistribution and clearance of $\sim$20 nm AuNPs coated with citrate or cysteine-alanine-leucine-asparagine-asparagine pentapeptide (CALNN) was evaluated after a single intravenous injection of 0.7 mg Au $\cdot$ kg^{-1} in rats. Both NP formulations were rapidly cleared out of the bloodstream and accumulated in the liver and spleen, where they were still detected 28 days after the injection.[9] Depending on their surface coating, they induced spleen atrophy and anaemia, and AuNPs coated with CALNN were more toxic than AuNPs coated with citrate.[9] This report underlines the impact of AuNP coating on their toxicity, but not on their biodistribution, which is similar with the two AuNP formulations. The biodistribution of larger AuNPs in rats (10–250 nm) also depends on their size. NPs with 10 nm diameters, as a suspension in phosphate buffer saline (PBS) where they agglomerated, distributed in the blood, liver, spleen, kidneys, testis, thymus, heart and brain, 24 h after injection. Conversely, 50–250 nm NPs were mainly accumulated in the liver and spleen. In addition, the smallest NPs (10 nm) were shown to pass through the blood-brain barrier (BBB), which is a highly selective barrier that allows the passage of only necessary biomolecules from the bloodstream to the brain extracellular fluid. These NPs may thus have reached brain cells.[10] This distribution pattern was also observed in mice after intravenous injection of 15–200 nm NPs (at $1 \, \mathrm{g \cdot kg^{-1}}$ of body weight, suspended in alginate where they did not agglomerate) and 15–50 nm NPs also passed the BBB while larger ones did not.[11] Accumulation in the liver, lung and spleen increased when AuNP diameter increased. Evidence of passage through the BBB was also shown *in vitro*, in cultures of primary human endothelial cells exposed to 4 nm AuNPs coated with glucose.[12] More important passage through brain endothelium was observed, as compared to passage through non-brain endothelium. Moreover, subsequent passage through the extracellular matrix and accumulation in the nucleus of astrocytes was also observed.[12] Passage through the BBB may occur via the paracellular route, since AuNPs were also recently shown to alter the endothelial tight junctions and consequently to increase paracellular permeability of an *in vitro* endothelial model.[13] Passage through the BBB may cause highly detrimental consequences on brain cells. Among them, intraperitoneal injection of citrate-coated AuNPs at the dosage of $70 \, \mu\mathrm{g \, Au \cdot kg^{-1}}$ b.w. was shown to induce DNA damage in the brain of exposed rats, although Au accumulation in brain cells was not specifically identified; genotoxic effect may thus be indirect.[14] AuNPs with a diameter of 30 nm induced more important damage than AuNPs with a diameter of 10 nm when injected at the same concentration, and acute exposure to a single dose of NPs induced more damage than chronic exposure with a single, daily injection of AuNPs during 28 days.[14] Still some studies also report absence of passage though the BBB of citrate-coated, 20 nm AuNPs after intravenous injection in rats, certainly because the physico-chemistry of these NPs differs from that of NPs that were shown to pass through.[9]

This biodistribution pattern, i.e. preferential storage in liver, spleen and lung, is classical for colloidal materials; it is explained by their uptake by the mononuclear phagocyte system (MPS), namely resident immune cells of the liver and spleen that subsequently differentiate into macrophages. Strategies to avoid NP uptake by the MPS include their coating with a layer of amphiphilic polymer chains such as polyethylene glycol (PEG). This protective cover renders NPs more hydrophilic, and finally confers them a longer circulation time in the blood, which is necessary in therapeutic applications. PEG-5000 grafting to gold nanorods was achieved by Niidome *et al.*, who introduced PEG during the synthesis process.[15] The resulting nanorods had different zeta potentials: 41 mV for the original nanorods and –0.5 mV for the pegylated ones. The authors observed that 54% of the coated nanorods still circulated in the bloodstream 30 minutes after injection, whereas most non-PEGyltaed nanorods were immediately detected in the liver. This proportion then decreased with time, in favour of preferential accumulation in the liver, but 5% of the initial dose was still circulating 24 h after injection. The same group then injected gold nanorods grafted with different quantities of PEG in tumour-bearing mice.[16] It was concluded from this study that a PEG:gold molar ratio of 1.5 was sufficient for MPS avoidance by AuNPs. Nanoparticles finally accumulated in the liver, and when liver storage capacity was overwhelmed, AuNPs distributed in other tissues and among them in tumours.[16] This strategy would then be interesting in cancer treatment strategies, although it would certainly cause adverse effects on hepatic function.

To conclude, these data tend to demonstrate that intravenously injected AuNPs are rather eliminated through the MPS, i.e. in faeces, when their diameter is above 15–20 nm. When their diameter is below 15–20 nm, a proportion of the NPs are eliminated through the urinary tract. Moreover, blood retention of PEG-grafted nanorods is longer than that of plain gold nanorods.

17.2.3 *Translocation of Gold Nanoparticles through Physiological Barriers*

External exposure to NPs through inhalation, ingestion or skin contact may lead to their internalisation and subsequent redistribution in the body. This transfer through physiological barriers is classically termed translocation. Although translocation is required when AuNPs are intended to be used for therapeutic purposes, it can also cause harmful adverse effects both on the directly exposed organs and on secondary target organs where NPs accumulate after translocation. A series of studies were conducted with a series of sulphonated triphenylphosphine (S-TPP)-coated AuNPs with diameters 1.4, 5, 18, 80 and 200 nm, and with a positively charged (coated with cysteamine) and a negatively charged (coated with thioglycolic acid) 2.8 nm AuNP. These AuNPs were neutron-activated, i.e. radiolabelled with ^{198}Au, with a very high specific activity, which enabled the administration of low doses of NPs and the measurement of a few nanograms of Au in rat tissues and organs without causing any acute radio-toxic effect. These NPs were instilled either intraoesophageally or intratracheally in rats, or injected intravenously in pregnant rats to assess translocation through the gut, the respiratory epithelium and the placenta, respectively.

In these studies, AuNPs were shown to translocate through the gastrointestinal barrier[17,18] and after 24 h, the smallest AuNPs passed from the gut lumen to the body more efficiently than larger AuNPs.[19] Surprisingly 18 nm NPs were more efficiently absorbed than 5 nm AuNPs,[19] which suggests that even if the NP size is an important determinant of NP absorption through the gut, absorption does not evolve linearly with respect to particle size. The authors hypothesised that this specific behaviour of 18 nm AuNPs could be related to their specific curvature and surface structure, which would lead to the absorption of specific proteins on their surface, particularly proteins with increased ability to penetrate the epithelium. They also noted that these 18 nm AuNPs accumulated more heavily in the brain of exposed rats than the others,[19] possibly for the same reason. Moreover, this study shows that absorption also depends on NP surface charge, negatively charged NPs being more efficiently absorbed than positively charged ones.[19] Other reports demonstrated that smallest AuNPs (4 nm), coated with citrate and tannic acid, translocated more efficiently through the gut than larger ones and then accumulated in the kidneys, liver, spleen, lungs and brain.[17] Oral administration induced toxicity, especially oxidative stress and DNA damage.[18]

The same series of [198]AuNPs was intratracheally instilled in rats by the same team, in order to probe their translocation through the air–blood barrier in lungs.[20] Again, the smallest AuNPs translocated more efficiently than the other NPs, suggesting that size was a major determinant of air–blood barrier translocation. Again, 200 nm AuNPs translocated more efficiently than 80 nm AuNPs[20] showing that no general trend can be established. Moreover, again in this model, negatively charged 2.8 nm AuNPs more efficiently translocated through the air–blood barrier than positively charged ones.[20] Regarding the biodistribution of these AuNPs in the tissues and organs 24 h after instillation, a major conclusion was the high retention in the carcass of animals, possibly in the bone marrow. The finest AuNPs (1.4 and 2.8 nm in diameter) were mainly retained in kidneys while larger ones accumulated more in the liver.[20] The influence of NP size on their translocation was further demonstrated in another study, showing small NPs (1.4 nm) were efficiently transferred, whereas 18 nm NPs were almost entirely retained in the lungs after instillation in the trachea of the rats.[8] One day after instillation, 8.5% of the 1.4 nm NPs were found in secondary target organs, mainly the liver and kidneys.[8] When administrating AuNPs via inhalation rather than instillation, translocation of AuNPs through the air–blood barrier was demonstrated, with again smaller NPs (7 nm) being more efficiently translocated than larger ones (20 nm).[21] The target organs where NPs accumulated were the lungs and blood, liver, kidneys, small intestine as well as the brain.[21] AuNPs were suspected to be cleared out of the body via macrophage-mediated mucociliary escalation, while hepatobiliary elimination of translocated NPs was a minor mechanism of clearance.[21] Finally, translocation through the air–blood barrier had also been demonstrated on an *in vitro* model composed of epithelial cells, macrophages and dendritic cells.[22] The authors showed that 25 nm gold NPs were accumulated in cells as non-agglomerated particles and were not grouped in intracellular vesicles. AuNP accumulation triggered inflammatory response in cells, through the release of tumour necrosis factor-α.[22]

Regarding other exposure routes, translocation of 15, 102 and 198 nm AuNPs through the skin and intestine was also reported, *ex vivo*.[11] Two

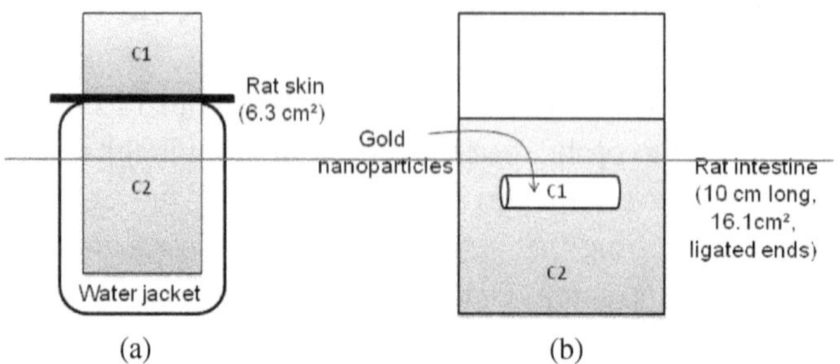

Figure 17.3   Schematic representation of Franz diffusion cell and intestinal sac method, employed to study NP permeation through skin or intestine. Adapted from Ref. 11.

ex vivo models were used, namely Franz diffusion cell for permeation through the skin, and the 'intestinal sac' method for assessing intestinal permeation (Figure 17.3).

Permeation was demonstrated to be size dependent; it decreased as NP diameter increased. Permeation through the intestine was higher than permeation through the skin, where a time lag was observed before NPs diffused through the epithelium. In addition, NPs of smaller size penetrated deeper in the skin, whereas larger NPs were retained in the epidermis and dermis regions.[11]

Another important question is the translocation from mother to foetus, i.e. passage of NPs through the placenta. Absence of translocation of 15 and 30 nm AuNPs was demonstrated in an *ex vivo* human placenta.[23] Translocation was not evidenced through the placenta of mice exposed to citrate-stabilised 20 nm and 50 nm AuNPs: Au was detected in the placenta but not in foetuses.[24] Still recently Semmler-Benke and collaborators demonstrated the translocation of 1.4, 18 and 80 nm, negatively charged, sulphonated triphenylphosphine (S-TPP)-coated [198]AuNPs through the placenta of pregnant rats, with 1.4 nm being much more translocated than the largest NPs, and with 1.4 nm and 18 nm AuNPs being subsequently detected in foetuses.[25] The amounts of Au detected in fetuses were very low but significant, with 30 and 0.1 ng detected in foetuses from pregnant mice exposed to 1.4 and 18 nm AuNPs, respectively.[25] Only radiolabelling enabled the measurement of so low quantities of Au, and this certainly explains the discrepancies between the conclusions of the articles by Semmler-Benke *et al.*,[25] Myllynen *et al.*[23] and Rattanapinyopituk *et al.*[24]

Finally, the blood–retinal barrier has received attention as it may be a route to achieve systemic treatment of eye pathologies, particularly those related to pathological angiogenesis.[26] Permeation of AuNPs through this barrier occurred with small NPs including 20 nm NPs, while larger ones (100 nm) did not cross the barrier. Transferred NPs were then taken up by neurons, endothelial cells and peri-endothelial glial cells, without causing any adverse effect.[26]

To conclude, the translocation of small AuNPs is likely to occur through most of the physiological barriers, including lungs, intestine, skin, placenta, blood–brain barrier and blood–retinal barrier. Currently, lacking in the literature are studies that would consider populations with pathologies, i.e. with altered barrier integrity. For instance, patients suffering from asthma or lung obstructive disorders, or from inflammatory bowel disease, may respond differently to AuNPs since the permeability of these barriers might be impacted by pathological conditions.

17.2.4 *Cellular Toxicity, In Vitro Studies*

Cellular toxicity of AuNPs has been described to be low, which is why they are attractive candidates for therapeutic applications. Several studies attempted to identify the influence of the physico-chemical properties of AuNP, such as size, shape and agglomeration state, on the biological response of exposed cells, but failed to provide any clear conclusion. Cytotoxicity, genotoxicity and other toxicological responses to gold nanoparticles have been demonstrated in a number of studies and recently reviewed by Hadrup *et al.*[27] Regarding toxicity assessment, one must keep in mind that most NPs might interfere with most of the commercial toxicity assays due to their inherent properties such as absorption, fluorescence or simply due to the adsorption of dyes on their surface. Results from *in vitro* studies should thus be interpreted with caution, especially if the authors do not describe interference checking. The main conclusions of these studies are described and commented thereafter.

Most studies show that AuNPs are generally accumulated in cells, *in vitro*. Their uptake does not necessarily correlate with cytotoxicity, for instance Connor *et al.* showed that AuNPs with very low toxicity were internalised and localised in intracytoplasmic vesicles.[28] On the other hand, NPs interaction with cell membrane, without any internalisation, may also be the cause of toxic events. Nevertheless, AuNP internalisation was shown to depend upon their size,[29] with 50 nm NPs being more efficiently internalised than their smaller or larger counterparts. The uptake mechanism is suspected to be through interaction of AuNP surface with serum proteins, which, in turn, are actively transported through the cell membrane.[29]

Generally, 15–100 nm AuNPs have been described to cause only a low decrease of cell viability,[29–31] independently of their surface coating. For instance, biotin- and citrate-coated AuNPs were shown to display very low toxicity. Conversely, residues from synthesis processes such as cetyltrimethylammonium bromide (CTAB) greatly increased cell death.[28] Consequently while in some reports AuNPs trigger cell death, they were not inherently toxic to human cells but surface pollutants caused toxicity. This statement was also true for gold nanorods, also synthesised with CTAB, since washing nanorods or overcoating them with polymers or polyelectrolytes drastically reduced their cytotoxicity.[15,30,32] Overcoating was presumed to prevent the desorption of contaminant molecules from nanorods.[30]

On the other hand, Pernodet *et al.* showed that 13 nm citrate-coated AuNPs accumulated in human dermal fibroblasts from primary cultures, which in consequence showed altered cell division and changes in cell morphology.[33] These modifications were attributed to a modification of actin structure and they were observed only after long-term exposure, i.e. four or six days. Cell growth and differentiation, cell morphology, cell cycle regulation and cellular function and maintenance were also among the pathways showing altered protein contents in mouse Balb/3T3 fibroblasts exposed to 5 and 15 nm, citrate-coated AuNPs, in a proteomic study conducted by Gioria *et al.*[34] In addition, oxidative stress and inflammatory response also showed altered response in cell exposed to these NPs.[34] At the gene expression level, Bajak *et al.* showed up- or down-regulation of genes involved in response to metal exposure, particularly metallothioneins, oxidative stress signalling and selenium homeostasis, in Caco-2 enterocytes exposed to 300 μM of 5 nm citrate-coated AuNPs for 72 h.[35] Although the exposure dose is very high, these results show interesting insights into the mechanisms by which

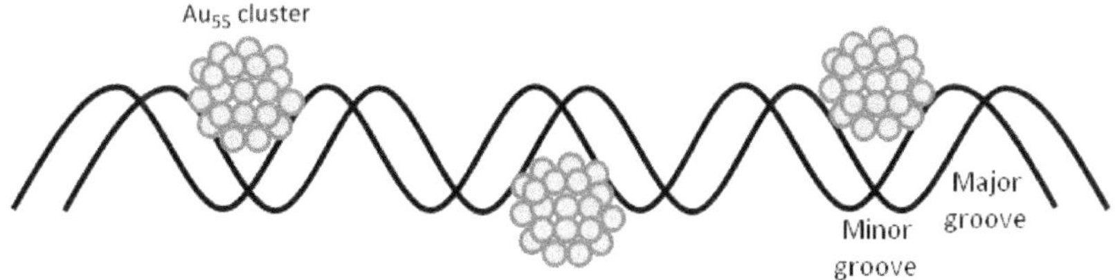

Figure 17.4   Au$_{55}$ cluster inserted in the major groove of DNA.

AuNP may cause toxicological outcomes. In particular the up-regulation of metallothionein-encoding genes suggest the presence of Au ions in the exposed cells, which may either be impurities remaining from the AuNP synthesis process or result from AuNP dissolution in the intracellular compartment. Although AuNP dissolution is improbable due the chemistry of these NPs, it is possible that their sequestration in highly acidic intracellular compartments (the pH in lysosomes is close to 4) may result in desorption of loosely bond Au ions remaining at the surface of NPs after the synthesis process.

A distinct impact was noticed for 1.4 nm Au$_{55}$ nanoclusters (Au$_{55}$(Ph$_2$PC$_6$H$_4$SO$_3$H)$_{12}$Cl$_6$). These AuNPs have been shown to interact in a unique manner with DNA: due to their specific size, these clusters have been shown to fit into the major groove of DNA where they irreversibly linked (Figure 17.4).[36]

Whether this interaction with DNA and their cellular impact are related or not, these nanoclusters have been shown to be highly toxic, *in vitro*, to 11 cancerous or normal cell lines. The authors assumed that cytotoxicity was not due to contaminant molecules. Moreover, 42.5% of the gold was detected in cell nuclei, and 21% of this proportion was shown to be bound to DNA.[37] Accumulation of NPs in cell nuclei may trigger their interaction with DNA, and possibly cause genotoxicity. Genotoxicity may also be indirectly caused by NPs, and especially AuNPs, for instance, if NPs impair the cell's ability to repair damaged DNA, or if NPs deplete cells from antioxidant molecules. Indication of AuNP genotoxicity, their direct or indirect, have been clearly shown in several *in vitro* studies (for review, see Ref. 27).

17.3 Environmental Impact of Gold Nanoparticles

17.3.1 *What Can Make Nanoparticles Toxic for the Environment?*

NPs are at present produced by the ton and used in many commercial products (see Chapter 18), which suggests that they will certainly be released into the environment.[38] As discussed in the previous section dealing with human toxicology, the risk (R) is the potential that a substance causes harm. R is a function of two elements: the extent of exposure (E) and the type of harm, or hazard (H).[39] Information related to NP concentration in the environment is almost entirely absent, and consequently data concerning exposure (E) are lacking. This is due to the absence of a valuable and precise analytical method, which would enable the correct quantification of traces of NPs in the environment. Every part of the environment already contains high concentrations of natural nanosized matter; it is thus not possible to distinguish this background from traces of the so-called engineered NPs. Today, the only way to have access to exposure data is modelling and the reader must keep in mind that the parameters used for modelling are often very uncertain. For instance, estimations of worldwide NP production are not clear, and the behaviour of NPs in the environment is poorly known, particularly regarding their potential dissolution. So far, two recent modelling studies predict TiO_2^-, ZnO^-, AgNPs and carbon nanotube concentrations in the different parts of the environment.[40,41] Moreover, the concentration of AuNPs occurring in the environment was predicted to reach 0.14 μg/L in water and 5.99 μg/kg in soils, as anticipated from AuNP use in consumer products.[38] In all the environmental reservoirs, NP concentration is predicted to exponentially increase in the next years.[40] For example, the concentration of AgNPs in U.S. sediments was predicted to rise from 0.7 to 2.2 μg $\cdot$ kg^{-1} between 2008 and 2012.[40] In the same conditions, the concentration of TiO_2NPs would rise from 0.2 to 0.6 mg $\cdot$ kg^{-1}. It is considered likely that AuNP agglomeration and/or mobilisation in environmental conditions will occur (Figure 17.5).

AuNPs agglomerate when ionic strength of the liquid medium increases: it is immediate when ionic strength reaches 0.1 M.[42] Conversely, when naturally occurring organic substances such as humic substances are added, the colloidal stability of AuNP suspension increases, i.e. NP agglomeration is hindered.[42] In conclusion, NP fate in the environment, and consequently

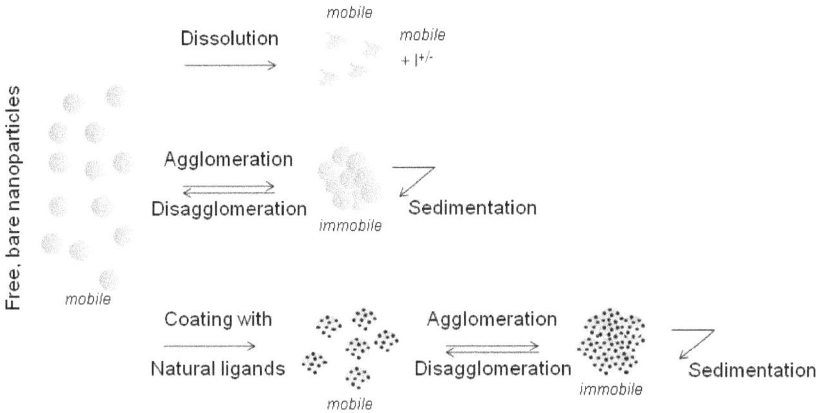

Figure 17.5   The fate of gold nanoparticles' fate in the environment.

NP exposure, is greatly governed by agglomeration–sedimentation and/or disagglomeration-mobilisation processes, themselves governed by changes of pH, ionic strength and the presence of organic substances.

Concerning data related to AuNP hazard (H), most of the studies published to date report their potential impact on living species, from unicellular organisms to animal and vegetal species. As already discussed in the human toxicology section, noxious NP effects may be direct, i.e. derived from the particle itself, but can also be indirect since NPs may mobilise pollutants by adsorbing them on their surface. Most of the studies related to AuNP toxicity towards environmental species have been, up to now, done on fresh water species, using standardised assays validated for regulatory toxicology, for instance OECD, ISO and DIN guidelines (for assays based on algae: ISO 8692, OECD 201, DIN 38412-33; for assays on daphnids: ISO 6341, OECD 202, DIN 38412-30). As underlined previously, these protocols are not yet validated for nanotoxicology applications, and some of the results described herein should be considered with caution.

17.3.2 *Impact of Gold Nanoparticles on Unicellular Organisms: Bacteria and Algae*

Bacteria and microalgae constitute the lowest level of the ecosystems. They may therefore be a route of entrance of NPs to the food chain. Moreover, some NPs like AgNPs show efficient bactericidal and antifungal effects, which have justified their use in first aid bandages, soap and self-cleaning textiles but also in consumers products such as washing machines or fridges. Due to these properties, they might also be lethal for some environmental bacteria, leading to ecotoxicological issues upon release in the environment. These properties have been attributed to various mechanisms, and among them the release of toxic Ag ions, the deficiency of proper regulation of transport through bacterial membranes due to NP adsorption, or interaction with membrane proteins or with DNA, which may induce bacterial cell death (for review, see Ref. 43). Considering these effects of AgNPs, a common view is that they may be extrapolated to AuNPs. However, most of the studies published to date instead show that AuNPs are quite inert for environmental species, certainly because they do not dissolve.

Firstly, some articles gave indirect evidence that AuNPs did not cause adverse effects when interacting with bacteria. Lin *et al.* reported that mannose-coated AuNPs (6 nm in diameter, $\sim$200 mannose per nanoparticle) efficiently bound to *E. coli* type I pili, through specific interaction of mannose with FimH adhesin of bacterial pili.[44] These NPs would then be used as efficient labelling probe. The authors did not mention any toxicity. Then AuNPs have been used in several studies for photothermal killing of bacteria. Pristine AuNPs are always shown to be non-toxic, but the surface coating that is added to NP in a view to trigger their interaction with bacterial cell wall sometimes renders them toxic. It should be mentioned here that the surface charge of AuNPs, dependent on their synthesis procedure, does not always allow their interaction with bacterial cell wall. Negatively charged AuNPs, for instance those synthesised using citrate as a reducing agent, do not adhere to the negatively charged bacterial cell wall (Figure 17.6(a)). Conversely, when synthesising AuNPs with $NaBH_4$ and cysteamine, AuNPs are positively charged and strongly adsorb onto bacterial cell wall (Figure 17.6(b)).

Exposure of *Staphylococcus aureus* bacteria to AuNP-anti-protein A antibody conjugates (NP diameter was 10–40 nm) leads to efficient adhesion of nanoconjugates on the bacterial surface.[45] Then irradiation with focused laser pulses induces over-heating of AuNPs (see Chapter 4), causing

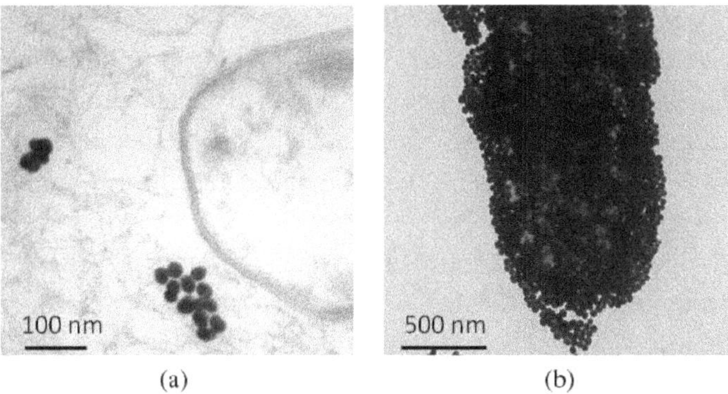

(a)   (b)

Figure 17.6 Interaction between bacterial cell wall and AuNPs, depending on their surface charge. AuNPs have been synthesised either with citrate (a) or with NaBH4 and cysteamine (b), resulting in negatively charged (a) or positively charged (b) NP surfaces. This image was kindly provided by C. Sicard-Roselli, LPC, Univ. Paris Sud, France.

bubbling on bacterial surfaces, which physically damages them, leading to their complete disintegration at the higher laser energies. Without any laser irradiation, gold nanoconjugates cause a slight increase in bacterial death, whereas pristine AuNPs are non-toxic.[45] In another report, polygonal, 30 nm, vancomycin-coated AuNPs, were shown to induce the death of several bacterial strains, Gram-positive or Gram-negative, antibiotic-sensitive or resistant, after irradiation in the near-infrared region (808 nm). Without irradiation, the toxicity of vancomycin-coated AuNPs was lower but still significant. Conversely pristine AuNPs were not toxic.[46] Finally, gold nanorods (68 nm × 18 nm, synthesised with CTAB) coated with an anti-*Pseudomonas aeruginosa* antibody, and irradiated in the near-infrared region (~785 nm), were shown to efficiently kill a multi-antibiotic-resistant bacterial strain of *Pseudomonas aeruginosa*, sampled in the upper respiratory tract of sinusitis patients. Nanorods alone did not inhibit bacterial growth over a 24 h period, suggesting that they were not inherently toxic to the cells.[47] In these three studies, the toxicity of pristine AuNPs is shown to be null, that of coated AuNPs is slightly higher and a combination of AuNPs with irradiation (which would improbably occur in natural environments) causes cell death.

A direct proof of AuNPs harmlessness was published by Williams *et al.*[18] These authors exposed *E. coli* bacteria in normal growth conditions to several types of inorganic NPs, and among them PEG-coated, 30 nm-diameter AuNPs. They observed that these NPs caused no inhibition of bacterial cell multiplication. A possible reason to explain absence of impact on bacteria is that NPs agglomerate. Indeed, if agglomerated, NPs would settle down, bacterial exposure would be hindered. In this study, AuNPs were not agglomerated.

Lastly, several microorganisms have been reported to produce biogenic AuNPs from gold ions ($AuCl_4^-$). For instance, in the metal-reducing bacterium *Shewanella algae*, gold deposits were formed in less than 30 minutes upon exposure to $AuCl_4^-$. After 90 minutes almost all the bacteria contained gold deposits of 5–15 nm, without evidence that these deposits induced any lethality.[49]

Among the unicellular organisms, algae are also one of the first steps in the food chain in aquatic ecosystems. A recent review summarises the results obtained up to now on marine microalgae exposed to Ag and AuNPs.[50] Again on microalgae, some articles report absence of toxicity of AuNPs, such as the article by Moreno-Garrido *et al.*, where the lethal doses of gold ions and AuNPs coated with citrate were compared, leading to the conclusion that the LD50 (i.e. the concentration leading to 50% of cell death) of AuNPs could not be measured because even at the highest exposure concentration the cell mortality never exceeded 50% of the population.[51] Conversely, in a study by Renault *et al.*, exposure for 24 h of the green algae *Scenedesmus subspicatus* to amine-coated 10 nm AuNPs induced a dose-dependent mortality, reaching 20–50% of the algae population.[52] AuNPs strongly adsorbed on algae cell wall, but they were not observed in the cell cytoplasm. However, this experiment was carried out at very high cell densities, which is known to directly influence the outcome of toxicology assays.[51] A more recent report shows that the impact of AuNPs, on the same species of microalgae, depends on surface coating: AuNPs were coated with an amphiphilic polymer consisting of a hydrophobic part (dodecylamine) and a hydrophilic part (poly-(isobutylene-*alt*-maleic anhydride)) and then grafted with polyethyleneglycol 10,000 (PEG) were less toxic than AuNPs coated with the amphiphilic polymer only.[53] These AuNPs were not shown to accumulate in the microalgae.[53]

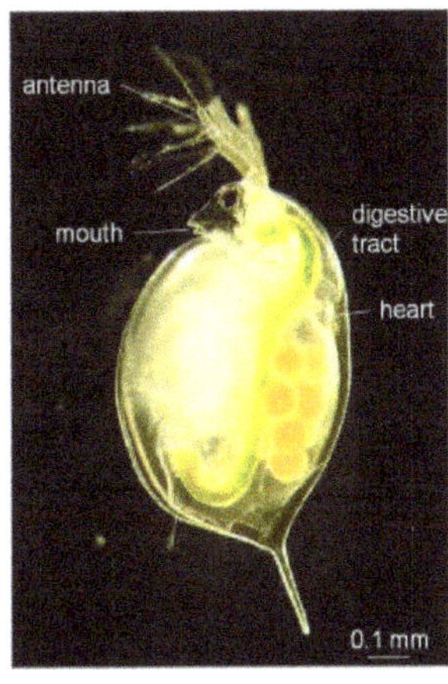

Figure 17.7 (Left) a daphnid; (right) a male zebrafish. These images were kindly provided by M. Floriani, LRE, IRSN, France.

17.3.3 *Impact of Gold Nanoparticles on Aquatic Organisms: Daphnids, Bivalves, Fishes*

Daphnids (*Daphnia magna*) (Figure 17.7) are largely used in ecotoxicity tests due to their sensitivity to chemical toxicants. They are considered to be a valid test model species prior to mammalian testing.

When applied to daphnids at sub-lethal concentration, AuNPs (17–23 nm, coated with tannic acid and citrate, 500 ppb) distributed through the gut, with highest gold accumulation after 12 h of exposure. AuNPs were observed in the lumen, but did not enter tissues.[54] A higher content of AuNPs was found in the mouth as compared to the tail region, and AuNPs were gradually excreted by the animal when placed in a NP-free environment.[54] The authors did not report any toxic effect of AuNPs in these conditions.

More recently, exposure of daphnids to spherical, 15 nm citrate-coated-AuNPs was achieved by Li *et al.*[55] who proved that 65–75 mg L^{-1} of AuNPs caused the death of 50% of the population in 48 h. Daphnid motility was gradually reduced upon NP exposure, the animals finally settled down and died. Daphnids exposed for eight days at sub-lethal dosage (10 mg L^{-1}) did not show any decreased birth rates or embryo development, but their death rate was higher than that of control samples. The authors observed that AuNPs adhered to external appendages of daphnids, impairing their movements. These appendages have two functions for daphnids. Their movement creates a water column around the animal, which brings food close to their mouth for their feeding, but also brings oxygen for their respiration. Inhibiting appendage movement would then cause starvation and oxygen deprivation. The authors also suggest that AuNPs likely inhibited food uptake by concentrating in the gut, as previously reported by Lovern *et al.*[54] In comparison, AgNPs were much more toxic, with a LD_{50} value $10^3 \times$ lower (concentration of AuNP leading to the death of 50% of daphnid population).[55]

The impact of AuNPs was tested on two bivalves, *Corbicula fluminea* and *Mytilus edulis*. Exposure to AuNPs of the blue mussel, *M. edulis*, caused oxidative stress in several organs. Thirteen-nanometre NPs, coated with citrate, led to the increase of catalase activity and protein ubiquitination in the digestive gland, and protein carbonylation in gills.[56] Five-nanometre NPs induced a weak oxidative stress, but impaired thiol antioxidant defences and oxidised protein thiols in the digestive gland, where they mainly accumulated.[57] This modest effect may impact the long-term ability to cope with traditional chemical pollutants. The report by Renault *et al.* shows the trophic transfer of AuNPs from green algae to *C. fluminea*. In their experiments, *C. fluminea* bivalves were exposed to green algae, themselves having been previously exposed to AuNPs. As described in Section 17.3.2, AuNPs rapidly adsorbed onto the algae's cell wall. Two bivalves were then fed with these algae cells. Such exposure did not cause any structural disturbance of bivalves' branchial and digestive tissues, but AuNPs were shown to penetrate into the cells of these epithelia. They caused a three-fold increase of metallothionein content in the gills and visceral mass at the highest exposure concentration. In addition, in the visceral mass, the expression of metallothionein and several markers of response to oxidative stress (superoxide dismutase, glutathione S-transferase, cytochrome *c* oxidase, catalase) were modulated by AuNP exposure.[52]

These data thus prove that AuNPs display pro-oxidant properties, which may alter the function of filter-feeding organisms, but also that they may be transferred through the trophic chain, and may thus reach the food chain if disseminated in the environment.

Lastly, the impact of AuNPs has also been studied on the zebrafish (*Danio rerio*) (Figure 17.7). This fish is classically used as a model to study environmental toxicity, but also as an alternative model to study development and physiology of higher organisms, including humans. The first article reporting the impact of AuNPs on this species showed almost no mortality, and no sub-lethal effects of the NPs on zebrafish embryos at 120 h postfertilisation, whatever their size (3, 10, 50, 100 nm, coated either with citrate or with citrate and tannic acid), when exposed to concentrations reaching 250 μM. Conversely, numerous embryo malformations were observed after exposure to AgNPs, even if low mortality is reported.[58]

Another article reports that AuNPs (11 nm) caused a low mortality and an insignificant amount of deformed embryos after chronic exposure (120 h, 0.025–1.2 nM).[59] Again, AuNPs are much more biocompatible than AgNPs. The authors show that AuNPs are accumulated in embryos, where they passively diffuse following a random route when embryonic cells divide, causing stochastic events altering embryo development.[59] In another report, 5 nm AuNPs were microinjected in zebrafish embryos whose development was then followed. After 7 days, normal-looking larval fishes were observed, with ability to feed and responsive nervous systems. Most of their organs functioned normally (cardiovascular system, eye mobility, etc.) showing that physiological functions of embryos were not affected. Neither was affected the gene expression profile.[60] Finally, zebrafish embryos were used for large-scale screening of nanoparticle toxicity, with a battery of NPs varying in their diameter and surface coating, with a view to developing a reliable hazard ranking method for NPs. Among the tested NPs, 16 different types of AuNPs were chosen and the results show that their toxic effect highly depends on both diameter and surface chemistry, with positively charged AuNPs much more toxic than neutral or negatively charged AuNPs, which may be explained by more or less intense interaction between these different AuNPs and zebrafish embryos.[61]

17.3.4 *Impact of Gold Nanoparticles on Plants*

Very few assays report the impact of AuNPs on plants. The earliest published article on this topic shows that exposure to 10 nm citrate-coated AuNPs

$(62\,\text{g.l}^{-1})$ significantly increases plant germination and root elongation of cucumber and lettuce. The authors attributed this effect to the presence of citrate during the exposure (remaining from nanoparticle synthesis), which is also known as a food additive. Root elongation was induced, but root weight was not increased significantly. The hypothesis is that it might be an avoidance mechanism of the seed to a stress factor produced by the presence of NPs. Conversely, silver and Fe_3O_4 NPs caused the reduction of seed germination and root elongation.[62] Leaf necrosis was observed in tobacco plants exposed to citrate-coated 3.5 nm AuNPs but not 18 nm AuNPs. It was correlated with accumulation in root cells in the case of 3.5 nm AuNPs, while 18 nm AuNPs only adsorbed on the root surface.[63] A recent study shows that citrate-coated, 2–19 nm diameter AuNPs do not affect plant germination but accumulate in plant roots and dramatically affect root and shoot growth.[64] Accumulation in the roots, translocation to leaves and toxic effect of AuNPs thus depend on plant species as well as AuNP diameter.

17.4 Conclusions

Most of the studies reported in this chapter have been achieved in acute exposure conditions, i.e. living organisms have been exposed to very high concentrations of AuNPs, during short incubation times. The obtained data prove that the toxicity of AuNPs *per se* is relatively low. When lethality is observed, most of the time it may be related to toxic substances adsorbed onto NP surface. However, the sublethal effects of these NPs have been poorly reported. Poorly reported is also their long-term and chronic effects, which may occur if AuNPs persisted in living organisms after translocation/accumulation. These exposure conditions are more realistic and representative of environmental exposure, and future research in these areas would be required. Finally, even if a fair amount of work has been done in recent years, research into nanotoxicology and particularly nano-ecotoxicology, is still at an early stage, and more data is needed before gold nanoparticles can be used safely for diagnostic or therapeutic purposes.

References

1. M. E. Lane, *J. Microencapsul.* **28**, 709–716 (2011).
2. M. P. Monopoli, A. S. Pitek, I. Lynch *et al.*, *Method Molec. Biol.* (Clifton, NJ) **1025**, 137–55 (2013).
3. R. Liu, W. Jiang, C. D. Walkey *et al.*, *Nanoscale* **7**, 9664–9675 (2015).
4. I. A. Khalil, K. Kogure, H. Akita et al., *Pharmacol. Rev.* **58**, 32–45 (2006).
5. J. M. Worle-Knirsch, K. Pulskamp, and H. F. Krug, *Nano Lett.* **6**, 1261–1268 (2006).
6. T. Pfaller, R. Colognato, I. Nelissen *et al.*, *Nanotoxicology* **4**, 52–72 (2010).
7. J. F. Hainfeld, D. N. Slatkin, and H. M. Smilowitz, *Phys. Med. Biol.* **49**, N309–N315 (2004).
8. M. Semmler-Behnke, W. G. Kreyling, J. Lipka *et al.*, *Small* **4**, 2108–2111 (2008).
9. S. Fraga, A. Brandao, M. E. Soares *et al.*, *Nanomedicine* **10**, 1757–1766 (2014).
10. W. H. De Jong, W. I. Hagens, P. Krystek *et al.*, *Biomaterials* **29**, 1912–1919 (2008).
11. G. Sonavane, K. Tomoda, A. Sano *et al.*, *Colloids Surf B Biointerfaces* **65**, 1–10 (2008).
12. R. Gromnicova, H. A. Davies, P. Sreekanthreddy *et al.*, *PloS One* **8**, e81043 (2013).
13. C. H. Li, M. K. Shyu, C. Jhan *et al.*, *Toxicol. Sci.* (*Soc. Toxicol.*) **148**, 192–203 (2015).
14. E. Cardoso, G. T. Rezin, E. T. Zanoni *et al.*, *Mutation Res.* **766–767**, 25–30 (2014).
15. T. Niidome, M. Yamagata, Y. Okamoto *et al.*, *J Control Release* **114**, 343–347 (2006).
16. Y. Akiyama, T. Mori, Y. Katayama *et al.*, *J Control Release* **139**, 81–84 (2009).
17. J. F. Hillyer and R. M. Albrecht, *J. Pharmaceut. Sci.* **90**, 1927–1936 (2001).
18. R. Shrivastava, P. Kushwaha, Y. C. Bhutia *et al.*, *Toxicol. Industrial Health* **32**, 1391–1404 (2016).
19. C. Schleh, M. Semmler-Behnke, J. Lipka *et al.*, *Nanotoxicology* **6**, 36–46 (2012).
20. W. G. Kreyling, S. Hirn, W. Moller *et al.*, *ACS Nano* **8**, 222–233 (2014).
21. S. K. Balasubramanian, K. W. Poh, C. N. Ong *et al.*, *Biomaterials* **34**, 5439–5452 (2013).
22. B. Rothen-Rutishauser, C. Muhlfeld, F. Blank *et al.*, *Part Fibre Toxicol.* **4**, 9 (2007).
23. P. K. Myllynen, M. J. Loughran, C. V. Howard *et al.*, *Reproduc. Toxicol.* (Elmsford, N.Y.) **26**, 130–137 (2008).
24. K. Rattanapinyopituk, A. Shimada, T. Morita *et al.*, *J. Veterinary Med. Sci.* **76**, 377–387 (2014).
25. M. Semmler-Behnke, J. Lipka, A. Wenk *et al.*, *Part Fibre Toxicol.* **11**, 33 (2014).
26. J. H. Kim, K. W. Kim, M. H. Kim *et al.*, *Nanotechnology* **20**, 19 (2009).
27. N. Hadrup, A. K. Sharma, M. Poulsen *et al.*, *Regul. Toxicol. Pharmacol. RTP* **72**, 216–221 (2015).
28. E. E. Connor, J. Mwamuka, A. Gole *et al.*, *Small* **1**, 325–327 (2005).
29. B. D. Chithrani, A. A. Ghazani, and W. C. Chan, *Nano Lett.* **6**, 662–668 (2006).
30. A. M. Alkilany, P. K. Nagaria, C. R. Hexel *et al.*, *Small* **5**, 701–708 (2009).
31. J. A. Khan, B. Pillai, T. K. Das *et al.*, *Chembiochem* **8**, 1237–1240 (2007).
32. T. S. Hauck, A. A. Ghazani, and W. C. Chan, *Small* **4**, 153–159 (2008).
33. N. Pernodet, X. Fang, Y. Sun *et al.*, *Small* **2**, 766–773 (2006).
34. S. Gioria, H. Chassaigne, D. Carpi *et al.*, *Toxicol. Lett.* **228**, 111–126 (2014).

35. E. Bajak, M. Fabbri, J. Ponti *et al.*, *Toxicol. Lett.* **233**, 187–199 (2015).
36. Y. Liu, W. Meyer-Zaika, S. Franzka *et al. Angew. Chem. Int. Ed.* **42**, 2853–2857 (2003).
37. M. Tsoli, H. Kuhn, W. Brandau *et al.*, *Small* **1**, 841–844 (2005).
38. K. Tiede, M. Hassellov, E. Breitbarth *et al.*, *J. Chromatogr. A* **1216**, 503–509 (2009).
39. G. Oberdorster, V. Stone, and K. Donaldson, *Nanotoxicology* **1**, 2–25 (2007).
40. F. Gottschalk, T. Sonderer, R. W. Scholz *et al.*, *Environ. Sci. Technol.* **43**, 9216–9222 (2009).
41. N. C. Mueller and B. Nowack, *Environ. Sci. Technol.* **42**, 4447–4453 (2008).
42. S. Diegoli, A. L. Manciulea, S. Begum *et al.*, *Sci. Total Environ.* **402**, 51–61 (2008).
43. M. Farre, K. Gajda-Schrantz, L. Kantiani *et al.*, *Anal. Bioanal. Chem.* **393**, 81–95 (2009).
44. C. C. Lin, Y. C. Yeh, C. Y. Yang *et al.*, *J. Am. Chem. Soc.* **124**, 3508–3509 (2002).
45. V. P. Zharov, K. E. Mercer, E. N. Galitovskaya *et al.*, *Biophys. J.* **90**, 619–627 (2006).
46. W. C. Huang, P. J. Tsai, and Y. C. Chen, *Nanomedicine* **2**, 777–787 (2007).
47. R. S. Norman, J. W. Stone, A. Gole *et al.*, *Nano Lett.* **8**, 302–306 (2008).
48. D. N. Williams, S. H. Ehrman, and T. R. Pulliam Holoman, *J. Nanobiotechnol.* **4**, 3 (2006).
49. Y. Konishi, T. Tsukiyama, N. Saitoh *et al.*, *J. Biosci. Bioeng.* **103**, 568–571 (2007).
50. I. Moreno-Garrido, S. Perez, and J. Blasco, *Mar. Environ. Res.* **111**, 60–73 (2015).
51. I. Moreno-Garrido, L. M. Lubian, and A. M. Soares, *Ecotoxicol. Environ. Safety* **47**, 112–116 (2000).
52. S. Renault, M. Baudrimont, N. Mesmer-Dudons *et al.*, *Gold Bull.* **41**, 116–126 (2008).
53. K. Van Hoecke, K. A. De Schamphelaere, Z. Ali *et al.*, *Nanotoxicology* **7**, 37–47 (2013).
54. S. B. Lovern, H. A. Owen, and R. Klaper, *Nanotoxicology* **2**, 43–48 (2008).
55. T. Li, B. Albee, M. Alemayehu *et al.*, *Anal. Bioanal. Chem.* **398**, 689–700 (2010).
56. S. Tedesco, H. Doyle, G. Redmond *et al.*, *Mar. Environ. Res.*, **66**, 131–133 (2008).
57. S. Tedesco, H. Doyle, J. Blasco *et al.*, *Comp. Biochem. Physiol. C Toxicol. Pharmacol.* **151**, 167–174 (2010).
58. O. Bar-Ilan, R. M. Albrecht, V. E. Fako *et al.*, *Small* **5**, 1897–1910 (2009).
59. L. M. Browning, K. J. Lee, T. Huang *et al.*, *Nanoscale* **1**, 138–152 (2009).
60. Y. Wang, J. L. Seebald, D. P. Szeto *et al.*, *ACS Nano* **4**, 4039–4053 (2010).
61. B. Harper, D. Thomas, S. Chikkagoudar *et al.*, *J. Nanoparticle Res.* **17**, 1–12 (2015).
62. R. Barrena, E. Casals, J. Colon *et al.*, *Chemosphere* **75**, 850–857 (2009).
63. T. Sabo-Attwood, J. M. Unrine, J. W. Stone *et al.*, *Nanotoxicology* **6**, 353–360 (2012).
64. N. S. Feichtmeier, P. Walther, and K. Leopold, *Environ. Sci. Pollut. Res. Int.* **22**, 8549–8558 (2015).

Chapter 18

Technological Applications of Gold Nanoparticles

Michael Cortie

Institute for Nanoscale Technology, University of Technology Sydney, Australia

18.1 Introduction

This chapter surveys the existing and potential commercial applications of AuNPs. This is an area in which there is currently rapidly developing activity, with about 19% growth in sales predicted recently.[1] This newfound interest is ironic, given that gold was historically the first of the metals to be exploited[2] and that AuNPs themselves were first used over 1500 years ago.[3] These earlier applications are discussed in greater detail in Chapter 1. Here I consider only the modern applications. The chapter starts with a brief review of why AuNPs are useful building blocks in various nanotechnologies, before considering in turn their electronic, optical, catalytic, decorative, biotechnological and medical applications.

The ever-expanding applications of nano-particulate gold are due to gold in this form possessing a unique cluster of desirable material properties: it is metallic and relatively inert, the optical properties are conducive to a plasmon resonance in the visible region of the electromagnetic spectrum (see Chapter 3), there is a useful tendency to bond selectively to sulphur and some other chalcogens, it is an excellent electrical conductor and, very importantly, AuNPs can be readily manufactured in a wide range of shapes and sizes (see Chapters 6 and 7). These attributes are exploited either individually or in combinations in the various technological applications to be discussed here.

The commercial market for AuNPs is satisfied by a number of vendors, who can supply either the basic colloidal gold nanospheres or AuNps that have been functionalised in various ways. Very small nanoparticles, more akin to clusters because they contain less than 100 gold atoms, are also available for specialised applications. Gold nanorods also have become available from suppliers. In general, the commercial customers for these products are in the decorative or biomedical industries for, as we shall see, these are areas in which the technological uses of AuNPs are best established. However, there is also some uptake in the electronics industry for use in conductive inks. Of course, part of the 'market' for AuNPs is not strictly commercial in nature and I refer here to the sale of nanoparticles to researchers at universities and elsewhere for use in pursuits of a rather scientific nature. Nevertheless, as will be shown in this chapter, a great diversity of other potential technological applications for AuNPs are being researched, and it seems likely that some of these will also become genuine commercial products in due course. Therefore they are described here too.

18.2 Electronic and Opto-electronic Applications

Thin films, wires and electronic and soldering pastes, which are not the subject of this book, are important industrial uses of gold in the electronics industry, normally consuming about of the order of a hundred tons of the element per year.[4] Here, however, we consider only the technological applications of discrete AuNPs. The market for these latter products currently consumes a rather smaller mass of Au, of the order of several tons, almost all of which is currently used in decorative inks or glazes, but the added value is extremely high. A nominal breakdown of the industrial uses of gold at time of writing is provided in Figure 18.1.

18.2.1 *Applications of the Optical and Electronic Properties of Gold*

The physical and optical properties of gold and AuNPs were discussed in Chapters 3 and 4. The electronic and optoelectronic applications of AuNPs rely mainly on only two of these properties: the dielectric function (i.e. the optical properties or refractive index as a function of frequency) and the DC electrical conductivity. Of course, other metals are electrically

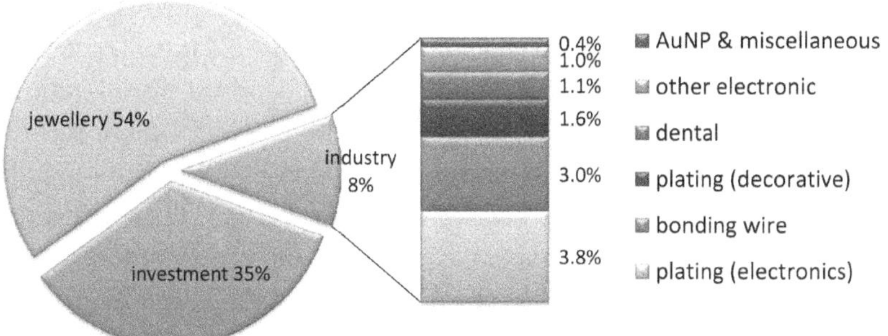

Figure 18.1 Approximate proportion of gold consumed in various applications in 2014. Data drawn from diverse Internet sources (including www.wgc.org[5]). About 3900 tons of gold were 'consumed' in 2014 but most of this was for investment and jewellery. Only 389 tons of gold was consumed for 'technology' applications. 'Other electronic' includes sputtering targets and pastes. The importance of electroplating as an end use of gold within the 'industry' sector is obvious. The author estimates that about 15 tons of the total gold consumed in 2014 were in the high tech areas ('AuNPs and miscellaneous').

conductive too, but, with the exception of platinum, nanoscale structures of these other elements would be rapidly destroyed by oxidation under ambient atmospheric conditions. Similarly, although it can be argued that the dielectric function of some other elements, such as Al, Ag, K, Na or Li, are superior to Au for the development of localised plasmon resonances at most wavelengths,[6] the oxidation resistance of the alternative elements is either relatively poor or non-existent, which complicates or prevents their use in technological applications.

18.2.2 *Sinter Inks*

The oxidation resistance of AuNPs makes them very suitable for the manufacture of conductive inks. The principle here is that an electric circuit can be 'printed' or 'painted' using readily available digital printing technologies, using an ink that is loaded with an appreciable volume fraction of AuNPs. The AuNPs are prevented from agglomerating prematurely because they are individually coated with some suitable surface ligand, such as a thiol, mercaptide or alkyl amine.[7] The potential flexibility in manufacturing that can be achieved is obvious: besides purely decorative applications, these types of inks have been used to print radiofrequency identification (RFID) tag antennas, circuits and circuit components like transistors.[7,8] However, in general, a thermal treatment ('metallising' or 'sintering') is required to destroy the protective ligand so that the AuNPs can weld together, an essential prerequisite to yield the desired electrical conductivity or reflective optical properties. In the past, this has limited the application of this technology to substrates capable of withstanding at least 200°C. Ways to induce sintering without appreciably heating the substrate have been investigated and include laser, microwave or UV irradiation, and chemical treatment.[9,10] Of course, conductive inks based on copper and silver are also used in the field, but gold has the advantage of providing the ultimate in oxidation resistance and durability.

Recently, improved formulation of the gold-based inks has led to the 'sintering' temperature being routinely brought to between 120°C and 180°C[7,11] (Figure 18.2), while sintering temperatures of below 100°C have been demonstrated under laboratory conditions.[10] The rapid low-temperature sintering of AuNPs is driven by their enormous surface energy. New formulations of inks containing AuNPs now make it possible to even coat polymer artefacts with conductive gold coatings by this technique. (Note, however, that electroless deposition of Au films or nanoparticles from solution[12] is a competing technology.) Inks of this general type, possibly containing co-additions of Cu or Ag nanoparticles, can also be selectively

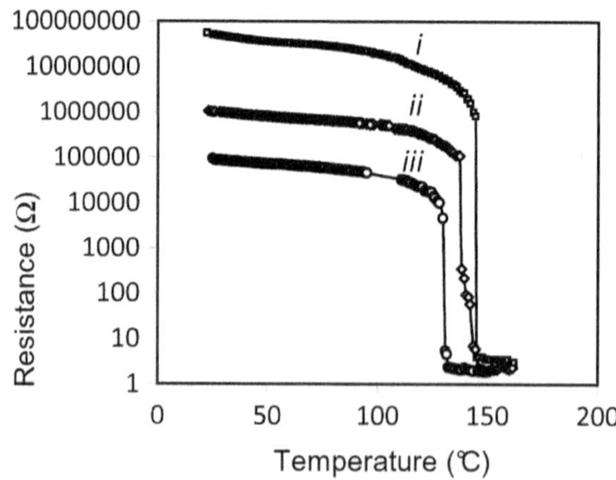

Figure 18.2   Sintering of films made with AuNP inks that have been stabilised with various ligands. (i) 1-hexanethiol, (ii) 1-butanethiol and (iii) 2-propanethiol. (Data courtesy of M. Coutts, University of Technology, Sydney. More information on these inks may be found in Ref. 10)

sintered by laser,[13] opening the way to quite flexible manufacturing option (Figure 18.3).

18.2.3 *Spectrally Selective Coatings*

Continuous coatings of gold of 20–50 nm thickness are relatively transparent to light in the visible part of the electromagnetic spectrum but are highly reflective in the infrared region. Therefore, the application of such coatings onto glass or another transparent substrate produces a spectrally selective surface, which blocks infrared radiation. This has been used in the past to produce energy-efficient windows, but gold has now been surpassed in this application by other materials, such as multilayer stacks comprising silver and transparent conducting oxides (TCOs). However, gold coatings are still used in some specialised optical applications.

More recently, the possibility of using coatings of discrete AuNPs in a spectrally selective role has been considered.[4,12,14] These block light by absorption (rather than by reflection) and the position of the absorption maximum corresponds to the position of the plasmon resonance of the nanostructure. A reasonable proportion of solar infrared can be attenuated by this means, as seen in Figure 18.4, with the grey-shaded region of the spectrum almost totally removed. In the case of glass coated with gold nanorods, the pronounced extinction in the upper visible and near-infrared between 750 and 1000 nm blocks a significant proportion of incoming solar heat. The advantage for architectural use would be that AuNPs can be readily deposited onto glass by wet chemical means, rather than by the expensive vacuum coating technologies currently used for Ag/TCO stacks. The potential reduction in processing cost appears to be large enough to compensate for the cost of the gold used.[12] Gold nanorods should give even better performance than nanospheres because their resonances can be tuned into the near-infrared region of the spectrum[15] but, so far, trials on prototype glass panes have yielded only modest solar screening figures-of-merit.[16]

18.2.4 *Nonlinear Optical Applications*

The fluctuation of the electric field of light with time, $E(t)$, induces a fluctuating dipole moment in the material through which the light is passing.

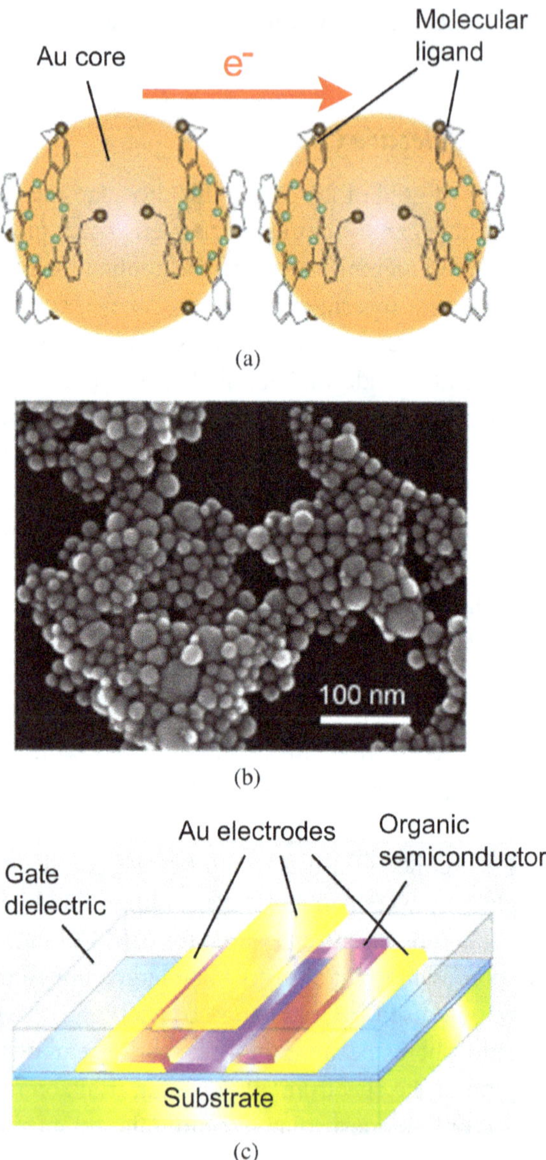

Figure 18.3   (a) Schematic illustration of π-junction AuNPs used for one type of sinter ink. The metal core is surrounded by aromatic molecular ligands. (b) Scanning electron microscope image of Au NPs deposited on a substrate. (c) Structure of OTFT devices fabricated by the printing process at room temperature. Reprinted from Ref. 8 by permission of John Wiley and Sons.

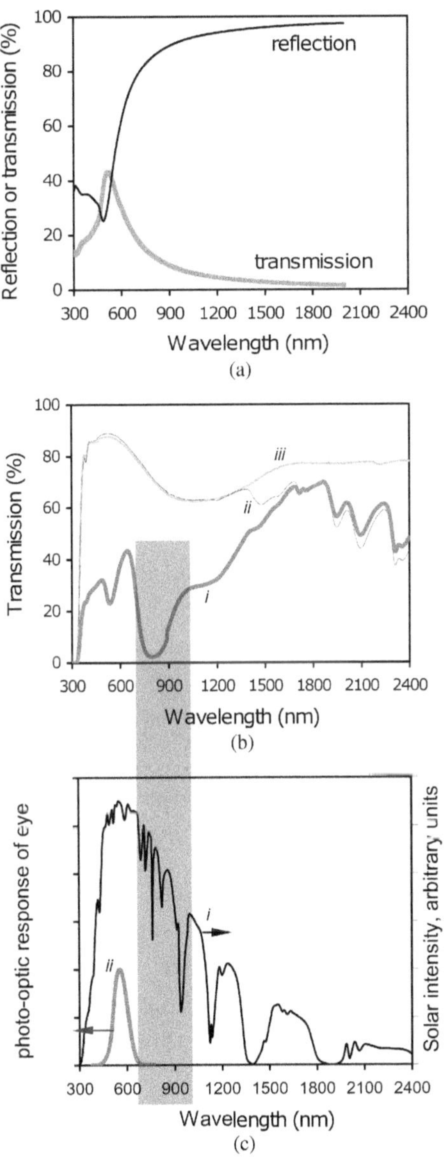

Figure 18.4   (a) Simulated transmission and reflection spectra of glass coated with a continuous nanoscale film of 28 nm thick gold (b) (i) Measured transmission of 3 mm window glass coated with a dispersion of gold nanorods in poly(vinyl alcohol) (PVA). (ii) Plain glass coated with PVA film only. (iii) Plain glass only. (c) (i) Energy distribution in standard solar spectrum, (ii) photo-optic response of the human eye. The shaded grey block indicates the region of the incoming solar spectrum that is most effectively blocked by this coating. (Figure 3(b) modified from Ref. 16 and reproduced here with kind permission from Springer Science+Business Media.)

At low intensities the relationship between electric field and dielectric polarisation, P, is linear. However, accurate description of the dielectric polarisation at high intensities of illumination requires the introduction of higher order terms, so that

$$P(t) \propto \chi^{(1)} E(t) + \chi^{(2)} \cdot (E(t))^2 + \chi^{(3)} \cdot (E(t))^3 + \cdots,$$

where the material parameters $\chi^{(1)}$, $\chi^{(2)}$ and $\chi^{(3)}$ are the first-, second- and third-order electric susceptibilities respectively. $\chi^{(2)}$ has non-zero values only for some materials and is not usually of interest in the context of AuNPs. On the other hand, values of $\chi^{(3)}$ in the range of 1×10^{-6} to 1×10^{-13} e.s.u are reported for various materials containing AuNPs,[17–22] although it should be noted that the technological figure-of-merit for exploitation of these phenomena also requires the lowest possible linear absorption coefficient too[18] (which is problematic for metallic nanoparticles). Nonlinear optical properties have existing or potential applications in devices like holography, frequency converters, optical limiters, signal processing and phase conjugators.[18] There is a degree of nonlinearity in the optical properties of AuNPs,[17] and in particular the $\chi^{(3)}$ parameter of composites containing AuNPs has attracted attention.[17–22] So far, however, there seem to have been few or no actual practical applications for AuNPs in this domain.

18.2.5 *Data Storage*

There are two means by which AuNPs can be used to store digital data. The first is capacitive: an isolated AuNP is used to store or release electric charge in the manner of the now familiar 'flash memory'.[23,24] Nanoparticles produced by the citrate method are suitable because they are in the optimum 10–20 nm diameter size range for this application. In contrast, very small particles ($<$5 nm) are subject to quantum size effects, which would complicate the operation of the device.

The other storage scheme makes use of the anisotropic optical properties of gold nanorods, and their dependence on both the aspect ratio of the nanoparticle and the polarisation of the light. This mode could find application in the CD or DVD type of recordable storage medium, although only in 'Write Once Read Many' (WORM) mode. Considerable density of data storage can be achieved by using more than one layer of nanorods of different aspect ratios embedded in a polymer matrix, with so-called five-dimensional optical recording having been demonstrated.[25,26]

Neither technology has been commercialised yet but they remain under development.

18.2.6 *Single-Electron Conductivity and Quantum Devices*

Capacitance is a measure of the ease with which electric charge can be stored on or in an object. Because AuNPs are so small, adding an extra electron causes a relatively significant increase in Coulomb repulsion between the free electrons on the particle.[27] There is therefore a significant energy cost associated with such a charge transfer process. The capacitance of an individual AuNP varies from between about 1×10^{-18} F for a 2 nm particle to about 2×10^{-16} F for an 11 nm particle.[28] Because charge itself is discretised into units of about 1.6×10^{-19} C, and the voltage on a capacitor is given by Q/C, where Q is the charge and C the capacitance, the transfer of a single quantum of charge will cause a voltage step of 0.16 V over a 1×10^{-18} F capacitor.[28] This means that the voltage on a sufficiently small metallic nanoparticle changes in a step-wise fashion as it is charged or discharged. Once one unit of charge is placed on the nanoparticle, no further addition of charge can take place until the applied potential difference is further increased to exceed the voltage step. Only then can a second unit of charge be forced onto the particle. This phenomenon is termed the Coulomb blockade. The step-wise change in voltage that accompanies a series of charge transfer events is termed a Coulomb staircase.

A single-electron transistor can be obtained by exploiting this phenomenon, using a third electrode, which applies an external electric field to modulate the step size of the Coulomb blockade. In general, these devices need to operate at very low temperatures because thermal noise at room temperature readily obscures the phenomenon. On the other hand, devices of this type should be scalable to extremely small size. A recent publication provides insights into the state of the art.[29]

18.3 Catalytic Applications

The scientific basis of catalysis by AuNPs has been described in Chapters 2 and 9. Here, some technological examples of these principles will be presented. Although the oxidation of carbon monoxide by AuNPs has received the largest share of scientific attention, the first commercial application of AuNPs as a catalyst seems to have been as a toilet deodoriser in 1992.[30] In principle, however, the lower operating temperature of AuNP-based catalysts should make their use for CO oxidation in automobile exhaust catalysts very attractive. Unfortunately, there have been only a few transfers of this knowledge to industry, so far. One issue is that the potentially high temperature of exhaust gases in a petrol-fuelled automobile running at full power will sinter and destroy a catalyst system based on AuNPs. Diesel-fuelled vehicles, which have a lower exhaust gas temperature, might be a more forgiving application and a commercial application for gold catalysts in this area was trialled by Fiat in 2011.[31] Since then, however, there have significant changes in the relative prices of Pt and Au, and it appears that Au-based catalysts are now less competitive in this application.[31] On the other hand, use of AuNP catalysts in respirators for use by emergency or mining personal remains attractive, because the gold-based units are much more tolerant of humidity and can operate for longer than the older, hopcalite-based units.[31,32] Gold catalysts have also reportedly been implemented for production of vinyl acetate monomer by BP Chemicals,[33] vinyl chloride monomer,[34] methyl methacrylate[35] and, at pilot plant scale, for the one-step direct production of methyl glycolate from ethylene glycol and methanol,[36] for the oxidation of mercury in the gas exhaust of coal-fired power stations[37] and for the collection of Hg in mercury analysers,[38] for the breakdown of toxic organic compounds such as vinyl chloride, trichloroethene and chloroform,[39] and in fuel cells as a means to selectively remove CO out of the gas stream.[40]

18.4 Decorative Applications

18.4.1 *Historic Uses in Ceramics and Glass*

The bright colour of gold itself, and of the dispersions of its nanoparticles, have been used in decorative applications for thousands of years. Besides the well-known red colour of colloidal suspensions of AuNPs, which have been used to colour glass as far back as Roman times, there is also a range of purple-hued pigments based on an intimate mixture of Au and tin oxide nanoparticles. These are known as the Purple of Cassius and have been used since 1679 to colour high-quality ceramic ware.[41] More information on the historical uses of gold may be found in Chapter 1.

18.4.2 *Colouring Textiles*

It is certainly possible to stain textiles with AuNPs, gold-silver nanoparticles or the Purple of Cassius-type nanoparticle, to produce a range of red to purple colours. The principles were already established in 1794.[42] In recent times, there has been some experimentation with dying silk or wool by these means[43–45] (see Figure 18.5) and there is currently some interest from the commercial sector.[4] Such a product would provide a means to add value to woollen fashion items intended for the high end of the market.

18.4.3 *Use in Paint and Polymers*

The vivid colours of colloidal gold are occasionally exploited in specialised paints or to colour polymers. A range of reddish to purple colours are

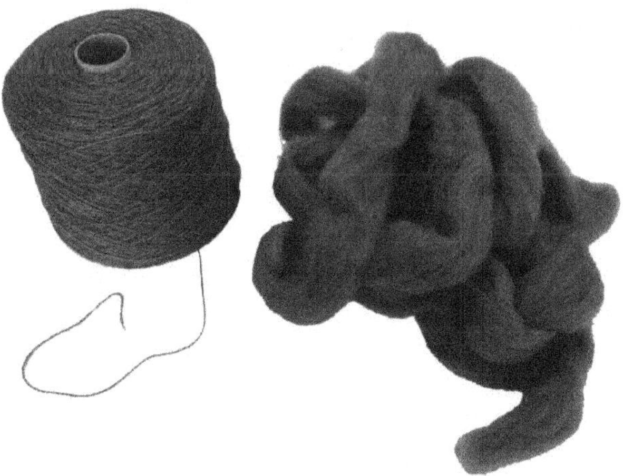

Figure 18.5   High-quality wool that has been dyed purple using AuNPs that have been grown *in situ* on the fibre. Photo courtesy of Prof. J. Johnston, Victoria University, New Zealand. Further detail may be found in Ref. 45.

obtained, with the exact colour depending on the angle of incidence of the viewer and light source.[46] Gold loading is high, however, so these materials are relatively expensive. A dichroic label for the packaging of, for example, high value consumer goods can be obtained by exploiting the fact that sufficiently large nanoparticles (>50 nm) will display a different colour in transmission than in reflection.[47]

18.5 Use in Sensors and Biomedical Diagnostics

Considerable information on this topic may be found in Chapters 14 and 16. Here only a rather applied summary of the topic is provided.

18.5.1 *Refractometric Sensors*

Refractometric sensors using gold are based on either of two related optical phenomena. In the older and more developed scheme, the characteristics of a surface plasmon polariton (SPP) propagating on the surface of a nanoscale gold film are monitored using suitable instrumentation. The details of the SPP can be tuned by controlling the average refractive index in the region of medium immediately adjacent to the surface of the gold. Generally, this is exploited to make a sensor that is selective to specific biological molecules or protein fragments. This is done by first coating the surface of the gold film with a molecule that will selectively bind to the analyte. When the analyte is present, it coats the film, thereby raising the average refractive index of the immediately adjacent medium and changing the critical angle and absorption maximum of the SPP. More information on this technique is available in Chapter 16. Several instruments that exploit this system are commercially available but, as the measurement does not usually involve nanoparticles, this scheme will not be discussed further here.

The other type of refractometric sensor exploits the localised surface plasmon resonance of AuNPs. The important point here is that the position of the resonance peak or peaks on the optical extinction spectrum of the sensor is influenced by the magnitude of the refractive index of the surrounding medium, in particular the resonance is shifted to longer wavelengths with an increase in the refractive index. This phenomenon can be exploited in two main ways. In the simplest version, the refractive index of a liquid medium can be sensed directly, or, similarly to the case of the SPP sensor described above, the surfaces of the AuNPs are functionalised so that they selectively bind to specific biological molecules or protein fragments. When the analyte is present, it attaches to the nanoparticle, thereby raising the average refractive index of the immediately adjacent medium and redshifting the particle's plasmon resonance. Generally the sensor system is designed to measure the magnitude of the redshift,[48,49] but a monochromatic interrogation of the optical absorbance also appears to be feasible too.[50] Although the scheme does work with gold nanospheres, greater sensitivity is obtained when using other shapes, such as nanorods or nano-bipyramids.[50–52] The capability to detect a single protein molecule by these means has recently been claimed.[53]

Of course, refractometric sensors can be constructed out of arrays of AuNPs too, with the advantage being that constructive inter-particle interactions can, in some cases, give sharper extinction peaks (and hence ultimately greater sensitivity) and the arrays can be used to collect and focus light onto a central 'hot spot'. However, precise positioning of the AuNPs is required for array devices and they are currently made by 'top down' lithographic techniques and, as such, do not involve the chemically produced AuNPs that are the focus of this book.

18.5.2 *Colorimetric Assays and Related Diagnostic Techniques*

A rather different platform for sensing an analyte using AuNPs is based simply on a colour change. This has the huge advantage that suitable test kits can be used in the field by a person with only basic training. The scheme is based either on a red-to-blue colour change that occurs when AuNPs aggregate together, or on the development of a red colour on a white background when one kind of Au nanoparticle binds to a white surface. Detailed information on the operation of these sensor schemes is available in the literature in Refs. 42, 54–58; here only a general overview will be presented.

In the first scheme, as the particles of Au approach closely to one another, the electric fields of their plasmon resonances interact, a factor that lowers (redshifts) their resonant frequency. This generates a new resonance peak in the optical extinction spectrum, which mainly absorbs the red wavelengths. The result is that only the blue wavelengths are transmitted and the colloidal suspension turns blue to the eye (Figure 18.6). A sensitive and selective sensor platform can be designed using this basic principle by arranging for the colloidal suspension to contain two populations of nanoparticles, which we will designate here as A and B. The 'A' and 'B' nanoparticles each bind to some different part of an analyte molecule, C. When C is added to the suspension, the A and B nanoparticles are brought into close proximity because both bind to the same C molecule, and a strongly redshifted plasmon resonance is developed. The concept was elegantly demonstrated for polynucleotides in the 1990s,[54] and has been developed for a wide variety of analytes since then. Recently, it has been shown that a one-pot detection of two target sequences can be achieved, which further extends the range of possibilities. A related type of assay is achieved by monitoring the aggregation of just one kind of Au nanoparticle. A high precision can be achieved with the correct functionalisation and control. An example of a system designed to detect Cd^{2+} ions[59] is shown in Figure 18.7.

On the other hand, in the so-called lateral flow sensor, a suspension of AuNPs to which a secondary antibody has been attached flow over a white surface, which has been functionalised to previously selectively collect the primary analyte. The AuNPs bind to the immobilised analyte (if it is present), with the appearance of a red line after ten minutes or so indicating a positive analysis. More information on this technique is available in Chapter 16.

A very large variety of colorimetric tests using Au nanoparticles has been successfully demonstrated, ranging from pregnancy tests, to tests for microbiological pathogens such as *Salmonella*, *Streptococcus*, viral agents such as herpes, specific genetic sequences, levels of specific human cell lines in the blood or the proteins that signal the onset of prostrate cancer.[4,60–62]

18.5.3 *Assays Based on Quartz Microbalance*

Because Au nanoparticles can be functionalised so that they will bind selectively to specific biological molecules, it follows that they can also be used to enhance the sensitivity of detection schemes based on the

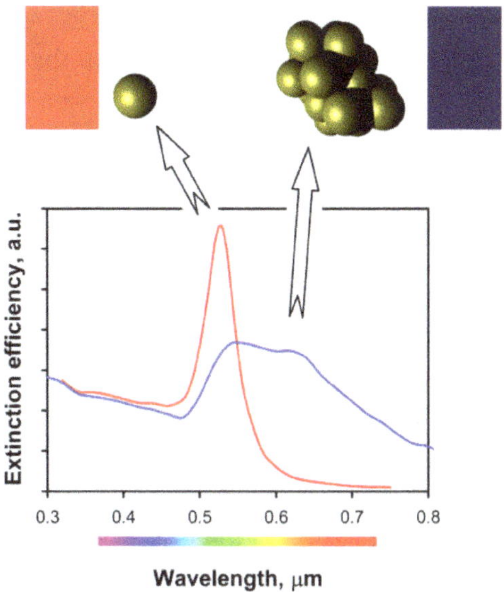

Figure 18.6   Isolated spherical AuNPs have a strong plasmon resonance with 520 nm light, causing extinction of green colours but transmission of red. When the AuNPs agglomerate, the plasmon resonance shifts to longer wavelengths and broadens, causing red and orange colours to be absorbed. Now the AuNPs seem blue in colour to the eye. Reprinted from Ref. 63, with permission from Elsevier.

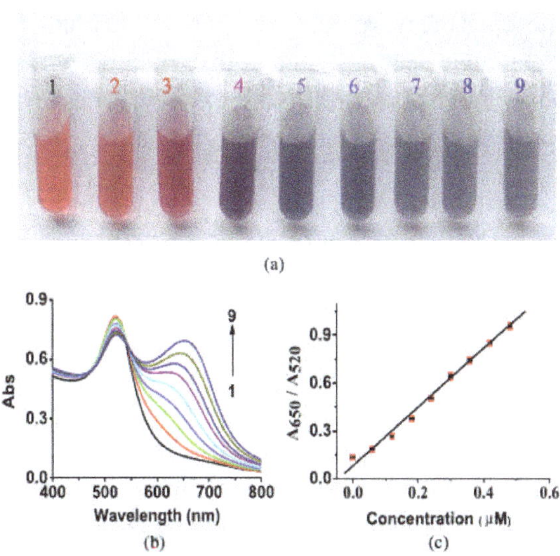

Figure 18.7   Example of a gold nanoparticle colorimetric sensor that can detect concentration of Cd^{2+} ions with a high degree of accuracy. (Reprinted from Ref. 59 with permission of Springer.)

piezoelectric quartz microbalance. In one version, the AuNPs are attached to the surface of the microbalance itself,[64] and in the other, they are in the solution.[65] In either event, a significant improvement in sensitivity is reported.

18.5.4 *Contrast Enhancement in Electron and Optical Microscopy*

There is a specialised market for AuNPs as a contrast enhancer for electron microscopy. Here the high electron density and chemical properties of Au are exploited. If the AuNPs are suitably functionalised to bind to a specific site or molecule, then they will facilitate the detection and imaging of such a site in a transmission electron microscope section due to their high atomic number. Further information is available in Chapter 15 and in a comprehensive text on the subject.[66] Resolution and functionality can be improved by use of very small nanoparticles, more of the nature of clusters. The NanoGold[TM] particle is an example of these and can be purchased pre-functionalised to bind to a wide range of targets.[67]

The use of AuNPs may also be advantageous in optical microscopy[68] and optical coherence tomography.[69] Even though the individual nanoparticles are much smaller than the wavelength of light used to illuminate the sample, they can, if big enough, scatter light and thereby modify the optical image in a beneficial manner. Enhanced detection of cancerous cells, e.g. Ref. 70, would be a typical objective of work in this domain.

18.5.5 *Bifunctional Metallo-dielectric Hybrids for Microscopy*

It is possible to prepare hybrid nanoparticles, consisting for example of a gold part and a second material, which exhibits some additional functionality. For example, a magnetic material (e.g. Fe_3O_4) or a fluorescent material (e.g. a semiconducting quantum dot) can be attached to an AuNP to produce a hybrid that has multiple functionalities. In this way, the electron density of Au can be combined with the light emission of a semiconductor, for example, to facilitate optical microscopy. Or, the presence of the magnetic material can be used to manipulate the position of the hybrid nanoparticle. Recent reviews on this topic provide more information; see Refs. 71 and 72.

18.5.6 *Surface-enhanced Raman Spectroscopy*

Although Ag is historically the better established substrate for SERS, there has also been interest from the very earliest days of SERS research in nanostructured Au substrates.[73] One reason for this is that silver nanostructures are susceptible to oxidation and this changes or compromises their performance. On the other hand, the plasmon resonances of the gold nanostructures are at lower energies (longer wavelengths) than those of Ag, and so somewhat different results are obtained. However, these older types of substrates, whether of Ag or Au, are produced by etching and are of an essentially random morphology. Interest in using geometrically well-defined configurations, such as nanoparticle dimers, for SERS enhancement is more recent and appears to date from 1999.[74] Good results are reported on specially prepared configurations of AuNPs, see for example Refs. 51, 75–77. Gold nanoshells, semi-shells or crescents seem most suitable.

18.5.7 *Two-photon Technologies*

Various types of nonlinear optical phenomena are possible on gold nanostructures but, of course, losses are high due to the opacity of metals. Four-wave mixing, for example, to produce 600 nm photons from a mixture of 800 and 1200 nm photons, is possible, as is broadband two photon luminescence (TPL). In general, however, complex geometries are needed, and the output from ordinary gold nanospheres is poor. Gold nanorods, on the other hand, have moderately good nonlinear properties and have been used in biomedical contexts as a type of two photon marker.[78] More information on this topic is available in Chapter 14.

18.6 Potential or Actual Therapeutic Applications

The reader is referred to publications on the use of AuNPs in medical contexts for more detailed information (see Refs. 63, 79–81, and other chapters in the present book). Despite occasional reports to the contrary,[82–85] it is generally believed that pure gold colloid itself has a negligible biological effect. This would be the expected situation considering the chemical unreactivity of AuNPs in physiological environments. Therefore, the use of AuNPs in medical contexts is more usually predicated upon the incorporation of some additional active functionality. Most commonly, this is the attachment of a pharmaceutically active molecule[86] or genetic material,[87] energisation of the colloid by plasmonic heating,[88] or, rarely, exploitation of the radioactive decay of ^{198}Au.[89] The presence of AuNPs can also improve the efficacy of X-ray radiotherapy by a sensitising mechanism (see Chapter 15).[90]

18.6.1 *Drug Delivery*

Delivery of a pharmaceutical compound can be, in principle, facilitated by conjugating it to the surface of a AuNP. The argument in favour depends on whether a more selective targeting of the compound to the desired region of the body can be obtained when using a nanoparticle vector. Certainly, there are some specific mechanisms by which such selective targeting can be achieved. For example, the vascular structure around rapidly growing tumours is relatively porous and so nanoparticles of less than 300 nm diameter diffuse out of the vascular system and collect in the tumour.[91–93] This mechanism of 'passive targeting' is known as extravasation. Delivery of a AuNP–drug conjugate by exploiting endocytosis is a less well-established process, but can certainly occur, in principle, because cells such as macrophages have evolved to engulf foreign particles in a certain size range. Furthermore, in principle, the drug component of the conjugate could be released or activated once in the desired location by the application of illumination and hence plasmonic heating, thereby invoking two functionalities simultaneously.

The argument against drug delivery by means of a nanoparticle vector is that only a minute dose can be delivered, of the order of a few tens of drug molecules per AuNP, because the drug necessarily forms only a surface coating on the nanoparticle core. Drug delivery by this means is therefore by necessity restricted to very potent substances, such as tumour necrosis factor (TNF)[86] or photodynamic sensitisers such as a phthalocyanine.[93] At least one of these prototype drug delivery schemes is now in Phase 1 clinical trials.[4]

Gold nanoparticles have also been used as adjuvants in vaccines. In this role the presence of the gold particles acts to enhance the effectiveness of the vaccine. A typical example is provided in a paper of Niikura *et al.*[94]

18.6.2 *Gene Therapy*

The penetration of an AuNP through a cell membrane, whether by endocytosis or some other means, provides a way to move foreign genetic material into a cell. Of course, the DNA or other payload must escape being destroyed in an endosome within the cell or degradation by a nuclease, and must somehow progress from there through the internal nuclear membrane and into the nucleus proper, before the transfective effect of its payload can be realised. Considerable progress has been made, and transfection using AuNPs as a vector has been frequently demonstrated.[87,95,96] Since an AuNP can undergo plasmonic heating, the additional possibility of controlling the release of genetic material by application of light has also been considered.[97–101] This provides an interesting and unique functionality, and permits better targeting since the genetic payload will be effectively unloaded in the illuminated cells only.

A related idea uses DNA-coated AuNPs of 2–3 μm diameter[102] to deliver a dose of genetic material that can (if successfully transfected) cause the recipient's body to express a target protein to which an immune response is desired.[103] This neatly overcomes the problem of the very small chemical payloads that are possible when using AuNPs because the patient's own body is manipulated into generating the desired protein. In this case, the particles are projected with high kinetic energy against the skin of the patient and carry their payload into the patient.

18.6.3 *Radiotherapy*

The [198]Au isotope, which has a half-life of 2.7 days, can be used in cancer therapy,[89,104] and has in the past been occasionally administered in the form of colloidal gold nanospheres. Targeting the radioactive material to the desired location is the key challenge, but can, in principle, be accomplished by extravasation, as discussed in Section 18.5.1. It is also known that the presence of AuNPs increases the efficacy of therapies using ionising electromagnetic radiation[90,105–107] but the reasons for this are, however, not yet well understood.

18.6.4 *Hyperthermal Techniques*

The energy of any light that is absorbed during a plasmon resonance in an AuNP (see Chapter 4) makes the particle a potent point source of heat when appropriately illuminated. Local temperature increases of a few tens of kelvin[108,109] are obtained with comparatively low-power laser diodes, quite adequate to kill mammal cells or single-cell pathogens. More intense illumination causes the AuNPs to melt or even vapourise, and the resulting pressure impulse can perforate a nearby cell membrane.[110] Gold nanospheres, nanoshells and nanorods are most frequently mentioned in respect of these proposed treatment schemes. Medical trials using nanoshells have progressed the furthest, and are now in Phase 1 clinical trials[4] for head and neck cancer. However, bacteria[63] and protozoa[111] may, in principle, also be targeted (Figure 18.8). It has been argued that nanorods would provide a similar or superior platform on which to develop technologies based on plasmonic heating than nanoshells.[112,113] In either case, however, it is desired that the plasmonic heating be applied by a laser located outside the body. Therefore, since the body is reasonably transparent in the near-infrared region of the spectrum (the so-called tissue window[114]), it is most helpful if the AuNP has a strong plasmon resonance in this part of the spectrum. This is readily achieved by adjusting the aspect ratio of gold nanoshells or nanorods. Ordinary gold nanospheres, on the other hand, absorb most strongly in the mid-visible, and would, at first sight, seem to be unsuitable for such an application. Nevertheless, it has been noted that usable optical extinction in the tissue window can be obtained from Au nanospheres provided that they are allowed to aggregate.[115,116] This collapses the single-particle plasmon and replaces it with a broad, redshifted resonance, which has an increased extinction cross-section at wavelengths within the tissue window.

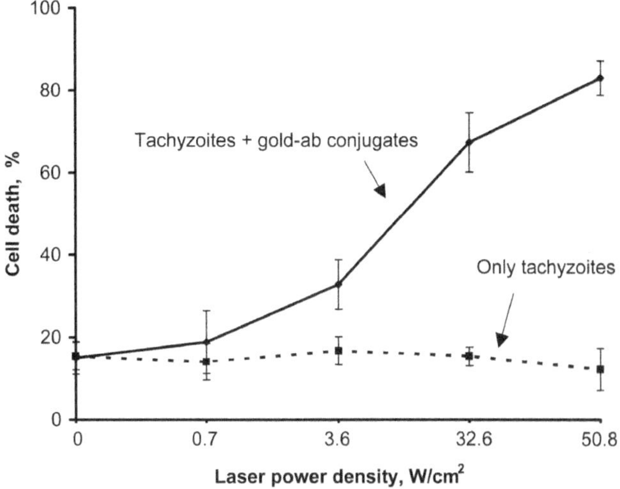

Figure 18.8 Death rates of *tachyzoites* of *Toxoplasmosis gondii* that had been labelled with gold nanorod–antibody conjugates compared to those of non-labelled *T. gondii tachyzoites* (a *tachyzoite* is one of the mobile stages in the pathogen's complex life cycle). Reprinted with permission from Pissuwan *et al.* Copyright 2007 American Chemical Society.

18.7 Environmental Remediation

The chemical affinity of AuNPs for elements such as sulphur and mercury have also suggested their use as an absorbent for specific pollutants. For example, it has been reported that Hg ions can be readily removed from water by exposure to Au colloid.[117] Of course, the colloid would have to be collected and purged of its mercury in a fully reversible process before such a scheme could be technologically feasible and, even then, the commercial feasibility of removing mercury contamination this way is doubtful. It has also been suggested that AuNPs could be useful in photocatalytic applications with the potential to remove some environmentally deleterious contaminant such a volatile organic compound.[118]

18.8 Conclusions and Outlook

AuNPs have unique and useful chemical and physical properties. For this reason, they are currently used in commercial quantities in biological diagnostic assays, as specialised stains for electron microscopy, as the colouring agent in niche glassware and ceramics, in high-end decorative inks and in some electronics as sinter inks. There are also, however, a very large range of potential applications under development, which this chapter has attempted to cover too. With so many examples under development it is likely that several will be successful, but of course, many others will fail to reach the level of commercial application. Only a very small quantity of gold is used in any individual device or kit in most of the applications or potential applications described here. Furthermore, in many cases, these applications are in the high-technology domain and have very high intrinsic added value. Therefore, the value of the gold that they contain is relatively insignificant as a proportion of the total cost.

References

1. Anon, Global Gold Nanoparticles Market 2015–2019, http://www.prnewswire.com/news-releases/global-gold-nanoparticles-market-2015-2019-with-bbi-solutions-cline-scientific-cytodiagnostics-goldsol-meliorum-technologies-nano-composix-sigma-aldrich-tanaka-technologies-dominating-300188136.html, Accessed 18 January 2016.
2. R. F. Tylecote *History of Metallurgy* (Institute of Materials, London, 1992).
3. I. Freestone, N. Meeks, M. Sax and C. Higgitt, *Gold Bull.* **40**, 270 (2007).
4. T. Keel, R. Holliday and T. Harper, *Gold for Good. Gold and Nanotechnology in the Age of Innovation* (World Gold Council, London, 2010).
5. L. Street, K. Gopaul, A. Hewitt and M. Grubb, *Gold Demand Trends. Full Year 2014* (World Gold Council, London, 2015).
6. M. G. Blaber, M. D. Arnold, N. Harris, M. J. Ford and M. B. Cortie, *Phys. B (Amsterdam, Neth.)*, **394**, 184 (2007).
7. P. T. Bishop, L. J. Ashfield, A. Berzins, A. Boardman, V. Buche, J. Cookson, R. J. Gordon, C. Salcianu and P. A. Sutton, *Gold Bull.* **43**, 181 (2010).
8. T. Minari, Y. Kanehara, C. Liu, K. Sakamoto, T. Yasuda, A. Yaguchi, S. Tsukada, K. Kashizaki and M. Kanehara, *Adv. Func. Mater.* **24**, 4886 (2014).
9. S. Sun, P. Mendes, K. Critchley, S. Diegoli, M. Hanwell, S. D. Evans, G. J. Leggett, J. A. Preece and T. H. Richardson, *Nano Lett.* **6**, 345 (2006).
10. M. Coutts, M. B. Cortie, M. J. Ford and A. M. McDonagh, *J. Phys. Chem. C* **113**, 1325 (2009).
11. T. Bakhishev and V. Subramanian, *J. Electronic Mater.* **38**, 2720 (2009).
12. X. Xu, M. Stevens and M. B. Cortie, *Chem. Mater.* **16**, 2259 (2004).
13. W. Zhao, T. Rovere, D. Weerawarne, G. Osterhoudt, N. Kang, P. Joseph, J. Luo, B. Shim, M. Poliks and C.-J. Zhong, *ACS Nano* **9**, 6168 (2015).

14. H. Chowdhury, X. Xu, P. Huynh and M. B. Cortie, *J. Solar Energ. Trans. ASME* **127**, 70 (2005).
15. X. Xu, T. Gibbons and M. B. Cortie, *Gold Bull.* **39**, 156 (2006).
16. N. Stokes, A. McDonagh and M. B. Cortie, *J. Nanopart. Res.* **12**, 2821 (2010).
17. M.-J. Kim, H.-J. Na, K. C. Lee, E. A. Yoob and M. Lee, *J. Mater. Chem.* **13**, 1789 (2003).
18. D. Compton, L. Cornish and E. van der Lingen, *Gold Bull.* **36**, 10 (2003).
19. M. Li, Z. S. Zhang, X. Zhang, K. Y. Li and X. F. Yu, *Optics Express* **16**, 14288 (2008).
20. E. Cattaruzza, G. Battaglin, F. Gonella, R. Polloni, B. F. Scremin, G. Mattei, P. Mazzoldi and C. Sada, *Appl. Surf. Sci.* **254**, 1017 (2007).
21. Y. Hamanaka, K. Fukuta, A. Nakamura, L. M. Liz-Marzán and P. Mulvaney, *Appl. Phys. Lett.* **84**, 4938 (2004).
22. C. Pecharromán, A. Esteban-Cubillo, H. Fernández, L. Esteban-Tejeda, R. Pina-Zapardiel, J. S. Moya, J. Solis and C. N. Afonso, *Plasmonics* **4**, 261 (2009).
23. J.-S. Lee, J. Cho, C. Lee, I. Kim, J. Park, Y.-M. Kim, H. Shin, J. Lee and F. Caruso, *Nature Nanotech.* **2**, 790 (2007).
24. L. D. Bozano, B. W. Kean, M. Beinhoff, K. R. Carter, P. M. Rice and J. C. Scott, *Adv. Funct. Mater.* **15**, 1933 (2005).
25. J. W. M. Chon, C. Bullen, P. Zijlstra and M. Gu, *Adv. Funct. Mater.* **17**, 875 (2007).
26. P. Zijlstra, J. W. M. Chon and M. Gu, *Nature* **459**, 410 (2009).
27. T. Laaksonen, V. Ruiz, P. Liljeroth and B. M. Quinn, *Chem. Soc. Rev.* **37**, 1836 (2008).
28. M. B. Cortie, M. H. Zareie, S. R. Ekanayake and M. J. Ford, *IEEE Trans. Nanotech.* **4**, 406 (2005).
29. L. Caillard, S. Sattayaporn, A.-F. Lamic-Humblot, S. Casale, P. Campbell, Y. J. Chabal and O. Pluchery, *Nanotechnology* **26**, 065301 (2015).
30. M. Haruta, *Catal. Today* **36**, 153 (1997).
31. M. Peplow, *Nature* **495**, S10 (2013).
32. Anon, in *CatGold News* (World Gold Council, London, p. 1, 2008).
33. Anon, in *CatGold News* (World Gold Council, London, p. 1, 2003).
34. P. Johnston, N. Carthey and G. J. Hutchings, *J. Am. Chem. Soc.* **137**, 14548 (2015).
35. License of PMMA manufacturing technology in Thailand, www.asahi-kasei. co.jp/ asahi/en/news/2013/e130424.html, Accessed 28 February 2016.
36. Anon, in *CatGold News* (World Gold Council, London, p. 1, 2006).
37. Anon, in *CatGold News* (World Gold Council, London, p. 1, 2005).
38. M. Taguchi, H. Kagoshima, T. Watanabe, R. Hasegawa, T. Toyoguchi and K. Tanida, in *10th International Conference on Mercury as a Global Pollutant*, ed. B. B. *et al.*, Halifax, Nova Scotia, Canada, 2011, p. 302.
39. D. A. Mann, DuPont plans environmental test in Louisville, http://www. bizjournals. com/louisville/news/2013/04/15/dupont-plans-environmental-test-in.html, Accessed 28 February 2016.
40. Anon, in *CatGold News* (World Gold Council, London, p. 1, 2004).
41. J. Carbert, *Gold Bull.* **13**, 144 (1980).
42. M.-C. Daniel and D. Astruc, *Chem. Rev.* **104**, 293 (2004).
43. Y. Nakao and K. Kaeriyama, *J. Appl. Polymer Sci.* **36**, 269 (1988).

44. M. Richardson and J. Johnston, *J. Colloid Interface Sci.* **310**, 425 (2007).

45. J. H. Johnston and K. A. Lucas, *Gold Bull.* **44**, 85 (2011).

46. A. Iwakoshi, T. Nanke and T. Kobayashi, *Gold Bull.* **38**, 107 (2005).

47. O. Pluchery, H. Remita Bosi and D. Schaming, French patent FR 15 56942, 2015.

48. A. J. Haes and R. P. V. Duyne, *Expert Rev. Mol. Diagn.* **4**, 527 (2004).

49. M. Himmelhaus and H. Takei, *Sens. Actuators, B* **63**, 24 (2000).

50. C. S. Kealley, M. D. Arnold, A. J. Porkovich and M. B. Cortie, *Sens. Actuators, B* **148**, 34 (2010).

51. K.-S. Lee and M. A. El-Sayed, *J. Phys. Chem. B* **110**, 19220 (2006).

52. H. Chen, X. Kou, Z. Yang, W. Ni and J. Wang, *Langmuir* **24**, 5233 (2008).

53. K. M. Mayer, F. Hao, S. Lee, P. Nordlander and J. H. Hafner, *Nanotechnol.* **21**, 255503 (2010).

54. R. Elghanian, J. J. Storhoff, R. C. Mucic, R. L. Letsinger and C. A. Mirkin, *Science* **277**, 1078 (1997).

55. E. Hutter and J. H. Fendler, *Adv. Mater.* **16**, 1685 (2004).

56. S.-Y. Lin, S.-H. Wu and C.-H. Chen, *Angew. Chem. Int. Ed.* **45**, 4948 (2006).

57. J. Zhang, L. Wang, D. Pan, S. Song, F. Y. C. Boey, H. Zhang and C. Fan, *Small* **4**, 1196 (2008).

58. N. T. K. Thanh and Z. Rosenzweig, *Anal. Chem.* **74**, 1624 (2002).

59. A.-J. Wang, H. Guo, M. Zhang, D.-L. Zhou, R.-Z. Wang and J.-J. Feng, *Microchim. Acta.* **180**, 1051 (2013).

60. E. I. Laderman, E. Whitworth, E. Dumaual, M. Jones, A. Hudak, W. Hogrefe, J. Carney and J. Groen, *Clin. Vaccine Immunol.* **15**, 159 (2008).

61. C. R. Martin and D. T. Mitchell, *Analytical Chemistry News & Features*, 1st May (1998) 322A.

62. A. M. Horgan, J. D. Moore, J. E. Noble and G. J. Worsley, *Trends Biotechnol.* **28**, 485 (2010).

63. D. Pissuwan, C. H. Cortie, S. M. Valenzuela and M. B. Cortie, *Trends Biotechnol.* **28**, 207 (2010).

64. L. Wang, Q. Wei, W. ChunSheng, Z. Hu, J. Ji and P. Wang, *Chin. Sci. Bull.* **53**, 1175 (2008).

65. H. Wang, C. Lei, J. Li, Z. Wu, G. Shen and R. Yu, *Biosens. Bioelec.* **19**, 701 (2004).

66. M. A. Hayat, *Colloidal Gold: Principles, Methods, and Applications* (Academic Press, San Diego, CA, 1989).

67. R. Powell and J. Hainfeld, *Micron* **42**, 163 (2011).

68. L. Tong, Q. Wei, A. Wei and J.-X. Cheng, *Photochem. Photobiol.* **85**, 21 (2009).

69. M. Hu, J. Chen, Z.-Y. Li, L. Au, G. V. Hartland, X. Li, M. Marquez and Y. Xia, *Chem. Soc. Rev.* **35**, 1084 (2006).

70. X. Huang, I. H. El-Sayed, W. Qian and M. A. El-Sayed, *J. Am. Chem. Soc.* **128**, 2115 (2006).

71. N. Sanvicens and M. P. Marco, *Trends Biotechnol.* **26**, 425 (2008).

72. M. B. Cortie and A. M. McDonagh, *Chem. Rev.* **111**, 3713 (2011).

73. C. G. Blatchford, J. R. Campbell and J. A. Creighton, *Surf. Sci.* **120**, 435 (1982).

74. H. Xu, E. J. Bjerneld, M. Käll and L. Börjesson, *Phys. Rev. Lett.* **83**, 4357 (1999).

75. C. L. Haynes, A. D. McFarland and R. P. V. Duyne, *Anal. Chem.* 338A (2005).

76. J.-F. Li, Z.-L. Yang, B. Ren, G.-K. Liu, P.-P. Fang, Y.-X. Jiang, D.-Y. Wu and Z.-Q. Tian, *Langmuir* **22**, 10372 (2006).
77. F. Le, D. W. Brandl, Y. A. Urzhumov, H. Wang, J. Kundu, N. J. Halas, J. Aizpurua and P. Nordlander, *ACS Nano* **2**, 707 (2008).
78. H. Wang, T. B. Huff, D. A. Zweifel, W. He, P. S. Low, A. Wei and J.-X. Cheng, *Proc. Natl. Acad. Sci.* **102**, 15752 (2005).
79. D. Pissuwan, T. Niidome and M. B. Cortie, *J. Controlled Release* **149**, 65 (2011).
80. D. Pissuwan, S. M. Valenzuela and M. B. Cortie, *Biotechnol. Genet. Eng. Rev.* **25**, 93 (2008).
81. J. Li, S. Gupta and C. Li, *Quant. Imaging. Med. Surg.* **3**, 284 (2013).
82. G. E. Abraham and P. B. Himmel, *Nutri. Env. Med.* **7**, 295 (1997).
83. N. Pernodet, X. Fang, Y. Sun, A. Bakhtina, A. Ramakrishnan, J. Sokolov, A. Ulman and M. Rafailovich, *Small* **2**, 766 (2006).
84. C. L. Brown, M. W. Whitehouse, E. R. T. Tiekink and G. R. Bushell, *Inflammopharmacology* **16**, 133 (2008).
85. C. L. Brown, G. Bushell, M. W. Whitehouse, D. Agrawal, S. Tupe, K. Paknikar and E. R. Tiekink, *Gold Bull.* **40**, 245 (2007).
86. G. F. Paciotti, L. Myer, D. Weinreich, D. Goia, N. Pavel, R. E. McLaughlin and L. Tamarkin, *Drug Delivery* **11**, 169 (2004).
87. T. Niidome, K. Nakashima, H. Takahashi and Y. Niidome, *Chem. Commun.* 1978 (2004).
88. D. Pissuwan, S. Valenzuela and M. B. Cortie, *Trends Biotechnol.* **24**, 62 (2006).
89. A. Sherman and M. Ter-Pogossian, *Cancer* **6**, 1238 (1953).
90. J. F. Hainfeld, D. N. Slatkin and H. M. Smilowitz, *Phys. Med. Biol.* **49**, N309 (2004).
91. D. P. O'Neal, L. R. Hirsch, N. J. Halas, J. D. Payne and J. L. West, *Cancer Lett.* **209**, 171 (2004).
92. S. M. Moghimi, A. C. Hunter and J. C. Murray, *FASEB J.* **19**, 311 (2005).
93. Y. Cheng, C. S. Anna, J. D. Meyers, I. Panagopoulos, B. Fei and C. Burda, *J. Am. Chem. Soc.* **130**, 10643 (2008).
94. K. Niikura, T. Matsunaga, T. Suzuki, S. Kobayashi, H. Yamaguchi, Y. Orba, A. Kawaguchi, H. Hasegawa, K. Kajino, T. Ninomiya, K. Ijiro and H. Sawa, *ACS Nano* **7**, 3926 (2013).
95. A. C. Bonoiu, S. D. Mahajan, H. Ding, I. Roy, K. T. Yong, R. Kumar, R. Hu, E. J. Bergey, S. A. Schwartz and P. N. Prasad, *Proc. Natl. Acad. Sci. USA* **106**, 5546 (2009).
96. N. L. Rosi, D. A. Giljohann, C. S. Thaxton, A. K. R. Lytton-Jean, M. S. Han and C. A. Mirkin, *Science* **312**, 1027 (2006).
97. Y. Niidome, T. Niidome, S. Yamada, Y. Horiguchi, H. Takahashi and K. Nakashima, *Mol. Cryst. Liq. Cryst.* **445**, 201/[491] (2006).
98. C. C. Chen, Y. P. Lin, C. W. Wang, H. C. Tzeng, C. H. Wu, Y. C. Chen, C. P. Chen, L. C. Chen and Y. C. Wu, *J. Am. Chem. Soc.* **128**, 3709 (2006).
99. H. Takahashi, Y. Niidome and S. Yamada, *Chem. Commun.* **17**, 2247 (2005).
100. A. Wijaya, S. B. Schaffer, I. G. Pallares and K. Hamad-Schifferli, *ACS Nano* **3**, 80 (2008).
101. S. E. Lee, G. L. Liu, F. Kim and L. P. Lee, *Nano Lett.* **9**, 562 (2009).

102. C. D. Medley, B. K. Muralidhara, S. Chico, S. Durban, P. Mehelic and C. Demarest, *Anal. Bioanal. Chem.* **398**, 527 (2010).
103. D. Tang, M. DeVit and S. A. Johnston, *Nature* **356**, 152 (1992).
104. P. Rubin and S. H. Levitt, *J. Nuclear Med.* **5**, 581 (1964).
105. P. Diagaradjane, A. Shetty, J. C. Wang, A. M. Elliott, J. Schwartz, S. Shentu, H. C. Park, A. Deorukhkar, R. J. Stafford, S. H. Cho, J. W. Tunnell, J. D. Hazle and S. Krishnan, *Nano Lett.* **8**, 1492 (2008).
106. A. Simon-Deckers, E. Brun, B. Gouget, M. Carrière and C. Sicard-Roselli, *Gold Bull.* **41**, 187 (2008).
107. S. Jain, D. G. Hirst and M. O'Sullivan, *Bri. J. Radiology* **85**, 101 (2012).
108. L. R. Hirsch, R. J. Stafford, J. A. Bankson, S. R. Sershen, B. Rivera, R. E. Price, J. D. Hazle, N. J. Halas and J. L. West, *Proc. Natl. Acad. Sci.* **100**, 13549 (2003).
109. D. Pissuwan, S. M. Valenzuela, M. C. Killingsworth, X. Xu and M. B. Cortie, *J. Nanopart. Res.* **9**, 1109 (2007).
110. C. M. Pitsillides, E. K. Joe, X. Wei, R. R. Anderson and C. P. Lin, *Biophys. J.* **84**, 4023 (2003).
111. D. Pissuwan, S. Valenzuela, C. M. Miller and M. B. Cortie, *Nano Lett.* **7**, 3808 (2007).
112. N. Harris, M. J. Ford, P. Mulvaney and M. B. Cortie, *Gold Bull.* **41**, 5 (2008).
113. P. K. Jain, K. S. Lee, I. H. El-Sayed and M. A. El-Sayed, *J. Phys. Chem. B.* **110**, 7238 (2006).
114. R. Weissleder, *Nature Biotechnol.* **19**, 316 (2001).
115. V. P. Zharov, E. N. Galitovskaya, C. Johnson and T. Kelly, *Lasers Surg. Med.* **37**, 219 (2005).
116. D. Pissuwan, S. M. Valenzuela, C. M. Miller, M. C. Killingsworth and M. B. Cortie, *Small* **5**, 1030 (2009).
117. K. P. Lisha, Anshup and T. Pradeep, *Gold Bull.* **42**, 144 (2009).
118. X. Chen, H.-Y. Zhu, J.-C. Zhao, Z.-F. Zheng and X.-P. Gao, *Angew. Chem. Int. Ed.* **47**, 5353 (2008).

Glossary

AAO: anodic aluminium oxide
AAPTS: 3-(2-aminoethylamino)propyltrimethoxysilane
AFM: atomic force microscopy
AES: Auger electron spectroscopy
APTES: 3-aminopropyltriethoxysilane
APTMS: 3-amino-propyltri-methoxysilane
AS: action spectrum
AuNP: gold nanoparticle
BBB: blood–brain barrier
BCE: Before the Common Era
BDE: ballistic diffusive equations
BEM: boundary element method
BPTS: (1,4)-bis(triethoxysilyl)propane tctrasufide (thioether)
BRIJ 30: poly(ethyleneglycol)-dodecylether
BSA: bovin serum albumin
BTE: Boltzmann transport equation
CB: conduction band
CNT: carbon nanotube
CPBr: cetylpyridinium bromide
CTAB: cetyltrimethylammonium bromide
CVD: chemical vapour deposition
DAPTS: N-[3-(trimethoxysilyl)propyl]ethylenediamine
DDA: 1,12-dodecanediamine
DFT: density functional theory
DLS: dynamic light scattering

DMF: dimethylformamide

DNA: deoxyribonucleic acid

DOS: local density of states

DSC: differential scanning calorimetry

DSSC: dye-sensitised solar cell

EBL: electron beam lithography

EDC: 1-ethyl-3-(3-dimethylaminopropyl) carbodiimide

EDX: energy dispersive X-ray spectroscopy

EGFR: epidermal growth factor receptor

EM: electromagnetic field

EMAs: effective medium approximations

ET: electron trap

Eugenol: 2-methoxy-4-allyl-phenol

FCC: face-centred cubic

FDTD: finite difference time domain

FPA: fluorescence polarisation anisotropy

FT-IR: Fourier transform infrared spectroscopy

FTO: fluorine doped tin oxide

GGA: generalised gradient correction

GFP: green fluorescent protein

HCAI: human carbonic anhydrase I

HR-PES: high-resolution photoelectron spectroscopy

HRTEM: high-resolution transmission electron microscopy

HSA: human serum albumin

Hydrotalcite: Mg-Al layered double hydroxide

ICG: indocyanine green

IET: inelastic electron tunnelling

IFE: inner-filter effect

IRAS: infrared reflection absorption spectroscopy

LBL: layer-by-layer

LED: light emitting diodes

LEISS: low-energy ion surface scattering

LEMF: local electromagnetic field transfer

LMs: longitudinal modes

LSP: localised surface plasmon

LSPR: localised surface plasmon resonance

MB: methylene blue (dye)

MCA: melamine cyanurate

MD: molecular dynamics

MG: Maxwell Garnett

MIP: molecular-imprinted polymer

MMP: multiple multipole

MOF: metal organic framework

MPCs: monolayer protected clusters

MPTS: 3-mercaptopropyltrimethoxysilane

MRI: magnetic resonance imaging

MTBE: methyl *tert*-butyl ether

MTMOS: methyltrimethoxysilane

MW: microwaves

NIR: near infrared

NP-5: polyethylene glycol mono-4-nonylphenyl ether

NPP: nanoparticle plasmon

NTCP: normal tissue complication probability

NV: nitrogen vacancy

ORR: oxygen reduction reaction

PAA: poly(acrylic acid)

PAEMH: poly(2-aminoethyl methacrylate hydrochloride)

PAMAM: poly(amidoamine)

PBS: phosphate buffer saline

PCE: power conversion efficiency

PCPs: porous coordination polymers

PCWC: probability of cure without complication

PEG: polyethylene glycol

PEO-b-P4VP: poly(ethylene oxide)-b-poly(4-vinylpyridine)

Pluronic P123: poly(ethylene glycol)-poly(propylene glycol)-poly(ethylene glycol) (triblock copolymer)

PolyHIPE: poly(styrene-co-divinylbenzene

PRET: plasmon energy transfer

PS-*b*-P2VP: poly(styrene)-block-poly(2-vinylpyridine) (diblock copolymer)

PS-b-P4VP: polystyrene-b-poly(4-vinylpyridine) (diblock copolymer)

PVA: poly(vinyl alcohol)

PVD: physical vapour deposition
PVDF: polyvinylidene difluoride
PVP: poly(vinyl pyrrolidone)
PVP-PVAc: poly(vinyl pyrrolidoneran-vinyl acetate)
PZC: point of zero charge
QE: quantum emitter
QSAR: quantitative structure activity relationship
RET: resonant energy transfer
RIU: refractive index unit
ROS: reactive oxygen species
SAED: selected electron diffraction
SAMs: self-assembled monolayers
SE: spontaneous emission
SEM: scanning electron microscopy
SERS: surface-enhanced Raman spectroscopy
SFG: sum frequency generation
SIMS: secondary ion mass spectrometry
SOMO: singly occupied molecular orbital
SP: surface plasmon
SPP: surface plasmon polariton
SPR: plasmon resonance at surfaces
SPN: surface plasmon on metal nanoparticles
STM: scanning tunnelling microscopy
TA: thermal analysis
TCP: tumour control probability
TDDFT: time-dependent density functional theory
TDLDA: time-dependent local density approximation
TEM: transmission electron microscopy
TEOS: tetraethoxysilane
TGA: thermogravimetric analysis
THPC: tetrakis (hydroxymethyl) phosphonium chloride
TIED: trapped-ion electron diffraction
TMs: transverse modes
TNF: tumour necrosis factor
TOAB: tetraoctylammonium bromide
TOF: turnover frequency

TOPO: *tri-n*-octylphosphine oxide
Triton-X 114: polyoxyethylene(7-8)octylphenyl ether
TTM: two temperature model
UHV: ultrahigh vacuum
UPD: underpotential deposition
UV: ultraviolet
VB: valence band
VDE: vertical detachment energy
WFT: wave function theories
XAFS: X-ray absorption fine structure
XRD: X-ray diffraction

Index

fluorescence, 101, 143, 373, 494, 547, 565
  fluorescence microscopy, 494
  fluorescence quenching, 143, 547–548
Fourier law, 92, 95, 111, 114–115, 117
Fourier transform infrared (FT-IR) spectroscopy, 170, 469
Fowler–Nordheim tunnelling phenomenon, 153
Fröhlich criterion, 136, 138
Frens method, 176
functional ligand, 209

G

G-value, 532
galvanic replacement, 188–189, 192–193, 258–259, 266
Geant4, 521, 525
gene gun, 566–567
gene therapy, 505, 558, 619
genotoxicity, 587, 589
glass, 7–12, 14–19, 22, 24–25, 52, 71, 171, 268, 270, 334, 382, 605, 610
Glauber, Rudolf Johann, 14–16, 19
gold cation, 235, 322, 339
gold chloride, 6, 10, 14, 23, 230
  AuCl, 38, 230, 267, 466
  $AuCl_4^-$, 168, 176, 179, 181, 255, 266, 454, 594
  tetrchloraoauric acid ($HAuCl_4$), 41, 173, 175, 230, 233, 236–237, 241–242, 244, 246–248, 252, 254–261, 263–265, 267, 269–270
gold coined, 3, 5
gold cyanide, 6, 38
gold film, 2, 7–8, 404, 412, 416, 424, 484–485, 604
gold leave, 6, 8, 10–11, 22
gold ruby (glass enamel), 9–25, 51
gold sol, 21–22, 230, 249
gold tetrammine nitrate, 239
gold(III) ethylenediamine, 239
gold–silver hollow structure, 193

H

hazard, 576, 578, 590–591, 597
heat transfer, 91–92, 96, 102, 109–111, 120
Heisenberg equations, 138
heteroepitaxial growth, 191
heterolytic dissociation, 307
high atomic number, 511, 519, 521, 533
high-energy electron, 372, 375, 379, 383–384, 386
high-resolution transmission electron microscopy (HRTEM), 170, 402
hole, 119, 319, 324–325, 328, 330, 351, 414, 490
hot electron, 326, 328–330, 334, 347–348, 356, 367–368
hot spot, 67, 336, 380, 487
hot-injection, 172
human health, 575–576, 578
hybrid, 132, 142, 154
  hybrid (particle, material), 101, 103, 211, 265, 353–355, 383–384, 548, 616
hybrid eigenmode, 144
hybrid mode, 143–144
hybridisation, 37, 41, 194, 348, 351, 355, 446, 469–470, 505
hydride transfer, 308
hydrogen oxidation, 292
hydrogen peroxide, 290–291, 299–300, 319, 330
hydrogenation reaction, 285–286, 295, 297–298, 304–306, 310, 393, 476
hydroperoxide, 301
hydrophilicity, 219, 250, 343, 359, 577
hydroxycarbonyl, 287–290
hydroxyl, 180, 210–212, 234, 287, 293, 297, 319, 324, 326–327, 346, 348, 412, 416, 512–514, 531–532
hydroperoxyl, 292, 297–298
hydroxyl radicals ($^\bullet$OH), 319
hyperthermia, 101–102, 557, 560–562, 568, 620

localised surface plasmon resonance
(LSPR) (*see also* plasmon and
Chapter 3), 21, 131, 168, 327, 371, 385,
483–484, 490, 492,
loss, 29, 54, 87, 97, 132, 134, 136, 142,
148, 154, 368, 370, 378, 396, 504,
531–532, 543, 565, 603, 612, 617
lustre, 7–8, 39
Lycurgus cup, 9–10, 21
lysine, 247–248, 251

M

magnetic dipole, 146–147, 369, 381
magnetic resonance imaging (MRI), 353,
494, 555
many-body quantum states, 132, 145
mass energy absorption coefficients,
519–520
Maxwell Garnett model (*see also* effective
medium approximation), 82
medical (treatment, uses), 39, 538–539,
556
medicine, 6–7, 218, 321, 483, 509, 537
metal-organic framework (MOF), 229,
241, 266–267
metallophilic attraction, 41
metallurgy in a beaker, 190
metamaterial, 131, 133, 144–146, 153, 489
micelle, 177, 215, 220, 248, 255, 269,
413–414
microwave, 171, 178, 181, 215–216, 501
microwave dielectric heating, 181
microwave irradiation, 179, 215,
243–244, 249, 604
microwave-assisted reduction, 257
Mie theory, 54, 60, 64, 68, 77, 93, 119,
171, 543
Mie, Gustav, 21, 53
mineralisation, 324, 334, 341, 343
mode volume, 142–143
morphology of nanoparticles, 25, 118,
166, 170, 182–183, 188, 192–193, 201,
320–321, 358, 399, 422
mosaic, 7–8, 10–11
multipolar modes, 68, 77, 119, 379–380

Murano, 13
mutagenesis/mutagenic, 511–512, 529

N

nanoalloy, *see* alloy
nanoantenna, 66–67, 380, 490
nanoelectronics, 268, 384, 437
nanoplatform, 565
nanorod, *see* shape of nanoparticles
nanosource of light, 377, 380–386
nanowire, *see* shape of nanoparticles
near-field approximation, 137
near-infrared laser irradiation, 179
Neri, Antonio, 13, 15–16, 18, 22
neutron-activated, 584
non-biofouling, 220
non-destructive, 542
non-toxic, 476, 491, 542, 592–593
nonlinear optics, 496, 605
normal tissue complication probability,
515
nuclear membrane, 528, 619
nuclearity, 291
nucleation, 24, 171–172, 182, 191, 202,
213, 244, 396, 404, 406, 416, 419

O

octahedron, *see* shape of crystalline
nanoparticles
oleylamine, 184, 190, 209, 244, 246, 248
on-chip applications, 384–385
on-chip information transfer, 384
one-pot preparation, 261
optical antenna, *see* nanoantenna
optical interconnect, 384
optical microscopy, 100, 170, 564, 616
optical properties, 21, 36, 53, 78, 87, 121,
146, 151, 182, 365, 440, 473, 537, 559,
568, 601–602, 604, 608
optical techniques, 483, 543
opto-acoustic-based imaging, 554
organogold, 7, 240
organometallic chemistry of gold, 39
Ostwald ripening, 172, 422
Otto, Andreas, 52